WM
AF684778

Paleozoic and Early Mesozoic Paleogeographic Relations; Sierra Nevada, Klamath Mountains, and Related Terranes

Edited by

David S. Harwood
U.S. Geological Survey
MS 975, 345 Middlefield Road
Menlo Park, California 94025

M. Meghan Miller
Jet Propulsion Laboratory 183-501
California Institute of Technology
4800 Oak Grove Drive
Pasadena, California 91109

THE GEOLOGICAL SOCIETY OF AMERICA
1888

SPECIAL PAPER

255

1990

© 1990 The Geological Society of America, Inc.
All rights reserved.

All materials subject to this copyright and included
in this volume may be photocopied for the noncommercial
purpose of scientific or educational advancement.

Copyright is not claimed on any material prepared
by government employees within the scope of their
employment.

Published by The Geological Society of America, Inc.
3300 Penrose Place, P.O. Box 9140, Boulder, Colorado 80301

Printed in U.S.A.

GSA Books Science Editor Richard A. Hoppin

Library of Congress Cataloging-in-Publication Data

Paleozoic and early Mesozoic paleogeographic relations : Sierra Nevada, Klamath Mountains, and related terranes / edited by David S. Harwood, M. Meghan Miller.

p. cm. — Special paper / Geological Society of America ; 255)

Based on papers from combined annual meetings of the Cordilleran and Rocky Mountain Sections of the Geological Society of America held in May 1989.

Includes bibliographical references and index.

ISBN 0-8137-2255-1

1. Paleogeography—Paleozoic. 2. Paleogeography—Mesozoic.
3. Paleogeography—Klamath Mountains (Calif. and Or.)
4. Paleogeography—Sierra Nevada Mountains (Calif. and Nev.)
5. Geology, Structural—Klamath Mountains (Calif. and Or.)
6. Geology, Structural—Sierra Nevada Mountains (Calif. and Nev.)
I. Harwood, David S. II. Miller, M. Meghan, 1957–
III. Geological Society of America. Cordilleran Section.
IV. Geological Society of America. Rocky Mountain Section.
V. Series: Special papers (Geological Society of America) ; 255.
QE501.4.P3P374 1990
557.94′21—dc20 90-19451
CIP

Cover photo: Sierra Buttes viewed to the southwest from Mtn. Mine Road, northern Sierra Nevada, California. Late Devonian volcanic center (top toward viewer) composed of quenched andesitic (dark) and felsic (light gray) rocks intrusive into submarine tuff, tuff breccia, and minor phosphatic chert. Photo by David S. Harwood.

10 9 8 7 6 5 4 3 2

Contents

Preface

Paleozoic and lower Mesozoic volcanic-arc sequences in the Klamath Mountains and northern Sierra Nevada formed outboard of the Cordilleran miogeocline and are now fault-bounded terranes with unique stratigraphic and structural histories. The tectonic evolution of these arc sequences has been the subject of debate since before the plate tectonic revolution, but the debate has intensified and broadened in scope during the past decade as the concept of accreted terranes has gained wider support. Although the accreted terrane concept was born in the Klamath Mountains, controversy still exists about the origin of those Paleozoic and lower Mesozoic volcanic rocks, their relation to similar rocks in the northern Sierra Nevada, and the paleogeographic affinity of these arc terranes to North America. The pivotal question in the debate is whether these arc terranes are far-travelled oceanic fragments unrelated to each other and exotic to North America or whether they formed along the active North American paleo-plate margin. Related controversy exists concerning the paleogeography of imbricated basinal terranes to the east, in Nevada, and fragmented oceanic and high-pressure metamorphic complexes to the west in the Klamath-Sierran belt. Determining relations between the basinal, arc, and subduction-related terranes is critical to understanding the broader framework of Paleozoic and early Mesozoic tectonism.

In June 1988, a Penrose Conference was convened to discuss new stratigraphic, structural, geophysical, geochemical, geochronological, and biogeographic data pertinent to Paleozoic and early Mesozoic paleogeographic relations between the Klamath Mountains, Sierra Nevada, and western North America. Subsequently, in May 1989, a theme session on the same topic was held at the combined annual meetings of the Cordilleran and Rocky Mountain Sections of the Geological Society of America. These meetings provided the stimulus for this volume.

We thank all of the contributors for their cooperation. In addition, we are particularly grateful to the reviewers for their time and efforts to improve this volume.

David S. Harwood
M. Meghan Miller

Geological Society of America
Special Paper 255
1990

Tectonic development of Cordilleran mid-Paleozoic volcano-plutonic complexes; Evidence for convergent margin tectonism

Charles M. Rubin
Division of Geological and Planetary Sciences, California Institute of Technology, Pasadena, California 91125
M. Meghan Miller
Jet Propulsion Laboratory, California Institute of Technology, 183-501, 4800 Oak Grove Drive, Pasadena, California 91109
George M. Smith
1027 Dellwood Avenue, Boulder, Colorado 80302

ABSTRACT

Devonian and Lower Mississippian volcano-plutonic complexes and related basinal strata that extend from California to British Columbia lie outboard of coeval North American shelf rocks. Remarkable similarities between these complexes indicate that they form the remnants of a mid-Paleozoic east Pacific fringing arc system. Arc basement was composite. It is characterized in part by high initial Sr ratios, radiogenic Pb, and $-\epsilon_{Nd}$ values, and in part by low initial Sr ratios, various initial Pb ratios, and $+\epsilon_{Nd}$ values. These distinctive isotopic signatures, together with the presence of Proterozoic average ages for detrital zircon in ultimately continent-derived sedimentary rocks and Proterozoic upper-intercept (inheritance) ages on zircon from arc- and rift-related magmatic rocks, imply that basement was composed of both continental crust and transitional or oceanic crust overlain by continent-derived sedimentary rocks. These data suggest that the arc system formed adjacent to a continental landmass. The initial geometry of the arc has been highly modified by subsequent crustal thickening, extension, associated uplift, and transport along strike-slip faults.

INTRODUCTION

A belt of Upper Devonian and Lower Mississippian volcanic and basinal rocks and coeval plutons extends for more than 5,000 km along the length of the North American Cordillera (Fig. 1) and lies outboard of contemporaneous shelf and platform strata. Regional lithologic and structural ties between these mid-Paleozoic volcanic and basinal assemblages and western North America have long been recognized (e.g., Burchfiel and Davis, 1972; Boucot and others, 1974; Monger and others, 1972), although some workers disagree. Recent isotopic and geochemical studies and geologic mapping, however, have substantiated and added to these paleogeographic constraints. A growing body of evidence, including new geochronologic, isotopic, and geochemical data, together with stratigraphic and structural studies, indicates that subduction was active west of the North American continental margin during mid-Paleozoic time. The arc remnants were originally underlain by a composite basement, consisting of attenuated continental crust, transitional or oceanic crust, and possibly rifted elements of North American continental crust. This chapter describes mid-Paleozoic volcanic and basinal strata, plutonic rocks, and their basement from several localities from northern California to northern Alaska (Fig. 1). We present a synthesis of geologic, isotopic, and geochemical data for the mid-

Rubin, C. M., Miller, M. M., and Smith, G. M., 1990, Tectonic development of Cordilleran mid-Paleozoic volcano-plutonic complexes; Evidence for convergent margin tectonism, *in* Harwood, D. S., and Miller, M. M., eds., Paleozoic and early Mesozoic paleogeographic relations; Sierra Nevada, Klamath Mountains, and related terranes: Boulder, Colorado, Geological Society of America Special Paper 255.

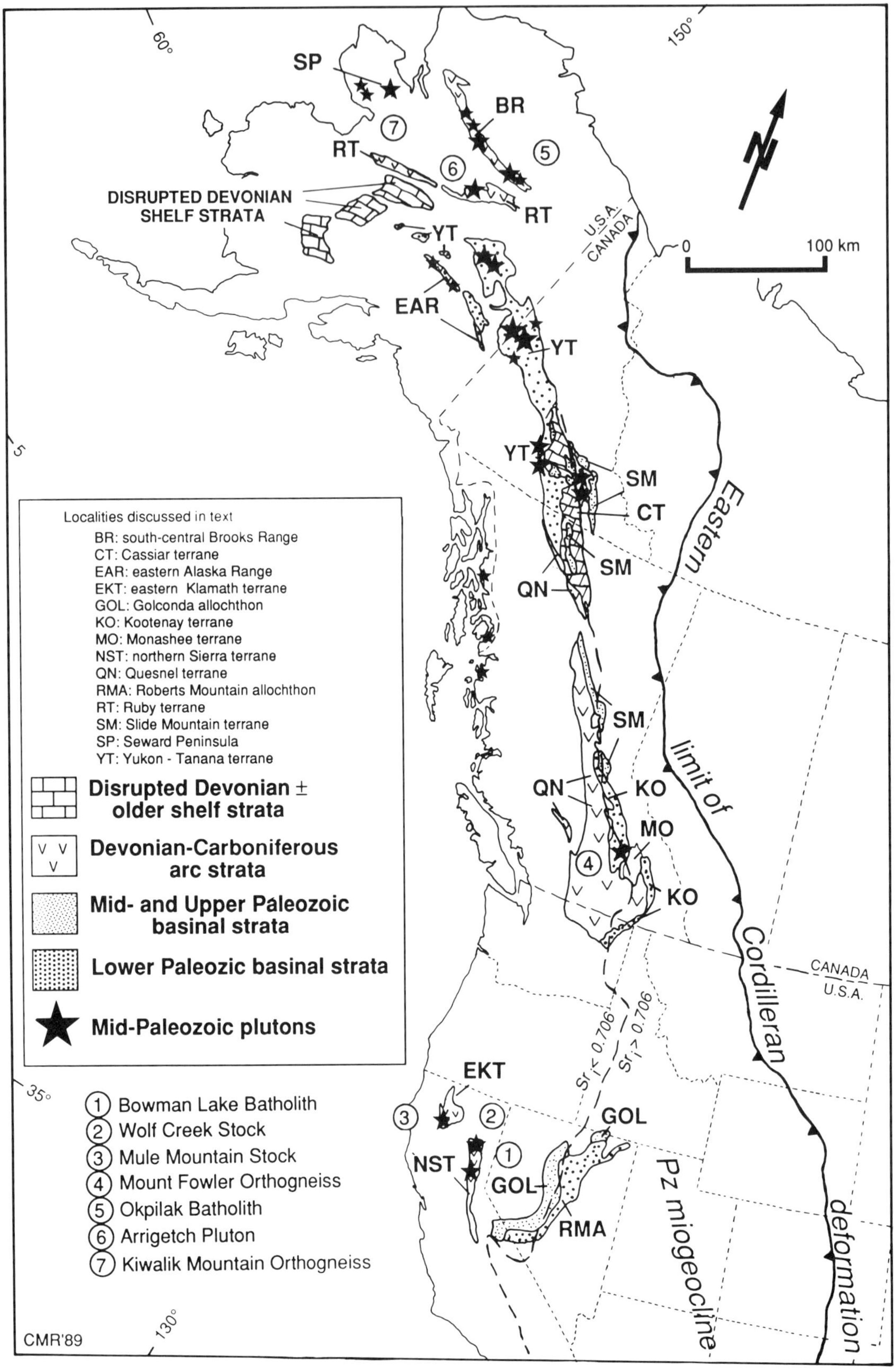

Figure 1. Generalized distribution of mid-Paleozoic volcanic and basinal rocks and coeval plutonic rocks in the western North American Cordillera. Modified from Monger and Berg (1987), Miller and others (1990), Silberling and others (1984), and Wheeler and others (1988).

Paleozoic volcano-plutonic belt, suggesting that disparate terranes have a common origin, and also that these terranes evolved adjacent to western North America.

GEOLOGIC FRAMEWORK

Mid-Paleozoic volcano-plutonic complexes and related basinal strata are recognized in many areas in the western Cordillera. These include the northern Sierra terrane, eastern Klamath terrane, Roberts Mountains and Golconda allochthons, Kootenay, Slide Mountain, Yukon-Tanana and Cassiar terranes, eastern Alaska Range, Ruby terrane, and the south-central Brooks Range (Fig. 1). Each of these areas contains several of four important lithologic components: (1) a sedimentary or metasedimentary basement complex indicative of a continent margin or slope and rise setting, (2) Devonian and Mississippian volcaniclastic strata and lava flows, with (3) intercalated basinal and epiclastic strata, and (4) intrusive rocks that are coeval with volcanic units. In addition to these lithologic characteristics, most of the areas contain evidence for mid-Devonian to earliest Mississippian tectonism. Deformation appears to have been diachronous throughout the belt, but in many areas it is merely poorly dated. The areas are discussed herein from south to north, and each lithologic and structural element, where present, is described.

Northern Sierra terrane

The northern Sierra terrane (Fig. 1) provides one of the most coherent records of mid-Paleozoic events in the western Cordillera. There, the lower Paleozoic Shoo Fly Complex forms depositional basement to the middle and upper Paleozoic volcanic, volcaniclastic, and basinal strata of the Taylorsville sequence (Fig. 2; Harwood, 1983; Girty and Schweickert, 1984). Deep-water siliciclastic sedimentary rocks, chert, carbonate, basaltic lavas, submarine tuff, and melange are the dominant lithologies within the Shoo Fly Complex and belong to four imbricate thrust plates: the Lang sequence, the Duncan Peak allochthon, the Culbertson Lake allochthon, and the Sierra City melange (Fig. 2; Schweickert and others, 1984). The Shoo Fly Complex is only sparsely fossiliferous and therefore poorly dated. Age constraints include Late Ordovician conodonts and megafossils from limestone blocks in the Sierra City melange (Harwood, 1988), late Middle to Late Ordovician conodonts from limestone interbedded in the Lang sequence (Harwood and others, 1988), and a Pb/U zircon age of 423 ± 8 Ma (Silurian) from a silicic aquagene tuff in the melange matrix (Saleeby and others, 1987). The four structural units were imbricated in a south-vergent thrust belt (Girty and others, this volume) and beveled by an angular unconformity prior to deposition of Upper Devonian (Frasnian-Famennian) rocks of the overlying Taylorsville sequence.

Three lines of evidence imply that the Shoo Fly Complex developed adjacent to the continent margin. Late Ordovician conodonts from blocks in the Sierra City melange have affinities to North American mid-continent faunas (A. Harris, *in* Hannah and Moores, 1986; Harwood, 1988). Petrographic studies on quartz arenite from the Lang sequence indicate a mature, monocrystalline-quartz–rich source, implying derivation from a continent (Bond and Devay, 1980). Detrital zircon from mature quartzose sandstone in the Culbertson Lake allochthon yields a Pb/U upper intercept age of 2.09 Ga, implying a Precambrian continental source for some terrigenous rocks (Girty and Wardlaw, 1984).

The lower part of the Taylorsville sequence unconformably overlies the imbricated Shoo Fly Complex. Upper Devonian (Frasnian-Famennian) and Lower Mississippian (Kinderhookian) volcanogenic rocks of the Taylorsville sequence consist of the Grizzly, Sierra Buttes, and Taylor Formations, and the lower member of the Peale Formation (Harwood, 1988). Areally restricted, coarse clastic rocks of the Grizzly Formation were derived from the Shoo Fly Complex, and are interbedded with quartzo-feldspathic sandstone. Volcanic strata within the overlying, regionally extensive Sierra Buttes Formation are predominantly rhyolitic to dacitic in composition, consisting of lapilli tuff, tuff breccia, and tuffaceous turbidites with interbedded phosphatic chert (Hanson and Schweickert, 1986). Intermediate and mafic submarine volcanic and volcaniclastic rocks of the Taylor Formation conformably overlie the Sierra Buttes Formation, and consist of augite-phyric basaltic to andesitic flows, breccia, and volcaniclastic turbidites. Continued volcanism is recorded by the lower part of the overlying Peale Formation, where massive latite flows, tuff, breccia, and tuffaceous epiclastic strata are present (Harwood, 1988). Epizonal trondhjemitic plutons (Figs. 1, 2; Bowman Lake batholith and Wolf Creek stock) represent the subvolcanic feeders to the Devonian arc strata, and yield Pb/U zircon ages of 375 ± 10 and 378 ± 5 Ma, respectively (Figs. 2, 3; Hanson and others, 1988; Saleeby and others, 1987). Complicated inheritance and disturbance patterns in zircon from the Bowman Lake batholith prevent exact resolution of the age of inheritance (Hanson and others, 1988). The volcanic strata that are dated by conodonts and ammonites and coeval plutonic rocks collectively record the initiation of Late Devonian arc magmatism in the northern Sierra terrane that continued into the Early Mississippian (Hanson and Schweickert, 1986; Hanson and others, 1988).

Stratigraphic, isotopic, and geochemical evidence point to a paleogeographic relation between the volcano-plutonic arc rocks and a continental landmass. The Late Devonian and Early Mississippian magmatic arc was constructed on the imbricated Shoo Fly Complex, which contains continental slope and rise deposits. Negative ϵ_{Nd} values for the Sierra Buttes Formation, together with light rare earth element (LREE) enrichment and strong negative Eu anomalies from both the Sierra Buttes Formation and Bowman Lake batholith, imply a contribution of continental crust to the evolving magmas (Rouer and others, 1989). High initial $^{87}Sr/^{86}Sr$ ratios from Sierra Buttes extrusive rocks (0.7088) appear to reflect interaction with sea water, whereas sills with intermediate values (0.7041) may reflect initial magmatic

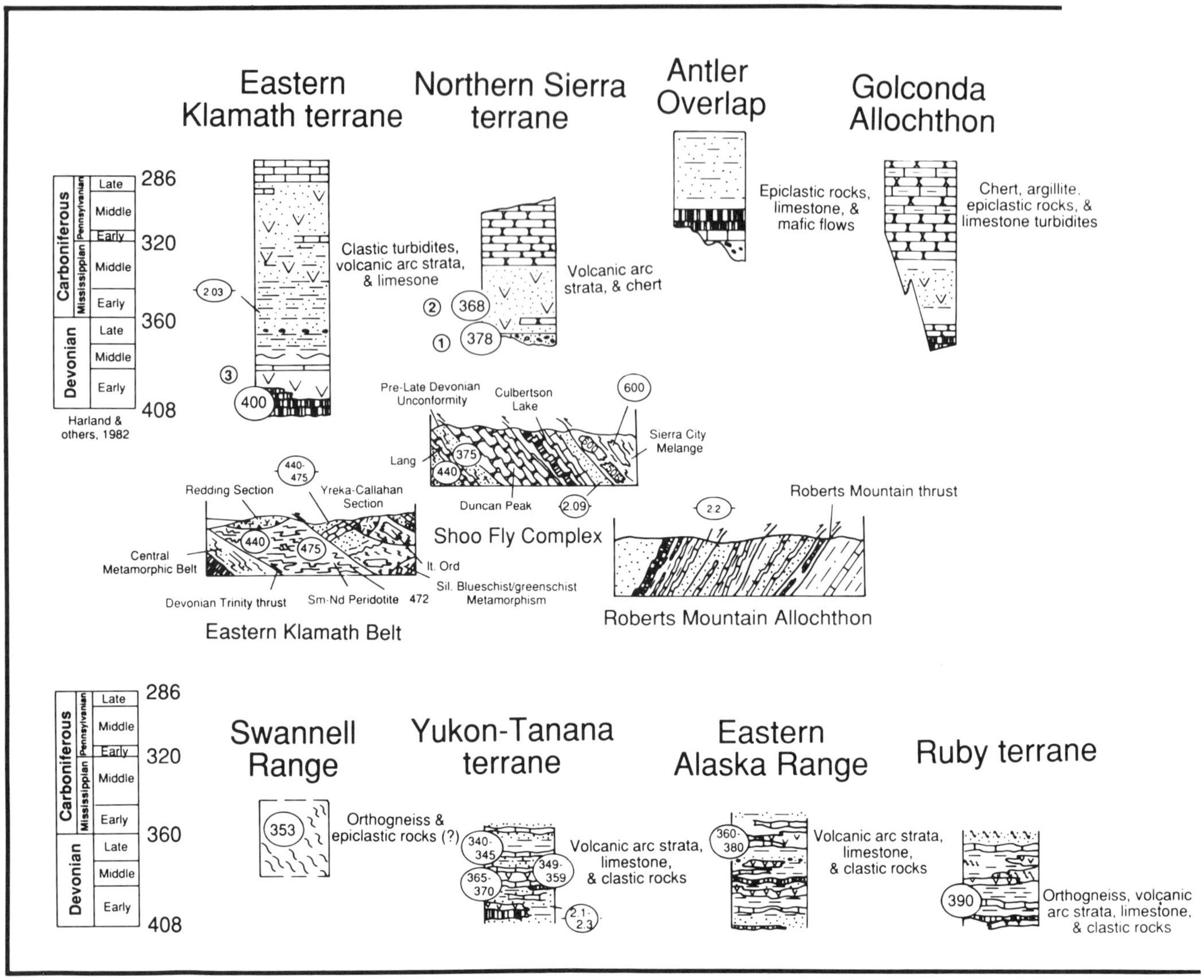

Figure 2. Correlation chart for mid-Paleozoic volcanic and basinal assemblages of the western North American Cordillera. Numbered circles indicate specific igneous bodies referred to in Figure 1 and text. References used in constructing this chart include Albers and Robertson (1961), Albers and others (1981), Aleinikoff (1984), Aleinikoff and others (1986), Dusel-Bacon and Aleinikoff (1985), Danner (1977), Dillon and others (1980), Dillon and Tilton (1985), Dillon and Bakke (1986), Gabrielse and others (1982), Girty and Wardlaw (1984, 1985), Girty and others (1985), Hanson and others (1988), Harms (1985), Harwood, (1983, 1988), Hotz (1977), Jacobsen and others (1984), Klepacki and Wheeler (1985), Little (1987), Madrid (1987), Miller and others (1984), Miller (1987, 1989), Miller and others (1988), Monger (1977, 1984), Mortensen and Jilson (1985), Nokleberg and others (1985), Okulitch and others (1975), Okulitch (1985), Patton and others (1987), Potter and others (1977), Quick (1981), Saleeby and others (1987), Schweickert and others (1984), Smith and others (1978), Speed (1979), Struik and Orchard (1985), Till (1983), and Wallin and others (1988).

values (Rouer and others, 1989), suggesting that the underlying crust was not highly evolved. The presence of phosphatic chert indicates deposition in a zone of upwelling adjacent to a continent, on the eastern margin of an open ocean basin (Varga, 1982). In summary, lower Paleozoic continent-margin–related basinal rocks were imbricated prior to Late Devonian time (mid-Famennian), forming basement for the Upper Devonian–lowermost Mississippian volcano-plutonic complex.

Eastern Klamath terrane

The eastern Klamath belt consists of three lithotectonic assemblages: (1) the structurally lowest and westernmost Ordovician Trinity complex, (2) the lower Paleozoic sedimentary and volcanic rocks of the Yreka terrane to the north, and (3) the Devonian to Middle Jurassic arc-related volcanogenic and sedimentary rocks of the eastern Klamath terrane (Redding section)

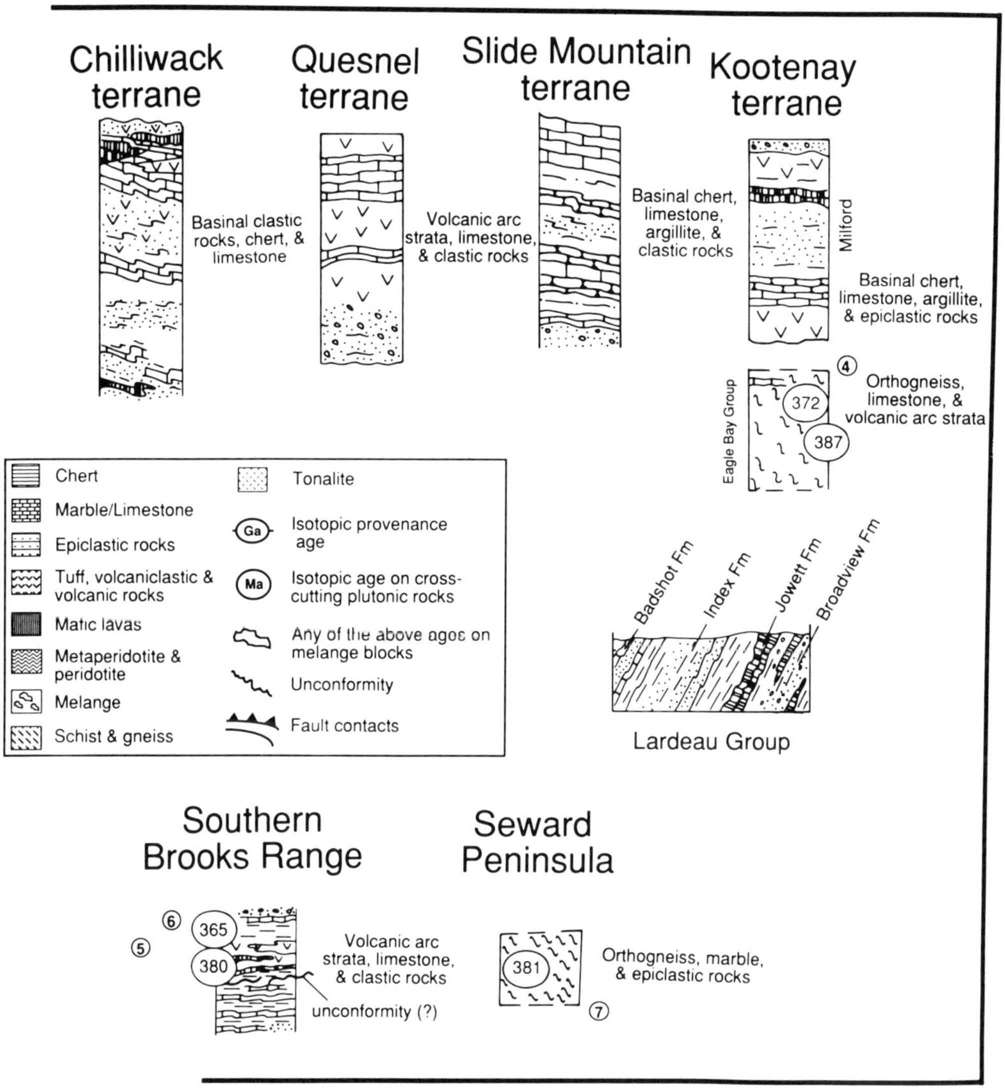

to the south (Fig. 2). Mid-Paleozoic volcanic arc strata are well known from the basal part of the Redding section, and possible correlative units are present in the Yreka terrane.

Complex petrologic and isotopic data from many parts of the Trinity complex suggest that it is a polygenetic body that is predominantly Ordovician in age, although younger (Silurian) mafic rocks crosscut the complex (Fig. 2; see review in Saleeby, 1990). Petrologic and isotopic data from the Trinity peridotite are not similar to those of typical ophiolites, although positive ϵ_{Nd} values imply evolution in the oceanic realm (Quick, 1981; Jacobsen and others, 1984). The ultramafic rocks are thought to represent a mantle diapir that formed beneath a volcanic highland (Quick, 1981) or at an attenuated continental margin (Boudier and others, 1989); although some workers have interpreted both the mafic and ultramafic rocks of the Trinity complex as cogenetic parts of an ophiolite (e.g., Brouxel and Lapierre, 1988).

The Trinity complex is overlain to the north by the Yreka terrane, which contains lower Paleozoic low-grade metasedimentary and metavolcanic rocks. Important components include volcaniclastic and quartz-rich sandstone, argillite, and limestone. Detrital zircons from various units in the Yreka terrane indicate an early Paleozoic crystalline (first cycle) source, as well as Proterozoic and Archean multicycle zircon populations (Wallin, 1989). Lithologic correlatives between the Yreka terrane and Shoo Fly Complex include monocrystalline-quartz–rich sedimentary rocks, volcaniclastic rocks derived from an Ordovician-Silurian source terrane, sedimentary matrix melange, and biostratigraphically similar Ordovician limestone (Bond and Devay, 1980; Wallin, 1989; Potter and others, this volume).

The eastern Klamath terrane overlies the Trinity complex to the south. The oldest rocks in the eastern Klamath terrane are Middle Devonian or older (Fig. 2), and thus predate arc-related sequences described elsewhere in this chapter. These rocks represent an older and paleotectonically distinct phase of magmatism, and thus—together with the Trinity complex—form depositional basement to Upper Devonian and Lower Carbonif-

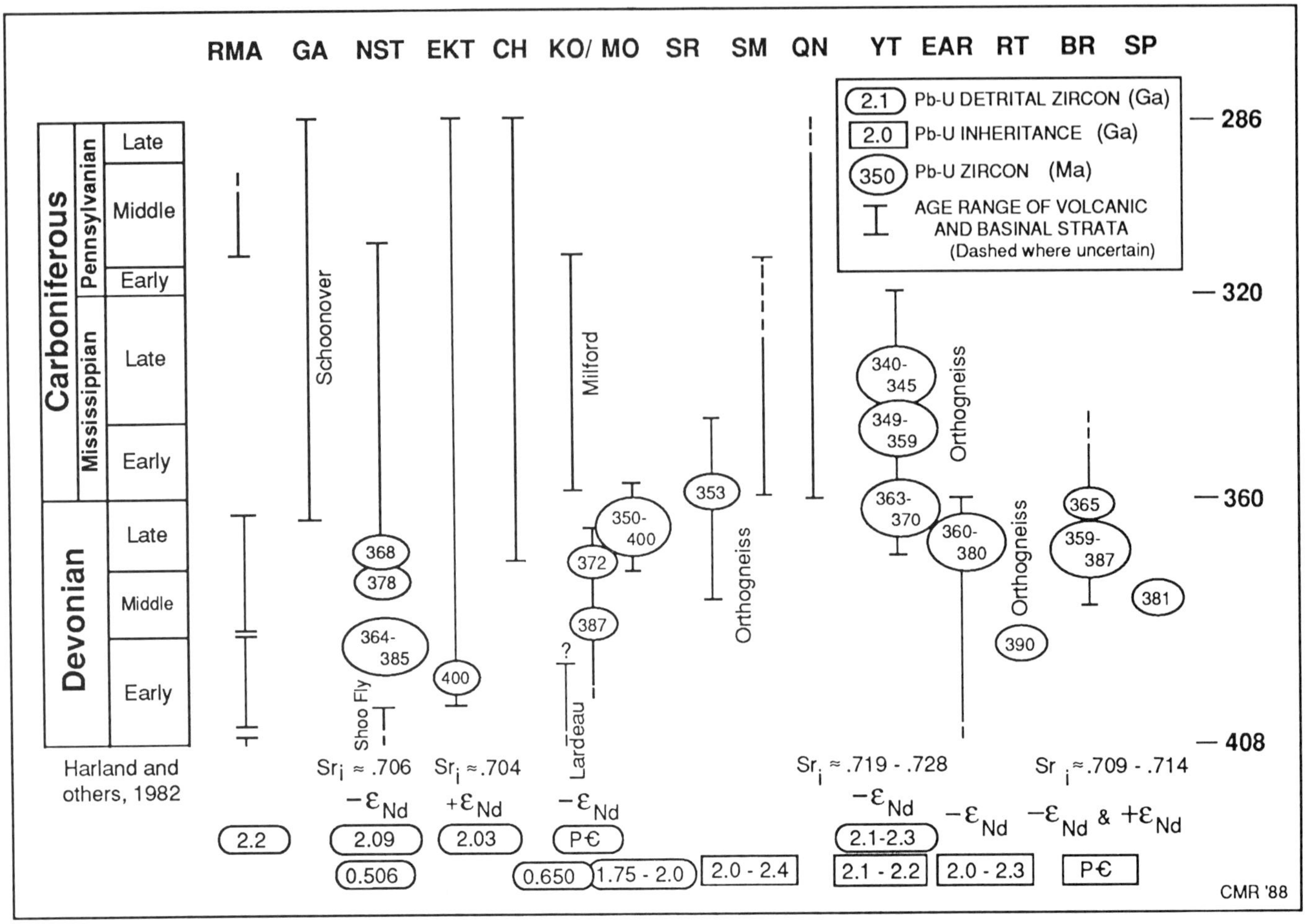

Figure 3. Summary of mid-Paleozoic zircon isotopic ages, detrital zircon average upper-intercept ages, zircon inheritance ages, initial Sr ratios, and ϵ_{Nd} values in the western North American Cordillera. RMA: Roberts Mountains allochthon; GA: Golconda allochthon; NST: northern Sierra terrane; EKT: eastern Klamath terrane; CH: Chilliwack; KO/MO: Kootenay-Monashee terranes; SR: Swannell Ranges; SM; Slide Mountain terrane; QN: Quesnel terrane; YT: Yukon-Tanana terrane; EAR: eastern Alaska Range; BR: Brooks Range; RT: Ruby terrane; SP: Seward Peninsula. References used on constructing this summary include Aleinikoff and others (1981), Aleinikoff (1984), Aleinikoff and others (1986), Dusel-Bacon and Aleinikoff (1985), Dillon and others (1980), Dillon and Bakke (1987), Gabrielse and others (1982), Girty and Wardlaw (1984, 1985), Girty and others (1985), Hanson and others (1988), Harwood (1988), Madrid (1987), McCulloch and Wasserburg (1978), Miller and others (1984), Miller (1987), Miller and others (1988), Monger (1977), Mortensen (1988), Nokleberg and others (1985), Nelson and others (1989), Okulitch and others (1975), Okulitch (1985), Parrish and others (1989), Patton and others (1987), and Saleeby and others (1987).

erous arc-related and basinal strata. The Copley Greenstone consists predominantly of basaltic flows and mafic pyroclastic debris (Lapierre and others, 1985a). The interfingering and overlying Balaklala "Rhyolite" is characterized by silicic aquagene tuff and flows that are associated with Kuroko-type massive sulfides, recording primitive volcanic-arc magmatism in an extensional setting (Lapierre and others, 1985b). Crystal-rich tuff in the Balaklala has yielded a Middle Devonian (Eifelian) fish plate (Boucot and others, 1974). LREE depletion and high ϵ_{Nd} values imply a depleted mantle source (Rouer and others, 1989). The trondhjemitic Mule Mountain stock (Fig. 1), which intrudes the Copley Greenstone and the basal member of the Balaklala Rhyolite, is thought to be cogenetic with the Balaklala Rhyolite (Barker and others, 1979). The Mule Mountain stock (Figs. 1, 2) yields a date of 400 Ma (Pb/U zircon, Albers and others, 1981; Fig. 2), although stratigraphic relations suggest that it may be slightly younger. These volcanic and plutonic rocks are overlain by basinal volcaniclastic strata and carbonate rocks of the Middle Devonian (Eifelian) Kennett Formation (Savage, 1976), recording the persistence of magmatism and onset of subsidence during early Middle Devonian time (Watkins and Flory, 1986).

The Middle Devonian and older(?) rocks are overlain by epiclastic and volcaniclastic strata of the Upper Devonian(?) and Lower Carboniferous Bragdon Formation, which is in part age correlative with arc and rift sequences discussed elsewhere in this chapter. The nature of the basal contact is disputed (see discussion in Miller and Harwood, this volume). The Bragdon Formation may overlap preexisting bathymetry of the older, subsided arc edifice with local conformity (Watkins and Flory, 1986), or it may unconformably overlie the imbricated Middle Devonian arc succession, Trinity peridotite, and central metamorphic belt to the west (Boucot and others, 1974; Charvet and others, 1989). The Bragdon Formation consists of a thick succession of turbidites and related channel-fill sequences (Watkins and Flory, 1986; Miller and Cui, 1989), with chert- and sedimentary-lithic–rich clastic rocks and penecontemporaneous, water-laid pyroclastic rocks (Miller and Cui, 1989). Pyroclastic rocks, which are lithologically similar to the Sierra Buttes Formation in the northern Sierra terrane, record the persistence of volcanism (Miller and Cui, 1989). Isotopic studies on detrital zircon from epiclastic rocks of the Bragdon Formation yield upper-intercept ages of ~2.03 Ga, and thus indicate an ultimate Precambrian source for some of the Bragdon detritus (Miller and others, 1988; Miller and Saleeby, 1990). They were derived from rocks lithologically and isotopically like those of the Shoo Fly Complex or Yreka terrane. The only direct age constraints on the relatively thick Bragdon Formation come from Late Mississippian fossils near the top of the formation (Watkins, 1986). If the basal contact is conformable and gradational, as suggested by Watkins (1986), the Bragdon Formation may span the late Middle Devonian to Early Carboniferous time. Alternatively, if the contact is an angular unconformity that onlaps the Trinity peridotite (Charvet and others, 1989), post–early Middle Devonian and pre–Early Carboniferous tectonism is implied, as originally suggested by Boucot and others (1974). No direct age constraints on the base of the formation exist.

In summary, although the initiation of mid-Paleozoic arc magmatism in the Klamath Mountains significantly preceded that in the northern Sierra terrane (Middle Devonian); contemporaneous (Late Devonian and Early Mississippian) magmatism is recorded by pyroclastic rocks in the Bragdon Formation. Some workers infer that the Middle Devonian and older arc complex and the Trinity complex, together with other older terranes in the Klamath Mountains, were imbricated in post–early Middle Devonian time, and were overlain by reworked sedimentary debris of the Bragdon Formation along an angular unconformity. Although the Middle Devonian volcanic and intrusive rocks were derived from a depleted mantle source, the presence of recycled quartzose debris and Precambrian zircon in the Bragdon Formation indicates an indirect tie to a continental source, and provenance ties to the Shoo Fly Complex or Yreka terrane.

Roberts Mountains allochthon

Middle Cambrian to Lower Mississippian strata of the Roberts Mountains allochthon consist of siliceous basinal and volcanic rocks that are exposed in northeastern Nevada (Fig. 1; Poole and Sandberg, 1977; Johnson and Pendergast, 1981; Poole and Claypoole, 1984). Depositional basement to the allochthon is unknown. Rocks in the allochthon include rhythmically bedded radiolarian chert, quartzose and feldspathic sandstone, siltstone, greenstone, tuff, limestone, and bedded barite (Madrid, 1987), which were deposited in a basinal continental slope-and-rise setting. Mafic volcanic rocks are predominantly alkalic in composition (Madrid, 1987), and volcanogenic debris flows interfinger with siliciclastic strata. Based on high Ti contents, LREE-enriched and HREE-depleted patterns, low Hf/Ta and Zr/Nb and high Nb/Y ratios, the mafic lavas were erupted within a continental intraplate setting, such as a rifted outer continental margin (Madrid, 1987). These alkalic mafic rocks are similar in composition to Devonian volcanic rocks in many continental margin sequences exposed in the northern Canadian Cordillera (e.g., Cecile, 1982) discussed below. Stratigraphic and petrologic studies indicate that some clastic debris within the Roberts Mountains allochthon was derived from an eastern continental source (Madrid, 1987). In contrast, feldspathic sandstones were probably derived from a western source, an interpretation corroborated by paleocurrent indicators (Madrid, 1987). Rocks within the Roberts Mountains allochthon were deformed and thrust over miogeoclinal rocks during Early Mississippian time (Nilson and Stewart, 1980) and were unconformably overlain by Upper Mississippian to Permian shelf deposits (Roberts and others, 1958; Little, 1987).

Stratigraphic, sedimentologic, and isotopic evidence (Fig. 3) suggest that basinal strata of the Roberts Mountains, allochthon were deposited adjacent to a continental landmass. Isotopic studies from quartz-rich sandstones yield a Pb/U detrital zircon upper-intercept age of 2.2 Ga (Girty and others, 1985), implying a continental source.

Golconda allochthon

The Golconda allochthon consists of middle and upper Paleozoic basinal rocks of the Havallah and Schoonover sequences (Miller and others, 1984; Murchey, this volume; Whiteford, this volume) that structurally overlie the Roberts Mountains allochthon (Fig. 1). Upper Devonian and Lower Mississippian rocks are present only in the Schoonover sequence and consist of ribbon chert, argillite, mafic to andesite volcaniclastic and volcanic rocks, lithic-rich turbidites, and debris flows (Fig. 2; Miller and others, 1984; Whiteford, this volume). Basinal deposition began in the latest Devonian (mid-Famennian), with the extrusion of basalt and accumulation of pelagic sediment, and was coeval with deposition of the youngest rocks in the Roberts Mountains allochthon (e.g., Poole and Sandberg, 1976; Wrucke and Jones, 1978; Johnson and Pendergast, 1981; Poole and Claypoole, 1984). Sedimentary-lithic–rich turbidites derived from the Roberts Mountains allochthon overlie the chert and volcaniclastic deposits (Miller and others, 1984). Turbidite deposition and volcanism continued into the Meramecian, which spans the time of emplacement of the Roberts Mountains allochthon (Miller and others, 1984).

Kootenay terrane

The Kootenay terrane contains deformed and variably metamorphosed sedimentary and volcanic rocks that extend from east-central Washington to east-central Alaska (Fig. 1; Wheeler and McFeely, 1987; Wheeler and others, 1988). In south-central British Columbia, the Kootenay terrane consists of four lithotectonic packages: (1) upper Proterozoic to lower Paleozoic miogeoclinal strata; (2) lower Paleozoic clastic and volcanic rocks of the Lardeau Group; (3) mid-Paleozoic volcanic, volcaniclastic, and associated plutonic rocks that form parts of the Eagle Bay assemblage and Fennell Formation; and (4) upper Paleozoic clastic and volcanic rocks of the Milford Group (Fig. 2). The Fennell Formation also contains upper Paleozoic rocks, which are excluded from this discussion.

The lower Paleozoic Lardeau Group forms depositional basement to the upper Paleozoic volcanic, volcaniclastic, and basinal strata of the Milford Group, as well as parts of the Eagle Bay assemblage and perhaps older portions of the Fennel Formation (Klepacki and Wheeler, 1985; Schiarizza, 1986; Schiarizza and Preto, 1987). The Lardeau Group consists of calcareous pelitic schist, phyllite, quartzite, coarse clastic rocks, mafic flows, and minor marble (Fig. 2; Klepacki and Wheeler, 1985; Read and Wheeler, 1976). Based on stratigraphic and lithologic similarities, the Covada Group, exposed in northeastern Washington, may be correlative with the Lardeau Group (Snook and others, 1981). The Covada Group consists of phyllite, coarse clastic rocks, metabasalt, and marble. The Lardeau Group is not fossiliferous; age constraints come from stratigraphic relations with adjacent units. The Lardeau Group overlies Lower Cambrian miogeoclinal strata and is unconformably overlain by Upper Mississsippian rocks of the Milford Group (Ross, 1970; Read, 1975; Klepacki and Wheeler, 1985). Early Ordovician conodonts have been reported from the correlative Covada Group, which is consistent with an early Paleozoic age for the Lardeau sedimentary rocks. Thus, the Lardeau Group is considered to be post–Early Cambrian and pre–Late Mississippian in age.

The Lardeau Group, and perhaps the correlative Covada Group, were deformed and structurally imbricated prior to deposition of Upper Mississippian rocks of the Milford Group (Klepacki and Wheeler, 1985; Gehrels and Smith, 1987; Read, 1975; Ross, 1970). Rocks in both groups were emplaced over miogeoclinal strata in the mid-Paleozoic (Smith and Gehrels, 1989). These structural and stratigraphic relations are analogous to those seen to the south in the Roberts Mountains allochthon (Fig. 2). Early Proterozoic intercept ages on detrital zircon from quartzite in the Lardeau Group imply a Precambrian continental source for some of the terrigenous metasedimentary rocks that form basement to the mid-Paleozoic volcanic and basinal strata (Smith and Gehrels, 1989).

Mid- and upper Paleozoic arc and basinal sequences of the Kootenay terrane unconformably overlie the deformed Lardeau Group. The arc or basinal strata include the Eagle Bay assemblage, parts of the Fennell Formation, the Davis and McHardy assemblages, and the lower part of the Keen Creek assemblage of the Milford Group (Klepacki and Wheeler, 1985). The Eagle Bay assemblage consists of mafic and felsic metavolcanic strata, clastic metasedimentary rocks, and marble (Fig. 2; V. Preto, *in* Brown and others, 1981; Schiarizza, 1986; Schiarizza and Preto, 1987). Felsic to intermediate volcanic and cogenetic intrusive rocks of the Eagle Bay assemblage collectively record arc magmatism (Toy, 1987; Schiarizza and Preto, 1987). High-grade metavolcanic and metasedimentary rocks of the Eagle Bay Group are intruded by the Mount Fowler orthogneiss (Fig. 1), which yields a 372 ± 6 Ma Pb/U zircon age (Fig. 2; Okulitch and others, 1975; Okulitch, 1985). Felsic metavolcanic rocks yield Pb/U zircon ages of 367 to 379 Ma (V. Preto, *in* Brown and others, 1981), indicating a Late Devonian age for both the felsic metavolcanic strata and the orthogneiss. These isotopic ages are similar to an upper-intercept crystallization age of 387 ± 4 Ma for felsic metavolcanic rocks of the Eagle Bay assemblage (Schiarizza and Preto, 1987), and most likely represent the protolith age for both the Eagle Bay and Fennell metavolcanic suites. Locally, a quartz-feldspar porphyry, which intrudes metabasalt that is intercalated with chert in the lower part of the Fennell Formation, yields a Pb/U zircon age of 370 to 380 Ma (V. Preto, *in* Brown and others, 1981), which suggests that parts of the Fennell Formation are probably age equivalent to the Eagle Bay assemblage. Based on trace-element geochemistry, the Fennell mafic lavas were erupted in an oceanic environment (Schiarizza and Preto, 1987).

Upper Paleozoic basinal sequences of the Milford Group consist of the Upper Mississippian Davis assemblage and the Upper Mississippian to Lower Pennsylvanian Keen Creek and McHardy assemblages (Klepacki and Wheeler, 1985). The Davis assemblage is dominated by clastic sedimentary rocks, chert, limestone, and mafic flows. It unconformably overlies the lower Paleozoic Lardeau Group (Fig. 2; Klepacki and Wheeler, 1985). The Keen Creek assemblage consists of predominantly tholeiitic basalt and limestone and rests unconformably over the Lardeau Group. Preliminary geochemical analyses suggest a rift-related setting for the volcanic rocks (Klepacki, 1985). The basinal rocks are age equivalent or slightly younger than the Eagle Bay and Fennell metavolcanic suite and are interpreted by Klepacki (1985) to have been deposited within an extensional setting based on rapid facies changes, syndepositional normal faulting, inferred basin geometry, and high Ti/Fe ratios in the metabasaltic lavas. The Upper Mississippian part of the McHardy assemblage consists of clastic sedimentary rocks, limestone, and minor volcanic rocks (Klepacki and Wheeler, 1985) and structurally overlies either the Davis or Keen Creek assemblages (Klepacki, 1985). Preliminary geochemical analyses indicate the McHardy assemblage was deposited on oceanic or transitional crust in an extensional setting (Klepacki, 1985) and is similar in composition and tectonic setting to other mid-Paleozoic rift-related volcanic rocks in outer continental margin sequences exposed in the Canadian Cordillera (Fig. 2).

Slide Mountain terrane

Mississippian and older(?) strata in the Slide Mountain terrane consist of clastic sedimentary rocks, chert, limestone, and volcanic rocks (Monger, 1977; Nelson and Bradford, 1989) that structurally overlie outer continental margin facies of North America (Nelson and Bradford, 1987). Sedimentary rocks include argillite, chert, limestone, sandstone, and conglomerate (Monger, 1977; Nelson and Bradford, 1989). The metavolcanic succession consists of mafic flows, breccia, and tuff (Fig. 2; Monger, 1977), with a basaltic protolith for the mafic lavas indicated by preliminary whole-rock geochemical analyses (Nelson and others, 1988). The mafic flows probably formed in an extensional environment, similar to the tectonic setting in other extension-related mid-Paleozoic basinal sequences in the northern Cordillera (Gordey and others, 1987; Turner and others, 1989). Extension may have occurred in an arc setting (Nelson and others, 1988).

Yukon-Tanana terrane: Yukon Territory and east-central Alaska

The Yukon-Tanana terrane (Fig. 1) consists of three lithotectonic assemblages: the Nisutlin, Anvil, and Simpson allochthons (Tempelman-Kluit, 1979) that structurally overlie Proterozoic(?) to lower Paleozoic rocks of the Nisling terrane (Wheeler and McFeely, 1987). Lower Paleozoic rocks of the Nasina and Nisling assemblages consist of carbonaceous and siliceous argillite, quartzite, and marble (Fig. 2), which are interpreted as continental shelf, slope, and rise deposits and are now assigned to the Nisling terrane (Wheeler and McFeely, 1987; Wheeler and others, 1988). The Nisling terrane likely formed basement to parts of the Kootenay and Yukon-Tanana terranes.

Mid-Paleozoic arc and basinal sequences are well represented in the Yukon-Tanana terrane (Fig. 2; Monger and Berg, 1987). A structural sequence of layered metavolcanic, metasedimentary, and metaplutonic rocks makes up the mid-Paleozoic part of the Yukon-Tanana terrane (Mortensen and Jilson, 1985). Metasedimentary rocks include quartzofeldspathic schist, pelitic and siliceous schist, marble, and quartzite (Fig. 2; Dusel-Bacon and Aleinikoff, 1985; Mortensen and Jilson, 1985). Gneissic metaplutonic rocks are predominantly granitic to quartz dioritic in composition, and locally are composed of quartz monzonite (Dusel-Bacon and Aleinikoff, 1985; Mortensen, 1986; Mortensen and Jilson, 1985). Metaplutonic rocks, exposed in southeast Yukon Territory, yield Pb/U zircon ages of 340 to 359 Ma (Mortensen and Jilson, 1985; Mortensen, 1986). Older zircon ages of 363 to 350 Ma were reported by Mortensen (1988) for the metaplutonic and metavolcanic rocks in the Yukon Territory. Felsic metavolcanic rocks, exposed in east-central Alaska, yield discordant Pb/U zircon ages of 360 to 380 Ma; metaplutonic rocks yield a Pb/U zircon age of 343 ± 3 Ma with an upper intercept age of 2.16 ± 0.03 Ga (Aleinikoff and others, 1986). Sm-Nd isotopic data yield a 2.09 ± 0.08 model age, which is in agreement with the upper intercept Pb/U age of 2.16 Ga (Aleinikoff and others, 1981). These isotopic ages are similar to a Nd model age of 2.33 ± 0.05 Ga for the Birch Creek Schist (McCulloch and Wasserburg, 1978), which is likely equivalent with the Yukon-Tanana terrane. Highly radiogenic initial Pb isotopic ratios indicate an Early Proterozoic source for some of these metaigneous rocks (Aleinikoff and others, 1987). LREE enrichment, negative Eu anomalies, and low Ti and Cr contents, together with limited isotopic data, suggest a continental arc origin for the metavolcanic and metaplutonic rocks (Dusel-Bacon and Aleinikoff, 1985; Aleinikoff and others, 1987; Mortensen, 1988).

Mid-Paleozoic ages on zircon from metaplutonic rocks exposed in southeast Alaska and western British Columbia (Fig. 1) were reported by Gareau (1989), G. E. Gehrels (written communication), and Saleeby and Rubin (1989). These mid-Paleozoic metaplutonic rocks intruded quartzite turbidites, marble, metabasalt, and metapelite that are correlative with the Yukon-Tanana terrane. Proterozoic average upper-intercept ages from zircon in quartzite turbidites (Gehrels and others, 1989; C. M. Rubin and J. B. Saleeby, unpublished data) imply a Precambrian continental source for part of the metasedimentary sequence. Sm-Nd isotopic data yield model ages as old as 2.4 Ga from quartz-rich metasedimentary rocks (Samson and others, 1989), which are broadly in agreement with the upper intercept Pb/U age 2.17 Ga (Gehrels and others, 1989).

The initiation of mid-Paleozoic magmatism in the Yukon-Tanana terrane was coeval with magmatism in correlative terranes to the south. Major-element geochemistry indicates that the mid-Paleozoic metaplutonic rocks record continental arc magmatism. The magmatic event was accompanied by coeval basinal sedimentation. Isotopic, stratigraphic, and geochemical evidence suggest a paleogeographic relation between the volcanoplutonic complexes and a continental landmass. High initial $^{87}Sr/^{86}Sr$ ratios, which range between 0.709 to 0.728, and radiogenic Pb isotopic values (Aleinikoff and others, 1987; Mortensen, 1988) indicate an old crustal component in the mid-Paleozoic metaplutonic rocks, as also suggested by negative ϵ_{Nd} values (Aleinikoff and others, 1980; Bennett and Hansen, 1988; McCulloch and Wasserburg, 1978). Detrital zircon from quartzite yields a Pb/U upper-intercept age of 2.2 Ga (Aleinikoff and others, 1981), which implies a Precambrian continental source for parts of the supracrustal sequences. These data suggest that the magmatic arc was likely constructed on continental margin and slope-and-rise deposits now represented by the Nisling terrane. Moderate initial $^{87}Sr/^{86}Sr$ ratios of <0.706 and minor Proterozoic inheritance within the 370- to 350-Ma Simpson Range plutonic suite (Simpson allochthon of Tempelman-Kluit, 1979) suggests that the plutonic suite did interact with older continental crust (e.g., Nisling terrane), but not to the extent that coeval parts of the Yukon-Tanana terrane did (Mortensen, 1988).

Cassiar terrane

The record of Devonian magmatism is found east of the Yukon-Tanana terrane in outer miogeoclinal rocks of the Cassiar terrane (Fig. 1), where coeval metavolcanic rocks, consisting of lapilli tuff, breccia, flows, volcaniclastic debris, and tuffaceous chert, are present (Mortensen, 1982). The metavolcanic rocks are interbedded with shale, chert, and conglomerate. The mid-Paleozoic felsic metavolcanic rocks are enriched in Nb, Zr, and Y, and have low abundances of Ti and P (Mortensen, 1982), indicating metaluminous and peralkaline affinities and a probable origin in an extensional setting. Stratigraphic, structural, and geochemical evidence points to an extensional origin for the mid-Paleozoic volcanic and basinal rocks of the Cassiar terrane (Gordey and others, 1987). Extension along the outer miogeocline probably occurred in an overall convergent-margin setting.

Eastern Alaska Range

The Windy, Hayes Glacier, and adjacent smaller terranes (Fig. 1) exposed in the eastern Alaska Range represent slivers of the mid-Paleozoic arc offset along the Denali fault system (Aleinikoff and Nokleberg, 1985; Nokleberg and others, 1985). These terranes are correlative with the Kootenay terrane (Wheeler and McFeely, 1987). The Windy terrane is composed of metavolcanic and metasedimentary rocks (Fig. 2) that are variably metamorphosed and deformed. The metavolcanic rocks consist of metaandesite and metadacite tuff, and interlayered flows that are interstratified with marble, pelitic schist, quartzite, and argillite (Nokleberg and others, 1985). Based on fossil ages, rocks in the Windy terrane are Devonian (W. A. Oliver and A. G. Harris, in Nokleberg and others, 1985). The Hayes Glacier and adjacent terranes consist of schist, amphibolite, metaandesite and rhyolite, and metaplutonic rocks (Aleinikoff and Nokleberg, 1985; Nokleberg and others, 1985). Metaigneous rocks within these smaller terranes yield Pb/U zircon ages of 360 to 380 Ma, with upper-intercept ages of 2.0 to 2.3 Ga (Fig. 3; Aleinikoff, 1984). Highly radiogenic initial Pb isotopic ratios in the metaigneous rocks indicate a Precambrian contribution for much of the arc sequence (Aleinikoff and others, 1987). The Windy, Hayes Glacier, and adjacent smaller terranes represent fragments of a Devonian volcanic arc and its plutonic roots (Aleinikoff and others, 1987; Nokleberg and others, 1985). Devonian igneous and fossil ages and Proterozoic upper intercept (average inheritance) ages in the metaplutonic rocks suggest that the mid-Paleozoic arc interacted with Precambrian basement. Based on similar Pb/U ages and geologic relations, the older portions of the adjacent Yukon-Tanana terrane (Nisling terrane of Wheeler and others, 1988) may form basement to the mid-Paleozoic arc sequences in the eastern Alaska Range.

Brooks Range and adjacent areas

Mid-Paleozoic volcanic and basinal strata exposed in the southern Brooks Range (Fig. 1) comprise felsic and mafic metavolcanic rocks, siliceous and graphitic schist, pelitic schist, and marble (Fig. 2; Hitzman and others, 1986; Mayfield and Tailler, 1978; Nelson and Grybeck, 1980). The metavolcanic rocks are predominantly massive mafic schist, mafic flows, felsic tuff, and breccia. The age of the sequence is constrained by both fossil ages and isotopic ages. Late Devonian to Mississippian(?) corals from marble have been reported by Nelson and Grybeck (1980) and Smith and others (1978). Pb/U ages on zircon from felsic metavolcanic rocks yield highly discordant ages of 365 ± 15 Ma (Dillon and others, 1980). Coeval metaplutonic rocks yield discordant Pb/U zircon ages of 380 ± 10 Ma for the Okpilak batholith (Fig. 1; Dillon and Bakke, 1987; Dillon and others, 1987a), and 360 ± 15 Ma for the Arrigetch pluton (Fig. 1; Dillon and others, 1980). Limited Pb/U zircon ages of 359 to 387 Ma for the Arrigetch pluton have been reported by Patton and others (1987) and are consistent with previous Pb/U zircon studies. The mid-Paleozoic metavolcanic and metasedimentary strata and coeval plutons are locally unconformably overlain by Mississippian rocks (Dillon and others, 1987a).

Major-element geochemical analyses on the metavolcanic and metaplutonic rocks exposed in the south-central Brooks Range suggest a tholeiitic basalt protolith for the mafic schist and a calc-alkaline rhyolite protolith for the felsic metavolcanic rocks (Hitzman and others, 1986). Based on inferred protoliths, these rocks have been interpreted to have formed in a rift setting. The major-element data are suspect, however, due to the likely mobility of large ion lithophile elements during alteration and subsequent metamorphism. Thus, the paleotectonic setting for the mid-Paleozoic volcanoplutonic belt is ambiguous. These data may equally imply a volcanic arc setting, based on the presence of calc-alkaline felsic and tholeiitic mafic rocks and abundant coeval granitic plutons and stocks. Devonian plutonism may have extended to the east-central Sewart Peninsula, where the Kiwalik Mountain orthogneiss (Fig. 1) yields a Pb/U zircon age of 381 Ma (Fig. 2; J. Aleinikoff, *in* Till, 1983).

Basement to the Devonian igneous sequence is unknown; however, rocks of Proterozoic to early Paleozoic age have been recognized across the Brooks Range (Dumoulin and Harris, 1987; Mayfield and others, 1982; Dillon and Tilton, 1985; Dillon and others, 1987a; Moore, 1987; Nelson and others, 1989). Sm-Nd isotopic data yield a 1.6- to 2.0-Ga model age for basement rocks consisting of metapelite and quartzite (Nelson and others, 1989). This Proterozoic to lower Paleozoic sequence might form basement to parts of the arc, which is consistent with high initial $^{87}Sr/^{86}Sr$ ratios (0.707 to 0.714) in the Devonian metaplutonic rocks (Silberman and others, 1979). Inheritance in zircon from the Ernie Lake pluton records an average age of 2.0 Ga for the basement contribution to the magmas (Dillon and Tilton, 1985), which indicates continent-derived sedimentary rocks beneath the Devonian arc remnants in the Brooks Range. This interpretation is consistent with 0.7- to 1.6-Ga Nd model age for the Devonian plutons in the Brooks Range (Nelson and others, 1989). Variable ϵ_{Nd} values (–6 to +3) for the Devonian plutons indicate a composite basement consisting of older crust and younger primitive material (Nelson and others, 1989). The geologic and isotopic relations within the arc rocks in the Brooks Range are broadly similar to those in arc and basement rocks exposed in the eastern Alaska Range and in east-central Alaska.

Ruby terrane

The record of Devonian magmatism is found south of the Brooks Range in the Ruby terrane (Fig. 1), where granite gneiss of the Ray Mountains yields a discordant Pb/U zircon age of 390 ± 25 Ma (Fig. 2; Patton and others, 1987). Proterozoic(?) to lower Paleozoic metavolcanic and metasedimentary rocks are present in this region (Fig. 2), which locally may be analogous to the mid-Paleozoic arc sequence exposed in the south-central Brooks Range. Proterozoic(?) to mid-Paleozoic metavolcanic and metasedimentary rocks are present in this region (Dover and Miyaoka, 1983). Elsewhere in the Ruby terrane, fossil ages suggest the presence of a mid-Ordovician shelf sequence and mid-Devonian through Mississippian metavolcanic and metasedimentary assemblage, with protoliths consisting of graywacke, siltstone, and slate with minor marble, chert, and mafic volcanic rocks (Smith and Puchner, 1985; Chapman and Patton, 1978).

DISCUSSION

Fragments of mid-Paleozoic volcano-plutonic complexes and associated basinal sequences that lie outboard of coeval shelf and platform strata are recognized throughout the western North American Cordillera. The mid-Paleozoic rocks were affected by subsequent deformation, and their original stratigraphic relations and paleogeographic distribution are obscured by their particular deformational histories. Thus, their current structural and tectonic setting with respect to the mid-Paleozoic continental margin is largely the result of later deformation. The mid-Paleozoic volcanoplutonic complexes and related basinal strata share many geologic, geochemical, and isotopic characteristics that suggest they have had a common tectonic history and shared paleogeographic ties to the continent.

In general, basement is composed of either sedimentary or metasedimentary strata that formed within a continent margin or slope-and-rise setting. Detrital zircon from quartzose sandstone in basement complexes that underlie the northern Sierra terrane, Kootenay, and Yukon-Tanana terranes yield Proterozoic Pb/U average upper-intercept ages (Fig. 3), implying an old continental source for some of the terrigenous rocks that underlie the mid-Paleozoic volcanic and related basinal strata. Negative ϵ_{Nd} values for the overlying volcanic rocks record a continental crustal contribution, which is consistent with relatively high initial $^{87}Sr/^{86}Sr$ ratios, where known (Fig. 3). In contrast, basement to the eastern Klamath terrane is characterized by low initial Sr ratios and positive ϵ_{Nd} values, implying the presence of transitional to oceanic crust underlying the mid-Paleozoic volcanic and sedimentary sequences in that terrane.

In regions with negative ϵ_{Nd} values and high initial Sr ratios (Fig. 3), basement complexes may represent attenuated North American continental crust or rifted fragments of North American crust, separated from their homeland during the late Precambrian or early Paleozoic. In addition, continental-derived slope-and-rise deposits that overlie oceanic crust may form parts of the mid-Paleozoic arc substrate. In this view, the oceanic substrate formed during or shortly after rifting along the western edge of North America. Regardless of local variations in basement, the presence of Proterozoic average upper-intercept ages on zircon from quartz sandstones in basement complexes and overlying epiclastic sequences demonstrates an ultimate continental tie.

The overlying mid-Paleozoic volcano-plutonic complexes and related basinal strata share many common elements. Similarities in detrital zircon provenance ages, the presence of Precambrian inheritance in arc-related intrusive rocks, and similarities in crystalline ages among the mid-Paleozoic volcanic rocks suggest they have a common geologic history (Fig. 3). In most areas, volcanism began in the Late Devonian. Upper Devonian rocks in the northern Sierra terrane, parts of the Eagle Bay and Milford Groups, the eastern Alaska Range and south-central Brooks Range record the early stages of arc activity. Although volcanism in the eastern Klamath terrane began earlier, Upper Devonian(?) to Mississippian volcanogenic rocks are present in the Bragdon Formation and are age-correlative with other volcanic sequences discussed here. The alkalic volcanic rocks in the Roberts Mountains allochthon and southeast portion of the Yukon-Tanana and Slide Mountain terranes and sporadic occurrences elsewhere record coeval extensional activity. Coeval intrusive rocks are present throughout the mid-Paleozoic sequences (Fig. 3). These intrusive rocks include the Bowman Lake batholith and Wolf Creek stock in the northern Sierra terrane, the Mount Fowler orthogneiss and related megaigneous rocks in the Kootenay terrane, the Upper Devonian and Mississippian metaigneous rocks in the Yukon-Tanana terrane, eastern Alaska Range, and the mid-Devonian plutons in the south-central Brooks Range (Fig. 3). Major- and trace-element data indicate that many of these igneous rock are arc-related, whereas a few (e.g., parts of the Slide Mountain and Yukon-Tanana terranes, Roberts Mountains allochthon, and Milford Group) formed inboard of the arc in an extensional tectonic setting.

Mid-Paleozoic volcano-plutonic sequences and associated basinal strata record an early period of convergent-margin tectonism and extension along the western margin of the North American Cordillera. Thick turbiditic and fine-grained clastic strata interfinger with volcanic rocks and represent intra-arc and locally back-arc basins that formed along the length of the North American Cordillera during the mid-Paleozoic. Extension in the mid- to late Paleozoic outer continental margin and adjacent basinal deposits of the North American Cordillera reflects continued continent-margin deformation.

Most basement complexes and overlying supracrustal sequences contain evidence for mid-Devonian to earliest Mississippian tectonism. A pre–middle Late Devonian unconformity is present in the northern Sierra terrane (Harwood, 1988), and similar deformation may have affected the eastern Klamath terrane during post–early Middle Devonian to pre–Late Mississippian time (Boucot and others, 1974; Charvet and others, 1989). Pre–Late Devonian unconformities are also present in the Brooks Range (Dillon and others, 1987a) and in outer continental margin deposits of the northern Canadian Cordillera (e.g., Gordey and others, 1987; Turner and others, 1989). Early Devonian contractional deformation is present in the north-central Brooks Range (Grantz and Moore, 1987) and to the southeast, where fine- to coarse-grained sediments were shed continentward into a foreland basin (Howell and Wiley, 1987). We suggest that Devonian foreland deformation in northern Alaska is directly related to the onset of continent-margin magmatism. Deformation elsewhere is younger. The Roberts Mountains allochthon was emplaced over the western North American continental margin strata during the Early Mississippian (Poole and Sandberg, 1977), whereas a Late Mississippian unconformity separates the Lardeau Group from the overlying Milford Group in the Kootenay terrane (Klepacki and Wheeler, 1985). These relations indicate that the basement and associated volcanic and basinal strata shared a similar structural history and were affected by continent-margin tectonism during the mid-Paleozoic.

There are several important differences and uncertainties in the structural history of the mid-Paleozoic terrane: (1) the exact age and initiation of deformation varies among terranes; (2) where dated, deformation in the arc sequences (e.g., the northern Sierra terrane) is distinctly older than that in eastward marginal basin terranes (e.g., the Roberts Mountains allochthon); (3) styles of deformation vary from contractional (e.g., northern Sierra terrane and Roberts Mountain allochthon) to extensional (e.g., Cassiar terrane) in different places—in some areas both styles are well documented; and (4) the mid-Paleozoic deformational history of the Yukon-Tanana terrane and adjacent terranes in the eastern Alaska Range is poorly known. These structural differences imply that deformation was diachronous throughout the arc and basinal sequences, and record both extensional and compressional modes of convergent-margin tectonism.

In summary, a discontinuous belt of mid-Paleozoic volcanic and basinal strata that extends from northern California to northern Alaska (Fig. 1) represents fragments of an east-Pacific fringing arc system. The exact tectonic setting and paleogeographic relations to the North American continental margin varied along strike, but all the terranes contain evidence for continental ties. The relation of younger marginal basins to the volcanic arc also varied along strike. For example, younger marginal basin deposits of the Schnoover and Havallah sequences lay inboard of the remnant mid-Paleozoic arc (northern Sierra terrane), whereas to the north, upper Paleozoic basinal rocks of the Slide Mountain terrane were deposited west of the older arc remnants (e.g., Kootenay terrane).

CONCLUSIONS

Fundamental similarities in isotopic ages and geologic relations between mid-Paleozoic arc sequences and their basement in the North American Cordillera suggest a common tectonic heritage. The volcanic arc and related marginal basins were constructed in part on older, previously deformed basement that has provenance links to both the North American continent and to the volcanic arc. Arc magmatism began in the middle Late Devonian and persisted for over 20 m.y. Devonian and Mississippian unconformities, orogenic clastic deposits, contractional and extensional structures, and associated tectonothermal events sporadically are present in many parts of the arc system and represent a protracted and diachronous Antler-age orogeny recorded throughout the western North American Cordillera.

ACKNOWLEDGMENTS

We thank Cynthia Dusel-Bacon, Dave Klepacki, Bill McClelland, Elizabeth Miller, Jason Saleeby, and Jim Wright for helpful discussions regarding mid-Paleozoic Cordilleran stratigraphy and tectonics, and Jim Mortensen and JoAnne Nelson for thorough and careful reviews of the manuscript. Some of these ideas were refined by discussions in the field with Jim Monger, JoAnne Nelson, Andy Okulitch, and Mike Orchard.

REFERENCES CITED

Albers, J. P., and Robertson, J. F., 1961, Geology and ore deposits of East Shasta copper-zinc district, Shasta County, California: U.S. Geological Survey Professional Paper 338, p. 1–107.

Albers, J. P., Kistler, R. W., and Kwak, L., 1981, The Mule mountain stock; An early Devonian pluton in northern California: Isochron/West, v. 31, p. 17.

Aleinikoff, J. N., 1984, Age and origin of metaigneous rocks from terranes north and south of the Denali fault, Mt. Hayes Quadrangle, east-central Alaska: Geological Society of America Abstracts with Programs, v. 16, p. 266.

Aleinikoff, J. N., and Nokleberg, W. J., 1985, Age of Devonian igneous-arc terranes in the northeast Mount Hayes Quadrangle, eastern Alaska Range, *in* Bartsch-Winkler, S., ed., The United States Geological Survey in Alaska; Accomplishments during 1984: U.S. Geological Survey Circular 962, p. 44–49.

Aleinikoff, J. N., Dusel-Bacon, C., Foster, H. L., and Futa, K., 1981, Proterozoic zircon from augen gneiss, Yukon-Tanana upland, east-central Alaska: Geology, v. 9, p. 649–473.

Aleinikoff, J. N., Foster, H. L., Nokleberg, W. J., and Dusel-Bacon, C., 1984, Isotopic evidence from detrital zircons from Early Proterozoic crustal material, east-central Alaska, *in* Coonard, W. L., and Elliott, R. L., eds., The United States Geological Survey in Alaska; Accomplishments during 1981: U.S. Geological Survey Circular 886, p. 43–45.

Aleinikoff, J. N., Dusel-Bacon, C., and Foster, H. L., 1986, Geochronology of augen gneiss and related rocks, Yukon-Tanana terrane, east-central Alaska: Geological Society of America Bulletin, v. 97, p. 626–637.

Aleinikoff, J. N., Dusel-Bacon, C., Foster, H. L., and Nokleberg, W. J., 1987, Lead isotopic fingerprinting of tectonostratigraphic terranes, east-central Alaska: Canadian Journal of Earth Sciences, v. 24, p. 2089–2098.

Barker, F., and Millard, H. T., Knight, R. J., 1979, Reconnaissance geochemistry of Devonian island-arc volcanic and intrusive rocks, west Shasta District, California, *in* Barker, F., ed., Trondjemites, dacites, and related rocks: Amsterdam, Elsevier, p. 531–545.

Bennett, V., and Hansen, V. L., 1988, Neodymium isotopic similarities between the Yukon-Tanana terrane, Yukon Territory and continental North America: Geological Society of America Abstracts with Programs, v. 20, p. A111.

Bond, G. C., and Devay, J. C., 1980, Pre-Upper Devonian quartzose sandstones in the Shoo Fly Formation, northern California; Petrology, provenance, and implications for regional tectonics: Journal of Geology, v. 88, p. 285–388.

Boucot, A. J., Dunkle, D. H., Potter, A., Savage, N. M., and Rohr, C., 1974, Middle Devonian orogeny in western North America?; A fish and other fossils: Journal of Geology, v. 82, p. 691–708.

Boudier, F., Le Sueur, E., and Nicolas, A., 1989, Structure of an atypical ophiolite; The Trinity complex, eastern Klamath Mountains, California: Geological Society of America Bulletin, v. 101, p. 820–833.

Brouxel, M., and Lapierre, H., 1988, Geochemical study of an early Paleozoic island-arc–back-arc basin system; Part 1, The Trinity ophiolite [sic] (northern California): Geological Society of America Bulletin, v. 100, p. 1111–1119.

Brown, R. L., and 6 others, 1981, Southern Cordilleran cross-section; Cranbrook to Kamloops, *in* Thompson, R. I., and Cook, D. G., Field guides to geologic and mineral deposits, Calgary '81 annual meeting: Calgary, Alberta, University of Calgary, p. 335–372.

Burchfiel, B. C., and Davis, G. A., 1972, Structural framework and evolution of the southern part of the Cordilleran orogen, western United States: American Journal of Science, v. 272, p. 97–118.

Cecile, M. P., 1982, The lower Paleozoic Misty Creek embayment, Selwyn Basin, Yukon and Northwest Territories: Geological Survey of Canada Bulletin 335, 78 p.

Chapman, R. M., and Patton, W. W., Jr., 1978, Two upper Paleozoic sedimentary rock units identified in southwestern part of the Ruby Quadrangle, *in* Johnson, K. M., and Williams, J. R., eds., The United States Geological Survey in Alaska; Accomplishments during 1978: U.S. Geological Survey Circular 804-B, p. 59–61.

Charvet, J., Lapierre, H., and Campos, C., 1989, Les effets de la phase Antler (Devonien supérieur-Carbonifère inférieur) dans les Klamath orientales (Californie, U.S.A.): Implications geodynamiques: Comptes Rendu Academie Science Paris, Serie II, p. 1629–1635.

Danner, W. R., 1977, Paleozoic rocks of northwest Washington and adjacent part of British Columbia, *in* Stewart, J. H., Stevens, C. H., and Fritsche, A. E., eds., Pacific Coast Paleogeographic Symposium 1; Paleozoic Paleogeography of the western United States: Society of Economic Paleontologists and Mineralogists, Pacific Section, p. 481–502.

Dillon, J. T., and Bakke, A. A., 1987, Evidence of Devonian ages of the Okpilak batholith, north-central Brooks Range, Alaska: Geological Society of America Abstracts with Programs, v. 19, p. 373.

Dillon, J. T., and Tilton, G. R., 1985, Devonian magmatism in Brooks Range, Alaska: American Association of Petroleum Geologists Bulletin, v. 69, p. 662.

Dillon, J. T., Pessel, G. H., Chen, J. H., and Veach, N. C., 1980, Middle Paleozoic magmatism and orogenesis in the Brooks Range, Alaska: Geology, v. 8, p. 338–343.

Dillon, J. T., Harris, A. G., and Dutro, J. T., Jr., 1987a, Preliminary description and correlation of Lower Paleozoic fossil-bearing strata inthe Snowden Mountain area of the south-central Brooks Range, *in* Tailleur, I. L., and Weimer, P., eds., Alaska North Slope geology: Society of Economic Paleontologists and Mineralogists, Pacific Section, and the Alaska Geological Society, Bakersfield and Anchorage, p. 337–345.

Dillon, J. T., Tilton, G. R., Decker, J., and Kelly, M. J., 1987b, Resource implications of magmatic and metamorhic ages from Devonian igneous rocks in the Brooks Range, *in* Tailleur, I. L., and Weimier, P., eds., Alaska North Slope geology: Society of Economic Paleontologists and Mineralogists, Pacific Section, and the Alaska Geological Society, Bakersfield and Anchorage, p. 713–723.

Dover, J. M., and Miyaoka, R. T., 1983, Major rock packages of the Ray Mountains, Tanana and Bettles Quadrangles, *in* Bartsch-Winker, S., and Reed, K. M., eds., The United States Geological Survey in Alaska; Accomplishments during 1983: U.S. Geological Survey Circular 945, p. 32–36.

Dumoulin, J. A., and Harris, A. G., 1987, Lower Paleozoic carbonate rocks of the Baird Mountains Quadrangle, western Brooks Range, Alaska, *in* Tailleur, I. L., and Weimer, P., eds., Alaska north slope geology: Society of Economic Paleontologists and Mineralogists, Pacific Section, and the Alaska Geological Society, Bakersfield and Anchorage, p. 311–336.

Dusel-Bacon, C., and Aleinikoff, J. N., 1985, Petrology and tectonic significance of augen gneiss from a belt of Mississippian granitoids in the Yukon-Tanana terrane, east-central Alaska: Geological Society of America Bulletin, v. 96, p. 411–425.

Gabrielse, H., Loveridge, W. D., Sullivan, R. W., and Stevens, R. D., 1982, U-Pb measurements on zircon indicate middle-Paleozoic plutonism in the Omineca crystalline belt, north-central British Columbia: Geological Survey of Canada Paper 82-1C, p. 139–146.

Gareau, S. A., 1989, Metamorphism, deformation, and geochronology of the Esctall-Quaal River Area, coast Plutonic Complex, British Columbia, *in* Current Research, Part E: Geological Survey of Canada Paper 89-1E, p. 155–162.

Gehrels, G. E., and Smith, M., 1987, "Antler" allochthon in the Kootenay arc?: Geology, v. 15, p. 769–770.

Gehrels, G. E., McClelland, W. C., Samson, S. D., Patchett, P. J., and Jackson, J. L., 1989, Ancient continental margin assemblage in the northern Coast Mountains, southeast Alaska and northwest Canada: Geology (in press).

Girty, G. H., and Schweickert, R. A., 1984, The Culbertson Lake Allochthon; A newly indentified structure within the Shoo Fly Complex, California; Evidence for four phases of deformation and extension of the Antler orogeny to the northern Sierra Nevada: Modern Geology, v. 8, p. 181–198.

Girty, G. H., and Wardlaw, M. S., 1984, Was the Alexander terrane a source of feldspathic sandstones in the Shoo Fly Complex, Sierra Nevada, California?: Geology, v. 12, p. 339–342.

——, 1985, Petrology and provenance of pre-Late Devonian sandstones, Shoo Fly Complex, northern Sierra Nevada, California: Geological Society of America Bulletin, v. 96, p. 516–521.

Girty, G. H., Reiland, D. N., and Wardlaw, M. S., 1985, Provenance of the Silurian Elder Sandstone, north-central Nevada: Geological Society of America Bulletin, v. 96, p. 925–930.

Gordey, S. P., Abbott, J. G., Tempelman-Kluit, D. J., and Gabrielse, H., 1987, "Antler" clastics in the Canadian Cordillera: Geology, v. 15, p. 103–107.

Grantz, A., and Moore, T. E., 1987, Crustal structure of northern Alaska along Geodynamics Continent/Ocean Transect (A-3): Geological Society of America Abstracts with Programs, v. 19, p. 383.

Hannah, J. L., and Moores, E. M., 1986, Age relationships and depositional environments of Paleozoic strata, northern Sierra Nevada: Geological Society of America Bulletin, v. 97, p. 787–797.

Hanson, R. E., and Schweickert, R. A., 1986, Stratigraphy of mid-Paleozoic island-arc rocks in part of the northern Sierra Nevada, Sierra and Nevada Counties, California: Geological Society of America Bulletin, v. 97, p. 986–998.

Hanson, R. E., Saleeby, J. B., and Schweickert, R. A., 1988, Composite Devonian island-arc batholith in the northern Sierra Nevada, California: Geological Society of America Bulletin, v. 100, p. 446–457.

Harland, W. B., and 5 others, 1982, A geologic time scale: Cambridge, England, Cambridge University Press, 131 p.

Harms, T. A., 1985, Cross-sections through the Sylvester allochthon and underlying Cassiar platform, northern British Columbia, *in* Current Research, Part B: Geological Survey of Canada Paper 85–1B, p. 341–346.

Harwood, D. S., 1983, Stratigraphy of upper Paleozoic volcanic rocks and regional unconformities in part of the northern Sierra terrane, California: Geological Society of America Bulletin, v. 94, p. 413–422.

——, 1988, Tectonism and metamorphism in the northern Sierra terrane, northern California, *in* Ernest, W. G., ed., Metamorphism and crustal evolution of the western United States; Rubey Volume 7: Englewood Cliffs, New Jersey, Prentice-Hall, p. 764–788.

Hitzman, M. W., Proffett, J. M., Schmidt, J. M., and Smith, T. E., 1986, Geology and mineralization of the Ambler District, northwestern Alaskaa: Economic Geology, v. 81, p. 1592–1618.

Hotz, P. E., 1977, Geology of the Yreka Quadrangle, Siskiyou County, California: U.S. Geological Survey Bulletin 1436, 72 p.

Howell, D. G., and Wiley, T. J., 1987, Crustal evolution of northern Alaska inferred from sedimentology and structural relations of the Kandik areas: Tectonics, v. 6, p. 619–631.

Jacobsen, S. B., Quick, J. E., and Wasserburg, G. J., 1984, A Nd and Sr isotopic study of the Trinity peridotite; Implications for mantle evolution: Earth and Planetary Science Letters, v. 68, p. 361–378.

Johnson, J. G., and Pendergast, A., 1981, Timing and mode of emplacement of the Roberts Mountains allochthon, Antler orogeny: Geological Society of America Bulletin, v. 92, p. 648–658.

Klepacki, D. W., 1985, Stratigraphy and structural geology of the Goat Range area, southeastern British Columbia [Ph.D. thesis]: Cambridge, Massachusetts Institute of Technology, 268 p.

Klepacki, D. W., and Wheeler, J. O., 1985, Stratigraphic and structural relations of the Milford, Kaslo, and Slocan Groups, Goat Range, Lardeau and Nelson Map areas, British Columbia, *in* Current Research, Part A: Geological Survey of Canada Paper 85-1A, p. 277–286.

Lapierre, H., Albarede, F., Albers, J., Coulon, C., and Cabanis, B., 1985a, Early Devonian volcanism in the eastern Klamath Mountains, California; Evidence for an immature island arc: Canadian Journal of Earth Sciences, v. 22, p. 214–227.

Lapierre, H., Cabanis, B., Coulon, C., Brouxel, M., and Albarede, F., 1985b, Geodynamic setting of Early Devonian Kuroko-type sulfide deposits in the eastern Klamath Mountains (northern California) inferred by the petrological and geochemical characteristics of the associated island-arc volcanic rocks: Economic Geology, v. 80, p. 2100–2113.

Little, T. A., 1987, Stratigraphy and structure of metamorphosed upper Paleozoic rocks near Mountain City, Nevada: Geological Society of America Bulletin, v. 98, p. 1–17.

Madrid, R., 1987, Stratigraphy of the Roberts Mountains allochthon in north-central Nevada [Ph.D. thesis]: Stanford, California, Stanford University, 336 p.

Mayfield, C. F., and Tailleur, I. L., 1978, Bedrock geology map of the Ambler River Quadrangle, Alaska: U.S. Geological Survey Open-File Report 78-120A, scale 1:250,000.

Mayfield, C. F., Silberman, M. L., and Tailleur, I. L., 1982, Precambrian metamorphic rocks from the hub Mountain terrane, Baird Mountain Quadrangle, Alaska, *in* Coonard, W. L., ed., The United States Geological Survey in Alaska; Accomplishments during 1980: U.S. Geological Survey Circular 844, p. 18–24.

McCulloch, M. T., and Wasserburg, G. J., 1978, Sm-Nd and Rb-Sr chronology of continental crust formation: Science, v. 200, p. 1003–10011.

Miller, E. L., Holdsworth, B. K., Whiteford, W. B., and Rodgers, D., 1984, Stratigraphy and structure of the Schoonover sequence, northeastern Nevada; Implications for Paleozoic plate-margin tectonics: Geological Society of America Bulletin, v. 95, p. 1063–1076.

Miller, E. L., Miller, M. M., Stevens, C. H., Wright, J. E., and Madrid, R., 1990, Late Paleozoic paleogeography and tectonic evolution of the western U.S. Cordillera, *in* Burchfiel, B. C., Lipman, P., and Zoback, M. L., eds., The Cordilleran Orogen; Conterminous U.S.: Boulder, Colorado, Geological Society of America, The Geology of North America, v. G-3 (in press).

Miller, M. M., 1987, Dispersed remnants of a northeast Pacific fringing arc–Upper Paleozoic island arc terrane of Permian McCloud faunal affinity, western U.S.: Tectonics, v. 6, p. 807–830.

——, 1989, Intra-arc sedimentation and tectonism; Late Paleozoic evolution, eastern Klamath terrane, California: Geological Society of America Bulletin, v. 101, p. 170–187.

Miller, M. M., and Cui, B., 1989, Submarine fan characteristics and dual sediment provenance, Lower Carboniferous Bragdon Formation, eastern Klamath terrane, California: Canadian Journal of Earth Sciences, v. 26, p. 927–940.

Miller, M. M., Saleeby, J. B., and Cui, B., 1988, Provenance of the Carboniferous Bragdon Formation, eastern Klamath terrane, California; Proterozoic detrital zircon and quartzose detritus as continental tracers in oceanic convergent margin systems, *in* Howell, D. G., and Wiley, T. J., eds., Provisional Proceedings of the Fourth International Tectonostratigraphic Terrane Conference, Nanjing University, September, 1988, Nanjing: U.S. Geological Survey, p. 67–69.

Miller, M. M., and Saleeby, J. B., 1990, Continental detrital zircon in Carboniferous ensimatic arc rocks, Bragdon Formation, eastern Klamath terrane, northern California: Geological Society of America Bulletin, in press.

Monger, J.H.W., 1977, Upper Paleozoic rocks of the western Canadian Cordillera and their bearing on Cordilleran evolution: Canadian Journal of Earth Sciences, v. 14, p. 1832–1859.

——, 1984, Cordilleran tectonics; A Canadian perspective: Geological Society of France Bulletin, v. 26, p. 255–278.

Monger, J.H.W., and Berg, H. C., 1987, Lithotectonic terrane map of western Canada and southeastern Alaska: U.S. Geological Survey Miscellaneous Field Studies Map MF-1874-B, scale 1:2,500,000.

Monger, J.H.W., Souther, J. G., and Gabrielse, H., 1972, Evolution of the Canadian Cordillera; A plate tectonic model: American Journal of Science, v. 272, p. 577–602.

Moore, T. E., 1987, Geochemistry and tectonic setting of some volcanic rocks of the Franklinian Assemblage, central and eastern Brooks Range, *in* Tailleur, I. L., and Weimer, P., eds., Alaska north slope geology: Society of Economic Paleontologists and Mineralogists, Pacific Section and the Alaska Geological Society, Bakersfield and Anchorage, p. 691–710.

Mortensen, J. K., 1982, Geological setting and tectonic significance of Mississippian felsic metavolcanic rocks in the Pelly Mountains, southeastern Yukon territory: Canadian Journal of Earth Sciences, v. 19, p. 8–22.

——, 1986, U-Pb ages for granitic orthogneiss from western Yukon Territory; Selwyn gneiss and Fiftymile batholith revisited, *in* Current Research, Part B: Geological Survey of Canada Paper 86-1B, p. 141–146.

——, 1988, Significance of episodic magmatism in Yukon-Tanana terrane, Yukon and Alaska: Geological Society of America Abstracts with Programs, v. 20, p. A111.

Mortensen, J. K., and Jilson, G. A., 1985, Evolution of the Yukon-Tanana terrane; Evidence from southeastern Yukon Territory: Geology, v. 13, p. 806–810.

Nelson, B. K., Nelson, S. W. and Till, A. B., 1989, Isotopic evidence of an Early Proterozoic crustal source for granites of the Brooks Range, Northern Alaska: Geological Society of America Abstracts with Programs, v. 21, p. A105.

Nelson, J. A., and Bradford, J., 1987, Geology of the area around the Midway deposit northern British Columbia, *in* Geologic fieldwork, 1986: British Columbia Ministry of Energy, Mines, and Petroleum Resources Paper 1987-1, p. 181–192.

——, 1989, Geology and mineral deposits of the Cassiar and McDame map areas, British Columbia, *in* Geologic fieldwork, 1988: British Columbia Ministry of Energy, Mines, and Petroleum Resources America Bulletin, v. 96, p. 1251–1270.

Nelson, J. A., Bradford, J. A., Green, K. C., and Marsden, H., 1988, Geology and patterns of mineralization, Blue Dome map area, Cassiar District, *in* Geologic fieldwork, 1987: British Columbia Ministry of Energy, Mines, and Petroleum Resources Paper 1988-1, p. 233–243.

Nokleberg, W. J., Jones, D. L., and Silberling, N. J., 1985, Origin and tectonic evolution of the Maclaren and Wrangellia terranes, eastern Alaska Range, Alaska: Geological Society of America Bulletin, v. 96, p. 1251–1270.

Okulitch, A. V., 1979, Lithology, stratigraphy, structure, and mineral occurrences of the Thompson-Shuswap-Okanagan area, British Columbia: Geological Survey of Canada Open-File 637, scale 1:250,000.

——, 1985, Paleozoic plutonism in southeast British Columbia: Canadian Journal of Earth Sciences, v. 22, p. 1409–1424.

Okulitch, A. V., Wanless, R. K., and Loveridge, W. D., 1975, Devonian plutonism in south-central British Columbia: Canadian Journal of Earth Sciences, v. 12, p. 1760–1769.

Parrish, R. R., Ross, G., Carr, S. D., and Coleman, V., 1989, Single grain U-Pb provenance ages of detrital zircons from metaquartzites of the southern Omineca Belt, British Columbia: Geological Society of America Abstracts with Programs, v. 10, p. 127.

Patton, W. W., Stern, T. W., Arth, J. G., and Carlson, C. L., 1987, New U/Pb ages from granite and granite gneiss in the Ruby geanticline and southern Brooks Range, Alaska: Journal of Geology, v. 95, p. 118–126.

Poole, F. G., and Claypoole, G. E., 1984, Petroleum source-rock potnetial and crude-oil correlation in the Great Basin, *in* Woodward, J., Meissner, F. F., and Clayton, G. E., eds., Hydrocarbon source rocks of the greater Rocky Mountain region: Denver, Colorado, Rocky Mountain Association of Geologists, p. 179–229.

Poole, F. G., and Sandberg, C. A., 1977, Mississippian paleogeography and tectonics of the western United States, *in* Stewart, J. H., Stevens, C. H., and Fritsche, A. E., eds., Paleozoic paleogeography of the western United States: Pacific Section, Society of Economic Paleontologists and Mineralogists Pacific Coast Paleogeography Symposium 1, p. 179–230.

Potter, A. W., Hotz, P. E., and Rohr, D. M., 1977, Stratigraphy and inferred tectonic framework of lower Paleozoic rocks in the eastern Klamath Mountains, northern California, *in* Stewart, J. H., Stevens, C. H., and Fritsche, A. E., eds., Paleozoic paleogeography of the western United States: Pacific Section, Society of Economic Paleontologists and Mineralogists Pacific Coast Paleogeographic Symposium 1, p. 421–440.

Quick, J. E., 1981, Petrology and petrogenesis of the Trinity peridotite; An upper mantle diapir in the eastern Klamath Mountains, northern California: Journal of Geophysical Research, v. 86, p. 11837–11863.

Read, P. B., 1975, Lardeau Group, Lardeau Map area, west ½ (82K west-half), British Colombia: Geological Survey of Canada Paper 75-1A, p. 29–30.

Read, P. B., and Wheeler, J. O., 1976, Geology, Lardeau west-half, British Columbia: Geological Survey of Canada Open-File Mapr 432, scale 1:125,000.

Roberts, R. J., Hotz, P. E., Gilluly, J., and Ferguson, H. G., 1958, Paleozoic rocks of northwest Nevada: American Association of Petroleum Geologists Bulletin, v. 42, p. 2813–2857.

Ross, J. V., 1970, Structural evolution of the Kootenay Arc, southeastern British Columbia, *in* Wheeler, J. O., ed., Structure of the southern Canadian Cordillera: Geological Association of Canada Special Paper 6, p. 53–65.

Rouer, O., Lapierre, H., Mascle, G., Coulon, C., and Albers, J., 1989, Geodynamic implications of Devonian silicic arc magmatism in the Sierra Nevada and Klamath Mountains, California: Geology, v. 17, p. 177–180.

Saleeby, J., 1990, Tectonic and petrogenetic settings of U.S. Cordillera ophiolites, *in* Burchfiel, B. C., Lipman, P., and Zoback, M. L., eds., The Cordilleran Orogen; Conterminous U.S.: Boulder, Colorado, Geological Society of America, The Geology of North America, v. G-3 (in press).

Saleeby, J. B., and Rubin, C. M., 1989, The western margin of the Coast Plutonic Complex (CPC) in southernmost SE Alaska: Geological Society of America Abstracts with Programs, v. 21, p. 139.

Saleeby, J., Hannah, J. L., and Varga, R. J., 1987, Isotopic age constraints on middle Paleozoic deformation in the northern Sierra Nevada, California: Geology, v. 15, p. 757–760.

Samson, S. D., McClelland, W. C., Patchett, P. J., and Gehrels, G. E., 1989, Nd isotopic evidence for ancient continental material between juvenile terranes of the Canadian Cordillera: EOS Transactions of the American Geophysical Union, v. 70, p. 485.

Savage, N. M., 1976, Middle Devonian species of *Polygnathus* from the Kennett Formation, northern California: Journal of Paleontology, v. 50, p. 374–379.

Schiarizza, P., 1986, Geology of the Eagle Bay Group between the Raft and Baldy batholiths (82M/5, 11, 12): British Columbia Ministry of Eenrgy, Mines, and Petroleum Resources Paper 1986-1, p. 89–94.

Schiarizza, P., and Preto, V. A., 1987, Geology of the Adams Plateau–Clearwater–Vavenby area: British Columbia Ministry of Energy, Mines, and Petroleum Resources Paper 1987-2, 88 p.

Schweickert, R. A., Harwood, D. S., Girty, G. A., and Hanson, R. E., 1984, Tectonic development of the northern Sierra terrane; An accreted late Paleozoic island arc and its basement, *in* Lintz, J., ed., Geological excursions; Geological Society of America Cordilleran Section Meeting Guidebook: Reno, Nevada, Mackay School of Mines, p. 1–65.

Silberling, N. J., Jones, D. L., Blake, M. C., and Howell, D. G., 1984, Lithotectonic terrane map of the western conterminous United States: U.S. Geological Survey Open-File Report 84-523, p. C1–C43.

Silberman, M. L., Brookins, D. G., Nelson, S. W., and Grybeck, D., 1979, Rubidium-strontium and potassium-argon dating of emplacement and metamorphism of the Arrigetch Peaks and Mount Igikpak plutons, Survey Pass Quadrangle, Alaska, *in* Johnson, K. M., and Williams, J. R., eds., The United States Geological Survey in Alaska; Accomplishments during 1978: U.S. Geological Survey Circular 804-B, p. 18.

Smith, G. M., and Puchner, C. M., 1985, Geology of the Ruby geanticline between Ruby and Poorman Alaska and the tectonic emplacement of the Ramparts Group: EOS Transactions of the American Geophysical Union, v. 66, no. 46, p. 1102.

Smith, M., and Gehrels, G. E., 1989, Lower Paleozoic eugeoclinal strata in the Kootenay arc, NE Washington and SE B.C.; A northern continuation of Roberts Mountains Allochthon-type outer continental margin facies: Geological Society of America Abstracts with Programs, v. 21, p. 146.

Smith, T. E., and 5 others, 1978, Evidence for a mid-Paleozoic depositional age of volcanogenic base-metal massive sulfide occurrences and enclosing strata, Ambler District, northwest Alaska: Geological Society of America Abstracts with Programs, v. 10, p. 148.

Snook, J. R., Ellis, M. A., Mills, J. W., and Watkinson, A. J., 1981, The Kootenay arc in N.E. Washington: Geological Society of America Abstracts with Programs, v. 14, p. 235.

Speed, R. C., 1979, Collided Paleozoic microplate in the western United States: Journal of Geology, v. 87, p. 279–292.

Struik, L. C., and Orchard, M. J., 1985, Upper Paleozoic conodonts from ribbon cherts indicate thrust imbrication of the Antler Formation of Slide Mountain terrane, central British Columbia: Geology, v. 13, p. 794–798.

Tempelman-Kluit, D. J., 1979, Transported cataclasite, ophiolite and granodiorite in Yukon; Evidence of arc-continent collision: Geological Survey of Canada Paper 79–14, 27 p.

Till, A. B., 1983, Granulite, peridotite, and blueschist; Precambrian to Mesozoic history of the Seward Peninsula: Journal of the Alaska Geological Society, v. 3, p. 59–66.

Toy, T., 1987, Alteration, chemistry, and tectonic setting of volcanogenic massive sulfide-barite deposits at Rea Gold and Homestake southeastern British Columbia (82M/4W): British Columbia Ministry of Energy, Mines, and Petroleum Resources, Exploration in British Columbia, p. B17–B19.

Turner, R.J.W., Madrid, R. J., and Miller, E. L., 1989, Roberts Mountains allochthon; Stratigraphic comparison with lower Paleozoic outer continental margin strata of the northern Canadian Cordillera: Geology, v. 17, p. 341–344.

Varga, R. J., 1982, Implications of Paleozoic phosphorites in the western Sierra Nevada Range: Nature, v. 297, p. 217–2202.

Wallin, E. T., 1989, Provenance of lower Paleozoic sandstones in the eastern Klamath Mountains and the Roberts Mountains allochthon, California and Nevada [Ph.D. thesis]: Lawrence, University of Kansas, 152 p.

Wallin, E. T., Mattinson, J. M., and Potter, A. W., 1988, Early Paleozoic magmatic events in the eastern Klamath Mountains, northern California: Geology, v. 16, p. 144–148.

Watkins, R., and Flory, R. A., 1986, Island arc sedimentation in the Middle Devonian Kennett Formation, eastern Klamath Mountains, California: Journal of Geology, v. 94, p. 753–761.

Wheeler, J. O., and McFeely, P., 1987, Tectonic assemblage map of the Canadian Cordillera: Geological Survey of Canada Open-File Report 1565, scale 1:2,000,000.

Wheeler, J. O., and 5 others, 1988, Terrane map of the Canadian Cordillera: Canadian Geological Survey Open-File Report 1894, scale 1:2,000,000.

Wricke, C. T., and Jones, D. L., 1978, Allochthonous Devonian chert in northern Shoshone Range, north-central Nevada, *in* Geological Survey Research 1977; U.S. Geological Survey Professional Paper 1100, p. 70–71.

MANUSCRIPT ACCEPTED BY THE SOCIETY MARCH 27, 1990

Printed in U.S.A.

Geological Society of America
Special Paper 255
1990

Provenance of selected lower Paleozoic siliciclastic rocks in the Roberts Mountains allochthon, Nevada

E. T. Wallin*
Department of Geology, University of Kansas, Lawrence, Kansas 66045-2124

ABSTRACT

Lower Paleozoic strata of the Roberts Mountains allochthon were derived both directly and indirectly from Precambrian basement rocks. The Upper Cambrian(?) Harmony Formation was derived directly from erosion of Proterozoic crystalline rocks along the western margin of North America. Zircon data from feldspar-bearing strata of the Vinini Formation (Lower Ordovician) suggest a provenance link with the Harmony Formation. Compositionally mature, Middle Ordovician eugeoclinal (Palmetto Formation) and miogeoclinal (Eureka Quartzite) quartzites exhibit similar isotopic signatures except that the miogeoclinal quartzite unequivocally contains an Archean component. Volcaniclastic sediment within the Slaven Chert (Middle Devonian) contains Precambrian zircon but no evidence of neovolcanic zircon. New U-Pb isotopic data for detrital zircon from the allochthon are most consistent with models of Antler orogenesis that do not require significant interaction (collision) of a fringing volcanic arc with the continental margin.

INTRODUCTION

Geologic setting

The Roberts Mountains allochthon in Nevada consists of complexly deformed, lower Paleozoic eugeoclinal deposits that were thrust eastward over the Cordilleran miogeocline during the Late Devonian–Early Mississippian Antler orogeny (Roberts and others, 1958; Nilsen and Stewart, 1980; Fig. 1). The allochthon extends southwestward from Nevada into central California and northward into Idaho, but the record in these areas is more fragmentary than in Nevada because of Mesozoic tectonic disruption (e.g., Schweickert and Lahren, 1987; Dover, 1980). Rocks of the Roberts Mountains allochthon make up sparsely fossiliferous siliciclastic, hemipelagic, pelagic, and volcanic deposits that range from Middle Cambrian to Late Devonian and perhaps Early Mississippian in age (Johnson and Pendergast, 1981). Between the 1940s and 1960s, a stratigraphic framework for the Roberts Mountains allochthon based on physical stratigraphy was erected by pioneering reconnaissance work (Merriam and Anderson, 1942; Roberts and others, 1958; Kerr, 1962; Roberts, 1964; Hotz and Willden, 1964; Kay and Crawford, 1964; Gilluly and Gates, 1965; Gilluly and Masursky, 1965). Although these tenuous lithostratigraphic correlations are supported by rather sparse biostratigraphic data (e.g., Ross, 1961, 1964; Ross and Berry, 1963), they constitute a vital background for further study.

Sedimentary rocks of the Roberts Mountains allochthon consist of chert, argillite, quartzite, quartzose sandstone, arkosic sandstone, siltstone, shale, greenstone, barite, and minor limestone. These strata occur within complexly deformed thrust nappes. Thicknesses of lithofacies assemblages are uncertain due to structural disruption, but estimates for the original stratigraphic thickness of the Roberts Mountains allochthon range from 810 m (Churkin and Kay, 1967) to 7,500 m (Gilluly and Gates, 1965). Even the minimum thickness estimates for the siliciclastic sequences record substantial erosion of lower Paleozoic paleogeographic elements. Yet, despite the paleogeographic significance of these deposits, their provenance is poorly understood. Knowledge of the provenance of these units is essential if we are to further evaluate competing models for the tectonic evolution of the western United States.

*Present address: Department of Geoscience, University of Nevada, Las Vegas, Nevada 89154

Wallin, E. T., 1990, Provenance of selected lower Paleozoic siliciclastic rocks in the Roberts Mountains allochthon, Nevada, *in* Harwood, D. S., and Miller, M. M., eds., Paleozoic and early Mesozoic paleogeographic relations; Sierra Nevada, Klamath Mountains, and related terranes: Boulder, Colorado, Geological Society of America Special Paper 255.

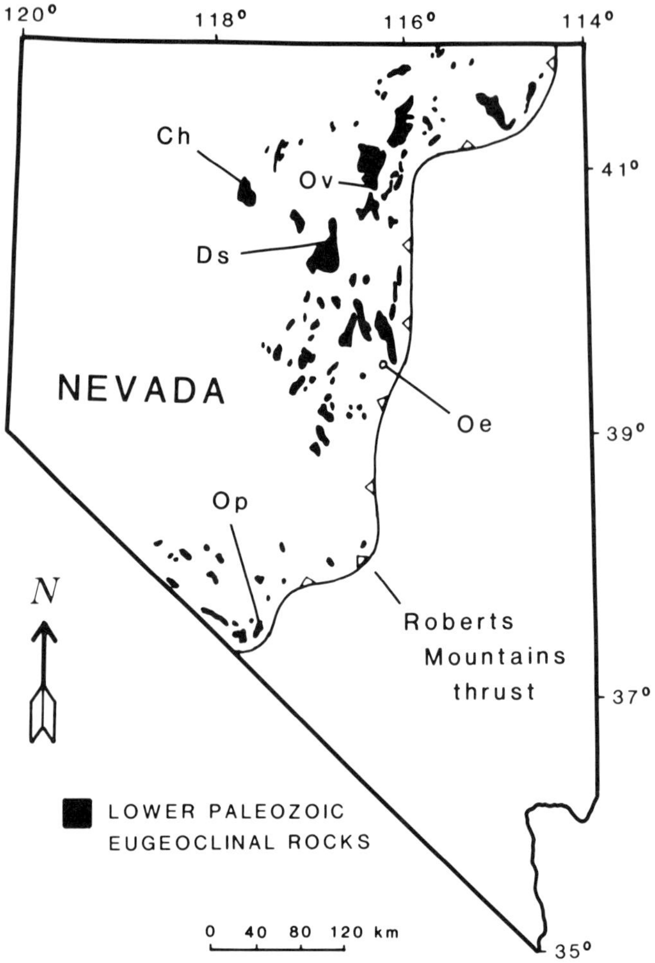

Figure 1. Sketch map illustrating the distribution of lower Paleozoic eugeoclinal rocks in Nevada. Ch = Upper Cambrian(?) Harmony Formation; Ov = Lower Ordovician Vinini Formation; Op = Middle Ordovician Palmetto Formation; Oe = Middle Ordovician Eureka Quartzite; Ds = Middle Devonian Slaven Chert. Modified after Schweickert, 1981.

Several different tectonic models have been proposed for the origin of the Roberts Mountains allochthon. Some workers view the Roberts Mountains allochthon as a closed back-arc basin that was associated with a west-facing volcanic arc (Burchfiel and Davis, 1972, 1975; Poole, 1974; Poole and Sandberg, 1977; Poole and others, 1977). Other workers view the Roberts Mountains allochthon as an accretionary prism that developed in front of an east-facing volcanic arc (Dickinson, 1977; Speed, 1979; Johnson and Pendergast, 1981; Speed and Sleep, 1982; Oldow, 1984). Recently, Burchfiel and Royden (1989) presented a model for the Antler Orogeny based on an analogy with the Apennine–East Carpathian orogen in Europe. Their model is appealing because it explains anomalous aspects of the Antler Orogeny such as the absence of widespread magmatism and regional metamorphism.

In addition to the uncertainty concerning the origin of the Roberts Mountains allochthon itself, paleogeographic relations between the Roberts Mountains allochthon and lower Paleozoic terranes in California remain controversial (Speed, 1979, 1983; Schweickert and Snyder, 1981; Turner and others, 1989). Consequently, the sources of voluminous clastic deposits within the Roberts Mountains allochthon have been the subject of considerable speculation in attempts to reconstruct early Paleozoic tectonics of the western United States. Plausible sources have been proposed for each of the major siliciclastic units, but often they are either difficult to reconcile with other geologic evidence, tenuous due to poor age control, or in the case of competing hypotheses, mutually exclusive. In this chapter, geochronology of detrital zircon is employed to constrain the provenance of four major clastic units within the Roberts Mountains allochthon. The large number of U-Pb zircon crystallization ages for the North American craton is used as a basis for comparison. New U-Pb isotopic data for detrital zircon from the Harmony, Vinini, and Palmetto Formations, and a sandstone within the Slaven Chert, are discussed below in the context of existing models for the evolution of sedimentation along the lower Paleozoic continental margin of the western United States.

Approach to analysis of detrital zircon

Many recent zircon geochronological studies have documented the potential complexity of both zircons and U-Pb zircon isotopic systematics. Isotopic analyses constitute averages of the isotopic composition of the analyzed material. In many samples, this isotopic averaging introduces sufficient geological uncertainty into the analysis as to preclude results that are geologically useful. Within-grain isotopic heterogeneity of zircon is well documented (e.g., Scharer and Allegre, 1982; Black and others, 1986) and must be considered when using standard thermal ionization mass spectrometry. However, useful isotopic analyses of <1-mg zircon samples can be performed using standard thermal ionization mass spectrometry, provided careful sample preparation is employed. Zircons in mixed populations can often be subdivided into species according to physical characteristics such as color, size, shape, roundness, magnetic susceptibility, clarity, core/overgrowth relations, and the nature of inclusions. Meticulous hand selection of zircon species for analysis can, in many instances, significantly reduce within-fraction age heterogeneity. The success with which isolation of different zircon species can be achieved varies from sample to sample. Small-grain sizes (silt to fine sand) usually require multiple-grain samples. When possible, single-grain analyses remove much of the geologic uncertainty associated with multiple-grain analyses.

In this study, only grains without visible cores were selected for analysis in an attempt to avoid the within-grain isotopic averaging that results from core/overgrowth relations. Due to the slight to moderate discordance of most of the analyzed fractions, the $^{207}Pb/^{206}Pb$ ages reported in Table 2 are interpreted here as the minimum age of the oldest component in each fraction. The certainty of different Pb-loss histories for the component grains of individual multiple-grain analyses precludes the use of linear regression to obtain meaningful upper intercept ages.

Several geologic inferences made below are based on the empirical observation that $^{207}Pb/^{206}Pb$ ages are often useful approximations of zircon crystallization ages. However, there are many exceptions to this generalization, and the reader is cautioned to interpret ages discussed below only as minimum ages when using them to solve or constrain geologic problems. This caveat and the considerations discussed above extend to all data presented in this chapter.

HARMONY FORMATION (UPPER CAMBRIAN?), HUMBOLDT COUNTY

The arkosic Harmony Formation is lithologically anomalous relative to other eugeoclinal deposits of the Roberts Mountains allochthon, which consist principally of chert, argillite, shale, quartzose siliciclastic rocks, and mafic volcanic rocks. The Harmony contains a tremendous volume of compositionally immature sediment that clearly represents substantial erosion of crystalline rocks. Due to its paleogeographic significance, the source of the Harmony has been the subject of considerable speculation (Palmer, 1971; Erickson and Marsh, 1974; Stewart and Poole, 1974; Stewart and Suczek, 1977; Suczek, 1977; Ketner, 1977; Schweickert and Snyder, 1981; Madrid, 1987). The Harmony is a distinctive unit that consists of feldspathic, poorly sorted grit, pebbly sandstone, sandstone, siltstone, shale, calcareous shale, and minor limestone (Roberts, 1964; Suczek, 1977; Madrid, 1987). Roberts (1964) did not attempt to measure a stratigraphic section because of structural complexity, but he estimated a total structural thickness of about 900 m for the Harmony in the Antler Peak Quadrangle in the Sonoma Range.

It is generally accepted that the Harmony Formation was derived from a crystalline igneous and metamorphic terrane and was deposited by sediment-water gravity flows along the western margin of North America. Petrographic data from the Harmony (Suczek, 1977) plot in the cratonic block field of the provenance discrimination diagrams of Dickinson and Suczek (1979). The relatively high angularity of most grains and the paucity of sedimentary- and volcanic-lithic fragments preclude the existence of abundant supracrustal rocks in the source region. Thus, the Harmony consists of first-cycle sediment derived from crystalline basement rocks. The problem is to locate possible sources that fit this description.

Faunal evidence indicates that the Harmony must have been derived, at least in part, from North America because trilobites from the Harmony display shallow-water, North American affinities (A. R. Palmer, *in* Hotz and Willden, 1964). The biogeographic tie to North America led workers to advocate the Salmon River Arch (Armstrong, 1975) in Idaho and Montana as a source of the Harmony (Suczek, 1977; Stewart and Suczek, 1977; Rowell and others, 1979). Selection of the Salmon River Arch as a source was based on the paleogeographic interpretation that the remainder of the craton in the western United States was undergoing marine carbonate sedimentation during the Late Cambrian and thus was unexposed and unable to yield terrigenous sediment. Rowell and others (1979) pointed out that evidence for the existence of a Late Cambrian positive feature along the Idaho-Montana border is equivocal but permissive. Suczek (1977) reviewed previously proposed sources of the Harmony and concluded that the only other reasonable possibility for a source was block faulting of Precambrian basement of the North American continental slope.

Analysis of detrital zircon from the Harmony Formation was undertaken to constrain the age of the supposed cratonic source of the Harmony. In addition, despite the biogeographic link with the North American miogeocline, no data exist that preclude partial or complete derivation of the siliciclastic component of the Harmony faleogeographic elements outboard of the miogeocline and craton. Study of the age of detrital zircon within the Harmony was undertaken also to test the hypothesis that the siliciclastic component was derived from the roots of a lower Paleozoic volcanic arc outboard of the lower Paleozoic continental margin.

Age of the Harmony Formation

Hotz and Willden (1964) collected trilobites from discontinuous limestone beds or lenses within sheared shale and sandstone at two exposures of the Harmony Foundation. Palmer (*in* Hotz and Willden, 1964) identified these trilobites as Late Cambrian (Franconian to early Trempealeauan) in age. It should be emphasized that the fossils that have been recovered from the Harmony constitute a derived fauna because they occur within resedimented limestone (Hotz and Willden, 1964; Suczek, 1977; Stewart and Suczek, 1977; Rowell and others, 1979) and therefore provide only a maximum age of deposition. Hotz and Willden (1964) interpreted an unexposed contact between the Harmony and the underlying Upper Cambrian (lower Dresbachian) Paradise Valley Chert as depositional on the basis of conformable dips and compatible age relations. Such an interpretation is equivocal, however, and some workers interpret this contact as a low-angle fault (Madrid, 1987). If the contact is indeed tectonic, then the Harmony would appear to be everywhere in fault contact with older rocks. A minimum age of deposition for the Harmony is provided by the lower Middle Pennsylvanian Battle Formation, which unconformably overlies the Harmony in the Sonoma Range (Roberts, 1964). Thus, the Harmony is constrained to have been deposited between the middle Late Cambrian and the early Middle Pennsylvanian. Madrid (1987) reported the occurrence of clasts in Lower Ordovician to Upper Devonian strata of the Roberts Mountains allochthon that are lithologically and petrographically indistinguishable from the Harmony Formation. His observations support the notion that the Harmony is Late Cambrian in age.

Geochronology of detrital zircon

Arkose from the Harmony was collected from a well-exposed, very coarse-grained, normally graded sandstone bed at

TABLE 1. SAMPLE LOCALITIES

Unit	Lat/Long	Township and Range	Description
Harmony Formation	41°56'10"N 117°37'29"W	SE1/4,NE1/4,SE1/4,Sec. 6,35N,39E	Well-exposed graded bed immediately east of road in Harmony Canyon at 5,840-ft contour (Winnemucca Quad)
Vinini Formation	40°54'17"N 116°17'7"W	Center N1/2,Sec.19,35N,51E	Outcrop of basal unit of tripartite sequence at 5,800-ft contour along north side of Carlin Mine Road (Simon Creek; Rodeo Creek NE Quad)
Palmetto Formation	32°32'8"N 117°28'50"W	Unsurveyed	Roadcut exposure at 6,300-ft contour on south side of road 0.5 km northwest of Railroad Pass (Montezuma Peak NW Quad)
Eureka Quartzite	39°35'6"N 116°16'18"W	Unsurveyed	8 m below top of Eureka at 6,680-ft contour on southwest flank of Lone Mountain (Bartine Ranch Quad)
Slaven Chert	40°27'12"W 116°45'48"W	Center S1/2,Sec.23,30N,46E	6,240-ft contour on south side of flat ridge (Mount Lewis Quad)

the type locality in the Sonoma Range (Fig. 1; Table 1). Approximately 40 kg yielded 4.8 g of zircon and other subordinate heavy minerals. Energy dispersive x-ray microanalysis of subhedral to anhedral heavy minerals of various colors (black, brown, yellow, green, and orange) indicated that they belong to the rutile, tourmaline, sphene, and monazite mineral groups.

Zircon grains exhibit considerable variability in size, color, roundness, clarity, and crystal form. Zircon grain size ranges from about 100 to 750 μm. Colors include dark purple, purple, pink, colorless, pearly white, and, rarely, brown. Zircons display the entire spectrum of roundness, from euhedral to very well rounded (Fig. 2A–C). Many sand-size euhedral zircons display brilliant crystal faces that exhibit no evidence of abrasion. These unabraded, euhedral grains occur in a variety of crystal forms that include both bipyramidal prisms and more complexly faceted types. Typically, most zircons are partially abraded. Very well-rounded grains occur in both elliptical and spherical shapes. Both multiple- and single-grain analyses were performed on handpicked fractions of these grain types selected from the bulk zircon population.

The analyses of detrital zircon from the Harmony reported here emphasize large, unabraded, euhedral, presumably first-cycle grains in an effort to date the crystalline source rocks. A bulk fraction of zircon containing all species yielded a discordant $^{207}Pb/^{206}Pb$ age of 1.63 Ga. Four relatively large fractions of handpicked, unabraded, euhedral, light pink zircon yielded $^{207}Pb/^{206}Pb$ ages between 1.74 and 1.84 Ga. Four smaller fractions of distinctive, handpicked, unabraded, complexly faceted, purple zircon yield a slightly narrower range of ages from 1.77 to 1.82 Ga. One fraction of well-rounded, purple zircon (Fig. 2c) yielded a $^{207}Pb/^{206}Pb$ age of 1.76 Ga (Fig. 3a).

Five fractions of distinctive, tabular, partially abraded, colorless zircon (Fig. 3b) yielded poorly constrained $^{207}Pb/^{206}Pb$ ages of 1.08 to 1.13 Ga. The mineralogy of five of these partially rounded grains was confirmed by energy dispersive x-ray microanalysis. A fraction of well-rounded colorless zircon yielded a $^{207}Pb/^{206}Pb$ age of 1.18 Ga (Fig. 3b). These data clearly record a distinct crystallization event that occurred roughly between 1.1 and 1.3 Ga.

Several analyses of single grains were performed to assess further the severity of zircon age heterogeneity. Two subhedral grains yielded $^{207}Pb/^{206}Pb$ ages of 1.75 and 1.77 Ga, which are similar to ages reported above. Two other single-grain analyses yielded younger $^{207}Pb/^{206}Pb$ ages of 1.34 and 0.69 Ga (Fig. 3c).

Discussion of results

The U-Pb zircon data indicate that feldspathic detritus of the Harmony Formation was derived from Precambrian rocks. $^{207}Pb/^{206}Pb$ ages for unabraded, euhedral, sand-size, presumably first-cycle zircon that range from 1.74 to 1.84 Ga provide a minimum age for the oldest component in individual analyses. These ages correspond to those of a well-documented, early Proterozoic basement terrane in the central United States (Fig. 4; Van Schmus and others, 1987). As indicated earlier, faunal evidence requires the Harmony to have been deposited within the North American trilobite biogeographic province. If we assume that the 1.74- to 1.84-Ga $^{207}Pb/^{206}Pb$ ages of first-cycle Harmony zircons approximate the crystallization age of zircon in the source region, then the Harmony is compatible with derivation from Precambrian basement in the west-central United States. Basement rocks with crystallization ages of 1.69 to 1.79 Ga crop out throughout Colorado, Utah, and northwestern Arizona, and presumably extend to the western limit of the craton in east-central Nevada (Fig. 4). The 1.34-Ga $^{207}Pb/^{206}Pb$ age is compatible with derivation from Middle Proterozoic anorogenic igneous rocks that intrude and overlie Early Proterozoic basement in the central United States (Thomas and others, 1984). The 691-Ma

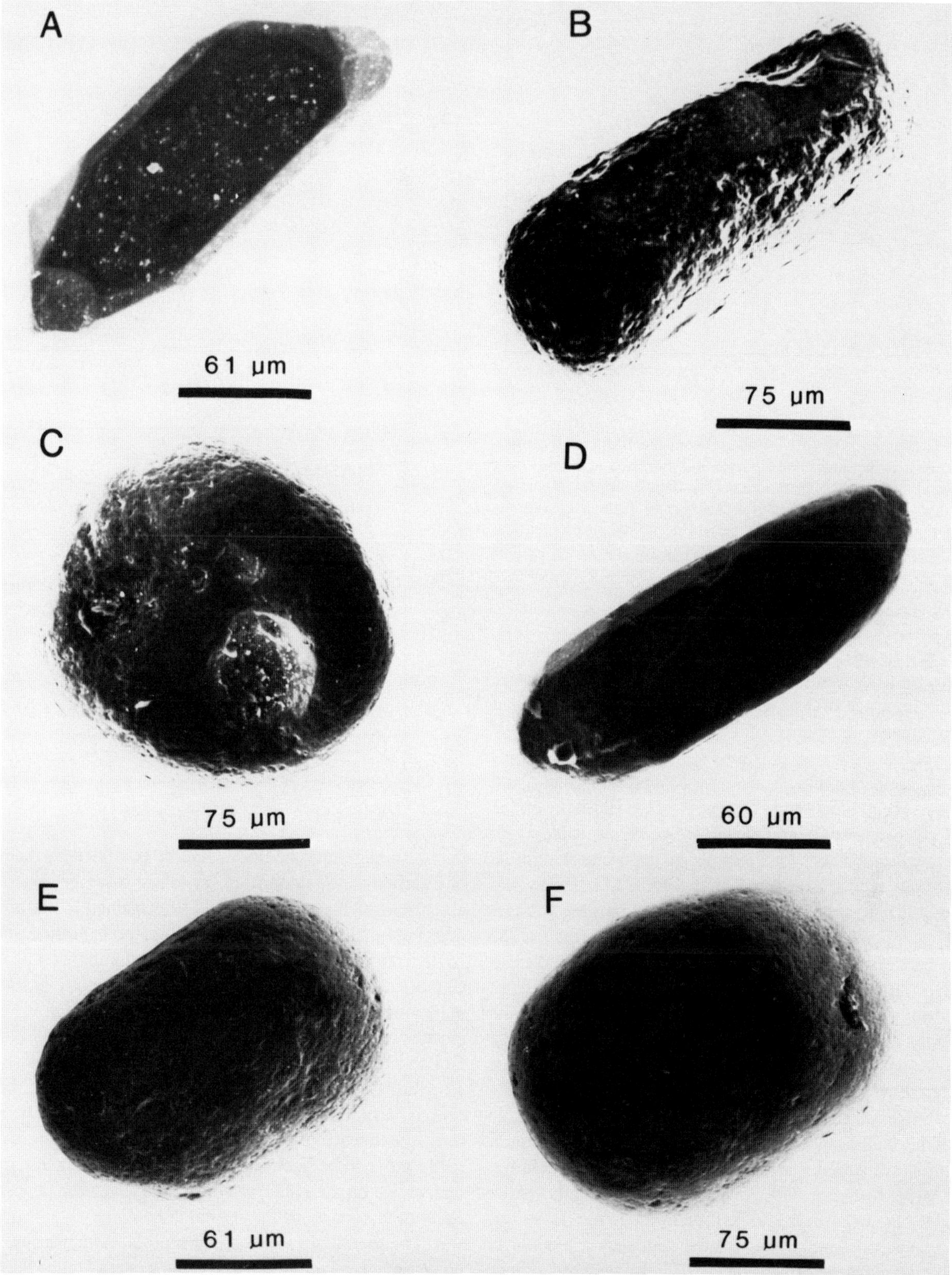

Figure 2. Scanning electron images of the shape of representative zircons. A, Harmony Formation, unabraded, euhedral purple zircon. B, Harmony Formation, abraded, colorless zircon with partially preserved crystal faces. C, Harmony Formation, well-rounded purple zircon. D, Vinini Formation, euhedral, unabraded purple zircon. E, Vinini Formation, well-rounded purple zircon. F, Palmetto Formation, very well-rounded purple zircon.

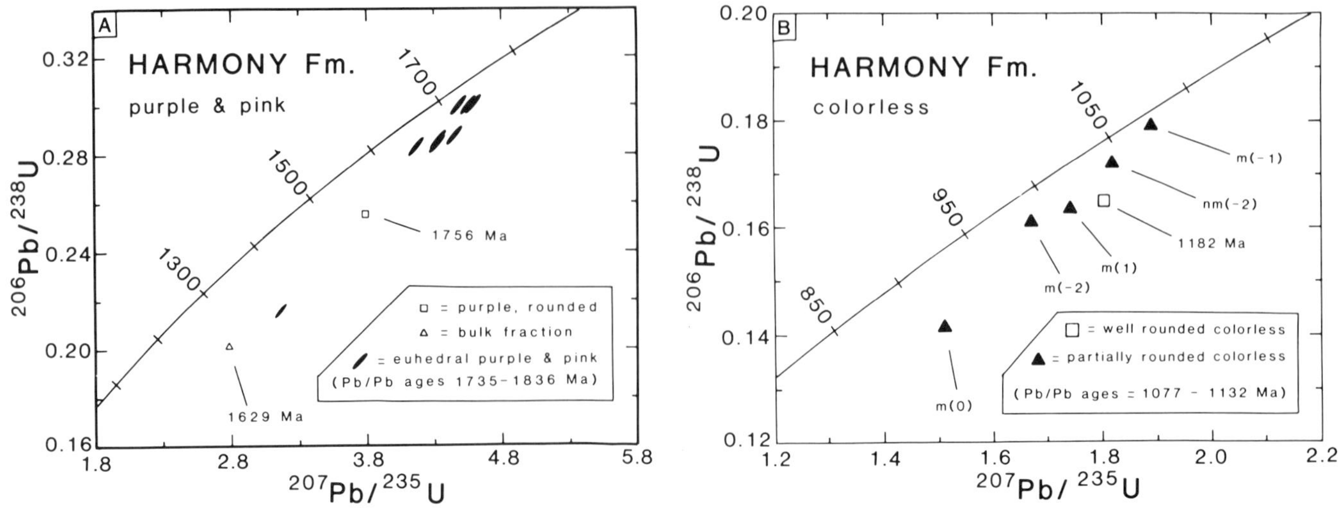

Figure 3. Concordia diagrams for zircon from the Harmony Formation. A, Multiple-grain analyses of purple and pink zircon; ellipses indicate nominal ±1 percent uncertainties. Uncertainties for the other two fractions are similar. B, Multiple-grain analyses of distinctive, colorless zircon with partially preserved crystal faces. Uncertainties are larger than indicated by triangles (see Table 2). C, Single-grain analyses of various species of zircon; uncertainties too small to show at scale of diagram.

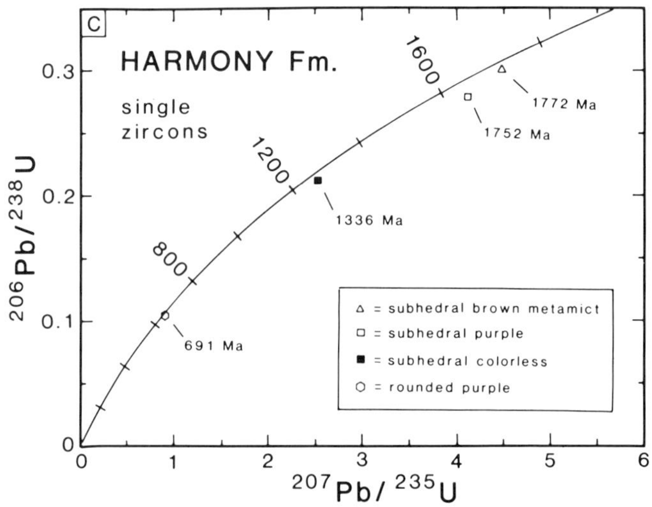

$^{207}Pb/^{206}Pb$ age may record a Late Proterozoic magmatic event. Given this evidence for older and relatively younger components, the bulk zircon age of 1.63 Ga is probably a hybrid age that reflects a greater abundance of 1.7- to 1.8-Ga zircon than of younger species.

As mentioned earlier, the Salmon River arch in Idaho and Montana has been proposed as the source of the Harmony on the basis of paleogeographic arguments. However, available U-Pb zircon crystallization ages for Precambrian exposures within the Salmon River arch cluster at 1370 and 1576 Ma (Evans and Zartman, 1981; Evans and Fischer, 1986) and thus are younger than the age of euhedral detrital zircon from the Harmony. The disparity between the age of first-cycle Harmony zircon and existing ages for the Precambrian of the Salmon River arch is difficult to reconcile. If the limited amount of available age data for the Precambrian of the Salmon River arch is representative of the age of the basement there, then we must look elsewhere for the source of the Harmony.

If we reject the Salmon River arch proper as a permissible source, block faulting of Precambrian basement along the western margin of North America is the most likely source, a conclusion reached earlier by Suczek (1977). The detrital zircon data restrict permissible basement sources to be Proterozoic in age. However, an important qualification is associated with such a model. As others have argued before (Ketner, 1977; Rowell and others, 1979), the lack of siliciclastic detritus in Upper Cambrian shelf strata is difficult to reconcile with the notion of significant lateral transport of arkosic sediment across the shelf from cratonic basement east of the shelf edge. In addition, there is no evidence for the unroofing sequence that would be generated by the removal of Cambrian shelf strata from crystalline basement, although the absence of such a sequence could be explained by the allochthonous nature of the Harmony. Therefore, this hypothetical faulting of the cratonic margin probably occurred at or west of the Late Cambrian shelf edge. Outcrops of the Harmony occur immediately west of the 0.706 isopleth of initial $^{87}Sr/^{86}Sr$ in Mesozoic igneous rocks (Kistler and Peterman, 1978; see also Farmer and DePaolo, 1983; Solomon and Taylor, 1989) that appears to delineate the western extent of the Precambrian cratonic margin in central Nevada. This spatial coincidence may simply be a fortuitous consequence of Phanerozoic tectonism, but also may eventually prove to be significant once these displacements are better understood because the Harmony could have been deposited as a result of faulting along this crustal discontinuity.

The source of 1.1-Ga zircon is problematic because there is no basement terrane of that age in the western United States, with the exception of the Pikes Peak batholith in Colorado (Scharer and Allegre, 1982). Given the absence of arkosic sediment within the Cordilleran miogeocline (Ketner, 1977), it is highly unlikely that the Pike's Peak batholith could have been a source of the

Harmony. The nearest documented extensive terrane of 1.0- to 1.3-Ga basement rocks occurs in the southwestern extension of the Grenville Province in Mexico (Hoffman, 1988). This latter observation suggests several intriguing possibilities concerning the provenance of the Harmony. The simplest explanation for the occurrence of 1.1-Ga zircon in the Harmony would be that rocks of that age reside (or did reside) in the western United States but are simply undocumented. Alternatively, detritus in the Harmony may have been transported northward longitudinally along the toe of the Cordilleran miogeocline from an area south of 30°N latitude (present day). Finally, it is conceivable that the Harmony could have been deposited far south of its present position and transported tectonically northward prior to its accretion to the miogeocline during the Antler Orogeny. These possibilities merit future consideration.

Ketner (1977) advocated a western source for the Harmony and other siliciclastic strata of the Roberts Mountains allochthon on the basis of sedimentological and paleogeographical arguments. He proposed that deposition of strata within the Roberts Mountains allochthon occurred in a structural trough in continental crust, and that siliciclastic detritus was shed eastward to form the Harmony and other units. However, as Suczek (1977) pointed out, (1) there is no evidence for the existence of such a cratonic block west of the Roberts Mountains allochthon today; and (2) if such a block did exist and was subsequently tectonically displaced, its location is unknown and its existence is completely speculative, given alternative explanations. Nevertheless, Ketner's (1977) arguments are based on regional field observations that must be explained by models of tectonic development and sedimentation. The work of Madrid (1987) appears to shed light on this problem. He reported the occurrence of sandstone clasts that are lithologically and petrographically indistinguishable from the Harmony within Lower Ordovician to Upper Devonian strata of the Roberts Mountains allochthon. If these clasts represent recycled Harmony Formation, then Madrid's observations are highly significant for two reasons. First, as mentioned earlier, they support the notion that the Harmony is Late Cambrian in age. Second, they suggest that the Harmony may have been the source of second-cycle feldspathic detritus that was deposited throughout the Roberts Mountains allochthon during Early Ordovician to Late Devonian time (Madrid, 1987). The significance of this concept for Ketner's (1977) model is that protracted uplift and erosion of an arkose would leave a geologic record similar in appearance to that of an exposed continental block.

Madrid (1987) similarly envisioned a western source for Ordovician to Devonian strata of the Roberts Mountains allochthon. He wrote that this source was a block-faulted uplift that exposed the Harmony Formation to erosion. The only significant difference between these models is that Madrid (1987) accepted Suczek's (1977) proposition that the Salmon River arch is the source of the Harmony, whereas Ketner (1977) invoked a western source for the Harmony. While the isotopic data reported here are compatible with derivation of the Harmony from Precambrian basement in the western United States, it should be emphasized that they do not preclude a cratonic block source that was outboard of the miogeocline. Additional information relevant to the idea that detritus from the Harmony may have been recycled into younger strata of the Roberts Mountains allochthon is discussed below in the section on the Vinini Formation.

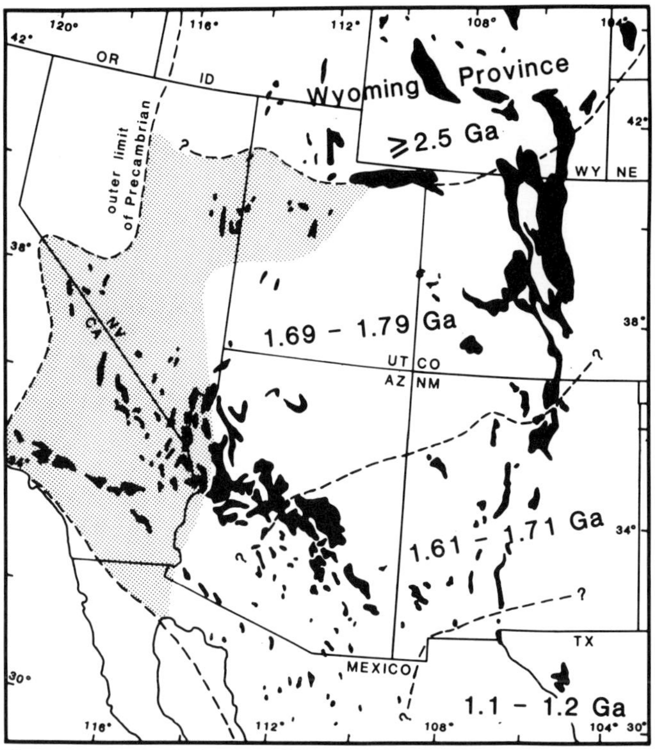

Figure 4. Generalized map of Precambrian crustal provinces of the southwestern United States. Outcrops of Precambrian rocks shown in black. Age ranges represent the dominant age of crust defined on the basis of U/Pb zircon and Rb/Sr isotopic ages. Stippled area indicates rocks that have Sm-Nd model ages between 2.0 to 2.3 Ga. Modified after Condie, 1982, and Hoffman, 1988.

ORDOVICIAN SANDSTONES OF THE ROBERTS MOUNTAINS ALLOCHTHON

Western Nevada contains chert, argillite, quartzite, sandstone, siltstone, shale, greenstone, and minor amounts of limestone and bedded barite that were deposited during all epochs of the Ordovician (Stewart, 1980). These strata consist of siliciclastic, hemipelagic, pelagic, and volcanic rocks that were deposited outboard of the coeval carbonate shelf to the east. Competing hypotheses concerning the source of Ordovician quartzose sandstones have been proposed. For example, Ketner (1977) suggested a western source, whereas Miller and Larue (1983) advocated a source to the east. Data presented below suggest that both hypotheses may have merit.

As a result of structural complexity, the concomitant uncertainty in lithostratigraphic correlations, and regional variability in sedimentation, a variety of local and regional stratigraphic nomenclature has been applied to the Ordovician System of the Roberts Mountains allochthon. Several workers (Schweickert and Snyder, 1981; Madrid, 1987) have pointed out that the term "formation" has been inappropriately applied to much of the Ordovician System in the past. Schweickert and Snyder (1981) emphasized the lithologic discontinuity and heterogeneity within the Valmy and Vinini Formations and suggested that, in many places, use of the term "formation" as a rock stratigraphic unit is inappropriate. Most of the Roberts Mountains allochthon consists of thrust nappes and thus contain strata that are disrupted and laterally discontinuous on regional and commonly local scales. Despite the structural complexity of the Roberts Mountains allochthon, Madrid (1987) found locally consistent stratigraphic sequences within both repeated and laterally adjacent thrust nappe exposures. He subdivided quartzose, Ordovician sandstones of the Roberts Mountains allochthon into feldspar-bearing and feldspar-free varieties and proposed different sources for each. Data presented below are consistent with his two-fold classification.

In the sections below, geochronological analyses of detrital zircon are presented to constrain the provenance of these Ordovician siliciclastic deposits. Lower and Middle Ordovician sandstones were selected from two widely spaced areas of the allochthon (Fig. 1). In addition, detrital zircon from Middle Ordovician quartzite of the shelf to the east has been analyzed similarly to evaluate the hypothesis that Middle Ordovician eugeoclinal quartzite contains detritus that was resedimented from coeval shelf sediments (Fig. 1).

VININI FORMATION (LOWER ORDOVICIAN), ELKO COUNTY

A fairly well-exposed, continuous sequence of Ordovician to Silurian Vinini Formation crops out along Simon Creek in the Tuscarora Mountains northwest of Elko, Nevada (Fig. 1; Table 1). The tripartite Ordovician sequence consists of mixed siliciclastic and carbonate rocks at the base, shale and quartzite in the middle, and black chert at the top (K. B. Ketner, unpublished data; author's unpublished data). Three limestone samples from an outcrop of the basal sequence yielded abundant conodonts that are early Early Ordovician in age (collected by K. B. Ketner, identified by J. Repetski: K. B. Ketner, written communication, 1985). Silurian chert and siltstone overlie the Ordovician strata along an apparently conformable contact. Rocks in the basal carbonate-siliciclastic sequence include thinly bedded, micritic and calcarenitic limestone, quartz-bearing limestone, calcareous shale, calcareous sandstone, pebbly sandstone, and both matrix- and clast-supported, flat limestone pebble breccia. Thin, centimeter- to decimeter-scale, structureless beds and abundant, thicker, decimenter- to meter-scale, normally graded beds are present. Clasts within the pebbly sandstone consist of limestone, dolomite, argillite, and siltstone. Petrographic study of this fine- to medium-grained, pebbly sandstone indicates that it consists principally of subangular to rounded monocrystalline quartz, extrabasinal carbonate lithic grains, and intrabasinal carbonate grains. Trace amounts of feldspar are extensively replaced by calcite. The presence of feldspar and subangular quartz in this rock indicates considerable textural and compositional immaturity relative to other Ordovician quartzose strata of the Roberts Mountains allochthon. The relative immaturity of this Lower Ordovician sequence suggests that it may have had a different source than highly quartzose, nonfeldspathic Ordovician strata.

Geochronology of detrital zircon

Forty kilograms of pebbly, medium-grained, quartz wacke yielded 10 mg of zircon with associated barite, sulfides, and subordinate tourmaline, rutile, monazite, sphene, and magnetite. Zircon grains are well sorted about a mean size of approximately 100 μm and consist of colorless, light pink, and purple varieties. The roundness of zircon varies considerably from very well rounded to euhedral (Fig. 2D–E). Sparse euhedral and subhedral zircons exhibit little evidence of abrasion.

Five discordant zircon analyses yield a variety of $^{207}Pb/^{206}Pb$ ages. A bulk fraction of zircon yields a discordant $^{207}Pb/^{206}Pb$ age of 1.73 Ga. Four handpicked fractions of different zircon species were also analyzed to identify within-population age heterogeneity. Rounded, spherical, purple zircon yielded a $^{207}Pb/^{206}Pb$ age of 2.28 Ga. Analyses of other species yielded younger $^{207}Pb/^{206}Pb$ ages ranging from 1485 to 1795 Ma (Table 2; Fig. 5a).

Discussion of results

These detrital zircon data are relatively discordant and should be interpreted with caution. Nevertheless, except for the 2.28-Ga fraction, the remaining ages are quite similar to those obtained from the Harmony Formation. This observation, coupled with the remarkable similarity of their heavy mineral suites, suggests the possibility of a provenance link between the Harmony and the Ordovician section in the Tuscarora Mountains. The composition and age of detrital zircon of this sandstone suggest two intriguing possibilities concerning its source. The first is that it may have been derived directly from the source of the slightly older(?) Harmony Formation. The second is that it may represent second-cycle detritus derived from uplift of the Harmony Formation itself. Ketner (1966, 1977) had advocated a western source for many siliciclastic units of the Roberts Mountains allochthon on the basis of regional sedimentological observations. As mentioned earlier, Madrid (1987) reported abundant Harmony-like detritus within Lower Ordovician to Upper Devonian strata of the Roberts Mountains allochthon. He advocated uplift and erosion of the Harmony as the source of feldspathic, second-cycle detritus in these younger strata. The new detrital zircon data are compatible with this hypothesis. Nonfeldspathic, quartz-rich Ordovician deposits in Nevada are discussed below.

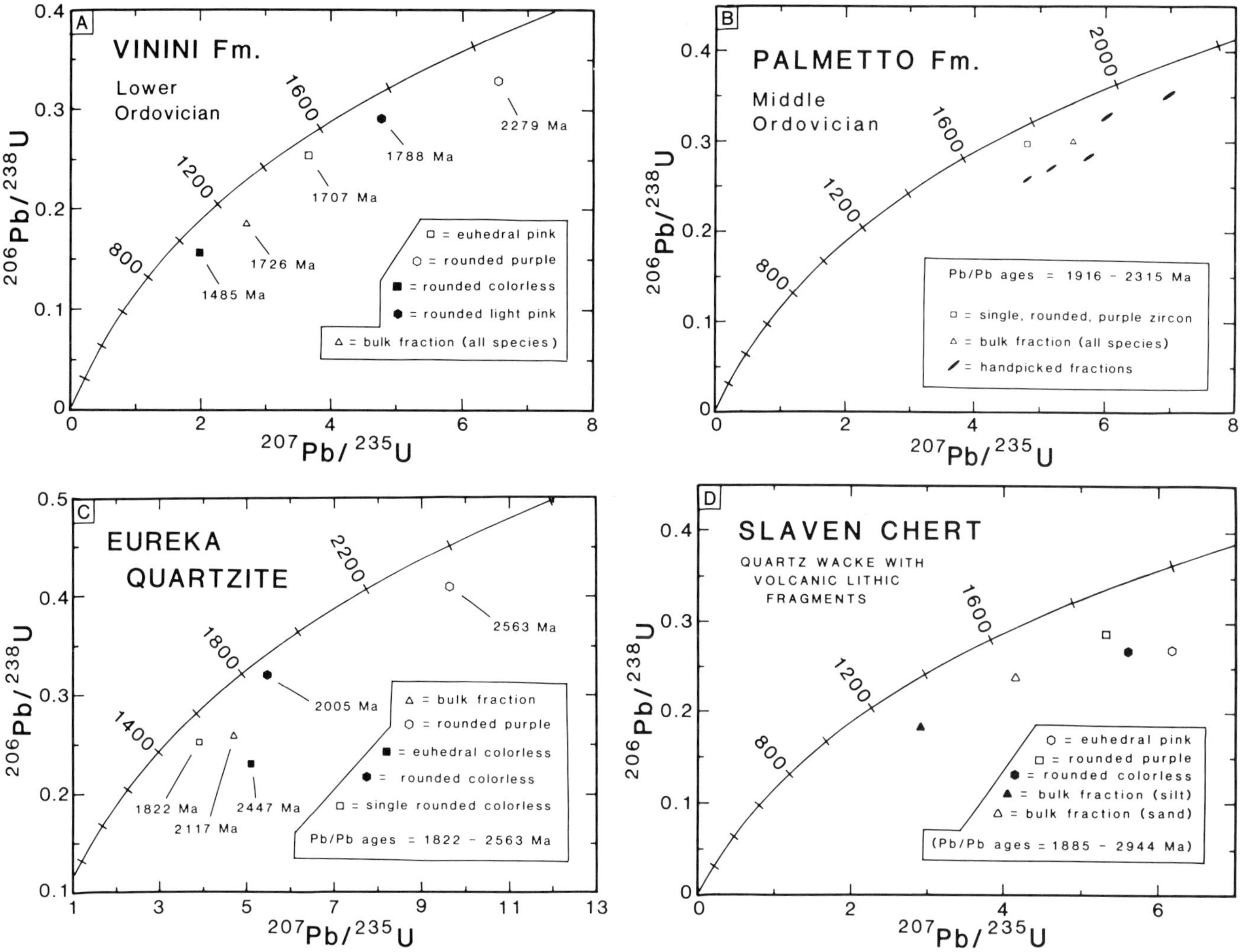

Figure 5. Concordia diagrams for Ordovician and Devonian units. Most calculated uncertainties are too small to show at scale of diagram. A, Vinini Formation. B, Palmetto Formation; ellipses indicate nominal (±1 percent uncertainties. C, Eureka Quartzite. D, Slaven Chert. See text for discussion.

PALMETTO FORMATION (MIDDLE ORDOVICIAN), ESMERALDA COUNTY

A well-exposed Middle Ordovician to Lower Silurian sequence of the Palmetto Formation (Albers and Stewart, 1972) crops out near Railroad Pass in Esmeralda County (Fig. 1; Table 1). This 80-m-thick sequence consists of interbedded quartzite and graptolitic shale at its base, interbedded Upper Ordovician black chert and shale in the middle, and lowermost Silurian colored chert at the top (K. B. Ketner, unpublished data). Middle Ordovician graptolites interbedded with the quartzites belong to the Zone of *Orthograptus Amplexicaulis* (identified by Claire Carter, written communication to K. B. Ketner, 1983). Quartzites are thickly bedded, gray, and medium- to fine-grained with minor amounts of carbonate cement. Framework grains are 99 percent very well-rounded, well-sorted, monocrystalline quartz, many of which exhibit partial abraded quartz overgrowths. The quartzite beds are structureless except for minor normal grading, which suggests that they were deposited below storm wave base by sediment gravity flows.

Forty kilograms of medium-grained quartzite from the Railroad Pass locality yielded 150 mg of virtually pure zircon. Zircon grains are well sorted about a mean size of approximately 160 μm. All zircon grains are very well rounded and highly spherical, and consist of colorless, light pink, and purple varieties (Fig. 2f). The high roundness and sphericity, coupled with the observed recycled sedimentary quartz, clearly indicate a polycyclic history for this zircon population.

$^{207}Pb/^{206}Pb$ ages obtained for multiple-grain fractions of these zircon species vary considerably, but all ages are older than 2.0 Ga. These analyses exhibit considerable nonsystematic scatter on the concordia diagram (Fig. 5b), which probably reflects thorough mixing of sediments from different sources by sediment transport and recycling. A bulk fraction of zircon yielded a

TABLE 2. U-Pb ISOTOPIC DATA AND APPARENT AGES

Fraction*	Sample Wt. (mg)	Concentrations† ^{238}U (ppm)	$^{206}Pb^*$ (ppm)	Measured§ Pb-204/Pb-206	Pb-207/Pb-206	Radiogenic Isotopic Ratios** Pb-207/Pb-206	Pb-206/U-238	Pb-207/U-235	Ages‡ Pb-206/U-238	Pb-207/U-235	Pb-207/Pb-206
Harmony Formation (Upper Cambrian(?)), Sonoma Range, Nevada											
					Multiple-grain analyses						
m(0) blk	3.592	356.1	62.00	0.000372(1)	0.105355(8)	0.10025	0.2013	2.782	1182	1350	1629 ± 3
nm(-2) lt pk eu- and sub	0.773	581.8	143.21	0.000368(5)	0.115271(22)	0.11027	0.2845	4.324	1614	1698	1804 ± 4
m(-2) lt pk eu- and sub	1.787	414.9	101.66	0.000124(4)	0.108431(17)	0.10674	0.2832	4.167	1607	1668	1745 ± 3
m(-1) lt pk eu- and sub	1.644	667.2	166.04	0.006656(3)	0.202357(30)	0.11226	0.2876	4.452	1630	1722	1836 ± 60
m(5) lt pk eu- and sub	0.463	401.3	74.97	0.000985(5)	0.119646(19)	0.10620	0.2159	3.162	1260	1448	1735 ± 8
nm(-2) pur eu	0.248	440.1	114.31	0.000894(1)	0.122149(7)	0.11001	0.3002	4.553	1692	1740	1800 ± 5
m(-2) pur eu	0.240	473.9	123.17	0.000934(1)	0.123626(6)	0.11096	0.3004	4.596	1693	1749	1815 ± 6
m(-1) pur eu	0.226	538.7	139.78	0.000412(1)	0.113994(6)	0.10839	0.2999	4.481	1691	1728	1773 ± 3
m(1) pur eu	0.136	554.2	137.14	0.000484(1)	0.116381(5)	0.10981	0.2860	4.330	1622	1699	1796 ± 3
m(-1) pur rnd	0.112	497.5	110.08	0.000230(3)	0.110313(21)	0.10739	0.2557	3.786	1468	1590	1756 ± 3
nm(-2) clr -100+240	0.129	376.9	56.11	0.006615(52)	0.170568(67)	0.07663	0.1721	1.818	1024	1052	1111 ±144
m(-2) clr -100+240	0.146	71.02	9.89	0.008659(46)	0.198596(72)	0.07531	0.1610	1.672	963	998	1077 ±112
m(-1) clr -100+240	0.120	131.7	20.41	0.009362(44)	0.209525(98)	0.07651	0.1791	1.889	1062	1077	1109 ±128
m(0) clr -100+240	0.100	140.0	17.15	0.013437(28)	0.268121(90)	0.07743	0.1416	1.512	854	935	1132 ±164
m(1) clr -100+240	0.139	128.1	18.12	0.007930(16)	0.189861(74)	0.07728	0.1635	1.742	976	1024	1129 ± 80
m(-1) clr rnd -100+240	0.428	136.8	19.50	0.000722(2)	0.089354(21)	0.07938	0.1647	1.803	983	1047	1182 ± 3
					Single-grain analyses						
nm(-2+100 brn mct sub	0.050	613.4	159.20	0.000210(1)	0.111114(28)	0.10386	0.2997	4.482	1690	1728	1772 ± 3
nm(-2)+100 pur sub	0.070	151.2	36.47	0.000613(1)	0.115159(31)	0.10715	0.2785	4.119	1584	1658	1752 ± 3
nm(-2)+10 clr sub	0.105	92.99	17.15	0.002259(8)	0.117265(34)	0.08591	0.2126	2.526	1243	1279	1336 ± 18
nm(-2)+100 pur rnd	0.065	165.1	15.01	0.003392(4)	0.110914(15)	0.06250	0.1050	0.9055	644	654	691 ± 24
Palmetto Formation (Middle Ordovician), Railroad Pass, Montezuma Range, Nevada											
					Multiple-grain analyses						
m(10) blk	6.055	186.1	48.20	0.000072(1)	0.134350(8)	0.13341	0.2995	5.508	1689	1902	2143 ± 2
nm(-2)1.7A pur rnd	0.173	334.7	94.91	0.000040(1)	0.133760(22)	0.13338	0.3278	6.029	1827	1980	2143 ± 3
m(-2) pur rnd	0.300	328.7	99.97	0.000076(2)	0.144899(24)	0.14404	0.3515	6.981	1942	2109	2276 ± 3
nm(-2)1.7A lt pk rnd	0.184	93.76	21.95	0.000384(2)	0.143027(22)	0.13876	0.2705	5.178	1543	1849	2212 ± 3
m(-2)1.7A lt pk rnd	0.148	107.3	26.27	0.000377(2)	0.151312(32)	0.14735	0.2829	5.752	1606	1939	2315 ± 3
m(-2) lt pk rnd	0.216	94.42	21.12	0.000924(5)	0.146575(31)	0.13507	0.2584	4.814	1482	1787	2165 ± 5

TABLE 2. U-Pb ISOTOPIC DATA AND APPARENT AGES (continued)

		Concentrations†		Measured§		Radiogenic Isotopic Ratios**			Ages‡		
Fraction*	Sample Wt. (mg)	^{238}U (ppm)	$^{206}Pb^{*}$ (ppm)	Pb-204/Pb-206	Pb-207/Pb-206	Pb-207/Pb-206	Pb-206/U-238	Pb-207/U-235	Pb-206/U-238	Pb-207/U-235	Pb-207/Pb-206
Palmetto Formation (Middle Ordovician), Railroad Pass, Montezuma Range, Nevada (continued)											
					Single-grain analyses						
m(-2) pur rnd	0.033	405.7	104.38	0.000388(3)	0.122062(26)	0.11737	0.2971	4.812	1677	1787	1916 ± 3
Vinini Formation (lower Lower Ordovician), Tuscarora Mountains, Nevada											
					Multiple grain analyses						
m(10) blk	0.998	344.1	55.34	0.000588(1)	0.113695(21)	0.10567	0.1859	2.708	1099	1330	1726 ± 5
m(-1) pk eu	0.121	324.4	71.42	0.001491(3)	0.124720(18)	0.10460	0.2544	3.670	1461	1565	1707 ± 12
M(-1) pur rnd	0.071	372.4	106.14	0.001590(3)	0.164303(19)	0.14425	0.3293	6.552	1835	2053	2279 ± 8
m(-1) lt pk rnd	0.151	232.4	57.74	0.001578(2)	0.130378(30)	0.10931	0.2870	4.327	1627	1699	1788 ± 11
M(-1) clr rnd	0.066	270.4	36.42	0.001200(3)	0.108952(16)	0.09284	0.1555	1.922	932	1112	1485 ± 9
Slaven Chert (Middle Devonian), northern Shoshone Range, Nevada											
					Multiple-grain analyses						
m(10)+240 blk	3.177	557.5	115.01	0.000271(1)	0.130041(16)	0.12643	0.2385	4.157	1379	1666	2049 ± 3
m(10)-240 blk	2.528	478.1	75.79	0.000268(1)	0.118936(19)	0.11533	0.1832	2.913	1085	1385	1885 ± 3
m(-2) pur rnd	0.171	483.3	120.10	0.000232(1)	0.137298(26)	0.13437	0.2872	5.321	1627	1872	2156 ± 2
m(-2) clr rnd	0.133	89.13	20.75	0.000700(2)	0.159713(41)	0.15133	0.2688	5.613	1535	1918	2361 ± 4
m(-2) pk eu	0.051	248.3	58.14	0.000634(2)	0.172722(50)	0.16562	0.2704	6.179	1542	2002	2514 ± 4
					Single-grain analyses						
nm(-2) clr rnd	0.048	95.13	20.31	0.002199(5)	0.236136(61)	0.21504	0.2463	7.318	1419	2151	2944 ± 5
Eureka Quartzite (Middle Ordovician), Lone Mountain, Nevada											
					Multiple-grain analyses						
m(10) blk	2.004	222.9	49.90	0.000580(1)	0.139058(6)	0.13141	0.2588	4.689	1484	1765	2117 ± 3
euhedral	0.120	223.9	44.87	0.000575(2)	0.165407(90)	0.15921	0.2314	5.085	1342	1834	2447 ± 3
nm(-1) pur rnd	0.132	238.9	84.71	0.000164(1)	0.172129(26)	0.17049	0.4096	9.632	2213	2400	2563 ± 3
nm(-1) clr rnd	0.660	193.3	53.75	0.000162(1)	0.125381(37)	0.12335	0.3213	5.465	1796	1895	2005 ± 3
					Single-grain analyses						
nm(-1) clr rnd	0.036	278.5	61.06	0.000639(1)	0.118946(17)	0.11139	0.2531	3.891	1455	1612	1822 ± 3

Notes: The isotopic composition of nonradiogenic Pb used to calculate radiogenic ^{207}Pb and ^{206}Pb was estimated using Stacey and Kramers' (1975) Pb evolution model and the $^{207}Pb/^{206}Pb$ age of individual analyses; analytical blanks ranged between 50 and 160 picograms total Pb.

*nm = nonmagnetic; m = magnetic; numbers in parentheses indicate side tilt used on Franz separator at 1.4 amps unless otherwise indicated; AA = air-abraded fraction; ±240 indicates mesh size fraction; blk = bulk; eu = euhedral; sub = subhedral; an = anhedral; rnd = rounded; lt = light; pk = pink; pur = purple; clr = colorless; brn = brown; mct = metamict.

†U corrected for analytical blank; Pb corrected for blank and nonradiogenic Pb.

§Corrected for mass fractionation using 0.12 percent bias per mass unit; uncorrected for blank.

**Pb corrected for blank and nonradiogenic Pb; U corrected for blank.

‡Ages in Ma; U decay constants from Jaffey and others (1971); uncertainties reported at 2σ level are based on precision of measured ratios and a nominal uncertainty of ±0.2 in initial $^{207}Pb/^{204}Pb$ and $^{206}Pb/^{204}Pb$.

$^{207}Pb/^{206}Pb$ age of 2.14 Ga. Five other fractions of two different zircon species yielded $^{207}Pb/^{206}Pb$ ages ranging between 2.14 and 2.32 Ga (Fig. 5b).

EUREKA QUARTZITE (MIDDLE ORDOVICIAN), EUREKA COUNTY

The Middle Ordovician Eureka Quartzite is a distinctive siliciclastic shelf deposit that crops out widely in eastern Nevada. Quartzite for detrital zircon analyses was collected from the type locality on the southwest flank of Lone Mountain in Eureka County (Fig. 1; Table 1), as redesignated by Kirk (1933), where Nolan and others (1956) reported a thickness of 107 m. Framework grains are 99 percent very well rounded, very well sorted, monocrystalline quartz. Some grains exhibit abraded quartz overgrowths.

Forty kilograms of quartzite yielded 140 mg of predominantly rounded zircon with trace amounts of euhedral unabraded zircon, apatite, rutile, tourmaline, and sphene. Abraded zircon grains are very well rounded and highly spherical, and are quite similar in character to zircon from the Palmetto Formation (Fig. 2f). Colors include purple, light pink, colorless, and, rarely, translucent pearly white.

A bulk fraction of zircon containing all species yielded a $^{207}Pb/^{206}Pb$ age of 2.12 Ga. Handpicked fractions of very well-rounded, purple and colorless zircon yielded $^{207}Pb/^{206}Pb$ ages of 2.56 and 2.05 Ga, respectively. A single, rounded, colorless grain yielded a $^{207}Pb/^{206}Pb$ age of 1.82 Ga (Fig. 5c). The trace amounts of euhedral zircon and other heavy minerals constitute a significant textural inversion relative to the rounded zircon. For this reason, the euhedral zircon was separated from the population by hand and dated to test the possibility that it was derived from pyroclastic airfall related to Middle Ordovician volcanism. These unabraded grains yielded a $^{207}Pb/^{206}Pb$ age of 2.45 Ga (Fig. 5c).

COMPARISON OF THE PROVENANCE OF MIDDLE ORDOVICIAN EUGEOCLINAL AND MIOGEOCLINAL QUARTZITES IN NEVADA

Extremely quartz-rich sand was deposited in both the Cordilleran miogeocline (Eureka Quartzite) and eugeocline (Palmetto Formation) during a short interval of the Middle Ordovician. It is generally accepted that Cordilleran miogeoclinal quartzites were derived from rocks of the North American craton to the east and north (present day) (Ross, 1964, 1977). Original stratigraphic relations between miogeoclinal and eugeoclinal quartzites in Nevada remain controversial. Ketner (1966) interpreted eugeoclinal quartzites to have been derived from a western source that was distinct from an eastern source for coeval miogeoclinal quartzites. His interpretation was based on subtle differences in sorting and grain size between miogeoclinal and eugeoclinal quartzites. Although similarities between miogeoclinal and eugeoclinal quartzites are remarkable, he believed that the subtle textural differences have paleogeographic significance. He concluded that the relatively coarse, more poorly sorted, eugeoclinal quartzites could not have been derived from spillage of coeval, finer grained, better sorted shelf sediment over the shelf edge.

In contrast, Miller and Larue (1983) found similarities between miogeoclinal and eugeoclinal quartzites to be compelling; they inferred an eastern cratonic source for eugeoclinal quartzites. Miller and Larue argued that the subtle textural differences reported by Ketner (1966) might be explained simply by differences in depositional mechanisms and by sediment bypassing on the continental shelf.

New U-Pb isotopic ages for multiple-grain fractions of detrital zircon from eugeoclinal and miogeoclinal quartzite presented above exhibit broad similarity with regard to both their magnitude and systematics. These data agree with previously determined ages of detrital zircon from both eugeoclinal and miogeoclinal quartzites (Ketner, 1977; K. B. Ketner, written communication, 1985). The range of $^{207}Pb/^{206}Pb$ ages for zircon from the eugeoclinal quartzite (2.14 to 2.32 Ga) raises an interesting question because it can be interpreted in three fundamentally different ways. The first is that the $^{207}Pb/^{206}Pb$ ages may simply be discordant age estimates of exclusively Archean zircon. This possibility is unlikely if the sediment was transported from the Canadian Shield as envisioned by Ross (1964, 1977). The second possibility is that the $^{207}Pb/^{206}Pb$ ages represent isotopic averages that result from sedimentary mixing of Archean and Proterozoic zircon. Ross (1964) documented the southward (present day) dispersal of quartz sand into Utah and Nevada during the Middle Ordovician. Presumably, this tremendous volume of sediment was derived from rocks of the Canadian Shield that contain both Archean and Proterozoic zircon. The third possible interpretation of the $^{207}Pb/^{206}Pb$ ages is that they may record fairly accurate crystallization-age estimates for 2.0- to 2.3-Ga basement source terrane. A fragmentary, but growing, body of evidence suggests that such a terrane, which has been documented in Canada (Parrish and Wheeler, 1983; Aleinikoff and others, 1986; Ross and others, 1988), may underlie Phanerozoic allochthons in the western United States (Fig. 4; Farmer and DePaolo, 1983, 1984; Bennett and DePaolo, 1987; Girty and Wardlaw, 1985; Girty and others, 1985; Bowring and Podosek, 1988).

Available data do not permit discrimination between the latter two hypotheses. However, regardless of which interpretation is correct, the isotopic ages and systematics provide a useful isotopic signature. The principal difference between detrital zircon age data from eugeoclinal and miogeoclinal quartzites is the absence of a documented Archean component in the eugeoclinal quartzite. It is uncertain whether this apparent lack of an Archean component reflects a real difference in provenance or whether it is an artifact of isotopic systematics related to discordance of the zircon ages or of insufficient sampling. Additional single-grain analyses may eventually reveal the presence of Archean zircon. Yet, until an unequivocal Archean zircon component can be demonstrated for eugeoclinal quartzite, the differences between the ages of zircon from eugeoclinal and miogeoclinal quartzites must be considered to be as compelling as the similarities.

SLAVEN CHERT (MIDDLE DEVONIAN), LANDER COUNTY

Dickinson (1977) pointed out that reliable documentation of Silurian or Devonian arc-derived detritus within the Roberts Mountains allochthon would lend much support to the notion that the allochthon represents a closed back-arc basin. A sample of Middle Devonian, volcanic lithic fragment–bearing quartz wacke was selected according to existing biostratigraphic control in an attempt to isolate and date any Paleozoic zircon component. This sample was collected from fossil locality F-29 of Gilluly and Gates (1965), which contains numerous ostracodes identified by Jean Berdan as Middle Devonian (probably Givetian) in age.

The Slaven Chert was named by Gilluly and Gates (1965) for exposures in the northern Shoshone Range (Fig. 1; Table 1). The west side of Slaven Canyon displays a siliciclastic sequence that is overlain along a poorly exposed contact by a sequence of black chert, cherty argillite, and subordinate, discontinuous, decimeter-scale limestone beds. The siliciclastic sequence contains a variety of lithologies, including moderately sorted, medium- to fine-grained, calcareous quartz wacke and arenite; poorly sorted, very fine-grained lithic wacke; pebbly grit; siltstone; cherty argillite; and minor, very fine-grained quartz arenite. Reconnaissance petrographic study of these rocks reveals that the quartzose sandstones consist principally of well-rounded, monocrystalline quartz, and a diverse assemblage of lithic fragments including a wide variety of carbonate fragments, cherty argillite, chert, quartzose sandstone, strongly foliated polycrystalline quartz, and sparse mafic and intermediate(?) volcanic lithic fragments. Feldspars and micas occur sparsely. The minor volcanogenic component consists of pervasively altered, microlitic and lathwork volcanic lithic fragments and devitrified vesicular pyroclastic fragments. The alteration consists mainly of calcite pseudomorphs after plagioclase microphenocrysts, chloritization of mafic minerals, and silicification of devitrified glass.

Geochronology of detrital zircon

Eighty kilograms of the moderately sorted quartz wacke described above yielded a 200-mg heavy mineral separate that consists principally of zircon, tourmaline, rutile, magnetite, sphene, and authigenic sulfides. The heavy-mineral suite was examined carefully in an unsuccessful attempt to isolate neovolcanic or other zircon of probable early Paleozoic age. The apparent absence of such zircon is supported by isotopic data discussed below.

Zircon roundness varies from euhedral (sparse) to very well rounded (abundant), and grains have a mean size of approximately 80 μm. Zircons occur in purple, pink, colorless, and metamict white varieties. Six isotopic analyses of detrital zircon yield moderately discordant, early Proterozoic and Archean $^{207}Pb/^{206}Pb$ ages (Fig. 5d; Table 2). Two bulk fractions of zircon split on the basis of size yield $^{207}Pb/^{206}Pb$ ages of 1.89 and 2.05 Ga for the fine and coarse splits, respectively. $^{207}Pb/^{206}Pb$ ages for handpicked splits of different zircon species are all greater than 2.1 Ga.

Discussion of results

The ages of detrital zircon discussed above provide no isotopic evidence for a lower Paleozoic volcanogenic component within this Middle Devonian siliciclastic interval of the Slaven Chert. The high degree of roundness of many of the zircons suggests that they have a polycyclic history. The wide variability in the roundness of the zircons, coupled with the considerable scatter in the isotopic ages, suggests that the Precambrian detritus was derived from many sources. The sparse volcanogenic component of this sandstone must have been derived from either or both of two possible sources. It may have been derived from Precambrian volcanogenic rocks and carried into the basin with the quartzose component. Alternatively, and perhaps more likely, it may have been derived from zircon-poor, lower Paleozoic mafic volcanogenic rocks of the allochthon itself. Greenstones of subaqueous origin are common within Middle Cambrian to Upper Devonian strata of the Roberts Mountains allochthon. Madrid (1987) reported field and geochemical evidence indicating that these greenstones formed in an extensional, marginal basin tectonic setting rather than in an arc environment.

REGIONAL PALEOGEOGRAPHIC RELATIONS

Available data indicate that sandstones of the Roberts Mountains allochthon were derived both directly and indirectly from Precambrian rocks (Fig. 6). The apparent absence of coarse, clastic, arc-derived detritus in strata of the Roberts Mountains allochthon permits several possible interpretations of early Paleozoic paleogeography. Burchfiel and Royden (1989) recently presented a model for the Antler Orogeny that invoked simple rollback of a subduction-zone hinge to explain the emplacement of the Roberts Mountains allochthon (cf. Malinverno and Ryan, 1986). According to such a scenario, a fringing volcanic arc in close proximity to the continental margin is unnecessary for paleogeographic reconstruction. The detrital zircon data presented here are most consistent with their model.

However, there is abundant, albeit fragmentary, evidence in California that fringing volcanic arcs may have existed along the continental margin during much of the early and middle Paleozoic (Burchfiel and Davis, 1972; Girty and Wardlaw, 1984; Hanson and others, 1988; Harwood and Murchey, this volume; Saleeby and others, 1987; Wallin and others, 1988; Wallin, 1989). If arc environments did exist offshore during much of the early Paleozoic, then the lack of arc-derived detritus within the allochthon can be explained by two scenarios. First, this absence may be due simply to substantial paleogeographic separation between the arc environment and the Cordilleran eugeocline. Alter-

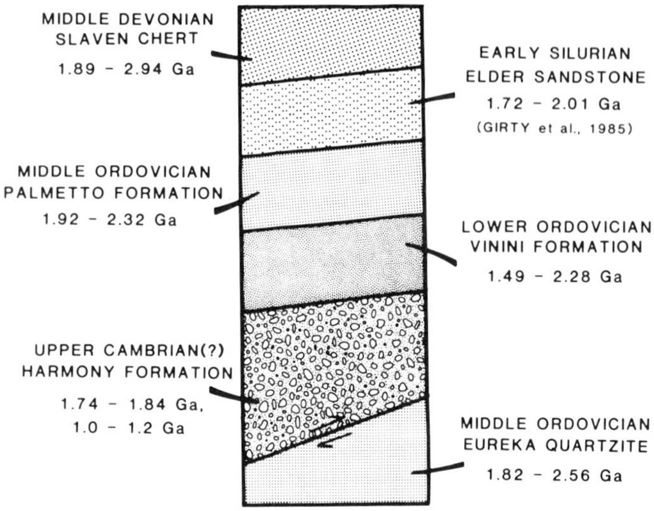

Figure 6. Pseudostratigraphic column of selected terrigenous strata of the Roberts Mountains allochthon. Column is highly schematic because the units occur in different mountain ranges. Depositional ages based on existing biostratigraphic control, but stacking order shown does not imply physical contact of units. Numbers indicate the range of $^{207}Pb/^{206}Pb$ ages for detrital zircon from each unit.

natively, such an absence might be the result of some physiographic barrier to the eastward dispersal of coarse detritus (e.g., a trench or an uplifted back-arc–basin ridge). Given this uncertainty, it appears that we must investigate fine-grained tuffaceous deposits of the Roberts Mountains allochthon if we are to evaluate further the timing of early Paleozoic volcanism and its proximity to the western margin of central North America.

Regional stratigraphic and age relations between poorly understood lower Paleozoic strata within allochthonous terranes of Nevada and California remain controversial. Reconnaissance studies of detrital zircon in these areas (Girty and Wardlaw, 1985; Girty and others, 1985; Wallin, 1988, 1989, this chapter) provide essential data for evaluation of tectonic models. Lower Paleozoic strata in the Roberts Mountains allochthon, the Shoo Fly Complex, and the Klamath Mountains exhibit both similarities and differences in the age of detrital zircon from potentially correlative units. For example, Schweickert and Snyder (1981) proposed a tectonostratigraphic correlation of the Harmony Formation and feldspathic sandstones in the Shoo Fly Complex (Poison Canyon and Red Hill Formations). Despite the fact that these rocks have remarkably similar compositions (Suczek, 1977; Girty and Wardlaw, 1985), analyses of detrital zircon indicate that they were derived from different sources (Girty and Wardlaw, 1985; Wallin, 1989). Even though these units have different sources, it should be emphasized that each unit does consist of voluminous, feldspathic, Precambrian detritus that may simply have been derived from different point sources. This comparison suggests that, although specific correlation of the Harmony and Shoo Fly sandstones as lithostratigraphic units is probably incorrect, the idea that these units were part of a broadly coeval apron of sediment along a continental margin may still have merit. Indeed, Schweickert and Snyder's (1981) general hypothesis is consistent with the observation that detrital zircons from Ordovician and Silurian strata of the Roberts Mountains allochthon are similar to those of the Shoo Fly Complex in that they yield early Proterozoic, but not Archean, $^{207}Pb/^{206}Pb$ ages.

CONCLUSIONS

The principal conclusions of this study are:

1. Siliciclastic rocks of the Roberts Mountains allochthon were derived both directly and indirectly from Precambrian crystalline rocks.

2. The arkosic Harmony Formation (Upper Cambrian(?) represents first-cycle detritus derived principally from early and middle Proterozoic crystalline rocks.

3. Quartzose sandstone (Lower Ordovician) of the Vinini Formation yields $^{207}Pb/^{206}Pb$ ages similar to those of the Harmony Formation, which suggests a similar ultimate source for both units.

4. Middle Ordovician eugeoclinal and miogeoclinal quartzites exhibit similar, but not identical, detrital zircon isotopic signatures that permit but do not require the interpretation that they have a similar source.

5. Middle Devonian, volcanic lithic fragment–bearing quartz wacke from the Slaven Chert contains no evidence of neovolcanic zircon. The Slaven contains polycyclic Precambrian zircon that was probably derived principally from erosion of older miogeoclinal strata.

ACKNOWLEDGMENTS

I thank Gary Girty, Raul Madrid, Gerry Ross, Rich Schweickert, and Chris Suczek for stimulating discussions. Keith Ketner provided unpublished biostratigraphic control early in the study. Thoughtful reviews by Gerry Ross, Ken Coles, and Meghan Miller improved the manuscript significantly. I am grateful for financial support provided by the Department of Geology and the Isotope Geochemistry Laboratory of the University of Kansas, the Geological Society of America, the American Association of Petroleum Geologists, and Sigma Xi, The Scientific Research Society.

Geological Society of America
Special Paper 255
1990

Biogeography of the Upper Ordovician Montgomery Limestone, Shoo Fly Complex, northern Sierra Nevada, California, and comparisons of the Shoo Fly Complex with the Yreka terrane

A. W. Potter
Department of Geosciences, Oregon State University, Corvallis, Oregon 97331-5506
R. Watkins
Geology Department, Milwaukee Public Museum, 800 West Wells Street, Milwaukee, Wisconsin 53233
A. J. Boucot
Department of Zoology, Oregon State University, Corvallis, Oregon 97331-2914
R. J. Elias
Department of Geological Sciences, University of Manitoba, Winnipeg, Manitoba R3T 2N2, Canada
R. A. Flory
Department of Geology, California State University, Chico, California 95929
J. K. Rigby
Department of Geology, 258 ESC, Brigham Young University, Provo, Utah 84602

ABSTRACT

The Ashgillian (Upper Ordovician) Montgomery Limestone occurs as slide blocks in melange of the Shoo Fly Complex, northern Sierra Nevada, northern California. Brachiopods and sphinctozoan sponges from the Montgomery Limestone have closest biogeographic ties to coeval faunas of the eastern Klamath Mountains (Yreka terrane), and in the case of the brachiopods, to east-central Alaska (Jones Ridge). The latter was part of North America in the Ordovician. A small collection of Montgomery rugose corals yielded one species that is known elsewhere only in the Yreka terrane and in northern Maine. Montgomery tabulate corals have affinities with contemporaneous faunas of the Yreka terrane, northern Europe/Asia, Australia, and eastern North America. The apparent absence of similar tabulate taxa in western North America may be an artifact of incomplete collecting.

As a whole, the biogeographic data indicate that the Montgomery Limestone was deposited close enough to Ordovician North America for faunal interchange to occur, and during its deposition was probably relatively near that continent.

A comparison of lithologic units of the Shoo Fly Complex with those of the Yreka terrane indicates that some units in each area have no counterparts in the other (e.g., schist of Skookum Gulch in the Yreka terrane), and other units have general similarities in age (where known) and lithology, but differ in detail. The Yreka terrane has been interpreted as the remnants of an Early Cambrian arc and Ordovician-Devonian arc–fore-arc–accretionary prism, and the Shoo Fly Complex as a fragment of a Devonian or older accretionary wedge. Available biogeographic and stratigraphic data can be reasonably explained, as has been done by earlier authors, by a paleogeography in

Potter, A. W., Watkins, R., Boucot, A. J., Elias, R. J., Flory, R. A., and Rigby, J. K., 1990, Biogeography of the Upper Ordovician Montgomery Limestone, Shoo Fly Complex, northern Sierra Nevada, California, and comparisons of the Shoo Fly Complex with the Yreka terrane, *in* Harwood, D. S., and Miller, M. M., eds., Paleozoic and early Mesozoic paleogeographic relations; Sierra Nevada, Klamath Mountains, and related terranes: Boulder, Colorado, Geological Society of America Special Paper 255.

which the Yreka terrane and Shoo Fly Complex were parts of the same arc-trench system but were situated at different points along the strike of the arc. Lateral changes along strike in tectonic conditions and source areas could account for the observed disparities.

INTRODUCTION

The Late Devonian or older Shoo Fly Complex is a regionally extensive rock assemblage in the northern Sierra Nevada that holds many keys to understanding the Paleozoic geologic history of northern California. Hannah and Moores (1986) characterized it as chiefly a coherent unit of phyllite, quartzose sandstone, and chert, and a melange unit. Unfortunately, the complex has produced relatively few fossils, so that ages and stratigraphic/paleogeographic relations of its components are imperfectly known. Only one unit, the Montgomery Limestone, has produced a fauna with enough diagnostic elements to provide some information on its biogeographic affinities (Boucot and Potter, 1977; A. G. Harris, *in* Hannan and Moores, 1986, and *in* Harwood, 1988). In this respect we update our earlier discussion of the biogeographic relations of the macrofossils. We also compare and contrast certain elements of the Shoo Fly Complex with possible lithologic correlatives in the nearby Yreka terrane of the eastern Klamath belt in order to gain more insight on the early Paleozoic paleogeographic relations of these two terranes.

MONTGOMERY LIMESTONE

The Montgomery Limestone was named by Diller (1908) for several limestone lenses in the Taylorsville area (Fig. 1). McMath (1966) correlated the Montgomery Limestone with part of the Shoo Fly Complex. The Montgomery limestone bodies were interpreted as "mega-clasts" in melange of the Shoo Fly Complex by D'Allura and others (1977). Work by Watkins (this report) indicates that the Montgomery limestone lenses make up an aligned series of discontinuous outcrops ranging from less than 1 m to 61 m in length and as much as 18 m thick, within a highly deformed, fine-grained siliciclastic matrix.

One 10-m-long, 2-m-thick outcrop (Fig. 2) consists of massive skeletal lime mudstone containing locally silicified fossils and terrigenous material common along stylolites. The limestone is bounded above and below by intervals about 1 to 1.4 m thick of highly sheared limestone breccia that has angular limestone pebbles and cobbles usually supported in a matrix of reddish or gray to brown siliceous argillite and argillaceous chert. Stratigraphically above and below the breccia is highly fractured argillaceous

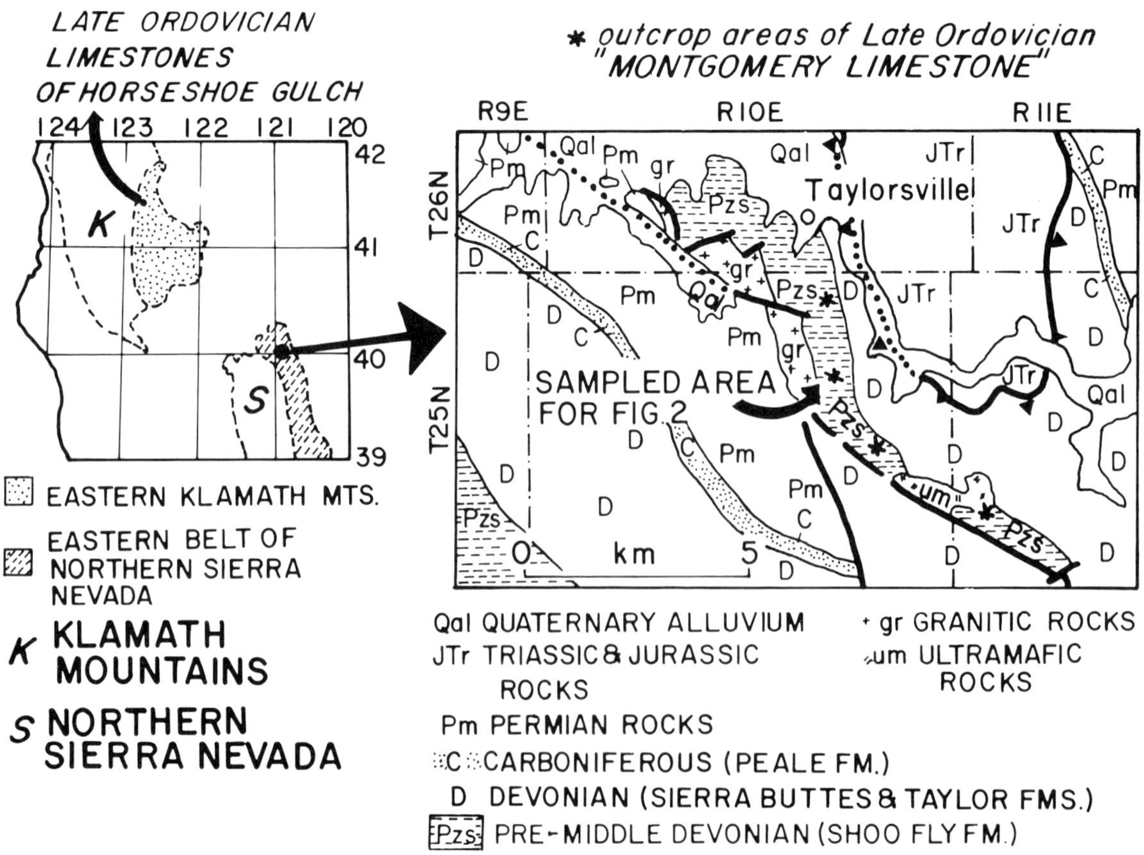

Figure 1. Index map of eastern Klamath Mountains and northern Sierra Nevada and simplified geologic map of Taylorsville area showing outcrops of Montgomery Limestone.

chert that appears largely homogeneous and massive, with some suggestion of tabular beds ⩽10 cm thick. The 10-m-long limestone outcrop is interpreted as a large clast emplaced in a silica-rich slope environment by gravity sliding.

Age

Fossils recovered from the Montgomery Limestone include brachiopods, conodonts, rugose and tabulate corals, and sphinctozoan sponges. Partial faunal lists are given in the section on biogeography below. The brachiopods indicate an Ashgillian (Late Ordovician) age (Boucot and Potter, 1977), based on the co-occurrence of *Dicoelosia* and *Phragmorthis,* and the conodonts a middle Maysvillian through Gamachian age (Harris, *in* Hannah and Moores, 1986; latter half of the Ashgillian *in* Ross and others, 1982, sheet 1).

Comparison with Horseshoe Gulch unit of Yreka terrane

Most of the brachiopod, coral, and sponge genera found in the Montgomery Limestone also are present in Ashgillian limestone of member 5 of the Horseshoe Gulch unit in the Yreka terrane of the eastern Klamath belt (Fig. 1). Therefore, it is of interest to compare the lithologies and stratigraphic settings of the Montgomery and Horseshoe Gulch units.

The Horseshoe Gulch unit has nine members: Member 1, the lowest, comprises Middle Ordovician or older quartz keratophyre, silicic volcaniclastic rocks, intermediate-mafic volcanic/dike rocks, and perhaps terrigenous sedimentary rocks. Members 2 through 9 contain shale, mudstone, siltstone, sandstone, conglomerate, and limestone of Middle Ordovician to Late Silurian (and possibly younger) age (details in Potter, 1990; see Fig. 2 of Potter and others, this volume).

Lindsley-Griffin (1982) and Lindsley-Griffin and Griffin (1983) considered the Horseshoe Gulch unit a melange (as used by Silver and Beutner, 1980) and part of their melange of Gregg Ranch. Potter (1990) and Potter and others (this volume) interpreted it to have been originally a stratigraphically coherent unit, now disrupted to a broken formation by faulting and associated tectonic deformation. For comparison to the Montgomery Limestone and its surrounding rocks, only members 4, 5, 7, and 8 of the Horseshoe Gulch unit are of concern here.

Member 4 consists of mudstone, shale, siltstone, sandstone, conglomerate, and one intercalated lens of limestone containing scattered gastropods. The terrigenous rocks generally weather to shades of brown, green, and gray. Thickness of member 4 is estimated at 40 m from map across sections.

The overlying member 5 bears limestone that weathers light to medium gray and ranges from massive to bedded-stylobedded, lenticular-nodular, or stylobreccia with a fitted fabric (terminology of Logan and Semeniuk, 1976). The limestone is commonly skeletal wackestone-packstone. Macrofossils are absent to abundant, and include corals, sphinctozoan sponges, calcareous algae, gastropods, brachiopods, and bivalves. Thickness is estimated at

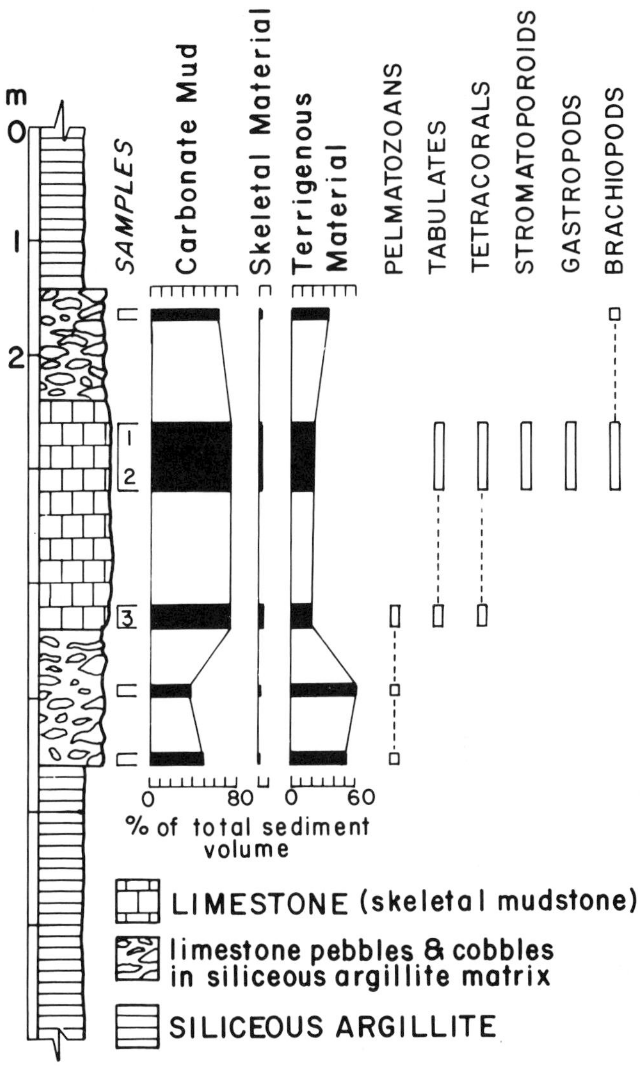

Figure 2. Measured section of Ashgillian Montgomery Limestone showing sample horizons studied in detail. Numbered horizons are within massive limestone.

50 m. The lenticular-nodular rocks and stylobreccia are inferred to be products of differential pressure solution ±early preferential diagenetic cementation and local fracturing (Potter, 1989).

Contacts between members 4 and 5 are usually faulted or concealed. In south Horseshoe Gulch a sheared contact is present between green-brown–weathering shale of member 4 and massive nodular limestone and limestone stylobreccia of member 5; the latter has a fitted fabric and irregular lenses of green-brown mudstone 2.5 to 15 cm long and ⩽6 mm thick that become thinner at 20 cm and higher above the contact. The shale of member 4 beneath the contact bears scattered limestone bodies up to cobble size that may be clasts, fault slices, or faulted interbedded lenses. The green-brown mudstone lenses in member 5 are taken as evidence that the contact between members 4 and 5 was originally depositional; it is now sheared because of minor faulting or ductility contrasts between mudstone and limestone during deformation.

Member 7 of the Horseshoe Gulch unit contains shale, mudstone, siltstone, sandstone, conglomerate, and limestone of Late Silurian age. The limestone is generally light gray weathering and bedded to massive; tan- to yellowish tan–weathering silty and muddy lenses and patches are locally common and due in part to bioturbation. Nonsilicified pelmatozoan debris and other fossils are locally abundant. In north Horseshoe Gulch, limestone of member 7 unconformably overlies massive Ashgillian limestone of member 5 at a well-exposed, unsheared, indisputably depositional contact. The unconformity itself is undulatory, and irregularities in it are filled by limestone pebble conglomerate with a tan matrix.

Member 8 depositionally overlies member 7 in two places and is commonly a distinctive limestone breccia with dark gray–weathering clasts in a light gray–weathering matrix; silicified tabulate corals are locally abundant.

While the Montgomery Limestone and member 5 of the Horseshoe Gulch unit themselves are not strikingly different lithologically, the common locally fossiliferous Upper Silurian limestones of members 7 and 8 have no reported counterparts in rocks associated with the Montgomery Limestone, nor have Silurian limestone or macrofossils yet been recorded from elsewhere in the northern Sierra Nevada.

Depositional setting

In view of the faunal similarities (enumerated below) between the Montgomery Limestone and member 5 of the Horseshoe Gulch unit, the depositional environments of the two units were probably analogous. Therefore, the Montgomery Limestone probably accumulated at about 30 to 90 m depth in warm, normal-marine salinity, relatively low-energy waters, the environment inferred by Rigby and Potter (1986) for member 5 of the Horseshoe Gulch unit.

Potter (1990) proposed that Upper Ordovician and Silurian rocks of the Horseshoe Gulch unit were deposited in a basin in the outer fore arc or near the outer ridge of the eastern Klamath arc (terminology of Silver and Reed, 1988; see Potter and others, this volume, Fig. 3). A fore arc–outer ridge setting seems possible for the original site of formation of the Montgomery Limestone as well, prior to its sliding and incorporation into a melange during Late Ordovician to Late Devonian time.

Biogeographic relations

A. G. Harris (*in* Hannah and Moores, 1986) reported four species of conodonts from the Montgomery Limestone: *Belodina* sp., *Drepanoistodus suberectus* Branson and Mehl, *Panderodus gracilis* Branson and Mehl, and *Protopanderodus insculptus* Branson and Mehl; she indicated that this faunule is widespread at equatorial latitudes. A comparison of this list with that given by Bergström (*in* Potter and others, this volume) for member 5 of the Horseshoe Gulch unit indicates that two of the above species may also occur in member 5.

The biogeographic affinities of macrofossils from the Montgomery Limestone are discussed in the following sequence: brachiopods by Potter, solitary rugose corals by Elias, tabulate corals by Flory, and sphinctozoan sponges by Rigby.

The brachiopod faunal list published by Boucot and Potter (1977) is here revised and expanded, based on study of additional material collected by V. McMath in 1955–1956. Taxa present include *Bimuria* sp.; *Diambonia* sp.; *Dicoelosia* sp.; *Diochthofera* (new genus of Potter, 1990); cf. *Elasmothyris* sp.; *Epitomyonia*? sp.; *Phragmorthis* aff. *buttsi,* Cooper, 1956, morphotype 1 of Potter (1990); *Ptychopleurella (Glypterina)* aff. *uniplicata* Cooper, 1956; *Skenidioides* sp.; and a bizarre new orthide genus with a transverse, emarginate outline.

All the taxa listed are present also in member 5 of the Horseshoe Gulch unit, and except for *Epitomyonia*? sp., in the upper member of the Jones Ridge Limestone, Jones Ridge, east-central Alaska and Yukon Territory (Potter, 1990; Potter and others, this volume, Fig. 4, and Rigby and others, 1988, gave locality data and maps). Churkin and others (1980), Payne and Allison (1981), and Rigby and others (1988) all concluded that the Jones Ridge area was part of North America in the late Proterozoic and Paleozoic, because it contains rock units of these ages that have lithologic counterparts in autochthonous northwest Canada.

Of this assemblage, the taxa with restricted geographic distributions are *Diochthofera* sp., cf. *Elasmothyris* sp., and the new genus of transverse, emarginate orthide. The geographic ranges of *Diochthofera* and *Elasmothyris* were discussed by Potter and others (this volume), and the new transverse, emarginate orthide has been found elsewhere only in member 5 of the Horseshoe Gulch unit and in the upper member of the Jones Ridge Limestone.

The brachiopods from the Montgomery Limestone have the same biogeographic relations as the brachiopods from member 5 of the Horseshoe Gulch unit discussed by Potter and others (this volume); therefore, faunal communication with Ashgillian western North America (e.g., Jones Ridge) is indicated, and in at least two cases was at the species level.

McMath collected a small number of rugose corals from the Montgomery Limestone in 1955–1956. They are generally poorly preserved. Some specimens could represent *Streptelasma* if they are solitary forms, or *Palaeophyllum* if they are isolated corallites from phaceloid colonies. Also present is the solitary rugosan species *Grewingkia penobscotensis* Elias, 1982. The latter taxon is of biogeographic significance. It is recorded elsewhere only in member 5 of the Horseshoe Gulch unit in the Yreka terrane (Elias and Potter, 1984), and in an unnamed Ashgillian formation in Penobscot County, northern Maine (Elias, 1982; see also Neuman, 1980, Fig. 3, and Neuman, 1984, p. 1197), which may have been close to or part of Ashgillian North America (Elias and Potter, 1984). *Grewingkia penobscotensis* is inferred to have been widely distributed in tectonically and volcanically active areas situated relatively near the North American continental margin during Ashgillian time.

Tabulate corals from the Montgomery Limestone are *Paleofavosites* sp., *Propora* sp., and *Quepora* sp. *Paleofavosites* is widespread in Late Ordovician time. *Propora* occurs elsewhere in member 5 of the Horseshoe Gulch unit (same species as in the Montgomery Limestone); Anticosti Island, Quebec, Canada (part of North America during the Ordovician); northern Europe; and Asia. *Quepora* is present elsewhere in member 5 of the Horseshoe Gulch unit (same species as in the Montgomery Limestone), Altai Mountains, Mongolia–south-central Siberia; and New South Wales, Australia. In summary, the tabulates indicate faunal communication with the Yreka terrane, northern Europe–Asia, Australia, and eastern North America. As noted by Potter and others (this volume), Ordovician tabulates from western North America are little-studied, so the seeming absence of *Propora* and *Quepora* in that region may be only an artifact of incomplete sampling.

The sphinctozoan sponges *Amblysiphonella* sp., *Pseudoimperatoria mega*(?) (Rigby and Potter, 1986), *Pseudoimperatoria*(?) sp., *Corymbospongia* cf. *adnata* Rigby and Potter, 1986, *Cystothalamiella* sp., *Girtyocoelia* cf. *epiporata* Rigby and Potter, 1986, *Girtyocoelia* sp., and *Rigbyetia obconica* (Rigby and Potter, 1986) are all present in collections from the Montgomery Limestone. The forms here referred to *Pseudoimperatoria* are similar to those described as *Imperatoria* by Rigby and Potter (1986, p. 23–28), but are somewhat more fragmented than specimens in the Klamath faunas. Senowbari-Daryan and Rigby (1988, p. 195–196) reported that sections of types of *Imperatoria* De Gregoria, 1903, clearly show that this Permian genus is not a sphinctozoan and that the Ordovician species are not related to the Sicilian genus. They placed these Ordovician species in the new genus *Pseudoimperatoria*. Similarly, sponges designated by Rigby and Potter (1986) as *Cliefdenella obconica* have since been placed by Webby and Lin (1988) in the new genus, *Rigbyetia*.

Sphinctozoan sponges in the Montgomery Limestone have their greatest taxonomic similarity to contemporaneous assemblages described from the Yreka terrane by Rigby and Potter (1986). All six of the Montgomery genera and one to four of the four identifiable Montgomery species also occur in the Yreka terrane.

Collections from Alaska and Yukon Territory described by Rigby and others (1988) are too small for adequate comparison with the Montgomery fauna. Nevertheless, the genus *Pseudoimperatoria* is shared by the Montgomery Limestone with an Ashgillian locality at Jones Ridge, Yukon Territory (it was the only sphinctozoan reported there by Rigby and others, 1988). *Cystothalamiella* was one of three genera identified by Rigby and others (1988) at an Ashgillian site in the White Mountain area of the Nixon Fork terrane, west-central Alaska, and the species *Rigbyetia obconica* and, possibly, *Girtyocoelia epiporata* are present in the same area in Llandeilian to Caradocian deposits (Rigby and others, 1988). The latter two species are known elsewhere only in the Yreka terrane. Blodgett and Clough (1985) inferred that the Nixon Fork terrane was part of North America in the Ordovician.

The biogeographic relations of the macrofossils discussed above demonstrate that the Montgomery Limestone was close enough to western North America and to the Yreka terrane in Late Ordovician time for genetic interchange to occur, in several cases so frequently that the same species existed in all or two of the three areas.

As with the Klamath Ordovician faunas discussed by Potter and others (this volume), the high faunal similarities between Montgomery and Jones Ridge faunas and apparent absence of similar faunas in coeval rocks of western Canada and points south could indicate that the Montgomery Limestone was near Alaska in Late Ordovician time. The alternative scenarios suggested for the Klamath faunas apply to the Montgomery Limestone fauna as well: (1) the Montgomery Limestone was near western Canada–Idaho–Nevada, and Klamath-Montgomery–type faunas are in these regions but have not yet been found due to insufficient sampling; (2) the Montgomery Limestone was near western Canada–Idaho–Nevada, but current circulation patterns or differing water-temperature regimes created barriers that prevented faunal interchange with North America except near Alaska-Yukon. Thus, we cannot specify the exact location of the Montgomery Limestone during Late Ordovician time; nonetheless, we do infer from the biogeographic evidence that it was probably relatively near North America and the Yreka terrane at that time and likely no farther from North America than was the Yreka terrane.

COMPARISONS BETWEEN THE SHOO FLY COMPLEX AND YREKA TERRANES

The Montgomery Limestone was contrasted with parts of the Horseshoe Gulch unit above. In this section, other rock units of the Yreka terrane are compared with parts of the Shoo Fly Complex, based on published descriptions of the latter. Figure 2 of Potter and others (this volume) provides columnar sections of many of the rock units in the Yreka terrane and northern Sierra Nevada mentioned in this discussion.

As described above, the Montgomery Limestone is encased in argillaceous chert and siliceous argillite of melange of the Shoo Fly Complex, the so-called Sierra City melange of Schweickert (1974) and Schweickert and others (1984). Hannah and Moores (1986) described this melange as containing blocks of limestone, greenstone, gabbro, ultramafic rocks, and exotic sandstone slices in a matrix of serpentinite or argillaceous sandstone. Chert occurs as blocks and matrix in places (Schweickert and others, 1984); blocks of plagiogranite (Mount and Schweickert, 1986) and additional lithologies also are present (Girty and Pardini, 1987). The melange was considered to have had a sedimentary origin in an accretionary prism by D'Allura (1977, *in* Hannah and Moores, 1986), Mount and Schweickert (1986), and Girty and Pardini (1987), either as debris shed from an advancing thrust sheet (D'Allura, *in* Hannah and Moores, 1986) or as a subaqueous gravity slide(s) (Girty and Pardini, 1987).

The Shoo Fly Complex is overlain unconformably by the Upper Devonian (Frasnian to mid-Famennian) Grizzly Formation (Hanson and Schweickert, 1986), on which was deposited the mid- and late? Famennian Sierra Buttes Formation (Hanson and Schweickert, 1986). Saleeby and others (1987) measured U-Pb ages on zircons of 423 +5/−15 Ma from a tuff within the Shoo Fly Complex and of 378 ± 5 Ma from a stock intruding both the Shoo Fly Complex and Sierra Buttes Formation. They concluded that the Shoo Fly Complex was deformed between these two isotopic dates (i.e., between Late Silurian and late Middle Devonian [Givetian] time on the scale of Harland and others, 1989). Presumably, the Sierra City melange also formed in this time interval.

Wallin (*in* Wallin and others, 1988) speculated that the melange of the lower parts of members 1 and 3 of the Gazelle Formation of Potter and others (1977) is "tectonostratigraphically equivalent" to the Sierra City melange. These parts of the Gazelle Formation bear limestone lenses and clasts in which Late Silurian to Early (and, at one location, Middle) Devonian macrofossils are locally common and in which distinctive chert-lithic–rich sandstone, breccia, and conglomerate are widespread, and so appear to differ lithologically from the Sierra City melange in these respects. Also, no phosphatic nodules have been reported in the Gazelle Formation or elsewhere in the Yreka terrane; they do occur in the Shoo Fly Complex (Bond and DeVay, 1980; Hannah and Moores, 1986, and references therein).

The Moffett Creek Formation, which structurally (and perhaps depositionally in one place) underlies the melange of lower Gazelle member 1, is areally extensive in the Yreka terrane and disrupted to melange over wide areas (Potter and others, 1977; Hotz, 1977; Potter, 1990). Again, it differs lithologically from the Sierra City melange in that serpentinite, ultramafic rocks, and gabbro are absent or so rare as to be not yet recognized in it as matrix or inclusions. Blocks of volcanic rocks are also very rare.

Although they differ lithologically from the Sierra City melange, the melanges of the Moffett Creek Formation and lower Gazelle member 1 may have formed in a manner similar to, and at approximately the same time as, the Sierra City melange. Potter (1990) inferred that the Moffett Creek and lower Gazelle member 1 melanges contain olistostromes (megamictites of Sundell and Fisher, 1985) that were also tectonically deformed. Lindsley-Griffin and Griffin (1983) reached a similar conclusion. Most of the melange of these two units is estimated to have been produced in Late Silurian to Middle or Late Devonian time (Potter and others, this volume).

The Sierra City melange is commonly considered to have formed in a subduction complex (Girty and Pardini, 1987, and references therein), and Girty and Guthrie (1989) inferred westward accretion. Generation of the Moffett Creek and lower Gazelle member 1 melanges probably took place both on the accretionary complex/outer ridge and in the forearc of a west-to-northwest–facing arc (Potter and others, this volume; arc polarity from Potter, 1990, using present-day directions). Lindsley-Griffin and Griffin (1983) hypothesized an accretionary-wedge origin for the melange of Gregg Ranch, which includes the melange of lower Gazelle member 1 of this paper.

TABLE 1. AVERAGE CHEMICAL ANALYSES OF ALTERED BASALTIC ROCKS FROM DUZEL ROCK, YREKA TERRANE, AND BULLPEN LAKE SEQUENCE, NORTHERN SIERRA NEVADA

	Duzel Rock Samples	Bullpen Lake Sequence Samples
SiO_2	49.90	48.79
TiO_2	3.23	3.38
Al_2O_3	15.13	12.25
FeO*	12.44	11.88
MgO	5.06	8.38
CaO	9.44	12.74
Na_2O	4.15	2.09
K_2O	0.65	0.49
Total	100.00	100.00
Y	37	34
Zr	215	284
Nb	26	20-60

Note: Oxides are in weight percent, recalculated to 100 percent total on a volatile-free basis. Total Fe as FeO*. Y, Zr, and Nb in ppm. Averages include three Duzel Rock samples from Potter and others (1975; unpublished data; one sample provided by N. Lindsley-Griffin) and three Bullpen Lake sequence samples from Schweickert and others (1984, samples 82-11B, 82-1, and 82-5).

Although the melanges of the Yreka terrane seem to differ lithologically from the Sierra City melange, some units in the Yreka terrane do resemble rocks of the Shoo Fly Complex. Altered pillow basalt and overlying laminated oolitic and pisolitic limestone and sponge-spicule-radiolarian chert of the Bullpen Lake sequence of Schweickert and others (1984) and Girty (1985) have a general similarity to altered alkalic pillow basalts at Duzel Rock and nearby White's Gulch Limestone of Nachman (1977). Samples of the pillow basalts of Duzel Rock contain uncommon phenocrysts and microphenocrysts of plagioclase, clinopyroxene, metallic opaques, and olivine pseudomorphs in a groundmass of metallic opaques, plagioclase microlites, scattered clinopyroxene, and secondary minerals (Potter and others, 1975, and unpublished data). Girty (1985) inferred a similar phenocryst assemblage in Bullpen Lake sequence basalts, although their original mineralogy has been replaced. A comparison of average chemical analyses of Duzel Rock and Bullpen Lake sequence basalts (Table 1; sources of data given in note to table) shows that the former are higher in Al_2O_3, and lower in MgO and Zr (ignoring the more mobile elements Ca, Na, and K). Nachman (1977) regarded the pillow basalt unit as about 100 to 300 m thick, fault-bounded, and structurally overlying the Moffett Creek Formation; its age is unknown, but is speculated to be early Paleozoic.

Nachman (1977) described the White's Gulch Limestone as 354 m thick. The lower part is 195 m of interbedded argillaceous, locally laminated platy micrites and massive silty micrites and oosparites. It is intercalated with altered basalt near its base. The middle part is 50 m of bedded laminated green cherts that lack radiolarians and sponge spicules. The chert is depositionally overlain by 109 m of massive, recrystallized, locally laminated micrites and oomicrites. The White's Gulch Limestone is undated, fault-bounded, and interpreted to be thrust over the Moffett Creek Formation (Nachman, 1977).

Chert in the White's Gulch unit differs from Bullpen Lake sequence chert of the Shoo Fly Complex in lacking radiolarians and sponge spicules.

The best-documented lithologic correlation between the Shoo Fly Complex and Yreka terrane comes from the work of Bond and DeVay (1980), who pointed out the close resemblance of some quartzose sandstones of the Shoo Fly Complex to the Antelope Mountain Quartzite of the Yreka terrane. Both units are imprecisely dated, however, and chronologic correlation has yet to be demonstrated.

As noted by previous authors, phyllite, quartzose sandstone, chert, and limestone occur in both the Shoo Fly Complex (Schweickert and Snyder, 1981; Schweickert and others, 1984) and in the Duzel Phyllite and Schulmeyer Gulch sequence of Hotz (1977); unfortunately, there is insufficient data available to confirm either lithologic or chronologic correlation between these Shoo Fly and Klamath units.

Wallin (1989) determined $^{207}Pb/^{206}Pb$ ages of 432 to 455 Ma (U-Pb ages are nearly concordant) on detrital zircon from feldspathic sandstone of the Sissel Gulch Graywacke (Yreka terrane) and Pb/Pb detrital zircon ages of 415 to 474 Ma from silicic volcaniclastic sandstone of member 2 of the Gazelle Formation. Wallin and others (1988) reported U-Pb upper intercept ages of 469 $\pm$ 21 Ma and 475 $\pm$ 10 Ma on tonalites intruding the Trinity complex, a 470-Ma Pb/Pb zircon age on gabbro of the Trinity complex, and a 440-Ma zircon age on a tonalite clast at Lovers Leap, Yreka terrane (the last two ages were recalculated from Mattinson and Hopson, 1972). Wallin and others (1988) also published two concordant zircon ages of 412 Ma for two pegmatitic trondhjemite dikes intruding the Trinity complex. Zircons from a pegmatitic gabbroid pluton that intrudes the Trinity complex yielded Pb/Pb ages of 414 to 416 Ma (U-Pb ages nearly concordant; Wallin and others, 1990).

Girty and Wardlaw (1984) measured Pb/Pb ages of 411 to 486 Ma on fractions of detrital zircons from a sample of arkose from the Sierra City melange; eight of ten of their zircon fractions had Pb/Pb ages in the range 448 to 469 Ma; a concordia upper-intercept age of 506 $\pm$ 22 Ma was determined for this sample. As noted above, a U-Pb zircon age of 423 +5/−15 Ma on silicic tuff from the Shoo Fly complex was reported by Saleeby and others (1987).

The 412- to 416-Ma ages on pegmatitic rocks in the Trinity terrane are within analytical uncertainty of the Shoo Fly tuff age of Saleeby and others (1987). Some of the Pb/Pb detrital zircon ages of Wallin (1989) and the tonalite-gabbro ages of Wallin and others (1988) are close to or within the 448- to 469-Ma range of Pb/Pb ages of Girty and Wardlaw (1984). These comparisons suggest a gross similarity in ages of magmatism and of detrital-zircon source areas for the Yreka-Trinity terranes and Shoo Fly Complex.

CONCLUSIONS

Several rock units of the Shoo Fly Complex thus have broadly similar lithologic counterparts in the Yreka terrane, yet each differs from its proposed correlative in detail, except possibly for Shoo Fly quartzose sandstones studied by Bond and DeVay (1980) and the Antelope Mountain Quartzite. Other rock units of the Yreka terrane not discussed here (e.g., schist of Skookum Gulch) have no known counterparts at all in the Shoo Fly Complex, and vice versa (Duncan Peak allochthon of Schweickert and others, 1984). Nevertheless, there seems to be a general correspondence in lithology, age, inferred setting of formation, and in timing of magmatism that suggests a common paleogeography for the Yreka terrane and Shoo Fly Complex. In such a paleogeography the Yreka terrane represents the remnants of an Early Cambrian arc and Ordovician-Devonian arc-trench system that faced west to northwest in the Late Ordovician and Devonian (Potter, 1990; Potter and others, this volume; Potter and Grunder, 1989) and that contains parts of the accretionary complex, fore arc, and arc. The Shoo Fly Complex contains a fragment of the accretionary complex from the same system, but formed elsewhere at an unknown distance along the strike of the arc. Burchfiel and Davis (1972, 1975) and Churkin (1974) earlier inferred that the eastern Klamath belt and northern Sierra Nevada were parts of one arc complex. Differences in lithology, age, and mode of formation of units in the two terranes can be explained by lateral changes in tectonic conditions, exposed source areas of detritus, and nature of basement and magmatic source regions along regional strike (Hamilton, 1988, gave recent examples). Similarly, disparities of the Duzel Rock pillow basalts and nearby White's Gulch limestones and cherts from the Bullpen Lake sequence might be attributed to lateral variation within different parts of one seamount chain that was offscraped into the accretionary prism of a Klamath-northern Sierra Nevada arc.

The above paleogeographic relations suggest an explanation for the enigmatic episode of uplift and erosion of the Shoo Fly Complex in Late Silurian to Middle Devonian time (dates based on the work of Saleeby and others, 1987, discussed above). Potter and others (this volume) inferred that the parts of the accretionary prism in the Yreka terrane that correspond to the Shoo Fly Complex (e.g., Duzel Phyllite and Antelope Mountain Quartzite) were uplifted in Middle Devonian to Late Mississippian time. This uplift was attributed to backthrusting in the fore-arc region that emplaced and elevated these units near the arc's magmatic axis. In view of the possible similarity in timing of uplift of the Shoo Fly Complex and its Yreka terrane counterparts, the interpretation that uplift of the Complex resulted from backthrusting of it into the arc-axis region should be investigated.

ACKNOWLEDGMENTS

Reviews by J. A. D'Allura, M. M. Miller, and F. G. Poole were very helpful. National Science Foundation Grant EAR 7919993 (A.J.B.) provided support for this work.

REFERENCES CITED

Blodgett, R. B., and Clough, J. G., 1985, The Nixon Fork terrane; Part of *in situ* peninsular extension of the Paleozoic North American continent: Geological Society of America Abstracts with Programs, v. 17, p. 342.

Bond, G. C., and DeVay, J. C., 1980, Pre-Upper Devonian quartzose sandstones in the Shoo Fly Formation, northern California; Petrology, provenance, and implications for regional tectonics: Journal of Geology, v. 88, p. 285–308.

Boucot, A. J., and Potter, A. W., 1977, Middle Devonian orogeny and biogeographical relations in areas along the North American Pacific rim, *in* Murphy, M. A., Berry, W.B.N., and Sandberg, C. A., eds., Western North America; Devonian: Riverside, University of California, Museum Contribution, v. 4, p. 210–219.

Burchfiel, B. C., and Davis, G. A., 1972, Structural framework and evolution of the southern part of the Cordilleran orogen, western United States: American Journal of Science, v. 272, p. 97–118.

—— , 1975, Nature and controls of Cordilleran orogenesis, western United States: Extensions of an earlier synthesis: American Journal of Science, v. 275-A, p. 363–396.

Churkin, M., Jr., 1974, Paleozoic marginal ocean basin-volcanic arc systems in the Cordilleran foldbelt, *in* Dott, R., Jr., and Shaver, R., eds., Modern and ancient geosynclinal sedimentation: Society of Economic Paleontologists and Mineralogists Special Publication 19, p. 174–192.

Churkin, M., Jr., Carter, C., and Trexler, J. H., Jr., 1980, Collision-deformed Paleozoic continental margin of Alaska; Foundation for microplate accretion: Geological Society of America Bulletin, part 1, v. 91, p. 648–654.

Cooper, G. A., 1956, Chazyan and related brachiopods: Smithsonian Miscellaneous Collections, v. 127, pt. 1, p. 1-1024, pt. 2, p. 1025–1245.

D'Allura, J. A., 1977, Stratigraphy, structure, petrology, and regional correlations of metamorphosed upper Paleozoic volcanic rocks in portions of Plumas, Sierra, and Nevada Counties, California [Ph.D. thesis]: Davis, University of California, 338 p.

D'Allura, J. A., Moores, E. M., and Robinson, L., 1977, Paleozoic rocks of the northern Sierra Nevada; Their structural and paleogeographic implications, *in* Stewart, J. H., Stevens, C. H., and Fritsche, A. E., eds., Paleozoic paleogeography of the western United States: Pacific Section, Society of Economic Paleontologists and Mineralogists Pacific Coast Paleogeography Symposium 1, p. 395–408.

Diller, J. S., 1908, Geology of the Taylorsville region, California: U.S. Geological Survey Bulletin 353, 128 p.

Elias, R. J., 1982, Latest Ordovician solitary rugose corals of eastern North America: Bulletins of American Paleontology, v. 81, no. 314, 116 p.

Elias, R. J., and Potter, A. W., 1984, Late Ordovician solitary rugose corals of the eastern Klamath Mountains, northern California: Journal of Paleontology, v. 58, p. 1203–1214.

Girty, G. H., 1985, An accreted seamount; A working hypothesis for the Bullpen Lake sequence, Sierra Nevada, California: California Geology, v. 38, p. 123–128.

Girty, G. H., and Guthrie, L., 1989, A westward verging imbricate structure, Shoo Fly Complex, Sierra Nevada, California: Geological Society of America Abstracts with Programs, v. 21, p. 84.

Girty, G. H., and Pardini, C. H., 1987, Provenance of sandstone inclusions in the Paleozoic Sierra City melange, Sierra Nevada, California: Geological Society of America Bulletin, v. 98, p. 176–181.

Girty, G. H., and Wardlaw, M. S., 1984, Was the Alexander terrane a source of feldspathic sandstones in the Shoo Fly Complex, Sierra Nevada, California?: Geology, v. 12, p. 339–342.

Hamilton, W. B., 1988, Plate tectonics and island arcs: Geological Society of America Bulletin, v. 100, p. 1503–1527.

Hannah, J. L., and Moores, E. M., 1986, Age relationships and depositional environments of Paleozoic strata, northern Sierra Nevada, California: Geological Society of America Bulletin, v. 97, p. 787–797.

Hanson, R. E., and Schweickert, R. A., 1986, Stratigraphy of mid Paleozoic island-arc rocks in part of the northern Sierra Nevada, Sierra and Nevada Counties, California: Geological Society of America Bulletin, v. 97, p. 986–998.

Harland, W. B., Armstrong, R. L., Cox, A. V., Craig, L. E., Smith, A. G., and Smith, D. G., 1989, A geologic time scale 1989: Cambridge, Cambridge University Press, 256 p.

Harwood, D. S., 1988, Tectonism and metamorphism in the northern Sierra terrane, northern California, *in* Ernst, W. G., ed., Metamorphism and crustal evolution of the western United States; Rubey Volume 7: Englewood Cliffs, New Jersey, Prentice-Hall, p. 764–788.

Hotz, P. E., 1977, Geology of the Yreka Quadrangle, Siskiyou County, California: U.S. Geological Survey Bulletin 1436, 72 p.

Lindsley-Griffin, N., 1982, Structure, stratigraphy, petrology, and regional relationships of the Trinity ophiolite, eastern Klamath Mountains, California [Ph.D. thesis]: Davis, University of California, 445 p.

Lindsley-Griffin, N., and Griffin, J. R., 1983, The Trinity terrane; An Early Paleozoic microplate assemblage, *in* Stevens, C. H., ed., Pre-Jurassic rocks in western North American suspect terranes: Pacific Section, Society of Economic Paleontologists and Mineralogists, p. 63–75.

Logan, B. W., and Semeniuk, V., 1976, Dynamic metamorphism; Processes and products in Devonian carbonate rocks, Canning Basin, Western Australia: Geological Society of Australia Special Publication 6, 138 p.

Mattinson, J. M., and Hopson, C. A., 1972, Paleozoic ophiolitic complexes in Washington and northern California: Carnegie Institution of Washington Yearbook, v. 71, p. 578–583.

McMath, V. E., 1966, Geology of the Taylorsville area, northern Sierra Nevada, California: California Division of Mines and Geology Bulletin 190, p. 173–183.

Mount, C. M., and Schweickert, R. A., 1986, Early Paleozoic orogeny in the Shoo Fly; Evidence from clastic rocks in the Sierra City melange, California: Geological Society of America Abstracts with Programs, v. 18, p. 162.

Nachman, D. A., 1977, Geology of the Duzel rock area, Yreka quadrangle, California [M.S. thesis]: Corvallis, Oregon State University, 153 p.

Neuman, R. B., 1980, Trip B-4, Itinerary; The core of the Weeksboro–Lunksoos Lake Anticline, and the Ordovician, Silurian, and Devonian rocks on its northwest flank, *in* Roy, D. C., and Naylor, R. S., eds., A guidebook to the geology of northeastern Maine and neighboring New Brunswick: 72nd Annual Meeting, New England Intercollegiate Geological Conference, p. 114–126.

—— , 1984, Geology and paleobiology of islands in the Ordovician Iapetus Ocean; Review and implications: Geological Society of America Bulletin, v. 95, p. 1188–1201.

Payne, M. W., and Allison, C. W., 1981, Paleozoic continental-margin sedimentation in east-central Alaska: Geology, v. 9, p. 274–279.

Potter, A. W., 1990, Middle and Late Ordovician brachiopods from the eastern Klamath Mountains, northern California, part 1: Palaeontographica, Abteilung A, v. 212, p. 31–158.

Potter, A. W., and Grunder, A. L., 1989, Inferred tectonic settings of Early Cambrian and Ordovician intrusive rocks, eastern Klamath terrane (EKT), northern California: Geological Society of America Abstracts with Programs, v. 21, p. 130.

Potter, A. W., Scheidegger, K. F., Corliss, J. B., and Dasch, E. J., 1975, Magma types present in Paleozoic keratophyres and spilites from the Gazelle area, eastern Klamath Mountains, northern California: Geological Society of America Abstracts with Programs, v. 7, p. 1251.

Potter, A. W., Hotz, P. E., and Rohr, D. M., 1977, Stratigraphy and inferred tectonic framework of Lower Paleozoic rocks in the eastern Klamath Mountains, northern California, *in* Stewart, J. H., Stevens, C. H., and Fritsche, A. E., eds., Paleozoic paleogeography of the western United States: Pacific Section, Society of Economic Paleontologists and Mineralogists Pacific Coast Paleogeography Symposium 1, p. 421–440.

Rigby, J. K., and Potter, A. W., 1986, Ordovician sphinctozoan sponges from the eastern Klamath Mountains, northern California: Paleontological Society Memoir 20 (Journal of Paleontology, v. 60, no. 4, supp.), 47 p.

Rigby, J. K., Potter, A. W., and Blodgett, R. B., 1988, Ordovician sphinctozoan sponges of Alaska and Yukon Territory: Journal of Paleontology, v. 62, p. 731–746.

Ross, R. J., Jr., and 27 others, 1982, The Ordovician System in the United States: International Union of Geological Sciences Publication 12, 73 p.

Saleeby, J., Hannah, J. L., and Varga, R. J., 1987, Isotopic age constraints on middle Paleozoic deformation in the northern Sierra Nevada, California: Geology, v. 15, p. 757–760.

Schweickert, R. A., 1974, Probable late Paleozoic thrust fault near Sierra City, California: Geological Society of America Abstracts with Programs, v. 6, p. 251.

Schweickert, R. A., and Snyder, W. S., 1981, Paleozoic plate tectonics of the Sierra Nevada and adjacent regions, *in* Ernst, W. G., ed., The geotectonic development of California; Rubey Volume I: Englewood Cliffs, New Jersey, Prentice-Hall, p. 182–201.

Schweickert, R. A., Harwood, D. S., Girty, G. H., and Hanson, R. E., 1984, Tectonic development of the northern Sierra terrane; An accreted Late Paleozoic island arc and its basement (Field Trip 5), *in* Lintz, J. Jr., ed., Western geological excursions; Geological Society of America Annual Meeting guidebook: Reno, University of Nevada Department of Geology, p. 1–65.

Senowbari-Daryan, B., and Rigby, J. K., 1988, Upper Permian segmented sponges from Djebel Tebaga, Tunisia: Facies, v. 19, p. 171–250.

Silver, E. A., and Beutner, E. C., 1980, Melanges: Geology, v. 8, p. 32–34.

Silver, E. A., and Reed, D. L., 1988, Backthrusting in accretionary wedges: Journal of Geophysical Research, v. 93, p. 3116–3126.

Sundell, K. A., and Fisher, R. V., 1985, Very coarse-grained fragmental rocks; A proposed size classification: Geology, v. 13, p. 692–695.

Wallin, E. T., 1989, Reconstruction of the timing and sequence of early Paleozoic sedimentation in the Yreka-Callahan terrane, eastern Klamath Mountains, California: Geological Society of America Abstracts with Programs, v. 21, p. 155.

Wallin, E. T., Mattinson, J. M., and Potter, A. W., 1988, Early Paleozoic magmatic events in the eastern Klamath Mountains, northern California: Geology, v. 16, p. 144–148.

Wallin, E. T., Lindsley-Griffin, N., and Potter, A. W., 1990, Trinity ultramafic complex, Klamath Mountains, California: Cambro-Ordovician ophiolite or composite terrane?: Geological Society of America Abstracts with Programs, v. 22, p. 91–92.

Webby, B. D., and Lin Baoyu, 1988, Upper Ordovician cliefdenellids (Porifera: Sphinctozoa) from China: Geological Magazine, v. 125, p. 149–159.

Manuscript Accepted by the Society March 27, 1990

Printed in U.S.A.

Geological Society of America
Special Paper 255
1990

Pre–Late Devonian geochemical, stratigraphic, sedimentologic, and structural patterns, Shoo Fly Complex, northern Sierra Nevada, California

Gary H. Girty, Kenneth C. Gester, and Joe B. Turner
Department of Geological Sciences, San Diego State University, San Diego, California 92182

ABSTRACT

In the northern Sierra Nevada, California, the pre–Upper Devonian Shoo Fly Complex has been subdivided into the following, from structurally lowest to structurally highest: (1) Lang sequence, (2) Duncan Peak chert, (3) Culbertson Lake allochthon, and (4) Sierra City melange. Detailed studies have been completed recently on the rocks in the Culbertson Lake allochthon in the Bowman Lake/Culbertson Lake area. The results of geochemical, sedimentologic, stratigraphic, and structural studies of rocks in the Culbertson Lake allochthon suggest that: (1) the allochthon is composed of a lower fault-bounded succession of basaltic rock and a complexly imbricated upper sandstone–dominated sequence; (2) the structurally lowest unit in the allochthon contains alkalic basalts that formed in a seamount or ocean island setting; (3) sedimentologic and stratigraphic patterns in the structurally higher units may be the result of sedimentation and deformation in a trench setting; (4) the allochthon is composed of a complex of northeast-dipping, thrust-fault–bounded slices, and is an imbricate structure; and (5) the imbricate structure, in terms of present-day geographic coordinates, formed as a result of generally westward-directed translations (i.e., northwest, west, or southwest) prior to the Late Devonian and probably after the Cambrian.

Data presented here generally support models that portray the Shoo Fly Complex as having formed in a subduction complex setting that developed near enough to a continental landmass to receive sand-sized detritus derived from it. However, current data do not uniquely constrain the location of the continental landmass that provided sand-sized detritus to the Shoo Fly Complex, nor do they rule out the possibility that the sand-rich portions of the Shoo Fly Complex were deposited as part of a continental margin submarine fan system that was subsequently accreted and imbricated within a subduction complex. The continental landmass supplying continental detritus to the Shoo Fly Complex may have been located in western North America, or it may have been located somewhere in Panthalassa.

INTRODUCTION

The Shoo Fly Complex, a metamorphosed assemblage of eugeoclinal rocks of pre–Late Devonian age, crops out in eastern California in a belt that is ~20 km wide and ~300 km long. It extends from approximately 37°30′N to beyond 40°N (Fig. 1; Hannah and Moores, 1986; Merguerian and Schweickert, 1987; Girty and Pardini, 1987; Harwood, 1988). North of 39°15′N, the eastern boundary of the complex is an angular unconformity that separates the Shoo Fly from a thick section of Upper Devonian through Permian volcanic and basinal rocks (Fig. 1; Varga and Moores, 1981; Schweickert, 1981; Harwood, 1988). Devonian and Permian volcanic rocks have been interpreted to represent magmatic arcs that, along with the Shoo Fly Complex, were

Girty, G. H., Gester, K. C., and Turner, J. B., 1990, Pre–Late Devonian geochemical, stratigraphic, sedimentologic, and structural patterns, Shoo Fly Complex, northern Sierra Nevada, California, *in* Harwood, D. S., and Miller, M. M., eds., Paleozoic and early Mesozoic paleogeographic relations; Sierra Nevada, Klamath Mountains, and related terranes: Boulder, Colorado, Geological Society of America Special Paper 255.

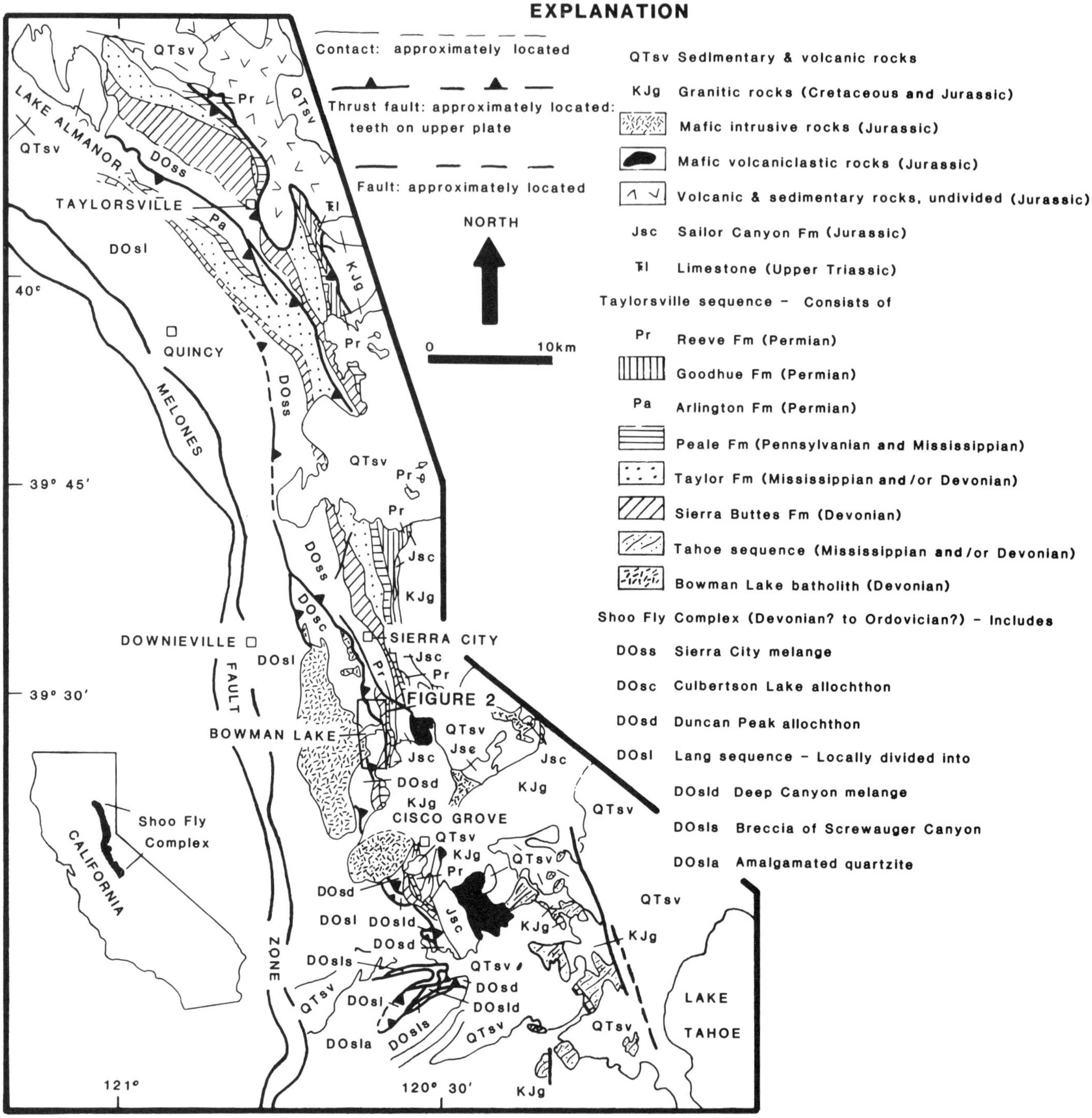

Figure 1. Generalized geologic map of a portion of the northern Sierra Nevada showing location of Figure 2. Map is modified from Figure 1 in Harwood (1988).

accreted to western North America during the Sonoman orogeny in latest Permian to early Triassic time (Burchfiel and Davis, 1975; D'Allura and others, 1977; Speed, 1979; Schweickert and Snyder, 1981). However, the pre-accretionary history of the Shoo Fly Complex is poorly understood and controversial. For example, some investigators argue that the Shoo Fly Complex developed adjacent to North America (D'Allura and others, 1977; Schweickert and Snyder, 1981; Varga, 1982; Girty and Wardlaw, 1984; Hannah and Moores, 1986), whereas others view the complex as a suspect or exotic terrane with respect to North America (Speed, 1979; Coney and others, 1980; Girty and Pardini, 1987).

In the area shown in Figure 1, the Shoo Fly Complex has been subdivided informally into four units by Schweickert and others (1984a) and Harwood (1988). In ascending structural order, these units include: (1) the Lang sequence, (2) the Duncan Peak chert or allochthon, (3) the Culbertson Lake allochthon, and (4) the Sierra City melange. Here we present a summary of recently completed geochemical, sedimentological, stratigraphic, and structural studies of units mapped within the Culbertson Lake allochthon in the Bowman Lake/Culbertson Lake area (Figs. 1, 2) (Girty, 1983). The results of our work support models that imply that the Shoo Fly Complex formed as a result of successive accretion associated with early Paleozoic subduction (e.g., D'Allura and others, 1977; Schweickert and Snyder, 1981; Schweickert and others, 1984a; Hannah and Moores, 1986; Girty and Pardini, 1987). However, our results do not uniquely constrain the position of the Shoo Fly Complex within the Paleo-Pacific Ocean, Panthalassa. Readers interested in the general geology of the Lang sequence, Duncan Peak chert, and Sierra City melange are referred to Taylor (1986), Girty and Pardini (1987), and Harwood (1988). Although rocks in the Culbertson Lake allochthon are generally metamorphosed to albite-epidote hornfels or greenschist facies, original volcanic and clastic textures are moderately to well preserved, and identification of protoliths is not difficult (Girty, 1983; Turner, 1986; Taylor, 1986; Girty and Pardini, 1987; Harwood, 1988). For ease of discussion, the prefix meta- is not used in this chapter.

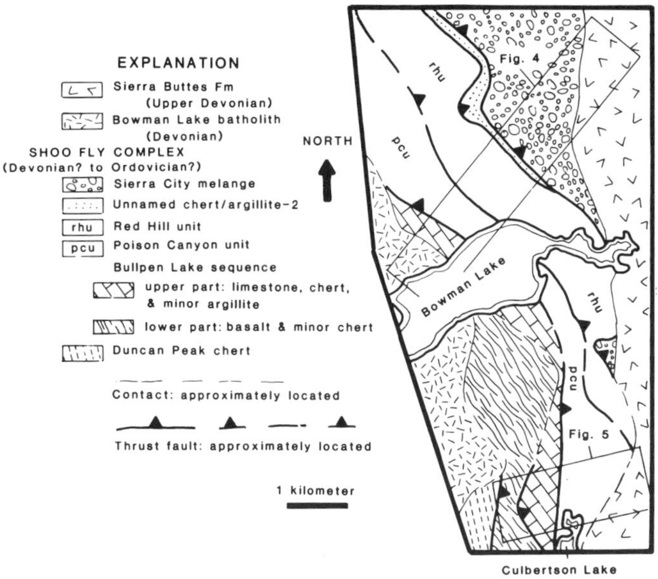

Figure 2. Simplified geologic map of the Bowman Lake/Culbertson Lake area showing locations of Figures 4 and 5.

STRATIGRAPHY/SEDIMENTOLOGY/ GEOCHEMISTRY

Structural relations in the Culbertson Lake allochthon are complex, but detailed mapping of folds and faults at a scale of 1:12,000 indicates that the allochthon consists of, in ascending structural order, the following informal map units (Fig. 2): (1) the Bullpen Lake sequence; (2) the Poison Canyon unit; (3) an unnamed ~50-m-thick unit of rhythmically bedded chert and argillite too thin to be shown in Figure 2; (4) the Red Hill unit; and (5) an unnamed 0- to ~100-m-thick interval of rhythmically bedded chert and argillite. In earlier papers by Girty and Schweickert (1984), Girty and Wardlaw (1984), and Girty and others (1984), the Poison Canyon unit was referred to as the "Poison Canyon formation" and the Red Hill unit as the "Red Hill formation." These older informal designations do not follow the recommendations of the North American Commission on Stratigraphic Nomenclature (1983), and as a result, are not used here. In the following sections, geochemical, stratigraphic, or sedimentologic data are briefly described from units mapped in the Bowman Lake/Culbertson Lake area.

BULLPEN LAKE SEQUENCE

The Bullpen Lake sequence crops out along the western margin of the Culbertson Lake allochthon (Fig. 2). The lower part of the sequence is intruded by the Bowman Lake batholith, or is cut out by a steep, northeast-dipping fault that forms the western boundary of the allochthon (Fig. 2). Thus, its stratigraphic thickness is unknown, but it attains a maximum outcrop thickness of 1,200 m (Turner, 1986).

Girty (1983, 1985) and Turner (1986) subdivided the Bullpen Lake sequence into mappable lower and upper parts (Fig. 2). The lower part consists mostly of basaltic pillow lava, pillow breccia, tuffaceous breccia, and massive basaltic lava that are locally interstratified with rhythmically bedded radiolarian chert. This latter stratigraphic characteristic, in conjunction with the presence of pillow lava and quench textures, indicates that the basalts were emplaced subaqueously (Bryan, 1972; Girty, 1985; Turner, 1986).

Lower Bullpen Lake basalts were analyzed for major and selected trace elements, and in some cases for selected rare earth elements (REE) following standard analytical procedures (Turner, 1986). The resulting analyses are listed and discussed in detail in Turner (1986), and are summarized in Table 1, together with the average compositions of alkalic basalts from the oceanic islands of the Marquesas Archipelago and Ponape Island (Dixon and others, 1984; Liotard and others, 1986). Uncertainties in oxide values are less than ±1 percent. Typical uncertainties in REE and trace-element values can be found in Nelson and others (1986).

Basalts from the Bullpen Lake sequence are similar in composition to unmetamorphosed alkalic basalts from the Marquesas Archipelago and Ponape Island (Table 1). This chemical similarity is underscored by the chondrite-normalized REE profiles for basalts from the Bullpen Lake sequence, which are like those characteristic of alkalic basalts (Fig. 3). In addition, Turner (1986) has shown that minor and trace-element concentrations in basalts from the Bullpen Lake sequence plot in the alkalic or ocean-island fields on: (1) the Zr vs. P_2O_5 and TiO_2 vs. Zr/P_2O_5

TABLE 1. AVERAGE ANALYSES OF BASALTS FROM THE BULLPEN LAKE SEQUENCE COMPARED TO AVERAGE ANALYSES OF ALKALIC BASALTS FROM PONAPE ISLAND AND THE MARQUESAS ARCHIPELAGO

	Bullpen Lake Sequence (n = 25)			Marquesas Archipelago* (n = 11)			Ponape Island† (n = 11)		
	X	S		X	S		X	S	
Wt. %									
SiO_2	45.15	2.93		44.03	2.33		44.12	1.00	
TiO_2	3.42	1.08		3.17	0.44		3.55	0.31	
Al_2O_3	12.65	2.49		12.76	2.03		13.99	1.17	
Fe_2O_3§	13.90	3.53		12.97	1.14		13.79	0.83	
MgO	8.13	3.38		9.96	4.01		7.88	2.00	
MnO	0.24	0.07		0.19	0.03		0.19	0.02	
CaO	10.56	4.53		9.81	1.16		10.57	0.07	
Na_2O	1.87	1.38		2.50	0.67		3.16	0.68	
K_2O	1.02	1.02		1.21	0.68		1.27	0.26	
P_2O_5	1.19	0.60		0.65	0.17		0.73	0.24	
	n = 7			n = 10			n = 11		
ppm									
Cr	423	370		442	281	(n = 11)	246	196	
Rb	29	25	(n = 25)	39	27	(n = 11)	33	11	
Sr	400	89	(n = 25)	870	408	(n = 11)	795	162	(n = 10)
Zn	122	53		122	15	(n = 11)	....	....	
V	190	71		292	46	(n = 11)	....	....	
Ni	196	246		270	176	(n = 11)	194	96	(n = 10)
Y	39	4.5		30	2.9		....	....	
Zr	243	62	(n = 25)	269	47		290	108	
Nb	28	14		43	16		....	....	
Sm	13	3.0		8.0	1.5		10.5	2.9	
Eu	4.1	1.0		2.6	0.4	(n =9)	3.3	0.8	
Lu	0.31	0.06		0.29	0.03		0.30	0.06	
Ba	548	606	(n = 25)	478	197		428	176	
U	2.7	1.2		....	....		1.0	0.2	(n = 5)
Th	5.3	1.4		5.7	2.4		4.6	2.0	
Yb	2.5	0.4		2.0	0.2		2.2	0.3	
Nd	63	17		43	12	(n = 8)	....	....	
La	62	20		41	15		43	15	
Ce	128	37		86	28		95	36	
Tb	2.3	0.6		1.0	0.1		1.7	0.4	
Hf	7.3	2.0		5.6	0.7	(n = 9)	7.7	2.4	
Cs	1.6	2.7		....	....		0.5	0.3	(n = 10)
Sc	24	5		22	5		20	4	
Ta	12	9		....	....		....	....	
CO	66	29		57	16		....	....	

Notes: *Liotard and others (1986).
†Dixon and others (1984).
§Fe_2O_3 = Fe total as Fe_2O_3.
X = Mean value.
S = One standard deviation.
n = Number of samples analyzed.

diagrams of Winchester and Floyd (1976), (2) the Ti/100-Zr-YX3 discrimination diagram of Pearce and Cann (1973), (3) the TiO_2 vs. Zr discrimination diagram of Gale and Pearce (1982), and (4) the MnOX10-TiO_2-P_2O_5X10 discrimination diagram of Mullen (1983). Thus, geochemical data indicate that basalts in the Bullpen Lake sequence are derived from alkalic magmas that are like those in modern-day intraplate seamounts or ocean islands.

Limestones in the upper part of the Bullpen Lake sequence are generally finely laminated at the centimeter scale, but locally they display graded beds or Bouma-like T_{bcd}, T_{bc}, and T_{cd} beds that range in thickness from 10 to 20 cm. North of Bowman Lake, they are in contact with the Bowman Lake batholith, and are more than 250 m thick (Figs. 2, 4). South of Bowman Lake, the same limestones depositionally overlie basalts in the lower Bullpen Lake sequence and are only ~50 m thick (Fig. 2). The same limestones are in fault contact with the lower Bullpen Lake sequence west of Culbertson Lake, where they are more than 300 m thick (Figs. 2, 5), and appear to be interstratified with a thin (~10 to ~25 m thick) sequence of bedded chert and a section of argillite that is ~10 to ~25 m thick (Fig. 4).

As described above, the apparent stratigraphic thickness of the limestones in the upper part of the Bullpen Lake sequence varies from ~50 to >300 m. The transition from a thick to a thin section of limestone is poorly exposed. The cause of thickness variations within this unit is unknown and may be due to either structural or depositional processes.

Limestones that overlie alkalic basalts in the lower Bullpen Lake sequence are interpreted to have been deposited in a sub-wave–base environment above the Paleozoic carbonate compensation depth (CCD) (Girty, 1985; Turner, 1986). In the Bowman Lake/Culbertson Lake area, the limestones are stratigraphically overlain by a unit of rhythmically bedded pelagic chert and argillite (bls[ch]) of variable thickness (Figs. 4, 5). The deposition of the siliceous sediment now represented by the pelagic chert and argillite may reflect subsidence of the limestones below the paleo-CCD (Girty, 1985; Turner, 1986), or the drowning of the source of carbonate detritus (D. S. Harwood, 1989, personal communication).

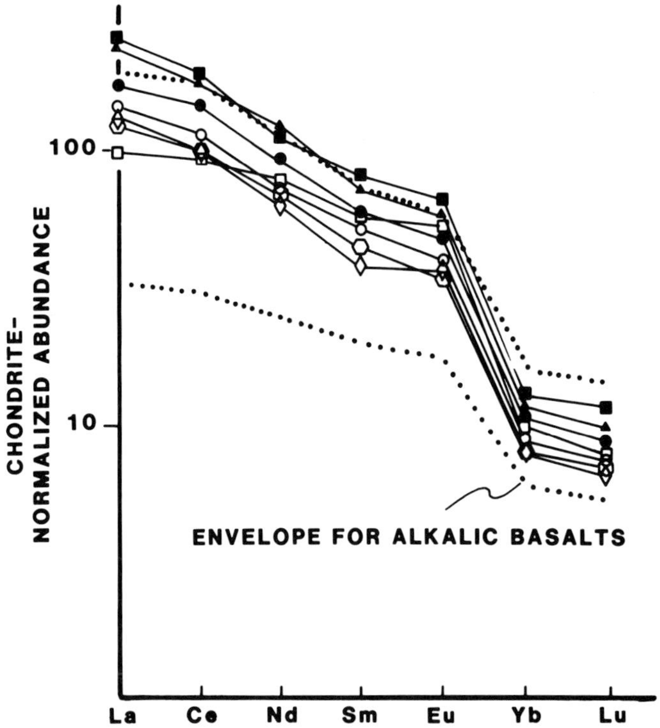

Figure 3. Chondrite-normalized REE profiles for basalts from the Bullpen Lake sequence. See Turner (1986) for tabulation of analyses, descriptions of samples analyzed, and locations of each sample. Dotted lines represent envelope of alkalic basalts from Ponape Island, the Marquesas Archipelago, and the Hawaiian Islands (Basaltic Volcanism Study Project, 1981, p. 173). Lines connecting symbols represent profiles for samples from the Bullpen Lake sequence.

POISON CANYON AND RED HILL UNITS

Seven representative stratigraphic sections were measured in both the Poison Canyon and Red Hill units (Gester, 1987). Bed thicknesses were measured in thick, relatively coherent, upright limbs of folds described later in this chapter. A representative section is shown in Figure 6, and characteristic bed forms and sedimentary structures observed in the measured sections are shown schematically in Figure 7.

As shown in the latter illustration, sandstone beds contain primary depositional features that are customarily attributed to deposition by highly concentrated turbidity currents and other forms of sediment gravity flow (e.g., Lowe, 1982; Nelson and Nilsen, 1984). Mudstone/argillite beds, in turn, contain primary depositional features that suggest deposition from dilute turbidity currents or debris flows (Girty, 1983; Gester, 1987). The sedimentary structures illustrated in Figure 7 also define four distinct sedimentary facies that are described in detail in the following sections.

Massive sandstone facies

Between 91 and 98 percent of the sandstone beds in the seven measured sections were assigned to the massive sandstone facies (Gester, 1987). Detritus ranges from medium-grained sand to granule conglomerate, with coarse- to very coarse-grained sand most common. Both graded and non-graded sandstone beds occur in the massive sandstone facies (Fig. 7).

Most (>50 percent) of the graded and non-graded massive sandstone beds range in thickness from 0.1 to 3.0 m. Thicker graded and non-graded beds reach a maximum thickness of 6 m, but they make up a smaller percentage (5 to 15 percent) of the sandstone beds present in any given section (Gester, 1987).

Within the measured sections, amalgamated massive sandstone intervals (2 to more than 130 m thick) make up 14 to 28 percent of the sandstone beds. Some amalgamated intervals are

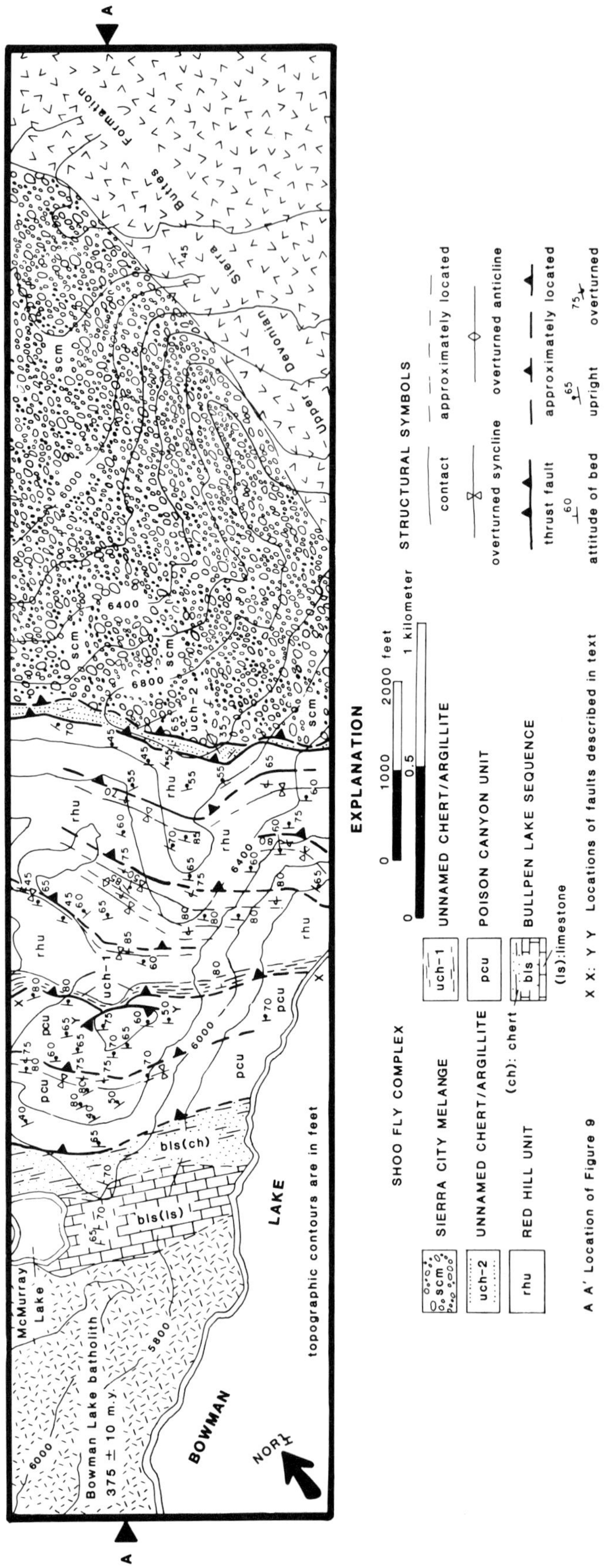

Figure 4. Structural map of a moderately to well-exposed portion of the Shoo Fly Complex north of Bowman Lake.

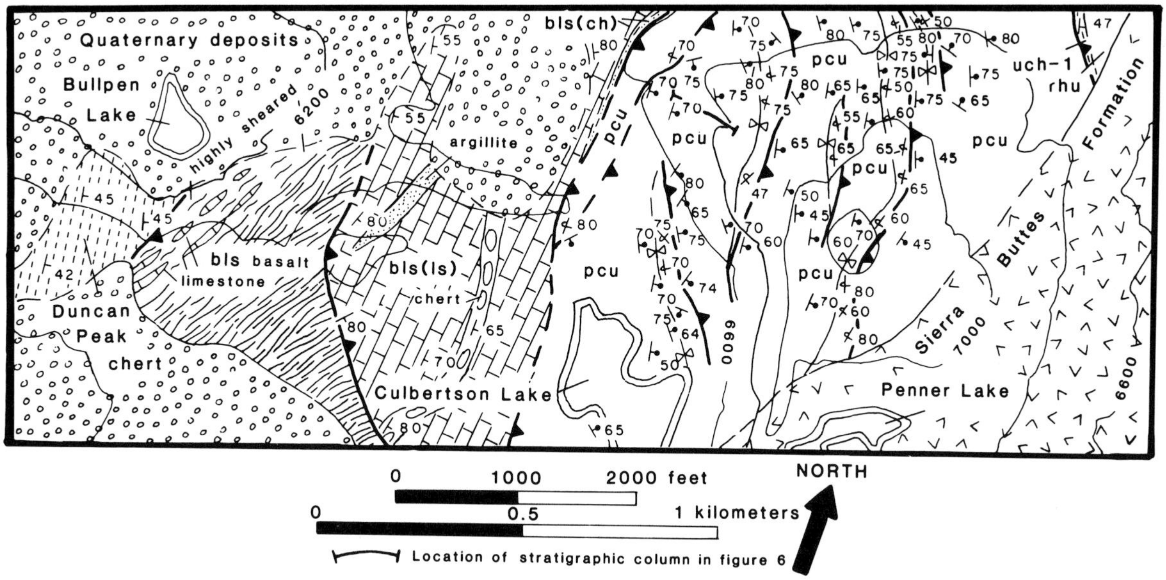

Figure 5. Structural map of a moderately to well-exposed portion of the Shoo Fly Complex south of Bowman Lake. See Figure 4 for explanation of map units and structural symbols.

clearly associated with the base of channelized deposits; in others, sedimentary relations are obscured by structural complications such as boudinage and extensive layer-parallel fragmentation of bedding.

T_a-present turbidite facies

T_a-present and T_a-absent turbidites are described using the terminology of Bouma (1962). In the measured sections, only 2 to 9 percent of the sandstone beds can be described by Bouma terminology (Gester, 1987). Of these beds, T_a-present turbidites account for more than 90 percent; remaining sandstone beds were assigned to the T_a-absent facies. T_a-present turbidites range in thickness from 0.1 to 2.34 m, but 75 percent are between 0.1 and 0.7 m thick; only 3 percent are more than 1.2 m thick (Gester, 1987).

Approximately 80 percent of T_a-present sandstone beds display T_{ab} or T_{abc} Bouma intervals. T_{abc} beds composed of very fine-grained to medium-grained sandstone and siltstone are an uncommon form of T-present turbidites in the Bowman Lake/Culbertson Lake area. Most T_a-present turbidites have either coarse-grained or medium-grained bases that grade upward to fine-grained sandstone or siltstone tops (Fig. 7). Coarse tail grading in the T_a interval is common; faint fluid-escape structures are uncommon.

T_a-absent turbidite facies

Beds composed of T_b, T_{bc}, or T_{cd} Bouma intervals are characteristic of the T_a-absent turbidite facies (Fig. 7). T_a-absent turbidite beds are commonly 9 to 20 cm thick. T_b intervals are, on average, 7.5 cm thick. They are mostly medium-grained sandstones that display thinly (1 to 2 cm) to widely spaced (5 to 10 cm) parallel laminations. T_c intervals range in thickness from <1 to 20 cm. They are typically ripple cross-laminated, very fine-grained to fine-grained sandstones and siltstones. Sets of climbing ripple cross-laminations, or in some instances, starved ripples, make up some T_c Bouma intervals.

Mudstone/argillite facies

The mudstone/argillite facies consists of beds of pebbly or sandy mudstone and mudstone/argillite (Fig. 7). Pebbly mudstones occur throughout the Poison Canyon and Red Hill units but are volumetrically insignificant. They occur in association with sandy mudstones that are interstratified with the massive sandstone facies. Beds of pebbly or sandy mudstone are generally 25 cm to several meters thick, and are composed of sand-sized quartz grains or centimeter-scale clasts of siltstone or sandstone enclosed in a mudstone matrix. In contrast, ubiquitous beds (<50 cm thick) of hemipelagic mudstone/argillite are either ripple cross-laminated, parallel laminated, homogeneous, or mixtures of these three varieties (Fig. 7).

INTERPRETATION OF STRATIGRAPHY AND SEDIMENTOLOGY

Pre–Upper Devonian sedimentologic and stratigraphic patterns identified in the Poison Canyon, Red Hill, and adjacent units are similar to those inferred to occur as a result of sedimentation and deformation in a trench setting (Piper and others, 1973; von Huene, 1974; Moore and Karig, 1976; Lash, 1985, 1986). For example, Lash (1985) has suggested that accreted slices of trench sediment should consist of a series of fault-

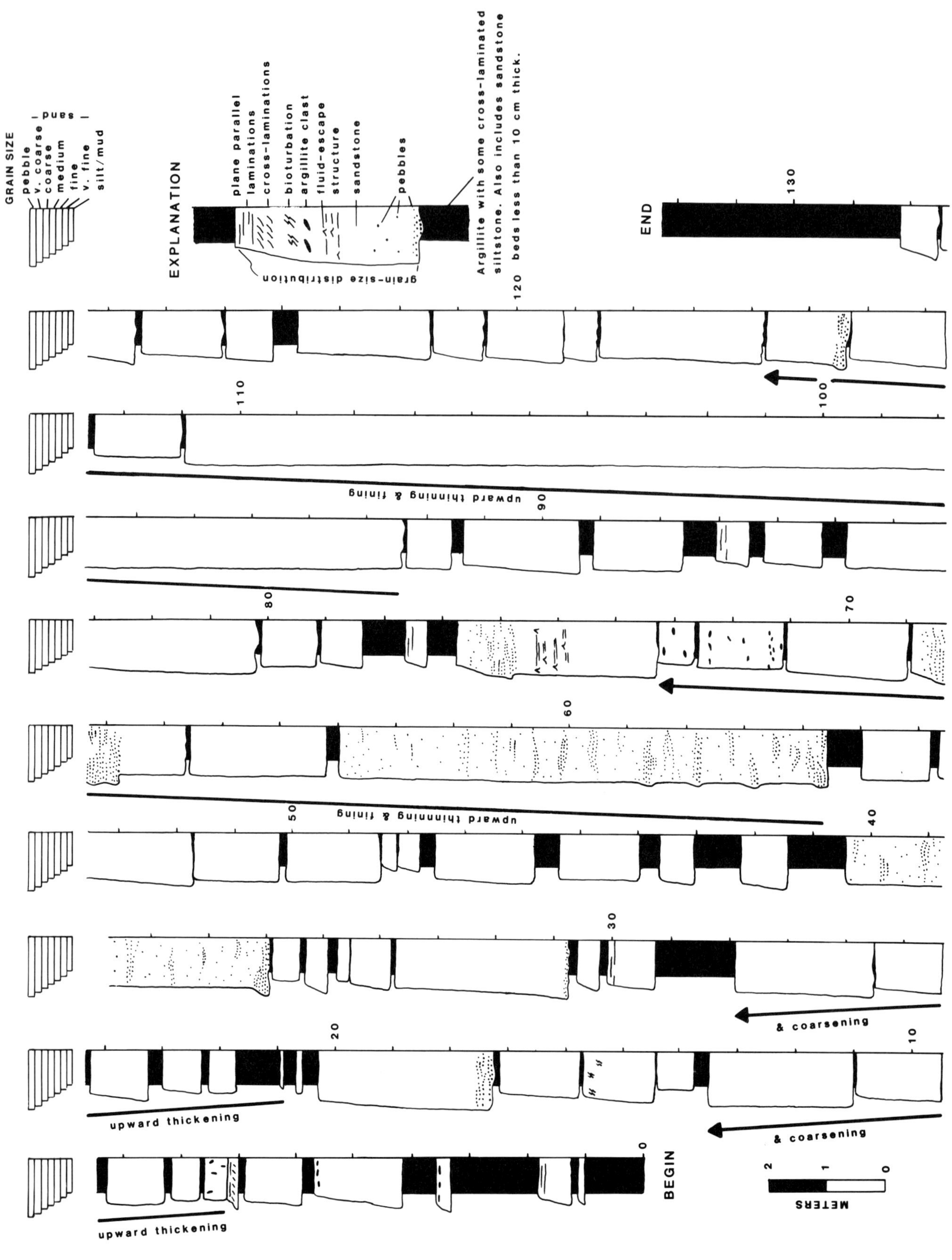

Figure 6. Detailed stratigraphic column for a small portion of the lower Poison Canyon unit. Location of column is shown in Figure 5.

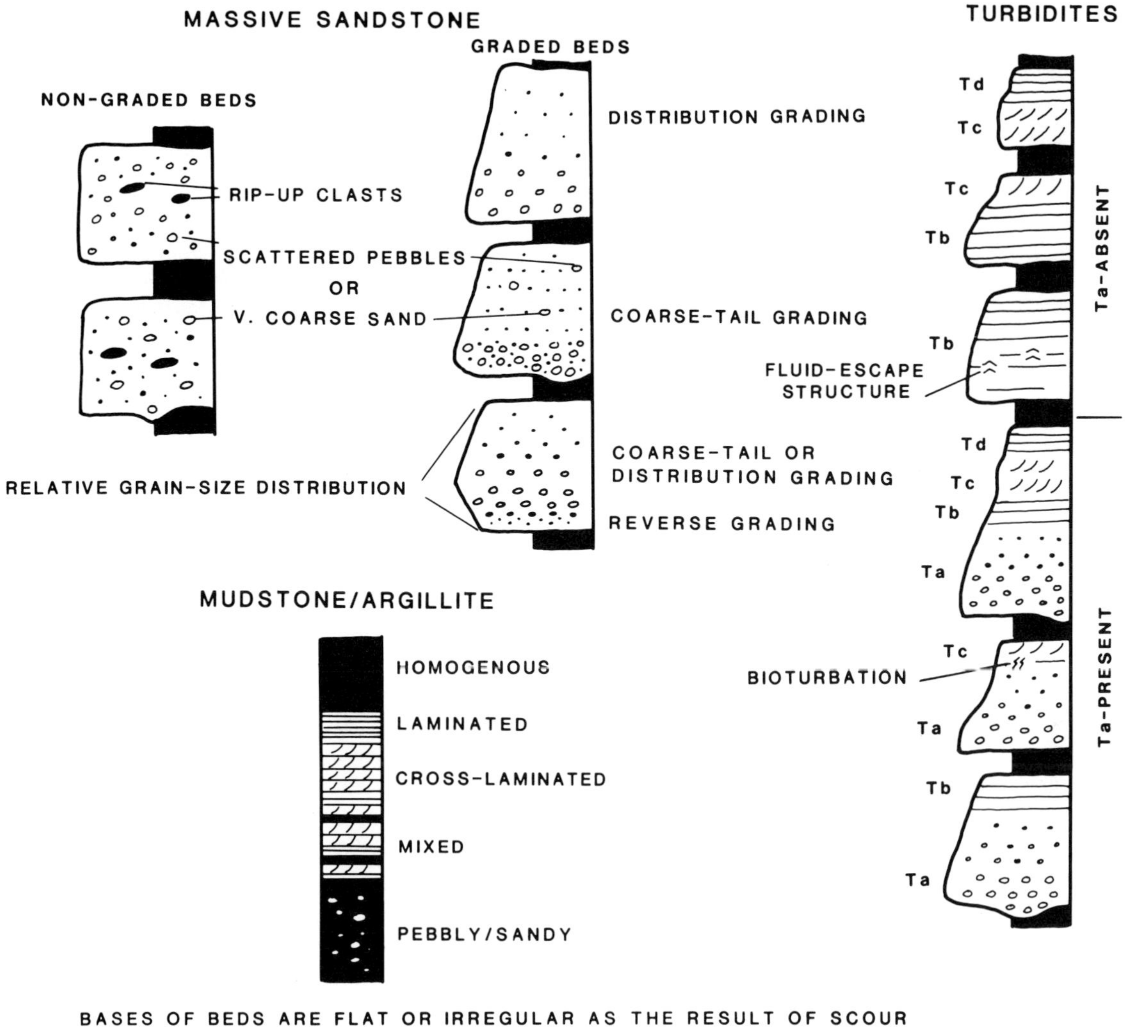

Figure 7. Sedimentologic characteristics of sandstone and mudstone/argillite beds, and sedimentary facies commonly observed in the Poison Canyon and Red Hill units.

bounded, coarsening-upward sequences. Each ideal coarsening-upward sequence is generally thought to begin with pelagic deposits (e.g., radiolarian cherts) overlain by "near-trench" siltstones and hemipelagic mudstones. The latter deposits are overlain, in turn, by "trench-fill" sandstones derived from sediment gravity flows. This ideal upward-coarsening sequence may be incomplete in each accreted slice as the result of faulting or the removal and subduction of stratigraphically lower intervals (Lash, 1985).

Stratigraphic and sedimentologic patterns observed in the Poison Canyon and Red Hill units are summarized in Figure 8. Note that they resemble those patterns attributed by Lash (1985) to accreted trench sediments. For example, observations of the structurally lower 150 m of the Poison Canyon and Red Hill units indicate that a few (two to four) laterally discontinuous 5- to 10-m-thick, upward-thickening and upward-coarsening sequences composed of four or more massive sandstone beds are locally well developed in the lower 25 to 30 m of the sand-rich upper portion of each formation (e.g., Figs. 6, 8). These upward-thickening and upward-coarsening sequences are inferred to represent "trench-fill" progradational lobes. Significantly, the inferred progradational lobes commonly overlie a structurally complex interval (~10 to ~40 m thick) composed primarily of hemipelagic mudstone/argillite that is interpreted to represent "near-trench" deposits (Lash, 1985) (Fig. 8). The hemipelagic deposits depositionally or structurally overlie a thin unit of pelagic radiolarian-bearing chert, which probably was derived from abyssal plain sediments (Fig. 8).

Stratigraphically above the lower progradational lobes in the westernmost or structurally lower fault-bounded slices of the Poison Canyon and Red Hill units is a thick section (locally on the order of ~100 to ~110 m thick) that contains channelized, ~15 to ~25 m thick, upward-thinning and upward-fining sequences composed almost entirely of four or more massive sandstone beds (e.g., Figs. 6, 8). Structurally higher or more eastern fault-bounded slices tend to be composed of randomly dispersed,

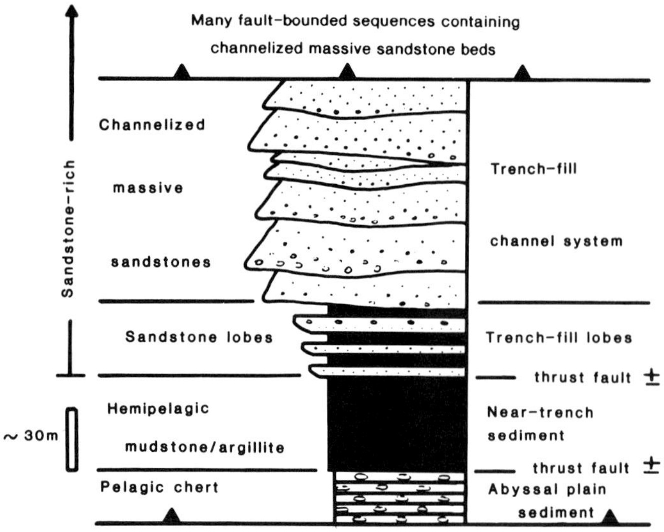

Figure 8. General sedimentologic/stratigraphic/structural patterns for the Poison Canyon and Red Hill units. Illustrated thicknesses were estimated from structurally coherent sections and are considered to be minimum values.

thick-bedded massive sandstones separated by centimeter-thick layers of hemipelagic mudstone/argillite, and/or channelized, upward-thinning and upward-fining sequences composed of four or more massive sandstone beds. As shown in Figure 8, the channelized massive sandstone beds in the complexly imbricated sand-rich upper portions of the Poison Canyon and Red Hill units are inferred to represent channelized "trench-fill" deposits. Hence, the above data support the hypothesis that the Poison Canyon and Red Hill units represent the remnants of a sand-rich depositional system that was accreted to the lower trench slope during pre–Late Devonian subduction. The question remains, however, are pre–Late Devonian structures inthe Bowman Lake/Culbertson Lake area consistent with this interpretation?

STRUCTURE

Recently published structural data and seismic reflection profiles suggest that modern and ancient subduction complexes are composed of oceanward-verging, thrust-fault–bounded packages of accreted sediment, and in some instances, bathymetrically high areas such as seamounts (e.g., Moore and others, 1985; Davey and others, 1986; Westbrook and others, 1988). Although the above description generally fits the rocks in the Bowman Lake/Culbertson Lake area, structures that can be used to determine the kind of displacement associated with faults that bound map units are uncommon. North of Bowman Lake, however, fault X in Figures 4 and 9 is well exposed at the base of the unnamed chert/argillite unit below the Red Hill unit (e.g., uch-1 in Fig. 4). The fault surface is sharp, dips ~70° northeastward, and is slightly polished. Asymmetric, drag-related folds in the hanging wall have axes that plunge shallowly to moderately to the northwest or southeast (Fig. 10). Folds that plunge to the northwest exhibit sinistral asymmetry, whereas southeastward-plunging folds display dextral asymmetry (Fig. 10). These structural relations indicate that fault X is a reverse or oblique-reverse fault, in which the eastern hanging wall moved generally westward (i.e., northwest, west, or southwest) over the western footwall. Other unit-bounding faults mapped in the Bowman Lake/Culbertson Lake area predate the unconformity at the base of the Upper Devonian Sierra Buttes Formation, and are parallel to subparallel to fault X (Figs. 2, 4, 5, 9). Hence, they are geometrically similar and probably formed in a similar manner to fault X. As mentioned previously, the Upper Devonian Sierra Buttes Formation unconformably overlies the Shoo Fly Complex east of Bowman Lake. In this area, bedding in the Sierra Buttes Formation generally dips 10 to 25° less than the pre–Late Devonian faults in the Shoo Fly Complex. This observation suggests that the pre–Late Devonian faults in the Shoo Fly Complex may have dipped about 10° to 25° prior to post–Sierra Buttes Formation tilting. They are thus inferred to be thrust or oblique-thrust faults.

Within the Poison Canyon and Red Hill units, many overturned synclines and thrust faults, and to a lesser extent overturned anticlines, also have been mapped (Figs. 4, 5). These structures were identified in the field by mapping the stratigraphic facing direction (>500 determinations) as indicated by graded beds (e.g., Girty, 1983; Girty and Schweickert, 1984). The results of this exercise indicate that overturned synclines generally are composed of a thick, southwestern, stratigraphically intact, long, upright limb, and a much thinner and shorter overturned northeastern limb composed of highly attenuated and broken beds (Figs. 4, 5, 9). The thickness of the upright southwestern limb varies between structures, but generally ranges from ~75 to ~300 m. In a similar manner, the overturned northeastern limb also varies in thickness, but generally ranges from ~15 to ~75 m. Observations of complete synclinal closures are uncommon because of poor exposure, or more often, because of the highly extended and broken character of strata. Nevertheless, strata in the overturned northeastern limb commonly terminate over a short distance (~2 to ~5 m) perpendicular to the strike of bedding against a poorly exposed fault at the base of the next thick, stratigraphically intact succession of upright beds. Based on the direction of overturning, and the westward-verging asymmetry of observed folds, the fault at the base of the next upright succession of beds is interpreted to be a thrust fault (Figs. 4, 9). These observations and interpretations suggest that the overturned folds and faults within the Poison Canyon and Red Hill units were produced by generally westward-directed translations (i.e., NW, W, or SW), relative to current geographic coordinates. Thus, our data are compatible with the idea that the Shoo Fly Complex in the Bowman Lake/Culbertson Lake area represents a remnant of a generally westward-verging subduction complex.

The timing of formation of the folds and thrust faults described above is constrained by the following relations.

(1) Some thrust faults that bound major tectonostratigraph-

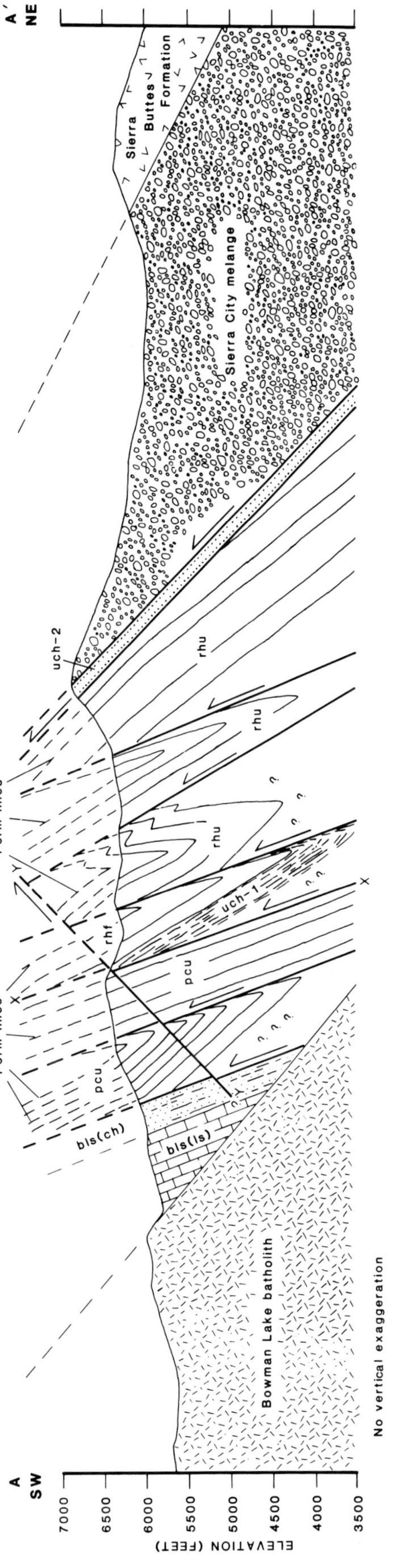

Figure 9. Structural cross section for area shown in Figure 4. See Figure 4 for explanation of mapped units.

ic units in the Shoo Fly Complex are cut by the Bowman Lake batholith (Figs. 1, 2), which has a poorly constrained U-Pb zircon age of 375 ± 10 Ma (Hanson and others, 1988).

(2) The overturned folds and thrust faults in the Shoo Fly Complex described in this chapter do not extend into the unconformably overlying Upper Devonian to Mesozoic volcanic and sedimentary rocks that form a moderately northeast-dipping (~45–~60°) and northeast-facing homocline in the Bowman Lake/Culbertson Lake area (Hanson and Schweickert, 1986).

(3) Radiolaria, which could not be identified as to genera, are common in the cherts in the Bowman Lake/Culbertson Lake area (Girty and Wardlaw, 1985; Turner, 1986; Girty and Pardini, 1987). Radiolaria are sparse in Cambrian rocks on a worldwide basis, but are relatively common in younger Paleozoic strata (White, 1986), suggesting a post-Cambrian age for chert in the Culbertson Lake allochthon. These temporal relations suggest that the imbricate structure in the Bowman Lake/Culbertson Lake area probably formed sometime after the Cambrian and before the Late Devonian.

Three phases of Mesozoic deformation produced structures that, at the scale of Figures 4 and 5, weakly distort the pre–Upper Devonian structures described above (Girty and Schweickert, 1984; Schweickert and others, 1984b; Girty and others, 1984). For example, the mapped traces of the pre–Upper Devonian folds and thrust faults are curvilinear and relatively continuous in the Bowman Lake/Culbertson Lake area (Figs. 4, 5). The curvilinear character of the mapped traces of the pre–Upper Devonian structures is due primarily to the widely spaced second- and third-phase Mesozoic folds described by Girty and Schweickert (1984) and Schweickert and others (1984b). First-phase Mesozoic folds are developed only locally at the mesoscopic scale, and are approximately coaxial with pre–Upper Devonian structures. In order to emphasize Paleozoic relations, the traces of Mesozoic folds are not shown in Figures 4 and 5. However, fault Y in Figures 4 and 9 may be a northeast-directed thrust fault of Mesozoic age. The northeastward direction of movement across fault Y was indicated by the "smooth" direction of fibrous slickensides that plunge down-dip of the exposed southwest-dipping fault surface. Exposures of fault Y are limited, and those that exist do not provide adequate timing constraints. Thus, we cannot rule out the possibility that fault Y is related to the development of the imbricate structure.

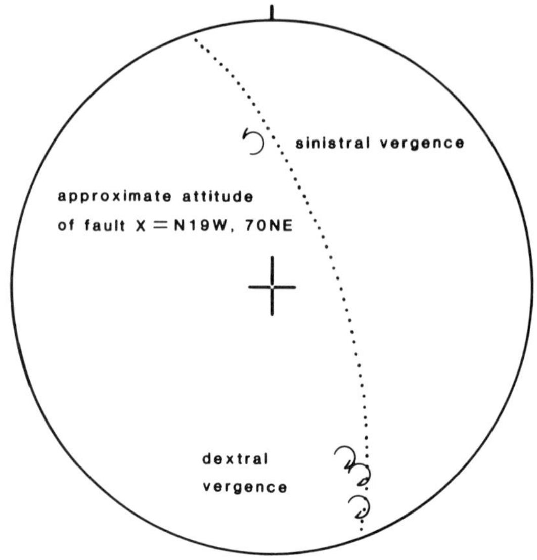

Figure 10. Equal area stereonet plot of the attitudes of hinge lines of asymmetric drag folds in the hanging wall of fault X in Figures 4 and 9.

CONCLUSIONS

Data presented here are consistent with the idea that the pre–Upper Devonian rocks in the Bowman Lake/Culbertson Lake area are the remnants of an early Paleozoic imbricate structure that probably formed in a subduction complex setting. The imbricate structure contains a structurally lower fault-bounded unit composed of the remnants of an oceanic intraplate seamount or ocean island, and structurally higher units are composed of the remnants of a sand-rich submarine depositional system that may have prograded into a trench setting. If the latter statement is correct, then the sand-rich submarine depositional system may have formed a trench fan or fans (e.g., Lash, 1985, 1986). Alternatively, the sand-rich depositional system may have been a part of a continental-margin submarine fan system that was later accreted and imbricated within a subduction complex (e.g., Schweickert and Snyder, 1981; Girty, 1983).

Girty and Wardlaw (1985) presented petrologic data that suggest that sandstones in the Poison Canyon and Red Hill units contain detritus derived from a continental landmass. This interpretation was supported by U-Pb detrital zircon data that indicate that the continental landmass contains Proterozoic rocks. These data imply that the Poison Canyon and Red Hill units were deposited close enough to some continent or continental fragment to receive detritus derived from it. Although we favor the interpretation that the continental landmass was located somewhere in western North America, there is nothing in the geochemical, stratigraphic, sedimentologic, structural, or petrologic data discussed in this chapter that precludes a setting either in or along other parts of the margin of Panthalassa. Nevertheless, detailed studies, such as the ones summarized here, offer the only means of unraveling the complex history of deposition and accretion along the long-lived southwestern Cordilleran margin. We hope that our work will stimulate additional studies in the perplexing pre–Upper Devonian rocks of the northern Sierra Nevada and elsewhere along the margin of Panthalassa. Only through continued detailed field-oriented studies of the various terranes scattered about the margin of Panthalassa will we be able to reconstruct their origins.

ACKNOWLEDGMENTS

The contents of this chapter were constructively and thoroughly reviewed by E. L. Miller, M. M. Miller, D. S. Harwood, W. D. Sharp, and R. J. Varga; we gratefully acknowledge their efforts. Work reported here was partially funded by a Summer Faculty Fellowship from San Diego State University and by National Science Foundation grant EAR-8902382 (to G.H.G.).

REFERENCES CITED

Basaltic Volcanism Study Project, 1981, Basaltic volcanism on the terrestrial planets: New York, Pergamon Press, 1,286 p.

Bouma, A. H., 1962, Sedimentology of some flysch deposits: Amsterdam, Elsevier, 163 p.

Bryan, W. B., 1972, Morphology of quench crystals in submarine basalts: Journal of Geophysical Research, v. 77, p. 5812–5819.

Burchfiel, B. C., and Davis, G. A., 1975, Nature and controls of Cordilleran orogenesis, western United States; Extensions of an earlier synthesis: American Journal of Science, v. 275A, p. 363–396.

Coney, P. J., Jones, D. L., and Monger, J.W.H., 1980, Cordilleran suspect terranes: Nature, v. 288, p. 329–333.

D'Allura, J. A., Moores, E. M., and Robinson, L., 1977, Paleozoic rocks of the northern Sierra Nevada; Their structural and paleogeographic implications, *in* Stewart, J. H., Stevens, C. H., and Fritsche, A. E., eds., Paleozoic paleogeography of the western United States: Pacific Section, Society of Economic Paleontologists and Mineralogists Pacific Coast Paleogeography Symposium 1, p. 395–408.

Davey, F. J., and 5 others, 1986, Structure of a growing accretionary prism, Hikurangi margin, New Zealand: Geology, v. 14, p. 663–666.

Dixon, T. H., Batiza, R., Futa, K., and Martin, D., 1984, Petrochemistry, age and isotopic composition of alkali basalts from Ponape Island, western Pacific: Chemical Geology, v. 43, p. 1–28.

Gale, G. H., and Pearce, J. A., 1982, Geochemical patterns in Norwegian greenstones: Canadian Journal of Earth Sciences, v. 19, p. 385–397.

Gester, K. C., 1987, Stratigraphic evidence for a Paleozoic submarine fan, Shoo Fly Complex, northern Sierra Nevada, California [M.S. thesis]: San Diego, California, San Diego State University, 236 p.

Girty, G. H., 1983, The Culbertson Lake allochthon, a newly identified structural unit within the Shoo Fly Complex; Sedimentologic, stratigraphic, and structural evidence for extension of the Antler orogenic belt to the northern Sierra Nevada, California [Ph.D. thesis]: New York, Columbia University, 155 p.

—— , 1985, An accreted seamount; A working hypothesis for the Bullpen Lake sequence, Sierra Nevada, California: California Geology, v. 38, p. 123–128.

Girty, G. H., and Pardini, C. H., 1987, Provenance of sandstone inclusions in the Paleozoic Sierra City melange, Sierra Nevada, California: Geological Society of America Bulletin, v. 98, p. 176–181.

Girty, G. H., and Schweickert, R. A., 1984, The Culbertson Lake allochthon, a newly identified structure within the Shoo Fly Complex, California; Evidence for four phases of deformation and extension of the Antler orogeny to the northern Sierra Nevada: Modern Geology, v. 8, p. 181–198.

Girty, G. H., and Wardlaw, M. S., 1984, Was the Alexander terrane a source of feldspathic sandstones in the Shoo Fly Complex, Sierra Nevada, California?: Geology, v. 12, p. 339–342.

—— , 1985, Petrology and provenance of pre–Late Devonian sandstones, Shoo Fly Complex, northern Sierra Nevada, California: Geological Society of America Bulletin, v. 96, p. 516–521.

Girty, G. H., Wardlaw, M. S., Schweickert, R. A., Hanson, R. E., Bowring, S., 1984, Timing of pre-Antler deformation in the Shoo Fly Complex, Sierra Nevada, California: Geology, v. 12, p. 673–676.

Hannah, J. L., and Moores, E. M., 1986, Age relationships and depositional environments of Paleozoic strata, northern Sierra Nevada, California: Geological Society of America Bulletin, v. 97, p. 787–797.

Hanson, R. E., and Schweickert, R. A., 1986, Stratigraphy of mid-Paleozoic island-arc rocks in part of the northern Sierra Nevada, Sierra and Nevada Counties, California: Geological Society of America Bulletin, v. 97, p. 986–998.

Hanson, R. E., Saleeby, J. B., and Schweickert, R. A., 1988, Composite Devonian island-arc batholith in the northern Sierra Nevada, California: Geological Society of America Bulletin, v. 100, p. 446–457.

Harwood, D. S., 1988, Tectonism and metamorphism in the northern Sierra terrane, northern California, *in* Ernst, W. G., ed., Metamorphism and crustal evolution of the western United States; Rubey volume 7: Englewood Cliffs, New Jersey, Prentice-Hall, p. 765–788.

Lash, G. G., 1985, Recognition of trench fill in orogenic flysch sequences: Geology, v. 13, p. 867–870.

—— , 1986, Sedimentology of channelized turbidite deposits in an ancient (early Paleozoic) subduction complex, central Appalachians: Geological Society of America Bulletin, v. 97, p. 703–710.

Liotard, J. M., Barsczus, H. G., Dupy, C., and Dostal, J., 1986, Geochemistry and origin of basaltic lavas from Marquesas Archipelago, French Polynesia: Contributions to Mineralogy and Petrology, v. 92, p. 260–268.

Lowe, D. R., 1982, Sediment gravity flows; 2, Depositional models with special references to the deposits of high-density turbidity currents: Journal of Sedimentary Petrology, v. 52, p. 279–297.

Merguerian, C., and Schweickert, R. A., 1987, Paleozoic gneissic granitoids in the Shoo Fly Complex, central Sierra Nevada, California: Geological Society of America Bulletin, v. 99, p. 699–717.

Moore, J. C., and Karig, D. E., 1976, Sedimentology, structural geology, and tectonics of the Shikoku subduction zone, southwestern Japan: Geological Society of America Bulletin, v. 87, p. 1259–1268.

Moore, J. C., Cowan, D. S., and Karig, D. E., 1985, Penrose Conference report; Structural styles and deformation fabrics of accretionary complexes: Geology, v. 13, p. 77–79.

Mullen, E. D., 1983, $MnO/TiO_2O_2/P_2O_5$; A minor element discriminant for basaltic rocks of oceanic environments and its implications for petrogenesis: Earth and Planetary Science Letters, v. 62, p. 53–62.

Nelson, C. H., and Nilsen, T. H., 1984, Modern and ancient deep-sea fan sedimentation: Society of Economic Paleontologists and Mineralogists Short Course No. 14, 404 p.

Nelson, K. L., Collinsworth, B. D., and Bell, R. W., 1986, Determination of intraflow element variation in a thick flow of the Crossen Trachyte, west Texas: Texas Journal of Science, v. 38, p. 71–80.

North American Commission on Stratigraphic Nomenclature, 1983, North American Stratigraphic Code: American Association of Petroleum Geologists Bulletin, v. 67, p. 841–875.

Pearce, J. A., and Cann, J. R., 1973, Tectonic setting of basic volcanic rocks determined using trace element analyses: Earth and Planetary Science Letters, v. 19, p. 290–300.

Piper, D.J.W., von Huene, R., and Duncan, J. R., 1973, Late Quaternary sedimentation in the active eastern Aleutian trench: Geology, v. 1, p. 19–22.

Schweickert, R. A., 1981, Tectonic evolution of the Sierra Nevada, *in* Ernst, W. G., ed., The geotectonic development of California; Rubey volume 1: Englewood Cliffs, New Jersey, Prentice-Hall, p. 88–131.

Schweickert, R. A., and Snyder, W. S., 1981, Paleozoic plate tectonics of the Sierra Nevada and adjacent regions, *in* Ernst, W. G., ed., The geotectonic development of California; Rubey volume 1: Englewood Cliffs, New Jersey, Prentice-Hall, p. 182–201.

Schweickert, R. A., Harwood, D. S., Girty, G. H., and Hanson, R. E., 1984a, Tectonic development of the northern Sierra terrane; An accreted late Paleozoic island arc and its basement *in* Lintz, J., ed., Geological excursions; Geological Society of America Annual Meeting Guidebook: Reno, University of Nevada Department of Geology, 50 p.

Schweickert, R. A., Bogen, N. L., Girty, G. H., Hanson, R. E., and Merguerian,

C., 1984b, Contrasting styles of deformation during the Nevadan orogeny, Sierra Nevada, California: Geological Society of America Bulletin, v. 95, p. 967–979.

Speed, R. C., 1979, Collided Paleozoic microplate in the western United States: Journal of Geology, v. 87, p. 279–292.

Taylor, G. W., 1986, Structural, sedimentological, and petrological setting of the Lang-Halsted sequence and Duncan Peak chert, lower Shoo Fly Complex, northern Sierra Nevada, California [M.S. thesis]: San Diego, California, San Diego State University, 110 p.

Turner, J. B., 1986, Petrography and geochemistry of basalts from the Bullpen Lake sequence, northern Sierra Nevada, California; A test of a working hypothesis [M.S. thesis]: San Diego, California, San Diego State University, 105 p.

Varga, R. J., 1982, Implications of Paleozoic phosphorites in the northern Sierra Nevada range: Nature, v. 297, p. 217–220.

Varga, R. J., and Moores, E. M., 1981, Age, origin, and significance of an unconformity that predates island-arc volcanism in the northern Sierra Nevada: Geology, v. 9, p. 512–518.

Von Huene, R., 1974, Modern trench sediments, *in* Burk, C. A., and Drake, C. L., eds., The geology of continental margins: New York, Springer-Verlag, p. 261–283.

Westbrook, G. K., Ladd, J. W., Buhl, P., Bangs, N., and Tiley, G. J., 1988, Cross section of an accretionary wedge; Barbados Ridge complex: Geology, v. 16, p. 631–635.

White, R. D., 1986, Cambrian radiolaria from Utah: Journal of Paleontology, v. 60, p. 778–780.

Winchester, J. A., and Floyd, P. A., 1976, Geochemical magma type discrimination; Applications to altered and metamorphosed basic igneous rocks: Earth and Planetary Science Letters, v. 28, p. 459–469.

Manuscript Accepted by the Society March 27, 1990

Printed in U.S.A.

Geological Society of America
Special Paper 255
1990

Early Paleozoic stratigraphic, paleogeographic, and biogeographic relations of the eastern Klamath belt, northern California

A. W. Potter
Department of Geosciences, Oregon State University, Corvallis, Oregon 97331-5506
A. J. Boucot
Department of Zoology, Oregon State University, Corvallis, Oregon 97331-2914
S. M. Bergström
Department of Geology and Mineralogy, Ohio State University, Columbus, Ohio 43210
R. B. Blodgett
U.S. Geological Survey, MS 970, National Center, Reston, Virginia 22092
W. T. Dean
Department of Geology, University College, P.O. Box 78, Cardiff CF1 1XL, United Kingdom
R. A. Flory
Department of Geology, California State University, Chico, California 95929
A. R. Ormiston
Amoco Production Company, Box 591, Tulsa, Oklahoma 74102
A.E.H. Pedder*
Geological Survey of Canada, 3303-33rd Street NW, Calgary, Alberta T2L 2A7, Canada
J. K. Rigby
Department of Geology, 258 ESC, Brigham Young University, Provo, Utah 84602
D. M. Rohr
Department of Geology, Sul Ross State University, Alpine, Texas 79832
N. M. Savage
Department of Geology, University of Oregon, Eugene, Oregon 97403

ABSTRACT

The eastern Klamath belt contains the fault-bounded Yreka, Trinity, and eastern Klamath terranes. The Yreka terrane comprises Lower Cambrian to Middle Devonian or younger igneous, metamorphic, and sedimentary rocks. The Trinity terrane consists of the Trinity ultramafic-mafic complex of Ordovician to Silurian age and an amphibolitic gabbro unit of Early Cambrian age. The eastern Klamath terrane bears igneous and sedimentary rocks of Early Devonian to Middle Jurassic age.

Stratigraphic and intrusive relations imply that the Yreka and Trinity terranes were amalgamated by Early to Middle? Devonian time and that the Trinity terrane was the basement on which the eastern Klamath terrane formed.

The lower Paleozoic rocks of the Yreka and eastern Klamath terranes are interpreted to represent the remnants of an Early Cambrian arc overlain by part of an Ordovician to Devonian arc-trench complex that faced west to northwest (present coordinates)

*Publishes with permission of the Director General, Sedimentary and Marine Geoscience, Geological Survey of Canada.

Potter, A. W., Boucot, A. J., Bergstrom, S. M., Blodgett, R. B., Dean, W. T., Flory, R. A., Ormiston, A. R., Pedder, A.E.H., Rigby, J. K., Rohr, D. M., and Savage, N. M., 1990, Early Paleozoic stratigraphic, paleogeographic, and biogeographic relations of the eastern Klamath belt, northern California, *in* Harwood, D. S., and Miller, M. M., eds., Paleozoic and early Mesozoic paleogeographic relations; Sierra Nevada, Klamath Mountains, and related terranes: Boulder, Colorado, Geological Society of America Special Paper 255.

in Late Ordovician and Early to Middle Devonian time. The Trinity complex may have formed in a marginal or back-arc basin northeast to southeast of the Lovers Leap–Gregg Ranch portion of the arc in Ordovician to Silurian? time (prior to existence of the eastern Klamath terrane).

The biogeographic affinities of eight groups of early Paleozoic fossils, taken as a whole, demonstrate that the eastern Klamath belt was close enough to North America in the Middle Ordovician to Middle Devonian for faunal communication to occur, in some cases at the species level. This evidence and the presence in the belt of some coarse-grained strata possibly derived from a continent indicate that the belt may have been relatively near to North America in Silurian or Devonian time, yet its location is obscure.

INTRODUCTION

The eastern Klamath belt contains Cambrian to Devonian rocks that provide important information about the early Paleozoic history of western North America. The location of the belt in that time interval is poorly known; some authors regard it as allochthonous to North America. In this chapter, we update the stratigraphy of selected lower Paleozoic units of the eastern Klamath belt, describe the evidence bearing on its paleolatitude and the biogeographic affinities of its faunas during the early Paleozoic, and then discuss the implications of this data for the early Paleozoic geographic and biogeographic relations of the belt.

The eastern Klamath belt has been subdivided into three tectonostratigraphic terranes by Silberling and others (1987); from north to south they are: the Yreka terrane; the Trinity terrane, composed of the Trinity ultramafic-mafic complex; and the eastern Klamath terrane, containing the Redding section (Fig. 1).

STRATIGRAPHY

Yreka terrane

Space limitations prevent a unit-by-unit discussion of the stratigraphy of the Yreka terrane in this chapter. Hotz (1977), Lindsley-Griffin and Griffin (1983), and Potter and others (1977) provided maps and descriptions of the major rock units in the terrane; other relevant work includes that of Cashman (1980), Lindsley-Griffin (1977, 1982), Lindsley-Griffin and Fisher (1989), Lindsley-Griffin and Rohr (1977), Potter (1977, 1982, 1990), Rigby and Potter (1986), Savage (1976b, 1977), Wallin (1988, 1989), and Wallin and others (1988). We have summarized schematically in Figure 2 the descriptions, inferred ages, and contact relations of the units pertinent to this report, and comment below only on some points that need elaboration or emphasis. We have adhered to the stratigraphic nomenclature of Potter and others (1977), even though such units as the Gazelle "Formation" are in obvious need of revision. Modifications to the present nomenclature will be made elsewhere.

In the Yreka terrane, Lower Cambrian to Middle Devonian igneous, metamorphic, and sedimentary rocks are imbricated in a series of thrust slices on the Trinity complex. The oldest dated rocks are in the Lovers Leap–Gregg Ranch area, where the generalized structural sequence is (in ascending order): (1) Ordovician layered gabbro and ultramafic units of the Trinity complex; (2) Lower Cambrian (Wallin and others, 1988) plagiogranite of Gregg Ranch; (3) Ordovician or older keratophyre of Lovers Leap; (4) Ordovician to Devonian? Kangaroo Creek Formation and Gregg Ranch unit; and (5) Devonian portions of the Gazelle Formation. The Kangaroo Creek Formation contains nine numbered members (Lindsley-Griffin and Rohr, 1977), the Gregg Ranch unit is made up of two numbered members, and the Gazelle "Formation" bears three numbered members (Potter and others, 1977). Based on work by Lindsley-Griffin and Griffin (1983) and Lindsley-Griffin and Fisher (1989), Gazelle member 1 is here subdivided into a structurally lower unit of melange and a structurally higher bedded unit, with a thrust fault intervening.

In the Bonnet Rock to Antelope Mountain tract, the structural sequence is (from lowest to highest): Moffett Creek Formation, Gazelle Formation, Sissel Gulch Graywacke, Duzel Phyllite, and Antelope Mountain Quartzite. Another unit, the Schulmeyer Gulch sequence, is of uncertain structural position. At Horseshoe Gulch, the Horseshoe Gulch unit, with nine numbered members, and the schist of Skookum Gulch are present.

Many of the rock units in the Yreka terrane have been referred to as melanges, and interpretation of these units and their contact relations is controversial. In this chapter, the terms *melange* and *broken formation* are used as defined in Silver and Beutner (1980), and *olistostrome* as defined in Bates and Jackson (1987); mappable submarine slide, slump, and debris-flow deposits are here regarded as olistostromes. In particular, Lindsley-Griffin (1982) and Lindsley-Griffin and Griffin (1983) included the Kangaroo Creek Formation, Gregg Ranch unit, Horseshoe Gulch unit, and structurally lower parts of members 1 and 3 of the Gazelle Formation of Potter and others (1977) in their melange of Gregg Ranch, and they inferred that it formed by sedimentary and tectonic processes acting over a long period of time. We agree with Lindsley-Griffin and Griffin (1983) that the lower part of member 1 of the Gazelle Formation and member 2 of the Gregg Ranch unit are best interpreted as melanges. In the other units listed above, some original depositional contacts between different subunits are preserved or can be readily inferred within individual fault blocks or olistoliths, and the original stratigraphic relations can be reconstructed or partly reconstructed. Therefore, at least members 1 through 4 of the Kangaroo Creek Formation, member 1 of the Gregg Ranch unit, the Horseshoe

Gulch unit, and member 3 of the Gazelle Formation are here interpreted as broken formations.

The remainder of this section elaborates on selected features of the units and stratigraphic relations depicted in Figure 2, in the order of units listed above.

In the Lovers Leap and Gregg Ranch areas, the only contacts found between the Trinity gabbro and plagiogranite of Gregg Ranch or between plagiogranite and keratophyre of Lovers Leap are faults. However, the keratophyre may intrude the plagiogranite unit at one point.

Member 1 of the Gregg Ranch unit depositionally overlies the keratophyre of Lovers Leap. Member 1 is dominantly volcaniclastic breccia, conglomerate, and sandstone that is locally bedded and bears planar laminations and trough cross-laminations. These deposits are inferred to have been laid down by mass flows (breccia and conglomerate) and tractive currents (sandstone). Member 1 is here correlated lithologically and chronologically with member 1 of the Kangaroo Creek Formation.

Member 2 of the Gregg Ranch unit includes breccia, conglomerate, sandstone, mudstone, limestone, limestone breccia (locally red), and dikes ± volcanic rocks (keratophyre-spilite) that clearly intrude sandstone, conglomerate, and limestone of member 2 and underlying plagiogranite of Gregg Ranch. The contact between pebbly quartz-rich sandstone of member 2 of the Gregg Ranch unit and the plagiogranite of Gregg Ranch is considered to have been originally an unconformity. Member 2 is provisionally (owing to very poor exposure) interpreted to be a tectonized sedimentary melange. The contacts between the Gregg Ranch unit and overlying Gazelle Formation are faults or are concealed.

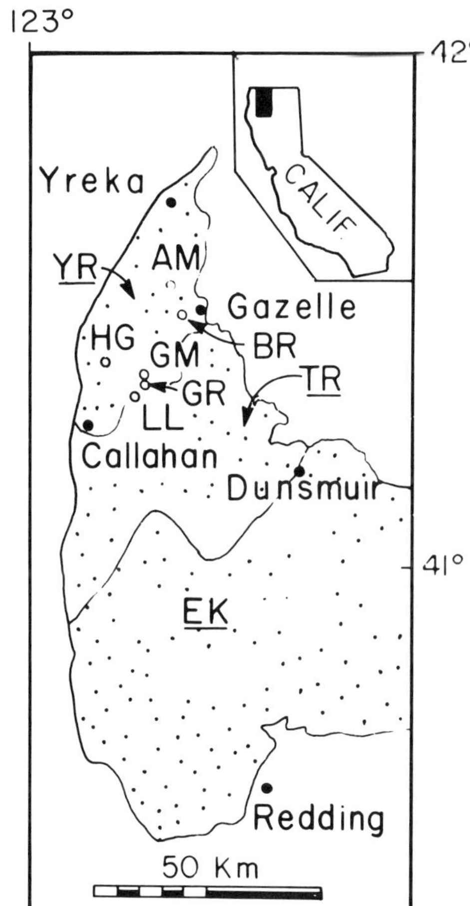

Figure 1. Index map showing the eastern Klamath belt subdivided into the Yreka (*YR*), Trinity (*TR*), and eastern Klamath (*EK*) terranes; terrane boundaries after Silberling and others (1987). AM = Antelope Mountain; BR = Bonnet Rock; GM = Gazelle Mountain; GR = Gregg Ranch; HG = Horseshoe Gulch; LL = Lovers Leap.

The melange of lower Gazelle member 1 contains a foliated matrix of mudstone and wacke that bears pebble-size to possibly 1.2-km-size inclusions of many rock types, comprising several varieties of fine- to coarse-grained clastic rocks, limestone, chert, and uncommon to rare ultramafic to granitic plutonic rocks, volcanic rocks (rare for >cobble size), and metamorphic rocks (description modified from Lindsley-Griffin and Griffin, 1983, and Lindsley-Griffin and Fisher, 1989). Chert and mudstone detritus is widespread in the clastic rocks. This melange contains tectonized olistostromes in places; its origin is being studied by Lindsley-Griffin and coworkers.

Distinctive inclusion lithologies within Gazelle member 1 melange are Upper Silurian fossil-bearing limestone and bedded to massive chert-lithic deposits of sandstone, breccia, and conglomerate. These limestone and chert-lithic deposits occur within the same inclusion in some cases, either in depositional contact or at nonadjacent stratigraphic horizons.

The age of the melange of Gazelle member 1 is partly delimited by a block within it that consists of limestone and limestone breccia with slump-folded beds of chert-lithic sandstone and chert-lithic breccia (A. W. Potter and R. Herbert, unpublished data). The limestone is undated, but its distinctive association with the chert-lithic deposits suggests that it is Late Silurian; this block is therefore taken as evidence that some slumping and, possibly, melange formation occurred in the Late Silurian. The youngest fossils in Gazelle member 1 melange were collected from limestone float inferred to have once been clasts in member 1 debris flows. One piece of float included conodonts of probable Eifelian age (Savage, this work). Two other pieces of float nearby produced probable Middle Devonian corals, one of which is early Givetian? (*Jiangzhaiphyllum* sp.; Pedder, this work). Therefore, Gazelle member 1 melange is in part Middle Devonian or younger. As no fossils younger than Givetian have been reported in the Yreka terrane, the melange is assigned a provisional minimum age of Middle or Late Devonian, although it could be younger. Our speculative estimate for the time of formation of the melange of Gazelle member 1 is thus Late Silurian (based on slump folds) to Middle or Late Devonian.

The upper portion of Gazelle member 1 is bedded and consists of siltstone, mudstone, shale, lithic sandstone, and local lithic granule and pebble breccia. Clasts are chiefly mudstone and siliceous mudstone. The upper part of member 1 is Early Devonian (*dehiscens*-zone age; Savage, 1977) at one locality and struc-

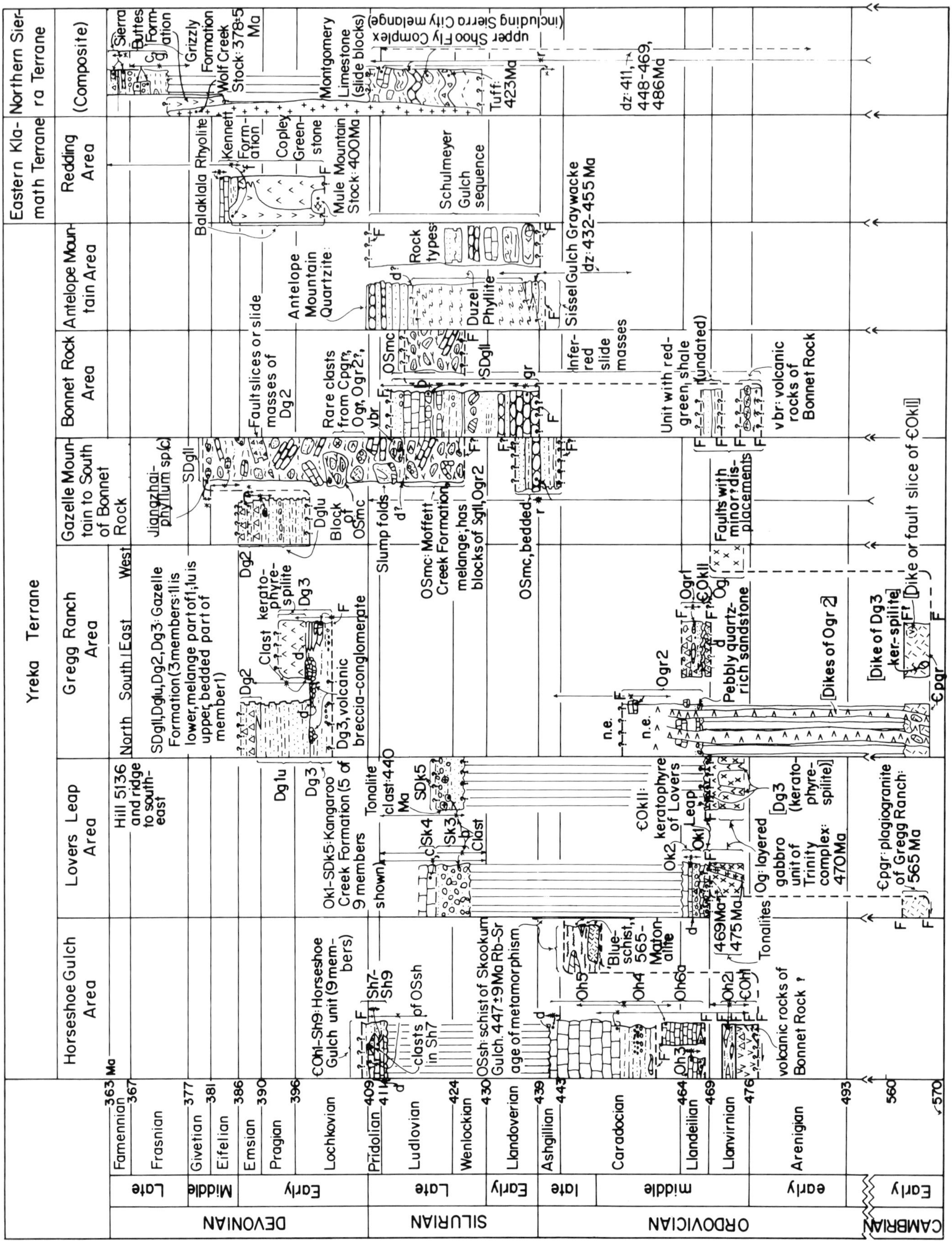

Yreka Terrane
Eastern Klamath Terrane
Northern Sierra Terrane
Horseshoe Gulch Area
Lovers Leap Area
Gregg Ranch Area
Gazelle Mountain to South of Bonnet Rock
Bonnet Rock Area
Antelope Mountain Area
Redding Area
(Composite)
DEVONIAN
SILURIAN
ORDOVICIAN
CAMBRIAN

turally overlies the melange of lower member 1 along a thrust contact (Lindsley-Griffin and Griffin, 1983).

Gazelle member 2 is bedded and chiefly siltstone, mudstone, and shale with intervals of quartz-bearing volcaniclastic siltstone, mudstone, and sandstone. It depositionally overlies the upper, bedded part of member 1, and is Early Devonian at one locality (upper *gronbergi-* to lower *inversus-* zone age; Savage, this work).

The upper, bedded parts of Gazelle member 1 and member 2 were deposited in large part by sediment gravity flows. East of Gregg Ranch, both units become thinner toward the south, suggesting the presence of a local high in that area in the Early Devonian.

Member 3 of the Gazelle Formation was described by Potter (1977) and Potter and others (1977). It contains units of volcanic breccia-conglomerate and overlying keratophyre-spilite that are of critical importance for dating juxtaposition of both local units and regional terranes, and are dated themselves from the following evidence: The volcanic breccia-conglomerate is closely related chronologically and, locally, lithologically to the keratophyre-spilite, and was deposited contemporaneously with volcanism (Potter, 1977). It pinches out in one area, where the keratophyre-spilite rests depositionally on a unit of black mudstone. Trenching and a thin section clearly show local soft-sediment deformation of the mudstone unit at the contact; hence, the mudstone was unlithified during deposition of the keratophyre. The mudstone unit contains lenses or slide blocks of limestone and mass-flow deposits of limestone breccia that bear Lochkovian conodonts and corals (Savage and Pedder, this work). The top of the keratophyre-spilite unit is not preserved. However, the upper, bedded portion of Gazelle member 1, dated elsewhere as Early Devonian, overlies the volcanic breccia-conglomerate of Gazelle member 3 at an unquestionably depositional contact. Mudstone and siltstone of member 1 fill interstices between volcanic clasts of member 3 volcanic breccia-conglomerate at the contact, and therefore are only slightly younger than the breccia-conglomerate. These relations indicate that the volcanic breccia-conglomerate is Early Devonian and that the keratophyre-spilite is most likely Early Devonian; however, a Middle Devonian age cannot be completely discounted for the keratophyre-spilite.

Dikes of the keratophyre-spilite intrude the layered gabbro and pegmatitic gabbro units of the Trinity complex, the plagiogranite of Gregg Ranch, and the keratophyre of Lovers Leap.

The Moffett Creek Formation ranges from well bedded to melange. It is predominantly shale, mudstone, siltstone, and sandstone, with minor bedded chert, siliceous mudstone, and in melange, limestone.

Melange of the Moffett Creek Formation structurally underlies the melange of lower Gazelle member 1 in the area south of Bonnet Rock. The contact is usually faulted or concealed, but may be depositional at one trenched point (Potter, 1990). In this area the Moffett Creek melange is considered to be one or more tectonized olistostromes (Lindsley-Griffin and Griffin, 1983; Potter, 1990; Potter and others, 1977). Some clasts in the Moffett

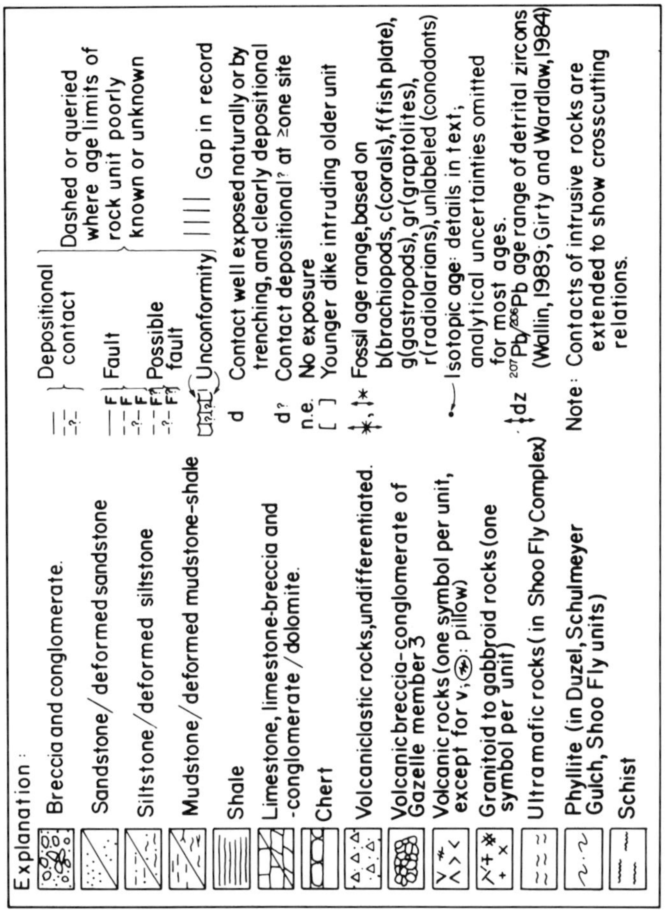

Figure 2. Columnar sections of selected areas in the eastern Klamath belt and northern Sierra Nevada. Those in the Yreka terrane were updated from Potter and others (1977, Fig. 4), based on post-1977 work by authors listed in the section of the text on the stratigraphy of the Yreka terrane and by Cotkin and Armstrong (1987), Haessig and others (1987), and Wallin (oral communication, 1988). Area locations are shown in Figure 1. The column for the eastern Klamath terrane is adapted from Albers and Bain (1985, Fig. 3). The northern Sierra Nevada column is a pastiche from Saleeby and others (1987, Fig. 3), Hanson and Schweickert (1986, Fig. 3), and Girty and Wardlaw (1984). Each column is a diagrammatic and generalized summary of stratigraphic relations from several areas. The effects of tectonic shearing of melanges and broken formations are not shown. Names of the various rock units are written out where they first occur, from left to right, and abbreviations are used elsewhere. Because the ordinate is the time scale of Harland and others (1989) rather than stratigraphic thickness, and because minor lithologies and small inclusions and clasts have been relatively greatly enlarged for illustrative purposes, proportions of both lithologic units and of rock types within units are highly distorted from actual values. The age limits of melanges and most other units are poorly known or unknown, hence are arbitrarily placed.

Creek melange south of Bonnet Rock are chert-lithic sandstone and chert-pebble breccia presumed to be Late Silurian because they strongly resemble chert-lithic sandstone and chert-pebble breccia associated with Upper Silurian limestone *within* blocks in melange of lower Gazelle member 1 (Potter, 1990). Therefore, the Moffett Creek melange in the Bonnett Rock area is considered to be Late Silurian or younger. The minimum age is most likely Middle or Late Devonian, if the contact with overlying melange of lower Gazelle member 1 is depositional, but could be younger.

The Moffett Creek Formation is bedded in some places. Irwin and others (1978) reported Ordovician-Silurian radiolarians from a chert interbed in undisrupted Moffett Creek Formation.

The schist of Skookum Gulch and the Horseshoe Gulch unit are fault-bounded units exposed in the Horseshoe Gulch area. The former is a tectonic melange containing blueschist (Hotz, 1977; Haessig, 1988, 1989). The age of blueschist metamorphism was inferred to be 447 ± 9 Ma by Cotkin and Armstrong (1987), based on a Rb-Sr phengite–whole-rock date measured on co-occurring quartz-albite schist. The Horseshoe Gulch unit ranges in age from Middle Ordovician or older to Late Silurian and possibly younger. Upper Silurian conglomerate within it contains clasts of siliceous schist from the schist of Skookum Gulch and tonalite clasts from fault blocks of Lower Cambrian tonalite within the schist.

The Sissel Gulch Graywacke, Duzel Phyllite, Antelope Mountain Quartzite, and Schulmeyer Gulch sequence occur in the Antelope Mountain area and are all assigned tentative depositional ages of Ordovician? to Silurian, based on stratigraphic relations and dating of detrital zircons by Wallin (1989). The Schulmeyer Gulch sequence ranges from coherent and deformed (S. Ott, unpublished mapping, 1983) to melange (J. R. Griffin, oral communication, 1989). The Moffett Creek, Duzel, and Antelope Mountain units contain Precambrian detrital zircons derived from a cratonal source (Wallin, 1989).

Trinity terrane

Mattinson and Hopson (1972), Lindsley-Griffin (1977, 1982), Brouxel and Lapierre (1988), and Boudier and others (1989) concluded that the Trinity complex is a partially preserved ophiolite. Lindsley-Griffin (1977, 1982), Cotkin and Armstrong (1987)," and Brouxel and Lapierre (1988) inferred its formation in a marginal basin, and Boudier and others (1989) suggested that it was produced by slow spreading at an oceanic ridge or in a marginal sea.

Isotopic ages from rocks of the Trinity complex were reviewed by Wallin and others (1988). They include a Sm-Nd mineral isochron age of 472 ± 32 Ma on plagioclase lherzolite (Jacobsen and others, 1984), a 470-Ma $^{207}Pb/^{206}Pb$ age (U-Pb ages nearly concordant) on zircons from a layered gabbro (recalculated from Mattinson and Hopson, 1972), and concordant U-Pb ages of 412 Ma on zircons from each of two pegmatitic trondhjemite dikes that intrude gabbroic rocks. Lindsley-Griffin (1982) included the 412-Ma dikes in her pegmatitic gabbro unit. Wallin and others (1990) recorded $^{207}Pb/^{206}Pb$ ages of 414 to 416 Ma (U-Pb ages nearly concordant) on zircon fractions from a pegmatitic gabbroid collected by Potter from Lindsley-Griffin's pegmatitic gabbro unit.

Wallin and others (1990) also reported $^{207}Pb/^{206}Pb$ ages of 552 to 580 Ma (U-Pb ages are slightly discordant) on zircon fractions from an amphibolitic gabbro collected by Potter from the amphibolitic gabbro unit of Lindsley-Griffin (1982). Lindsley-Griffin (1982) concluded that the amphibolitic gabbro unit is intruded by the pegmatitic gabbro unit and faulted against other units of the Trinity complex. Samples of the amphibolitic gabbro chemically resemble Fe-Ti basalts, and may have formed in a spreading-center or within-plate setting, possibly during arc-related rifting (Potter and Grunder, 1989).

Data compiled by Taylor and Karner (1983, Table 1) suggest that the maximum age difference between the oldest and youngest crust created in any one present-day circum-Pacific or western Mediterranean marginal basin is about 20 m.y. Applying this time span to the information on the Trinity complex reviewed above suggests that two or more Ordovician-Silurian episodes of magmatism of back-arc basin or other type are represented in the Trinity complex.

Potter and Grunder (1989) proposed that the tonalites and gabbro dated at 469, 475, and 470 Ma, respectively, by Wallin and others (1988) originated in an arc on the basis of trace-element contents. However, if those rocks are genetically related to the sheeted dike complexes described by Boudier and others (1989) or Brouxel and Lapierre (1988), formation in a marginal basin (or fore- or intra-arc basin) would be more probable.

Eastern Klamath terrane

Units of the eastern Klamath terrane considered here are, in ascending order, the Copley Greenstone, Balaklala Rhyolite, and Kennett Formation. The Copley Greenstone is inferred to be faulted against the subjacent Trinity complex (Schweickert and Irwin, 1989), to consist mainly of altered basaltic-andesitic (commonly keratophyre-spilite) flows, tuffs, and breccias, and to be about 1,800 m thick. It is overlain by the Balaklala Rhyolite along a contact that is locally gradational (Kinkel and others, 1956; Albers and Bain, 1985). The Balaklala Rhyolite is mostly altered silicic (commonly quartz keratophyre) flows, dikes, sills, tuffs, and breccias. It is as much as 1,100 m thick, and in the eastern part of the map area it lenses out and locally interfingers with the greenstone (Kinkel and others, 1956; Albers and Bain, 1985). The Kennett Formation depositionally overlies the Balaklala Rhyolite, and a tuff or tuffaceous sandstone lies transitionally between these two units (Albers and Bain, 1985). Watkins and Flory (1986) described the Kennett Formation as being 120 m thick and consisting of a lower part made up of volcaniclastic sandstone to mudstone and siliceous mudstone and siltstone; an upper part is composed of lenses of limestone (wackestone to packstone) inferred to have been deposited by sediment gravity flows. The Copley and Balaklala are intruded by the Mule Mountain stock, which was dated at 400 Ma by U-Pb and K-Ar methods (Albers and others, 1981) and considered to be cogenetic with the Balaklala Rhyolite (Barker and others, 1979; Lapierre and others, 1985a, b). On the time scale of Harland and others (1989), 400 Ma is Early Devonian (Lochkovian).

Tuff in the upper part of the Balaklala Rhyolite yielded a fish plate of Middle to Late Devonian age (Boucot and others, 1974, p. 696–699). Savage (1976a) recovered conodonts and brachiopods of late Eifelian age from the upper part of the Kennett Formation; the conodonts were from limestone, and the brachiopods from limestone and sandstone to mudstone. The Copley Greenstone has not produced fossils.

From the above stratigraphic, biostratigraphic, and isotopic-age evidence, the Kennett Formation is here assigned an age of Eifelian, the Balaklala Rhyolite is considered to be Early to Middle Devonian, and the Copley Greenstone is most likely Early to Middle Devonian.

The Copley Greenstone and Balaklala Rhyolite represent altered arc low-K tholeiitic-series rocks (Potter and others, 1976; Barker and others, 1979; Lapierre and others, 1985a, b). The Kennett Formation is interpreted to be an intra-arc deposit formed adjacent to volcanic islands (Watkins and Flory, 1986).

PALEOLATITUDE EVIDENCE

The only paleomagnetic studies of the eastern Klamath belt were carried out on Devonian and younger rocks in the Redding section (Mankinen and others, 1988, 1989; Renne and Scott, 1988; Renne and others, 1988; and references therein). These investigations indicate no significant latitudinal displacement of the sample sites relative to the North American craton since the Permian. Mankinen and others (1989) postulate about 106 ± 12° of clockwise rotation for Permian and Triassic beds of the eastern Klamath terrane with respect to cratonic North America. Renne and others (1988), however, concluded that no significant rotation occurred and that rotations inferred by other authors are probably artifacts of oroclinal bending of structural domains about vertical axes.

Carbonate deposits of Early Ordovician to Middle Devonian age occur in the Yreka and eastern Klamath terranes (Fig. 2). The limestones contain locally diverse and abundant macrofaunas of Middle and Late Ordovician, Late Silurian, and Early and Middle Devonian ages. Porter (1974) reported ooids in the massive limestone member of the Facey Rock Limestone, which is probably Arenigian (Early Ordovician; Bergström and others, 1980), and thus is the oldest dated sedimentary rock in the Yreka terrane. Abundant dasycladacean algal debris occurs in Ashgillian limestone of member 5 of the Horseshoe Gulch unit (G. F. Elliott, *in* Rigby and Potter, 1986), and probable calcareous algae (R. B. Blodgett, this work) and locally common oncoids occur in Early Devonian limestone of member 3 of the Gazelle Formation. This evidence indicates that the Yreka terrane was situated in warm, tropical to subtropical latitudes from Ordovician through Devonian time, as was the western margin of North America.

PALEOBIOGEOGRAPHIC RELATIONS

Paleobiogeographic relations of eight fossil groups are examined briefly here. Bergström is responsible for material on Early to Late Ordovician conodonts; Dean, Middle Ordovician trilobites; Potter, Middle to Late Ordovician brachiopods; Rohr, Middle Ordovician to Early Devonian gastropods; Berry, Middle Ordovician graptolites; Rigby, Middle to Late Ordovician sphinctozoan sponges; Flory, Late Ordovician to Middle Devonian tabulate corals; Elias, Late Ordovician tetracorals; Boucot, Silurian brachiopods; Potter and Boucot, Early Devonian brachiopods from Gazelle member 3; Savage, Early and Middle Devonian conodonts; Pedder, Early and Middle Devonian tetracorals; Ormiston, Early Devonian trilobite. Results are summarized in Table 1.

The oldest fossils in the eastern Klamath belt are early Whiterockian or slightly older (Arenigian, Early Ordovician) conodonts from the massive limestone member of the Facey Rock Limestone (shown in Potter and others, 1977, Fig. 2), including an *Oepikodus* close to *O. evae* and a primitive-looking *Periodon*. Middle Whiterockian (Llanvirnian, Middle Ordovician) conodonts occur in member 2 of the Horseshoe Gulch unit; *Histiodella holodontata* is present in this assemblage. Probable Llandeilian (Middle Ordovician) conodonts in member 3 of the Horseshoe Gulch unit include *Periodon aculeatus,* typical *Belodina,* and *Belodella nevadensis,* an association similar to coeval assemblages in the Great Basin and the eastern Appalachian Mountains. Llandeilian conodonts from member 1 of the Gregg Ranch unit include *Pygodus anserinus* and other taxa widely distributed along the margins of the North American plate as well as in the Baltic area, the Girvan district of Scotland, and southern China. Ashgillian (probably early to middle Ashgillian, based on co-occurring conodonts and brachiopods) conodonts from member 5 of the Horseshoe Gulch unit include *Drepanoistodus* cf. *suberectus, Panderodus* (either *gracilis* or *unicostatus*?), *Protoprioniodus* (either *liripipus* or *insculptus*), and a coniform species that may be identical to a species in the Upper Ordovician of northwest Europe that is still unnamed. In summary, Ordovician conodonts from the Yreka terrane are relatively cosmopolitan, and are not very diagnostic for biogeographic/paleogeographic purposes. However, they exhibit similarity to those in the Great Basin and do not resemble coeval faunas in northern China/Korea or on the Siberian Platform.

Llanvirnian graptolites from member 2 of the Horseshoe Gulch unit most likely belong to the Pacific (or Australian-American) Faunal Province of Berry (Berry and others, 1973). They are similar to coeval faunas from the Yukon and Nevada and differ from British faunas of this age.

Llanvirnian trilobites from member 2 of the Horseshoe Gulch unit are poorly preserved, but include *Geragnostus* sp.,

TABLE 1. SUMMARY OF THE ORDOVICIAN TO MIDDLE DEVONIAN BIOGEOGRAPHY OF THE EASTERN KLAMATH BELT

Fossil Group	Areas with faunas that most closely resemble coeval eastern Klamath faunas
Early Ordovician (Arenigian)	
Conodonts	Not diagnostic of one area
Middle Ordovician (Llanvirnian)	
Conodonts	Not diagnostic
Graptolites	Pacific Province (includes Yukon, Nevada)
Trilobites	Inner slope of eastern and western North America
Middle Ordovician (Llandeilian)	
Brachiopods	Southern Appalachians, west-central Alaska; *Xenambonites* and *Craspedelia* are important
Conodonts	Margins of North America; Baltic; Girvan, Scotland; southern China
Sphinctozoans	In Middle and Late Ordovician time, known elsewhere only in: Australia; Alaska–Yukon Territory; and northern Sierra Nevada, California. Four of eleven species in Alaska occur in eastern Klamath Mountains
Middle Ordovician (early Caradocian)	
Trilobites	Southern Appalachians; Girvan; Kazakhstan, USSR
Late Ordovician (Ashgillian)	
Brachiopods	West-central Alaska, east-central Alaska–Yukon Territory, Northwest Territories, northern Sierra Nevada. Three to eight species shared with east-central Alaska
Conodonts	Not diagnostic
Gastropods	Not diagnostic; *Siskiyouspira* occurs elsewhere only in Alaska and Wyoming
Solitary Rugosa	Genera not diagnostic; one species occurs elsewhere only in northern Maine and northern Sierra Nevada
Sphinctozoans	See Llandeilian above
Tabulates	Eurasia, Australia; one species resembles a Uralian species. (Incomplete sample?)
Late Silurian	
Brachiopods	Uralian-Cordilleran Region (western and northwestern North America, Urals, parts of northern Africa and southern Europe, eastern Australia, Malaysia)
Tabulates	Eastern Europe and Asia. (Incomplete sample?)
Early Devonian	
Brachiopods	Cordilleran Region (Nevada, northern British Columbia, Yukon, Canadian Arctic)
Conodonts	General affinities to Cordilleran Region
Rugosan Corals	Strong affinities with Cordilleran Region; seven species shared with Great Basin–western Canada.
Tabulates	Nevada; Five of ten species shared with Nevada.
Middle Devonian	
Yreka terrane	
Rugosan Corals	Old World Realm (includes Europe, Asia, Australia, arctic and western North America); two of four genera unknown in either Great Basin or northwest Canada; *Jiangzhaiphyllum*, one of the four genera, known elsewhere only in southern China
Eastern Klamath terrane	
Tabulates	Nevada; one of three species shared with Nevada

Triarthrus sp., *Telephina* sp., and *Carolinites* sp. nov. The assemblage exhibits affinities with Whiterockian faunas from inner-slope environments in both eastern and western North America. All four genera have a long range within the Ordovician, but *Carolinites* sp. nov. may best be compared with an unnamed species from the Middle Table Head Formation of western Newfoundland.

Early Caradocian trilobite fragments are numerous but very poorly preserved in thin layers of decalcified, silty mudstone of member 2 of the Gregg Ranch unit. The faunal list includes: *Ampyxina* sp.; *Bronteopsis* cf. *scotica* Nicholson and Etheridge, 1879; calymenid? gen. et sp. undet.; *Encrinuroides* sp.; *Pseudosphaerexochus* sp.; *Robergia* sp.; *Sphaerocoryphe* sp.; *Tretaspis* sp.; and *Yumenaspis* cf. *virginica* (B. N. Cooper, 1953). Affinities are with the so-called "Scoto-Appalachian" faunas in the southern Appalachians, the Girvan district of Scotland, and Kazakhstan, USSR. A previous assessment of a Llandeilian age in Boucot and others (1973) was based on the record at Gregg Ranch of *Taimyraspis,* a genus said then to be from the Llandeilian of Kazakhstan. Both *Taimyraspis* and *Effnaspis* (type species *Isbergia virginica,* from the Effna Limestone of Virginia) are now placed tentatively in synonymy with the Chinese genus *Yumenaspis* Chang and Fan, 1960. *Yumenaspis* is known from the Llandeilian and Caradocian, but the California material most closely resembles a species from the Effna Limestone of Virginia, a unit placed in the lowest Caradocian by Ross and others (1982). Further support for an early Caradocian age is provided by *Bronteopsis* cf. *scotica,* a species from the Balclatchie Group of Scotland, but recorded also in Kazakhstan. In a modern list of trilobites from the Balclatchie Group (Tripp, 1980, p. 132), the binomen *Bronteopsis scotica* has been replaced by its senior subjective synonym *B. concentrica* (Linnarsson, 1869), a species described from the Middle Ordovician of Sweden. *Tretaspis* is not reported from pre-Caradocian rocks (Hughes and others, 1975).

Biogeographic relations of Ordovician brachiopods are documented in detail in Potter (1990), and are summarized briefly here. Llandeilian brachiopods from Gregg Ranch unit member 1 most closely resemble broadly coeval faunas in the Effna, Athens, and Pratt Ferry rock units of Cooper (1956) in Virginia, Tennessee, and Alabama, respectively, and in the uOl unit of Gilbert (1981) in the White Mountain tract of west-central Alaska (Potter, 1990) (see Fig. 3). Genera with limited geographic ranges include *Xenambonites* and *Craspedelia.* In the Middle Ordovician, *Xenambonites* is limited to the eastern Klamath belt, the Girvan area of Scotland, the Pratt Ferry beds of Alabama, and the uOl unit of White Mountain area of west-central Alaska. In the Ashgillian, it occurs in the Jones Ridge Limestone, east-central Alaska. *Craspedelia* has a similarly restricted geographic distribution in the Llandeilian-Ashgillian, except that it occurs also in Kazakhstan in the Middle Ordovician.

Ashgillian brachiopods from Horseshoe Gulch unit member 5 most closely resemble coeval assemblages in unit uOll of Gilbert (1981), White Mountain area, west-central Alaska; upper member of the Jones Ridge Limestone, east-central Alaska and

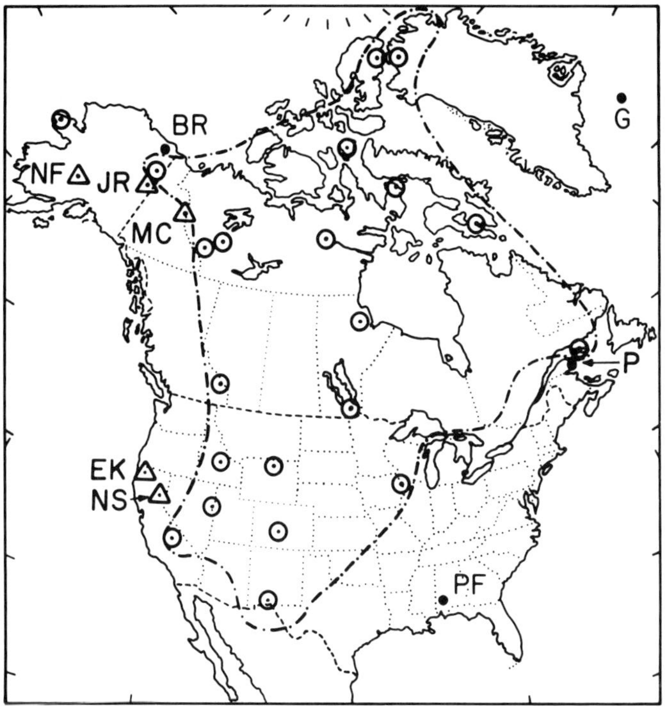

Figure 3. Map of North America showing essentially cratonic distribution of the Late Ordovician coral *Bighornia* (at dots in circles; from Elias, 1983) and present-day locations (except for Girvan) of areas containing Ordovician fossils similar to those of the eastern Klamath belt (at dots in triangles) and other areas mentioned in text (large dots). The location of the Girvan area is its approximate position relative to North America in the Ordovician, as modified from Williams (1973). BR = northeast Brooks Range; EK = eastern Klamath belt; G = Girvan area, Scotland; JR = Jones Ridge; MC = Misty Creek Embayment; NF = White Mountain area, Nixon Fork terrane; NS = northern Sierra terrane; P = Percé, Québec; PF = Pratt Ferry.

Yukon Territory, Canada; transitional Mount Kindle Formation, Misty Creek Embayment, Northwest Territories, Canada; and Montgomery Limestone, northern Sierra Nevada, northern California (see Potter and others, this volume). Genera with limited geographic distributions include: (1) *Diochthofera* Potter, 1990, which occurs in member 5 and all four other sites listed above but has not been reported elsewhere in the world to date; (2) *Elasmothyris* Cooper, 1956, found only in member 5, the Montgomery Limestone, and in the Jones Ridge upper member in the Ashgillian, and only in the Llandeilian Pratt Ferry unit in Alabama at any other time; (3) *Eripnifera* Potter 1990, present only in member 5 and unit uOll, White Mountain area. More detailed comparisons at the species level indicate that member 5 of the Horseshoe Gulch unit shares three to eight species with the Jones Ridge upper member (Potter, 1990).

The areas listed above with faunas similar to those of Gregg Ranch member 1 and Horseshoe Gulch unit member 5 were either part of North America in the Ordovician (the appropriate portions of Virginia, Tennessee, and Alabama [Walker and others, 1989]; Jones Ridge, east-central Alaska–Yukon Territory [Churkin and others, 1980; Payne and Allison, 1981; Rigby and others, 1988]; Misty Creek Embayment, Northwest Territories [Cecile, 1982]) or are inferred to have been relatively close to North America in the Ordovician (Girvan area, Scotland [Williams, 1973, p. 263]; White Mountain area, west-central Alaska [Blodgett and Clough, 1985]).

Rohr (1980) studied Ordovician through Devonian gastropods of the Callahan-Gazelle tract of the eastern Klamath belt and described 71 species representing 40 genera. While early Paleozoic gastropods are not well known, the Klamath taxa appear to be relatively cosmopolitan, with just five of the genera known only from the Klamath Mountains. *Siskiyouspira,* a distinctive Ordovician pleurotomariacean, has been found only in the Nixon Fork terrane, northeast Brooks Range of Alaska, Bighorn Mountains of Wyoming, and Yreka terrane. Ordovician gastropod genera common to both Alaska and the Klamath Mountains (but not restricted to these two areas) include *Maclurites, Ophiletina, Trochonemella,* and *Euomphalopterus*?. Ordovician gastropod faunas from the Great Basin are very poorly known except for the very abundant *Maclurites* and *Palliseria. Palliseria* is not present in the Klamath Mountains, but this may be due to the absence of rocks corresponding to its relatively limited stratigraphic and paleoecologic range.

Middle and Late Ordovician sphinctozoan sponges are at present recorded only in Australia, the eastern Klamath belt, and Alaska–Yukon Territory portion of western North America (see Rigby and others, 1988, for discussion). They occur in Llandeilian limestone of member 1 of the Gregg Ranch unit, Ashgillian Horseshoe Gulch unit member 5, and in Ordovician limestone clasts from Kangaroo Creek Formation member 3. This distribution is likely an artifact of incomplete sampling, because Ordovician sphinctozoans were not known until comparatively recently. Even so, it is significant that four of the eleven sphinctozoan *species* described by Rigby and others (1988) from Middle and Upper Ordovician rocks of Alaska are also present in the eastern Klamath belt.

Ashgillian tabulates from the Horseshoe Gulch unit member 5 contain *Plasmoporella* cf. *typylensis* Ivanov (*in* Ivanov and Miagkova, 1950), *Propora* sp., *Quepora* sp., and *Tetradium* sp. (Table 2). *Plasmoporella* occurs elsewhere in Norway, the Urals, central Asia, and Tasmania; *P. typylensis* is present in the Urals. *Propora* is found elsewhere in northern Europe; Asia; and Anticosti Island, Quebec. *Quepora* is reported elsewhere from Kazakhstan and New South Wales, Australia. Therefore, eastern Klamath Ashgillian tabulates are most like those of Eurasia and Australia. However, these faunal distributions may be artifacts of incomplete sampling on a global basis (see Discussion below).

Elias (in Elias and Potter, 1984) studied Ashgillian solitary rugose corals from member 5 of the Horseshoe Gulch unit. Of three species described, one, *Grewingkia penobscotensis* Elias, 1982, is present elsewhere only in the Montgomery Limestone of the northern Sierra Nevada (Potter and others, this volume) and in northern Maine in an area considered to have been near or part of North America in Ashgillian time (*in* Elias and Potter, 1984).

TABLE 2. GEOGRAPHIC DISTRIBUTIONS OF ORDOVICIAN TO DEVONIAN TABULATE CORALS FROM THE EASTERN KLAMATH BELT

Late Ordovician

Genera	Horseshoe Gulch	Montgomery Limestone	Appl.	Eur.	Urals	Sib.	NE USSR	Altai	China	Aust.
Tetradium	X		X		X	X		X		X
Quepora	X	X						X		X
Propora	X	X		X	X	X	X	X		
Plasmoporella	X				X		X	X	X	X
Chaetetella	X						X			
Paleofavosites		X	X	X	X	X	X	X	X	X

Silurian (Wenlockian)

Genera	Gazelle Fm.	Kangaroo Creek Fm.	Appl.	Eur.	Urals	Sib.	NE USSR	Altai	China	Aust.
Favosites		X	X	X	X	X	X	X	X	X
Squameo-favosites	X	X						X	X	
Syringopora		X	X	X	X	X	X	X	X	X
Heliolites		X	X	X	X	X	X	X	X	X
Multisolenia		X	X	X	X	X	X	X	X	X
Halysites		X	X	X	X	X	X	X	X	X

Silurian (Ludlovian)

Genera	Horseshoe Gulch	Appl.	Eur.	Urals	Sib.	NE USSR	Altai	China	Aust.
Favosites	X	X	X	X	X	X	X	X	X
Squameofavosites	X		X			X	X		
Thamnopora	X		X				X		

Early Devonian

Species	Grouse Creek Unit	Gazelle Fm.	Eureka, Nevada
Favosites sp. A		X	
Favosites sp. B		X	X
Favosites sp. C		X	
Squameofavosites sp. A		X	X
Squameofavosites sp. B		X	X
Squameofavosites sp. C	X	X	
Thamnopora sp. A	X		
Alveolites sp. A		X	X
Syringopora sp. A		X	
Syringopora sp. B	X	X	
Heliolites sp. A		X	X

Middle Devonian

Species	Kennett Fm.	Suplee, Oregon	Eureka, Nevada
Favosites sp. D	X		
Squameofavosites sp. D	X		
Heliolites sp. C	X	X	X

Note: Coeval occurrences are shown. The Montgomery Limestone is in the northern Sierra Nevada terrane. Appl. = Appalachians; Sib. = Siberia; Altai. = Altai Mountains in Mongolia, northwest China, and south-central Siberia; Aust. = Australia; X = present in that area. From Flory and Watkins (1983), after Hill (1981).

Wenlockian to Ludlovian (Late Silurian) tabulates occur in Members 4, 5, and 7 of the Kangaroo Creek Formation; Ludlovian to Pridolian tabulates are present in Members 7 and 8 of the Horseshoe Gulch unit (Table 2). Most are cosmopolitan; *Squameofavosites* characterizes Silurian faunas of Asia and eastern Europe (Flory and Watkins, 1983), but this distribution may be an artifact (see Discussion, below).

Ludlovian brachiopods from olistoliths in Gazelle member 1, studied by Boucot, include *Aenigmastrophia, Atrypoidea,* and *Cymbidium* and belong to the Uralian-Cordilleran Region of Boucot (1975).

Early Devonian brachiopods from Gazelle member 3 are listed in Boucot and Potter (1977, p. 211). They are of Cordilleran Region type and are similar to contemporary faunas of Nevada, northern British Columbia, the Yukon, and Canadian Arctic; all four of the latter geographic regions belonged to Early Devonian North America (Boucot and Potter, 1977, p. 215).

A conodont fauna from limestone of the Grouse Creek unit of Boucot and others (1974, p. 695) is of *delta* zone age (Lochkovian, Early Devonian) (Savage, 1976a). It contains *Ozarkodina remscheidensis, Ozarkodina excavata excavata, Icriodus angustoides bidentatus, Pedavis pesavis,* and *Ozarkodina* n. sp. E of Savage. The first four of these taxa are cosmopolitan, and the fifth is known elsewhere only from Nevada.

Conodonts from a concretion in the upper, bedded part of Gazelle member 1 are of *dehiscens* zone age (Emsian, Early Devonian) (Savage, 1977) and include *Polygnathus pireneae, Polygnathus dehiscens, Pelekysgnathus klamathensis* and *Pandorinellina* cf. *optima* (=*Pandorinellina* n. sp. O of Klapper and Johnson, 1980). All four are characteristic of the Cordilleran Region.

In summary, the Devonian conodonts from the Callahan-Gazelle area show general Cordilleran Region affinities.

Lochkovian (Early Devonian) tabulate faunas from Gazelle member 3, studied by Flory, share five of the ten species present with coeval faunas of Eureka County, Nevada (Table 2), which was part of Devonian North America.

Early Devonian tetracorals from limestone clasts in Gazelle member 3 include *Pseudamplexus* cf. *bohemicus* (Pocta), *Stylopleura* cf. *julli* Pedder, *S.* sp. nov., *Australophyllum johnsoni* (Merriam), *Carlinastraea halysitoides* (Etheridge), *Lyrielasma* 2 spp. nov., *Neomphyma greggi* (Merriam), and *Stauromatidium* cf. *marylandicum* (Swartz) sensu Pedder and Oliver, 1982. This assemblage is unequivocally Lochkovian and shows strong biogeographic affinities with the Great Basin and western Canada. Seven of the nine identified species in the Gazelle Formation are shared with the combined faunas of the other two regions (Pedder, 1985a, b). In particular, the new species of *Stylopleura* also occurs in middle Lochkovian strata of northern Yukon Territory (Jackson and others, 1978, Plate 27, Figs. 7, 9), and the coral identified as *Stauromatidium* cf. *marylandicum* (Swartz) is conspecific with an unnamed species occurring in the Lockkovian part of the Roberts Mountains Formation and the Lochkovian Windmill Limestone of Elko and Eureka counties, Nevada (Pedder and Oliver, 1982, p. 5, Plate 6, Figs. 1–8). Both species are unusual and distinctive, and are unknown from any other region of the world.

Two clasts from debris flows or olistostromes in member 1 of the Gazelle Formation yielded species of *Microplasma*?, *Lekanophyllum, Jiangzhaiphyllum,* and a phaceloid analogue of *Endophyllum.* This fauna evidently derives from some part of the Middle Devonian Old World Realm, but none of the species and only two of the genera have been found in the Great Basin and the Mackenzie Province of western and northern Canada. *Jiangzhaiphyllum* is known elsewhere only in the lower Givetian Songjiaqiao Member of the Dushan Formation in southeastern Guizhou Province, China (Wang, 1983). Its presence in the Yreka terrane is anomalous and cannot be adequately explained at this time. However, it should be pointed out that the Givetian coral faunas of Cordilleran North America are poorly known.

The trilobite *Warburgella rugulosa* subsp. indet. was found in the limestone of the Grouse Creek unit by Blodgett and identified by Ormiston. It indicates an early Lochkovian age for the limestone. While this species is relatively widespread geographically, the different subspecies have restricted geographic ranges; unfortunately, the material is too distorted to identify to subspecies.

Conodonts of late Eifelian (Middle Devonian) age from the Kennett Formation in the Redding section include *Polygnathus kennettensis* and *Polygnathus* cf. *trigonicus.* The first is not definitely known outside the Klamath area, and the second is cosmopolitan. Eifelian tabulates of the Kennett Formation share one of the three species present with Eureka County, Nevada (Table 2).

DISCUSSION

Relations among Yreka, Trinity, and eastern Klamath terranes

The boundary between the Yreka and Trinity terranes in this report is taken as the contact between the plagiogranite of Gregg Ranch and the structurally subjacent Ordovician to Upper Silurian intrusive rocks of the Trinity complex (Fig. 2). This contact is commonly a fault, but has not been completely mapped. The plagiogranite of Gregg Ranch is included in the Yreka terrane rather than in the Trinity terrane because it is depositionally overlain by sedimentary rocks of the Yreka terrane (pebbly quartz-rich sandstone to conglomerate of member 2 of the Gregg Ranch unit), and is inferred to be genetically unrelated to the Ordovician and younger rocks of the Trinity complex.

Brouxel and others (1988) pointed out similarities in mineralogy, geochemistry, and Sm-Nd isotopic characteristics between the Copley Greenstone and their "Lovers Leap low-K tholeiites." The LREE—depleted samples of their Lovers Leap low-K tholeiites are here correlated lithologically and chronologically with the Lower to Middle? Devonian keratophyre-spilite of member 3 of the Gazelle Formation in the Gregg Ranch area. Therefore, the keratophyre-spilite of Gazelle member 3 probably correlates lithologically and chronologically with the LREE-depleted parts of the Copley Greenstone. The intrusive relation of the Gazelle keratophyre-spilite into the Trinity complex also implies that the latter was the basement on which the Copley Greenstone and overlying units were deposited.

Schweickert and Irwin (1989) proposed that the eastern Klamath terrane may be part of an allochthon transported from the Trinity complex toward the southeast along a low-angle detachment fault in Tertiary time. If such is the case, the Yreka terrane may be part of the breakaway zone that originally was much closer to the eastern Klamath terrane (Schweickert and Irwin, 1989). Such a suggestion is compatible with the relations outlined above between the keratophyre-spilite of Gazelle member 3, Copley Greenstone, and Trinity complex, and with the relation proposed below between the Yreka terrane and Bragdon Formation of the eastern Klamath terrane.

Another possible tie between the Yreka and eastern Klamath terranes might exist between the silicic volcaniclastic beds in member 2 of the Gazelle Formation, which are, at least in part, Emsian and similar to quartz keratophyre in chemical composition (A. W. Potter, unpublished data), and the Balaklala Rhyolite of the eastern Klamath terrane. However, Wallin (1989) reported Pb/Pb ages of detrital zircons from these beds of 415 to 474 Ma, suggesting derivation of at least the zircons from older rocks, such as silicic extrusive differentiates of tonalites that intrude the Trinity complex, rather than from the Balaklala Rhyolite.

Paleogeography

From the data of Potter and others (1977), Brouxel and others (1988), and Potter and Grunder (1989), it is proposed that the Yreka terrane contains remnants of an Early Cambrian arc overlain by parts of an episodically active Ordovician to Devonian arc-trench system. Owing to gaps in the rock record, it is unknown whether these Cambrian and Ordovician to Devonian vestiges are fragments of one long-lived, episodically active arc complex. Potter (1990) inferred that in the Late Ordovician the subduction zone beneath this arc dipped toward the east to southeast (present-day coordinates), based on the geographic position of the blueschist-bearing schist of Skookum Gulch relative to roughly coeval igneous rocks considered to be of arc or suprasubduction-zone type in the Lovers Leap–Gregg Ranch area to the southeast. Inferred arc-type rocks include the Middle Ordovician or older keratophyre of Lovers Leap (part of the "Lovers Leap Butte Unit" of Brouxel and others, 1988) and the probably Middle to Upper? Ordovician dike unit of member 2 of the Gregg Ranch unit. Suprasubduction-zone-type rocks (i.e., those formed in an arc or in a basin situated within or close to an arc) are found in tonalite and gabbro that formed as part of the Trinity complex or intruded it later: 435 ± 21-Ma microgabbro dike (Jacobsen and others, 1984); 469 ± 21-Ma and 475 ± 10-Ma tonalites and 470-Ma gabbro (Potter and Grunder, 1989); the 440-Ma tonalite boulder in conglomerate at Lovers Leap (age recalculated by Wallin and others, 1988); and the tonalite and gabbro studied by Brouxel and Lapierre (1988). The structural positions of the keratophyre of Lovers Leap and the dike unit of member 2 of the Gregg Ranch unit above the plagiogranite of Gregg Ranch and the apparent absence of intrusive contacts between them and the Trinity complex suggest that they are unrelated to the Trinity complex. From this information and the chemical and other data discussed by Potter and others (1977), Lapierre and others (1987), and Brouxel and others (1988), it is inferred that the keratophyre of Lovers Leap and the dike unit of member 2 are most likely arc magmatic rocks. Cotkin and Armstrong (1987) also proposed eastward subduction.

Obviously, one problem with the above interpretation is that the presently preserved Cambrian-Ordovician igneous rocks inferred to be of arc type are of small volume. However, nearly all of the original arc magmatic rock may have been removed by erosion and large-scale tectonic transport (e.g., rifting of the arc during formation of the Trinity complex in Ordovician to Silurian time, or strike-slip faulting along the arc during oblique convergence). Nevertheless, the Ordovician blueschist in the schist of Skookum Gulch, which is itself not particularly widespread, is strong evidence that such an arc complex existed. Two lines of evidence indicate that the schist of Skookum Gulch is not exotic to the Yreka terrane: (1) fault slices of Lower Cambrian tonalite included in it resemble in age and lithology the plagiogranite of Gregg Ranch (Haessig and others, 1987), which is depositional basement near Gregg Ranch; and (2) the clasts of schist of Skookum Gulch in Upper Silurian conglomerate in Horseshoe Gulch.

Accepting the interpretations cited above that the Trinity complex was generated in a marginal basin (intra- or back-arc basins are other possibilities), we infer that much of it formed at about 470 Ma (age of layered gabbro unit) northeast to southeast of the Lovers Leap-Gregg Ranch arc-magmatic units. Later strike-slip faulting along or oblique to the arc may have juxtaposed the Yreka and Trinity terranes and cut out most of any previously intervening volcanoplutonic arc rocks. Younger ages of 440, 435, and 412 Ma reported by Wallin and others (1988) and Jacobsen and others (1984) could represent later episodes of spreading or arc magmatism. The rocks studied by Brouxel and Lapierre (1988) near Dunsmuir (Fig. 1) suggest a spreading event of uncertain age. By Early Devonian time (prior to intrusion of keratophyre-spilite dikes of Gazelle member 3), the plagiogranite of Gregg Ranch and the overlying keratophyre of Lovers Leap, Kangaroo Creek Formation, and Gregg Ranch unit had all been emplaced structurally above the plutonic rocks of the Trinity complex, possibly by eastward to southeastward overthrusting. We surmise that the dike and volcanic units of the Trinity complex, now missing for the most part in the Lovers Leap–Gregg Ranch area, were removed during thrusting, and that the unconformity between Ordovician and Upper Silurian rocks that is recorded in both the Kangaroo Creek and Horseshoe Gulch units resulted from uplift and erosion of the arc system during or after this overthrusting. The Lower Cambrian amphibolitic gabbro unit was faulted against Trinity plutonic rocks before intrusion of

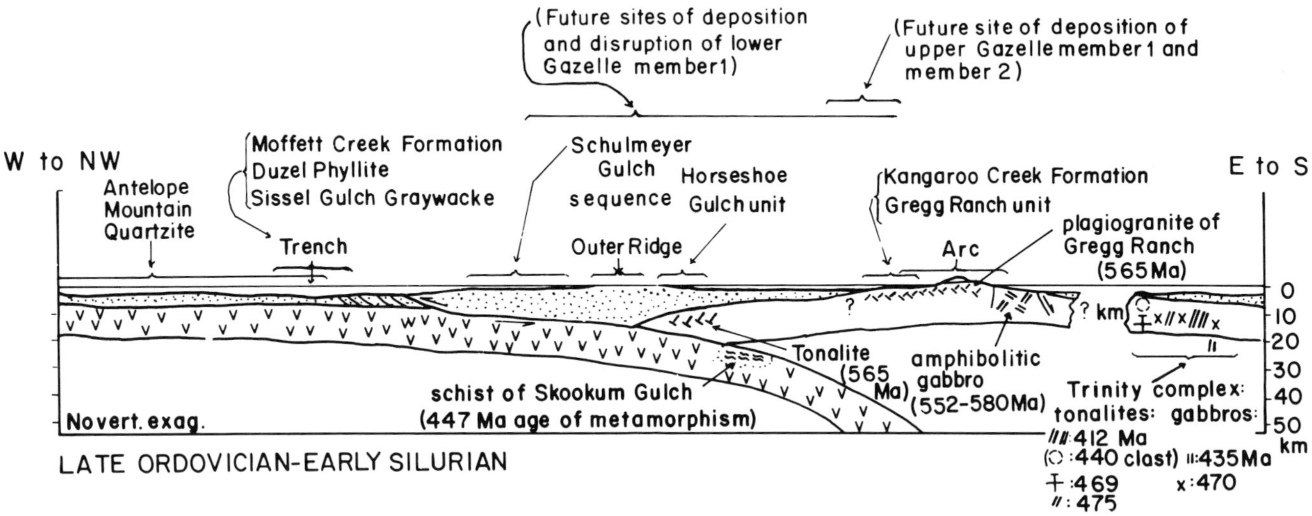

Figure 4. Cross section of hypothetical arc-trench complex showing the settings in which various rock units of the Yreka terrane are inferred to have formed in Late Ordovician to Early Silurian time. Several older and younger units are depicted as well. Ages of units are shown in Figure 2. Isotopic ages of igneous rocks are from Wallin and others (1988) and Jacobsen and others (1984). V pattern = oceanic crust; dot pattern = sedimentary deposits. Other symbols are explained in figure. Although projected onto one transverse section here for purposes of illustration, the Horseshoe Gulch, Gregg Ranch–Kangaroo Creek, and Schulmeyer Gulch units, and the lower, disrupted part of Gazelle member 1 are thought to have been deposited in different basins along the length of the arc. In this configuration the arc faces away from the North American continent, and continent-derived detritus arrives at the appropriate sites of deposition by transport subparallel to the trend of the arc, as discussed in text.

the pegmatitic gabbro unit in Late Silurian time (Lindsley-Griffin, 1982).

Burchfiel and Davis (1975) and Cashman (1980) concluded from their work on the rocks of the central metamorphic belt and Duzel Phyllite that subduction was eastward during the period of Devonian metamorphism that affected these units. Hotz (1977) reported K-Ar hornblende ages of 390 ± 12 and 399 ± 12 Ma for amphibolite from the central metamorphic belt near Yreka (analytical uncertainties from Cashman, 1980). These ages and the 370 ± 20-Ma time of metamorphism inferred by Cashman (1980) for the Duzel Phyllite are within analytical uncertainty of the Early to Middle Devonian ages of keratophyre-spilite of Gazelle member 3 and the Copley-Balaklala arc-volcanic units. The present location of the latter east of the central metamorphic belt is compatible with eastward subduction during that time.

In the context of the above model, the Gregg Ranch unit and Kangaroo Creek Formation are considered to have been deposited in an intra-arc or fore-arc basin near the magmatic axis, and members 2 through 9 of the Horseshoe Gulch unit are inferred to have formed in a basin in the outer fore arc to outer ridge area (Potter, 1990; see Fig. 4). These interpretations are based on the following geologic relations: (1) member 1 of the Gregg Ranch unit was deposited on the keratophyre of Lovers Leap, and member 1 of the Kangaroo Creek Formation is inferred to have been deposited originally on that same keratophyre; (2) member 2 of the Gregg Ranch unit is intruded by the dikes belonging to member 2 itself; and (3) the Horseshoe Gulch unit contains clasts from the schist of Skookum Gulch, presumed to have been uplifted near the outer ridge or outer fore arc by Late Silurian time. Additional corroborating evidence for these inferences is given in Potter (1990).

From the work of Hotz (1977) and Wallin (1988) and the discussion in Potter and others (1977), it is reasonable to surmise that the Moffett Creek Formation, Sissel Gulch Graywacke, and Duzel Phyllite were deposited in the trench of the arc complex, and that the Antelope Mountain Quartzite was deposited in the trench or on the sea floor in front of the trench. In such a framework, arc-derived volcanic detritus could reach the site of deposition of the Sissel Gulch and Moffett Creek units, both of which contain volcanic lithic grains (Potter and others, 1977; Wallin, 1988), and Precambrian zircons of continental origin reported by Wallin (1988) in the Moffett Creek, Duzel, and Antelope Mountain units could be transported to the sites of deposition of those units. (We are assuming here that these Precambrian zircons were derived from a continent or at least another terrane, and not from the arc complex itself.) The clast of 440-Ma tonalite at Lovers Leap (Wallin and others, 1988) implies that magmatic rocks of appropriate age and composition to act as a source for the Sissel Gulch Graywacke once existed near Lovers Leap. The Schulmeyer Gulch sequence may have been deposited on the trench inner slope at least locally above carbonate compensation depth, with uplifted fault blocks of quartz sandstone-siltstone acting as sources for its terrigenous deposits and sites of deposition for its limestone lenses.

We surmise that the lower part of Gazelle member 1 was originally deposited in a different basin that lay between the upper inner slope of the trench and the outer fore arc. Uplifted fault blocks presumably were again present as sources of slides and other mass-flow deposits and, in Late Silurian time, as loci of relatively shallow carbonate sedimentation. The upper, bedded part of Gazelle member 1 and members 2 and 3 were deposited near the magmatic axis, because, as described above, the upper, bedded part of member 1 overlies the volcanic breccia-conglomerate of Gazelle member 3 along a clearly depositional contact. Potter and others (1976, 1977) concluded that the volcanic rocks of Gazelle member 3 are of arc type. As discussed earlier in this report, the member 3 volcanic rocks have lithologic and chronologic counterparts in the much thicker arc rocks of the Copley Greenstone; based on the work of Schweickert and Irwin (1989), they were probably spatially close to the Copley Greenstone in Devonian time. (These close associations with arc volcanic rocks make unlikely the inference by other authors that the bedded part of Gazelle member 1 and member 2 were deposited in a trench-slope basin.)

The Horseshoe Gulch, Gregg Ranch–Kangaroo Creek, and Schulmeyer Gulch units, and the lower part of Gazelle member 1 are inferred to have been deposited in different local basins along the longitudinal axis of the arc in order to account for faunal and lithologic differences among them. Each of these basins was topographically complex and tectonically unstable. These units could then have been juxtaposed during later oblique convergence and faulting subparallel to the long axis of the arc.

The Moffett Creek, Sissel Gulch, Duzel, Antelope Mountain, and Schulmeyer Gulch units are envisioned to have been incorporated into an accretionary prism and uplifted from the subduction zone by later underthrusting of other sediment packets.

Redeposition of large parts of the Moffett Creek Formation and lower Gazelle member 1 to form olistostromes (megamictites of Sundell and Fisher, 1985) and tectonic disruption of these units probably occurred both on the accretionary complex/outer ridge (as also envisioned by Lindsley-Griffin and Griffin, 1983, for their melange of Gregg Ranch) and in the fore arc. In the latter case, some analogies exist with the mode of formation of the Lichi melange of Taiwan (Page and Suppe, 1981).

The presence of soft-sediment deformation in some clasts in olistostromes of the Moffett Creek Formation (Potter and others, 1977) suggests that some slumping and disruption of the Moffett Creek unit took place prior to lithification, presumably while it was part of the accretionary complex. On the other hand, two clasts, 2 and 3 m in maximum dimension, of volcanic or dike rock similar to that in member 2 of the Gregg Ranch unit near Gregg Ranch, occur in Moffett Creek melange south of Bonnet Rock. The large size of these clasts and their source in the inferred magmatic arc suggest that part of the Moffett Creek melange formed in the fore-arc region rather than far from the arc axis in the accretionary prism.

The presence of blocks of Gazelle member 2 as much as 0.6 km in maximum dimension in the melange of lower Gazelle member 1 (A. W. Potter, unpublished mapping) and the inferred site of deposition of member 2 near the magmatic axis are taken as evidence that at least the later stages of formation of the melange of lower Gazelle member 1 occurred in the inner fore-arc region.

Following metamorphism of the Duzel Phyllite, the Moffett Creek, Sissel Gulch, Duzel, Antelope Mountain, and Schulmeyer Gulch units and the lower part of Gazelle member 1 are inferred to have been transported east to southeast toward the magmatic arc during backthrusting of the accretionary prism in the fore arc (Potter, 1990). The eastward to southeastward vergence of this thrusting was inferred from map-scale structures by Potter and others (1977). Nias Island in the Sunda arc, Indonesia, is one of many sites where such backthrusting has been documented (Karig and others, 1980; Silver and Reed, 1988, and references therein).

The likelihood that parts of the melanges of the Moffett Creek Formation and lower Gazelle member 1 formed in the fore-arc region suggests that backthrusting also took place before or during the later stages of melange formation; it may have occurred episodically over a relatively long time span in the Silurian to Devonian. Detailed structural studies are needed to verify this hypothesis. In any case, the final episode of backthrusting most likely occurred in Middle to Late Devonian time. It postdated formation of the youngest units involved in thrusting, which are the melange of lower Gazelle member 1 (minimum age, early Givetian or younger) and the Duzel Phyllite (metamorphosed at 370 ± 20 Ma; Cashman, 1980). It probably preceded deposition of the sandstones with quartzose and sedimentary/metasedimentary-lithic detritus that occur in the Bragdon Formation of the eastern Klamath terrane. These sandstones are late Middle Devonian to Late Mississippian (Watkins, 1986), and were inferred by Miller and Cui (1989) and Miller and Saleeby (1989) to have been derived from a source lithologically and isotopically similar to the Shoo Fly Complex or the Yreka terrane. Because lithologic and chronologic correlatives of the Copley Greenstone occur in the Yreka terrane (see discussion above) and have not been reported in the northern Sierra, we infer that the Yreka terrane was closer to the eastern Klamath terrane in the Devonian than was the Shoo Fly Complex, and was more likely to be the source. (Paleocurrent data of Watkins, 1986, are too sparse to define the source area.)

We suggest that, after their emplacement near the arc by backthrusting, which telescoped or completely closed the fore-arc basin, the Antelope Mountain Quartzite, Moffett Creek Formation, and Duzel Phyllite (and, possibly, chert of Gazelle member 1 and siliceous schist and quartzite of the Skookum Gulch unit) supplied the chert, quartz, metasedimentary-lithic fragments, and Precambrian zircons reported by Miller and Saleeby (1989) in Bragdon sandstones. Loading of the thrust allochthons on the arc crust may have contributed to subsidence of the arc and formation of the basin in which Bragdon strata were deposited.

The orientation of the eastern Klamath arc-trench system

relative to the North American continent in the Cambrian to Devonian is unknown. Schweickert and Snyder (1981, p. 197) pictured the eastern Klamath arc as facing toward the continent. Perhaps it rotated into its present orientation in Devonian or later time. Alternatively, it could have faced away from or have been at high angles to the continent, and received continent-derived detritus by longitudinal transport along the trench. This is the configuration shown in Figure 4.

Paleobiogeographic Relations

Faunal evidence presented above indicates that the eastern Klamath belt was close enough to North America in the Middle Ordovician to Middle Devonian for faunal migration to occur from one to the other. For some genera of Ordovician brachiopods, Ordovician sphinctozoans, and Early and Middle Devonian tabulates, tetracorals, and conodonts, migration took place often enough that the same species could exist in the eastern Klamath belt and North America. These brachiopod, sphinctozoan, and coral species probably had relatively long-lived larvae.

The Late Ordovician tabulate evidence, showing affinities with Asia and the Baltic, contrasts with the brachiopod and sphinctozoan *species*-level affinities with North America, and may be an artifact of comparative lack of study of tabulates in North America. This caveat applies to the Silurian tabulate data as well, because they appear to be inconsistent with the better-documented evidence from Silurian brachiopods.

The presence of *Craspedelia, Phragmorthis,* and other typically "North American" brachiopods and trilobites in Kazakhstan in the Ordovician seems anomalous, yet a similar pattern of intermixing of "North American," "Asiatic," and even some "Tasman" elements in Kazakhstan and nearby areas persisted in Cambrian through Devonian time (Boucot and Gray, 1979). Boucot and Gray (1979) explained this phenomenon as a result of long-distance transport of teleplanic larvae by surface currents.

The relatively strong resemblance of eastern Klamath Ordovician brachiopods and sphinctozoans to coeval faunas in Alaska and the seeming absence of similar Ordovician faunas in British Columbia to Nevada could indicate that the eastern Klamath belt was near Alaska in the Ordovician. Source rocks exist in Alaska and western Canada that could have supplied the detrital zircons found in the Antelope Mountain Quartzite and other units (Hansen, 1990, and references therein; Ross and Bowring, 1989). However, the absence of Klamath-type macrofaunas in British Columbia to Nevada in the Ordovician may also be an artifact of inadequate sampling of appropriate environments there, or of current-circulation or temperature regimes that prevented faunal communication. Madrid (1987) postulated that an uplifted platform lay off the continental margin in Nevada in the early to middle Paleozoic and provided a westward source of sediment to the strata of the Roberts Mountains allochthon. This platform was probably not extensive enough, however, to block macrofaunal interchange between the eastern Klamath belt and the western United States and Canada outside of Nevada.

By and large, the Ordovician conodont faunas from the eastern Klamath belt are similar to those in Nevada, in apparent contrast to the brachiopod and sphinctozoan faunas. The Devonian tabulate data suggest that the eastern Klamath belt could have been relatively near to Nevada by Early Devonian.

In summary, the faunal evidence indicates that the eastern Klamath belt was probably relatively near to western North America throughout the Middle Ordovician to Middle Devonian time span. The exact location of the belt during this time cannot be specified with available data.

A speculative estimate of the distance between the eastern Klamath belt and North America at one time in the Silurian to Devonian can be made from the sediment-gravity–flow deposits of medium- to coarse-grained quartz wacke and some pebble conglomerate reported by Klanderman (1977) in the Antelope Mountain Quartzite, if this material came from North America. The maximum distance of transport of coarse sand and gravel by sediment gravity flows that we have found in published reports is about 700 km, measured from the edge of the continental shelf, for Pleistocene deposits along the axis of the Cascadia Deep-Sea Channel that were studied by Griggs and Kulm (1970). Using a distance of 1,000 km to reduce the chance of underestimation, we speculate that the eastern Klamath arc was less than 1,000 km from the edge of the North American continental shelf at the time in the Silurian to Devonian when the Antelope Mountain unit was incorporated into the accretionary complex of the arc. We reiterate that this estimate applies only if the Antelope Mountain Quartzite detritus under consideration was transported directly from North America.

ACKNOWLEDGMENTS

Thorough reviews by S. C. Finney, D. S. Harwood, and J. G. Johnson are greatly appreciated. National Science Foundation Grant EAR-7919993 (to A.J.B.) provided support for this work.

REFERENCES CITED

Albers, J. P., and Bain, J. H., 1985, Regional setting and new information on some critical geologic features of the West Shasta District, California: Economic Geology, v. 80, p. 2072–2091.

Albers, J. P., Kistler, R. W., and Kwak, L., 1981, The Mule Mountain stock; an early Middle Devonian pluton in northern California: Isochron/West, v. 31, p. 17.

Barker, F., Millard, H. T., and Knight, R. J., 1979, Reconnaissance geochemistry of Devonian island arc volcanic and intrusive rocks, West Shasta district, California, *in* Barker, F., ed., Trondhjemites, dacites, and related rocks: Amsterdam, Elsevier, 660 p.

Bates, R. L., and Jackson, J. A., eds., 1987, Glossary of geology, 3rd ed.: Alexandria, Virginia, American Geological Institute, 788 p.

Bergström, S. M., Potter, A. W., Porter, R. W., Boucot, A. J., and Rohr, D. M., 1980, Biostratigraphic and biogeographic significance of Ordovician conodonts in the eastern Klamath Mountains, northern California: Geological Society of America Abstracts with Programs, v. 12, p. 219.

Berry, W.B.N., Lindsley-Griffin, N., Potter, A. W., and Rohr, D., 1973, Early Middle Ordovician graptolites from the eastern Klamath Mountains, Siskiyou County, California: Geological Society of America Abstracts with Programs, v. 5, p. 11.

Blodgett, R. B., and Clough, J. G., 1985, The Nixon Fork terrane; Part of *in situ* peninsular extension of the Paleozoic North American continent: Geological Society of America Abstracts with Programs, v. 17, p. 342.

Boucot, A. J., 1975, Evolution and extinction rate controls: New York, Elsevier, 427 p.

Boucot, A. J., and Gray, J., 1979, Epilogue; A Paleozoic Pangaea?, *in* Gray, J., and Boucot, A. J., eds., Historical biogeography, plate tectonics, and the changing environment: Corvallis, Oregon State University Press, p. 465–482.

Boucot, A. J., and Potter, A. W., 1977, Middle Devonian orogeny and biogeographical relations in areas along the North American Pacific rim, *in* Murphy, M. A., Berry, W.B.N., and Sandberg, C. A., eds., Western North America; Devonian: Riverside, University of California, Museum Contribution, v. 4, p. 210–219.

Boucot, A. J., and 7 others, 1973, Biogeographic relations of the pre-Late Middle Devonian of the eastern Klamath belt, northern California: Geological Society of America Abstracts with Programs, v. 5, p. 14.

Boucot, A. J., Dunkle, D., Potter, A., Savage, N. M., and Rohr, D., 1974, Middle Devonian orogeny in western North America?; A fish and other fossils: Journal of Geology, v. 82, p. 691–708.

Boudier, F., Le Sueur, E., and Nicolas, A., 1989, Structure of an atypical ophiolite; The Trinity complex, eastern Klamath Mountains, California: Geological Society of America Bulletin, v. 101, p. 820–833.

Brouxel, M., and Lapierre, H., 1988, Geochemical study of an early Paleozoic island-arc–back-arc basin system; Part 1, The Trinity ophiolite (northern California): Geological Society of America Bulletin, v. 100, p. 1111–1119.

Brouxel, M., Lapierre, H., Michard, A., and Albarede, F., 1988, Geochemical study of an early Paleozoic island-arc–back-arc basin system; Part 2, Eastern Klamath early to middle Paleozoic island-arc volcanics (northern California): Geological Society of America Bulletin, v. 100, p. 1120–1130.

Burchfiel, B. C., and Davis, G. A., 1975, Nature and controls of Cordilleran orogenesis, western United States; Extensions of an earlier synthesis: American Journal of Science, v. 275-A, p. 363–396.

Cashman, S. M., 1980, Devonian metamorphic event in the northeastern Klamath Mountains, California: Geological Society of America Bulletin, v. 91, p. 453–459.

Cecile, M. P., 1982, The Lower Paleozoic Misty Creek Embayment, Selwyn Basin, Yukon and Northwest Territories: Geological Survey of Canada Bulletin 335, 78 p.

Chang, Went'ang, and Fan, Chiasung, 1960, Class Trilobita of the Ordovician and the Silurian Periods of the Ch'i-Lien Mountains, *in* Yin, Tsanhsun, ed., Geological gazetteer of the Ch'i-Lien Mountains 4 (1, Paleontological description . . .): Peking, Science Press, p. 83-147. (In Chinese; English translation by U.S. Department of Commerce National Technical Information Service, p. 246–430.)

Churkin, M., Jr., Carter, C., and Trexler, J. H., Jr., 1980, Collision-deformed Paleozoic continental margin of Alaska; Foundation for microplate accretion: Geological Society of America Bulletin, part 1, v. 91, p. 648–654.

Cooper, B. N., 1953, Trilobites from the Lower Champlainian formations of the Appalachian Valley: Geological Society of America Memoir 55, p. 11–69.

Cooper, G. A., 1956, Chazyan and related brachiopods: Smithsonian Miscellaneous Collections, v. 127, pt. 1, p. 1-1024, pt. 2, p. 1025–1245.

Cotkin, S. J., and Armstrong, R. L., 1987, Rb/Sr age, geochemistry, and tectonic significance of blueschist from the schist of Skookum Gulch, eastern Klamath Mountains, California; Introducing the Callahan event: Geological Society of America Abstracts with Programs, v. 19, p. 367.

Elias, R. J., 1982, Latest Ordovician solitary rugose corals of eastern North America: Bulletins of American Paleontology, v. 81, 116 p.

——, 1983, Late Ordovician solitary rugose corals of the Stony Mountain Formation, southern Manitoba, and its equivalents: Journal of Paleontology, v. 57, p. 924–956.

Elias, R. J., and Potter, A. W., 1984, Late Ordovician solitary rugose corals of the eastern Klamath Mountains, northern California: Journal of Paleontology, v. 58, p. 1203–1214.

Flory, R. A., and Watkins, R., 1983, Biogeographic relations of Lower Paleozoic tabulate corals from northern California: Geological Society of America Abstracts with Programs, v. 15, p. 409.

Gilbert, W. G., 1981, Preliminary geologic map and geochemical data, Cheeneetnuk River area, Alaska: Alaska Division of Geological and Geophysical Surveys Open-File Report 153, p. 1–10.

Girty, G. H., and Wardlaw, M. S., 1984, Was the Alexander terrane a source of feldspathic sandstones in the Shoo Fly Complex, Sierra Nevada, California?: Geology, v. 12, p. 339–342.

Griggs, G. B., and Kulm, L. D., 1970, Sedimentation in Cascadia Deep-Sea Channel: Geological Society of America Bulletin, v. 81, p. 1361–1384.

Haessig, P. A., 1988, Structure and petrography of the schist of Skookum Gulch, Callahan-Yreka area, eastern Klamath Mountains, northern California [M.S. thesis]: Corvallis, Oregon State University, 169 p.

——, 1989, Field relations and structure of the schist of Skookum Gulch, an Early Paleozoic melange, Callahan-Yreka area, eastern Klamath Mountains: Geological Society of America Abstracts with Programs, v. 21, p. 88.

Haessig, P. A., Potter, A. W., Wallin, E. T., and Mattinson, J. M., 1987, Early Cambrian tonalite-plagiogranite from the Yreka-Callahan-Gazelle area, eastern Klamath Mountains, northern California: Geological Society of America Abstracts with Programs, v. 19, p. 689.

Hansen, V. L., 1990, Yukon-Tanana terrane; A partial acquittal: Geology, v. 18, p. 365–369.

Hanson, R. E., and Schweickert, R. A., 1986, Stratigraphy of mid-Paleozoic island-arc rocks in part of the northern Sierra Nevada, Sierra and Nevada Counties, California: Geological Society of America Bulletin, v. 97, p. 986–998.

Harland, W. B., and 5 others, 1989, A geologic time scale 1989: Cambridge, Cambridge University Press, 256 p.

Hill, D., 1981, Rugosa and Tabulata, *in* Moore, R. C., ed., Treatise on invertebrate paleontology; Part F, Coelenterata, Supplement 1: Boulder, Colorado, Geological Society of America (and University of Kansas Press), v. 2, p. F430–F669.

Hotz, P. E., 1977, Geology of the Yreka Quadrangle, Siskiyou County, California: U.S. Geological Survey Bulletin 1436, 72 p.

Hughes, C. P., Ingham, J. K., and Addison, R., 1975, The morphology, classification, and evolution of the Trinucleidae (Trilobita): Philosophical Transactions of the Royal Society of London, Series B, v. 272, p. 537–607.

Irwin, W. P., Jones, D. L., and Kaplan, T. A., 1978, Radiolarians from pre-Nevadan rocks of the Klamath Mountains, California and Oregon, *in* Howell, D. G., and McDougall, K. A., eds., Mesozoic paleogeography of the western United States: Pacific Section, Society of Economic Paleontologists and Mineralogists Pacific Coast Paleogeography Symposium 2, p. 303–310.

Ivanov, A. N., and Miagkova, E. I., 1950, Opredelitel' fauny ordovika zapadnogo sklona srednego urala: Trudy Gorno-Geologiska in-ta Ural'skogo Filia, An SSSR, v. 18, p. 3–32.

Jackson, D. E., Lenz, A. C., and Pedder, A.E.H., 1978, Late Silurian and Early Devonian graptolite, brachiopod, and coral faunas from northwestern and Arctic Canada: Geological Association of Canada Special Paper 17, 159 p.

Jacobsen, S. B., Quick, J. E., and Wasserburg, G. E., 1984, A Nd and Sr isotopic study of the Trinity peridotite; Implications for mantle evolution: Earth and Planetary Science Letters, v. 68, p. 361–378.

Karig, D. E., Lawrence, M. B., Moore, G. F., and Curray, J. R., 1980, Structural framework of the forearc basin, N.W. Sumatra: Journal of the Geological Society of London, v. 137, p. 77.

Kinkel, A. R., Jr., Hall, W. E., and Albers, J. P., 1956, Geology and base metal deposits of West Shasta copper-zinc district, Shasta County, California: U.S. Geological Survey Professional Paper 285, 156 p.

Klanderman, D. S., 1977, Stratigraphy, structure, and depositional environments of the Antelope Mountain Quartzite, Yreka, California [M.S. thesis]: Corvallis, Oregon State University, 115 p.

Klapper, G., and Johnson, J. G., 1980, Endemism and dispersal of Devonian conodonts: Journal of Paleontology, v. 54, p. 400–455.

Lapierre, H., Albarede, F., Albers, J., Cabanis, B., and Coulon, C., 1985a, The Early Devonian volcanism in the eastern Klamath Mountains, California; Evidence for an immature island-arc: Canadian Journal of Earth Sciences, v. 22, p. 214–227.

Lapierre, H., Cabanis, B., Coulon, C., Brouxel, M., Albarede F., 1985b, Geodynamic setting of Early Devonian Kuroko-type sulfide deposits in the eastern Klamath Mountains (northern California) inferred by the petrological and geochemical characteristics of the associated island-arc volcanic rocks: Economic Geology, v. 80, p. 2100–2113.

Lapierre, H., and 7 others, 1987, Paleozoic and Lower Mesozoic magmas from the eastern Klamath Mountains (North California) and the geodynamic evolution of northwestern America: Tectonophysics, v. 140, p. 155–177.

Lindsley-Griffin, N., 1977, The trinity ophiolite, Klamath Mountains, California, *in* Coleman, R. G., and Irwin, W. P., eds., North American ophiolites: Oregon Department of Geology and Mineral Industries Bulletin 95, p. 107–120.

——, 1982, Structure, stratigraphy, petrology, and regional relationships of the Trinity ophiolite, eastern Klamath Mountains, California [Ph.D. thesis]: Davis, University of California, 445 p.

Lindsley-Griffin, N., and Fisher, M., 1989, Significance of the tectonic contact between Lower Paleozoic melange and shales of the Gazelle Formation, Yreka terrane, Klamath Mountains, California: Geological Society of America Abstracts with Programs, v. 21, p. 107.

Lindsley-Griffin, N., and Griffin, J. R., 1983, The Trinity terrane; An Early Paleozoic microplate assemblage, *in* Stevens, C. H., ed., Pre-Jurassic rocks in western North American suspect terranes: Pacific Section, Society of Economic Paleontologists and Mineralogists, p. 63–75.

Lindsley-Griffin, N., and Rohr, D. M., 1977, Lovers Leap; A geologic puzzle, *in* Lindsley-Griffin, N., and Kramer, J. C., eds., Geology of the Klamath Mountains, northern California: Sacramento, California, Geological Society of America Cordilleran Section meeting guidebook, p. 47–69.

Linnarsson, J.G.O., 1869, Om Vestergötlands Cambriska och Siluriska Aflagringar, Kongliga Svenska Vetenskaps: Akademiens Handlingar, Bandet 8, no. 2, p. 1–89.

Madrid, R.J.J., 1987, Stratigraphy of the Roberts Mountains allochthon in north-central Nevada [Ph.D. thesis]: Palo Alto, Stanford University, 340 p.

Mankinen, E. A., Irwin, W. P., and Gromme, C. S., 1988, Paleomagnetic results from the Shasta Bally plutonic belt in the Klamath Mountains Province, northern California: Geophysical Research Letters, v. 15, p. 56–59.

——, 1989, Paleomagnetic study of the eastern Klamath terrane, California, and implications for the tectonic history of the Klamath Mountains Province: Journal of Geophysical Research, v. 94, p. 10444–10472.

Mattinson, J. M., and Hopson, C. A., 1972, Paleozoic ophiolite complexes in Washington and northern California: Carnegie Institution of Washington Yearbook, no. 71, p. 578–583.

Miller, M. M., and Cui, B., 1989, Submarine-fan characteristics and dual sediment provenance, Lower Carboniferous Bragdon Formation, eastern Klamath terrane, California: Canadian Journal of Earth Sciences, v. 26, p. 927–940.

Miller, M. M., and Saleeby, J. B., 1989, Proterozoic detrital zircon from the Carboniferous Bragdon Formation, E. Klamath terrane, California; Continent-influenced arc sedimentation: Geological Society of America Abstracts with Programs, v. 21, p. A347.

Nicholson, H. A., and Etheridge, R., 1879, A monograph of the Silurian fossils of the Girvan district in Ayrshire: Edinburgh and London, v. 1, fasciculus II, p. i-vi, 138–236.

Page, B. M., and Suppe, J., 1981, The Pliocene Lichi Melange of Taiwan; Its plate-tectonic and olistostromal origin: American Journal of Science, v. 281, p. 193–227.

Payne, M. W., and Allison, C. W., 1981, Paleozoic continental-margin sedimentation in east-central Alaska: Geology, v. 9, p. 274–279.

Pedder, A.E.H., 1985a, Lower Devonian rugose corals of Lochkovian age from Yukon Territory, *in* Current Research, Part A: Geological Survey of Canada Paper 85-1A, p. 587–602.

——, 1985b, Lochkovian (Early Devonian) rugose corals from Prince of Wales and Baillie Hamilton islands, Canadiaan Arctic Archipelago, *in* Current Research, Part B: Geological Survey of Canada Paper 85-1B, p. 285–301.

Pedder, A.E.H., and Oliver, W. A., Jr., 1982, *Stauromatidium* and Stauromatidiidae, new genus and family of Upper Silurian and Lower Devonian rugose corals: Geological Survey of Canada Bulletin 352, 43 p.

Porter, R. W., 1974, Geology of the Facey Rock area, Etna Quadrangle, California [M.S. thesis]: Corvallis, Oregon State University, 87 p.

Potter, A. W., 1977, Stratigraphy of Moffett Creek and Gazelle Formations northeast Gregg Ranch, eastern Klamath Mountains, northern California, *in* Lindsley-Griffin, N., and Kramer, J. C., eds., Geology of the Klamath Mountains, northern California: Sacramento, California, Geological Society of America Cordilleran Section meeting guidebook, p. 80–103.

——, 1982, Klamath Mountains, *in* Ross, R. J., Jr., and 27 others, The Ordovician System in the United States: International Union of Geological Sciences Publication 12, p. 30–33.

——, 1990, Middle and Late Ordovician brachiopods from the eastern Klamath Mountains, northern California, part 1: Palaeontographica, Abteilung A, v. 212, p. 31–158.

Potter, A. W., and Grunder, A. L., 1989, Inferred tectonic settings of Early Cambrian and Ordovician intrusive rocks, eastern Klamath terrane (EKT), northern California: Geological Society of America Abstracts with Programs, v. 21, p. 130.

Potter, A. W., Scheidegger, K. F., and Corliss, J. B., 1976, Magma types of Early Paleozoic altered volcanic rocks of the eastern Klamath Mountains, northern California; Further results: EOS American Geophysical Union, v. 57, p. 1023.

Potter, A. W., Hotz, P. E., and Rohr, D. M., 1977, Stratigraphy and inferred tectonic framework of Lower Paleozoic rocks in the eastern Klamath Mountains, northern California, *in* Stewart, J. H., Stevens, C. H., and Fritsche, A. E., eds., Paleozoic paleogeography of the western United States: Pacific Section, Society of Economic Paleontologists and Mineralogists Pacific Coast Paleogeography Symposium 1, p. 421–440.

Renne, P. R., and Scott, G. R., 1988, Structural chronology, oroclinal deformation, and tectonic evolution of the southeastern Klamath Mountains, California: Tectonics, v. 7, p. 1223–1242.

Renne, P. R., Scott, G. R., and Bazard, D. R., 1988, Multicomponent paleomagnetic data from the Nosoni Formation, eastern Klamath Mountains, California; Cratonic Permian primary directions with Jurassic overprints: Journal of Geophysical Research, v. 93, p. 3387–3400.

Rigby, J. K., and Potter, A. W., 1986, Ordovician sphinctozoan sponges from the eastern Klamath Mountains, northern California: Paleontological Society Memoir 20 (Journal of Paleontology, v. 60, no. 4, suppl.), 47 p.

Rigby, J. K., Potter, A. W., and Blodgett, R. B., 1988, Ordovician sphinctozoan sponges of Alaska and Yukon Territory: Journal of Paleontology, v. 62, p. 731–746.

Rohr, D. M., 1980, Ordovician-Devonian Gastropoda from the Klamath Mountains, California: Palaeontographica, Abteilung A, v. 171, p. 141–199.

Ross, G. M., and Bowring, S. A., 1989, Detrital zircon geochronology of the Windermere Grits (Late Proterozoic, southern Canadian Cordillera); Implications for terrane boundaries: Geological Society of America Abstracts with Programs, v. 21, p. 136.

Ross, R. J., Jr., and 27 others, 1982, The Ordovician System in the United States: International Union of Geological Sciences Publication 12, 73 p.

Saleeby, J., Hannah, J. L., and Varga, R. J., 1987, Isotopic age constraints on middle Paleozoic deformation in the northern Sierra Nevada, California:

Geology, v. 15, p. 757–760.

Savage, N. M., 1976a, Middle Devonian species of *Polygnathus* from the Kennett Formation, northern California: Journal of Paleontology, v. 50, p. 374–379.

——, 1976b, Lower Devonian (Gedinnian) conodonts from the Grouse Creek area, Klamath Mountains, northern California: Journal of Paleontology, v. 50, p. 1180–1190.

——, 1977, Lower Devonian conodonts from the Gazelle Formation, Klamath Mountains, northern California: Journal of Paleontology, v. 51, p. 57–62.

Schweickert, R. A., and Irwin, W. P., 1989, Extensional faulting in southern Klamath Mountains, California: Tectonics, v. 8, p. 135–149.

Schweickert, R. A., and Snyder, W., 1981, Paleozoic plate tectonics of the Sierra Nevada and adjacent regions, *in* Ernst, W. G., ed., The geotectonic development of California; Rubey Volume 1: Englewood Cliffs, New Jersey, Prentice-Hall, p. 182–201.

Silberling, N. J., Jones, D. L., Blake, M. C., Jr., and Howell, D. G., 1987, Lithotectonic terrane map of the western conterminous United States, *in* Folio of the lithotectonic terrane maps of the North American Cordillera: U.S. Geological Survey Miscellaneous Field Studies Map MF–1874–C, scale 1:2,500,000.

Silver, E. A., and Beutner, E. C., 1980, Melanges: Geology, v. 8, p. 32–34.

Silver, E. A., and Reed, D. L., 1988, Backthrusting in accretionary wedges: Journal of Geophysical Research, v. 93, p. 3116–3126.

Sundell, K. A., and Fisher, R. V., 1985, Very coarse-grained fragmental rocks; A proposed size classification: Geology, v. 13, p. 692–695.

Taylor, B., and Karner, G. D., 1983, On the evolution of marginal basins: Reviews of Geophysics and Space Physics, v. 21, p. 1727–1741.

Tripp, R. P., 1980, Trilobites from the Ordovician Balclatchie and lower Ardwell Groups of the Girvan district, Scotland: Transactions of the Royal Society of Edinburgh, Earth Sciences, v. 71, p. 123–145.

Walker, K. R., Read, J. F., and Hardy, L. A., 1989, Cambro-Ordovician carbonate banks and siliciclastic basins of the United States Appalachians, 28th International Geological Congress Field Trip Guidebook T161: American Geophysical Union, 88 p.

Wallin, E. T., 1988, Evolution of Lower Paleozoic terrigenous clastic sedimentation in the Yreka-Callahan area, eastern Klamath Mountains, California: Geological Society of America Abstracts with Programs, v. 20, p. A231.

——, 1989, Reconstruction of the timing and sequence of Early Paleozoic sedimentation in the Yreka-Callahan terrane, eastern Klamath Mountains, California: Geological Society of America Abstracts with Programs, v. 21, p. 155.

Wallin, E. T., Mattinson, J. M., and Potter, A. W., 1988, Early Paleozoic magmatic events in the eastern Klamath Mountains, northern California: Geology, v. 16, p. 144–148.

Wallin, E. T., Lindsley-Griffin, N., and Potter, A. W., 1990, Trinity ultramafic complex, Klamath Mountains, California; Cambro-Ordovician ophiolite or composite terrane?: Geological Society of America Abstracts with Programs, v. 22, p. 91–92.

Wang, Zhi-ping, 1983, Two genera of Middle Devonian rugose corals from Guizhou: Acta Palaeontologica Sinica, v. 22, p. 66–67 (Chinese with English summary).

Watkins, R., 1986, Late Devonian to Early Carboniferous turbidite facies and basinal development of the eastern Klamath Mountains, California: Sedimentary Geology, v. 49, p. 51–71.

Watkins, R., and Flory, R. A., 1986, Island arc sedimentation in the Middle Devonian Kennett Formation, eastern Klamath Mountains, California: Journal of Geology, v. 94, p. 753–761.

Williams, A., 1973, Distribution of brachiopod assemblages in relation to Ordovician palaeogeography, *in* Hughes, N. F., ed., Organisms and continents through time, Special Papers in Palaeontology 12, Systematics Association Publication 9: Palaeontological Association, p. 241–269.

Manuscript Accepted by the Society March 27, 1990

Printed in U.S.A.

Geological Society of America
Special Paper 255
1990

Comparison of the Central Metamorphic Belt and Trinity terrane of the Klamath Mountains with the Feather River terrane of the Sierra Nevada

Bradley R. Hacker*
Department of Earth and Space Sciences, University of California, Los Angeles, California 90024-1567
Simon M. Peacock
Department of Geology, Arizona State University, Tempe, Arizona 85287-1404

ABSTRACT

The Central Metamorphic Belt and Trinity terrane of the Klamath Mountains and the Feather River terrane of the northern Sierra Nevada share strong similarities in their protolith types, metamorphic parageneses, structures, ages of formation and metamorphism, and relations with surrounding units. These terranes consist of ductilely deformed ultramafic to mafic plutonic and volcanic rocks and minor oceanic sedimentary rocks interpreted as oceanic lithosphere. Ultramafic rocks in the Trinity terrane and the Feather River terrane were hydrated and metasomatized under lower amphibolite-facies conditions. Mafic and sedimentary rocks in the Central Metamorphic Belt and the Feather River terrane contain upper greenschist to amphibolite-facies parageneses formed during Devonian time. Phase assemblages and mineral chemistries indicate peak *P-T* conditions of 500° to 650 ± 50°C and 500 ± 300 MPa for both terranes. The similarities imply that the Klamath Mountains and Sierra Nevada share a common early to middle Paleozoic history. The Trinity terrane/Central Metamorphic Belt represents an arc basement/subduction zone couple; the Feather River terrane may also represent such a couple.

INTRODUCTION

The Klamath Mountains and the Sierra Nevada of the North American Cordillera contain a variety of rocks that are interpreted as magmatic arcs, oceanic crust, and subduction complexes, and formed during different phases and styles of ocean/continent interaction. This chapter summarizes the results of detailed studies of the Central Metamorphic Belt and Trinity terrane of the Klamath Mountains, and the Feather River terrane of the Sierra Nevada, to illustrate the similarities and differences between early Paleozoic rocks of the Klamath Mountains and Sierra Nevada. Textural relations, determined by back-scattered electron and optical microscopy, constrain the sequence of deformational and metamorphic events. Phase relations and mineral compositions determined by electron probe microanalysis constrain the *P-T* conditions of metamorphism. Analytical techniques used in this study are discussed in Peacock and Norris (1989). Our observations, combined with those of other authors, permit correlations of the Central Metamorphic Belt and Trinity terrane of the Klamath Mountains with the Feather River terrane of the Sierra Nevada, and provide the petrogenetic framework in which to discuss the early to middle Paleozoic history of this area.

To understand the geologic relations among the Central Metamorphic Belt, the Trinity terrane, the Feather River terrane, and their surrounding units, the following section summarizes the early to middle Paleozoic rocks in both mountain belts. The descriptions of the Central Metamorphic Belt, the Trinity terrane,

*Present address: Department of Geology, Stanford University, Stanford, California 94305-2115.

Hacker, B. R., and Peacock, S. M., 1990, Comparison of the Central Metamorphic Belt and Trinity terrane of the Klamath Mountains with the Feather River terrane of the Sierra Nevada, *in* Harwood, D. S., and Miller, M. M., eds., Paleozoic and early Mesozoic paleogeographic relations; Sierra Nevada, Klamath Mountains, and related terranes: Boulder, Colorado, Geological Society of America Special Paper 255.

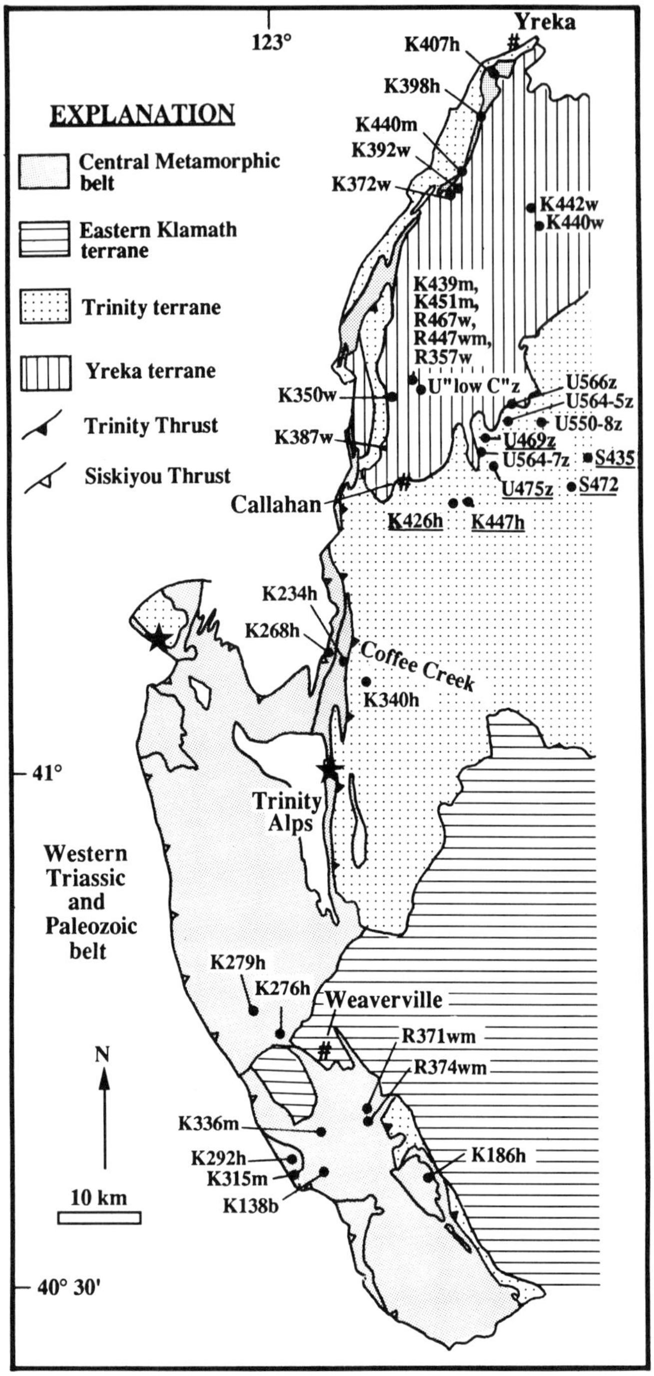

Figure 1. Geologic map of the eastern Klamath Mountains modified from Strand (1962), Davis and others (1965), Irwin (1981), Wagner and Saucedo (1987). The western Paleozoic and Triassic belt crops out west of the area shown. Radiometric abbreviations: A = Ar/Ar; K = K/Ar; R = Rb/Sr; S = Sm/Nd; U = U/Pb; m = mica; w = whole rock; h = hornblende; z = zircon (e.g., K389h indicates a 389-Ma hornblende dated by the K/Ar method. The locations of the six samples for the Sm/Nd isochron by Brouxel and Lapierre (1988) are not shown. For clarity, all age determinations of the Trinity terrane are underlined; ages that are not underlined but appear within the Trinity terrane correspond to outcrops of the Yreka terrane or Central Metamorphic Belt that are too small to show at this scale. Locations of samples used for geothermometry are indicated by stars.

and the Feather River terrane are based principally on our own observations. Radiometric ages have been recalculated as necessary, using the data of Dalrymple (1979).

KLAMATH MOUNTAINS REGIONAL GEOLOGY

The major lithostratigraphic units of the Klamath Mountains are the western Jurassic belt, the western Triassic and Paleozoic belt, the Central Metamorphic Belt, the Yreka terrane, the Trinity terrane, and the Eastern Klamath terrane (Fig. 1) (Irwin, 1960a, 1977; Silberling and others, 1987). The contacts between all units are faults (Irwin, 1960a; Lindsley-Griffin and Griffin, 1983; Renne and Scott, 1988; Schweickert and Irwin, 1989), except that the Eastern Klamath terrane may locally depositionally overlie the Trinity terrane (Brouxel and others, 1988).

Yreka terrane

The Yreka terrane consists of poorly understood, coherent, and disrupted lower Paleozoic sedimentary rocks, the more well known of which include the Duzel Phyllite and the Antelope Mountain Quartzite. The rocks form an imbricate stack of nappes in low-angle fault contact with the underlying Trinity terrane to the east and underlying unnamed amphibolite and ultramafic rocks to the west. Several of the units are inferred to have been deposited as turbidites and debris flows derived from quartzose, volcanoplutonic, and ophiolitic sources of Precambrian to Late Silurian age (Hotz, 1977; Potter and others, 1977; Lindsley-Griffin and Griffin, 1983; Wallin, 1989). Many of the sedimentary units are broken formation or melange that contain a wide variety of Early Cambrian to Early Devonian blocks, including blueschist, peridotite, and serpentinite, set in an unmetamorphosed to greenschist-facies Ordovician-Silurian sedimentary matrix. Some of these units may represent accretionary wedges (Potter and others, 1977).

Most of the sedimentary rocks in the Yreka terrane are chlorite grade or lower. Several of the units, however, contain Late Ordovician greenschist (Table 1, a and b) and Middle to Late Ordovician blueschist-facies rocks (Table 1, c and d). Near the Central Metamorphic Belt the isotopic systematics of these same units were altered, and now yield Devonian to Mississippian ages (Table 1, e and f). Moreover, these reheated rocks contain isoclinal folds that are coplanar to structures in the adjacent unnamed amphibolite and ultramafic rocks. The formation of the Yreka terrane may be related to the Trinity terrane and the Central Metamorphic Belt, as discussed later.

Trinity terrane

The Trinity terrane is an east-dipping sheet of mafic and ultramafic rocks (Davis and others, 1965; Irwin, 1966; La Fehr, 1966) overlain by thrust nappes of the Yreka terrane to the northwest and by Devonian to Jurassic rocks in the eastern Klamath terrane to the southeast, and underlain along the Trinity thrust by the Central Metamorphic Belt (Fig. 1). The Trinity terrane has been interpreted as an immature magmatic arc (Brouxel and Lapierre, 1988; Boudier and others, 1989) intruded into mid-ocean ridge–type ultramafic rocks (Jacobsen and others, 1984). The ultramafic rocks include predominantly plagioclase lherzolite, harzburgite, and dunite (Boudier and others, 1989). One sample of ultramafic rock yielded an Ordovician Sm/Nd age (Table 1, g), and a Sm/Nd isochron derived from gabbros and basalts is also Ordovician (Table 1, h). Older gabbros also yield Ordovician U/Pb ages (Table 1, i and j), whereas Silurian gabbros, diorites, and trondhjemites intrude the peridotite and older gabbros (Goullaud, 1977; Lindsley-Griffin, 1977; Quick, 1981) (Table 1, k through n). Nd isotopes indicate that the gabbros intruding the peridotite were not derived from partial melting of the peridotite, but crystallized from an isotopically distinct depleted mantle source (Jacobsen and others, 1984). It is plausible that the gabbros represent cumulate residue from arc magmas (Jacobsen and others, 1984; Brouxel and Lapierre, 1988).

Early deformation in the Trinity terrane produced isoclinal mesoscopic folds in the peridotite prior to intrusion of the later gabbros. Then, serpentine + magnetite veins formed in the peridotite synchronous with macroscopic warping (F_1 and F_2 of Goullaud, 1977). Adjacent to the Trinity thrust, the Trinity terrane was later foliated (Goullaud, 1977; Cannat and Boudier, 1985) and serpentinized by fluids derived from prograde reactions in the underlying Central Metamorphic Belt (Peacock, 1987a). The serpentinized peridotites were later folded together with the Central Metamorphic Belt (Davis and others, 1965; Goullaud, 1977).

Unnamed ultramafic rocks north of Callahan contain similar lithologies, metamorphic parageneses, structures, and ages of metamorphism (Hotz, 1977; Cashman, 1980), and in this chapter, they are considered equivalent to the Trinity terrane. However, these ultramafic rocks are not contiguous with the Trinity terrane (Hotz, 1977; Cashman, 1980) and are not in "normal" structural sequence with respect to the Yreka terrane and the amphibolite (Irwin, 1977). Ultramafic and mafic rocks of the Trinity terrane may have correlatives in the Feather River terrane of the Sierra Nevada.

Central Metamorphic Belt

The Central Metamorphic Belt consists chiefly of amphibolite structurally overlain by micaceous and feldspathic quartz

TABLE 1. ISOTOPIC AGES CITED

Yreka terrane
- a. 440 ± 11 and 442 ± 12 Ma, K/Ar whole-rock (Cashman, 1980)
- b. 440 ± 11 and 440 ± 13 Ma, K/Ar muscovite (Hotz, 1977; Hotz, reported in Cashman, 1980)
- c. 447 ± 9 and 467 ± 45 Ma, Rb/Sr phengite–whole-rock (Cotkin and Armstrong, 1987)
- d. 439 ± 13 and 451 ± 14 Ma, K/Ar white mica (Potter and others, 1981)
- e. 392 ± 11, 387 ± 10, 372 ± 9, and 350 ± 10 Ma, K/Ar whole-rock (Cashman, 1980)
- f. 357 ± 18 Ma, Rb/Sr whole-rock and minerals (Cotkin and Armstrong, 1987)

Trinity terrane
- g. 472 ± 32 Ma, Sm/Nd plagioclase and clinopyroxene (Jacobsen and others, 1984)
- h. 452 ± 40 Ma, Sm/Nd isochron on 6 samples (Brouxel and Lapierre, 1988)
- i. 460 and 470 Ma, U/Pb zircon (Mattinson and Hopson, 1972a, b)
- j. 475 ± 10 and 469 ± 21 Ma, U/Pb zircon (Wallin and others, 1988)
- k. 418 ± 17, 426 ± 17, 439 ± 18, and 447 ± 18 Ma, K/Ar hornblende (Lanphere and others, 1968)
- l. 412 Ma, U/Pb zircon (Potter and others, 1977)
- m. 435 ± 21 Ma, Sm/Nd plagioclase-clinopyroxene (Jacobsen and others, 1984)
- n. 412 ± 10 and 414–419 ± 15 Ma Pb/Pb zircon (Wallin and others, 1988)

Central Metamorphic Belt
- o. 371 ± 11 and 374 ± 10 Ma, Rb/Sr whole-rock and mica (Lanphere and others, 1968)
- p. 138 Ma, K/Ar biotite; 186 ± 10, 234 ± 30, 276, 279, 292 Ma, K/Ar hornblende; 268, 315, and 336 Ma, K/Ar muscovite (Lanphere and others, 1968)

Unnamed amphibolite west of the Yreka terrane
- q. 398 ± 12 and 407 ± 12 Ma, K/Ar hornblende (Hotz, 1974)

Mule Mountain stock
- r. 400 Ma, U/Pb, Pb/Pb, and K/Ar (Albers and others, 1981)

Bowman Lake batholith
- s. 364–385 Ma, U/Pb zircon (Hanson and others, 1988)

Wolf Creek stock
- t. 378 ± 5 Ma, U/Pb zircon (Saleeby and others, 1987)

Feather River area of Feather River terrane
- u. 306–324 Ma, U/Pb zircon (Saleeby and others, 1989)
- v. 241 ± 4 Ma, K/Ar hornblende (Weisenberg, 1979)

Devils Gate area of Feather River terrane
- w. 248 Ma, K/Ar hornblende (Hietanen, 1981)
- x. 272 ± 6 Ma, Ar/Ar hornblende (Standlee, 1978)
- y. 395 ± 13 Ma, Ar/Ar hornblende (Standlee, 1978)

Yuba River area of Feather River terrane
- z. 322 ± 27 and 345 ± 9 Ma, K/Ar hornblende (Böhlke and McKee, 1984)
- aa. "Devonian," U/Pb zircon (Saleeby and others, 1989)
- cc. 273 ± 5 Ma, K/Ar muscovite (Bohlke and McKee, 1984)
- dd. 285 ± 8 Ma, K/Ar hornblende (Hietanen, 1981)

Feather River terrane
- dd. ~300 Ma; Sm/Nd isochron on 4 samples (Saleeby and others, 1989)

schist and siliceous marble (Irwin, 1960b; Davis and Lipman, 1962; Davis and others, 1965). It generally crops out west of and structurally below the Trinity terrane along the Trinity thrust (Irwin, 1981). A seismic refraction study by Zucca and others (1986) suggested that the Central Metamorphic Belt extends eastward beneath the Trinity terrane for more than 100 km. The Central Metamorphic Belt overlies the Western Triassic and Paleozoic Belt along the Siskiyou thrust (Davis and others, 1965; Goodge, 1990).

Mafic rocks in the Central Metamorphic Belt are well-foliated and well-lineated amphibolite with pegmatitic, centimeter-thick quartzofeldspathic lenses (Davis and others, 1965; Hotz, 1977; Cashman, 1980; Peacock, 1985). Metamorphic recrystallization destroyed most primary igneous and sedimentary textures. Foliation is generally parallel to the Trinity thrust, and rare isoclinal fold hinges are parallel to the foliation (Davis and others, 1965; Peacock, 1985). Hornblende lineation patterns in the Central Metamorphic Belt are complex (Davis and others, 1965; Davis, 1968), but close to the Trinity thrust, lineations plunge moderately eastward.

Several observations strongly suggest that metamorphism of the Central Metamorphic Belt took place concurrent with emplacement of the Central Metamorphic Belt beneath the Trinity terrane along the Trinity thrust: (1) peak metamorphic temperatures and grain size generally increase structurally upward toward the Trinity thrust (Davis and others, 1965; Holdaway, 1965; Peacock and Norris, 1989); (2) the Trinity thrust, metamorphic foliation in the Central Metamorphic Belt, and mylonitic foliation at the base of the Trinity terrane are all subparallel (Lipman, 1964; Cannat and Boudier, 1985; Peacock, 1987a); and (3) extensive serpentinization near the base of the Trinity terrane occurred at the same time but at slightly lower temperature than the metamorphism of the underlying Central Metamorphic Belt (Peacock, 1987a). The sense of shear along the Trinity thrust, inferred from lineations and the lattice orientations of olivine and quartz crystals, is top to the west (Cannat and Boudier, 1985).

Unnamed amphibolite west of the Yreka terrane contains similar lithologies, metamorphic parageneses, and ages of metamorphism (Hotz, 1977), and in this chapter, they are considered equivalent to the Central Metamorphic Belt, as proposed by Cashman (1980). However, these rocks are not contiguous with the type Central Metamorphic Belt farther south (Hotz, 1977; Cashman, 1980), are not in "normal" structural sequence with respect to the Yreka terrane and the ultramafic sheet (Irwin, 1977), and contain slightly higher grade metamorphic mineral assemblages than the type Central Metamorphic Belt farther south (Peacock, 1985).

The oldest radiometric dates on rocks of the Central Metamorphic Belt are Middle or Late Devonian Rb/Sr whole-rock and mica isochrons (Table 1, o). Subsequent heating events affected these rocks, producing K/Ar ages ranging from 138 to 336 Ma (Table 1, p). The unnamed amphibolite west of the Yreka terrane has yielded Late Silurian to Middle Devonian K/Ar hornblende ages (Table 1, q). The Central Metamorphic Belt has probable correlatives in the Feather River terrane of the Sierra Nevada.

Eastern Klamath terrane

The eastern Klamath terrane contains more than 10 km of regionally east-dipping Devonian to Jurassic sedimentary and volcanic strata formed in an intra-oceanic arc that was active in Devonian, Permo-Triassic, and Jurassic time (Irwin, 1977; Eastoe and others, 1987; Miller, 1989). An Early Devonian basalt/rhyolite volcanoplutonic suite is represented by the Copley Greenstone, the Balaklala Rhyolite, and the Mule Mountain stock (Kinkel and others, 1956). The Balaklala Rhyolite was intruded into and erupted onto the Copley Greenstone (Albers and Bain, 1985; Lapierre and others, 1985a). The 400-Ma (Table 1, r) Mule Mountain stock is cogenetic with the Balaklala Rhyolite, and coeval with the Copley Greenstone (Barker and others, 1979; Albers and others, 1981; Lapierre and others, 1985b). REE patterns and Nd isotopes suggest that these volcanoplutonic rocks represent an immature intra-oceanic arc (Brouxel and others, 1988).

SIERRA NEVADA REGIONAL GEOLOGY

The Sierra Nevada of northern California have been divided into seven major lithostratigraphic units (Edelman and others, 1989). These are, from west to east, the Smartville, Slate Creek, Fiddle Creek, Calaveras, Red Ant, Feather River, and Northern Sierra terranes (Fig. 2). The latter five units were juxtaposed along early, low-angle, west-directed, large-displacement faults described by Edelman and others (1989), and all are separated by later, high-angle, small-displacement faults of the Foothills fault system (Clark, 1960). This chapter discusses only the Northern Sierra and Feather River terranes.

Northern Sierra terrane

The Northern Sierra terrane comprises regionally east-dipping Ordovician to Jurassic sedimentary and volcanic rocks. The Middle Ordovician to Devonian Shoo Fly Complex is the oldest unit in the Sierra Nevada, and forms the basement for magmatic arcs that were active in Devonian, mid-Permian, and Jurassic time (Harwood, 1988; Harwood and others, 1988). The Shoo Fly Complex consists of melange and structurally underlying deep-water siliciclastic rocks with minor chert, carbonate, and subaqueous tuff, and is interpreted as a west-vergent accretionary wedge thrust over a passive-margin sequence (D'Allura and others, 1977; Schweickert and others, 1984; Hannah and Moores, 1986; Saleeby and others, 1987; Girty and Guthrie, 1989).

The bulk of the Late Devonian volcanic sequence is represented by the Sierra Buttes and Taylor Formations, which consist principally of silicic to andesitic subaerial(?) and subaqueous vent complexes and flows grading to a distal submarine volcanic

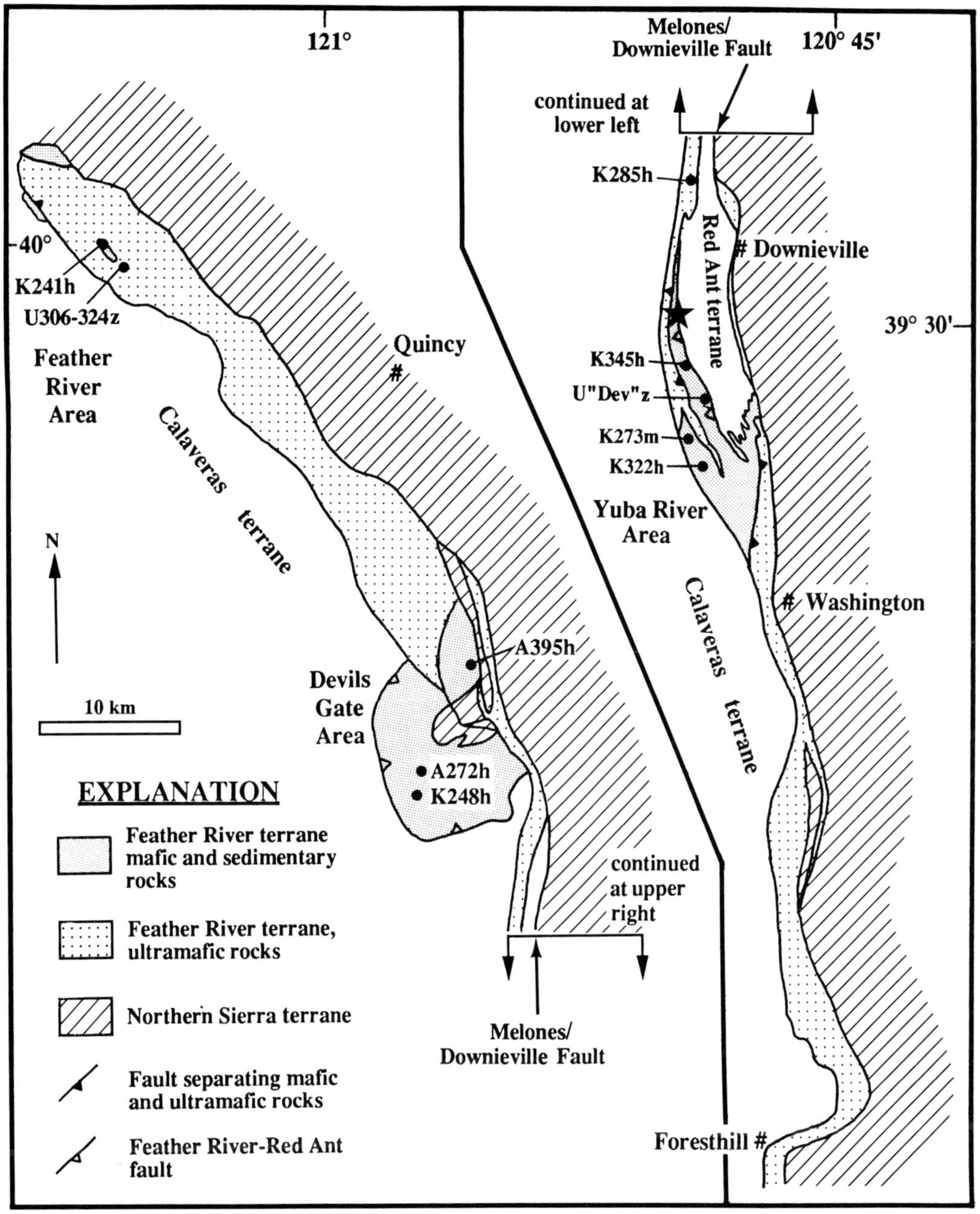

Figure 2. Geologic map of the northern (left) and southern (right) parts of the Feather River terrane and surrounding parts of the Sierra Nevada, modified from Ferguson and Gannett (1932), Lydon and others (1960), Ehrenberg (1975), Standlee (1978), Weisenberg (1979), Hietanen (1981), Edelman and others (1989), B. R. Hacker (unpublished map), and unpublished Chico Quadrangle geologic map of California. The Smartville, Slate Creek, and Fiddle Creek terranes crop out west of the area shown. Abbreviations same as in Figure 1. The locations of the five samples for the Sm/Nd isochron by Saleeby and others (1989) are not shown. Location of samples used for geothermometry are indicated by a star.

apron, also deposited unconformably on the Shoo Fly Complex (Varga and Moores, 1981; Brooks and others, 1982; Hanson and Schweickert, 1986; Hanson and others, 1988).

Several silicic intrusions, including the Middle to Late Devonian Bowman Lake batholith (Table 1, s) and the Middle Devonian Wolf Creek stock (Table 1, t) cut the Shoo Fly Complex and feed volcanic layers in the Sierra Buttes Formation (Hanson and Schweickert, 1986). Studies of trace-element (Brooks and Coles, 1980), REE (Rouer and others, 1989), and Pb isotope (Saleeby and others, 1987) abundances indicate that the Devonian to Mississippian arc comprises calc-alkaline rocks erupted through oceanic rocks.

Feather River terrane

The Feather River terrane is a heterogeneous assemblage of rocks no more than 10 km wide, but stretching for more than 150 km southward from the northern end of the Sierra Nevada. It has previously been interpreted as a partial (Moores, 1970) and complete (Edelman and others, 1989) ophiolite, although it does not represent normal mid-ocean ridge lithosphere. A broad spectrum of ages, from Devonian to Triassic, suggests that the Feather River terrane is an amalgam of rock packages whose interrelations are incompletely understood. The Feather River terrane contains all the requisite ultramafic, mafic, and sedimentary units for an ophiolite. Contacts between rock units include faults and primary igneous contacts. The Feather River terrane overlies the Red Ant and Calaveras terranes along low-angle, west-directed thrust faults, and is in contact with the northern Sierra terrane principally along the high-angle Downieville fault (Edelman and others, 1989). Much of the Feather River terrane remains unexplored, but three regions have been investigated in some detail: the Feather River area (studied by Hietanen, 1973, 1981; Ehrenberg, 1975; Weisenberg, 1979), the Devils Gate area (Standlee, 1978; Hietanen, 1981; Edelman and others, 1989), and the Yuba River area (Ferguson and Gannett, 1932; this study) (Fig. 2).

Ultramafic rocks are dominant in the Feather River terrane, and include chiefly dunite, harzburgite, and lherzolite. Primary phases of the peridotite are replaced by tremolite, chlorite, serpentine, and talc. In the large body of ultramafic rocks in the Feather River area, metamorphic minerals are abundant near faults, whereas alteration of the peridotite is nearly pervasive in the smaller ultramafic bodies in the Devils Gate and Yuba River areas. The metamorphic minerals locally define a northwest-striking foliation that is subparallel to the faults bounding the terrane, and subparallel to the foliation in the mafic rocks. This metamorphic foliation overprints and is discordant to chromite, spinel, and pyroxene layers, which strike ENE to NNE and dip steeply east and west in all three areas (Standlee, 1978; Hietanen, 1981; Edelman and others, 1989; this study).

Bodies of hornblende schist, micaceous and feldspathic quartz schists, and quartzite, ranging in size from square meters to tens of square kilometers, crop out in the Feather River terrane in all three areas; they are most abundant in the Devils Gate and Yuba River areas (Fig. 2). The quartzite occurs in ubiquitous, but volumetrically minor, isolated lenses approximately 1 m in length and 10 to 30 cm thick, and may be metamorphosed chert (Ehrenberg, 1975; Standlee, 1978; this study). The micaceous and feldspathic schists are also minor constituents with protoliths that range from arkosic to calcareous arenite. Garnet-bearing schist was mapped by Ferguson and Gannett (1932) as granite and aplite, but rare, partially unrecrystallized layers with detrital textures indicate that the protolith was sedimentary. Abundant almandine suggests that some sedimentary protoliths were ferruginous.

Hornblende schist is the most common nonultramafic rock. Igneous features such as centimeter-scale compositional layering, dikes, and pillowed flows, as well as pegmatitic, ophitic, and porphyritic textures, are sporadically preserved. Plutonic rocks range from massive to pegmatitic biotite-hornblende tonalite to massive and layered hornblende gabbro, pyroxene gabbro, hornblendite, hornblende clinopyroxenite, and clinopyroxenite. Chlorite pseudomorphs are interpreted to indicate that orthopyroxene too was once present (Hietanen, 1981; this study). P, Ti, Y, and Zr concentrations indicate that the rocks are compositionally tholeiitic basalts (Hietanen, 1981).

Locally, mafic rocks grade into ultramafic rocks (Ferguson and Gannett, 1932; Standlee, 1978; Hietanen, 1981), and many porphyritic diabase dikes cut the mafic and ultramafic rocks (Ehrenberg, 1975; Standlee, 1978; Hietanen, 1981; this study). Some dikes have chilled margins, indicating that intrusion occurred when the host rocks were relatively cool.

Foliation within the mafic and sedimentary rocks strikes NNW, dips steeply east and west, and is strongly disrupted by boudins, folds, and faults. Hornblende lineations plunge at moderate to shallow angles NNW and SSE. The folds are open to isoclinal, with amplitudes and wavelengths of millimeters to meters. Fold hinge surfaces and axes are subparallel to the foliation and lineation, respectively (Ehrenberg, 1975; Weisenberg, 1979; this study).

Feather River area. Most of the Feather River terrane in the Feather River area is altered peridotite, but there are three small exposures of mafic rocks (Fig. 2). The northern mafic body is greenschist-facies layered gabbro that grades into the ultramafic rocks. The western mafic body consists of epidote-amphibolite-facies mafic rocks, pelitic rocks, and chert. It contains a foliation that is parallel to the foliation in the structurally overlying peridotite (Ehrenberg, 1975; Weisenberg, 1979). A several-hundred-meter-wide body of amphibolite-facies mafic rock also crops out within the ultramafic rocks; zircons from interlayered plagiogranite yield a Carboniferous U/Pb age (Table 1, u), and hornblendes yield a Triassic K/Ar age (Table 1, v). The plagiogranite intrudes the ultramafic rocks (Saleeby and others, 1989), and provides a minimum age for the ultramafic rocks.

Devils Gate area. Massive gabbro, sheeted dikes, and pillowed flow rocks crop out in the Devils Gate area in the shape of a dome, with volcanic rocks near the perimeter of the mafic body and gabbro near the center (Edelman and others, 1989). Hornblende from the sheeted dikes yields Permian K/Ar and Ar/Ar ages (Table 1, w and x). Low-angle faults separate the mafic rocks from the underlying Calaveras and Red Ant terranes and from the overlying Northern Sierra terrane. The ultramafic body stretching southward from the Feather River area is separated from the Devils Gate mafic rocks by a fault. A fault also separates the Devils Gate mafic rocks from a sliver of chiefly ultramafic rock that continues south into the Yuba River area. This ultramafic sliver contains mafic and ultramafic cumulate rocks intruded by porphyritic diabase dikes with chilled margins. Hornblende from one dike yielded an Early to Middle Devonian Ar/Ar age (Table 1, y). Dikes, gabbros, and pyroxenite from the Devils Gate area contain Sm and Nd isotopic similarities with the plagiogranite in the Feather River area, and all are ~300 m.y. old (Saleeby and others, 1989).

Yuba River area. The Feather River terrane in the Yuba River area contains dunite, harzburgite, and pyroxenite that are mostly serpentinized, and amphibolite-facies layered gabbro, massive gabbro, and volcanic rocks ("Alleghany amphibolite" of Saleeby and others, 1989). The mafic igneous rocks yielded two Carboniferous K/Ar hornblende ages (Table 1, z), and are intruded by the Devonian Oriental Mine granite, which is pre- or syn-amphibolite-facies metamorphism (Coveney, 1981) (Table 1, aa). A Permian K/Ar age was obtained on white mica in a nearby micaceous schist (Table 1, bb). Hornblende from a small gabbro body yielded a Permian K/Ar age (Table 1, cc). Ultramafic rocks are gradational with the mafic rocks (Ferguson and Gannett, 1932), indicating that the ultramafic rocks are also pre-Devonian.

Age summary. The age data indicate that Carboniferous mafic rocks and pre-Carboniferous ultramafic rocks occur in the Feather River area. Metamorphism of the mafic rocks may have been Triassic. The Devils Gate area contains Carboniferous mafic rocks, probably metamorphosed in Permian time. The sliver of chiefly ultramafic rocks between the Devils Gate and Yuba River areas is pre-Devonian. Mafic and ultramafic rocks in the Yuba River area crystallized and were metamorphosed in Devonian time. Permian metamorphic ages are also recorded in the Yuba River area. The K/Ar ages are of questionable value for interpreting the ages of metamorphism because many of these amphiboles contain core and rim compositions that are radically different, but Ar/Ar and U/Pb ages indicate that crystallization of mafic igneous rocks in the Feather River terrane spanned Devonian to Carboniferous time.

PETROLOGY AND MINERAL CHEMISTRY

Ultramafic rocks

The primary minerals in ultramafic rocks of the Trinity and Feather River terranes are olivine, enstatite, Cr-diopside, and spinel (Table 2); the Trinity terrane also contains plagioclase. The compositions of olivine, enstatite, and Cr-diopside in both terranes are compared in Table 3. There are few differences, and the minerals of both terranes are similar in composition to primary minerals in most Alpine-type ultramafic bodies. Enstatite in the Feather River terrane is more aluminous, and some Cr-diopside in the Trinity terrane contains more chromium.

The anhydrous primary minerals in the ultramafic rocks are replaced by metamorphic serpentine, carbonate, chlorite, diopside, talc, tremolite, andradite, hornblende, and magnetite (Table 4). Alteration is locally extensive, obliterating all primary minerals and textures. Antigorite, lizardite, and chrysotile have been identified in both terranes. Lizardite is widespread, chrysotile occurs dominantly in veins, and antigorite in the Trinity terrane is generally restricted to exposures within 1 to 2 km of the Trinity thrust (Peacock, 1987a). Carbonate minerals are common; calcite, dolomite, and magnesite have all been observed. Chlorite commonly forms coronas around chromite in the Trinity and Feather River terranes. Both terranes contain metamorphic diopside of similar composition (Table 3). In contrast to primary Cr-diopside, which contains 1 to 5 wt% Al_2O_3, metamorphic diopside contains <0.5 wt% Al_2O_3. Talc occurs throughout the Feather River and Trinity terranes between olivine grains, as rims on tremolite and enstatite crystals, and as fine-grained masses and veins in serpentinite. Tremolite is a common minor phase that is particularly abundant near the margins of ultramafic bodies in the Feather River terrane, and near the Trinity thrust in the Trinity terrane. Tremolite crystals are grown on diopside in the Feather River terrane and replaced by serpentine in the Trinity and Feather River terranes. Magnetite is ubiquitous in both terranes, and is concentrated along the centers of serpentine "channels," and as remnants of altered spinel. Andradite occurs along enstatite and Cr-diopside cleavage planes in the Feather River terrane. The minimum variance assemblages in ultramafic rocks are listed in Table 5. Ehrenberg (1975) reported anthophyllite near the North Feather River in the Sierra Nevada, but Weisenberg (1979) specifically stated that anthophyllite does not occur there. Coleman and others (1988) listed anthophyllite as a metamorphic mineral in the Trinity terrane, but the references they cited do not. Anthophyllite was not observed in the Feather River terrane or Trinity terrane in this study.

Mafic rocks

Mafic rocks of the Central Metamorphic Belt and Feather River terrane have been metamorphosed in the greenschist, albite-epidote amphibolite, and amphibolite facies. In the Central Metamorphic Belt, metamorphic grade decreases structurally downward from amphibolite facies adjacent to the Trinity thrust to greenschist facies near the Siskiyou thrust (Peacock and Norris, 1989). Metamorphic grade also varies in the Feather River terrane, but the metamorphic zonation is poorly resolved because of smaller exposures, more pronounced deformation, and later metamorphism. Representative electron microprobe analyses of silicate phases (Table 6) illustrate the variations in mineral chemistry among the different facies.

Amphibolite facies. Amphibolite-facies mafic rocks occur in the Feather River and Yuba River areas of the Feather River terrane. They occur in the Central Metamorphic Belt north of Callahan (Cashman, 1980), and are inferred to have existed in the Trinity Alps area prior to Mesozoic(?) retrogression. The minimum-variance assemblages are listed in Table 5.

Clinopyroxene, interpreted as both igneous and metamorphic in origin, is found in the Feather River and Yuba River areas of the Feather River terrane. Pink salite ($Wo_{53}En_{32}Fs_{16}$ to $Wo_{51}En_{28}Fs_{22}$) with 3.8 to 6.3 wt% Al_2O_3, rich in TiO_2 and Na_2O, and coexisting with ferro-kaersutite or ferroan pargasite is interpreted as igneous. Colorless augite and diopside ($Wo_{44}En_{35}Fs_{21}$ to $Wo_{45}En_{45}Fs_{10}$) contain 0.6 to 3.5 wt% Al_2O_3, and are interpreted as metamorphic (Hietanen, 1981; this study). Clino-

TABLE 2. REPRESENTATIVE ANALYSES OF PRIMARY MINERALS OF THE FEATHER RIVER TERRANE*

	Olivine			Enstatite			Cr-Diopside			Spinel		Diopside		Hornblende	
Sample	5A	85	85	86	86	86	7A	7A	A4	1B	5A	40	6C	40	6C
SiO_2	41.50	41.32	40.69	55.72	55.78	55.06	53.66	51.44	51.98	0.00	0.00	48.19	48.70	39.18	39.82
Al_2O_3	b.d.	b.d.	b.d.	3.78	4.62	4.44	2.37	3.83	4.70	33.86	3.34	6.18	6.26	13.92	14.43
TiO_2	b.d.	0.05	0.03	0.09	0.10	0.15	0.24	0.74	0.52	0.10	0.53	1.67	1.43	5.08	3.04
$FeO^{\dagger}$	9.44	9.03	9.30	5.52	5.42	5.27	3.54	3.01	3.84	17.01	31.90	11.91	9.10	18.25	15.27
Cr_2O_3	b.d.	0.03	0.02	0.57	0.67	0.69	0.46	0.75	0.64	34.44	57.25	b.d.	0.05	b.d.	0.17
MnO	0.17	0.13	0.13	0.11	0.15	0.13	0.13	0.21	0.13	0.22	0.65	0.22	0.15	0.20	0.20
MgO	49.35	50.41	50.16	33.65	33.18	33.00	15.98	15.69	15.83	14.93	3.08	8.92	10.21	6.87	8.83
CaO	b.d.	b.d.	b.d.	0.67	0.66	0.61	23.82	23.82	23.12	b.d.	b.d.	22.52	23.26	11.56	11.93
Na_2O	b.d.	b.d.	b.d.	b.d.	b.d.	b.d.	0.42	0.41	0.38	b.d.	b.d.	1.07	0.80	2.14	2.44
K_2O	b.d.	b.d.	b.d.	b.d.	b.d.	b.d.	b.d.	b.d.	b.d.	b.d.	b.d.	b.d.	b.d.	1.95	1.39
Total	100.46	100.97	100.34	100.11	100.57	99.35	100.62	99.89	101.14	100.59	96.75	100.68	99.95	99.15	97.52
Cations	4 oxygens			6 oxygens			6 oxygens			4 oxygens		6 oxygens		23 oxygens	
Si	1.00	1.00	0.99	1.90	1.91	1.91	1.95	1.89	1.88	b.d.	b.d.	1.82	1.83	5.99	6.06
Al^{IV}	b.d.	b.d.	b.d.	0.08	0.09	0.09	0.05	0.11	0.12	0.00	0.00	0.18	0.17	2.01	1.94
Al^{VI}	n.a.	n.a.	n.a.	0.07	0.09	0.09	0.05	0.05	0.08	1.16	0.15	0.10	0.11	0.49	0.65
Ti	b.d.	0.00	0.00	0.00	0.00	0.00	0.01	0.02	0.01	0.00	0.01	0.05	0.04	0.58	0.35
Cr	b.d.	0.00	0.00	0.02	0.02	0.02	0.01	0.02	0.02	0.79	1.69	b.d.	0.00	b.d.	0.02
Fe	0.19	0.18	0.19	0.16	0.15	0.15	0.11	0.09	0.12	0.41	1.00	0.38	0.28	2.33	1.95
Mn	0.00	0.00	0.00	0.00	0.00	0.00	0.00	0.01	0.00	0.01	0.02	0.01	0.00	0.03	0.03
Mg	1.77	1.82	1.82	1.71	1.69	1.70	0.87	0.86	0.85	0.65	0.17	0.50	0.57	1.56	2.01
Ca	b.d.	b.d.	b.d.	0.03	0.02	0.02	0.93	0.94	0.90	b.d.	b.d.	0.91	0.94	1.89	1.95
Na	b.d.	b.d.	b.d.	b.d.	b.d.	b.d.	0.03	0.03	0.03	b.d.	b.d.	0.08	0.06	0.53	0.67
K	b.d.	b.d.	b.d.	b.d.	b.d.	b.d.	b.d.	b.d.	b.d.	b.d.	b.d.	b.d.	b.d.	0.38	0.27

*See Quick (1981) and Peacock (1987a) for analyses of similar minerals of the Trinity terrane. b.d. = below detection.

TABLE 3. COMPOSITIONS OF MINERALS IN ULTRAMAFIC ROCKS OF THE TRINITY AND FEATHER RIVER TERRANES

	Trinity Terrane	Feather River Terrane
Primary olivine	Fo_{88-93}	Fo_{88-92}
Primary enstatite	$Wo_{01}En_{91}Fs_{08}$ to $Wo_{04}En_{87}Fs_{09}$ <3.5 wt.% Al_2O_3	$Wo_{01}En_{91}Fs_{08}$ to $Wo_{02}En_{90}Fs_{08}$ >3.5 wt.% Al_2O_3
Primary Cr-diopside	$Wo_{41}En_{51}Fs_{05}$ to $Wo_{48}En_{49}Fs_{03}$ ≤1.4 wt.% Cr_2O_3	$Wo_{41}En_{54}Fs_{08}$ to $Wo_{49}En_{48}Fs_{03}$ ≤0.8 wt.% Cr_2O_3
Metamorphic Cr-diopside	$Wo_{46}En_{53}Fs_{01}$ to $Wo_{50}En_{39}Fs_{11}$	$Wo_{46}En_{53}Fs_{01}$ to $Wo_{50}En_{39}Fs_{11}$

*Trinity terrane analyses from Lindsley-Griffin (1977) and Quick (1981); Feather River terrane analyses from Ehrenberg (1975), Hietanen (1973), and this study.

pyroxene occurs only in contact aureoles in the Central Metamorphic Belt.

Ferroan pargasite and ferro-kaersutite in the Feather River and Yuba River areas of the Feather River terrane are interpreted as igneous (Table 2). Metamorphic amphiboles in both terranes are tschermakitic (Table 6). Hornblende in the Central Metamorphic Belt contains >0.75 wt% TiO_2; hornblende in quartz-bearing and quartz-absent rocks of the Feather River terrane contains 0.50 to 2.6 wt% TiO_2, and up to 5.1 wt% TiO_2, respectively.

Plagioclase is oligoclase to andesine, An_{15-36} (Ehrenberg, 1975; this study) in the Feather River terrane, and averages $\sim An_{30}$ in the highest grade Central Metamorphic Belt rocks. Garnet occurs in one mafic sample of the Feather River terrane of the Yuba River area. It ranges from $Alm_{62}Gro_{28}Pyr_{03}Spe_{01}$ to $Alm_{59}Gro_{36}Pyr_{04}Spe_{01}$ (Table 6). Two mafic samples contain garnet ranging from $Alm_{49}Gro_{26}Pyr_{18}Spe_{07}$ to $Alm_{54}Gro_{26}Pyr_{14}Spe_{03}$ in the Central Metamorphic Belt.

Albite-epidote amphibolite facies. The bulk of the Central Metamorphic Belt, small portions of the Feather River terrane in the Feather River area, and most of the Devils Gate area of the Feather River terrane, are composed of albite-epidote

TABLE 4. REPRESENTATIVE ANALYSES OF METAMORPHIC MINERALS OF THE FEATHER RIVER TERRANE*

	Diopside		Tremolite		Andradite		Serpentine		Chlorite	
Sample	5A	7A	5A	85	7A	86	85	85	85	85
SiO_2	54.65	54.27	57.45	57.26	36.56	33.77	43.81	43.39	40.70	33.78
Al_2O_3	0.54	0.03	0.80	0.68	9.53	1.52	0.21	0.25	7.02	13.82
TiO_2	0.11	0.03	0.09	0.08	4.33	0.21	0.05	0.03	0.07	0.06
FeO*	0.61	6.27	1.58	1.60	8.26	24.87	1.24	1.37	2.14	2.97
Cr_2O_3	0.25	0.04	0.52	0.39	3.69	0.51	b.d.	0.06	1.01	2.40
MnO	0.15	1.13	0.06	0.06	0.08	0.08	0.02	0.08	b.d.	b.d.
MgO	19.99	14.16	24.07	24.28	1.32	0.12	40.62	40.53	35.25	34.98
CaO	23.86	25.41	12.53	12.38	33.62	34.04	0.30	0.11	2.15	b.d.
Na_2O	b.d.	b.d.	0.29	0.26	b.d.	b.d.	b.d.	b.d.	b.d.	b.d.
K_2O	b.d.	b.d.	b.d.	b.d.	b.d.	b.d.	b.d.	b.d.	b.d.	b.d.
Total	100.16	101.34	97.39	96.99	97.39	95.13	86.25	85.82	88.34	88.01
Cations	6 oxygens		23 oxygens		12 oxygens		14 oxygens		14 oxygens	
Si	1.97	2.00	7.87	7.88	2.96	2.88	4.13	4.11	3.82	3.17
Al^{IV}	0.03	0.00	0.13	0.11	0.04	0.12	0.00	0.00	0.18	0.83
Al^{VI}	0.00	0.00	0.00	0.00	0.86	0.04	0.02	0.03	0.60	0.69
Ti	0.00	0.00	0.01	0.01	0.24	0.01	0.00	0.00	0.01	0.00
Cr	0.01	0.00	0.02	0.01	0.24	0.03	b.d.	0.00	0.08	0.18
Fe	0.02	0.19	0.18	0.19	0.56	1.94	0.10	0.11	0.17	0.23
Mn	0.00	0.04	0.01	0.01	0.01	0.01	0.00	0.01	b.d.	b.d.
Mg	1.07	0.78	4.91	4.98	0.16	0.02	5.71	5.73	4.93	4.89
Ca	0.92	1.00	1.84	1.83	2.91	3.12	0.03	0.01	0.22	0.00
Na	b.d.	b.d.	0.08	0.07	b.d.	b.d.	b.d.	b.d.	b.d.	b.d.
K	b.d.	b.d.	b.d.	b.d.	b.d.	b.d.	b.d.	b.d.	b.d.	b.d.

*See Peacock (1987a) for analyses of similar minerals from the Trinity terrane. b.d. = below detection.

TABLE 5. METAMORPHIC ASSEMBLAGES

Ultramafic Rocks
(1) Ol + Tr + Atg + Chl + Mt (Peacock, 1987a; this study)
(2) Tc + Tr + Srp + Mt (Lipman, 1964; Hietanen, 1973; Ehrenberg, 1975; this study)
(3) Di + Tr + Srp + Mt (Ehrenberg, 1975; this study)

Assemblages (1) through (3) occur in the Feather River terrane; (1) and (2) have been reported from the Trinity terrane.

Rodingites
(1) And + Di + Chl + Ido (Hietanen, 1981; this study)
(2) And + Di + Ido + Pg-Hb (this study)
(3) Gro + Ido + Zo + Tr + Cal (Hietanen, 1973)

Rodingites have been recognized only in the Feather River terrane.

Mafic Rocks
(1) Di + Hb + Olg + Bio + Ilm + Qtz (this study)
(2) Hb + Olg + Bio + Ilm + Qtz (Hietanen, 1981; Edelman and others, 1989)
(3) Sal + Hb + Ab + Gar + Spn + Qtz + Rut (Ehrenberg, 1975)
(4) Di + Hb + Olg + Ep + Ep + Chl (Holdaway, 1965; Ehrenberg, 1975)
(5) Hb + Olg + Ep + Spn + Qtz (Davis and others, 1965; Holdaway, 1965; Hietanen, 1981; Edelman and others, 1989; this study)
(6) Act + Ab + Ep + Qtz + Chl + Spn (Hotz, 1977; Standlee, 1978; Peacock, 1985; Edelman and others, 1989)
(7) Hb + Ab + Ep + Qtz + Chl + Spn (Hotz, 1977; Standlee, 1978; Cashman, 1980; Hietanen, 1981; Peacock, 1985; Edelman and others, 1989; this study)
(8) Hb + Olg + Ep + Chl + Qtz + Bio + Gar (Holdaway, 1965)
(9) Hb + Ab + Ep + Chl + Gar + Spn + Qtz (Holdaway, 1965)

Assemblages (5) through (9) are found in the Central Metamorphic Belt, and (1) through (7) in the Feather River terrane.

Metasedimentary Rocks
(1) Qtz + Olg + Gar + Mus or Bio + Chl (Ehrenberg, 1975; this study)
(2) Qtz + Ab + Gar + Tr + Ep + Cal + Spn + Chl (Peacock, 1987)
(3) Qtz + Ab + Mus + Bio + Gar + Ido + Cal + Spn (Davis and others, 1965)

Assemblage (1) occurs in the Feather River terrane, and assemblages (2) and (3) are found in the Central Metamorphic Belt.

Notes: Ab = albite; Act = actinolite; And = andradite; Atg = antigorite; Bio = biotite; Cal = calcite; Chl = chlorite; Di = diopside; Ep = epidote; Gar = garnet; Gro = grossularite; Hb = hornblende; Ido = idocrase; Ilm = ilmenite; Mt = magnetite; Mus = muscovite; Ol = olivine; Olg = oligoclase; Pg-hb = pargasitic hornblende; Qtz = quartz; Rut = rutile; Sal = salite; Spn = sphene; Srp = serpentine; Tr = tremolite; Tc = talc; Zo = zoisite.

amphibolite-facies rocks (Table 5). Hornblende typically contains 0.25 to 1.0 wt% TiO_2. Epidote is commonly zoned, with $Fe^{3+}/(Al+Fe^{3+})$ increasing from core to rim in most samples, although reverse zoning is also observed. Albite is typically An_{01-05} in the Central Metamorphic Belt and An_{01-11} in the Feather River terrane (Ehrenberg, 1975; this study). Several samples in the Central Metamorphic Belt contain peristerite pairs (An_{09-12} and An_{20-30}), suggesting metamorphic conditions transitional to amphibolite facies. Chlorite, interpreted as a stable prograde phase in some samples, occurs as elongate pods oriented parallel to the foliation.

Greenschist facies. Greenschist-facies mafic rocks occur around the border zone of the Devils Gate mafic body in the Feather River terrane and in the Central Metamorphic Belt near the basal Siskiyou thrust. Local green biotite indicates upper greenschist facies (Table 5). Hornblende rims on actinolite grains suggest that at least some of the Central Metamorphic Belt greenschists were not formed by retrogression, but rather represent the highest metamorphic grade attained.

Rodingites. Gabbroic dikes and inclusions within the ultramafic rocks in the Feather River terrane are commonly altered to rodingites (Table 5). When strongly altered, the entire rock may be replaced by magnesite and talc, but more typically, the rodingites are composed of andradite alterations of plagioclase, and idocrase + diopside + chlorite intergrowths replacing pyroxene. The rodingite and host ultramafic rock are usually separated by a centimeter-scale zone of intergrown Mg-Fe-, Fe-Mg-chlorite, and magnetite. Rodingites have not been recognized in the Central Metamorphic Belt.

Later metamorphism. All the Feather River terrane rocks in the Yuba River area were metamorphosed under pumpellyite-actinolite facies conditions during Jurassic(?) time (Hacker, 1988). Many of the minerals diagnostic of this metamorphic event have been recognized throughout the Feather River terrane and Central Metamorphic Belt, but the grade of metamorphism has always been termed greenschist facies (Davis and others, 1965; Ehrenberg, 1975; Standlee, 1978; Cashman, 1980; Hietanen, 1981; Peacock and Norris, 1989). Pumpellyite was only recognized in the Yuba River area by back-scattered electron microscopy, and it may occur in the Central Metamorphic Belt and in other parts of the Feather River terrane.

Sedimentary rocks

Mineral assemblages in quartz-bearing sedimentary rocks in the Central Metamorphic Belt and Feather River terrane record garnet-zone metamorphic conditions (Table 5). Amphibole in calcareous units is actinolite or hornblende, perhaps reflecting the original Al_2O_3 content of the protolith. Garnet compositions range from $Alm_{68}Pyr_{28}Gro_{03}Spe_{01}$ to $Alm_{58}Pyr_{35}Gro_{06}Spe_{01}$ in the Feather River terrane (Table 5) and $Alm_{66}Pyr_{04}Gro_{27}Spe_{02}$ to $Alm_{50}Pyr_{10}Gro_{05}Spe_{35}$ in the Central Metamorphic Belt. Muscovite in the Feather River terrane and Central Metamorphic Belt contains 10 percent celadonite component.

TABLE 6. REPRESENTATIVE ANALYSES OF METAMORPHIC MINERALS IN MAFIC ROCKS OF THE FEATHER RIVER TERRANE*

	Hornblende					Garnet	Pyroxene			
Sample	26	105	27	6E	126	126	26	105	27	6E
SiO_2	48.47	49.58	43.22	45.49	41.75	38.27	52.27	53.84	52.41	53.34
Al_2O_3	6.96	7.95	11.48	10.68	15.32	21.61	1.23	0.74	1.45	1.27
TiO_2	1.12	1.21	1.88	2.29	1.16	0.02	0.30	0.10	0.18	0.20
FeO*	13.56	9.41	16.50	9.19	16.59	27.47	11.02	7.33	10.63	5.65
Cr_2O_3	0.04	0.22	0.12	0.15	b.d.	b.d.	b.d.	0.06	0.05	0.10
MnO	0.27	0.17	0.25	0.15	0.13	1.78	0.36	0.27	0.35	0.17
MgO	13.60	16.04	10.30	14.69	8.77	3.64	13.29	14.93	12.20	15.40
CaO	11.87	11.84	11.31	12.19	11.25	7.60	21.87	22.87	22.52	23.65
Na_2O	1.15	0.98	1.98	1.50	1.89	b.d.	0.30	0.20	0.35	0.26
K_2O	0.46	0.57	0.33	1.04	0.76	b.d.	b.d.	b.d.	b.d.	b.d.
Total	97.50	97.97	97.37	97.37	97.62	100.39	100.64	100.34	100.14	100.04
Cations	23 oxygens					12 oxygens	6 oxygens			
Si	7.05	7.01	6.41	6.60	6.19	3.01	1.96	1.99	1.97	1.97
Al^{IV}	0.95	0.99	1.59	1.40	1.81	0.00	0.04	0.01	0.03	0.03
Al^{VI}	0.24	0.33	0.42	0.42	0.86	2.00	0.01	0.02	0.04	0.02
Ti	0.12	0.13	0.21	0.25	0.13	0.00	0.01	0.00	0.01	0.01
Cr	0.00	0.02	0.01	0.02	b.d.	b.d.	b.d.	0.00	0.00	0.00
Fe	1.65	1.12	2.05	1.12	2.06	1.81	0.35	0.27	0.33	0.17
Mn	0.03	0.02	0.03	0.02	0.02	0.12	0.01	0.01	0.01	0.01
Mg	2.95	3.38	2.28	3.18	1.94	0.43	0.74	0.82	0.68	0.85
Ca	1.85	1.79	1.80	1.89	1.79	0.64	0.88	0.90	0.91	0.93
Na	0.32	0.27	0.57	0.43	0.54	b.d.	0.02	0.01	0.03	0.02
K	0.09	0.10	0.06	0.19	0.14	b.d.	b.d.	b.d.	b.d.	b.d.

*See Peacock and Norris (1989) for analyses of similar minerals from the Central Metamorphic belt. b.d. = below detection.

METAMORPHIC HISTORY

Ultramafic rocks

As a first approximation, metamorphism of ultramafic rocks can be discussed in terms of the system MgO-SiO_2-H_2O. An internally consistent set of *P-T* reaction positions for the MgO-SiO_2-H_2O system generated from the thermodynamic data of Robinson and others (1982) was presented by Peacock (1987a). Antigorite is stable to temperatures as high as 570°C for P_{H_2O} = 500 MPa. The lack of antigorite + brucite constrains the temperature to >425°C. Coexisting talc and tremolite indicate temperatures of >475°C at 500 MPa. Actual temperatures were probably slightly lower because of dilution of the fluid by CO_2 (Trommsdorff and Evans, 1977).

Four lines of evidence strongly suggest that the sources of fluid responsible for the antigorite serpentinization were dehydration reactions in the underlying Central Metamorphic Belt during thrusting: (1) the spatial restriction of antigorite to within 1 to 2 km of the Trinity thrust; (2) the presence of talc and tremolite formed by metasomatism at the Trinity thrust; (3) metamorphic δD values indicate that nonmeteoric water was involved; and (4) the foliation in the antigorite schist, the Trinity thrust, and the foliation in the Central Metamorphic Belt are all subparallel (Peacock, 1987a).

Mafic rocks

Metamorphic temperatures and pressures in the Central Metamorphic Belt and Feather River terrane can be estimated from the observed mineral assemblages, mineral chemistry, and limited thermometry of the mafic rocks. Relevant experimentally determined reactions in basaltic systems were presented by Peacock and Norris (1989). At P = 500 MPa and f_{O_2} defined by the QFM buffer, the amphibolite facies occurs at T >650°C, the albite-epidote amphibolite facies at ~500° to 650°C, and the greenschist facies at T <550°C. The presence of hornblende + calcic plagioclase in the mafic rocks indicates that metamorphic conditions locally reached the lower amphibolite facies. Locally developed hornblende + clinopyroxene assemblages in the Feather River terrane indicate peak metamorphic temperatures in the upper amphibolite facies. Maximum metamorphic temperatures close to the Trinity thrust were probably in the range of 600° to 700°C, whereas upper greenschist facies assemblages at

TABLE 7. THERMOMETRY FOR THE FEATHER RIVER TERRANE

Graham and Powell (1984)

Garnet				Hornblende		T(°C)
X(Fe)	X(Mg)	X(Ca)	X(Mn)	Fe	Mg	
1.9	0.42	0.55	0.12	2.02	1.94	621

Hodges and Spear (1982)

Garnet				Biotite						T(°C)
X(Fe)	X(Mg)	X(Ca)	X(Mn)	Al^{VI}	Ti	Cr	Fe	Mn	Mg	
1.9	0.40	0.54	0.15	0.54	0.08	0.00	1.09	0.01	1.29	648

Garnet-biotite temperature calculated for 500 MPa pressure.

the base of the Central Metamorphic Belt record peak metamorphic temperatures of 450° to 550°C.

Only two samples each in the Feather River terrane and Central Metamorphic Belt were suitable for quantitative thermometry (see Fig. 2 for sample locations). Propagation of a ±2 percent uncertainty in the microprobe analyses through the thermometer equations results in an analytical uncertainty of approximately ±25°C. Calibration uncertainties are much larger, perhaps in excess of ±100°C for some geothermometers (Hodges and McKenna, 1987). Graham and Powell (1984) calibrated Fe/Mg partitioning between garnet and hornblende against the garnet-clinopyroxene geothermometer of Ellis and Green (1979). The Central Metamorphic Belt sample yields a metamorphic temperature of 640°C (Peacock and Norris, 1989), and the Feather River terrne sample yields 621°C (Table 7). Fe-Mg exchange between garnet and biotite is also temperature sensitive and has been calibrated as a geothermometer. For the Hodges and Spear (1982) calibration, garnet-biotite rim compositions yield average temperatures of 695° and 648°C, respectively, for the Central Metamorphic Belt (Peacock and Norris, 1989) and Feather River terrane samples (Table 7).

Mineral assemblages constrain metamorphic pressures to lie between approximately 200 and 900 MPa. At pressures less than ~200 MPa, albite + hornblende is unstable with respect to oligoclase + actinolite (e.g., Grapes and Graham, 1978; Liou and others, 1985). At pressures greater than ~900 MPa, mafic rocks should contain sodic amphibole (Liou and others, 1985), which is not observed in the Central Metamorphic Belt or Feather River terrane.

Amphibole chemistry can be a useful indicator of metamorphic pressure (e.g., Laird and Albee, 1981). In a comprehensive study of mafic schists from different metamorphic terranes, Laird and Albee (1981) demonstrated that the edenite and tschermakite contents of amphibole increase with increasing metamorphic grade in mafic rocks that contain a "common" mineral assemblage. Laird and Albee (1981) and Hynes (1982) presented discriminant diagrams that distinguish among amphiboles from high- and medium- to low-pressure facies series metamorphism. Amphibole compositions from the Central Metamorphic Belt and the Feather River terrane plot consistently within the moderate- to low-pressure facies series field (Fig. 3).

The rodingite assemblage andradite + diopside + chlorite + idocrase found in the Feather River terrane could have formed at temperatures of 300° to 650°C at 500 MPa in equilibrium with an H_2O-rich fluid (X_{CO_2}<0.03) (Rice, 1983).

Sedimentary rocks

Spear and Cheney (1989) calculated the maximum thermal stability of garnet + biotite + chlorite + quartz + muscovite in the system KFMASH as ~575°C at 500 MPa. Rocks with this assemblage occur in the Feather River terrane, but none was suitable for thermobarometry because of retrogression. Calc-schists in the Central Metamorphic Belt containing tremolite + calcite + quartz formed at temperatures about 550° to 600°C (Metz and Trommsdorff, 1968) and below ~650°C (Turner, 1981, p. 164) at 500 MPa, those containing diopside + quartz + tremolite formed above 600° to 650°C (Turner, 1981, p. 164). Diopside was found in one limestone block in the Feather River terrane (Ferguson and Gannett, 1932), suggesting temperatures of >600°C at pressures of 500 MPa if the diopside formed by the breakdown of quartz and dolomite (Turner, 1981, p. 164).

Summary of metamorphic P-T conditions

Based on the observed mineral assemblages, mineral chemistry, and limited thermometry, peak metamorphic temperatures in the Central Metamorphic Belt decrease from 650 ± 50°C at the Trinity thrust to 500 ± 50°C at the base of the Central Metamorphic Belt. Intercalated sedimentary rocks indicate peak metamorphic temperatures of 600 ± 50°C. Concomitant metamorphic pressures are loosely constrained as 500 ±0 MPa. Peak metamorphic temperatures and pressures for mafic rocks in the Feather River terrane were roughly 600° to 650 ± 50°C and 500 ± 300 MPa, and intercalated sedimentary rocks were metamorphosed at temperatures of 550° to 600°C. Mineral assemblages indicate that metamorphism in ultramafic rocks of the Feather River terrane and the Trinity terrane occurred at peak temperatures of 525 ± 50°C.

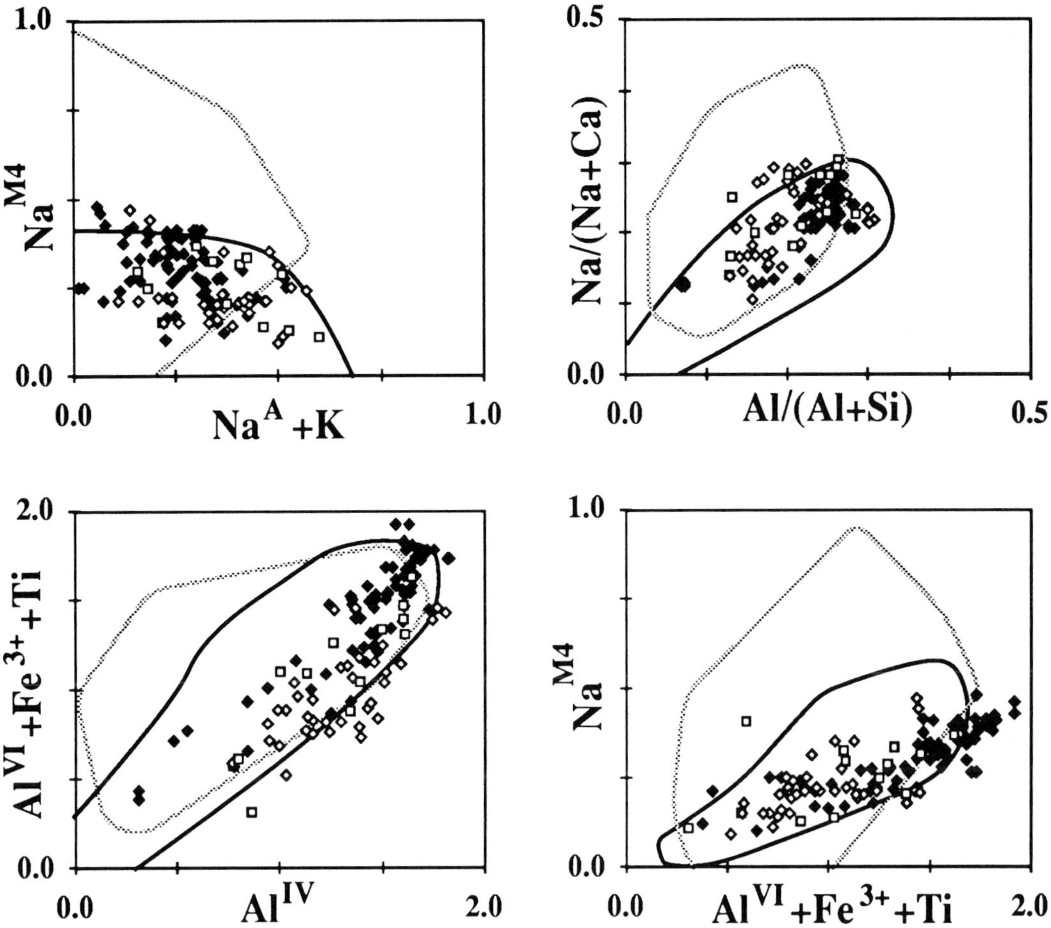

Figure 3. Amphibole analyses from the Central Metamorphic Belt (filled diamonds) and the Feather River (open squares) and Yuba River (unfilled diamonds) areas of the Feather River terrane. The field for high-pressure facies series amphiboles (pale line) is after Laird and Albee (1981, p. 140–141), and the medium- to low-pressure field (heavy line) is after Laird and Albee (1981) and Hynes (1982). The "medium-pressure" Haast schists compositions of Cooper and Lovering (1970) plotted by Laird and Albee (1981) are not shown in the figure because recent study of the Haast schists (Yardley, 1982) reveals rather different amphibole compositions than those of Cooper and Lovering, and reveals that the schists were metamorphosed under high-pressure conditions prior to medium-pressure metamorphism. Amphibole microprobe analyses were recalculated using stoichiometric constraints of Laird and Albee (1981), which generally result in maximum permissible Fe^{3+}.

DISCUSSION

Comparison of the Sierra Nevada and Klamath Mountains

The first attempt at correlations between the Sierra Nevada and Klamath Mountains was by Davis (1969), who correlated the Trinity thrust with the Melones (= Downieville) Fault in the northern Sierra Nevada. More recent correlations by Schweickert and Snyder (1981) and Sharp (1988) are discussed below. Based on the descriptions of lithologies, structures, and metamorphic parageneses in this chapter, we consider four potential correlations between the Trinity terrane and Central Metamorphic Belt of the Klamath Mountains with the Feather River terrane of the Sierra Nevada. References for data restated in this section have been cited earlier.

Feather River terrane ultramafic rocks = Trinity terrane ultramafic rocks. Ophiolitic rocks in the Klamath Mountains and Sierra Nevada have undergone multiple episodes of magmatism and metamorphism, some of which were coincident in both mountain belts. The ultramafic rocks in the Trinity terrane are Ordovician, and the dated ultramafic rocks in the Feather River terrane are pre-Carboniferous or pre-Devonian. Thus, the Trinity and Feather River terranes both contain ultramafic rocks that may be correlative. The ultramafic rocks in both mountain belts contain similar protoliths, although plagioclase lherzolite has not been recognized in the Feather River terrane. Perhaps the most compelling link is that both the Trinity and Feather River terrane ultramafic rocks were metamorphosed at ~525°C in Devonian time, and then cooled through greenschist-facies conditions. High ϵ_{Nd} initial values reported for the Trinity terrane (Jacobsen and others, 1984) have been measured in the Kings River ophiolite of the southern Sierra Nevada (Shaw and others, 1987); peridotite in the Sierra Nevada have not yet been analyzed. Metamorphism of the Trinity terrane occurred coincident with metamorphism of the Central Metamorphic Belt and movement along the Trinity thrust. Perhaps the metamorphism in the Feather River terrane occurred by similar means.

Feather River terrane mafic rocks = Trinity terrane mafic rocks. It is probable that some of the mafic igneous rocks in the Trinity and Feather River terranes are correlative. Mafic rocks of the Trinity terrane are Ordovician gabbros, dikes, and basalts, and Silurian gabbros and diorites. Some mafic rocks in the Sierra Nevada preserve igneous textures and minerals, and are gradational with the ultramafic rocks. These include the northernmost exposure of pre-Carboniferous mafic rocks in the Feather River area, the sliver of pre-Devonian ultramafic and mafic rock between the Devils Gate and Yuba River areas, and pre-Devonian mafic rocks in the Yuba River area. Pre-Devonian mafic rocks in the Kings River ophiolite of the southern Sierra Nevada are also potentially correlative with the Trinity terrane (Shaw and others, 1987).

Feather River terrane mafic and sedimentary rocks = Central Metamorphic Belt mafic and sedimentary rocks. Schweickert and Snyder (1981) suggested that the Central Metamorphic Belt is correlative with portions of the Shoo Fly Complex. Sharp (1988) suggested correlation of the Central Metamorphic Belt with mafic rocks in the Feather River terrane. Our work suggests that the Central Metamorphic Belt is correlative with mafic rocks in the Feather River terrane, and not with part of the Shoo Fly Complex. Both the Central Metamorphic Belt and the Feather River terrane in the Yuba River area were metamorphosed at greenschist to amphibolite-facies conditions in Devonian time, contain similar protolith lithologies and textures, underlie serpentinized ultramafic rocks with subparallel foliations, and overlie lower Mesozoic blueschist-facies rocks (Hacker and Goodge, this volume). The correlation would be strengthened if a systematic inverted metamorphic gradient could be demonstrated in the Feather River terrane. The most remarkable difference between the Central Metamorphic Belt and the Feather River terrane mafic rocks is that igneous activity continued during Carboniferous and Permian time in the Feather River terrane, but ceased by Devonian time in the Trinity terrane. Perhaps the late Paleozoic North Fork terrane ophiolite (Ando and others, 1983) shares affinities with the Carboniferous and Permian rocks in the Sierra Nevada.

Melones Fault = Trinity thrust. Davis's (1969) correlation of the Trinity thrust with the Melones Fault requires modification in light of our studies. The Trinity thrust separates the ultramafic rocks of the Trinity terrane from the dominantly mafic rocks of the underlying Central Metamorphic Belt. In contrast, the Melones Fault separates the melange and sedimentary allochthons of the Shoo Fly Complex from the underlying mafic/ultramafic rocks of the Feather River terrane. The more likely equivalent of the Trinity thrust in the Sierra Nevada is the unnamed contact between the ultramafic and mafic rocks within the Feather River terrane exposed in the Yuba River area.

Implications for the early to middle Paleozoic evolution of the Cordillera

These possible correlations allow more accurate reconstruction of the early to middle Paleozoic evolution of the Klamath Mountains and Sierra Nevada. In Late Proterozoic (~850 Ma; Dickinson, 1977) through Cambrian time (~650 Ma; Stewart and Suczek, 1977), rifting produced a passive margin along the western edge of the North America craton. From then until Devonian time, central Nevada was the site of an east-to-west transition from continental shelf and slope deposition adjacent to the craton, to eugeosynclinal deposition farther west (Stewart, 1980). The mafic section of the Trinity terrane formed during Ordovician time (~450 Ma).

Ages near the Ordovician/Silurian boundary (~440 Ma) for blueschist and greenschist-facies rocks in the Yreka terrane and magmatic arc plutons intruded into the Trinity terrane indicate subduction at that time. These features may be related to one intraoceanic subduction zone, but whether the subduction zone dipped east or west is not known. More plutons intruded the Trinity terrane in Late Silurian time (~415 Ma), suggesting continued subduction. Perhaps the tectonic setting was a Marianas-type intraoceanic arc, with active volcanoes built on multiply intruded basement separated from one or more remnant arcs by a back-arc basin. Inception of the Feather River terrane may have occurred in a similar environment. At this time, sediments in the Yreka terrane and the Shoo Fly Complex were derived from Precambrian cratonic and early Paleozoic volcanoplutonic sources (Girty and Wardlaw, 1984, 1985; Wallin, 1989). Sediments derived from similar sources were being deposited in central Nevada (Gilluly and Gates, 1965; Stewart, 1980; Girty and Wardlaw, 1984, 1985).

Plutonic and volcanic rocks were erupted in the eastern Klamath terrane (and possibly in the Trinity and Feather River terranes) in Early Devonian time (~400 Ma), suggesting continued arc magmatism. During Middle Devonian time (~380 Ma) an inverted metamorphic gradient developed in the Central Metamorphic Belt, coincident with serpentinization of the Trinity terrane, and portions of the Yreka terrane were metamorphosed. A critical constraint on the tectonism responsible for these events is that the rocks thrust over the Central Metamorphic Belt must have been at temperatures >650°C at a depth of 15 km (Peacock, 1987b). The Trinity terrane was probably cooler than this, however, because it was ~90 m.y. old at the time, and only oceanic lithosphere younger than 8 m.y. old is hotter than 650°C at 15 km (Parsons and Sclater, 1971). Three possible ways that the Central Metamorphic Belt could have been thrust beneath hot material associated with the Trinity terrane are: (1) the Central Metamorphic Belt was thrust beneath a portion of the Trinity terrane heated by arc magmatism (Fig. 4A); (2) the Central Metamorphic Belt was initially thrust beneath young, hot,

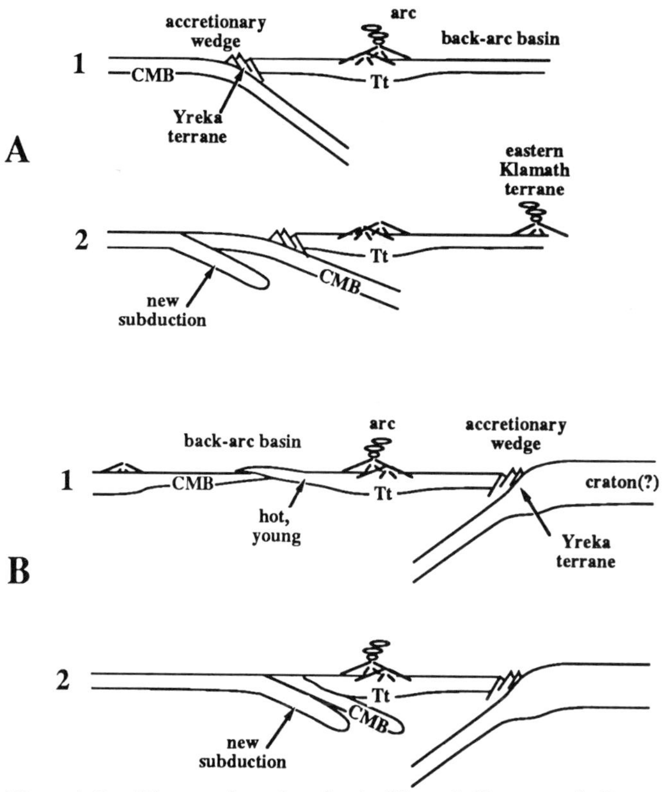

Figure 4. Possible tectonic settings for the Klamath-Sierran arc in Devonian time. This figure is labeled with rock units of the Klamath Mountains, but may apply to the Sierra Nevada as well. A, The Central Metamorphic Belt is thrust beneath a portion of the Trinity terrane that has been heated by arc magmatism. Perhaps the angle of subduction shallowed. B, Back-arc closure thrusts Central Metamorphic Belt rocks beneath hot, young back-arc lithosphere. This is similar to the situation envisioned by Boudier and others (1988) for the Oman ophiolite. This thrusting may have begun because continental material such as the Yreka terrane entered the old subduction zone. Metamorphism of the Central Metamorphic Belt occurred while the Central Metamorphic Belt was subducted beneath young back-arc lithosphere that has since been eroded. Further closure thrust the Trinity terrane over the Central Metamorphic Belt, causing serpentinization of the Trinity terrane. Isostatic uplift of the partially subducted craton may have allowed emplacement of the continentally derived Yreka terrane onto the Trinity terrane by low-angle normal faulting—analogous to the emplacement of the continentally derived Batinah allochthons onto the Oman ophiolite (Woodcock and Robertson, 1982). CMB = Central Metamorphic Belt; Tt = Trinity terrane.

back-arc lithosphere and only later beneath the Trinity terrane (Fig. 4B); or (3) the Central Metamorphic Belt was thrust beneath the Trinity terrane in Silurian time, and the Devonian metamorphic ages reflect later cooling. In case 2, thrusting may have begun along a spreading center between two young oceanic back-arc plates. Refrigeration of the Central Metamorphic Belt necessary for preservation of its inverted metamorphic gradient was accomplished by subduction of more material beneath the Central Metamorphic Belt. All three possibilities require that more than one subduction zone was active.

CONCLUSIONS

Detailed optical microscopy, back-scattered electron microscopy, and electron microprobe studies reveal that the Central Metamorphic Belt and Trinity terrane in the Klamath Mountains and the Feather River terrane of the Sierra Nevada contain similar protoliths and metamorphic parageneses. These data, combined with similar ages of metamorphism and relations with surrounding units, permit potential correlations between the Sierra Nevada and Klamath Mountains. The Trinity terrane is a multiply intruded intraoceanic arc composed of Ordovician to Devonian(?) ultramafic and mafic rocks. The Feather River terrane is also a polygenetic ophiolite with pre-Devonian to pre-Carboniferous ultramafic and pre-Devonian to Triassic mafic rocks. Both were metamorphosed during Devonian time at peak temperatures of 475° to 570°C. The Central Metamorphic Belt is a slice of oceanic crust accreted to the base of the Trinity terrane; the Feather River terrane also contains oceanic crust faulted against ultramafic rocks. Mafic and sedimentary rocks in both mountain belts were metamorphosed at temperatures of 500° to 650 ± 50°C and pressures of 500 ± 300 MPa. An inverted metamorphic gradient formed in the Central Metamorphic Belt during subduction beneath the Trinity terrane. Our correlation of the Central Metamorphic Belt and Trinity terrane of the Klamath Mountains and the Feather River terrane of the Sierra Nevada emphasizes large-scale similarities, and is intended to stimulte investigations of smaller scale differences that will ultimately provide a more detailed description of the early to middle Paleozoic evolution of the Klamath Mountains and Sierra Nevada.

ACKNOWLEDGMENTS

We thank H. W. Day, D. S. Harwood, and J. B. Saleeby for thoughtful reviews, W. G. Ernst for financial support of the Klamath Mountains studies (through National Science Foundation Grant EAR 80-17295 and 82-116257), and R. Jones for his help with the electron microprobe. Acknowledgment is made to the Donors of The Petroleum Research Fund, administered by the American Chemical Society, for the partial support of this research. Preparation of this chapter was stimulated by the Penrose Conference organized by Dave Harwood and Meghan Miller.

REFERENCES CITED

Albers, J. P., and Bain, J.H.C., 1985, Regional setting and new information on some critical geologic features of the West Shasta District, California: Economic Geology, v. 80, p. 2072–2091.

Albers, J. P., Kistler, R. W., and Kwak, L., 1981, The Mule Mountain stock; An early Middle Devonian pluton in northern California: Isochron/West, v. 31, p. 17.

Ando, C. J., Irwin, W. P., Jones, D. L., and Saleeby, J. B., 1983, The ophiolitic North Fork terrane in the Salmon River region, central Klamath Mountains, California: Geological Society of America Bulletin, v. 94, p. 236–252.

Barker, F., Millard, H. T., Jr., and Knight, R. J., 1979, Reconnaissance geochemistry of Devonian island-arc volcanic and intrusive rocks, west Shasta district, California, *in* Barker, F., Trondhjemites, dacites, and related rocks: Developments in Petrology, v. 6, p. 531–545.

Bohlke, J. K., and McKee, E. H., 1984, K-Ar ages relating to metamorphism, plutonism, and gold-quartz vein mineralization near Alleghany, Sierra County, California: Isochron/West, v. 39, p. 3–7.

Boudier, F., Ceuleneer, G., and Nicolas, A., 1988, Shear zones, thrusts, and related magmatism in the Oman ophiolite; Initiation of thrusting on an oceanic ridge: Tectonophysics, v. 151, p. 275–296.

Boudier, F., Le Sueur, E., and Nicolas, A., 1989, Structure of an atypical ophiolite; The Trinity complex, eastern Klamath Mountains, California: Geological Society of America Bulletin, v. 101, p. 820–833.

Brooks, E. R., and Coles, D. G., 1980, Use of immobile trace elements to determine original tectonic setting of eruption of metabasalts, northern Sierra Nevada, California: Geological Society of America Bulletin, v. 91, p. 665–671.

Brooks, E. R., Wood, M. M., and Garbutt, P. L., 1982, Origin and metamorphism of peperite and associated rocks in the Devonian Elwell Formation, northern Sierra Nevada, California: Geological Society of America Bulletin, v. 93, p. 1208–1231.

Brouxel, M., and Lapierre, H., 1988, Geochemical study of an early Paleozoic island arc-back arc system; Part 1, The Trinity Ophiolite (northern California): Geological Society of America Bulletin, v. 100, p. 1111–1119.

Brouxel, M., Lapierre, H., Michard, A., and Albarede, F., 1988, Geochemical study of an early Paleozoic island arc-back arc system; Part 2, Eastern Klamath Mountains, early to middle Paleozoic island-arc volcanic rocks (northern California): Geological Society of America Bulletin, v. 100, p. 1120–1130.

Cannat, M., and Boudier, F., 1985, Structural study of intra-oceanic thrusting in the Klamath Mountains, northern California; Implications on accretion geometry: Tectonics, v. 4, p. 435–452.

Cashman, S. M., 1980, Devonian metamorphic event in the northeastern Klamath Mountains, California: Geological Society of America Bulletin, v. 91, p. 453–459.

Clark, L. D., 1960, Foothills fault system, western Sierra Nevada, California: Geological Society of America Bulletin, v. 71, p. 483–496.

Coleman, R. G., Manning, C. E., Donato, M. M., Mortimer, N., and Hill, L. B., 1988, Tectonic and regional metamorphic framework of the Klamath Mountains and adjacent Coast Ranges, California and Oregon, *in* Ernst, W. G. , ed., Metamorphism and crustal evolution of the western United States: Englewood Cliffs, New Jersey, Prentice-Hall, p. 1061–1097.

Cooper, A. F., and Lovering, J. F., 1970, Greenschist amphiboles from Haast River, New Zealand: Contributions to Mineralogy and Petrology, v. 27, p. 11–24.

Cotkin, S. J., and Armstrong, R. L., 1987, Rb/Sr age, geochemistry, and tectonic significance of blueschist from the schist of Skookum Gulch, eastern Klamath Mountains, California; Introducing the Callahan event: Geological Society of America Abstracts with Programs, v. 19, p. 367–368.

Coveney, R. M., Jr., 1981, Gold quartz veins and auriferous granite at the Oriental Mine, Alleghany District, California: Economic Geology, v. 76, p. 2176–2199.

D'Allura, J. A., Moores, E. M., and Robinson, L., 1977, Paleozoic rocks of the northern Sierra Nevada; Their structural and paleogeographic implications, *in* Stewart, J. H., Stevens, C. H., and Fritsche, A. E., eds., Paleozoic paleogeography of the western United States: Pacific Section, Society of Economic Paleontologists and Mineralogists Pacific Paleogeography Symposium p. 395–408.

Dalrymple, G. B., 1979, Critical tables for conversion of K-Ar ages from old to new constants: Geology, v. 7, p. 558–560.

Davis, G. A., 1968, Westward thrusting in the south-central Klamath Mountains, California: Geological Society of America Bulletin, v. 79, p. 911–933.

——— , 1969, Tectonic correlations, Klamath Mountains and western Sierra Nevada, California: Geological Society of America Bulletin, v. 80, p. 1095–1108.

Davis, G. A., and Lipman, P. W., 1962, Revised structural sequence of pre-Cretaceous metamorphic rocks in the southern Klamath Mountains, California: Geological Society of America Bulletin, v. 73, p. 1547–1552.

Davis, G. A., Holdaway, M. J., Lipman, P. W., and Romey, W. D., 1965, Structure, metamorphism, and plutonism in the south-central Klamath Mountains, California: Geological Society of America Bulletin, v. 76, p. 933–966.

Dickinson, W. R. , 1977, Paleozoic plate tectonics and the evolution of the Cordilleran continental margin, *in* Stewart, J. H., Stevens, C. H., and Fritsche, A. E., eds., Paleozoic paleogeography of the western United States: Pacific Section, Society of Economic Paleontologists and Mineralogists Pacific Coast Paleogeography Symposium 1, p. 137–156.

Eastoe, C. J., Gustin, M. M., and Nelson, S. E., 1987, Problems of recognition of olistostromes; An example from the lower Pit Formation, eastern Klamath Mountains, California: Geology, v. 15, p. 541–544.

Edelman, S. H., and 5 others, 1989, Structure across a Mesozoic ocean-continent suture zone in the northern Sierra Nevada, California: Geological Society of America Special Paper 224, 56 p.

Ehrenberg, S. N., 1975, Feather River ultramafic body, northern Sierra Nevada, California: Geological Society of America Bulletin, v. 86, p. 1235–1243.

Ellis, D. J., and Green, D. H., 1979, An experimental study of the effect of Ca upon garnet-clinopyroxene exchange equilibria: Contributions to Mineralogy and Petrology, v. 71, p. 13–22.

Ferguson, H. G., and Gannett, R. W., 1932, Gold quartz veins of the Alleghany District, California: U.S. Geological Survey Professional Paper 172, 139 p.

Gilluly, J., and Gates, O., 1965, Tectonic and igneous geology of the northern Shoshone Range, Nevada: U.S. Geological Survey Professional Paper 465, 153 p.

Girty, G. H., and Guthrie, L., 1989, A westward verging imbricate structure, Shoo Fly Complex, Sierra Nevada, California: Geological Society of America Abstracts with Programs, v. 21, p. 84.

Girty, G. H., and Wardlaw, M. S., 1984, Was the Alexander terrane a source of feldspathic sandstones in the Shoo Fly Complex, Sierra Nevada, California?: Geology, v. 12, p. 339–342.

——— , 1985, Petrology and provenance of pre-Late Devonian sandstones, Shoo Fly Complex, northern Sierra Nevada, California: Geological Society of America Bulletin, v. 96, p. 516–521.

Goodge, J. W., 1990, Convergent margin setting and petrotectonic evolution of a coherent late Triassic subduction complex, Stuart Fork terrane, Klamath Mountains, northern California: Geological Society of America Bulletin, v. 102, p. 86–101.

Goullaud, L., 1977, Structure and petrology in the Trinity mafic-ultramafic complex, Klamath Mountains, northern California, *in* Lindsley-Griffin, N., and Kramer, J. C., eds., Geology of the Klamath Mountains, Northern California; 73rd Annual Meeting Cordilleran Section Geological Society of America: p. 112–133.

Graham, C. M., and Powell, R., 1984, A garnet-hornblende geothermometer; Calibration, testing, and application to the Pelona Schist, Southern California: Journal of Metamorphic Geology, v. 2, p. 13–31.

Grapes, R. H., and Graham, C. M., 1978, The actinolite-hornblende series in metabasites and the so-called miscibility gap; A review: Lithos, v. 11, p. 85–97.

Hacker, B. R., 1988, Multiple metamorphic and deformation events in the Central and Feather River belts, northern Sierra Nevada: Geological Society of America Abstracts with Programs, v. 10, p. 165.

Hannah, J. L., and Moores, E. M., 1986, Age relationships and depositional environments of Paleozoic strata, northern Sierra Nevada, California: Geological Society of America Bulletin, v. 97, p. 787–797.

Hanson, R., and Schweickert, R. A., 1986, Stratigraphy of mid-Paleozoic island-arc rocks in part of the northern Sierra Nevada, Sierra and Nevada Counties, California: Geological Society of America Bulletin, v. 97, p. 986–998.

Hanson, R., Saleeby, J., and Schweickert, R. A., 1988, Composite Devonian island-arc batholith in the northern Sierra Nevada, California: Geological Society of America Bulletin, v. 100, p. 446–457.

Harwood, D. S., 1988, Tectonism and metamorphism in the Northern Sierra Terrane, northern California, *in* Ernst, W. G., ed., Metamorphism and crustal evolution of the western United States: Englewood Cliffs, New Jersey, Prentice-Hall, p. 764–788.

Harwood, D. S., Jayko, A. S., Harris, A. G., Silberling, N. J., and Stevens, C. H., 1988, Permian-Triassic rocks slivered between the Shoo Fly Complex and the Feather River peridotite belt, northern Sierra Nevada, California: Geological Society of America Abstracts with Programs, v. 20, p. 167–168.

Hietanen, A., 1973, Geology of the Pulga and Bucks Lake Quadrangles, Butte and Plumas Counties, California: U.S. Geological Survey Professional Paper 731, 66 p.

——, 1981, Geology west of the Melones fault between the Feather and North Yuba Rivers, California: U.S. Geological Survey Professional Paper 920, 30 p.

Hodges, K. V., and McKenna, L. W., 1987, Realistic propagation of uncertainties in geologic thermobarometry: American Mineralogist, v. 72, p. 671–680.

Hodges, K. V. and Spear, F. S., 1982, Geothermometry, geobarometry, and the Al_2SiO_5 triple point of Mt. Moosilauke, New Hampshire: American Mineralogist, v. 67, p. 1118–1134.

Holdaway, M. J., 1965, Basic regional metamorphic rocks in part of the Klamath Mountains, northern California: American Mineralogist, v. 50, p. 953–977.

Hotz, P. E., 1974, Preliminary geologic map of the Yreka Quadrangle, California: U.S. Geological Survey Miscellaneous Field Studies Map MF-568, scale 1:62,500.

——, 1977, Geology of the Yreka Quadrangle, Siskiyou County, California: U.S. Geological Survey Bulletin 1436, 72 p.

Hynes, A., 1982, A comparison and amphiboles from medium- and low-pressure metabasites: Contributions to Mineralogy and Petrology, v. 81, p. 119–125.

Irwin, W. P., 1960a, Geologic reconnaissance of the northern Coast Ranges and Klamath Mountains, California, with a summary of the mineral resources: California Division of Mines and Geology Bulletin, 179, 80 p.

——, 1960b, Relations between the Abrams Mica Schist and Salmon Hornblende Schist in Weaverville Quadrangle, California: U.S. Geological Survey Professional Paper 400-B, p. 315–316.

——, 1966, Geology of the Klamath Mountains Province, California: Division of Mines and Geology Bulletin, v. 190, p. 19–38.

——, 1977, Review of Paleozoic rocks of the Klamath Mountains, *in* Stewart, J. H., Stevens, C. H., and Fritsche, A. E., eds., Paleozoic paleogeography of the western United States: Pacific Section, Society of Economic Paleontologists and Mineralogists Pacific Coast Paleogeography Symposium 1, p. 441–454.

——, 1981, Tectonic accretion of the Klamath Mountains, *in* Ernst, W. G., ed., The geotectonic development of California: Englewood Cliffs, New Jersey, Prentice-Hall, p. 29–49.

Jacobsen, S. B., Quick, J. E., and Wasserburg, G. J., 1984, A Nd and Sr isotopic study of the Trinity peridotite; Implications for mantle evolution: Earth and Planetary Science Letters, v. 68, p. 361–378.

Kinkel, A.R.J., Hall, W. E., and Albers, J. O., 1956, Geology of base metal deposits of West Shasta copper-zinc district, Shasta County, California: Geological Society of America Bulletin, v. 94, p. 413–422.

La Fehr, T. R., 1966, Gravity in the eastern Klamath Mountains, California: Geological Society of America Bulletin, v. 77, p. 1177–1190.

Laird, J., and Albee, A. L., 1981, Pressure, temperature, and time indicators in mafic schist; Their application to reconstructing the polymetamorphic history of Vermont: American Journal of Science, v. 281, p. 127–175.

Lanphere, M. A., Irwin, W. P., and Hotz, P. E., 1968, Isotopic age of the Nevadan orogeny and older plutonic and metamorphic events in the Klamath Mountains, California: Geological Society of America Bulletin, v. 79, p. 1027–1052.

Lapierre, H., Albarede, F., Albers, J., Cabanis, B., and Coulon, C., 1985a, Early Devonian volcanism in the eastern Klamath Mountains, California; Evidence for an immature island arc: Canadian Journal of Earth Sciences, v. 22, p. 214–227.

Lapierre, H., Cabanis, B., Coulon, C., Brouxel, M., and Albarede, F., 1985b, Geodynamic setting of Early Devonian Kuroko-type sulfide deposits in the eastern Klamath Mountains (northern California) inferred by the petrological and geochemical characteristics of the associated island-arc rocks: Economic Geology, v. 80, p. 2100–2113.

Lindsley-Griffin, N., 1977, The Trinity ophiolite, Klamath Mountains, California: Oregon Department of Geology and Mineral Industries Bulletin, v. 95, p. 107–120.

Lindsley-Griffin, N., and Griffin, J. R., 1983, The Trinity terrane; An early Paleozoic microplate assemblage, *in* Stevens, C. H., ed., Pre-Jurassic rocks in western North American suspect terranes: Pacific Section, Society of Economic Paleontologists and Mineralogists, p. 63–75.

Liou, J. G., Maruyama, S., and Cho, M., 1985, Phase equilibria and mineral paragenesis of metabasites in low-grade metamorphism: Mineralogical Magazine, v. 49, p. 321–333.

Lipman, P. W., 1964, Structure and origin of an ultramafic pluton in the Klamath Mountains, California: American Journal of Science, v. 262, p. 199–222.

Lydon, P. A., Gay, T. E., Jr., and Jennings, C. W., 1960, Geologic map of California, Westwood Sheet: California Division of Mines and Geology, scale 1:250,000.

Mattinson, J. M., and Hopson, C. A., 1972a, Paleozoic ages of ophiolite complexes in Washington and northern California: EOS Transactions of the American Geophysical Union, v. 53, p. 543.

——, 1972b, Paleozoic ophiolite complexes in Washington and northern California: Carnegie Institution of Washington Yearbook, v. 71, p. 578–583.

Metz, P., and Trommsdorff, V., 1968, On phase equilibria in metamorphosed siliceous dolomites: Contributions to Mineralogy and Petrology, v. 18, p. 305–309.

Miller, M. M., 1989, Intra-arc sedimentation and tectonism; Late Paleozoic evolution of the eastern Klamath terrane, California: Geological Society of America Bulletin, v. 101, p. 170–187.

Moores, E. M., 1970, Ultramafics and orogeny, with models of the U.S. Cordillera and the Tethys: Nature, v. 228, p. 837–842.

Parsons, B. P. and Sclater, J. G., 1971, An analysis of the variation of ocean floor bathymetry and heat flow with age: Journal of Geophysical Research, v. 82, p. 803–827.

Peacock, S. M., 1985, Thermal and fluid evolution of the Trinity thrust system, Klamath province, northern California; Implications for the effect of fluids in subduction zones [Ph.D. thesis]: Los Angeles, University of California, 328 p.

——, 1987a, Serpentinization and infiltration metasomatism in the Trinity peridotite, Klamath province, northern California; Implications for subduction zones: Contributions to Mineralogy and Petrology, v. 95, p. 55–70.

——, 1987b, Creation and preservation of subduction-related inverted metamorphic gradients: Journal of Geophysical Research, v. 92, p. 12763–12781.

Peacock, S. M., and Norris, P. J., 1989, Metamorphic evolution of the Central Metamorphic Belt, Klamath province, California; An inverted metamorphic gradient beneath the Trinity peridotite: Journal of Metamorphic Geology, v. 7, p. 191–209.

Potter, A. W., Hotz, P. E., and Rohr, D. M., 1977, Stratigraphy and inferred tectonic framework of lower Paleozoic rocks, the eastern Klamath Mountains, northern California, *in* Stewart, J. H., Stevens, C. H., and Fritsche, A. E., eds., Paleozoic paleogeography of the western United States: Pacific Section, Society of Economic Paleontologists and Mineralogists Pacific Coast Paleogeography Symposium 1, p. 421–440.

Potter, A. W., Hotz, P. E., and Lanphere, M., 1981, Evidence of Ordovician-Silurian subduction and Silurian or older igneous units of possible magmatic arc origin, eastern Klamath Mountains, northern California: Geological Society of America Abstracts with Programs, v. 13, p. 101.

Quick, J. E., 1981, Petrology and petrogenesis of the Trinity peridotite, an upper mantle diapir in the eastern Klamath Mountains, northern California: Journal of Geophysical Research, v. 86, p. 11837–11863.

Renne, P. R., and Scott, G. R., 1988, Structural chronology, oroclinal deformation, and tectonic evolution of the southeastern Klamath Mountains, California: Tectonics, v. 7, p. 1223–1242.

Rice, J. M., 1983, Metamorphism of rodingites; Part 1, Phase relations in a portion of the system CaO-MgO-Al_2O_3-SiO_2-CO_2-H_2O: American Journal of Science, v. 283-A, p. 121–150.

Robinson, G. R., Jr., Hoas, J. L., Jr., Schafer, C. M., and Haselton, H. T., Jr., 1982, Thermodynamic and thermophysical properties of selected phases in the MgO-SiO_2-H_2O-CO_2, CaO-Al_2O_3-SiO_2-H_2O-CO_2, and Fe-FeO-Fe_2O_3-SiO_2 chemical systems with special emphasis on the properties of basalts and their mineral components: U.S. Geological Survey Open-File Report 83–79, 429 p.

Rouer, O., Lapierre, H., Mascle, G., Coulon, C., and Albers, J., 1989, Geodynamic implications of Devonian silicic arc magmatism in the Sierra Nevada and Klamath Mountains, California: Geology, v. 17, p. 177–180.

Saleeby, J. B., Hannah, J. L., and Varga, R. J., 1987, Isotopic age constraints on middle Paleozoic deformation in the northern Sierra Nevada, California: Geology, v. 15, p. 757–760.

Saleeby, J. B., Shaw, H. F., Moores, E. M., and Edelman, S. H., 1989, U/Pb, Sm/Nd, and Rb/Sr geochronological and isotopic study of northern Sierra Nevada ophiolitic assemblages, California: Contributions to Mineralogy and Petrology, v. 102, p. 205–220.

Schweickert, R. A., and Irwin, W. P., 1989, Extensional faulting in southern Klamath Mountains, California: Tectonics, v. 8, p. 135–149.

Schweickert, R. A., and Snyder, W. S., 1981, Paleozoic plate tectonics of the Sierra Nevada and adjacent regions, *in* Ernst, W. G., ed., The geotectonic development of California: Englewood Cliffs, New Jersey, Prentice-Hall, p. 183–201.

Schweickert, R. A., Harwood, D. S., Girty, G. H., and Hanson, R. H., 1984, Tectonic development of the Northern Sierra terrane; An accreted late Paleozoic island arc and its basement, *in* Lintz, J., ed., Geological excursions; Geological Society of America Cordilleran Section Meeting Guidebook: Reno, Nevada, Mackay School of Mines, p. 1–65.

Sharp, W. D., 1988, Pre-Cretaceous crustal evolution in the Sierra Nevada region, California, *in* Ernst, W. G., ed., Metamorphism and crustal evolution of the western United States: Englewood Cliffs, New Jersey, Prentice-Hall, p. 823–864.

Shaw, H. F., Chen, J. H., Wasserburg, G. J., and Saleeby, J. B., 1987, Nd-Sr-Pb systematics and age of the Kings River ophiolite, California; Implications for depleted mantle evolution: Contributions to Mineralogy and Petrology, v. 96, p. 281–290.

Silverling, N. J., Jones, D. L., Blake, M. C., Jr., and Howell, D. G., 1987, Lithotectonic terrane map of the western conterminous United States, *in* Folio of the lithotectonic terrane maps of the North American Cordillera: U.S. Geological Survey Miscellaneous Field Studies Map MF-1874-C, scale 1:2,500,000.

Spear, F. S., and Cheney, J. T., 1989, A petrogenetic grid for pelitic schists in the system SiO_2-Al_2O_3-FeO-MgO-K_2O-H_2O: Contributions to Mineralogy and Petrology, v. 101, p. 149–164.

Standlee, L. A., 1978, Geology of the northern Sierra Nevada basement rocks, Quincy-Downieville area, California [Ph.D. thesis]: Houston, Texas, Rice University, 221 p.

Stewart, J. H., 1980, Geology of Nevada: Nevada Bureau of Mines and Geology Special Publication 4, 136 p.

Stewart, J. H., and Suczek, C. A., 1977, Cambrian and latest Precambrian paleogeography and tectonics in the western United States, *in* Stewart, J. H., Stevens, C. H., and Fritsche, A. E., eds., Paleozoic paleogeography of the western United States: Pacific Section, Society of Economic Paleontologists and Mineralogists, Pacific Coast Paleogeography Symposium 1, p. 1–17.

Strand, R. G., 1962, Geologic map of California, Redding Sheet: California Division of Mines and Geology, scale 1:250,000.

Trommsdorff, V., and Evans, B. W., 1977, Antigorite-ophicarbonates; Phase relations in a portion of the system CaO-MgO-SiO_2-H_2O-CO_2: Contributions to Mineralogy and Petrology, v. 60, p. 39–56.

Turner, F. J., 1981, Metamorphic petrology: New York, McGraw-Hill, 524 p.

Varga, R. J., and Moores, E. M., 1981, Age, origin, and significance of an unconformity that predates island-arc volcanism in the northern Sierra Nevada: Geology, v. 9, p. 512–518.

Wagner, D. L., and Saucedo, G. J., 1987, Geologic map of California, Weed Quadrangle: California Division of Mines and Geology, scale 1:250,000.

Wallin, E. T., 1989, Reconstruction of the timing and sequence of early Paleozoic sedimentation in the Yreka-Callahan terrane, eastern Klamath Mountains, California: Geological Society of America Abstracts with Programs, v. 21, p. 155.

Wallin, E. T., Mattinson, J. M., and Potter, A. W., 1988, Early Paleozoic magmatic events in the eastern Klamath Mountains, northern California: Geology, v. 16, p. 144–148.

Weisenberg, C. W., 1979, Structural development of the Red Hill portion of the Feather River ultramafic body, Plumas County, California [Ph.D. thesis]: Houston, Texas, Rice University, 181 p.

Woodcock, N. H., and Robertson, A.H.F., 1982, The upper Batinah Complex, Oman; Allochthonous sediment sheets above the Semail ophiolite: Canadian Journal of Earth Sciences, v. 19, p. 1635–1656.

Yardley, B.W.D., 1982, The early metamorphic history of the Haast Schists and related rocks of New Zealand: Contributions to Mineralogy and Petrology, v. 81, p. 317–327.

Zucca, J. J., Fuis, G. S., Milkereit, B., Mooney, W. D., and Catchings, R. D., 1986, Crustal structure of northeastern California: Journal of Geophysical Research, v. 91, p. 7359–7382.

Manuscript Accepted by the Society March 27, 1990

Printed in U.S.A.

Geological Society of America
Special Paper 255
1990

Geochronological and tectonostratigraphic framework of Sierran-Klamath ophiolitic assemblages

Jason B. Saleeby
Division of Geological and Planetary Sciences, California Institute of Technology, Pasadena, California 91125

ABSTRACT

Through the work of a number of investigators, a tremendous amount of geochronological data with excellent geological control exists on the numerous ophiolitic assemblages of the Sierran-Klamath orogen. Distinct ophiolitic assemblages of latest Precambrian, Ordovician-Silurian, Carboniferous-Permian, Late Triassic–Early Jurassic and late Middle–Late Jurassic (Callovian-Oxfordian) ages are recognized. The geochronological data, in conjunction with structural and petrologic data, facilitate a relatively detailed analysis of the petrotectonic and structural-stratigraphic development of the orogen from the perspective of ophiolite basement geology. Insights into the plate tectonic settings of ophiolite genesis and emplacement are offered by an abundance of information that now exists on active oceanic island arcs and related environments. Remnants of abyssal lithosphere, boundary transform melange, extensional fore-arc igneous sequences, and interarc basin lithosphere are all preserved within the Sierran-Klamath assemblages. In general, the older assemblages or the older elements of polygenetic assemblages have greater affinities to abyssal lithosphere, whereas the younger assemblages or elements reflect suprasubduction zone environments of genesis. The upper Paleozoic through Jurassic assemblages occur in regional belts, and their tectonic, as well as petrogenetic, histories may be related in detail with some confidence to the tectonic development of the Cordilleran margin.

INTRODUCTION

Over the past decade, a significant body of geochronological data has been generated in conjunction with structural and petrologic studies of the wide array of ophiolitic assemblages that constitute the basement framework of the Klamath Mountains and the northern and western Sierra Nevada. Various aspects of this work are presented in numerous publications, but a broad synthesis has yet to be presented. The integrated body of data has some very interesting implications on the nature and temporal relations of crustal genesis, paleogeography, and deformation along the Sierran-Klamath orogen. This chapter provides a review of the geochronologic and tectonic relations that are well established for key ophiolitic assemblages. Generalized interpretations of the petrotectonic significance of the relations are also presented in light of a rapidly growing body of data on the structure and petrogenesis of oceanic lithosphere in and adjacent to modern island arcs. The most pertinent aspects of this growing body of data are reviewed in Saleeby (1990). In this synthesis, the emphasis is on the critical relations within the Sierran-Klamath ophiolitic assemblages.

A number of plate tectonic models have been offered for the Paleozoic and Mesozoic development of the Sierran-Klamath orogen. Most of these models do not draw heavily on the wealth of information from the petrogenetic and structural-stratigraphic relations of the Sierran-Klamath ophiolitic assemblages. The simplistic view that ophiolites represent collisional sutures, subduction complexes, or terrane boundaries is outmoded. This is demonstrated from what we currently know about oceanic arcs, fore-arcs, interarc basins, fracture zones, and boundary transforms. It is not the intent of the interpretive aspects of this synthesis to develop, support, or critique plate tectonic models for

Saleeby, J. B., 1990, Geochronological and tectonostratigraphic framework of Sierran-Klamath ophiolitic assemblages, *in* Harwood, D. S., and Miller, M. M., eds., Paleozoic and early Mesozoic paleogeographic relations; Sierra Nevada, Klamath Mountains, and related terranes: Boulder, Colorado, Geological Society of America Special Paper 255.

the orogen; rather, specific plate tectonic environments that render distinct ophiolitic assemblages are merely identified from the relations discussed. These may be used as modules in existing or future plate tectonic models (e.g. Saleeby, 1990; Saleeby, Busby-Spera and others, 1990).

The discussion of the geochronological and tectonostratigraphic relations of the Sierran-Klamath ophiolites centers on Figure 1, which shows a generalized tectonic map of the orogen and diagrammatic stratigraphic columns for key assemblages. The columns also depict the tectonic relations between ophiolitic and adjacent terranes, as well as the pertinent age data. U/Pb zircon ages are used extensively in this synthesis. Where uncertainties are small (<2 percent), the absolute age is shown in the Figure 1 columns. Where uncertainties are large, or where other techniques, including biostratigraphy, are used, the relative ages are shown based on the DNAG (Decade of North American Geology) time scale (Palmer, 1983).

TECTONIC FRAMEWORK

The Figure 1 map emphasizes the major metamorphic units and pre-Cretaceous intrusive rocks of the Klamath Mountains and western Sierra Nevada (modified after Harper and others, 1985). As noted by Irwin (1981), the Klamath Mountains consist of a series of ensimatic terranes, bounded by east-dipping, west-vergent thrust faults, which in general, get older to the east. Similar regional units extend southward into the northern Sierra Nevada (Davis, 1969), although the prevailing fault geometry changes to predominantly steep, west-directed reverse faults with a number of east-vergent thrust faults (Clark, 1964; McMath, 1966; Day and others, 1985; Edelman and others, 1989). Jurassic plutons cut, and are cut by, major faults of the Klamath Mountains and Sierra Nevada (Davis and others, 1978; Saleeby, 1982; Snoke and others, 1982; Sharp, 1988; Wright and Fahan, 1988; Saleeby and others, 1989a, b). Lower Cretaceous plutons of the composite Sierra Nevada batholith crosscut the major faults of the western Foothills (Tobisch and others, 1989). The 140-Ma (earliest Cretaceous) Grants Pass pluton of southern Oregon crosscuts the youngest internal thrust system of the Klamath Mountains (Harper and others, in press).

Sierran-Klamath ophiolitic assemblages fall into four tectonostratigraphic regimes. Grouping or splitting the assemblages into tectonostratigraphic terranes is a purely operational exercise. The regimes are as follows:

1. Throughout much of the North American Cordillera, there is a repeated association of structurally complex lower Paleozoic continental rise strata with lower to middle Paleozoic arc plutonic and volcanic rocks (Rubin and others, this volume), and adjacent, well-stratified, mid-Paleozoic to lower Mesozoic island-arc sequences with uncertain basement relations. The upper Paleozoic arc strata are distinguished by the presence of the McCloud fauna, which formed in carbonate reefs that covered a fractured volcanic ridge during the Permian (Miller, 1987). The continental rise–McCloud arc association is considered the remnants of a major northeast Pacific fringing arc system that developed outboard of the Cordilleran miogeocline. The Yreka, eastern Klamath, and northern Sierra terranes are typical of this association, and the Trinity ophiolitic complex appears to be a basement assemblage within the Yreka, and perhaps eastern Klamath terranes.

2. A highly disrupted belt, characterized by upper Paleozoic ophiolitic fragments, upper Paleozoic to lower Mesozoic chert-argillite deposits, upper Paleozoic Tethyan-affinity limestone and associated mafic blocks, and local Triassic blueschist slabs and blocks, lies to the west of these terranes (Davis and others, 1978). The Jurassic cherty strata commonly have continent-derived epiclastic and tuffaceous admixtures. This association is referred to as the Central belt in the western Sierra Nevada, and terranes of the western Paleozoic and Triassic belt in the Klamath Mountains. Such rocks can be traced along much of the northern Cordillera, outboard of McCloud arc rocks. The regional belt appears to have formed by a combination of subduction and transform tectonics. The North Fork, Feather River, Bear Mountains, and Kings-Kaweah ophiolitic assemblages are important Sierran-Klamath members of this belt.

3. Submarine lava flow, pyroclastic, and hypabyssal complexes with overlying cherty, tuffaceous, and epiclastic strata, which range from Upper Triassic to Lower and locally lower Middle Jurassic, are intimately interwoven with the western margin of the Central belt rocks (Saleeby, 1979, 1982; Wright, 1981, 1982; Wright and Wyld, 1985; Sharp, 1988; Edelman and others, 1989; Saleeby and others, 1989a; Dilek and others, this volume). These volcanic-sedimentary accumulations are commonly associated with multi-stage serpentinite melange, and are locally shown to have been built on upper Paleozoic ophiolitic basement of the Central belt. These complex accumulations are interpreted as extensional fore-arc sequences (Saleeby, 1990). Important examples are the Rattlesnake Creek terrane, Fiddle Creek and Slate Creek complexes, Penon Blanco Formation, and western facies of the Kings sequence.

4. Along the western margin of the Sierran-Klamath orogen run a number of relatively intact Callovian to Kimmeridgian mafic submarine volcanic and slaty flysch sequences. These strata were deposited on a composite basement consisting of the second and third regimes listed above. In general, the volcanic strata represent island-arc sequences (Schweickert and Cowan, 1975; Garcia, 1979, 1982; Menzies and others, 1980), although some may also represent extensional fore-arc eruptions (Saleeby, 1990). Swarms of sheeted dikes and shallow-level intrusions associated with such eruptives in the northern Sierra constitute the Smartville complex, and give the overall assemblage an ophiolitic character (Menzies and others, 1980; Day and others, 1985; Beard and Day, 1987; Saleeby and others, 1989a). Within the western Klamaths, a complete spreading-ridge crust-and-mantle sequence constitutes the Josephine ophiolite, which formed in close spatial and temporal association with arc eruptive products (Harper, 1984). Extensive stratigraphic units that constitute the arc and/or fore-arc volcanic rocks include the Rogue Formation

and western Hayfork terrane of the Klamath Mountains, and the Logtown Ridge Formation and the Copper Hill, Gopher Ridge, and Browns Valley Ridge volcanic units of the Sierran Foothills.

The age and structural relations of the Sierran-Klamath ophiolitic assemblages and adjacent rocks record sequential westward growth of the Cordilleran plate edge by the accretion of progressively younger ensimatic assemblages. Igneous accretion appears to have been an important phenomenon, in addition to tectonic accretion. The critical ophiolitic relations on which this synthesis is based are presented below in a temporal sequence starting with the oldest events reaching back to the inception of the Cordilleran continental margin.

THE PASSIVE TO ACTIVE MARGIN TRANSITION

The earliest signs of Cordilleran active-margin phenomena are recorded in structurally complex lower Paleozoic continental rise strata that are intruded by, and interstratified with, lower to middle Paleozoic arc-derived rocks (Rubin and others, this volume). Such assemblages may be traced southward from the central Alaska–Yukon region to the eastern Klamath belt and northern Sierra terranes. Within these terranes lie the oldest ophiolitic remnants presently recognized in the North American Cordillera. They offer a glimpse into the onset of active-margin tectonism along the continental rise outboard of the Cordilleran miogeocline.

The oldest rocks in the Cordillera of probable oceanic crustal origin are within the Shoo Fly Complex of the northern Sierra Nevada, and of probable island-arc origin within the Yreka terrane of the eastern Klamath belt. The uppermost Precambrian and Lower Cambrian ophiolitic rocks of these areas and intimately associated metaclastic units were incorporated into a major subduction-arc system prior to the Devonian. The inclusion of continental rise strata within the ophiolitic structural sequences implies an east Pacific (parautochthonous) plate tectonic setting, although current distribution patterns of the Sierran relative to the Klamath remnants may not reflect original geometric relations.

Shoo Fly Complex

The Shoo Fly Complex consists of a southwest-vergent thrust sequence of mainly siliciclastic, argillaceous, and cherty strata of Ordovician-Silurian and possibly older age (Varga and Moores, 1981; Schweickert and others, 1984; Saleeby and others, 1987; Harwood, 1988; Girty and Guthrie, 1989). In general, the Shoo Fly Complex is considered to be imbricated North American continental rise strata built outboard of the miogeocline (Bond and DeVay, 1980). The structurally highest thrust sheet consists of the Sierra City melange. The melange contains blocks of serpentinized peridotite, gabbro, pillowed and massive basalt, chert, limestone, and sandstone, all in a sheared sedimentary matrix. Most of the blocks appear to represent the disrupted vestiges of an ophiolite sequence that was fragmented and then mixed into a sedimentary matrix by a combination of soft sediment slide and tectonic deformation mechanisms.

Ophiolitic fragments within the Sierra City melange appear to have been derived from late Precambrian oceanic crust. A rare plagiogranite block yields a Pb/U zircon age of 600 ± 10 Ma, and its initial lead isotopic compositions are indicative of a late Precambrian depleted mantle source (J. B. Saleeby and R. A. Schweickert, unpublished data). Additional isotopic studies on adjacent mafic blocks are in progress in order to test the consanguinity of the ophiolitic blocks. Oceanic crust ~600 m.y. old is likely to have existed as basement for continental rise strata built off the Cordilleran passive margin (Bond and Kominz, 1984; Bond and others, 1984). Remnants of such basement are also likely to have been incorporated within the initial thrust wedge during the onset of subduction in the continental rise environment.

Sedimentary petrography and detrital zircon data of selected feldspathic sandstone blocks in the Sierra City melange record the onset of arc activity during the Cambro-Ordovician (Girty and Wardlaw, 1984). The first direct signs of arc activity in the Shoo Fly complex are recorded by mid-Silurian tuff beds (Saleeby and others, 1987). Imbrication of the Shoo Fly complex and the incorporation of the melange were completed during the Devonian (Varga and Moores, 1981; Schweickert and others, 1984). Significant similarities have been noted by numerous authors between the Shoo Fly Complex and the Yreka terrane of the eastern Klamaths. The Yreka terrane most clearly records the earliest stages of Cordilleran subduction and arc activity.

Yreka terrane

The Yreka terrane consists of a series of thrust sheets that structurally overlie the northwest margin of the Trinity ophiolite. Between the ophiolite and the main structural sequence lies a disrupted zone that contains Early Cambrian plutonic rocks, pre–Middle Ordovician keratophyric volcanics and Ordovician mafic

Figure 1. Tectonic map of central Sierra Nevada through Klamath Mountains region emphasizing ophiolitic assemblages and adjacent terranes (modified after Harper and others, 1985). Also shown are reconstructed diagrammatic columnar sections for key assemblages and diagrammatic relations with adjacent terranes. Locations of columns are keyed to map. Indicated isotopic ages represent high-precision U/Pb zircon analyses (references given in text). Other indicated age constraints reflect a combination of isotopic and biostratigraphic techniques. Columns do not reflect stratigraphic thicknesses.

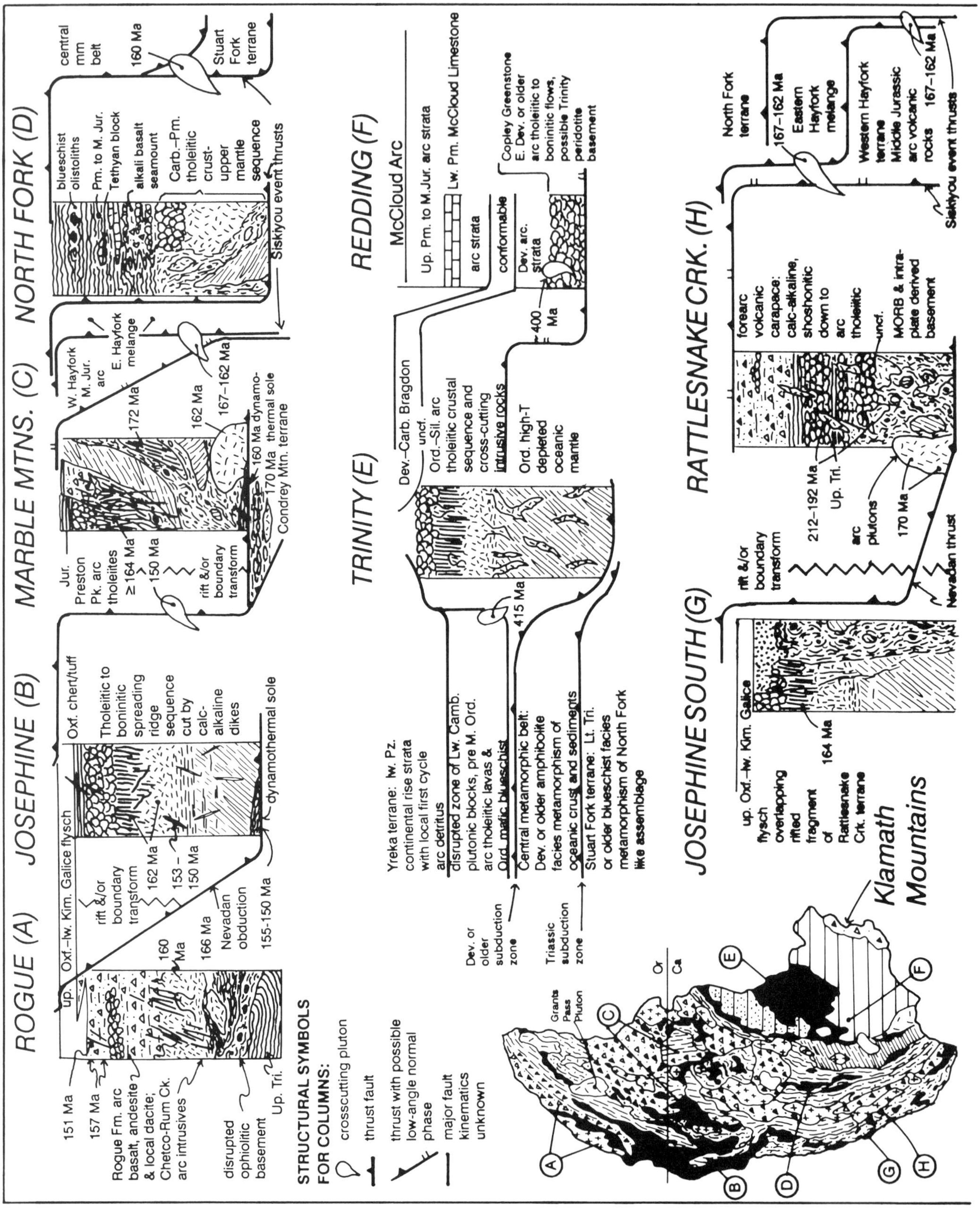
ROGUE (A)
JOSEPHINE (B)
MARBLE MTNS. (C)
NORTH FORK (D)
151 Ma
157 Ma
Rogue Fm. arc basalt, andesite & local dacite; Chetco-Rum Ck. arc intrusives
disrupted ophiolitic basement
Up. Tri.
up. Oxf.–lw. Kim. Galice flysch
160 Ma
166 Ma
rift &/or boundary transform
Nevadan obduction
155-150 Ma
162 Ma
153 – 150 Ma
Oxf. chert/tuff
Tholeiitic to boninitic spreading ridge sequence cut by calc-alkaline dikes
dynamothermal sole
Jur. Preston Pk. arc tholeiites
≥164 Ma
150 Ma
rift &/or boundary transform
172 Ma
162 Ma
167–162 Ma
160 Ma dynamo-
170 Ma thermal sole
Condrey Mtn. terrane
W. Hayfork M. Jur. arc
E. Hayfork melange
blueschist olistoliths
Pm. to M. Jur. Tethyan block
alkali basalt seamount
Carb.–Pm. tholeiitic crust-upper mantle sequence
Siskiyou event thrusts
central mm belt
160 Ma
Stuart Fork terrane
STRUCTURAL SYMBOLS FOR COLUMNS:
crosscutting pluton
thrust fault
thrust with possible low-angle normal phase
major fault kinematics unknown
TRINITY (E)
REDDING (F)
Yreka terrane: lw. Pz. continental rise strata with local first cycle arc detritus
disrupted zone of Lw. Camb. plutonic blocks, pre M. Ord. arc tholeiitic lavas & Ord. mafic blueschist
Central metamorphic belt: Dev. or older amphibolite facies metamorphism of oceanic crust and sediments
Stuart Fork terrane: Lt. Tri. or older blueschist facies metamorphism of North Fork like assemblage
Dev. or older subduction zone
Triassic subduction zone
415 Ma
Dev.–Carb. Bragdon
uncf.
Ord.–Sil. arc tholeiitic crustal sequence and cross-cutting intrusive rocks
Ord. high-T depleted oceanic mantle
McCloud Arc
Up. Pm. to M.Jur. arc strata
Lw. Pm. McCloud Limestone
arc strata
conformable
Dev. arc. strata
400 Ma
Copley Greenstone E. Dev. or older arc tholeiitic to boninitic flows, possible Trinity peridotite basement
JOSEPHINE SOUTH (G)
RATTLESNAKE CRK. (H)
up. Oxf.–lw. Kim. flysch overlapping rifted fragment of Rattlesnake Crk. terrane
Galice
164 Ma
rift &/or boundary transform
Nevadan thrust
212–192 Ma
Up. Tri.
arc plutons
170 Ma
forearc volcanic carapace: calc-alkaline, shoshonitic down to arc tholeiitic
uncf.
MORB & intra-plate derived basement
North Fork terrane
167–162 Ma
Eastern Hayfork melange
Western Hayfork terrane Middle Jurassic arc volcanic rocks
167–162 Ma
Siskiyou event thrusts
Klamath Mountains
Grants Pass Pluton
Or
Ca
A
B
C
D
E
F
G
H

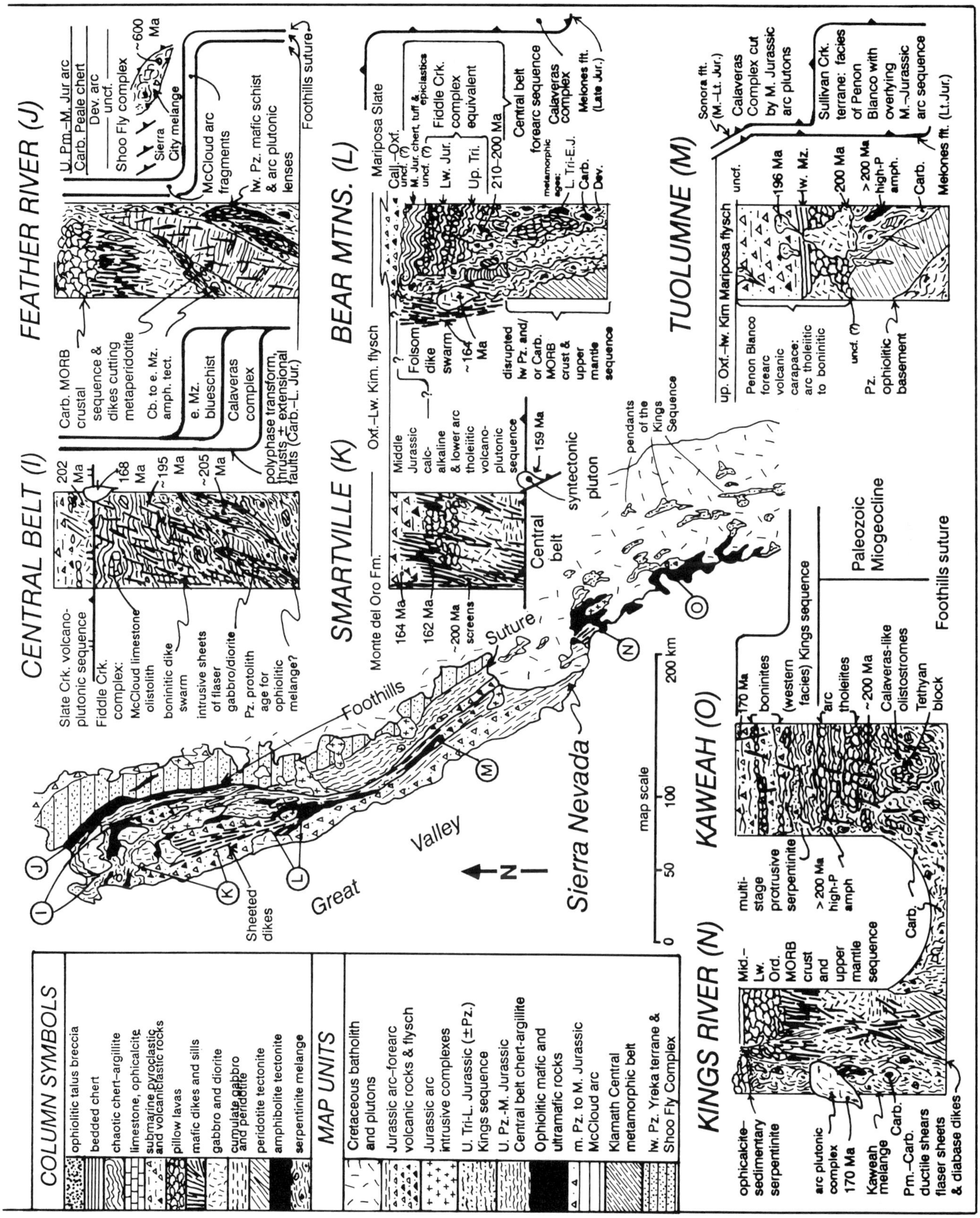

COLUMN SYMBOLS
ophiolitic talus breccia
bedded chert
chaotic chert–argillite
limestone, ophicalcite
submarine pyroclastic and volcaniclastic rocks
pillow lavas
mafic dikes and sills
gabbro and diorite
cumulate gabbro and peridotite
peridotite tectonite
amphibolite tectonite
serpentinite melange
MAP UNITS
Cretaceous batholith and plutons
Jurassic arc–forearc volcanic rocks & flysch
Jurassic arc intrusive complexes
U. Tri-L. Jurassic (±Pz.) Kings sequence
U. Pz.-M. Jurassic Central belt chert-argillite
Ophiolitic mafic and ultramafic rocks
m. Pz. to M. Jurassic McCloud arc
Klamath Central metamorphic belt
lw. Pz. Yreka terrane & Shoo Fly Complex
CENTRAL BELT (I)
FEATHER RIVER (J)
SMARTVILLE (K)
BEAR MTNS. (L)
TUOLUMNE (M)
KINGS RIVER (N)
KAWEAH (O)
Great Valley
Sierra Nevada
Foothills Suture
Sheeted dikes
map scale
0 50 100 200 km
Paleozoic Miogeocline
Foothills suture
Kings sequence
Central belt
syntectonic pluton
pendants of the Kings Sequence

blueschist (Potter and others, 1977; Cotkin and Armstrong, 1987; Wallin and others, 1988). The diversity of rock types and metamorphic grades represented in the disrupted zone indicates a polyphase development that could have continued to as late as Cretaceous time.

The Early Cambrian plutonic rocks of the disrupted zone have been loosely considered ophiolitic in character owing to their spatial association with the Trinity complex. They are generally tonalitic to plagiogranitic in composition, but with their original host rocks yet to be determined. Pb/U zircon ages cluster around 560 Ma (Wallin and others, 1988). Lithologic homogeneity in conjunction with local entrainment of Proterozoic xenocrystic zircon (Wallin, 1989a) suggests an arc plutonic origin adjacent to the Cordilleran margin rather than genesis as part of the Trinity oceanic lithosphere. Proterozoic xenocrystic zircon may have been entrained from continental rise sediments or foundered sialic blocks that were stranded off the rifted Cordilleran margin.

In addition to the Lower Cambrian plutonic blocks, the earliest phases of arc activity are recorded by pre–Middle Ordovician low-K tholeiitic (keratophyric) flows and hypabyssal rocks and a 480-Ma quartz diorite lens within the disrupted zone (Mattinson and Hopson, 1972; Potter and others, 1977; Brouxel and others, 1988). Upper Ordovician first-cycle volcanic sandstones occur within structurally higher thrust sheets above the disrupted zone (Wallin, 1989b) and, as in the Shoo Fly Complex, are mixed and structurally imbricated with detritus and strata of North American cratonic provenance. Subduction is recorded by blueschist blocks within the disrupted zone. Isotopic, geochemical, and petrographic data suggest Late Ordovician subduction zone metamorphism of a protolith sequence consisting of Middle to Early Ordovician ocean-floor basalt with superjacent arc and/or continent-derived clastic strata (Cotkin and Armstrong, 1987). Locally, blueschist blocks, as well as Lower Cambrian plutonic blocks and arc-derived strata, were intermixed in melange units of the disrupted zone and then unconformably overlapped by Upper Silurian conglomeratic strata (Haessig, 1989).

Despite the complex and fragmentary geologic history of the Yreka terrane, there is a clear implication of earliest Paleozoic subduction imbrication and arc activity along the Cordilleran continental rise. The different structural units of the Yreka terrane, as well as those of Shoo Fly Complex, probably represent the tectonic accumulation of a laterally vast expanse of strata. The generation and emplacement of the Trinity ophiolitic complex within, or adjacent to, this depositional system punctuated the tectonic history and probably contributed to much of the structural complexity of the eastern Klamath belt.

Trinity ophiolitic complex

The Trinity complex (Fig. 1E) consists of more than 1,600 km^2 of variably serpentinized peridotite with subordinate mafic intrusive and volcanic rock (Lindsley-Griffin, 1977; Quick, 1981; Brouxel and Lapierre, 1988; Boudier and others, 1989). The ophiolite is a semi-intact 2- to 4-km-thick, east-dipping thrust sheet (Irwin and Lipman 1962; Zucca and others, 1986). Mafic members are concentrated toward the east, suggesting that deeper levels are exposed to the west. The Yreka terrane is in thrust contact with the ophiolite to the north, and the Redding section sits structurally above the ophiolite to the southeast. To the west, the Central Metamorphic belt is thrust eastward beneath the Trinity peridotite.

The Trinity peridotite consists of a variable mixture of depleted harzburgite-lherzolite and relatively nondepleted plagioclase lherzolite, with subordinate dunite, and scattered clinopyroxene-rich dikes (Quick, 1981; Boudier and others, 1989). The plagioclase lherzolites appear to be concentrated near the structurally deeper levels of the massif (Brouxel and Lapierre, 1988). The peridotites exhibit high-temperature ductile deformation fabrics throughout, except for local clinopyroxene-rich dikes and dunite cumulate lenses. Mafic plutonic rocks consisting of a wide spectrum of cumulate, isotropic, and pegmatitic gabbroids crosscut the high-temperature fabrics (Quick, 1981). The upper levels of the complex are only locally preserved in the northeast and consist of sheeted and isolated diabasic dikes, sills, and local deeper-level plutonic rocks, and an overlying pillow flow sequence (Lindsley-Griffin, 1977; Brouxel and Lapierre, 1988; Boudier and others, 1989).

The Trinity complex was constructed from multiple depleted mantle sources over a time span of perhaps as long as 50 m.y. The earliest petrogenetic events recorded in the peridotite are the upper mantle ascent of hot, partially depleted lherzolite from greater than 30-km depths through the spinel and into the plagioclase lherzolite stability fields (Quick, 1981). During ascent, the lherzolite underwent local decompression and partial melting, yielding a basaltic magma. Remnants of the partial melt product were trapped in the peridotite and equilibrated as plagioclase lherzolite. Remnants of mafic crustal rocks that may have been generated during ascent and partial melting are not preserved. Ascent of the peridotite presumably represents mantle diapirism, and the thermal gradient required to produce plagioclase lherzolite implies diapirism beneath an oceanic spreading center. An abyssal versus interarc basin setting cannot be distinguished with the existing data. Mineral and bulk-rock Sm/Nd isochron-age data on the plagioclase lherzolite indicate an equilibration age of 472 ± 32 Ma, which approximates the spreading center residence of the peridotite (Jacobsen and others, 1984).

Subsequent to its spreading center residence, the peridotite was stripped of any crustal rocks that may have formed at the ridge and then it became the host and basement for the currently exposed mafic crustal sequence. Mafic magmas derived from a mantle source beneath and distinct from the Trinity peridotite rose through the peridotite and coalesced into thin mafic crust. Relative age relations are clearly displayed by the mafic rocks crosscutting the high-temperature mantle fabrics that formed during the earlier partial melting event, and by the presence of trapped olivine and clinopyroxene cumulates within the depleted peridotites (Quick, 1981).

A broad spectrum of age data exists for the mafic crustal sequence. A representative suite of crustal-level samples defines a 452 ± 20 Ma Sm/Nd bulk-rock isochron (Brouxel and Lapierre, 1988) and overlaps a 435 ± 21 Ma Sm/Nd isochron derived from microgabbro and pyroxenite dikes that crosscut the mantle peridotite. These samples form a distinct data array from that which yields the 472 ± 32 Ma age for the earlier mantle melting event (Jacobsen and others, 1984). K/Ar age data indicating cooling ages as old as 447 Ma are reported for crosscutting dikes within the peridotite (Lanphere and others, 1968), and zircon ages of ~415 Ma and ~430 Ma are reported for younger intrusives as well (Mattinson and Hopson, 1972; Wallin and Lindsley-Griffin, 1990).

The geochemical affinities of the Trinity complex crustal sequence are that of an immature island arc, fore-arc, or interarc basin (Brouxel and Lapierre, 1988; Brouxel and others, 1988). Mineralogical, trace-element, and Nd isotopic data indicate a low-K tholeiitic series, with low Ti and REE contents, and the crystallization sequence is more typical of oceanic arcs than MORB (mid-ocean ridge basalt) (Coleman, 1986). Initial ϵ_{Nd} values range from +6.6 to +7.5 and are clearly distinct from the $+10.4 \pm 0.4$ value reported for the underlying plagioclase lherzolite (Jacobsen and others, 1984). Boudier and others (1989) constructed a detailed structural model of the ophiolite, relating the thin mafic section, the presence of plagioclase lherzolite, and the high-temperature fabrics in the peridotite to a single consanguineous slow-spreading ridge sequence. These workers did not address the age and clear petrochemical contrasts between the lherzolites and mafic section.

A number of geologic relations suggest that the Trinity complex existed as basement for, or was adjacent to, Silurian and younger rocks of the eastern Klamath Mountains. Sparse ophiolitic debris, presumably derived from the ophiolite, is present in Silurian conglomeratic strata of the Yreka terrane (Mattinson and Hopson, 1972; Lindsley-Griffin, 1977; Potter and others, 1977). These data suggest that the Trinity complex was at least within the same subduction-arc system as represented by the Yreka terrane for the Silurian. Stronger ties are suggested between the ophiolite and the Redding section. Rocks of the Redding section lie structurally above the peridotite with general concordance (Fig. 1F), although a depositional contact has not been documented. The base of the Redding section consists of the pre- to Early Devonian Copley Greenstone, a pillowed mafic volcanic unit similar in a number of aspects to Trinity basalts exposed in the northeast part of the massif. The Copley Greenstone could have broad facies relations with the Trinity basalts, although existing age constraints suggest differences. The Upper Devonian(?)–Carboniferous Bragdon Formation is conformably within the Redding section, but unconformably above Trinity basalts in the northeast (Brouxel and others, 1988). Thus, a spatial tie between the Copley and Trinity basalts is clear by the end of the Devonian. The Redding section consists of one of the longest records of east-Pacific fringing arc volcanism and sedimentation preserved in the U.S. Cordillera (Miller, 1987, 1989). Starting with the Copley Greenstone, a near-continuous stratigraphic record is preserved spanning the pre- or earliest Devonian to the Middle Jurassic. The Redding section contains the McCloud limestone and thus may be considered the type McCloud arc (Miller, 1987).

The Copley Greenstone consists of pillowed and massive basalt to andesite flows and subordinate hyaloclastites with local volcanogenic sedimentary rocks (Lapierre and others, 1986). It is overlain conformably by the Lower Devonian Balaklala Rhyolite. The rhyolite is considered cogenetic with the ~400-Ma Mule Mountain Stock (Albers and others, 1981), which intrudes the Copley Greenstone. The Copley Greenstone belongs to the low-K tholeiite series like the Trinity mafic rocks, but also includes high-Mg andesitic flows and dikes of boninite affinity (Brouxel and others, 1988). An igneous fore-arc petrogenetic setting is implied for the Copley Greenstone (Saleeby, 1990). The Balaklala Rhyolite exhibits low-K tholeiitic geochemical features, which are also common in subordinate silicic rocks of fore-arc regions (Rouer and others, 1989). Volcanism and sedimentation recorded in the Redding section subsequent to deposition of the Copley and Balaklala are more typical of a mature axial arc environment. Deposurface relations at the base of the Bragdon Formation imply a Devonian disturbance that affected the Trinity ophiolite but apparently not the Copley-Balaklala section. Such a disturbance appears to correspond in time with thrusting of the Trinity complex over the Central Metamorphic belt, or alternatively, uplift and cooling of the Central belt–Trinity thrust stack.

Tectonics of the Trinity complex

The polygenetic development of the Trinity complex and its current structural setting attest to a complex tectonic history. Shortly after the diapiric rise and partial melting event that the peridotite underwent beneath a spreading ridge, the new oceanic lithosphere was apparently stripped of its basaltic crustal sequence. The only viable alternative to this view is that the diapiric rise occurred at a spreading center–transform intersection and was continuous with protrusion and exposure of the mantle rocks, with little or no superjacent crustal sequence having been generated. Regardless, the Early to Middle Ordovician oceanic mantle fragment was incorporated into a Middle Ordovician to Silurian oceanic arc, where it hosted a low-K arc tholeiitic magmatic series that was derived from a distinct underlying mantle source. The spreading center under which the peridotite rose may have been an interarc basin, in which case only a change in mantle magma source is needed to explain the petrogenetic sequence. Alternatively, if the peridotite rose under an abyssal spreading ridge, a distinct lithosphere detachment and subduction zone ± transform accretion event is required. Loss of possible abyssal crustal remnants may have occurred during such an event.

Subsequent to the construction of the arc tholeiitic crustal sequence, the Trinity complex was thrust westward over the Central Metamorphic belt in Devonian or perhaps earlier times (Lipman, 1964; Lanphere and others, 1968; Cannat and Boudier, 1985). The Central Metamorphic belt consists of an inverted metamorphic gradient ranging from upper amphibolite-facies rocks adjacent to the Trinity thrust to upper greenschist-facies rocks at its structural base (Peacock and Norris, 1989). Thrusting is presumed to have occurred in an east-dipping subduction zone, with residual heat in the peridotite inhibiting the development of high-P/T conditions. Bulk-rock and muscovite Rb/Sr ages of ~375 Ma (Lanphere and others, 1968) have been widely interpreted as dating metamorphism and underthrusting of the Central Belt. These ages, however, may reflect unroofing of the metamorphic terrane, with underthrusting and peak metamorphism having occurred substantially earlier. Such a possibility is appealing if Ordovician-to-Silurian primary heat from the Trinity peridotite is considered to have affected the metamorphic environment.

The Central Metamorphic belt appears to have been derived from a relatively intact fragment of oceanic crust with overlying siliceous and calcareous sediments. Remnants of an accretionary prism between the metamorphic belt and the Trinity peridotite are lacking. Medial structural-level thrust sheets of the Yreka terrane have been suggested as correlatives to the Central Metamorphic belt (Cashman, 1980), implying a very complex kinematic pattern for the construction of the terrane and its emplacement over the Trinity complex.

The geometry of the Central Metamorphic belt, Trinity ophiolite, and Redding section offers some constraints on the tectonic setting of petrogenesis for the Trinity mafic crustal section. The westward thrusting of the ophiolite over the Central Metamorphic belt is well documented, although the timing is in serious question. Geophysical data (Zucca and others, 1986) indicate that the Central Metamorphic belt, Trinity complex, and Redding section all extend to considerable depths as east-dipping slabs. The implied westward transport of the fore-arc–affinity Copley Greenstone over the Trinity complex requires a fore-arc setting for the complex in Silurian to Devonian time. Arrival at such a position from an interarc basin site of petrogenesis cannot be geometrically substantiated from the current structural relations, although it is possible with a hypothetical arc reversal. Generation of the Trinity crustal sequence as an in situ fore-arc igneous complex, however, is geometrically the simplest model.

The absence of an accretionary prism at the base of the Trinity thrust suggests an attritionary inner trench wall for Devonian and possibly earlier times. Such conditions are typical of extensional fore-arcs (Bloomer, 1983). The petrogenesis of the Trinity crustal sequence and the Copley-Balaklala sequence may represent one or more phases of fore-arc extensional magmatism. In this regard, the subduction zone "metamorphic events" recorded isotopically as Late Ordovician in the Yreka terrane blueschist (Cotkin and Armstrong, 1987), and as Devonian in the Central Metamorphic belt (Lanphere and others, 1968) may actually represent extensional disruption and denudation events of subduction-zone metamorphic environments temporally related to the extensional magmatic events.

Despite the large surface area of the Trinity ophiolite and its relatively intact state, some of the more fundamental tectonic questions it poses cannot be adequately addressed since it does not extend as a regional belt like many of the younger ophiolite terranes of the U.S. Cordillera. The Trinity peridotite does, however, carry unique geochemical fingerprints that can be used to trace its likely relatives. Such fingerprints tie the peridotite and its relatives to a unique oceanic lithosphere domain.

EARLY PALEOZOIC PANTHALASSA LITHOSPHERE REMNANTS

The Sm/Nd isotopic signature of the Trinity plagioclase lherzolite is unique and of broad geodynamic significance. It may also prove to be of great regional significance. The initial ϵ_{Nd} of +10.4 ± 0.4 recorded in the 472 ± 32 Ma upper mantle equilibrium assemblage, in conjunction with major- and trace-element data, indicates initial depletion of the protolherzolite substantially earlier in Earth history than the early Paleozoic partial melting event, perhaps as early as ~3.4 Ga (Jacobsen and others, 1984). This domain of the Earth's mantle remained undisturbed and chemically distinct until the diapiric rise and partial melting of the peridotite beneath a Panthalassa spreading ridge. The Trinity peridotite is not unique in this geochemical signature. The Kings River ophiolite of the southwestern Sierra Nevada Foothills carries the same geochemical signature, but within a complete ridge crest crust and upper mantle sequence (Saleeby, 1978; Shaw and others, 1987). The Trinity peridotite and the Kings River ophiolite are the only confirmed early Paleozoic ridge crest lithosphere remnants in the entire Cordillera. The fact that they share the same unique geochemical characteristics suggests that a significant domain of the early Paleozoic Panthalassa lithosphere also carried such characteristics.

Kings River ophiolite

The Kings River ophiolite is exposed over an area of ~150 km^2 along the western edge of the southern Sierra Nevada batholith (Fig. 1N). It consists of a complete ophiolite sequence with >6 km of tectonitic harzburgite with minor dunite and lherzolite, ~3 km of gabbroic and minor wehrlitic cumulates, ~0.7 km of sheeted basalt and diabase dikes, and ~1.8 km of basaltic pillow lava (Saleeby, 1978). Metalliferous chert blocks as thick as 20 m are associated with the ophiolite, but these are dislodged from the pillow lavas, and otherwise there is no sedimentary record. Felsic igneous rocks are exceedingly rare, except as distinctly younger crosscutting intrusives. Unlike most Cordilleran ophiolites, the Kings River section offers no geologic signatures suggesting generation within or adjacent to an island arc, and the thickness of its mafic crustal sequence compares favorably with abyssal crust.

The ophiolite sequence is preserved within several large, intact, tectonically bounded slabs that were incorporated into an upper Paleozoic regional ophiolitic melange interpreted as a transform system (Saleeby, 1977, 1978, 1979, 1990). In addition to domainal greenschist to amphibolite-facies dynamothermal metamorphism suffered during melange disruption and also in Jurassic intraarc shear zones, the ophiolite was contact-metamorphosed during Early Cretaceous batholith emplacement. Primary textures, fabrics, and structures are well preserved throughout the ophiolite, but primary mineral assemblages are commonly pseudomorphed by contact assemblages. In this regard, major-element chemistry is suspect for the ophiolite, although analyses of basalts, diabases, and gabbros show MORB affinities.

Mineral and bulk-rock Sm/Nd isochron age data from a representative suite of pillow lavas, sheeted mafic and plagiogranite dikes, cumulate gabbros, and clinopyroxene-bearing peridotite yield an age of 483 ± 21 Ma (Shaw and others, 1987). Zircon from the only known plagiogranite from the entire ophiolite yields complex, multi-disturbed systematics with concordia patterns suggesting an early Paleozoic age (Saleeby and Sharp, 1980; Shaw and others, 1987). The 483 ± 21 Ma isochron age indicates spreading-ridge petrogenesis during the Early to Middle Ordovician. The initial ϵ_{Nd} for the Kings River section is ±10.7 ± 0.5, and the Trinity peridotite data lie precisely on the Kings River Nd isotopic evolution line (Shaw and others, 1987).

The distinct Nd isotopic signature of the Kings River ophiolite is complemented by extremely nonradiogenic initial Sr and Pb isotopic compositions (Shaw and others, 1987). Initial $^{87}Sr/^{86}Sr$ as low as 0.7023—perhaps lower, considering metamorphism and extremely retarded Pb compositions ($^{206}Pb/^{204}Pb$ = 17.14, $^{207}Pb/^{204}Pb$ = 15.37, $^{208}Pb/^{204}Pb$ = 36.80—indicate derivation from an extraordinarily depleted mantle source. The extremely low initial $^{207}Pb/^{204}Pb$ values indicate that the source acquired a low U/Pb ratio early in Earth history before significant decay of ^{235}U could occur. Depletion of Rb relative to Sr, and Nd relative to Sm, as indicated by the initial Sr and Nd data, is most likely to have occurred during the early stage U depletion event as well. Thus, the suggestion by Jacobsen and others (1984) that the most likely mechanism for the distinct isotopic and trace-element aspects of the Trinity peridotite was a very early-stage depletion of the proto-lherzolite, followed by long-term isolation in the mantle, is corroborated by the Kings River data (see Jacobsen and others, 1984, and Shaw and others, 1987, for alternative mechanisms).

The initial Nd and Pb isotopic data for the Kings River ophiolite and the Trinity peridotite lie just beyond or at the extremity of all data from depleted mantle-related samples acquired on ophiolites and MORB throughout the Earth. This indicates that very little of the mantle that has been, or is, involved in the creation of oceanic crust attained, or has retained, the state of depletion as that represented by the Trinity peridotite and the Kings River ophiolite. Clearly the identification of additional fragments of this lithosphere type carries broad geodynamic significance.

The distinct geochemical characteristics of the Trinity and Kings River Panthalassa lithosphere remnants should facilitate the identification of possible relatives elsewhere in the Cordillera or circum-Pacific orogen by isotopic and trace-element studies. Such identification should not be considered in the same sense as a stratigraphic correlation. As with the Trinity peridotite and Kings River ophiolite, what can be said with confidence is that they both resided at a Panthalassa spreading ridge at the same time, during the Early to Middle Ordovician, and that they were both derived from the same or very similar depleted mantle domain(s). It is quite unlikely that such spreading occurred in a suprasubduction-zone environment. The spatial association of the Trinity peridotite and Kings River ophiolite along the Sierran-Klamath orogen strongly suggests genesis along the same spreading ridge. Furthermore, as discussed below, a number of large ophiolitic fragments along the Sierran Central belt are possible remnants of the same or similar lithosphere. The tectonic history of these fragments, however, as well as that of the Kings River ophiolite, is highly divergent from that of the Trinity peridotite. This suggests that the Trinity and Kings River fragments may represent isolated samples from a large lithospheric plate, or that they were dispersed from one another shortly after their petrogenesis.

The tectonic and geographic setting of the spreading ridge genesis for the Kings River ophiolite is poorly constrained. As noted above, the ophiolite is unique in that it lacks any indication of being near an island arc early in its history. As with the Trinity peridotite, it may have formed in an abyssal spreading ridge or an interact basin spreading ridge, although the isotopic characteristics make the latter unlikely. If the Kings River ophiolite formed in an interarc basin, it is constrained to the extreme distal reaches of a large basin by the lack of any arc signature. Alternatively, the ophiolite may have resided in a distant abyssal plain for a substantial period of time, or could have been trapped abyssal crust within a fore arc or a large marginal basin offshore of the Cordillera continental margin. The first entry in its geologic history, following its early Paleozoic petrogenesis and modest pelagic sedimentation, is fragmentation and incorporation into a regional ophiolitic melange adjacent to the Cordilleran margin during the Carboniferous. This amounts to a greater than 150-m.y. gap in its geologic history.

UPPER PALEOZOIC REGIONAL OPHIOLITE MELANGE

Ophiolitic assemblages of the California–southern Oregon region of late Paleozoic to Jurassic age occur in semi-continuous regional belts (Fig. 1), in contrast to the scattered uppermost Precambrian and lower Paleozoic assemblages discussed above. Such regional map patterns, in conjunction with broad stratigraphic relations, facilitate the interpretation of the younger belts in the context of regional Cordilleran tectonics. A major turning point in the tectonic development of the southwest Cordillera was the disruption of Paleozoic northeast-southwest–trending facies patterns in the miogeocline and tectonic trends along the outer shelf. These features ran continuously from eastern Nevada through eastern California prior to their truncation to the southwest and the superpositioning of northwest-southeast structural trends and allochthonous tectonic belts along the truncation zone (Hamilton and Meyers, 1966; Burchfiel and Davis, 1972). The truncation zone appears to have run along the axial Sierra Nevada, and subsequently the California segment of the Cordilleran plate edge followed this trend (Saleeby and Busby-Spera, 1990). Stratigraphic and structural relations in regions southeast of the Sierra Nevada indicate a sinistral transform truncation pattern during Carboniferous to Permian time (Stone and Stevens, 1988; Walker, 1988). This corresponds in time with the formation of the regional ophiolitic melange, which constitutes basement for the Sierran-Klamath Central belt. The temporal relations, along with tectonic position and internal structural-stratigraphic relations, suggest formation of the melange along a major boundary transform related to the truncation event (Saleeby, 1977, 1979, 1990). The truncation locus in the western Sierra Nevada is termed the Foothills suture (Saleeby, 1981).

Upper Paleozoic transform ophiolites

Structurally complex ophiolitic rocks of the western Sierra Nevada are regionally interleaved with belts of lower Mesozoic island-arc–affinity strata and upper Paleozoic to lower Mesozoic chert-rich and continent-derived clastic basinal strata along the Central belt (Saleeby, 1982). In some localities, the ophiolitic rocks can be shown to have constituted basement for such strata and to have been metamorphosed and highly disrupted prior to unconformable overlap. In most cases, however, the contacts between basement and superjacent strata are high-angle faults (Clark, 1964, 1976). From south to north, the major ophiolitic exposures consist of the Kings-Kaweah belt, Tuolumne River belt, Bear Mountains belt, and Feather River belt. Additionally, there are numerous thin, discontinuous serpentinite slivers, which locally contain metabasite, metachert, and carbonate-rock blocks that are structurally encased in younger strata, and which were probably derived from the ophiolitic basement. A number of common features between these belts and slivers suggest that the entire western Foothills belt has a common but complex polygenetic ophiolitic basement. The transform fracture-zone affinity of this basement complex is most clearly displayed in the Kings-Kaweah belt.

Kings-Kaweah belt. The Kings-Kaweah ophiolite belt (Fig. 1N, O) is exposed for ~125 km along the southwestern Sierra Nevada foothills (Saleeby, 1977, 1978, 1979). The Lower Ordovician Kings River ophiolite occurs as several large tectonic slabs at the northern end of the belt. Smaller additional slabs and blocks of the ophiolite are dispersed southward along the belt within the Kaweah serpentinite melange. Narrow belts of Kaweah-type melange, as well as ductile shear zones, bound the Kings River slabs. Metamorphic rocks to the east of the ophiolite belt consist of the Paleozoic(?)–lower Mesozoic Kings sequence (Bateman and Clark, 1974; Saleeby and others, 1978). Lower Mesozoic Kings sequence fore-arc strata are partly depositional above the Kaweah melange (Saleeby and others, 1978; Saleeby, 1979).

Rocks of the Kings-Kaweah ophiolite belt exhibit all the major features typical of major oceanic transform fracture zones (Saleeby, 1977, 1978, 1979, 1984a). These include: (1) injection and solidification of tholeiitic magma into zones of high shear strain, with the production of flaser gabbros and metamorphic tectonites ranging from upper amphibolite to low greenschist (dynamically retrograded) assemblages; (2) development of widespread high-strain mylonitic fabrics superposed over high-temperature mantle flow fabrics in peridotites; (3) evidence for sea-water infiltration into lower crustal–upper mantle rocks during high-temperature deformation; (4) widespread serpentinization of mantle rocks under a wide range of physical conditions, yielding high- and low-temperature as well as highly schistose-to-isotropic serpentinites; (5) widespread diapiric rise of serpentinites and protrusive breachment onto the sea-floor, contemporaneous with basaltic volcanism and pelagic sedimentation; (6) a crisscross pattern in dikes, faults, shear zones, and ultramafic protrusions, which is a common structural feature of spreading ridge–transform intersection zones; (7) complex accumulations and surface remobilizations of polymict sedimentary serpentinite and ophicalite; and (8) chaotic soft sediment deformation of overlying pelagic sediments and continued upward rise and surface breachment of serpentinites, resulting in highly mixed assemblages.

Uranium/lead zircon ages on flaser diorite and plagiogranite dikes and Sm/Nd isochron ages on these dikes, as well as flaser gabbro and diabase dikes, indicate a Carboniferous to Permian age for the transform assemblage (Saleeby and Sharp, 1980; Shaw and others, 1987). In the Kings River area, the dikes, ductile shear zones, and narrow serpentinite melange belts are superposed over Ordovician ridge-crest ophiolite. Detailed resolution of the relative proportions of Ordovician versus Carboniferous to Permian rocks within the Kaweah melange to the south has not been attempted. The younger rocks of the Kings River area yield initial ϵ_{Nd} = +9.9 ± 1.1, initial $^{87}Sr/^{86}Sr$ = 0.703, and initial Pb: 206/204 = 18.04, 207/204 = 15.54, and 208/204 = 37.84 (Shaw and others, 1987). These values clearly reflect a MORB-type depleted mantle source, but are quite distinct from the Ordovician values. It is not clear how much new oceanic lithosphere was created during the Carboniferous to Permian transform event, as opposed to how much Ordovician mantle rock was uplifted to surface levels.

Fragmentation and incorporation of the Kings River ophiolite into the transform melange seemingly require the propagation of a major fracture zone and presumed spreading-ridge segments into the early Paleozoic Panthalassa crust, in conjunction with major transform displacement. As noted above, it may have been abyssal crust or a large marginal basin behind the McCloud arc. Fragments of McCloud arc rocks scattered elsewhere along the Foothills suture are consistent with the latter (Saleeby, 1990). The Kings-Kaweah segment of the transform suture cut into the Cordillera Paleozoic miogeocline, as shown by the juxtaposition of the ophiolitic melange with the Kings sequence. Slide blocks and turbidites of the Kings sequence mixed into chaotic chert and argillite depositionally above the Kaweah melange, indicating juxtaposition prior to the Jurassic (Saleeby and others, 1978). The Kings sequence is an Upper Triassic–Lower Jurassic shelfal and fore-arc basin assemblage built for the most part on the truncated edge of the miogeocline.

Central foothills. The regional upper Paleozoic ophiolitic melange is well exposed in two regions of the central foothills, the lower Toulumne River area (Fig. 1M) and, to the north, along the Bear Mountains ophiolite belt (Fig. 1J). The major exposures occur within cores of large Late Jurassic anticlines.

Serpentinized dunite-harzburgite tectonite slabs and blocks and bounding zones of serpentinite melange of the lower Tuolumne River area have strong similarities to parts of the Kaweah melange (Morgan, 1976; Saleeby, 1982; Sharp, 1988). The ophiolitic rocks occur at the base of Lower Jurassic fore-arc volcanic sequences. Polyphase deformation, metamorphism, and melange mixing predated overlap by the Lower Jurassic sequences. Blocks within serpentinite melange zones consist primarily of mafic tectonite of amphibolite to greenschist facies, metachert and serpentinized peridotite. Rare blocks retaining primary ophiolitic features such as pillows or gabbroic/diabasic textures also occur, as well as rare marble blocks. Very little in terms of a mafic oceanic crustal section remains.

Geochronological data from the Tuolumne area are sparse, yet significant. A plagiogranite melange block yields a Carboniferous to Permian U/Pb zircon age (Saleeby, 1982), and is typical of similar-age blocks and dikes within the Kaweah melange. K/Ar and Ar/Ar ages from mafic tectonite blocks yield cooling ages indicative of pre–Early Jurassic dynamic metamorphism (Morgan, 1976; Sharp, 1988). Similar blocks with similar cooling ages are present within all the major ophiolitic melange and tectonite belts of the western Sierra (Weisenberg and Ave Lallement, 1977; Behrman, 1978; Saleeby and Sharp, 1980; Saleeby, 1982, and unpublished data).

Narrow, discontinuous slivers of serpentinite and serpentinite melange extend northward along major fault zones and occur within cleaved Jurassic strata from the region of the Tuolumne River exposures. These slivers merge with and suggest basement continuity with the Bear Mountains belt to the north (Fig. 1L). This disrupted and polymetamorphosed ophiolite occurs as a major basement exposure for ~80 km along the Bear Mountains fault zone (Behrman, 1978; Saleeby, 1982). The sequence is broken into numerous polylithologic slabs and monolithologic blocks that are separated by an anastomosing network of schistose serpentinite and serpentinite matrix melange. The overall assemblage and its structural configuration are very similar to mafic block and slab-rich domains of the Kaweah melange. The main ophiolite belt is bounded to the east by an Upper Triassic–Lower Jurassic pillow lava–chert sequence, which is overlain by chaotic chert and epiclastic-rich strata that pass upward into Jurassic arc-related strata. There are numerous serpentinite and serpentinite melange exposures along faults and protrusions within these strata that expose the Paleozoic ophiolite basement. To the west, the ophiolite is structurally bounded by Jurassic arc-related rocks. To the north, Middle Jurassic layered gabbro, diorite, and pyroxenite, as well as a mafic-to-intermediate composition sheeted dike complex, crosscut and contact metamorphosed the disrupted ophiolite.

The structural sequence of the ophiolite belt suggests a generally east-facing section with blocks and slabs of serpentinized dunite-harzburgite concentrated to the west and layered igneous peridotite and gabbro progressing to massive and pillowed basalt, and then chert, argillite, and ophicalcite/sedimentary serpentinite toward the east. Major- and trace-element data from the main ophiolite belt and melange blocks to the east suggest a MORB association (Behrman, 1978; Rogers and others, 1984). Plagiogranite melange blocks and a lone dike within a large mafic slab yield U/Pb zircon ages indicating Carboniferous intrusion and local minor inheritance of early Paleozoic zircon (Saleeby, 1982, and unpublished data). The lone dike carries structural relations with its mafic host that indicate that it is relatively late-stage within the ophiolite sequence, but was clearly emplaced prior to melange disruption. Mafic blocks and slabs yield a spectrum of K/Ar and Ar/Ar ages indicating multiple amphibolite-facies events in pre–Early Jurassic, Permo-Carboniferous, and Devonian times (Behrman, 1978; Saleeby, 1982; Sharp, 1988; J. B. Saleeby and W. D. Sharp, unpublished data). One of the metamorphic blocks has eclogitic affinities, although barometric studies have not been conducted in the area.

The available geochronological data, as well as structural setting and lithologic association, provide a strong basis for broad basement correlation between the Bear Mountains, Tuolumne River, and Kings-Kaweah belts. What is not clear concerning the Bear Mountains belt is whether the main mafic crustal sequence was generated during the Carboniferous spreading/transform event, or whether it could be a fragment of older Kings River–affinity crust encased in and intruded by younger dikes, within the transform zone. A small amount of zircon inheritance within the plagiogranite dike indicates the presence of preexisting, early Paleozoic crustal rocks. Lead isotopic data on a Carboniferous melange block that lacks inherited zircon are comparable to data from the Kings-Kaweah Carboniferous to Permian dikes, indicative of a MORB-like source. The presence of a Devonian mafic amphibolite melange block (Sharp, 1988) indicates the presence of older crustal rocks as well.

Feather River belt. The Feather River belt (Fig. 1J) mostly consists of metaperidotite and serpentinite with subordinate mafic rocks that together occur as a fault-bounded belt ranging from 2 to 10 km wide and extending for ~150 km southward from the northern end of the western metamorphic belt (Clark, 1976; Ehrenberg, 1975; Edelman and others, 1989). It is a polygenetic assemblage with both lower and upper Paleozoic ophiolitic constituents, as well as fragments of lower Paleozoic arc-related rocks. The main peridotite massif is composed of metamorphosed harzburgite, lherzolite and dunite tectonite, and serpentinite (Hietanen, 1973; Ehrenberg, 1975). Remnants of high-temperature mantle fabrics trend at high angles to the belt and are commonly transposed by northwest-trending regional metamorphic fabrics. Regional metamorphism was in the low amphibolite-to-greenschist facies and appears to have been predated by a period of serpentinization. Additional episodes of serpentinization occurred during and following regional metamorphism. The high-temperature mantle fabrics are cut by regionally metamorphosed pyroxenite, microgabbro, diabase, and plagiogranite dikes. One of the plagiogranite dikes yields a Carboniferous U/Pb zircon age and initial Pb compositions similar to Carboniferous dikes of the Bear Mountains and Kings-Kaweah belts (Saleeby and others, 1989). The plagiogranite shows synplutonic relations with the mafic dikes, suggesting that the zircon age is applicable to much or all of the dike swarm. The main regional metamorphic event was superposed over the dike swarm and is constrained to have occurred in pre–Late Triassic time by Ar/Ar amphibole age data on an amphibolitic crosscutting dike (Weisenberg and Ave Lallement, 1977), and possibly as early as the Carboniferous by a K/Ar amphibole age (Hietanen, 1981).

The Carboniferous crosscutting dikes in the peridotite appear to have been related to the construction of a mafic crustal sequence (Edelman and others, 1989; Saleeby and others, 1989a). Structurally rooted into part of the western margin of the peridotite is a ~2-km-thick section of amphibolite-to-greenschist facies pillow lava with minor metachert, sheeted mafic dikes, and subordinate gabbro (Devils Gate ophiolite of Edelman and others, 1989). Trace-element data indicate a tholeiitic affinity for the lavas (Hietanen, 1981). Strong deformation accompanied metamorphism of the Devils Gate sequence, and intrusive relations within the sheeted dikes indicate a significant component of deformation and metamorphism during the igneous construction. Amphibole K/Ar and Ar/Ar ages suggest Permo-Carboniferous deformation and metamorphism (Hietanen, 1981; Standlee, 1978). Silicic rocks have not been discovered in the section, and thus, igneous age constraints are not well defined. A representative suite of flows and intrusive rocks from the Devils Gate section and the Carboniferous plagiogranite and related microgabbro dikes that crosscut the peridotite define a linear array on an Sm/Nd evolution diagram, suggesting a Carboniferous igneous age and a common depleted mantle source (Saleeby and others, 1989). The systematics are complex, possibly due to polymetamorphism and contamination, but these data, in conjunction with the plagiogranite zircon age, the Ar/Ar and K/Ar age data, and field relations (Edelman and others, 1989), strongly suggest primary relations between the Devils Gate section and the Carboniferous intrusions within the peridotite.

Relations between the Devils Gate section and the peridotite massif alone demonstrate the polygenetic nature of the Feather River belt. The peridotite represents a preexisting depleted mantle framework in which, and perhaps adjacent to which, the mafic crustal sequence was generated. Depleted mantle rocks genetically related to the crustal sequence have not been identified, although an abundance of harzburgite tectonite occurs west of the Devils Gate sequence structurally mixed with, and as older basement for, Mesozoic basinal and arc strata of the Central belt. The structural-stratigraphic settings of these ultramafic rocks resemble those of the Bear Mountains, Tuolumne, and Kings-Kaweah belts, which suggests broad basement correlation of disrupted Paleozoic upper mantle.

The question of whether the Feather River peridotite bears any relation to the Trinity peridotite is amplified by the presence of structural slices of Devonian and perhaps older arc-related rocks along the Feather River belt. An Early Devonian or older age for a crosscutting hornblende pegmatite within a layered gabbro fault slice is indicated by hornblende Ar/Ar age data (Standlee, 1978). The gabbro and pegmatite are grossly similar to plutonic rocks of the superjacent arc tholeiitic suite of the Trinity ophiolite. An additional structural slice of Devonian or older mafic schist occurs along the Feather River belt near the Alleghany mining district (Bohlke and McKee, 1984; Saleeby and others, 1989). The Alleghany amphibolite consists mainly of a mafic-to-intermediate volcanic and a plutonic protolith assemblage with minor interbedded clastic sedimentary rocks (Hacker, 1988). This assemblage underwent amphibolite-facies metamorphism during, or perhaps prior to, the Devonian. A synmetamorphic peraluminous granitoid (Oriental Mine) intruded the amphibolite and yields complex U/Pb zircon systematics, indicating the incorporation of Proterozoic zircon components during Devonian igneous petrogenesis (Saleeby and others, 1989). The granitoid, as well as a representative sample of mafic schist, yields Nd and Sr isotopic compositions indicative of significant continental components as well. The continental components are considered to have been introduced into the system as detritus. The protolith age of the amphibolite is unknown, although possible correlatives could be reasonably sought in the early Paleozoic rocks of the Yreka terrane and Central Metamorphic belt of the eastern Klamaths.

The relations discussed above suggest the possible presence of structural slices of lower Paleozoic eastern Klamath belt rocks along the Feather River belt. A middle-to-late Paleozoic stage for such tectonic mixing is suggested by relations in the Devils Gate section. Within the sheeted dike sequence there are mafic schist screens, not unlike the Alleghany amphibolite, which yield isotopic signatures clearly distinct from the depleted mantle signature of the main igneous sequence and the dikes that cut the adjacent peridotite. An ϵ_{Nd} (O) value of −20.6 measured on such a screen indicates a strong sialic component, as seen in the Alleghany amphibolite (Saleeby and others, 1989). Local minor contamination of the ophiolitic intrusives with such host materials is a likely mechanism for the complexity noted above for the Carboniferous Sm/Nd evolution array. It is clear that the Carboniferous phase of ophiolite crustal genesis was superposed over a structurally complex basement regime that consisted of a vast expanse of depleted mantle rocks, and structural or stratigraphic units of lower Paleozoic arc-related rocks that carried components of North American sial. The clearest analogue to this basement complex is the lower Paleozoic of the eastern Klamath Mountains.

The tectonics of the Feather River belt appears to be intimately related to that of the eastern Klamaths, although in a more complex setting than originally envisaged by Davis (1969). Disruption and dispersal of eastern Klamath rocks along the Feather River belt is suggested for Carboniferous to Permian time within a transform ± spreading-ridge environment (Saleeby, 1990). Surface-level disruption of the eastern Klamath McCloud arc is indicated in the Feather River area west of the main ophiolite belt by a Permian olistostrome complex that contains blocks of McCloud limestone (Watkins and others, 1987). The olistostromes, as well as possible tectonic fragments of McCloud arc rocks, are tectonically bounded by serpentinized peridotite and serpentitinite melange zones of the Central belt, which are typical of basement exposures woven throughout the northern Sierra west of the Feather River belt and merging southward with the Bear Mountains belt (Hietanen, 1981; Edelman and others, 1989; Saleeby and others, 1989; Dilek and others, this volume).

Much of the Foothills belt is envisaged here as being underlain by a regional ophiolitic melange. The main constituent of this melange is depleted oceanic mantle and its serpentinized equivalents. Part or much of the mantle may be of Trinity–Kings River ophiolite affinity, and part may have spawned Carboniferous mafic crustal rocks in a transform-dominated spreading environment. Encased within the northern segments of this melange are fragments of the McCloud arc probably derived from the eastern Klamaths arc. Coextensive with this melange are belts of Permo-Carboniferous to Triassic chaotic argillite and chert-rich olistostrome complexes. These are interpreted below as large sediment accumulations in bathymetric troughs that formed within the transform and subsequent subduction environments.

Chert-rich olistostrome belts

Distinct belt-like exposures of chaotic chert and argillite run the entire length of the western Sierra metamorphic belt and extend into the Klamath Mountains (Schweickert and others, 1977; Behrman, 1978; Irwin and others, 1978; Saleeby, 1979; Blome and Irwin, 1983; Edelman and others, 1989). The thicker, more continuous belts, along with the ophiolitic melange, constitute basement for the Sierran-Klamath Central belt. The main chert-argillite belt in the Sierra Nevada is the Calaveras complex, which runs semi-continuously along the western margin of the Feather River belt in the north and along the western margin of the southern Shoo Fly complex to the south (Fig. 1). Additional belts of Calaveras-like rocks occur to the west of the main belt and along the Kings-Kaweah ophiolite belt, and are interleaved with Paleozoic ophiolitic basement rocks and Jurassic fore-arc strata. The outstanding feature of the chert-argillite belts is widespread evidence for deformation and accumulation by massive olistostromal phenomena (Schweickert and others, 1977; Saleeby, 1979).

Structurally complex chert and argillite of the North Fork terrane in the central Klamaths strongly resemble rocks of the Calaveras complex (Cox and Pratt, 1973; Irwin and others, 1978; Ando and others, 1983). Radiolarian dates from cherts range from Permian to Middle Jurassic (Blome and Irwin, 1983), and upper Paleozoic ophiolitic basement rocks are exposed in the core of a large antiform. The ophiolite is highly disrupted, yields a Carboniferous to Permian U/Pb zircon age, and consists of both MORB and within-plate–type basaltic flows. The North Fork terrane has strong similarities to the belt of Carboniferous ophiolitic rocks and the coextensive Calaveras complex of the western Sierra. It lies to the west of, and structurally beneath, the Central Metamorphic belt and may, in part, represent the protolith of Triassic high-P/T metamorphic rocks of the Stuart Fork terrane (Hotz and others, 1977; Goodge, 1990). Ophiolitic basement rocks of the North Fork terrane are suggested here to be the northern continuation of the Foothills upper Paleozoic belt.

The generally chaotic chert-argillite belts are considered here to have never constituted coherent sedimentary sequences, nor to have even rested on a coherent basement. In contrast, a tectonically active environment of deposition is envisaged where pelagic and continent-derived clastics accumulated in and along the flanks of linear troughs where ophiolitic basement rocks were being laterally sheared, and differentially uplifted and downdropped by transform and subsequent subduction and fore-arc tectonics (Saleeby, 1990). During chert-argillite deposition, the transform system became the locus of a new subduction zone, resulting in its fore-arc accretion. Thereafter, the accreted fracture zone basement and its chaotic sedimentary belts served as fore-arc basement for a wide variety of Jurassic volcanic and sedimentary sequences.

JURASSIC OPHIOLITIC ASSEMBLAGES

Geochronologic and structural-stratigraphic relations indicate that two major pulses of submarine mafic volcanism and related intrusive activity occurred during the Jurassic. The first commenced in the Late Triassic and was most pronounced in the Early Jurassic; the second was concentrated in the Callovian-Oxfordian. Vestiges of stratigraphic contacts suggest that the Central belt Paleozoic ophiolitic framework was the basement for the Jurassic sequences, although related production of new lithosphere by spreading is also recorded in the Josephine ophiolite (Harper, 1984) and was probably widespread in the Great Valley and Coast Ranges to the west of the Sierra Nevada (Saleeby, 1990). The regional geologic relations of the Jurassic rocks, in conjunction with the petrologic associations, indicate that the two magmatic events reflect major fore-arc extension and interarc basin formation events (Saleeby, 1990). The two extensional events were apparently separated in time by major Middle Jurassic contractile deformation termed the Siskiyou event (Coleman and others, 1988; Saleeby and Busby-Spera, 1990).

Early Jurassic fore-arc volcanism

Distinct volcanic and shallow-level plutonic sequences were produced along the entire length of the Sierran-Klamath orogen during the Early Jurassic. Some of the sequences are presently situated as nappes totally detached from their basement rocks, although some still retain vestiges of primary relations with ophiolitic melange basement rocks. Well-studied sequences discussed below include, from north to south, the Rattlesnake Creek Terrane, the Fiddle Creek complex, the Penon Blanco Formation, and the western facies of the Kings sequence. Petrogenetic ages on key intervals of each of these sequences cluster near 200 ± 10 Ma, (Irwin and others, 1978; Wright, 1981; Saleeby, 1982, and unpublished data; Bateman and others, 1985; Sharp, 1988; Saleeby and others, 1989a). The sequences are characterized by mainly mafic submarine pillow flows and pyroclastic rocks, and shallow-level intrusive rocks ranging mainly from peridotite to diorite in composition. Sequences of radiolarian and tuffaceous chert are associated with pillow flows, but are rare or absent in thick pyroclastic intervals.

Rattlesnake Creek terrane. Relations in the Rattlesnake Creek terrane are best preserved in the southwestern Klamath Mountains (Fig. 1-H), where multi-stage serpentinite matrix melange is depositionally overlain by Upper Triassic–Lower Jurassic volcanic and sedimentary strata (Wright, 1981; Wright and Wyld, 1985). Both the melange and cover rocks are cut by 212- to 192-Ma shallow-level intrusives. The serpentinite melange contains large tracts of tectonitic harzburgite, and blocks of mafic volcanic rock, metagabbro, metachert, argillite, and amphibolite within a low-temperature serpentinite matrix. The metabasaltic blocks, which locally contain well-preserved pillows, exhibit trace-element characteristics diagnostic of both MORB and intraplate lavas. The only age constraints on the protolith of the melange are those constrained by the cover sequence. Considering the structural association and position of the melange, the most likely protolith analogue is the upper Paleozoic regional ophiolitic melange discussed above.

The cover sequence consists of two units, the oldest of which consists of arc tholeiitic basaltic pillow flows and breccias interlayered with terrigenous epiclastic strata, as well as chert, argillite, and minor limestone (Wright and Wyld, 1985). Cherts from this unit have yielded Late Triassic radiolarians (Irwin and others, 1978). Overlying this unit is a predominantly volcanic sequence composed of monolithologic calc-alkaline to shoshonitic basaltic breccias and intercalated hemipelagic and terrigenous epiclastic strata.

The serpentinite melange developed prior to the deposition of the cover sequence but was at least locally remobilized during and after deposition (Wright, 1981; Wright and Wyld, 1985). In addition to distinct lithologic assemblage and metamorphic-grade contrasts across the unconformity, basaltic feeder dikes for the cover sequence intrude the melange, and terrigenous epiclastic strata of the cover sequence contain clasts and detritus derived from the melange basement.

Northern Sierra central belt. Structurally complex lower Mesozoic assemblages similar to the Rattlesnake Creek terrane occur as major constituents of the Central belt in the northern Sierra. Various constituents have been termed the Fiddle Creek Jarbo Gap, and Slate Creek Complexes (Edelman and others, 1989; Dilek and others, this volume). The former two extend southward as far as the eastern margin of the Bear Mountains ophiolite belt (Behrman, 1978; Saleeby, 1982), and the latter has strong affinity to the Penon Blanco Formation farther to the south. The oldest rocks within the Central belt are ophiolitic melange zones and harzburgite tectonite slabs that are structurally interwoven and reworked as diapirs and olistostromes within generally chaotic chert and argillite of the Fiddle Creek Complex (Edelman and others, 1989; Saleeby and others, 1989a; Dilek and others, this volume). The ophiolitic slabs and melange zones merge southward with the Bear Mountains belt (Behrman, 1982; Saleeby, 1982), suggesting that they are part of the regional upper Paleozoic ophiolitic melange. Ophiolitic and chert-argillite assemblages of the lower Feather River area are cut by basalt and basaltic-andesite solitary, composite, and locally sheeted dikes, and by tabular intrusions of flaser gabbro and diorite (Jarbo Gap Complex). The dikes represent low-FeO–high-MgO and LIL-enriched liquids with boninitic affinities (Dilek and others, this volume). Pb/U zircon and Nd/Sm age data indicate emplacement of the dikes and flaser intrusives at ~200 Ma (Saleeby and others, 1989a). Proximal volcanic equivalents to the intrusives have not been identified, although pillowed basaltic to andesitic flows and local silicic volcanic lenses are widespread throughout the adjacent Fiddle Creek Complex (Edelman and others, 1989); some are known to be as young as Early Jurassic (Saleeby and others, 1989a). Central belt strata to the south, adjacent to the Bear Mountains ophiolite belt, contain abundant mafic pillow flows interbedded with Upper Triassic–Lower Jurassic radiolarian chert and are cut by 210- to 200-Ma hypabyssal intrusive rocks (Saleeby, 1982, and unpublished data).

Between the Bear Mountains and Feather River areas, the Central belt structurally encases an intact volcano-plutonic sequence named the Slate Creek Complex (Edelman and others, 1989). The sequence is ~2.5 km thick, with the lower kilometer consisting of wehrlite, gabbro, diorite, and massive greenstone intrusive rocks that grade upward into basalt to basaltic-andesite volcanic rocks. The volcanics are chiefly clinopyroxene and locally plagioclase phyric and consist of submarine tuffs and breccias. Phyric and aphyric pillow flows, volcaniclastic sandstones, and cherty tuffs also occur subordinately. Pb/U zircon data on the hypabyssal complex indicate a 202 ± 2 Ma age (Saleeby and others, 1989), and the Slate Creek Complex strongly resembles the Penon Blanco Formation of the central Sierra, which has yielded very similar ages (Saleeby, 1982). The Slate Creek complex rests as a nappe above Fiddle Creek strata and was emplaced prior to the intrusion of a ~168-Ma crosscutting pluton. Together, upper and lower plate rocks are cut by Late Jurassic (Nevadan) high-angle faults (Edelman and others, 1989; Saleeby and others, 1989a). The nappe is considered to have rooted to the west along an east-vergent thrust. Alternatively, the nappe may have originated as a low-angle extensional fault flake within the Central belt basinal system (Saleeby, 1990).

Penon Blanco Formation. The Penon Blanco Formation is an intact volcano-plutonic sequence that was built unconformably above the Tuolumne ophiolitic melange (Fig. 1K; Clark, 1964; Morgan, 1976; Saleeby, 1982; Bogen, 1985). The base of the Penon Blanco Formation consists of up to 1 km of aphyric basalt and basaltic-andesite pillow lava (Jasper Point Formation of Bogen, 1985). Preliminary geochemical data (J. B. Saleeby, unpublished data) suggest arc tholeiitic and transitional arc tholeiitic-boninitic magma series. The pillowed sequence is overlain depositionally by as much as 20 m of radiolarian, tuffaceous, and locally metalliferous chert. Radiolarians are of "early Mesozoic" age (D. L. Jones, personal communication, 1979). The chert is, in turn, overlain conformably by the main Penon Blanco submarine pyroclastic sequence.

The pyroclastic sequence consists of as much as ~5 km pyroxene ± olivine and, locally, plagioclase or hornblende phyric basalt to basaltic-andesite tuff, tuff breccia, tuffaceous turbidites, and local massive to pillowed flows. Much of the volcanic material has been referred to as ankaramite (Saleeby, 1982), but it also resembles clinopyroxene variants of boninites, referred to as marianites. The base of the pyroclastic sequence consists of pyroxene and hornblende diorite, microdiorite, and gabbroic and wehrlitic intrusives. Pb/U zircon and igneous hornblende dates on the intrusive rocks cluster around 200 Ma (Saleeby, 1982). These ages approximate the age of much of the pyroclastic sequence, as shown by intrusion of dikes up to high levels of the extrusive sequence, and by the inclusion of intrusive series diorite, gabbro, and wehrlite clasts within pyroclastic breccias.

A stratigraphic reconstruction of the Penon Blanco pyroclastic and intrusive sequence suggests that the formation represents a distinct submarine volcano with the intrusive bodies clustered near the main vent area (Bogen, 1985). The outcrop dimensions of the pyroclastic sequence and its turbiditic apron are of appropriate magnitude, compared to modern submarine volcanoes of oceanic island arcs. Outcrop dimensions of the Slate Creek complex in the northern Sierra are also comparable, suggesting a similar origin.

Northward along the Penon Blanco outcrop belt, argillite-rich olistostromes of the Central belt lie depositionally above outer apron turbidites (Schweickert and Bogen, 1983). Analogous stratigraphic relations are preserved eastward from the Penon Blanco vent complex across the Melones fault zone (Sharp, 1988). Here, Lower Jurassic aphyric pillow flows and clinopyroxene phyric pyroclastic rocks of the Sullivan Creek terrane are overlain depositionally by argillite and chert-argillite olistostromes (Bateman and others, 1985; Sharp, 1988). The lower volcanic sequence of the Sullivan Creek terrane may represent an additional structural slice of the Penon Blanco volcano, as well as its ophiolitic melange basement and its overlapping olistostromes, or alternatively, a fragment of another volcano similar to those represented by the Penon Blanco and Slate Creek sequences. The overlap relations of Central belt strata above the Penon Blanco and Sullivan Creek volcanic sequences are critical, and suggest that the Slate Creek complex, as well, sat depositionally within the broad tectonostratigraphic belt now defined by the Fiddle Creek complex and the Penon Blanco–Sullivan Creek volcanics. The environment of deposition is envisaged as an elongate, volcanically active extensional basin with scattered submarine pyroclastic cones (Saleeby and Busby-Spera, 1990).

Western-facies Kings sequence. Quartz-rich and silicic volcaniclastic turbidites and olistostromes of the western Kings sequence are interbedded with and chaotically mixed with chert-argillite and mafic volcanic rocks above the Kaweah serpentinite melange (Fig. 10; Saleeby and others, 1978; Saleeby, 1979). These western-facies strata appear to represent a fore-arc sedimentary prism with proximal arc–volcaniclastic strata occurring in the eastern facies. The fore-arc strata are shed across a thick chert-argillite olistostrome tongue within the Kaweah melange and across a lower Mesozoic pillow flow–chert carapace that was constructed across protrusive serpentinite melange. The chert-argillite olistostrome tongue resembles much of the Calaveras complex, while the pillow flow–chert carapace resembles the lower levels of the Penon Blanco Formation.

Age constraints on the Kaweah pillow-flow carapace are derived from an ~200-Ma zircon age on a peperitic feeder dike that cuts the underlying sedimentary serpentinite (Saleeby and Sharp, 1980). Petrochemical data on the lavas indicate an arc tholeiitic suite, ranging from basalt to basaltic andesite in composition. Stratigraphically higher within turbidites and olistostromes of the Kings sequence are lenses of pyroxene phyric pillow flow, broken pillow breccia, and hyaloclastite as much as several hundred meters in thickness. These lavas have strong boninitic affinities (J. B. Saleeby, unpublished data). Direct age data are lacking, but the volcanic lenses and host strata are crosscut by hornblende phyric calc-alkaline dikes of Middle Jurassic age (Saleeby and Sharp, 1980).

In spite of the complications posed by the more voluminous batholithic material in the southern Sierra Nevada, a number of critical and unique relations are preserved. These include evidence for the shedding of silicic pyroclastic and volcaniclastic material into the fore-arc setting, which was undergoing sporadic phases of island-arc tholeiitic and boninitic volcanism, and the occurrence of fore-arc volcanic and sedimentary rocks with strong similarities to both the Penon Blanco Formation and the Fiddle Creek complex. In conjunction with the occurrence of Jura-Trias arc rocks along the eastern Sierra Nevada and Klamaths (Saleeby and Busby-Spera, 1990), the entire belt of lower Mesozoic strata along the western Foothills and western Klamaths is considered a complex extensional fore-arc terrane built on the accreted Paleozoic ophiolitic melange (Saleeby, 1990).

Late Middle Jurassic fore-arc and interarc basin ophiolites

Deformation and metamorphism of the Middle Jurassic Siskiyou event resulted in the west-vergent imbrication of Lower

Jurassic fore-arc sequences and their ophiolitic melange basement complexes (Harper and Wright, 1984; Sharp, 1988; Wright and Fahan, 1988; Coleman and others, 1988; Saleeby and Busby-Spera, 1990). Partly coincident and during the waning phases of this orogeny, basalt and basaltic-andesite pillowed and highly porphyritic pyroclastic sequences were erupted over a broad region, and a related belt of peridotitic-to-dioritic intrusive complexes was emplaced (Harper and Wright, 1984; Sharp, 1988; Wright and Fahan, 1988; Saleeby and Busby-Spera, 1990). This Middle Jurassic volcano-plutonic belt represents a distinct tholeiitic-to-calcalkaline chain, which was localized along the axis of the Siskiyou deformation zone. Representative volcanic units include the western Hayfork terrane in the Klamath Mountains and the Logtown Ridge Formation in the western Sierra Nevada. During and at the close of the Callovian, the fore-arc margin of the Siskiyou deformation zone and superposed arc began to rift, and then a family of interarc basins rapidly formed. Rift complexes are preserved in the western Sierra Nevada and Klamath Mountains, and interarc basin and fringing arc or rifted fore-arc igneous complexes are preserved in the western Klamath Mountains and Coast Ranges (Saleeby, 1990).

Rift complexes. A distinct phase of rifting is recorded in the northwestern Sierra Nevada between 164 to 162 Ma by major swarms of basaltic-to-plagiogranite sheeted dikes and parallel tabular intrusions (Saleeby, 1982; Saleeby and others, 1989a). At the northern end of the Bear Mountains ophiolite belt, sheeted dikes and coeval gabbro and diorite lenticular stocks were injected along the preexisting ophiolitic melange structural fabric and were accompanied by dynamothermal contact aureoles (Springer, 1980; Saleeby, 1982). The Folsom dike swarm resembles a sheeted complex except for the fact that it is bounded by, and locally encases, screens of Paleozoic host rocks. The dike complex extends for more than 30 km along strike. To the south, it feathers into the Bear Mountains belt; to the north, it apparently feathers into the upper volcanic unit of the Smartville Complex. Early phases of low-angle normal faulting may have occurred along the Bear Mountains fault zone in conjunction with the extensional magmatism (Saleeby, 1990).

The Smartville Complex to the north is a well-preserved pseudostratigraphic sequence consisting of minor layered gabbro, larger gabbro-diorite, and tonalite tabular plutons, massive diabasic hypabyssal rock, pillow basalt, mafic and plagiogranite sheeted dikes, and submarine pillowed and volcaniclastic rocks of basaltic to locally dacitic composition (Xenophontos and Bond, 1978; Menzies and others, 1980; Beard and Day, 1987). Petrographic and field data along with rare earth and major-element chemical data indicate a tholeiitic to calc-alkaline evolution from early to late magmas and flows, which along with the dominance of porphyritic volcaniclastic rocks and stratabound sulfide ore deposits, indicate an island-arc origin. The early (lower) mafic flows are typically pillowed and aphyric, with later (upper) basaltic to andesitic pyroclastic rocks distinctly clinopyroxene phyric. The volcanic sequence is considered Middle Jurassic in age, based on several relations. First, a representative suite of bulk-rock and mineral separates from the volcanic and underlying intrusive rocks define a strong linear array on a Sm/Nd evolution diagram corresponding to an igneous age of 178 ± 21 Ma and an ϵ_{Nd} (t) = +9.2 ± 0.6 (Saleeby and others, 1989a). These data permit an Early Jurassic age, but a silicic tuff from the uppermost stratigraphic levels of the volcanic rocks yields a 164 ± 1 Ma Pb/U zircon age. Breaks in the volcanic section are not recognized, and thus, a Middle Jurassic eruptive age is most reasonable. Major swarms of sheeted dikes and tabular intrusions invaded the volcanic section at 162 ± 1 Ma (Day, 1977; Beard and Day, 1987; Saleeby and others, 1989a). The dikes range from basalt to sodic rhyolite in composition, and have both arc tholeiitic and calc-alkaline geochemical affinities (Day, 1977). The major swarms are parallel to and on strike with the Folsom swarm to the south.

The Smartville Complex appears to have developed within the Central belt basement framework. This is suggested by map relations with Central belt rocks (Jarbo Gap complex) at the northern end of the Smartville Complex (Dilek and others, this volume), the dike swarms and intrusions observed cutting the Bear Mountains belt to the south, and the local inclusion of ~200-Ma metaplutonic rock as screens within Smartville intrusives (Bickford and Day, 1988).

In the west-central Klamath Mountains to the north, the Rattlesnake Creek terrane in the area of Preston Peak and the Marble Mountains also records a phase of Middle Jurassic rifting (Fig. 1C). At Preston Peak, dikes, irregular hypabyssal bodies, near-vent breccias, and local pillow flows of an arc tholeiitic series are well preserved in a section that intrudes and unconformably overlies older serpentinite-matrix melange basement (Snoke and others, 1977; Snoke and Whitney, 1979; Saleeby and others, 1982). The melange basement contains mainly large blocks of tectonitic harzburgite and lherzolite with smaller mafic amphibolite blocks typical of laterally continuous Rattlesnake Creek terrane rocks to the south. A late-stage dike in the hypabyssal complex yields a 164 ± 1 Ma zircon age, and Jurassic radiolarians have been extracted from tuffaceous chert that overlies pillow flows (Saleeby and others, 1982; Harper and others, in press).

To the east in the Marble Mountains area, high-grade ophiolitic melange is considered equivalent to the Rattlesnake Creek terrane, but there it has undergone amphibolite-facies metamorphism en masse within the Siskiyou thrust belt (Hill, 1985; Donato, 1987; Coleman and others, 1988). The interlayered mafic amphibolites, metaperidotites, and siliceous schists represent metamorphosed ophiolitic melange. The remnants of a sheeted dike complex were recognized within one of the larger amphibolite layers (Hill, 1985), and a plagiogranite dike yields a 172 ± 2 Ma zircon age (Harper and others, 1990).

The Rattlesnake Creek terrane formed the western basement complex for the Middle Jurassic western Hayfork arc sequence (Wright and Fahan, 1988). From the above relations, it appears that the basement complex underwent a rifting phase between 172 and 165 Ma. Such rifting may correspond to the initial phases of interarc basin formation represented in the western Klamaths by the Josephine ophiolite (Snoke, 1977; Saleeby and others, 1982; Harper, 1984; Donato, 1987). The oldest vestiges of this basin system, or alternatively, the attenuated fore arc, that faced the basin, may be exposed in the Condrey Mountain window as the subcreted Condrey Mountain schist terrane (Klein, 1977; Harper and others, 1990). The schist terrane consists of an internally imbricated intra- or interarc basin volcanic-sedimentary sequence that was metamorphosed in transitional blueschist-greenschist facies during subcretion beneath the Rattlesnake Creek terrane (Helper, 1986; Coleman and others, 1988). A premetamorphic dioritic lens yields a zircon age of 170 ± 2 Ma which places a minimum age constraint on the upper-level basic schist unit, and syn-thrusting leucosomes at the structural base of amphibolite-facies upper plate rocks yield a zircon age of 160 ± 2 Ma (Harper and others, 1990). These relations indicate that the initial underthrusting of the Condrey Mountain terrane began virtually during, or at the cessation of, spreading within the Josephine basin.

Josephine ophiolite. Remnants of oceanic crust and upper mantle that formed in an interarc basin are preserved as the Josephine ophiolite, which runs semi-continuously along the western edge of the Klamath Mountains for more than 200 km (Snoke, 1977; Harper, 1980, 1984, Saleeby and others, 1982). The most complete section of the ophiolite lies immediately south of the Oregon-California border (Fig. 1B; Harper, 1984). Here, the ophiolite consists of harzburgite tectonite, cumulate ultramafites and gabbro, high-level gabbro and diorite, a 1.5-km-thick sheeted dike complex, and 0.4 km of pillow lava and breccia (Harper, 1980, 1984). Pillow lavas show geochemical features of arc tholeiites and boninites (Harper, 1988). Pb/U zircon ages of 157 ± 2 Ma were reported by Saleeby and others (1982), but subsequent studies have shown these ages to be slightly disturbed, and the true ages to be as old as 162 Ma (Saleeby, 1987; Harper and others, 1990). A zircon age as old as 164 ± 1 Ma was reported from a small remnant of the ophiolite from the southern end of the belt (Fig. 1G; Wright and Wyld, 1986). Entrainment of substantially older zircon at this locality suggests ophiolite generation within, or in immediate proximity to, older Klamath terranes, which commonly contain Proterozoic detrital zircon within epiclastic strata (Miller and Saleeby, 1987; Miller and others, 1988). Structural relations within the Josephine ophiolite suggest sea-floor spreading, plate transport, and thrust emplacement kinematics with strong tangential components relative to the Cordilleran plate edge (Harper and others, 1985). Such relations suggest that rift edges to the Josephine basin originally contained significant boundary transform segments.

The Josephine ophiolite is conformably overlain by the Galice Formation, consisting of a thin, siliceous pelagic sequence overlain by a thick sequence of turbidites having both volcanic arc and recycled continental sources. The ophiolite and overlying Galice Formation were intruded by numerous 151- to 150-Ma calc-alkaline dikes and sills. Relations along the base of the Galice are critical for the interpretation of the tectonic evolution of the Josephine ophiolite. Northward along the Galice outcrop belt, the substrate geology changes dramatically. In southern Oregon, the lower Galice is interstratified with pyroclastic apron deposits of the Upper Jurassic Rogue Formation (Fig. 1A), which represents the volcanic levels of a fringing arc built outboard of the Josephine basin. The Josephine ophiolite lies structurally between the Rogue arc sequence and the Rattlesnake Creek terrane along Late Jurassic (Nevadan) thrust faults. The thrusts appear to have nucleated along boundary transforms or rift edges between the Josephine interarc basin and adjacent rifted arc segments represented by the Rogue and Rattlesnake Creek terrane. This is suggested by the conformable overlap of the Galice Formation across both the ophiolite and the Rogue sequence, and by the presence of rifted screens of Rattlesnake Creek terrane within the southern Josephine ophiolite (Fig. 1G; Harper and others, 1985, 1990; Wright and Wyld, 1985). Furthermore, fragments of Rattlesnake Creek terrane–like rocks occur within the Rogue basement (Fig. 1A), suggesting that the Josephine basin opened within the confines of older western Klamath lithosphere, and then shortly thereafter the basin closed, leaving the relatively narrow ophiolite belt as its remnants (Saleeby and others, 1982; Harper and others, 1990).

The Rogue Formation consists of possibly as much as 5 km of basaltic-andesite submarine pyroclastic rocks with subordinate basaltic flows and dacitic tuffs and tuff breccias (Wells and Walker, 1953; Garcia, 1979; Saleeby, 1984). Phase chemistry and bulk-rock major-, minor- and trace-element data show island-arc affinity for the volcanic sequence and for spatially-related intrusives (Garcia, 1979, 1982). The upper part of the Rogue has yielded 157 ± 1 Ma Pb/U zircon ages from dacitic tuff and nearby dacitic breccia (Saleeby, 1984b), and as suggested by the gradational contact with the upper Oxfordian–lower Kimmeridgian Galice Formation, little or no time elapsed between deposition of the two. A thin hemipelagic and arc-derived tuffaceous interval (<50 m) intervenes between typical Galice turbidites and uppermost pillow flows of the Josephine ophiolite in northernmost California. This interval probably represents the distal tephra apron of the Rogue arc, suggesting that the arc eruptives, as well as the Galice flysch, lapped across the Rogue arc/Josephine basin boundary. An additional critical relation is that the Rogue volcanic sequence rests on a complex ophiolitic substrate composed of amphibolite-facies metabasite, metachert, serpentinite, and Upper Triassic radiolarian chert (Garcia, 1979, 1982; Roure and DeWever, 1983; Saleeby, 1984b). The basement complex strongly resembles the Rattlesnake Creek terrane or similar central belt rocks of the Sierran Foothills. Furthermore, the basement is cut by Middle and Upper Jurassic peridotitic-to-dioritic arc-related intrusives (Rum Creek–Chetco complexes) and swarms of ~160-Ma mafic and plagiogranite dikes (Hotz, 1971; Garcia, 1979, 1982; Saleeby, 1984b). Thus, a fragment of Rattlesnake Creek terrane and/or Sierran Central belt rocks, along with the roots of the western Hayfork arc, appear to have been rifted to the outer fringes of the Josephine basin to serve as fringing arc basement.

Imbrication of the Josephine ophiolite above the Rogue arc sequence and beneath the Rattlesnake Creek terrane proper occurred during the Late Jurassic Nevadan orogeny. The age relations and structural stacking arrangement suggest west and north-northwest–vergent collapse of a remnant arc (Rattlesnake Creek terrane and related Middle Jurassic arc rocks), interarc basin (Josephine ophiolite), and fringing arc (Rogue sequence) system. Detailed structural and geochronological studies demonstrate the oblique collapse of the interarc basin immediately following its formation (Harper and others, this volume, and 1990). Integrated $^{40}Ar/^{39}Ar$ and U/Pb zircon geochronological and structural studies, along the amphibolite sole of the Josephine ophiolite where it rests tectonically above fringing arc plutonic rocks, indicate that the ophiolite was detached at upper mantle levels within 10 m.y. of its formation, and obducted primarily northward over the arc terrane and overlapping Galice Formation (Harper and others, 1990). Such thrusting was completed by 150 Ma. Within the same time interval, 160 to 150 Ma, perhaps as much as 100 km of the Josephine basin was subducted eastward beneath the Rattlesnake Creek terrane (Klein, 1977; Helper, 1986; Harper and others, 1990). During and at the end of Josephine basin closure, numerous calc-alkaline stocks and dikes cut all major plates of the thrust system, suggesting the migration of arc magmatism to the locus of the closed interarc basin, presumably above an outer east-dipping subduction zone.

Jurassic tectonics

The Siskiyou and Nevadan deformation events are suggested to have been separated by a brief interlude of regional extensional tectonism represented in the Klamath Mountains by the formation of the Josephine ophiolite, and in the Sierra Nevada by emplacement of the Smartville dike complex (Harper and Wright, 1984; Wright and Fahan, 1988). This is supported in the Klamath Mountains by the ~170- to 161-Ma age constraints for Siskiyou deformation, the 164- to 162-Ma age of the Josephine ophiolite, and the >150-Ma Nevadan collapse of the Josephine basin. The ~160-Ma age for the initiation of Condrey Mountain terrane subcretion raises the possibility, however, that a near-continuum existed between Siskiyou and Nevadan thrusting. In such a case, Josephine and possibly Condrey Mountain terrane interarc basin crust was subducted eastward virtually as it formed. Such a plate kinematic pattern is not highly unlikely if spreading and transform displacements were to have occurred with high tangential components relative to the Cordilleran plate edge, as is in fact suggested by structural and paleobiogeographic data (Harper and others, 1985, 1989, this volume; Saleeby, 1990).

The Nevadan orogeny appears not to have involved subduction in the western Sierra Nevada. The main structural expressions are eastward overthrusting of the Smartville complex across the Central belt, followed by high-angle reverse faulting (Day and others, 1985; Edelman and others, 1989), folding, and at least local cleavage development in the Galice-equivalent Mariposa flysch (Schweickert and others, 1988; Tobisch and others, 1989), and a combination of west-directed reverse and sinistral faulting along the Foothills fault system (Newton, 1986; Schweickert and others, 1988; Gefell and others, 1989). Such deformation occurs primarily in rock assemblages whose counterparts in the Klamath Mountains rode in the upper plate of the Condrey Mountain–Josephine subduction zone. Nevadan-age west-dipping subduction may have occurred in the Great Valley along Sierra Nevada latitudes (Saleeby and others, 1986). Such a plate kinematic pattern requires a Nevadan-age transform tear between the southern Klamaths and the northern Sierras (Harper and others, 1985; Saleeby and Busby-Spera, 1990).

The distinction between Nevadan and Siskiyou deformation is obscured locally in the Sierran Foothills, as it is in the west-central Klamath Mountains. Along the northern end of the Bear Mountains ophiolite belt, ~164-Ma sheeted dikes and layered mafic intrusions were intensely deformed while still hot, and possibly while still partly magmatic. The ophiolitic melange wall rocks were metamorphosed en masse to high-temperature amphibolite facies, along with the synplutonic deformation (Springer, 1980; Saleeby, 1982). The timing and style of the deformation and metamorphism are very similar to that reported for the Siskiyou event in the west-central Klamath Mountains (Donato, 1987; Coleman and others, 1988). Immediately north of the Bear Mountains deformation zone, however, Smartville sheeted dikes were emplaced ~2 m.y. later, without any such deformation. The dike complex and its volcanoplutonic wall rocks were overthrust eastward by ~158 Ma (Saleeby and others, 1989a). Rapid sequential Nevadan contractile deformation, longitudinal sheeted dike emplacement, and then subsequent homoaxial contraction is recorded at ~150 Ma at the southern end of the Foothills belt as well (Tobisch and others, 1989; Saleeby and others, 1989b; Saleeby, 1990; Wolf and Saleeby, 1990). These complex patterns are also interpreted as an integral part of strong tangential plate motion components during Nevadan time (Saleeby and Busby-Spera, 1990).

CONCLUSIONS

Ophiolitic assemblages of the Sierran-Klamath orogen record a wealth of plate tectonic and petrogenetic phenomena that span latest Precambrian to Late Jurassic time. Remnants of oceanic lithosphere created as abyssal crust and as extensional fore-arc complexes are represented by the dispersed fragments of uppermost Precambrian and lower Paleozoic assemblages. Regionally extensive ophiolitic belts were developed in late Paleozoic through Jurassic time. These may be related with some confidence to major tectonic events along the Cordilleran margin, the oldest of which was the late Paleozoic transform truncation of the Paleozoic miogeocline and outboard orogenic and island-arc terranes, followed by the Triassic initiation of east-dipping subduction beneath southwestern North America. Accretion of a regional ophiolitic melange during the transform truncation to subduction transition established the fore-arc basement regime for the Cordilleran convergent plate edge. Multiple episodes of early Mesozoic fore-arc extension, interarc basin nucleation and growth, and contractile deformation characterize the regionally extensive fore-arc basement. The rocks of this regime are considered to be parautochthonous in a plate tectonic sense, although structurally, virtually all rock units are allochthonous.

ACKNOWLEDGMENTS

Support for work on U.S. Cordillera ophiolites from National Science Foundation Grants EAR-7708691, EAR-7925998, EAR-8206382, and EAR-8407788, and the Alfred P. Sloan Foundation is gratefully acknowledged. Lively discussions and field excursions with M. C. Blake, R. G. Coleman, P. F. Dobson, S. H. Edelman, B. R. Hacker, G. D. Harper, J. W. Hawkins, C. A. Hopson, D. L. Jones, J. M. Mattinson, E. M. Moores, A. Nicholas, R. A. Schweickert, W. D. Sharp, A. W. Snoke, J. E. Wright, and W. H. Wright III were of great assistance in my ophiolite research and in the construction of this synthesis. Helpful reviews of the initial manuscript were provided by R. G. Coleman, M. M. Miller, and E. M. Moores.

REFERENCES CITED

Albers, J. P., Kistler, R. W., and Kwak, L., 1981, The Mual Mountains Stock; An early Middle Devonian pluton in northern California: Isochron/West, no. 31, p. 17.

Ando, C. J., Irwin, W. P., Jones, D. L., and Saleeby, J. B., 1983, The North Fork ophiolitic terrane of the central Klamath Mountains, California: Geological Society of America Bulletin, v. 94, p. 236–252.

Bateman, P. C., and Clark, L. D., 1974, Stratigraphic and structural setting of the Sierra Nevada batholith, California: Pacific Geology, v. 8, p. 78–89.

Bateman, P. C., Harris, A. G., Kistler, R. W., and Krauskopf, K. B., 1985, Calaveras reversed: Westward facing is indicated: Geology, v. 13, p. 338–342.

Beard, J. S., and Day, H. W., 1987, The Smartville intrusive complex, Sierra Nevada, California; The core of a rifted volcanic arc: Geological Society of America Bulletin, v. 99, p. 779–791.

Behrman, P. G., 1978, Pre-Callovian rocks west of the Melones Fault Zone, central Sierra Nevada Foothills, *in* Howell, D. E., and McDougal, K. A., eds., Mesozoic paleogeography of the western United States: Pacific Section, Society of Economic Paleontologists and Mineralogists Pacific Coast Paleogeography Symposium 2, p. 337–348.

Bickford, M. E., and Day, H. W., 1988, Jurassic ages of arc-ophiolite complexes, northern Sierra Nevada; Implications for duration of the Nevadan "orogeny": Geological Society of America Abstracts with Programs, v. 20, p. A274.

Blome, C. D., and Irwin, W. P., 1983, Tectonic significance of late Paleozoic to Jurassic radiolarians from the North Fork terrane, Klamath Mountains, California, *in* Stevens, C. H., ed., Pre-Jurassic rocks in western North America suspect terranes: Pacific Section, Society of Economic Paleontologists and Mineralogists Special Publication.

Bloomer, S. H., 1983, Distribution and origin of igneous rocks from the landward slopes of the Mariana Trench; Implications for its structure and evolution: Journal of Geophysical Research, v. 88, no. B9, p. 7411–7428.

Bogen, N. L., 1985, Stratigraphic and sedimentologic evidence of a submarine island-arc volcano in the lower Mesozoic Penon Blanco and Jasper Point Formations, Mariposa County, California: Geological Society of America Bulletin, v. 96, p. 1322–1331.

Bohlke, J. K., and McKee, E. H., 1984, K/Ar ages relating to metamorphism, plutonism, and gold-quartz vein mineralization near Alleghany, Sierra County, California, Isochron/West, no. 39, p. 3–7.

Bond, G. C., and DeVay, J. C., 1980, Pre–Upper Devonian quartzose sandstones in the Shoo Fly Formation, northern California; Petrology, provenance, and implications for regional tectonics: Geology, v. 88, p. 285–308.

Bond, G. C., and Kominz, M. A., 1984, Construction of tectonic subsidence curves for the early Paleozoic miogeocline, southern Canadian Rocky Mountains; Implications for subsidence mechanisms, age of breakup, and crustal thinning: Geological Society of America Bulletin, v. 95, p. 155–173.

Bond, G. C., Nickerson, P. A., and Kominz, M. A., 1984, Breakup of a supercontinent between 625 Ma and 555 Ma; New evidence and implications for continental histories: Earth and Planetary Science Letters, v. 70, p. 325–345.

Boudier, F., Le Sueur, E., and Nicholas, A., 1989, Structure of an atypical ophiolite; The Trinity complex, eastern Klamath Mountains, California: Geological Society of America Bulletin, v. 101, p. 820–833.

Brouxel, M., and Lapierre, H., 1988, Geochemical study of an early Paleozoic island-arc–back-arc system; Part 1, The Trinity Ophiolite (northern California): Geological Society of America Bulletin, v. 100, p. 1111–1119.

Brouxel, M., Lapierre, H., Michard, A., and Albarede, F., 1988, Geochemical study of an early Paleozoic island-arc–back-arc basin system; Part 2, Eastern Klamath, early to middle Paleozoic island-arc volcanic rocks (northern California: Geological Society of America Bulletin, v. 100, p. 1120–1130.

Burchfiel, B. C., and Davis, G. A., 1972, Structural framework and evolution of the southern part of the Cordilleran orogen, western United States: American Journal of Science, v. 272, p. 97–118.

Cannat, M., and Boudier, F., 1985, Structural study of intra-oceanic thrusting in the Klamath Mountains, northern California; Implications for accretion geometry: Tectonics, v. 4, no. 5, p. 435–452.

Cashman, S. M., 1980, Devonian metamorphic event in the northeastern Klamath Mountains, California: Geological Society of America Bulletin, part 1, v. 91, p. 453–459.

Clark, L. D., 1964, Stratigraphy and structure of part of the western Sierra Nevada metamorphic belt, California: U.S. Geological Survey Professional Paper 410, 70 p.

——, 1976, Stratigraphy of the north half of the western Sierra Nevada metamorphic belt, California: U.S. Geological Survey Professional Paper 923, 26 p.

Coleman, R. G., Manning, C. E., Mortimer, N., Donato, M. M., and Hill, L. B., 1988, Tectonic and regional metamorphic framework of the Klamath Mountains and adjacent Coast Ranges, California and Oregon, *in* Ernst, W. G., ed., Metamorphism and crustal evolution of the western United States; Rubey Volume 7: Englewood Cliffs, New Jersey, Prentice-Hall, p. 1061–1097.

Cotkin, S. J., and Armstrong, R. L., 1987, Rb/Sr age, geochemistry, and tectonic significance of blueschist from the schist of Skookum Gulch, eastern Klamath Mountains, California; Introducing the Callahan event: Geological Society of America Abstracts with Programs, v. 19, p. 367–368.

Cox, D. P., and Pratt, W. P., 1973, Submarine chert-argillite slide breccia of Paleozoic age in southern Klamath Mountains, California: Geological Society of America Bulletin, v. 84, p. 1423–1438.

Davis, G. A., 1969, Tectonic correlations, Klamath Mountains and western Sierra Nevada, California: Geological Society of America Bulletin, v. 80, p. 1095–1108.

Davis, G. A., Burchfiel, B. C., and Monger, J.W.H., 1978, Mesozoic construction of the Cordilleran "collage," central British Columbia to central California, *in* Howell, D. G., and McDougall, K. A., eds., Mesozoic paleogeography of the western United States: Pacific Section: Society of Economic Paleontologists and Mineralogists Pacific Coast Paleogeography Symposium 2, p. 1–32.

Day, H. W., Moores, E. M., and Tuminas, A. C., 1985, Structure and tectonics of the northern Sierra Nevada: Geological Society of America Bulletin, v. 96, p. 436–450.

Day, S. D., 1977, The petrology of a mafic dike complex near Smartville, Yuba County, California [M.S. thesis]: Davis, University of California, 113 p.

Donato, M. M., 1987, Evolution of an ophiolitic tectonic melange, Marble Mountains, northern California Klamath Mountains: Geological Society of America Bulletin, v. 98, p. 448–464.

Edelman, S. H., and 5 others, 1989, Structure across a Mesozoic ocean-continent suture zone in the northern Sierra Nevada, California: Geological Society of America Special Paper 224, 56 p.

Ehrenberg, S. N., 1975, Feather River ultramafic body, northern Sierra Nevada, California: Geological Society of America Bulletin, v. 86, p. 1235–1243.

Garcia, M. O., 1979, Petrology of the Rogue and Galice Formation, Klamath Mountains, Oregon; Identification of a Jurassic island-arc sequence: Journal of Geology, v. 86, p. 29–41.

—— , 1982, Petrology of the Rogue River island-arc complex, southwest Oregon: American Journal of Science, v. 282, p. 783–807.

Gefell, M. J., Twiss, R. J., and Moores, E. M., 1989, Ductile and brittle shear sense for the "melones Fault zone," northern Sierra Nevada, California: Geological Society of America Abstracts with Programs, v. 21, p. 83.

Girty, G. H., and Guthrie, L., 1989, A westward verging imbricate structure, Shoo Fly Complex, Sierra Nevada, California: Geological Society of America Abstracts with Programs, v. 21, no. 5, p. 38.

Girty, G. H., and Wardlaw, M. S., 1984, Was the Alexander terrane a source of feldspathic sandstones in the Shoo Fly Complex, Sierra Nevada, California?: Geology, v. 12, p. 339–342.

Goodge, J. W., 1990, Tectonic evolution of a coherent Late Triassic subduction complex, Stuart Fort terrane, Klamath Mountains, northern California: Geological Society of America Bulletin, v. 102, p. 86–101.

Hacker, B. R., 1988, Multiple metamorphic and deformation events in the Central and Feather River Belts, northern Sierra Nevada: Geological Society of America Abstracts with Programs, v. 20, p. 165.

Haessig, P. A., 1989, Field relations and structure of the schist of Skookum Gulch, an early Paleozoic melange, Callahan-Yreka area, eastern Klamath Mountains: Geological Society of America Abstracts with Programs, v. 21, p. 88.

Hamilton, W., and Myers, W. B., 1966, Cenozoic tectonics of the western United States: Reviews in Geophysics, v. 4, p. 509–549.

Harper, G. D., 1980, The Josephine ophiolite; Remains of a late Jurassic marginal basin in northwestern California: Geology, v. 8, p. 333–337.

—— , 1984, The Josephine ophiolite, northwestern California: Geological Society of America Bulletin, v. 95, p. 1009–1026.

—— , 1988, Episodic magma chambers and amagmatic extension in the Josephine ophiolite: Geology, v. 16, p. 831–834.

Harper, G. D., and Wright, J. E., 1984, Middle to late Jurassic tectonic evolution of the Klamath Mountains, California-Oregon: Tectonics, v. 3, p. 759–772.

Harper, G. D., Saleeby, J. B., and Norman, E.A.S., 1985, Geometry and tectonic setting of a sea-floor spreading for the Josephine ophiolite, and implications for Jurassic accretionary events along the California margin, *in* Earth Science Series for Energy and Mineral Resources of Circum-Pacific: American Association of Petroleum Geologists, v. 1, p. 239–258.

Harper, G. D., Saleeby, J. B., Pessagno, E. A., Jr., and Heizler, M., 1989, The Josephine ophiolite; Generation, translation, and emplacement of Late Jurassic parautochthonous back-arc lithosphere, Klamath Mountains, California-Oregon: Geological Society of America Abstracts with Programs, v. 21, p. A28.

Harper, G. D., Saleeby, J. B., and Heizler, M., 1990, Isotopic age of emplacement of the Josephine ophiolite and overlying flysch during the Late Jurassic Nevadan orogeny, Klamath Mountains, Oregon-California: Journal of Geophysical Research (in press).

Harwood, D. S., 1988, Tectonism and metamorphism in the northern Sierra terrane, northern, California, *in* Ernst, W. G., ed., Tectonic and metamorphic evolution of the U.S. Cordillera; Rubey Volume 7: Englewood Cliffs, New Jersey, Prentice-Hall, p. 764–788.

Helper, M. A., 1986, Deformation and high P/T metamorphism in the central part of the Condrey Mountain window, north-central Klamath Mountains, California and Oregon, *in* Evans, B. W., and Brown, eds., Blueschists and eclogites: Geological Society of America Memoir 164, p. 125–141.

Hietanen, A., 1973, Geology of the Pulga and Bucks Lake Quadrangles, Butte and Plumas Counties, California: U.S. Geological Survey Professional Paper 731, 66 p.

—— , 1981, Petrologic and structural studies in the northwestern Sierra Nevada, California: U.S. Geological Survey Professional Paper 1226-A-C, 59 p.

Hill, L. B., 1985, Metamorphic, deformational, and temporal constraints on terrane assembly, northern Klamath Mountains, California, *in* Howell, D. G., ed., Tectonostratigraphic terranes of the circum-Pacific region: American Association of Petroleum Geologists Earth Science Series 1, p. 173–186.

Hotz, P. E., 1971, Plutonic rocks of the Klamath Mountains, California and Oregon: U.S. Geological Survey Professional Paper 684-B, 19 p.

Hotz, P. E., Lanphere, M. A., and Swanson, D. A., 1977, Triassic blueschist from northern California and north-central Oregon: Geology, v. 5, p. 659–663.

Irwin, W. P., 1981, Tectonic accretion of the Klamath Mountains, *in* Ernst, W. G., eds., The geotectonic development of California; Rubey Volume 1: Englewood Cliffs, New Jersey, Prentice-Hall, p. 19–49.

Irwin, W. P., and Lipman, P. W., 1962, A regional ultramafic sheet in eastern Klamath Mountains, California: U.S. Geological Survey Professional Paper 450-C, p. 18–21.

Irwin, W. P., Jones, D. L., and Kaplan, T. A., 1978, Radiolarians from pre-Nevadan rocks of the Klamath Mountains, California and Oregon, *in* Howell, D. G., and McDougall, K. A., eds., Mesozoic paleogeography of the western United States: Pacific Section, Society of Economic Paleontologists and Mineralogists Pacific Coast Paleogeography Symposium 2, p. 303–310.

Jacobsen, S. B., Quick, J. E., and Wasserburg, G. J., 1984, A Nd and Sr isotopic study of the Trinity peridotite; Implications for mantle evolution: Earth and Planetary Science Letters, v. 68, p. 361–378.

Klein, C. W., 1977, Thrust plates of the north-central Klamath Mountains near Happy Camp California: California Division of Mines and Geology Special Publication 28, p. 23–26.

Lanphere, M. A., Irwin, W. P., and Hotz, P. E., 1968, Isotopic age of the Nevadan orogeny and older plutonic and metamorphic events in the Klamath Mountains, California: Geological Society of America Bulletin, v. 79, p. 1027–1052.

Lapierre, H., and 5 others, 1986, The Paleozoic and Mesozoic geodynamic evolution of the eastern Klamath Mountains (North California) inferred from the geochemical characteristics of its magmatism: Geological Society of France Bulletin, v. 2, no. 6, p. 969–980.

Lindsley-Griffin, N., 1977, The Trinity ophiolite, Klamath Mountains, California, *in* Coleman, R. G., and Irwin, W. P., eds., North American ophiolites: Oregon Department of Geology and Mineral Industries Bulletin, v. 95, p. 107–120.

Lipman, P. W., 1964, Structure and origin of an ultramafic pluton in the Klamath Mountains, California: American Journal of Science, v. 262, p. 199–222.

Mattinson, J. M., and Hopson, C. A., 1972, Paleozoic ophiolitic complexes in Washington and northern California: Carnegie Institution of Washington Yearbook 71, p. 578–583.

McMath, V. E., 1966, Geology of the Taylorsville area, northern Sierra Nevada, California: California Division of Mines Bulletin 190, p. 173–183.

Menzies, M. D., Blanchard, D., and Xenophontos, C., 1980, Genesis of the Smartville arc-ophiolite, Sierra Nevada foothills, California: American Journal of Science, v. 208A, p. 329–344.

Miller, M. M., 1987, Dispersed remnants of a northeast Pacific fringe arc; Upper Paleozoic island-arc terranes of Permian McCloud faunal affinity, western U.S.: Tectonics, v. 6, p. 807–830.

—— , 1989, Intra-arc sedimentation and tectonism; Late Paleozoic evolution of the eastern Klamath terrane, California: Geological Society of America Bulletin, v. 101, p. 170–187.

Miller, M. M., and Saleeby, J. B., 1987, Detrital zircon studies of the Galice Formation; Common provenance of strata overlying the Josephine ophiolite and Rogue Island arc, western Klamath Mountains terrane: Geological Society of America Abstracts with Programs, v. 19, p. 772–773.

Miller, M. M., Saleeby, J. B., Harper, G. D., and Wright, J. E., 1988, Jurassic orogenesis in the Klamath Mountains; Episodic tectonism and arc superposition within an evolving convergent margin: Geological Society of America Abstracts with Programs, v. 20, p. A274.

Morgan, B. A., 1976, Geology of Chinese camp and Moccasin Quadrangles, Tuolumne County, California: U.S. Geological Survey Miscellaneous Field Studies Map MF—840, scale 1:24,000.

Newton, M. C., 1986, The southern part of the Bear Mountains fault zone, Foothills terrane, western Sierra Nevada, California: Geological Society of America Abstracts with Programs, v. 18, no. 2, p. 164.

Palmer, A. R., 1983, The decade of North American Geology 1983 geologic time scale: Geology, v. 11, p. 503–504.

Peacock, S. M., and Norris, P. J., 1989, Metamorphic evolution of the Central Metamorphic Belt, Klamath province, California; An inverted metamorphic gradient beneath the Trinity peridote: Journal of Metamorphic Geology (in press).

Potter, A.W., Hotz, P. E., and Rohr, D. M., 1977, Stratigraphy and inferred tectonic framework of lower Paleozoic rocks, the eastern Klamath Mountains, northern California, *in* Stewart, J. H., Stevens, C. H., and Fritsche, A. E., eds., Paleozoic paleogeography of the western United States: Pacific Section, Society of Economic Paleontologists and Mineralogists Pacific Coast Paleogeography Symposium 1, p. 421–440.

Quick, J. E., 1981, Petrology and petrogenesis of the Trinity peridotite; An upper mantle diapir in the eastern Klamath Mountains, northern California: Journal of Geophysical Research, v. 87, p. 3831–3848.

Rogers, J.J.W., Suayah, I. B., and Edwards, J. M., 1984, Trace elements in continental-margin magmatism; Part 4, Geochemical criteria for recognition of two volcanic assemblages near Auburn, western Sierra Nevadas, California: Geological Society of America Bulletin, v. 95, p. 1437–1445.

Roure, F., and DeWever, 1983, Decouverte de radiolarites du Trias dans l'unite occidentale des Klamath, sudperidotites de Josephine: Paris, Comptes Rendus de l'Academie des Sciences, v. 297, p. 161–164.

Rouer, O., Lapierre, H., Mascle, G., Coulon, C., and Albers, J., 1989, Geodynamic implications of Devonian silicic arc magmatism in the Sierra Nevada and Klamath Mountains, California: Geology, v. 17, p. 177–180.

Saleeby, J. B., 1977, Fracture zone tectonics, continental margin fragmentation, and emplacement of the Kings-Kaweah ophiolite belt, southwest Sierra Nevada, California, *in* Coleman, R. G., and others, eds., North American ophiolites: Oregon Department of Geological and Mineralogical Industries Bulletin 95, p. 141–1594.

——, 1978, Kings River ophiolite, southwest Sierra Nevada foothills, California: Geological Society of America Bulletin, v. 89, p. 617–636.

——, 1979, Kaweah serpentine melange, southwest Sierra Nevada foothills, California: Geological Society of America Bulletin, v. 90, p. 29–46.

——, 1981, Ocean floor accretion and volcano-plutonic arc evolution of the Mesozoic Sierra Nevada, California, *in* Ernst, W. G., ed., The geotectonic development of California; Rubey Volume 1: Englewood Cliffs, New Jersey, Prentice-Hall, p. 132–181.

——, 1982, Polygenetic ophiolite belt of the California Sierra Nevada; Geochronological and tectonostratigraphic development: Journal of Geophysical Research, v. 87, p. 1803–1824.

——, 1984a, Tectonic significance of serpentinite mobility and ophiolitic melange, *in* Raymond, L. A., ed., Melanges; Their nature, origin, and significance: Geological Society of America Special Paper 198, p. 153–168.

——, 1984b, Pb/U zircon ages from the Rogue River area, western Jurassic belt, Klamath Mountains, Oregon: Geological Society of America Abstracts with Programs, v. 16, p. 331.

——, 1987, Discordance patterns in Pb/U zircon ages of the Sierra Nevada and Klamath Mountains: EOS Transactions of the American Geophysical Union, v. 68, p. 32.

——, 1990, Petrotectonic and paleogeographic settings of U.S. Cordilleran ophiolites, *in* Burchfiel, B. C., Lipman, P. W., and Zoback, M. L., The Cordilleran orogen; Conterminous U.S.: Boulder, Colorado, Geological Society of America, The Geology of North America, v. G-3 (in press).

Saleeby, J. B., and Sharp, W. D., 1980, Chronology of the structural and petrologic development of the southwest Sierra Nevada foothills, California: Geological Society of America Bulletin, part 2, v. 91, p. 1416–1535.

Saleeby, J. B., Busby, C., Goodin, W. D., and Sharp, W. D., 1978, Early Mesozoic paleotectonic-paleogeographic reconstruction of the southern Sierra Nevada region, California, *in* Howell, D. G., and McDougall, K. A., eds., Mesozoic paleogeography of the western United States: Pacific Section, Society of Economic Paleontologists and Mineralogists Pacific Coast Paleogeography Symposium 2, p. 311–336.

Saleeby, J. B., Harper, G. D., Sharp, W. D., and Snoke, A. W., 1982, Time relations and structural-stratigraphic patterns in ophiolite accretion, west-central Klamath Mountains, California: Journal of Geophysical Research, v. 87, p. 3831–3848.

Saleeby, J. B., and 12 others, 1986, Monterey Bay offshore to the Colorado plateau: Boulder, Colorado, Geological Society of America, The Geology of North America, Transect C2, scale 1:500,000.

Saleeby, J. B., Hannah, J. L., and Varga, R. J., 1987, Isotopic age constraints on middle Paleozoic deformation in the northern Sierra Nevada, California: Geology, v. 15, p. 757–760.

Saleeby, J. B., Shaw, H. F., Niemeyer, S., and Moores, E. M., 1989a, U/Pb, Sm/Nd, and Rb/Sr geochronological and isotopic study of northern Sierra Nevada ophiolitic assemblages, California: Contributions to Mineralogy and Petrology, v. 102, p. 205–220.

Saleeby, J. B., Geary, E. E., Paterson, S. R., and Tobisch, O. T., 1989b, Isotopic systematics of Pb/U (zircon) and $^{40}Ar/^{39}Ar$ (biotite/hornblende) from rocks of the Central Foothills terrane, Sierra Nevada, California: Geological Society of America Bulletin, v. 101, p. 1481–1492.

Saleeby, J. B., and Busby-Spera, C., 1990, Early Mesozoic tectonic evolution of the western U.S. Cordillera, *in* Burchfiel, B. C., Lipman, P. W., and Zoback, M. L., The Cordilleran orogen; Conterminous U.S.: Boulder, Colorado, Geological Society of America, The Geology of North America, v. G-3 (in press).

Schweickert, R. A., and Bogen, N. L., 1983, Tectonic transect of Sierran Paleozoic through Jurassic accreted belts: Pacific Section, Society of Economic Paleontologists and Mineralogists Field Trip Guide, 22 p.

Schweickert, R. A., and Cowan, D. S., 1975, Early Mesozoic tectonic evolution of the western Sierra Nevada, California: Geological Society of America Bulletin, v. 86, p. 1329–1336.

Schweickert, R. A., Saleeby, J. B., Tobisch, O. T., and Wright, W. H., III, 1977, Paleotectonic and paleogeographic significance of the Calaveras Complex, western Sierra Nevada, California, *in* Stewart, J. H., Stevens, C. H., and Fritsche, A. E., eds., Paleozoic paleogeography of the western United States: Pacific Section, Society of Economic Paleontologists and Mineralogists Pacific Coast Paleogeography Symposium 1, p. 381–394.

Schweickert, R. A., Harwood, D. S., Girty, G. H., and Hanson, R. E., 1984, Tectonic development of the northern Sierra terrane; An accreted late Paleozoic island-arc and its basement, *in* Lintz, J., Jr., ed., Western geological excursions, v. 4: Reno, University of Nevada Mackay School of Mines, p. 1–65.

Schweickert, R. A., Bogen, N. L., and Merguerian, C., 1988, Deformational and metamorphic history of Paleozoic and Mesozoic basement terranes in the western Sierra Nevada metamorphic belt, *in* Ernst, W. G., ed., Metamorphism and crustal evolution of the western United States; Rubey Volume 7: Englewood Cliffs, New Jersey, Prentice-Hall, p. 789–822.

Sharp, W. D., 1988, Pre-Cretaceous crustal evolution in the Sierra Nevada region, California, *in* Ernst, W. E., ed., Metamorphism and crustal evolution of the western United States; Rubey Volume 7: Englewood Cliffs, New Jersey, Prentice-Hall, p. 823–865.

Shaw, H. F., Chen, J. H., Saleeby, J. B., and Wasserburg, G. J., 1987, Nd-Sr-Pb systematics and age of the Kings River ophiolite, California: Contributions to Mineralogy and Petrology, v. 96, p. 281–290.

Snoke, A. W., 1977, A thrust plate of ophiolitic rocks in the Preston Peak area,

Klamath Mountains, California: Geological Society of America Bulletin, v. 88, p. 1641–1659.

Snoke, A. W., and Whitney, S. E., 1979, Relict pyroxenes from the Preston Peak ophiolite, Klamath Mountains, California: American Mineralogist, v. 64, p. 865–873.

Snoke, A. W., Bowman, H. R., and Hebert, A. J., 1977, The Preston Peak ophiolite, Klamath Mountains, California; An ancient immature island arc; Petrochemical evidence: Short Contributions to California Geology Special Report 129, p. 67–79.

Snoke, A. W., Sharp, W. D., Wright, J. E., and Saleeby, J. B., 1982, Significance of mid-Mesozoic peridotitic to dioritic intrusive complexes, Klamath Mountains, western Sierra Nevada, California: Geology, v. 15, p. 160–166.

Springer, R. K., 1980, Geology of the Pine Hill intrusive complex; A layered gabbroic body in the western Sierra Nevada foothills, California; Summary 1: Geological Society of America Bulletin, v. 91, p. 381–385.

Standlee, L. A., 1978, Middle Paleozoic ophiolite in Melones fault zone, northern Sierra Nevada, California: Geological Society of America Abstracts with Programs, v. 10, p. 148.

Stone, P., and Stevens, C. H., 1988, Pennsylvanian and Early Permian paleogeography of east-central California: Implications for the shape of the continental margin and the timing of continental truncation: Geology, v. 16, p. 330–333.

Tobisch, O. T., Paterson, S., Saleeby, J. B., and Geary, E. E., 1989, Nature and timing of deformation in the Foothills terrane, central Sierra Nevada, California; Its bearing on orogenesis: Geological Society of America Bulletin, v. 101, p. 401–413.

Varga, R. J., and Moores, E. M., 1981, Age, origin, and significance of an unconformity that predates island-arc volcanism in the northern Sierra Nevada: Geology, v. 9, p. 512–518.

Walker, J. D., 1988, Permian and Triassic rocks of the Mojave Desert and their implications for timing and mechanisms of continental truncation: Tectonics, v. 7, p. 685–709.

Wallin, E. T., 1989a, Petrogenetic and tectonic significance of xenocrystic Precambrian zircon in a lower Cambrian tonalite, eastern Klamath Mountains, California: Geological Society of America Abstracts with Programs, v. 21, no. 5, p. 155.

—— , 1989b, Reconstruction of the timing and sequence of early Paleozoic sedimentation in the Yreka-Callahan terrane, eastern Klamath Mountains, California: Geological Society of America Abstracts with Programs, v. 21, no. 5, p. 155.

Wallin, E. T., and Lindsley-Griffin, N., 1990, Trinity ultramafic complex, Klamath Mountains, California; Cambro-Ordovician ophiolite or composite terrane?: Geological Society of America Abstracts with Programs, v. 22, p. 91.

Wallin, E. T., Mattinson, J. M., and Potter, A. W., 1988, Early Paleozoic magmatic events in the eastern Klamath Mountains, northern California: Geology, v. 16, p. 144–148.

Watkins, R., Reinheimer, C. E., Wallace, J. W., and Nestell, M. K., 1987, Paleogeographic significance of a Permian sedimentary megamictite in the Central Belt of the northern Sierra Nevada: Geological Society of America Abstracts with Programs, v. 19, no. 6, 771 p.

Weisenberg, C. W., and Ave Lallement, H., 1977, Permo-Triassic emplacement of the Feather River ultramafic body, northern Sierra Nevada Mountains, California: Geological Society of America Abstracts with Programs, v. 9, p. 525.

Wells, F. G., and Walker, G. W., 1953, Geology of the Galice Quadrangle, Oregon: U.S. Geological Survey Geological Quadrangle Map GQ-25, scale 1:62,500.

Wolf, M. S., and Saleeby, J B., 1990, Crustal extension during the Nevadan orogeny in the southwestern Sierra Nevada Foothills terrane (FT), California: Geological Society of America Abstracts with Programs, v. 22, no. 6 (in press).

Wright, J. E., 1981, Geology and Pb/U geochronology of the western Paleozoic and Triassic subprovince, Klamath Mountains, California [Ph.D. thesis]: Santa Barbara, University of California, 350 p.

—— , 1982, Permo-Triassic accretionary subduction complex, southwestern Klamath Mountains, northern California: Journal of Geophysical Research, v. 87, no. B5, p. 3805–3818.

Wright, J. E., and Fahan, M. R., 1988, An expanded view of Jurassic orogenesis in the western United States Cordillera; Middle Jurassic (pre-Nevadan) regional metamorphism and thrust faulting within an active arc environment, Klamath Mountains, California: Geological Society of America Bulletin, v. 100, p. 859–876.

Wright, J. E., and Wyld, S J., 1985, Multi-stage serpentinite matrix melange development; Rattlesnake Creek terrane, southwestern Klamath Mountains: Geological Society of America Abstracts with Programs, v. 17, p. 419.

—— , 1986, Significance of xenocrystic Precambrian zircon contained within the southern continuation of the Josephine ophiolite: Devils Elbow ophiolite remnant, Klamath Mountains, northern California: Geology, v. 14, p. 671–674.

Xenophontos, C., and Bond, G. C., 1978, Petrology, sedimentation, and paleogeography of the Smartville terrane (Jurassic); Bering on the genesis of the Smartville ophiolite, *in* Howell, D. E., and McDougall, K. A., eds., Mesozoic paleogeography of the western United States: Pacific Section, Society of Economic Paleontologists and Mineralogists Pacific Coast Paleogeography Symposium 2, p. 291–302.

Zucca, J. J., Fuis, G. S., Milkereit, B., Mooney, W. D., and Catchings, R. D., 1986, Crustal structure of northeastern California: Journal of Geophysical Research, v. 91, no. B7, p. 7359–7382.

Manuscript Received by the Society April 9, 1990

Printed in U.S.A.

Geological Society of America
Special Paper 255
1990

Paleogeographic setting of the Schoonover sequence, Nevada, and implications for the late Paleozoic margin of western North America

William B. Whiteford*
Department of Geology, Stanford University, Stanford, California 94305

ABSTRACT

Details of the stratigraphy, depositional setting, and clastic petrography of the upper Paleozoic Schoonover sequence in the Independence Mountains, northern Nevada, provide the basis for a better understanding of the paleogeography of the continental margin of western North America during the late Paleozoic.

The Schoonover sequence represents the northernmost exposures of the Golconda allochthon, which was thrust over the outer edge of the continental margin of western North America during the Permo-Triassic Sonoma orogeny. The Schoonover sequence, like the Havallah sequence and other units of the Golconda allochthon, is an imbricated assemblage of thrust-bound packages of radiolarian "ribbon" chert, basaltic greenstone, silty limestone turbidites, and siliciclastic sandstones that range in age from Late Devonian to Early Permian.

Detailed mapping, and facies and petrographic analysis of lithostratigraphic units, combined with paleontologic control, indicate that the Schoonover sequence consists of a coherent stratigraphic succession of basinal deposits with paleogeographic ties to the autochthonous shelf margin and to a volcanic arc. Latest Devonian- to earliest Mississippian-age basaltic and andesitic greenstones and tuffaceous sedimentary rocks form the stratigraphic base of the sequence and are succeeded by Lower Mississippian siliciclastic sandstones. The petrography of these sandstones indicates that they consist of mixed detritus derived from volcanic and continental shelf source terranes, indicating the proximity of an arc to the continental margin. In addition, stratigraphic relations between autochthon and allochthon indicate that the Schoonover basin evolved adjacent to the continental shelf throughout the late Paleozoic, its history punctuated by periods of basaltic volcanism and pulses of continent-derived clastic input. The onset of the Sonoma orogeny resulted in closure of the basin and emplacement of its deposits on the continental margin.

*Present address: 2530 South Garfield Street, Denver, Colorado 80210.

Whiteford, W. B., 1990, Paleogeographic setting of the Schoonover sequence, Nevada, and implications for the late Paleozoic margin of western North America, *in* Harwood, D. S., and Miller, M. M., eds., Paleozoic and early Mesozoic paleogeographic relations; Sierra Nevada, Klamath Mountains, and related terranes: Boulder, Colorado, Geological Society of America Special Paper 255.

INTRODUCTION

The Schoonover sequence in the northern Independence Mountains, Nevada (Fig. 1), is a complexly deformed assemblage of upper Paleozoic chert, argillite, quartzose and volcanogenic siliciclastic sandstones, silty limestones, and mafic and intermediate volcanic rocks. The Schoonover sequence is in part lithologically similar to and coeval with the Havallah sequence of the Golconda allochthon in central Nevada (Stewart, 1980; Miller and others, 1984). The Schoonover sequence is inferred to be part of the Golconda allochthon, which was thrust across the Paleozoic continental margin during the Sonoma orogeny in latest Permian to earliest Triassic time (Stewart, 1980). The structure and lithologies of the Golconda allochthon are similar to those of the lower Paleozoic Roberts Mountains allochthon, which was thrust onto the continental margin during the Early Mississippian Antler orogeny. Although the general character of deformation within both of these allochthons is well established, the paleogeographic and plate tectonic setting of the rocks within the allochthons continue to be topics of debate. In part, this is because the Paleozoic record to the west and along strike of the allochthons is masked by Mesozoic and Cenozoic cover and structures; also, interpretations based on the allochthons themselves are hampered by an incomplete understanding of the rocks and their stratigraphy, and by complex internal deformation.

The Schoonover sequence is an exception to this characterization, and therefore provides a better basis for speculating on the nature of the plate tectonic setting of the late Paleozoic margin. Miller and others (1984), using extensive biostratigraphic studies of radiolarian chert, were able to unravel the nature of the original stratigraphic succession that was imbricated to form the Golconda allochthon. The succession ranges in age from Late Devonian to Early Permian and is therefore coeval with the emplacement of the Roberts Mountains allochthon (the Antler orogeny) and spans the time preceding thrusting of the Golconda allochthon (the Sonoma orogeny). This chapter presents data on the depositional setting and provenance of the Schoonover sequence and discusses the implications of these data for the paleogeographic setting of the Antler orogeny and the continental margin during the late Paleozoic.

Before delving into the Schoonover sequence in detail, a brief summary of the Paleozoic regional geology of Utah and Nevada is presented, along with a sketch of tectonic models for the Paleozoic margin. With this background, the reader will be prepared to follow discussions of the stratigraphic, tectonic, and paleogeographic correlations between the Schoonover sequence and the continental shelf.

Paleozoic geology of Nevada and Utah

The oldest rocks related to the development of the continental margin in the U.S. Cordillera are miogeoclinal shelf quartzite and carbonate deposited after rifting in the late Precambrian (Stewart and Suczek, 1977; Stewart, 1980). Paleozoic rocks in

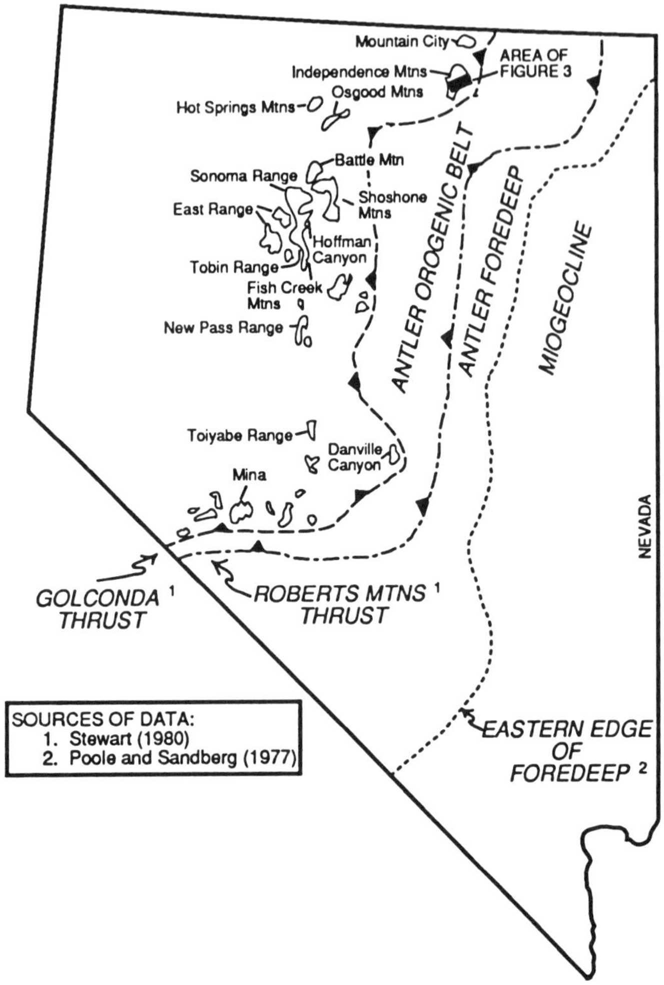

Figure 1. Paleozoic tectonic provinces of Nevada with exposures of the Golconda allochthon indicated.

Utah and eastern Nevada record relatively continuous shelf sedimentation under tectonically quiescent conditions, while structural and stratigraphic relations in central and northern Nevada reflect four deformational episodes along the outer shelf.

During the Early Mississippian Antler orogeny, mafic volcanic rocks, radiolarian chert, black shales, orthoquartzite, and feldspathic sandstone of the Roberts Mountains allochthon were transported eastward over the edge of the continental shelf (Roberts and others, 1958; see review *in* Stewart, 1980). The emplacement of the allochthon resulted in the creation of the "Antler highlands," a positive topographic feature that persisted along the outer continental shelf throughout the late Paleozoic. Coarse detritus poured eastward from the Antler highlands into a narrow, elongate foredeep basin (Fig. 1; Poole, 1974; Poole and Sandberg, 1977). By Pennsylvanian time, shallow-water carbonates were deposited across the filled foredeep and over portions of the Antler highlands (Rich, 1977; Stewart, 1980; Saller and Dickinson, 1982; Little, 1987). Stratigraphic evidence from northern Nevada indicates that the emplacement of the Roberts Mountains allochthon was followed by rifting and subsidence in the Antler highlands in Late Mississippian time (Little, 1987).

The third deformational event resulted in local uplift and tilting within the Antler orogenic belt and an increase in terrigenous sedimentation on the shelf in late Middle Pennsylvanian time. Ketner (1977) named this event the Humboldt orogeny.

Finally, rocks of the Golconda allochthon were emplaced over the shelf edge during the Permo-Triassic Sonoma orogeny (Silberling and Roberts, 1962; Silberling, 1973, 1975; review *in* Stewart and others, 1977). No coeval foredeep deposits are known, possibly because the orogen was never emergent. Subsequent to the Sonoma orogeny, western Nevada and California evolved into an active margin characterized by arc magmatism and episodic compressional tectonism during Mesozoic time.

The Antler and Sonoma orogenies were similar in many respects. Both events resulted in the emplacement of complexly folded and imbricately faulted oceanic volcanic and sedimentary rocks above less deformed miogeoclinal carbonate rocks along regionally extensive sole thrusts.

Review of tectonic models

Of the many tectonic-paleogeographic scenarios proposed for the Antler and Sonoma orogenies (see Dickinson, 1977; Nilsen and Stewart, 1980), back-arc thrusting and collision of "exotic" arcs are the two models most commonly invoked to explain the thrust emplacement of the allochthons. The two different models have important implications for interpretations of continental margin paleogeography during the Paleozoic.

The back-arc basin model postulates an offshore island arc, facing west and separated from the continental shelf by a marginal basin or basins. In this context, the Golconda and Roberts Mountains allochthons are viewed as marginal basin deposits that were obducted during arc-continent convergence (Burchfiel and Davis, 1972, 1975; Churkin, 1974; Poole and Sandberg, 1977; Miller and others, 1984). In the exotic arc model, the Roberts Mountains and Golconda allochthons are thought to represent the obducted accretionary prisms of arcs that collided with North America after the subduction of large tracts of paleo-Pacific oceanic crust. This model is premised in part on the similarity of structural style within the allochthons to that of known subduction complexes (Speed, 1977; Speed and Sleep, 1982; Snyder and Brueckner, 1983; Dickinson and others, 1983). In addition, Speed and Sleep (1982) have postulated that after the collision that was presumed to have caused the Antler orogeny, the arc thermally contracted and subsided to oceanic depths by Middle Pennsylvanian time.

SCHOONOVER SEQUENCE

The Schoonover sequence was first described by Fagan (1962), who assigned it a Carboniferous age based on sparse brachiopod collections. Poole and Sandberg (1977) confirmed this age with conodont collections. Subsequently, Miller and others (1981, 1984) recognized a latest Devonian (Famennian) to earliest Permian (Wolfcampian) succession based on more than 200 radiolarian ages. This succession is repeated in an imbricated stack of northwest-dipping thrust plates that structurally overlie largely undeformed Permian-age shallow-water limestones (Fig. 2). The age, lithology, and structural relations of the Schoonover sequence are similar to those of the Havallah sequence (Stewart, 1980; Miller and others, 1982). Below is a brief summary of the stratigraphic relations of the Schoonover lithostratigraphic map units based on the detailed biostratigraphy and mapping presented by Miller and others (1984) and Whiteford (1984), followed by a detailed discussion of lithofacies, depositional environment, and provenance of the Schoonover sequence.

Stratigraphy

The oldest dated unit in the Schoonover sequence consists of locally manganiferous chert-argillite of Famennian-Kinderhookian (DMc map unit; Fig. 3), which in some thrust plates overlies scraps of greenstone (Dg), in part andesitic (Dga; Miller and others, 1984). Kinderhook-age chert locally contains abundant tuffaceous interbeds. Overlying the chert unit is a siliciclastic sequence (Msv) of interbedded volcanogenic and quartzose sandstones and mudstones. This unit is succeeded by an interval of quartzose sandstones, conglomerates, mudstones, and pebbly mudstones (Mss) that locally contain interbeds of Kinderhookian-Osagian chert. The main distinction between these siliciclastic map units is the presence of interbedded quartzose and volcanogenic sandstones in the mixed Lower Mississippian siliciclastic unit (Msv) and their absence, together with local chert interbeds, in the quartzose siliciclastic unit (Mss). In structurally lower thrust plates, the Kinderhookian mixed unit (Msv) contains interbedded volcanogenic and quartzose sandstones and is much thinner than in the structurally highest positions (Figs. 3, 4). Abruptly above the Lower Mississippian clastic section is a distinctive chert-argillite unit (Mc) that locally contains laminae of sandstone and siltstone, and abundant knobby bedding-surface structures (diagenetic "penetration structures" of Fagan, 1962). This unit is early Meramec in age (earliest Late Mississippian). The chert units that bracket the clastic sections thus limit them to Kinderhook-Osage age.

In some thrust plates the Mississippian clastic section is missing (Figs. 2, 4), and a coeval Kinderhook-Meramec sequence of chert-argillite occurs instead (DMc overlain by CPcl or CPc). This succession is inferred to have been deposited in areas isolated from clastic deposition.

Along the basal thrust, Upper Mississippian pillow basalts form greenstone units (Mg), which to the southwest of the map area, locally contain quartzite boulder conglomerate horizons and lenses of limestone and chert (see map of Miller and others, 1984). Lower Meramecian chert generally overlies Mg greenstone units. Along the basal thrust, Mg greenstone units are quite thick, whereas in structurally higher thrust sheets, they are thin or absent.

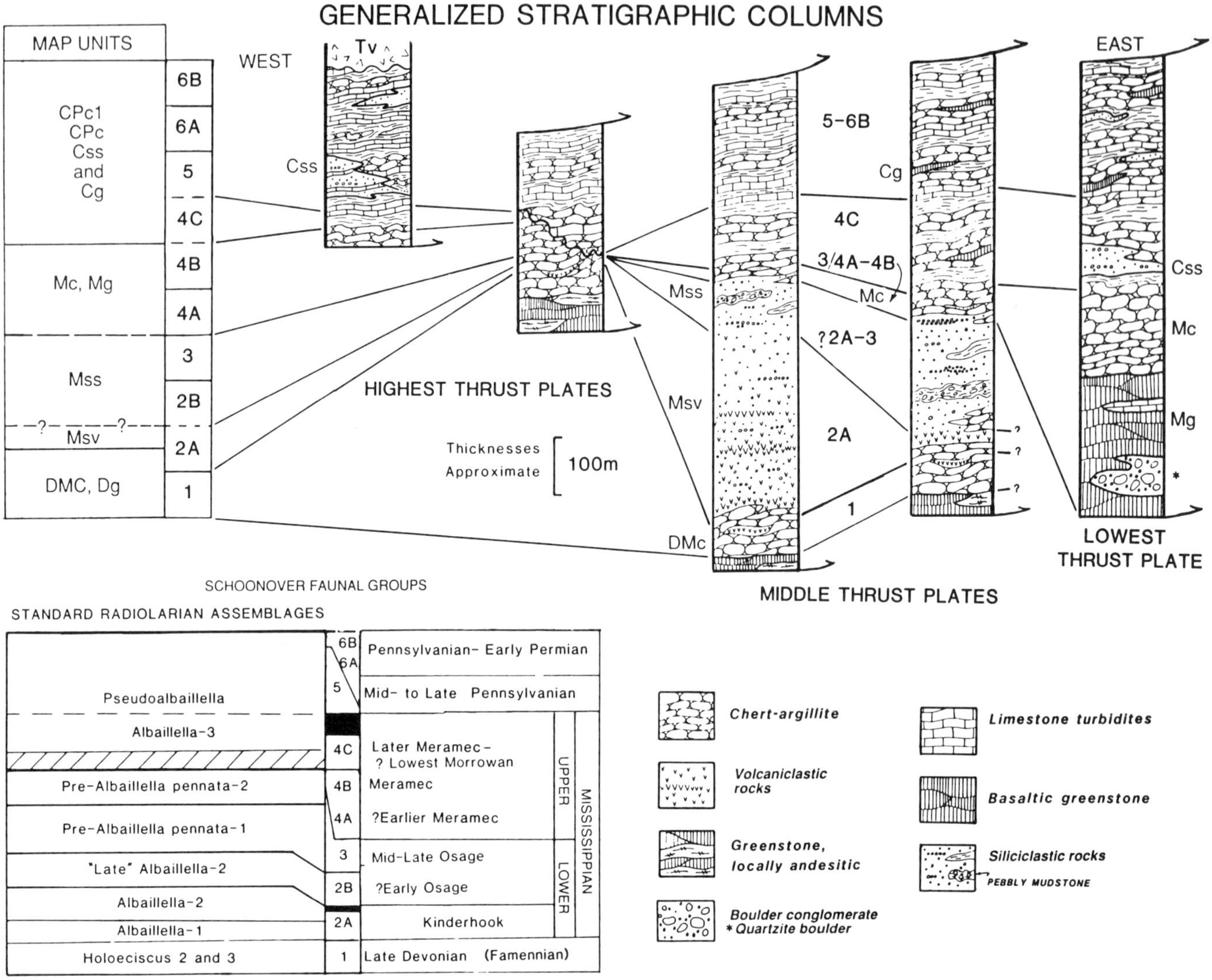

Figure 2. Generalized, schematic stratigraphic columns from the Schoonover sequence, based on composites of detailed stratigraphic sections and on biostratigraphic data presented in Miller and others (1984).

Meramecian chert is succeeded by an Upper Mississippian to Lower Permian section making up more than half the exposure of the Schoonover sequence and consisting of chert-argillite and silty limestone turbidites (CPcl) with relatively thin intervals of siliciclastic sandstone and conglomerate (Css) and rare, thin greenstone intervals composed of basaltic flows and hyaloclastites (Cg). The relative amount of chert-argillite versus limestone varies considerably throughout CPcl units; the calcareous rocks are more voluminous in the younger Pennsylvanian-Permian sections.

Depositional setting

Interpretation of the depositional environment of the Schoonover sequence is based on detailed facies analysis of measured sections (Fig. 4), which traverse each lithostratigraphic map unit at least once. Sections were located to best take advantage of rock exposure and paleontologic control. Measured sections do not include structurally higher plates because the rocks are poorly exposed and lithologically similar to the coeval units in sections D and F (Fig. 4). Although structural disruption and generally poor exposure hinder documentation of stratigraphic contacts in the Schoonover sequence, many exposures of depositional contacts, together with the consistent stratigraphic order preserved in different thrust plates, argue for intact, fault-bound stratigraphic sequences.

Sedimentary rocks of the Schoonover sequence are divided into two categories: (1) "starved-basin" sequences composed of chert-argillite and, in Pennsylvanian-Permian time, limestone turbidites; and (2) "submarine fan" sequences corresponding to the Mississippian and younger siliciclastic sequences (Msv, Mss, and Css map units). The siliciclastic sequences are further subdi-

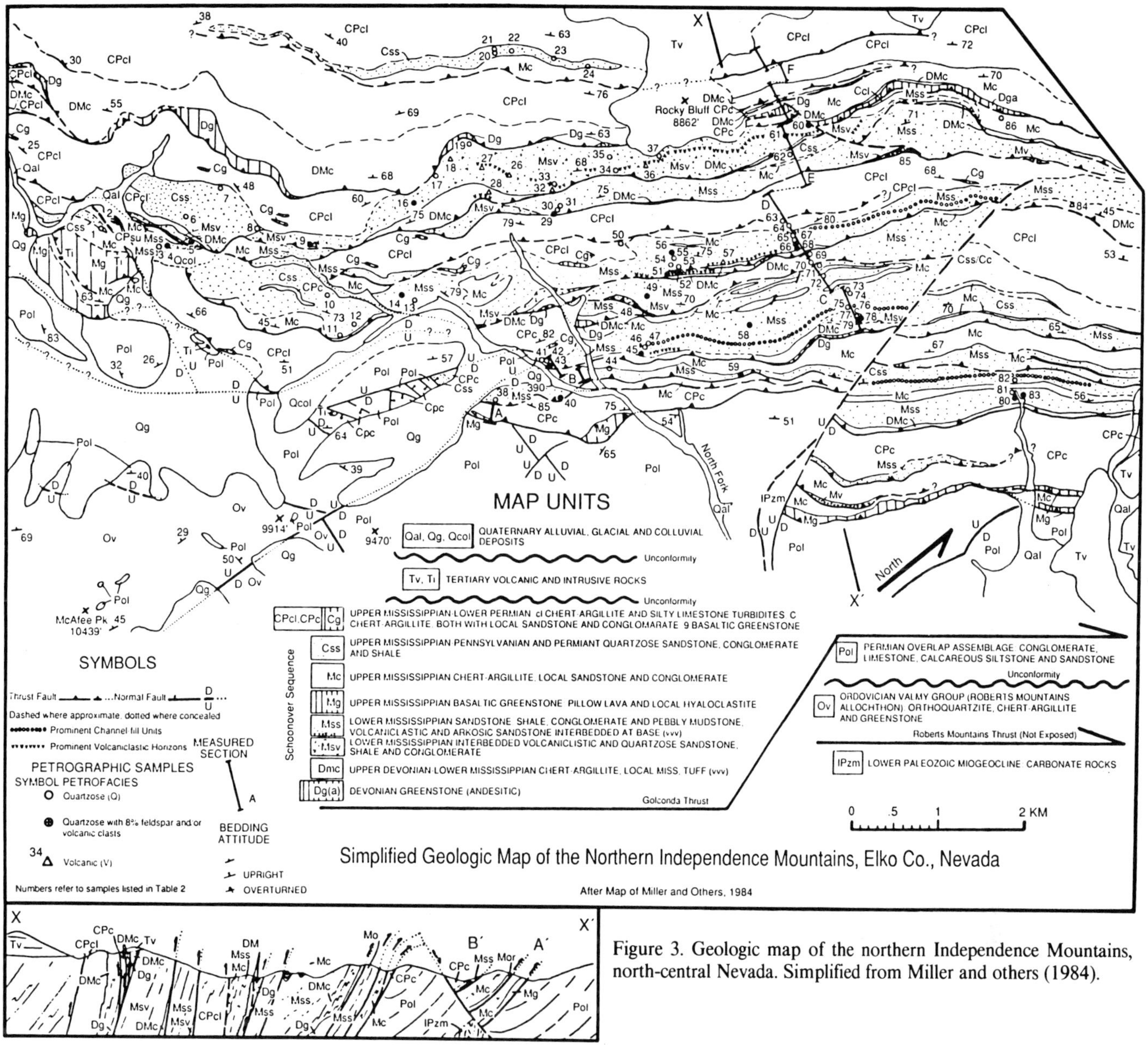

Figure 3. Geologic map of the northern Independence Mountains, north-central Nevada. Simplified from Miller and others (1984).

vided on the basis of sandstone compositions into volcanogenic and quartzose petrofacies, which are described in detail in the Petrography and Provenance section below. Rare bioclastic sandstones and limestone turbidites make up a calcareous petrofacies. The mixed Lower Mississippian siliciclastic unit (Msv map unit) contains both the volcanogenic and quartzose petrofacies, whereas other Carboniferous siliciclastic units (Mss and Css map units) contain only quartzose petrofacies. The calcareous petrofacies is present only in the Upper Mississippian to Permian sequences (CPcl map unit).

Submarine fan facies associations. Submarine fan facies associations consist of sandstone, mudstone, conglomerate, and pebbly mudstone. These lithofacies generally occur together as mappable lithostratigraphic bodies that form the Lower Mississippian (Msv, Mss) and Upper Mississippian and Pennsylvanian (Css) units. Pebbly mudstone only occurs within Mss intervals. The clastic units are interpreted as submarine fan deposits based on abundant evidence of gravity sedimentation and characteristic facies associations resembling those defined by Mutti and Ricci-Lucchi (1972). Lithologic descriptions of individual facies as they occur in the Schoonover sequence are summarized in Table 1.

Channel-fill sandstone and conglomerate. This facies association is characterized by massive or amalgamated thick beds of facies A conglomerate and pebbly sandstone and facies B sandstone. In the Mississippian section (Msv map unit), generally coarse-grained, locally pebbly sandstone is interbedded with facies A and B medium-grained sandstones (Fig. 4). Rare intercalations of facies C and D also occur, but in general the sandstone to

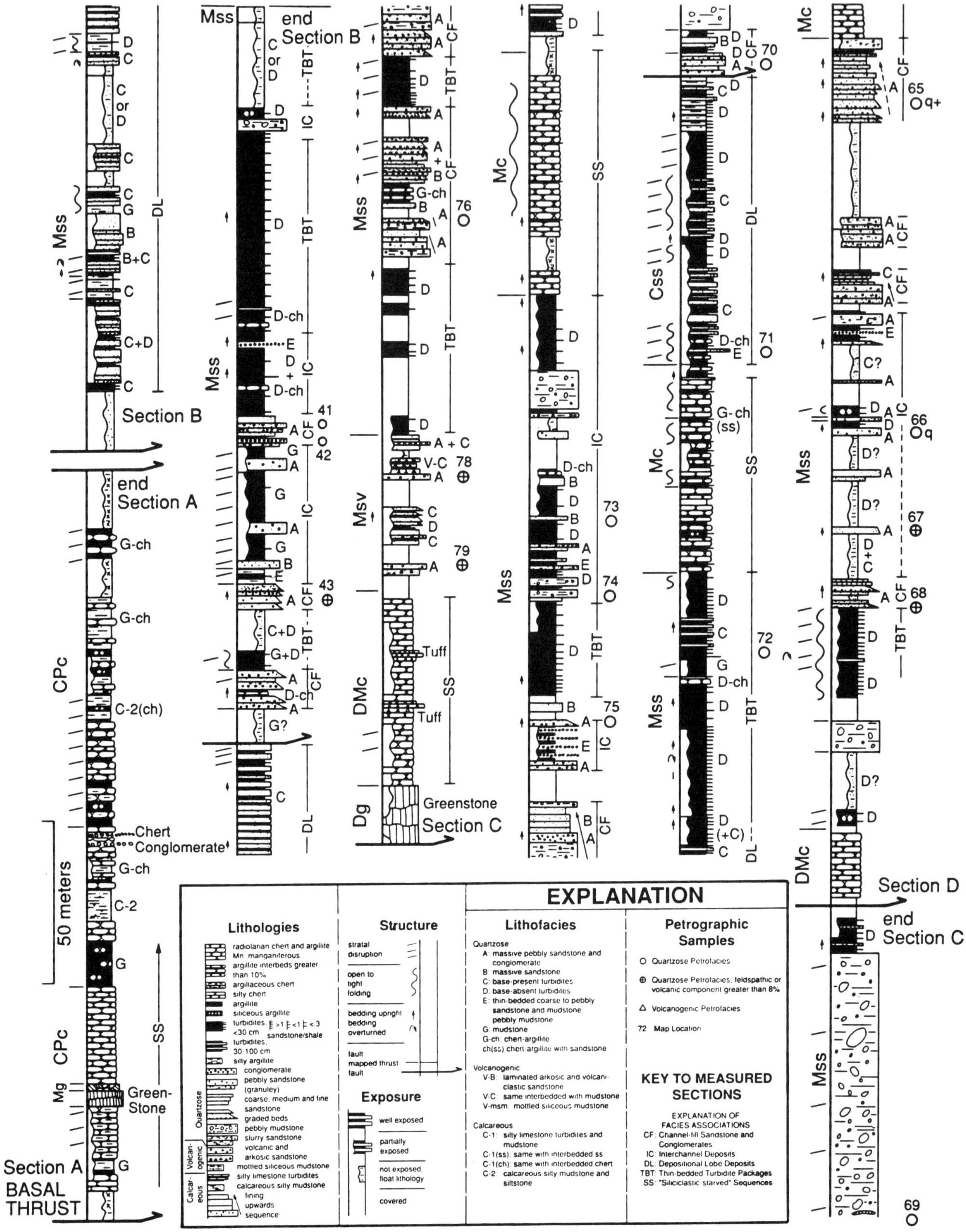

Figure 4. Detailed lithostratigraphic columns of measured sections from the Schoonover sequence, northern Independence Mountains, north-central Nevada. Locations of measured sections shown in Figure 3.

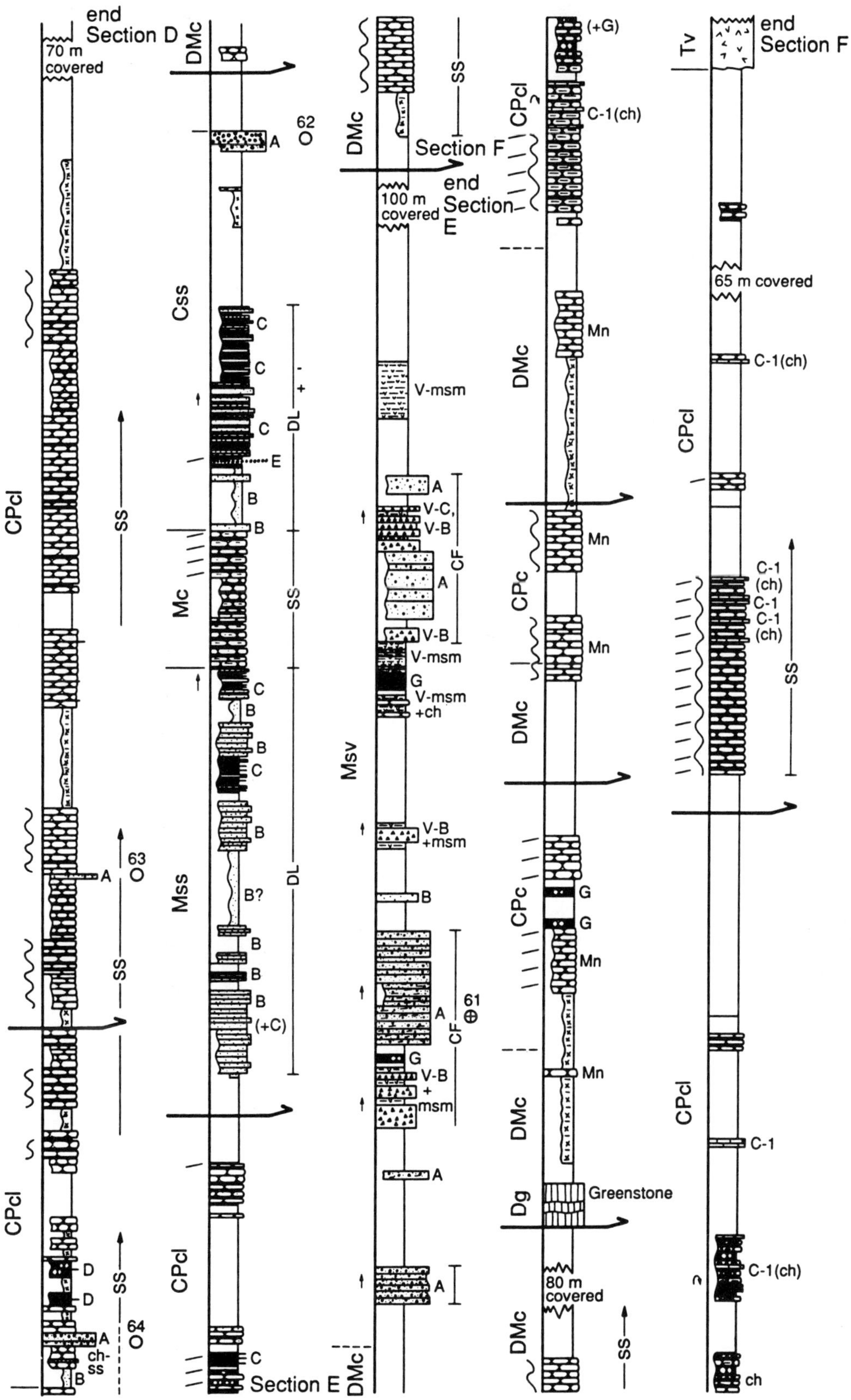
end
Section D
70 m
covered
DMc
Css
Mc
Mss
CPcl
Section E
end
Section E
100 m
covered
Section F
Msv
V-msm
V-C,
V-B
+msm
CF
DL
SS
Greenstone
Dg
CPc
Mn
80 m
covered
(+G)
C-1(ch)
Tv
end
Section F
65 m covered
C-1
(ch)
ch
62
63
64
61

TABLE 1. SUMMARIZED LITHOLOGIC DESCRIPTION OF DEPOSITIONAL FACIES

Facies*	Lithology	Bedding
A	Pebbly sandstone and conglomerate.	Massive or amalgamated intervals, 0.5 to 5 m thick.
B	Medium- to coarse-grained sandstone.	Massive or amalgamated intervals, 0.3 to 1 m thick.
C	Fine- to medium-grained sandstone.	Tabular 10 to 50 cm turbidite beds, Bouma Ta interval at base.
D	Very fine to fine-grained sandstone.	Turbidite beds, 2 to 15 cm thick, Bouma Ta intervals absent.
E	Interbedded coarse-grained to pebbly sandstone and mudstone.	Thin (<20 cm) tabular to lenticular sandstones; massive to finely laminated mudstone.

*Lettering follows scheme of Mutti and Ricci-Lucchi (1972).

mudstone ratio is very high (>15). The channel-fill sandstone and conglomerate facies form massive sandstone bodies 8 to 22 m thick, which, in several instances, can be traced up to 3.5 km along strike. Interpretation of the massive sandstone bodies as channel-fill deposits is based on three distinctive features: (1) massive sandstone bodies with a lenticular map pattern that thin laterally and pinch out in muddier facies; (2) demonstrably erosive contacts between the base of massive sandstone units and underlying mud-rich facies; and (3) fining-upward trends in many channel-fill intervals (Fig. 4, note key).

From these observations, massive sandstone bodies appear to be similar to channel-filling deposits described in ancient submarine fans (Mutti and Ricci-Lucchi, 1972; Walker and Mutti, 1973; Ingersoll, 1978; Hiscott, 1980; Link and Nilsen, 1980). Furthermore, in modern submarine fan and slope environments, coarse-grained deposits are generally confined to channels (Normark, 1978; Normark and Hess, 1980).

Interchannel deposits. This facies association is defined for intervals characterized by abundant mudstone with sporadic thin (<5 m) intervals of facies A, B, and/or E sandstones. Facies D turbidites predominate, although intervals of chert-argillite and, more rarely, facies C occur also. The coarse-grained beds are commonly lenticular in cross section, and are at least several meters thick and 8 m wide. Stratigraphic thicknesses of the interchannel facies association range from 8 to 22 m and are characterized by overall sandstone to mudstone ratios between 0.1 and 1.0.

The presence of thin, coarse-grained interbeds, and the association with channel-fill sequences suggest that the interchannel facies association represents levee or interchannel deposits. Following this interpretation, facies D turbidites were deposited from fine suspensates when large-volume turbidity currents overtopped channel margins, and coarser facies represent crevasse splay deposits accompanying the processes of aggradation and filling of active channels (Mutti and Ricci-Lucchi, 1972; Ricci-Lucchi, 1975). Alternatively, the interbedded coarser sandstones could be part of channel abandonment sequences, where up-fan channel avulsion left the former active channel to be filled with overbanked fines and rare coarse sands from the new active channel (e.g., Mutti, 1977).

Depositional lobe deposits. Depositional lobe deposits are represented by thick (50 to 135 m) packages of facies C and D turbidites with sandstone to mudstone ratios between 1 and 8 and bed thicknesses between 30 cm and 1.5 m. Facies B massive sandstone strata occur in a few 3- to 10-m-thick intervals. The lateral dimensions of this facies association appear to be generally extensive, but are uncertain due to poor exposure.

Based on their geometry, association with channelized facies, and on the abundance of turbidite beds, these sequences are interpreted to have been deposited down-fan from the channelized portions of the mid-fan (Mutti and Ricci-Lucchi, 1972; Normark, 1980), either in the outer mid-fan or proximal outer fan position.

Thin-bedded turbidite packages. Relatively thick packages of facies D turbidites, along with local chert-argillite, thin facies C turbidite and mudstone intervals, and an overall low sandstone to mudstone ratio (<1) characterize this facies association. Exposure of the mud-rich facies is rather poor, and lateral stratigraphic relations are thus poorly known. Tight folds are common in these sequences, although undeformed packages as thick as 30 m occur.

Thin-bedded turbidite sequences can occur in several fan settings (Nelson and others, 1978). Based on their association with the coarser facies associations, they probably were deposited in either interchannel or outer fan settings.

Pebbly mudstone units. Pebbly mudstone occurs as massive units that consist of 5 to 20 percent well-rounded to rounded pebble-, cobble-, and in places, boulder-size lithic clasts (compositions described in the section on Petrography and Provenance) supported by massive, dark gray, silty to sandy mudstone. Most clasts are from 1 to 14 cm in size, although boulders with diameters up to 8 m occur locally. Intervals of pebbly mudstone range from 1 to 10 m thick except for a single 87-m-thick interval. The contacts with other facies are seldom well exposed, and pebbly mudstone units commonly have irregular map geometries. In some instances, units are traceable for more than a kilometer along strike. These units only occur within the Lower Mississippian (Mss map unit) sections.

The mud-supported fabric and massive nature of the pebbly mudstone are typical of classic debris flow deposits (Middleton and Hampton, 1973), which have been shown to be common in deep-water, base-of-slope environments. Based on petrographic data, the pebbly mudstones appear to have been derived from a different point source than the sandstone facies. As discussed later, this point has important paleogeographic implications.

Starved-basin facies associations. Intervals of chert-argillite at the base and top of the Mississippian clastic section (DMc and Mc map units) and chert-argillite with sporadic interbeds or thick bundles (as much as 20 m) of silty limestone turbidites in the Upper Mississippian to Permian section (CPcl units) represent "starved-basin" facies associations. Intervals of facies D turbidites are a sparse component. These rocks belong to discrete stratigraphic time intervals and bound the Mississippian submarine fan sequences. Stratigraphic thicknesses of starved-basin units are difficult to establish because of mesoscopic scale folding and faulting, but are commonly greater than 30 m. These units compose more than 80 percent of the present structural thickness of the Schoonover sequence.

Chert-argillite. Chert-argillite facies consists of rhythmically interbedded radiolarian chert and gray- or green-hued siliceous to dark carbonaceous argillite. Chert beds are composed of recrystallized radiolaria tests, locally abundant sponge spicules, and quartz silt grains set in a microcrystalline to aphanitic quartz matrix with as much as 40 percent clay. Chert beds range in thickness from 1 to 25 cm and average 7 to 10 cm. The ratio of chert to argillite ranges from 20:1 to 1:1. High ratios are associated with relatively "pure" chert, whereas low ratios are associated with relatively silty and argillaceous chert. Most chert beds have sharp, flat upper and lower contacts with abundant knobby "penetration structures" (terminology of Fagan, 1962) and boudins. The latter are interpreted to have formed during diagenesis or tectonism (Fagan, 1962; Jenkyns and Winterer, 1982). Petrographic examination of individual chert beds indicates that radiolarian tests and silt particles are commonly normally graded by size and concentration, suggesting deposition by turbidity currents. The predominance of silt- and clay-size material in this facies association and the paucity of chert-argillite in the clastic sequences elsewhere in the Schoonover sequence imply a relatively quiet, low-energy setting in deep water for chert-argillite successions. Variations in the amount of clay and silt present within chert beds presumably reflect the relative flux of terrigenous material versus radiolarian production (Fagan, 1962).

Silty limestone turbidites. The silty limestone turbidites are composed mostly of silt- and clay-sized carbonate particles with as much as 30 percent quartz silt. Locally these beds have been partially or completely replaced by black, aphanitic chert. Beds range from 2 to 80 cm thick and are tabular with sharp, flat bases and sharp or graded tops. Interbeds of calcareous mudstone are common and locally compose as much as 75 percent of the section. The turbidite beds contain Tbc(e), Tcd(e), and Tbcd(e) Bouma sequences (Bouma, 1962), indicating deposition from low-density turbidity currents. Subhorizontal burrows and other trace fossils are common.

The limestone turbidites most likely represent distal deposits of a carbonate submarine fan that interfinger with the Upper Mississippian to Lower Permian basinal chert-argillite facies and intervals of coeval quartzose sandstone (Css map unit).

Summary. The facies associations present in the Lower Mississippian volcanogenic to siliciclastic units (Msv and Mss map units) and in the Upper Mississippian to Permian siliciclastic units (Css map unit) are interpreted as mid-fan deposits based on comparison with recent fan models. The abundance of mud-rich facies indicates a fan setting with a relatively high proportion of fine-grained sediment. These characteristics of the Schoonover siliciclastic intervals resemble those of "highly efficient" fans described in the Apennine Alps (Mutti and Ricci-Lucchi, 1972).

The uppermost Devonian to Mississippian, Upper Mississippian (Meramec), and Upper Mississippian to Permian chert-argillite units (DMc, Mc, and CPcl map units, respectively) that stratigraphically bound the clastic units are interpreted as "starved-basin" sequences deposited in the Schoonover basin. In this view, chert-argillite and, in the later half of the basin's evolution, limestone turbidites represent slow but steady background sedimentation that accumulated between relatively short-lived pulses of rapid siliciclastic deposition.

Combining this interpretation with age relations of the lithostratigraphic units, the following picture of the Schoonover depositional history emerges. The basin appears to have received one major clastic pulse in Early Mississippian (Kinderhook-Osage) time and three or four lesser pulses from Late Mississippian to Late Pennsylvanian time. The absence of clastic rocks in several thrust sheets (see Fig. 3), the presence of age-equivalent chert-argillite, and faunal evidence for reworking during periods of nondeposition or erosion (Miller and others, 1984) suggest that the Schoonover sequence was deposited in a bathymetrically complex basin. The intercalated basaltic greenstone units of Famennian, Meramecian, and mid-Pennsylvanian to Early Permian age suggest periods of extensional tectonism within the basin that may have caused the postulated complex bathymetry.

Petrography and provenance

The discussion of the petrography and provenance of the Schoonover sandstones is based on the study of more than 190 rock thin sections (Whiteford, 1984). Detrital modes (Table 2) for 64 thin sections of medium-grained sandstones were determined using the criteria of Dickinson (1970). In addition, "coarse clast modes" (Table 3) were determined for 61 coarse-grained to pebbly sandstones by counting approximately 100 clasts with diameters greater than 1.0 mm. The data are presented in ternary diagrams (Figs. 5, 6) that were plotted according to detrital parameters defined by Dickinson and Suczek (1979).

Three compositional suites are distinguished: a quartzose, a volcanogenic, and a calcareous petrofacies. The quartzose petrofacies consists primarily of sandstones that range from feldspathic litharenite to quartz arenite (terminology of Folk, 1974); such sandstones compose the bulk of the siliciclastic rocks. Pebbly mudstones and conglomerates also belong to the quartzose petrofacies. The composition of sandstones of the volcanogenic petrofacies range from arkose to feldspathic litharenite. The volcanogenic petrofacies also includes tuffaceous interbeds in the Devonian-Mississippian chert-argillite (DMc map unit). Volcanogenic sandstones are differentiated from quartzose petrofacies sandstones by distinctly higher abundance of feldspathic and lithic components. Silty limestone turbidites and rare bioclastic limestones make up a calcareous petrofacies.

Table 2. Sandstone Detrital Modes

				Grain Categories																					
				Quartzose							Feldspar		Lithic							Miscellaneous					
Sample	Map Location	Map Unit	Petrofacies	Qm	Qp	Recrystallized Chert	Sedimentary Chert	Quartzite	Metaquartzite	Foliate Quartz Aggregate	Plagioclase	Potassium Feldspar	Shale	Sandstone & Siltstone	Carbonate	Volcanic	Phyllite	Schist	Undifferentiated Lithic	Biotite	Opaques	Plutonic	Other*	Matrix & Cement	Total Grains
W-16	2	CPsu	q	149	74	29	34	3			27	4	37	6	1	33	3							68	400
W-18d	3	Mss	q	174	95	33	13	6		1	22		25	6		15	4	3				2		97	399
W-18c	4	Mss	q	186	76	13	2			1	41	1	38	2	9	17	8	5						161	399
W-19a	5	Msv	q	107	45	8	74	10			43	2	29	12		74			1	4				84	409
W-19b	6	Css	q	202	121		43	17		1			31	1										91	416
W-19c	7	Css	q	194	104	21	18	2			5	4	36	3		7	6							108	400
BW-56	9	Msv	v	152	31			1	2		179	6		2		13		4		4	2		4	190	400
BW-56 (2)	9	Msv	v	89	33						237	7	4			20	1	8					1	201	400
W-13e	11	Css	q	225	123	20					2		26	1		1					2			79	400
W-13f	12	Css	q	186	117	42	1	5			5		33	3		12								84	404
BW-68c	14	Mss	q	217	43	8				1	59	11	1			13	3	26	15		2		1	259	400
BW-47c	18	Msv	v	44	10	4			3		297	8	3	1		19		11						168	400
BW-46	19	Msv	q	153	121	30							30	2		9	1	1						53	347
Bw-82f	20	Css	q	267	103	16		2			4	1	5	2										96	400
BW-82e	21	Css	q	219	134	22		2			2		23	2		6	1	t						47	411
BW-82	22	Css	q	196	154	26		5			6	1	9	1	1	1								93	400
BW-81c	23	Css	q	239	109	21		2		1	4		15			8					1			112	400
BW-81c (1)	23	Css	q	265	105	12		8			2		3	2	3									91	400
BW-81h	24	Css	q	297	65	16		9	1		7		1	1	3								1	172	401
BW-81c (2)	23	Css	q	251	106	26		6					6		4					1				177	400
BW-49c	26	Msv	v	82	28			1		1	204	2		1		77	1	3						79	400
BW-49d (1)	27	Msv	v	45	20					1	277	11	1			38		4			1		2	122	400
BW-51	28	DMc	v	6							109	3				244†								72	362
BW-51 (2)	28	DMc	v	7							197	3	5			179							9	81	400
W-8a	29	Mss	q	97	61	21					55		32	6		39	4	11		1				73	327
W-8	30	Mss	q	177	95	28		3			8		17	5			4	5		t			6	76	348
W-8b	31	Mss	q	110	66	31					18		31	1		20	7	12		1	2		1	100	300
W-8c	32	Msv	v	9	4	1					170		2			54		3			3			154	246
W-8d	33	Msv	q	163	97	3				8	3		19			4	4	7		2	1		2	81	313
BW-26A	34	Msv	q	122	100	9				2	15		19	1		17	2	10		1	2		2	102	302
W-9	35	Msv	q	124	131	5				8	4		41			4		10			1			72	328
BW-5B	36	Msv	v	22	7			2			315	8	1	3		29		7			5		1	201	400
W-11d (2)	38	Mss	q	249	91	17	12					1	26	3		1								86	400
W-11d (3)	39	Mss	q	233	91	9	13				22	2	21	5	1	3					1			72	400
W-3	44	Mss	q	110	134	6		8					4			35	1	1						103	299
W-4	45	Mss^{b}	q	167	123	16	10	2		2	23	5	25	5		7	2	13				t		111	400
W-5a	46	Mss	q	95	52	61							71	6		1		1			11			102	298
W-5b	47	Mss	q	43	30	142		8					62	56			6				2			51	349
W-22a	48	Msv	q	185	74	20	3	2		3	22	1	31	1		53	1	2		1			1	108	400
W-22b	49	Msv	q	194	54	18	15	2			40	7	33	2		19	8	5				t	3	73	400
W-23 (1)	51	Mss^{b}	q	98	40	25	21	5			54	5	55	16		71	3	5			2			76	400
W-7	52	Mss	q	135	50	29	30	4			35	3	67	2		30	3				4			97	392
W-7a	53	Mss	q	210	79	12	10				22	3	43	10		5	2	1			1		2	89	400
W-23a	54	Mss	q	90	176	12	11	5			13	1	51	4	13	14		2		2	5		1	95	400
W-23c	56	Mss	q	183	87	8	10	3			39	2	16	3		41	1	2			3		2	132	400
81-44	57	Mss^{b}	v	1	1						209	2				185					1		1	185	400
LW-5-66	63	CP1	q	214	119	20		6		2	14	5	9	7		3	1							254	400
LW-4-1(1)	69	Mss	q	86	29	66	86	4	1		3	2	62	15	4	37					3			97	398
LW-3-99	70	Mss	q	197	76	29		1		1	17	3	23	5		31	5	10					1	239	399
LW-3-89	71	Css	q	202	102	19	5	2			13	6	28	6		15	1				3			114	402
LW-3-79	72	Mss^{u}	q	166	42	46	15	2		4	12	2	57	17		21	6	6			2	2		254	400
LW-3-57	73	Mss	q	138	162	20		7		3	5	1	29	8		23	8	1						87	405
LW-3-49	74	Mss^{u}	q	211	58	7	7	7		1	21	2	45	11		9	6	12		1	2			137	400
LW-3-10	77	Msv	v	65	16	4	1				78			1		200	1	4	28	2				67	400
LW-3-9	78	Msv	qf	177	83	8	26	3	1	1	34	3	43	8		8		2		2	1			82	400
LW-3-3	79	Msv	q	196	41	9				2	75	1	10			51	4	4		1	6			136	400
LW-3-3(1)	79	Mss^{b}	q	167	88	9	13	1			64	5	27	3		15	4	2		2				92	400
EW-2	80	Mss	q	187	80	20	9			5	21	9	35	3		6	14	10					1	108	400
EW-4	81	Mss	q	164	83	32	13	1			22	2	61	7	1	13		t			1			127	400
EW-5	82	Css	q	226	97	30	28	2	2	1	1		12	1										114	400
81-61	84	DMc	v	117	15	1					184	11	6			44	4	3	13		1		1	183	400
FC-2	§	Css	q	196	176	12					3		9	2	1	1					1			113	401
FC-6	§	Css	q	264	108	10		6			6		5			1								198	400
FC-7	§	Css	q	222	152	7		8			1		8			1					1			53	400
FC-8	§	Css	q	237	147	3		3					6	4										58	400

Note: Superscripts b and u respectively denote samples from basal and upper parts of units.

* Includes zircon, muscovite, and, in sample BW-51 (2), chlorite.

† Differentiated: 12 lathwork, 66 microlithic, 11 felsitic, 25 volcaniclastic, 130 undifferentiated.

§ Located in Css unit to northwest; see map of Miller and others, 1984.

Table 3. Coarse Clast Detrital Modes

				Clast Types: Quartzose					Sedimentary				Volcanic					Igneous				Miscellaneous						
Sample	Map Location	Map Unit	Petrofacies	Qm	Qp	Quartzite	Recrystallized Chert	Sedimentary Chert	Shale	Sandstone & Siltstone (Quartzose)	Sandstone & Siltstone (Feldspathic)	Carbonate	Lathwork Volcanic	Microlitic Volcanic	Felsitic Volcanic	Volcaniclastic & Pyroclastic	Undifferentiated Volcanic	Plagioclase	Potassium Feldspar	Gabbroic	Plutonic	Phyllite	Schist	Quartz-Mica Tectonite	Metaquartzite	Foliate Quartz Aggregate	Other*	Total Grains
W-18	1	Css	q	5	7	8	2	55	2	11																1		91
W-16	2	CPs^u	q	5	4	4	4	24	14	1		1	6	6	1	8	8	3								2	91	
W-16 (1)	2	CPs^u	q		3	1	9	31	5	6	2		19	11	1	11		1	1								101	
W-18d	3	Mss	q	2	4	3	14	42	14	8			3	1	1	1	2				1			1	2	2		101
W-19a	5	Msv	q	10	9	3	2	60	15	9	1		11	15			1	6			1							143
W-19b	6	Css	q	6	11	3	1	96	2	2																		121
BW-63c (1)	8	Css	q	2	9		30	38	18	3																		100
BW-56 (2)	9	Msv	v	20†					1				1	2			2	35	4					1				66
BW-56 (4)	9	Msv	q	5	8	1	12	27	18	5			9	1	1	4		2				1						94
W-14	10	Css	q	11	29	11	7	31	9	2																		100
BW-70b	13	Mss	q	2	5	4	11	27	11	10		22				3								1	4			100
BW-32	16	Msv	q	4	3	2	13	29	3	6			2	19	13	2		1				1				2		100
BW-43a	17	Msv	q	1	1	1	27	39	14	18																		101
BW-82f	20	Css	q	54	28	7	2	46	8				1							1					1			148
BW-81c	23	Css	q	10	17	1	2	8	3																			41
BW-81c (1)	23	Css	q	40	24	4	3	30	3			2													3			109
BW-81c (2)	23	Css	q	21	27	1	1	21		1		3													1			76
BW-81h	24	Css	q	40	17		1	24	4	1		2																89
BW-26B	34	Msv	q	4	24	2	5	7	2	1			1	2			1	1	1					3	5	32		91
BW-6	37	Msv	q	9	25		2	7	5				2		2	5						2		9		41		109
W-11d	40	Mss^b	q	1		2	18	20	9	5			8	19	6	11		1										100
W-11f	41	Mss	q	4	8	26	2	61	15	11													1				3	131
W-11e	42	Mss	q	6	14	6	25	34	7	2				1		2										1		98
LW-2-22	43	Mss	q	2	4	13	7	59	9	2		1	1				11						1			1		111
LW-2-22(1)	43	Mss	q	6	6	3	16	10	12	7		12	1	4			17						t		5			99
C-6A (2)	44	Mss	q	12	21		9	7	8	2			1	3	2	1		3						8		19		96
W-22a	48	Msv	q	5	6	2	19	30	7	1			6	14	1	3	4						2			1		101
W-22b	49	Mss	q	2	5	1	4	16	7	3	2		2	4			4	2					1			1		54
W-24c	50	CPc	q	31	47	7	3	20	1	1	2							1							3	1		117
W-23 (2)	51	Mss^b	q		4	1	11	23	3	6			9	8		10	19	5	1									100
W-7a	53	Mss	q	9	11	2	3	20	22	3			4			1						1	1					77
W-23b	55	Mss	q	4	4	10	12	29	19	10		9		1		2	6	1			1				2			110
W-23c	56	Mss	q	10	6	6	8	14	11	3			4	3			2	1								3		71
C-33	58	Mss	q		5	1	15	22	3	8			6	17	9	7		3				2		1		1		100
C-16	59	Mss	q	2	6		5	31	6	3			5	5		1		1					4					69
C-16A	59	Mss	q	10	19	2	21	30	10	3																	5	100
81-101	60	Msv	q	1	37		2	3	1	9			7	3	9	4		7						4		14		101
LW-6-58	61	Msv	q	4	15		9	3	5	4			10	1	4	1		10					4		8	23		101
LW-6-44	62	Css	q	28	30	1	4	11	10	1						1	3								1	10		100
LW-5-66	63	CPcl	q	25	18	1	1	7	1	1								1								1		56
LW-5-42	64	CPcl	q	19	48	7		26																				100
LW-5-35	65	Mss^u	q	2	8	9	19	36	11	14							1											100
LW-5-17	66	Mss	q	7	10	6	1	65	18	8	3																	118
LW-5-13	67	Mss	q	2	4	8	2	56	8	3			19	15			4	3										124
LW-5-9	68	Mss	q	11	10	2	1	58	7	3			4	10											1			107
LW-4-1 (1)	69	Mss	q	3	1	5	13	55	11	9			2				1											100
LW-4-1 (2)	69	Mss	q		1	4	6	51	23	10						1								2			2	100
LW-3-99	70	Mss	q	7	5		11	16	8	8			8		1	8												72
LW-3-79	72	Mss^u	q	11	7	6	16	17	30	7				1		1	3	1										100
LW-3-49	74	Mss^u	q	2	3	6	7	33	9	6						1	6											73
LW-3-40	75	Mss	q	5	5	5	15	28	26	11				2								1			2			100
LW-3-23	76	Mss	q	14	27	24	2	79	2	1																		149
LW-3-9	78	Msv	q	1	6	3	4	50	17	10			6	5			1	1							1			105
EW-3	81	Mss	q	3		1	9	24	19	2	2		9	16	3		2	10										100
EW-5	82	Css	q	13	20	6	13	23	18	1															4	2		100
A-59	83	Mss	q	1	3	1	12	25	12	6		1	5	16	13	5									1			101
81-85	82	Msv	v		1								6	46	10	12	17	22										114
81-79	86	Mc	q	16	28		5	23	4	3																		79
FC-1 (2)	§	Css	q	32	22	5	15	23	5																			102
FC-7	§	Css	q	29	24	3	1	14	8								1											80
BW-82	22	Css	q	30	28	3	4	28	4	6									1		1				1			106

Note: Superscripts b and u respectively denote samples from basal and upper parts of units.
* Opaques for sample LW-4-1 (2); undifferentiated lithics for sample W-11f.
† Includes 16 embayed, volcanic quartz grains.
§ Located in Css unit to northwest; see map of Miller and others, 1984.

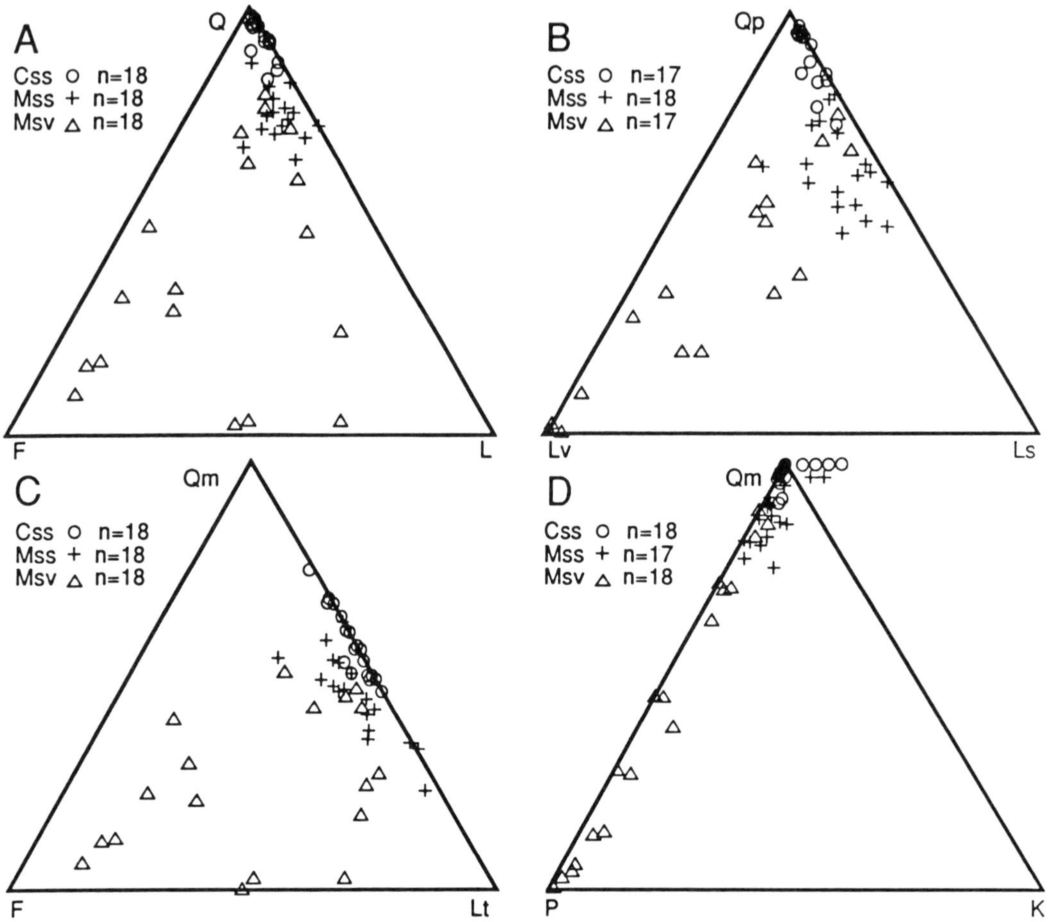

Figure 5. Detrital modes for sandstones of the Schoonover sequence. Point-count methods are after Dickinson (1970) and plotting methods follow the scheme of Dickinson and Suczek (1979); modes are plotted for sandstone with average grain size of 0.25 to 0.50 (medium grained) as estimated from thin sections and grain-count totals ≥400. F = P + K; P = plagioclase feldspar; K = alkali feldspar (orthoclase, microcline, and perthite); L = Ls + Lv; Ls = sedimentary lithic grains (shale, sandstone, siltstone, carbonate, phyllite, and schist); Lv = volcanic lithic grains; Lt = Ls + Lv + Qp; Q = Qm + Qp; Qm = monocrystalline quartz, Qp = polycrystalline quartz (Qp in Table 2, plus recrystallized chert, sedimentary chert, quartzite, metaquartzite, foliate quartz aggregate).

Petrography of quartzose petrofacies. *Sandstone and pebble conglomerate.* Sandstones and conglomeratic rocks of the quartzose petrofacies tend to be poorly to moderately sorted, and contain predominantly subangular to subrounded sand-size grains, with more rounded coarse- and pebble-size clasts.

Quartzose grains (Q) are the dominant constituent, accounting for more than 60 percent of all framework grains (Fig. 5A). Monocrystalline quartz is the most abundant grain type, and includes grains with straight or slightly undulatory extinction ("common plutonic" and "vein" quartz of Folk, 1974), strong undulatory extinction, and segmented undulatory extinction with subgrain boundaries (see Young, 1976). Many of the well-rounded "common plutonic" quartz grains have quartz overgrowths, indicating recycling from sedimentary rocks.

After monocrystalline quartz, microgranular to coarse quartz aggregate (Qp of Table 2) and radiolarian and recrystallized chert are the most common clast types (Fig. 7A). Chert clasts are particularly common in the coarse- to pebble-size fraction of pebbly sandstones and conglomerates. In quartz aggregate clasts, individual crystals are generally equidimensional and form a nonoriented, equigranular to seriate mosaic of crystals, with diameters generally less than 0.6 mm, but in places as great as 2 mm in coarse clasts. Quartz aggregate grains can be further subdivided into two categories: one in which the crystal-crystal boundaries are sutured or granulated, and both undulose and nonundulose extinction occur; and another in which polyhedral crystals with smooth crystal-crystal boundaries, interfacial angles of 120° and straight to slightly undulose extinction predominate. The latter type is composed of relatively coarse (>0.3 mm) individual crystals.

Orthoquartzite, metaquartzite, and foliate quartz aggregate clast types are common in many pebbly sandstones and conglomerates, but are generally rare in sandstones. Orthoquartzite clasts are composed of fine to coarse, well- to very well-rounded quartz grains (Fig. 7A, B); in larger clasts (>2 mm), the average grain size is medium, and sorting is moderate to good. Grain shapes are commonly masked by quartz overgrowth cement, resulting in a polygonal network of grains similar to that seen in many of the generic "quartz aggregate" clasts. Metaquartzite clasts appear to be slightly recrystallized and tectonized equivalents of the orthoquartzite clasts, similar in texture to foliate quartz aggregate clasts but containing relict detrital grains. Foliate quartz aggregate clasts are similar to the microgranular quartz aggregates with sutured crystal-crystal boundaries, except that the foliate type are composed of elongate and aligned crystals.

Feldspar grains in rocks of the quartzose petrofacies are mostly sand-size lath-shaped plagioclase grains that commonly exhibit partial replacement by clay and calcite. In some cases, the cores of grains are more altered than rims, suggesting that those plagioclase crystals were originally compositionally zoned. Alkali feldspars such as microcline and orthoclase are rare or absent (Fig. 5D).

A wide variety and abundance of labile lithic clasts occur in the quartzose petrofacies, and are most commonly argillaceous to siliceous, silty, and/or radiolarian-bearing mudstones ("shale" category in Tables 2, 3) and volcanic fragments. Based on clast texture, type, and abundance of phenocrysts, five types of volcanic clasts are recognized in very coarse to pebbly sandstones: lathwork, microlitic, felsitic, pyroclastic, and volcaniclastic. In

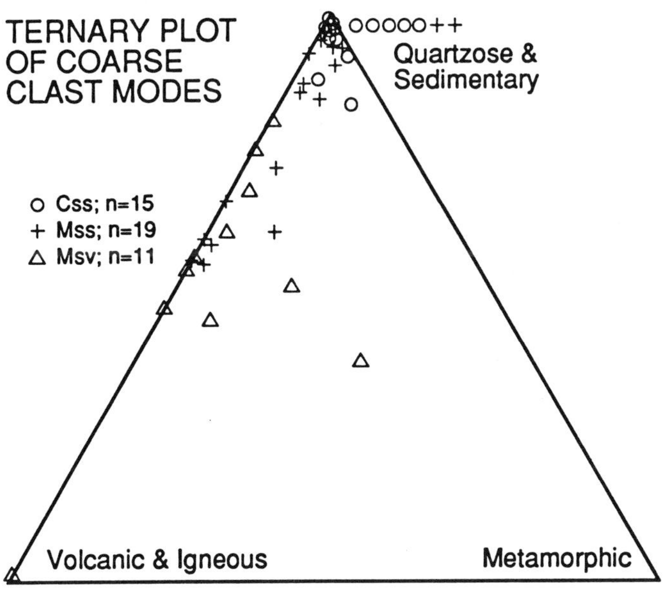

Figure 6. Ternary plot of coarse-grained sandstone from the Schoonover sequence. Only clasts >1 mm were counted. "Quartzose and Sedimentary" includes monocrystalline quartz, polycrystalline quartz, quartzite, recrystallized chert, sedimentary chert, shale, quartzose and feldspathic sandstone and siltstone, and carbonate fragments. "Volcanic and igneous" includes all volcanic clasts (see Table 3), plagioclase, alkali feldspar, and plutonic fragments. "Metamorphic" includes phyllite, schist, quartz-mica tectonite, metaquartzite, and foliate quartz aggregate.

many cases, however, alteration to clay, calcite, and rare chlorite has obscured the primary texture and composition of clasts; these are classified as "undifferentiated volcanics." In the quartzose petrofacies, lathwork and microlitic volcanic grains are abundant; felsitic and pyroclastic grains are only locally significant.

Other locally common clast types are quartzose siltstone/sandstone and, rarely, sparry recrystallized carbonate grains. Less abundant, but also locally common are: feldspathic sandstone clasts containing quartz, plagioclase, and microcline ± polycrystalline quartz, schist and "shale" grains; very rare plutonic fragments consisting of medium to coarse intergrown plagioclase ± quartz aggregates; and metasedimentary clasts. This last category includes phyllite, schist, and quartz-mica tectonite clasts, which in very coarse and pebble-size clasts commonly have elongated, sutured quartz grains with relict detrital textures (Fig. 7C). In schistose clasts, both biotite and muscovite are common, and define a micaceous foliation that is sometimes crenulated. The mineralogy and textures of these metamorphic clasts indicate derivation from a greenschist facies metasedimentary source terrane.

Sandstone composition is related to the age of the samples (Fig. 5A, C, and Fig. 6). The Lower Mississippian sandstones (Mss and Msv map units) are slightly to moderately feldspathic and usually have some volcanic and/or metamorphic detritus. The Upper Mississippian–Upper Pennsylvanian (Css map unit) sandstones, in contrast, contain only sparse feldspar and volcanic clasts, and lack metamorphic detritus, except for local clasts of foliate quartz aggregate, metaquartzite, and traces of phyllite/schist.

Pebby mudstone and boulder conglomerate. Clast compositions from pebbly mudstone units were tabulated in 100-point pebble counts at two outcrops (Table 4), while at numerous sites the types of clasts present and their relative abundance were noted. These data are similar to those gathered by Fagan (1962) at five localities. Nine clasts were examined in thin section.

Detrital modes of pebbly mudstones and the boulder conglomerates (from the Mississippian greenstone, Mg map unit along the basal thrust; Fig. 3) are somewhat similar to those of the sandstones and pebble conglomerates. The dominant clast lithologies are back to dark gray radiolarian chert and white fine- to coarse-grained, well-sorted orthoquartzite consisting entirely of well-rounded quartz grains (Fig. 7B). These latter clasts are similar to orthoquartzite clasts in sandstones. Other clast lithologies include argillite, siltstone, coarse-grained feldspathic sandstone, limestone that is locally bioclastic, and very rarely, mafic greenstone. Scattered silt and sand grains that are present in the mudstone matrix are mostly quartz, chert, and mudstone clasts.

Provenance of quartzose petrofacies. Although a broad range of lithic clasts and quartzose grains occur in sandstones of the quartzose petrofacies, the bulk of the detritus was derived from sedimentary lithologies. The abundance of radiolarian chert and argillaceous clasts points to the importance of "chert-argillite" sequences in the source terrane. While the quartzose siltstone and sandstone, orthoquartzite, and feldspathic sandstone

A

D

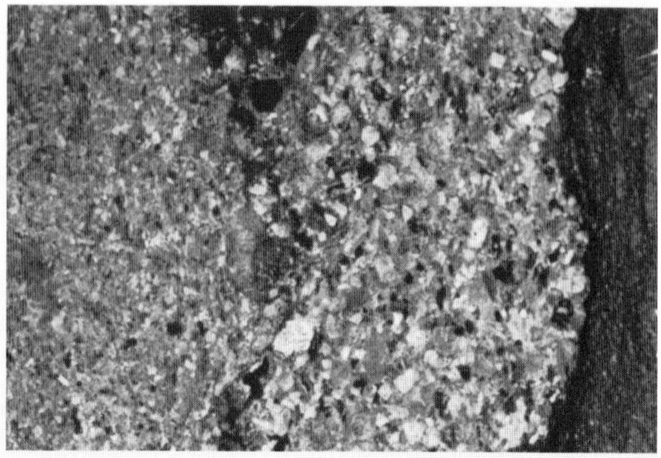

B

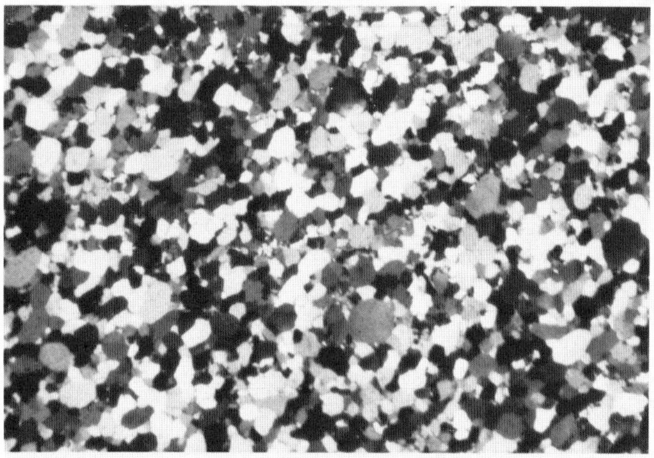

E

C

Figure 7. Photomicrographs of rocks from the Schoonover sequence; all image widths correspond to ~8 mm. A, Silty radiolarian chert and orthoquartzite lithic clasts in a pebbly sandstone from Lower Mississippian quartzose siliclastic rocks (Mss map unit). A carbonate clast is also present in the upper middle portion of the photograph. Note the well-rounded quartz grains in the orthoquartzite clasts. B, Orthoquartzite clast from pebbly mudstone unit in Mississippian siliciclastic sandstone (Mss map unit). A high degree of rounding of individual grains is masked by quartz overgrowth cement. C, Metasedimentary clasts in a Mississippian pebbly sandstone (Mss map unit). Across the middle of the photograph a metasandstone clast is visible. Only a hint of relict detrital quartz is visible; otherwise the clast has the texture of a quartz-mica tectonite. Parallel to this clast on the right is a dark biotite-muscovite schist clast. D, Fine-grained tuff composed primarily of altered plagioclase laths and tuffaceous material. The few opaque grains may be pseudomorphs of mafic phenocrysts. The tuff forms a graded bed (graded right to left) that is scoured into underlying mudstone layer. E, Subquartzose volcanogenic sandstone with example of embayed, euhedral quartz phenocryst.

clasts are much less prevalent, the abundant monocrystalline and polycrystalline quartz grains are most likely disaggregated fragments of these lithologies. That quartzose clastic sedimentary rocks and quartzite in particular were a principal component in the source terrane is supported by: (1) the presence of quartz overgrowths on monocrystalline grains, (2) the population of very well-rounded quartz grains that contrast with the general angularity of many coarser clasts; and (3) the paucity of alkali feldspars and of gneissic and granitic clasts, which would be expected if any significant proportion of the "generic" quartzose grains had been derived from a crystalline basement source.

The presence of as much as 18 percent volcanic clasts in some sandstone indicates the presence of a volcanic source. The volcanic-lithic and feldspathic components tend to occur together, suggesting that the plagioclase grains are derived from the volcanic source.

Finally, the locally significant phyllitic to schistose fragments, foliated quartzose grains, and metaquartzite clasts represent contributions from a metasedimentary source. These clasts are metasedimentary equivalents of the predominant chert-argillite-orthoquartzite-sandstone detritus. This observation, together with the relative abundance of coarse and pebbly labile lithic fragments and the angularity of many clasts, suggests derivation from an orogenic source consisting mainly of sedimentary rocks and their moderately metamorphosed counterparts, along with volcanic rocks of mafic to intermediate composition.

The provenance for the pebbly mudstone (in Mss map unit) and boulder conglomerate (in Cg map unit) appears to differ somewhat from that of the sandstones. Although chert-argillite-orthoquartzite-sandstone detritus is common to both, metamorphic clasts are absent and volcanic clasts are very rare in the former. This different composition and the high degree of clast roundness suggest derivation from a different part of the quartzose source terrane.

Petrography of volcanogenic petrofacies. *Tuffaceous rocks.* Thin beds of very poorly sorted tuffaceous rocks occur within Upper Devonian and Lower Mississippian chert units (DMc map unit). Tuffaceous rocks consist of 20 to 60 percent plagioclase laths, microlitic and vitroclastic lithic fragments, and rare broken hornblende and pyroxene(?) phenocrysts that range from 0.1 to 2.0 mm and are matrix-supported in a microcrystalline to finely crystalline mixture of altered plagioclase microlites, clay, and microcrystalline quartz (Fig. 7D). Rare euhedral and embayed quartz phenocrysts occur, as do traces of chert, shale, phyllite, and microgranular quartz. Commonly, pervasive alteration of these rocks to calcite, clay, chlorite, hematite, albite, and quartz obscures their original mineralogy and texture; however, the matrix-supported fabric, the poor sorting of framework grains, and the predominance of plagioclase laths and volcanic-lithic grains, together with rare mafic phenocrysts, suggest an originally tuffaceous matrix.

Volcanogenic sandstones. Volcanogenic sandstones occur as massive amalgamated beds that are intercalated with more quartzose, massive sandstones in the Lower Mississippian sequence (Msv map unit). In structurally lower thrust sheets, volcanogenic sandstones tend to occur as thinner amalgamated units or single beds within massive intervals of quartzose sandstone. Facies C and D turbidites with volcanogenic compositions are extremely rare, and occur within bundles of facies C and D quartzose turbidites.

TABLE 4. PEBBLE COUNTS FROM PEBBLY MUDSTONE UNITS*

Clast Type	Location 58†	Location 69
Radiolarian chert	52	45
Orthoquartzite	25	8
Feldspathic sandstone	2	4
Argillite/siltstone	11	8
Chert-bearing arenite	3	1
Calcareous siltstone	1	2
Limestone	1	1
Unknown	5	2
Total	100	71

*Data collected by Stanford University Field Study in 1980.
†See Simplified Geologic Map (Fig. 3).

Most volcanigenic sandstones are medium- to very coarse-grained and are composed of plagioclase and volcanic rock fragments with as much as 15 percent matrix and cement of clay and calcite. The lithic clasts are elongated and aligned, probably resulting from compaction, and are altered to clay, calcite, chlorite, and iron oxides. Grain boundaries are commonly obscured by alteration of lithic grains. Size sorting of the framework grains is moderate to good, suggesting that much of the observed "matrix" originated from the diagenetic destruction of volcanic grains.

The compositions of medium-grained volcanogenic sandstones (on which point counts were performed) vary from arkose to feldspathic litharenite, whereas coarser sandstones are lithic-rich (Fig. 5A). Minor amounts of Qm, Qp, and chert, and traces of shale, schist, muscovite, biotite, perthite, and microcline grains occur in many samples. Monocrystalline quartz grains in many cases are clearly of volcanic origin, as they are clear, euhedral, and embayed (Fig. 7E). Feldspar grains are commonly zoned, euhedral plagioclase laths that are altered to albite. The most abundant volcanic-lithic type is microlite (Table 3), exhibiting pilotaxitic or trachytic fabric. Such textures are characteristic of, though not unique to, intermediate composition volcanic rocks (Williams and McBirney, 1979). Also common are clasts with lathwork plagioclase in a holocrystalline groundmass, a texture typical of mafic volcanic rocks. Rare pyroclastic and felsic volcanic clasts can be distinguished in places by poorly preserved, welded fabrics or diagnostic phenocrysts of quartz or potassium feldspar. Also rare are volcaniclastic clasts with detrital grains of chert or quartz silt. Mafic phenocrysts in all varieties of volcanic clasts are almost always altered to opaques, although in a few sandstones, detrital biotite and hornblende are present.

Provenance of volcanogenic petrofacies. The abundance of volcanic clasts, broken mafic phenocrysts, and plagioclase microlites in the altered groundmass in tuffaceous beds are evidence for active pyroclastic volcanism during deposition of the tuffaceous chert unit. In the lithic-rich volcanogenic sandstones, the very high P/F and V/L ratios (Fig. 5), and low quartz contents are comparable to those of sandstones derived from undissected magmatic arcs (Dickinson and Suczek, 1979; Dickinson, 1982). The presence of locally common quartz and hornblende phenocrysts and the abundance of microlitic and felsitic types of volcanic grains suggest a source with intermediate to siliceous compositions, which is characteristic of magmatic arc terranes. The numerous arkosic compositions clearly represent sands derived from the same source area, as indicated by high P/F and V/L ratios (Fig. 5D); the predominance of plagioclase grains over volcanic clasts and the higher quartz (Q) content, however, indicate that they are compositionally more mature versions of the lithic-rich sandstone. Finally, the minor to locally common monocrystalline and polycrystalline quartz (Qm and Qp), chert, shale, and schist clasts in some volcanogenic sandstones are similar to the suite of clasts composing the bulk of the quartzose petrofacies.

Petrography of calcareous petrofacies. *Silty limestone turbidites.* Silty limestone turbidites typically have a silty to microcrystalline matrix of microspar and minor amounts of clay, with 5 to 30 percent silt-size monocrystalline quartz grains. Locally, the matrix is recrystallized to sparry calcite, or replaced by microcrystalline quartz. Sand-size grains of monocrystalline quartz, polycrystalline quartz, chert, and, more rarely, feldspar, are present locally.

Bioclastic limestones. Bioclastic limestones are composed of moderately sorted, medium-grained to pebble-size bioclasts and angular terrigenous grains in a cement of impure, sparry calcite and irregular masses of hematite. Bioclasts, though commonly recrystallized, include abundant crinoid debris, broken mollusc fragments, and minor poorly preserved clasts of coral. Terrigenous grains compose 10 to 15 percent of bioclastic limestones and include monocrystalline quartz grains with straight or undulose extinction, chert in minor amounts, and traces of feldspar and polycrystalline quartz.

Provenance of calcareous petrofacies. Silty limestone turbidites and the bioclastic limestones both contain commingled calcareous and quartzose detritus. This suggests that these rocks were derived from a source terrane where a carbonate depositional environment existed in proximity to a quartzose terrigenous supply.

Summary of petrography and provenance. The Schoonover sandstones fall into two petrofacies that represent two distinct source terranes. The quartzose petrofacies is interpreted to represent detritus derived from a chert-argillite-orthoquartzite-sandstone sedimentary source terrane, whereas volcaniclastic and arkosic sandstones of the volcanogenic petrofacies were largely derived from a magmatic arc terrane. Two aspects of the data from the Lower Mississippian unit (Msv map unit) suggest, however, that detritus from the two source terranes was mixed together before deposition in the Schoonover basin: (1) sandstones from both petrofacies are interbedded within single channel-fill units (Fig. 4), and (2) volcanogenic sandstones contain the same suite of quartzose and sedimentary-lithic clasts that are so abundant in the quartzose petrofacies (Fig. 5A, C; Table 2). Pebbly mudstone and quartzite-chert boulder conglomerate compositions from the younger Carboniferous units (Mss and Mg units, respectively) contain no evidence of source mixing and probably were derived solely from the quartzose sedimentary source.

Diminishing feldspar (F) and volcanic lithic (Lv) content in Lower Mississippian sandstones (Mss map unit) and deposition of quartz arenites by Late Mississippian (Meramec) time (Css map unit), indicate that the influx of volcanic detritus into the basin declined with time. The appearance of the silty limestone turbidites, also beginning in Late Mississippian time, marks the development of carbonate buildups, presumably in shallow water fringing the sedimentary source terrane; this is inferred from the quartzose component (Qm, chert, and rare Qp clasts) in the calcareous rocks. Viewed as a whole, the clastic deposits of the Schoonover sequence display an increase in mineralogic maturity with time and reflect sporadic deposition of penecontemporaneous carbonate in the source area.

Comparison of Schoonover clastic rocks with other units in the Golconda allochthon

The Schoonover sequence is considered part of the Golconda allochthon based on its stratigraphy and structural setting, as outlined by Miller and others (1984). As such, it represents the northernmost known exposure of the Golconda allochthon. The correlation is detailed below, in order to provide a basis for discussing the paleogeographic and tectonic setting of the allochthon.

The Upper Mississippian to Lower Permian section of the Schoonover sequence is considered correlative with the Havallah sequence as described in the Battle Mountain area and the Sonoma and Tobin Ranges (Fig. 1). In the Battle Mountain area, Upper Mississippian chert that depositionally overlies submarine basalt (Murchey, 1982) may be correlative in age with the sequence of Meramecian chert that depositionally overlies pillow basalts above the basal thrust fault in the Schoonover sequence.

The Havallah sequence consists of chert, silty limestone turbidites, and calcareous and argillaceous mudstones, with local scraps of greenstone, units of quartzose sandstones, and sandy bioclastic turbidites (Silberling and Roberts, 1962; Roberts, 1964; Miller and others, 1982; Snyder and Brueckner, 1983; Stewart and others, 1986). Because of the complex structure of the Golconda allochthon and limited fossil age data, however, the stratigraphy and ages of individual clastic units are commonly ambiguous. In addition, little systematic petrographic information on clastic units is available (cf. Turner, 1982; Tomlinson, 1988). Nevertheless, based on existing descriptions and age data, there are many similarities between the Schoonover and Havallah clastic rocks.

Compositions and ages of Golconda clastic units. In the Golconda allochthon in central Nevada, clastic units within the Havallah sequence include the Jory member of the Havallah Formation of Roberts (1964) in the Antler Peak area, the Farrel Canyon Formation in the Osgood Mountains (Hotz and Willden, 1964), and units from the Havallah sequence in the Tobin Range (Ferguson and others, 1951; Turner, 1982; Stewart and others, 1977, 1986; for localities, see Fig. 1).

Jory Member. The Jory Member was originally described and inferred to be of early mid-Pennsylvanian age, based on fusulinids present as bioclasts in calcareous sandstones (Roberts, 1964). Miller and others (1982), however, obtained a Late Mississippian age from interbedded radiolarian chert near the base of the Jory, so it is likely that this clastic interval may range in age from Late Mississippian to Pennsylvanian, and is coeval with carbonate and siliciclastic rocks (CPcl and Css map units) in the Schoonover sequence. Compositionally, the bulk of the Jory sandstones are described as quartz arenites containing monocrystalline quartz and lithic clasts of chert and quartzite. Locally, plagioclase, microcline, orthoclase, and biotite, muscovite, and/or collophane are present (Roberts, 1964). These sandstones are most similar to the quartzose sandstones from the Upper Mississippian and younger part of the Schoonover sequence, and possibly the slightly feldspathic quartzose sandstones from the older Mississippian siliciclastic units. Roberts (1964) also noted numerous sandy bioclastic limestones; this type of sandstone is only rarely present in the Schoonover sequence.

Farrel Canyon Formation. The Farrel Canyon Formation in the Osgood Mountains is an undated assemblage of quartzose sandstone, tuff, chert, shale, siltstone, and greenstone. Its structural relations with other rock units are unclear; however, it has a strong lithologic resemblance to the Havallah and Schoonover sequences and to an unnamed sequence in the adjacent Hot Springs Range (Fig. 1) in which Mississippian to Permian fossils have been found (Hotz and Willden, 1964; Jones, 1988). The quartzose sandstones consist mostly of quartz and chert, with lesser alkali and plagioclase feldspar and shale fragments. In pebbly sandstones, chert and quartzite clasts are abundant. Hotz and Willden (1964) have also described "crystal tuffs" consisting of quartz and potassium feldspar with relict vitroclastic textures, fragments of quartz, plagioclase, and sanidine(?), and clasts of chert, quartzite, and volcanic rock. Hotz and Willden inferred that these rocks, which may be analogous to the Schoonover volcaniclastic rocks, were derived from an andesitic to rhyolitic source.

Tobin Range sandstones. In the Tobin Range, sandstones of Late Mississippian–Early Permian age studied by Turner (1982) have mostly very quartzose compositions, although, as in the Jory sandstones, a gradation into sandy bioclastic limestones is observed. Coarse clast counts of pebbly sandstones include only chert, shale, siltstone, and carbonate lithic clasts; quartzite, volcanic, and metamorphic detritus are lacking, and feldspar is very minor. Detrital modes of the sandstones (Fig. 8) are similar to those obtained for many of the coeval quartzose sandstones in the Schoonover sequence (Css map unit).

Additional data. Sparse compositional data from undated sandstone and sandy calcarenite samples from the Tobin, Sonoma, and Independence Ranges (Dickinson and others, 1983; summarized in Fig. 8) indicates a range of compositions from lithic-rich sandstone to pure quartzose sandstone. Chert and shale are the dominant lithic clasts; feldspar and volcanic rock fragments are locally significant, but typically rare.

Volcaniclastic sandstones and breccias of Permian age occur in the Antler Peak area (Snyder and Brueckner, 1983; Miller and others, 1982), and southern Shoshone, New Pass, and southern Toiyabe Ranges (Tomlinson and Wright, 1986; Babaie, 1987; Tomlinson, 1988). Volcaniclastic rocks of unknown age and composition are also reported from the Sonoma Range (Snyder and Brueckner, 1983).

Depositional environment of siliciclastic rocks in the Golconda allochthon. Most workers have interpreted clastic sequences in the Golconda allochthon as deep-water fan deposits, based on bedding characteristics and their association with radiolarian chert, pillow basalt, and deep-water ichnofossils (Roberts, 1964; Hotz and Willden, 1964; Stewart and others, 1977). Turner (1982) recognized both middle and outer fan deposits in the Tobin Range. Coarse clastic rocks elsewhere in the Golconda allochthon, such as pebble and granule conglomerates, massive sandstones and pebbly sandstones, and facies C and D turbidites, are similar to strata in the Schoonover sequence. There are not sufficient data on the type and distribution of lithofacies in the Havallah sequence, however, to allow comparison and contrast of the depositional setting of the Havallah clastic rocks with that inferred for the Schoonover facies.

Summary of Schoonover-Havallah comparison. Although quantitative compositional data from the Golconda allochthon are sparse, the Upper Mississippian and Pennsylvanian siliciclastic rocks of the Schoonover sequence (Css map unit) are similar to many coeval clastic units elsewhere in the Golconda allochthon. The bulk of the sandstones in the allochthon are sublithic to pure quartzose sandstones (Fig. 8), containing as much as 10 percent feldspar (Dickinson and others, 1983). As in the Schoonover, the most abundant lithic clasts are chert, shale, and quartzite. Volcanic rock fragments are only locally significant, and metamorphic clasts are rare. This apparently reflects the probable Late Mississippian and younger age of quartzose sandstones in the Havallah sequence; coeval sandstones in the Schoonover sequence (Css map unit) are also mostly lacking in volcanic and metamorphic clasts. Volcaniclastic sandstones of Early Mississippian age are minor or absent in the Havallah sequence, while volcaniclastic rocks of Permian age are common in the Havallah, but not the Schoonover sequence. Equivalents to the Schoonover Devonian-Mississippian tuffaceous chert-argillite (DMc map unit) and Lower Mississippian volcaniclastic and quartzose turbidite sequences (Msv and Mss map units) have not been documented elsewhere in the Golconda allochthon.

Composition and age data suggest that the Upper Mississippian and Pennsylvanian Golconda clastic units and coeval clastic rocks of the Schoonover sequence were derived from a sedimen-

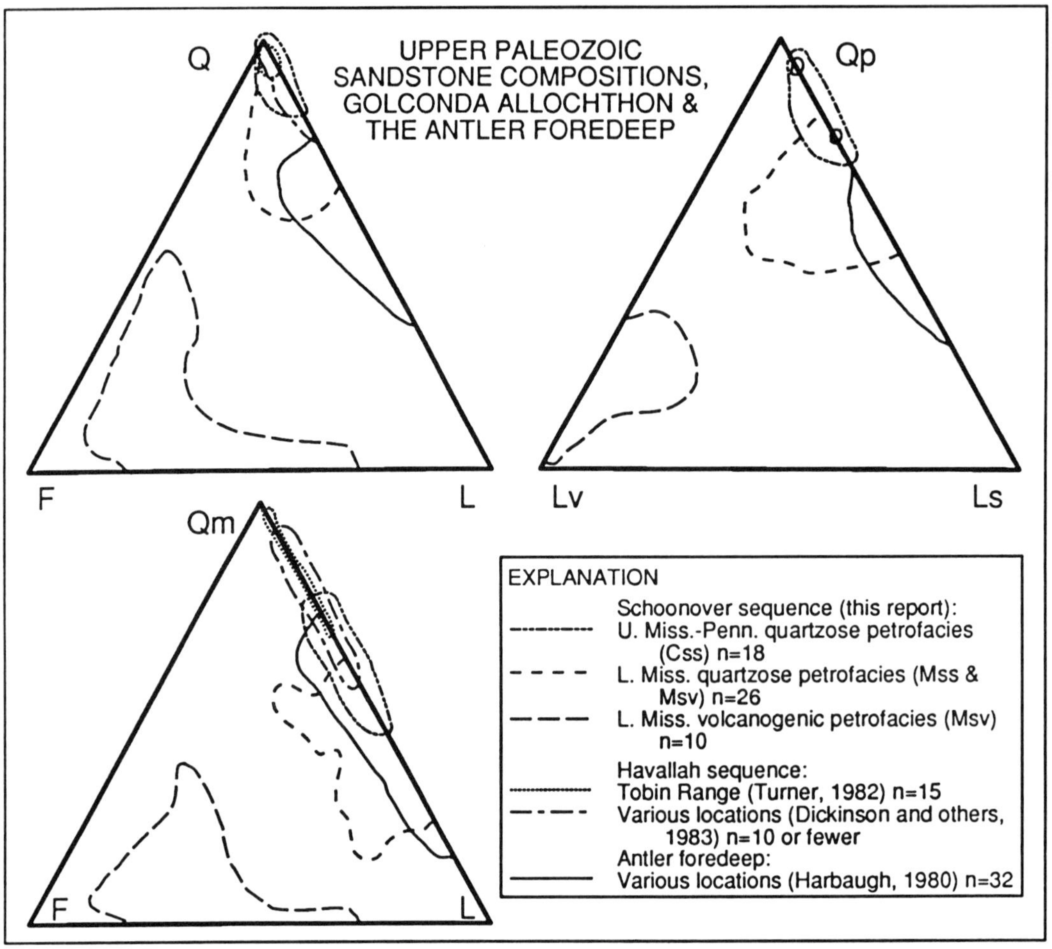

Figure 8. Summary of available data on composition of sandstones from the Golconda allochthon.

tary terrane composed of chert, shale, and quartzose clastic rocks, including quartzite. Thus, Upper Mississippian to Permian parts of the Havallah and Schoonover sequences appear to have been deposited in the same oceanic basin. Upper Devonian and Lower Mississippian rocks are apparently not present or are structurally omitted in the Havallah sequence. Deposition in the Schoonover part of the basin apparently ceased in Early Permian time, perhaps due to the onset of thrust faulting (Miller and others, 1984), while deposition continued into the Late Permian elsewhere in the Havallah basin.

DISCUSSION AND CONCLUSIONS

Paleogeography of Havallah-Schoonover basin

Given evidence for large-scale thrusting of the Havallah and Schoonover sequences over autochthonous Paleozoic continental margin sequences, some recent workers have cast the Golconda allochthon as a "suspect terrane" of uncertain paleogeographic origin (Speed, 1977; Dickinson and others, 1983; Snyder and Brueckner, 1983). The petrographic and stratigraphic data in this study, however, indicate that the Havallah-Schoonover basin evolved adjacent to North America and in association with a magmatic arc.

Geologic ties to North America. Primary evidence for paleogeographic ties between the Golconda allochthon and the North American continent lies within the clastic units of the Schoonover sequence. Roberts and others (1958) and Fagan (1962) noted that the most likely source for the orthoquartzite, radiolarian chert, argillite, and other sedimentary-lithic clasts within clastic units in the Schoonover sequence and elsewhere in the Golconda allochthon is the Roberts Mountains allochthon (Roberts and others, 1958; Hotz and Willden, 1964; Churkin and Kay, 1967; Gilluly and Gates, 1965; Stewart, 1980). Quartzite clasts from Lower Mississippian pebbly sandstone and pebbly mudstone in the Schoonover sequence are petrographically indistinguishable from the Valmy Formation orthoquartzite in the Roberts Mountains allochthon, distinctive for its purity and high degree of grain roundness (Ketner, 1966). The petrography of the feldspathic sandstone clasts in Schoonover pebbly mudstone units is similar to that of feldspathic sandstone of the Harmony formation from the Roberts Mountains allochthon (Suczek, 1977, oral communication, 1980). The Roberts Mountains allochthon was undergoing erosion throughout the late Paleozoic, after it was emplaced on the shelf during the Devonian-Mississippian Antler orogeny, based on the stratigraphy and petrography of rocks deposited in the foredeep to the east of the Antler highlands (Roberts and others, 1958; Poole and Sandberg, 1977; Stewart, 1980; Harbaugh, 1980; Dickinson and others, 1983). Carbonate, shale, and quartzite strata below the Roberts Mountains allochthon were locally exposed during the late Paleozoic in structural windows or fault blocks (Hotz and Willden, 1964; Miller and others, 1981) and may also have contributed detritus to the Havallah-Schoonover basin.

The Golconda allochthon sandstones can be linked to sandstones deposited in the Antler foredeep based on similar sedimentary-lithic clasts and similar detrital modes (Fig. 8). Furthermore, the onset of sandstone deposition in the Schoonover sequence was synchronous with the sandstone deposition in the Antler foredeep during Early Mississippian time (Poole, 1974). The higher level of mineralogic maturity for the Golconda sandstone is interpreted to reflect different dispersal and transport histories.

Other similarities between the stratigraphies of the continental shelf and the Havallah-Schoonover basin exist. Meramecian basalts in the Schoonover (Miller and others, 1984) and Havallah (Antler Peak area, Miller and others, 1982) sequences are coeval with basalt of the autochthonous Meramec-age Nelson Formation in the Mountain City area (Coats, 1969; Little, 1987; Fig. 1). Interbedded limestone and pillow basalts of Meramecian age also occur in the Goughs Canyon Formation in the Osgood Mountains (Fig. 1; Hotz and Willden, 1964), although there it is uncertain whether the sequence is part of the Golconda allochthon or is autochthonous. Finally, the beginning of limestone turbidite deposition in the Schoonover basin in latest Mississippian–Early Pennsylvanian time was coeval with the transgression of the Ely Limestone across the outer continental shelf (Rich, 1977). These stratigraphic ties to the Antler orogenic belt and the evidence for a deep-water depositional environment strongly suggest that the Havallah-Schoonover basin lay adjacent to the continental margin of North America throughout late Paleozoic time.

The Lower Mississippian tuff and volcaniclastic sandstones of the Schoonover sequence, and other arc-derived detritus of Permian or unspecified age elsewhere in the Golconda allochthon, indicate the proximity of a volcanic arc or arcs during the evolution of the Havallah-Schoonover basin. Mixture of intermediate volcanic detritus with chert, quartzite, and argillite clasts derived from the Roberts Mountains allochthon in the Schoonover Lower Mississippian sandstones suggests that the arc source(s) may have somewhere been attached to the continent (see Rubin and others, this volume). Metamorphic fabrics of Paleozoic age described in parts of the Roberts Mountains allochthon (Hotz and Willden, 1964; Roberts, 1964; Oldow, 1984; Madrid, 1987) perhaps support this interpretation. Volcanic arc detritus, however, has not been noted in the Antler highlands or foredeep rocks, and Mesozoic structures and sedimentary rocks obscure any relations outboard (westward) of the Golconda and Roberts Mountains allochthons and the Antler orogenic belt (see Fig. 1). Thus, the relation of the inferred Antler-age arc to the continental margin is uncertain. Potential Paleozoic arc sources are found in the northern Sierra Nevada and eastern Klamath Mountains of California (Schweickert and others, 1977; D'Allura and others, 1977; Harwood, 1983, 1988; Miller, 1987, 1989).

Given the constraints of Schoonover provenance and regional relations, a volcanic arc probably lay seaward and/or along strike of the Havallah-Schoonover basin, as depicted in Figure 9. If the arc lay between the basin and the Antler highlands, debris from the Roberts Mountains allochthon would have had to traverse or circumvent the arc, in which case the "pure" Antler-derived compositions found in the Schoonover pebbly mudstones and sandstones would not be expected. The mingled volcanic, metamorphic, and Roberts Mountains allochthon detritus in Schoonover Mississippian volcaniclastic and quartzose sandstones, as well as the local metamorphic fabrics in the Roberts Mountains allochthon, constitute evidence that an arc may have been joined to the continental margin at some point along strike of the Antler orogen, either to the north or south (Fig. 9).

Tectonic setting of the paleozoic continental margin

The paleogeographic relations discussed above suggest that the Schoonover-Havallah basin evolved in a back-arc setting. Late Devonian–age basalts suggest a rift origin for the Schoonover basin, coeval with Late Devonian extensional tectonism in the Roberts Mountains allochthon and northern Canadian Cordillera (Gordey and others, 1987; Turner and others, 1989). Andesitic greenstones overlain by Famennian–early Kinderhookian chert and tuff, and succeeded by Kinderhookian volcanogenic and quartzose sandstones imply that the Havallah-Schoonover basin experienced volcanism and tectonism adjacent to a volcanic arc immediately prior to the Antler orogeny and was also adjacent to the continent during emplacement of the Roberts Mountains allochthon. Episodes of basaltic volcanism and pulses of siliciclastic deposition suggest continued extensional or strike-slip tectonism along the continental margin during the late Paleozoic, before closure of the Havallah-Schoonover basin by back-arc thrusting or subduction in the Sonoma orogeny. A similar late Paleozoic history of extensional tectonics has been envisaged for the Canadian segment of the North American continental margin. There, extensional faulting, basaltic volcanism, and widespread deposition of chert-quartzite conglomerate are recorded (Eisbacher, 1983; Gordey and others, 1987; Turner and others, 1989).

This work on the Golconda allochthon supports and extends the suggestions of previous workers (e.g., Burchfiel and Davis, 1972; Churkin, 1974; Schweickert and Snyder, 1981; Miller and others, 1984) that the mid-Paleozoic western margin of North America was fringed by island arc(s) separated from the continent by marginal basins. Along strike, coeval arc rocks intruded

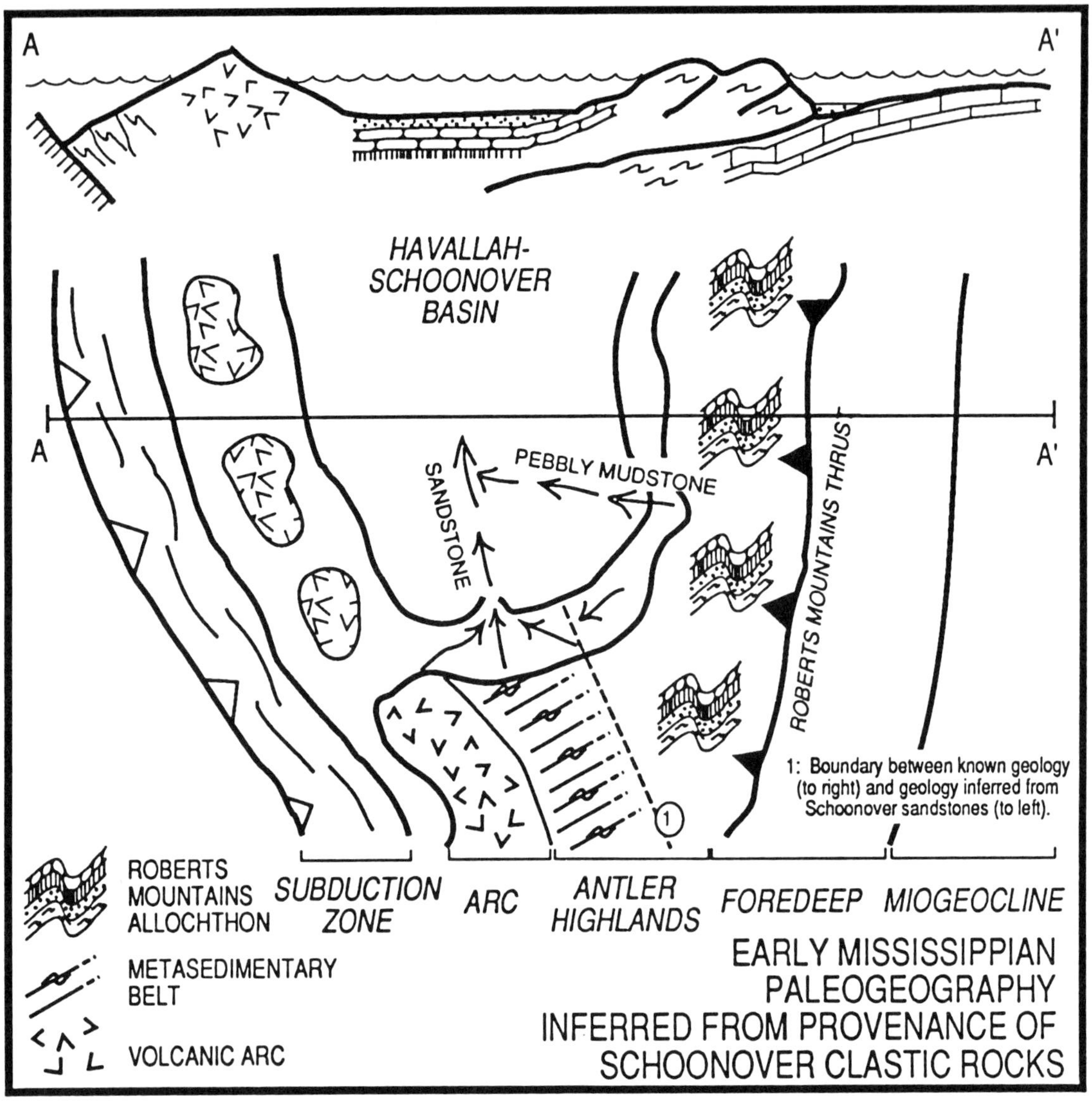

Figure 9. Schematic paleogeographic setting for the Havallah-Schoonover basin during Early Mississippian time, inferred from the provenance of clastic rocks in the Schoonover sequence.

the continental margin (e.g., Rubin and others, this volume). Imbrication and overthrusting of portions of intervening basins and/or arcs over the continental shelf may have been caused by short-lived episodes of arc-continent convergence due to changes in paleo-Pacific plate motions.

ACKNOWLEDGMENTS

Financial support for field and laboratory studies was provided by National Science Foundation Grant EAR-80-18763 to E. L. Miller and D. K. Larue. Field work benefited greatly from the assistance of Elizabeth Miller and David Larue. Dr. Larue, in particular, contributed to the gathering and analysis of sedimentologic data. Brian Holdsworth of the University of Keele, U.K., beyond his work on the Schoonover and Havallah biostratigraphy, contributed to discussion of the data presented here. E. L. Miller and A. J. Tomlinson generously reviewed the manuscript. Reviews by J. A. Dumoulin and R. J. Turner also improved the manuscript. Finally, the assistance of Meghan Miller made this report possible.

REFERENCES CITED

Babaie, H. A., 1987, Paleogeographic and tectonic implications of the Golconda allochthon, southern Toiyabe Range, Nevada: Geological Society of America Bulletin, v. 99, p. 231–243.

Bouma, A. H., 1962, Sedimentology of some flysch deposits: Amsterdam, Elsevier, 168 p.

Burchfiel, B. C., and Davis, G. A., 1972, Structural framework and evolution of the southern part of the Cordilleran orogen, western United States: American Journal of Science, v. 272, p. 97–118.

—— , 1975, Nature and controls of Cordilleran orogenesis, western United States; Extensions of an earlier synthesis: American Journal of Science, v. 275-A, p. 363–396.

Churkin, M., 1974, Paleozoic marginal ocean basin–volcanic arc systems in the Cordilleran fold belt, *in* Dott, R. H., and Shaver, R. H., eds., Modern and ancient geosynclinal sedimentation: Society of Economic and Petroleum Geologists Special Publication 19, p. 174–192.

Churkin, M., and Kay, M., 1974, Graptolite-bearing Ordovician siliceous and volcanic rocks, northern Independence Range, Nevada: Geological Society of America Bulletin, v. 78, p. 651–668.

Coats, R. R., 1969, Upper Paleozoic formations of the Mountain City area, Elko County, Nevada, *in* Cohee, G. V., Bates, R. G., and Wright, W. B., eds., Changes in stratigraphic nomenclature by the U.S. Geological Survey, 1967: U.S. Geological Survey Bulletin 1274-A, p. A22–A27.

D'Allura, J. A., Moores, E. M., and Robinson, C., 1977, Paleozoic rocks of the northern Sierra Nevada, California, *in* Stewart, J. H., Stevens, C. H., and Fritsche, A. E., eds., Paleozoic paleogeography of the western United States: Pacific Section, Society of Economic Paleontologists and Mineralogists, Pacific Coast Paleogeographic Symposium 1, p. 395–408.

Dickinson, W. R., 1970, Interpreting detrital modes of graywacke and arkose: Journal of Sedimentary Petrology, v. 40, p. 695–707.

—— , 1977, Paleozoic plate tectonics and the evolution of the Cordilleran continental margin, *in* Stewart, J. H., Stevens, C. H., and Fritsche, A. E., eds., Paleozoic paleogeography of the western United States: Pacific Section, Society of Economic Paleontologists and Mineralogists, Pacific Coast Paleogeographic Symposium 1, p. 137–155.

—— , 1982, Compositions of sandstones in Circum-Pacific subduction complexes and fore-arc basins: American Association of Petroleum Geologists Bulletin, v. 66, p. 121–137.

Dickinson, W. R., and Suczek, C. A., 1979, Plate tectonics and sandstone compositions: American Association of Petroleum Geologists Bulletin, v. 63, p. 2164–2182.

Dickinson, W. R., Harbaugh, D. W., Saller, A. H., Heller, P. L., and Snyder, W. S., 1983, Detrital modes of upper Paleozoic sandstones derived from Antler orogen in Nevada; Implications for nature of Antler orogeny: American Journal of Science, v. 238, p. 481–509.

Eisbacher, G. H., 1983, Devonian-Mississippian sinistral transcurrent faulting along the cratonic margin of western North America; A hypothesis: Geology, v. 11, p. 7–10.

Fagan, D. J., 1962, Carboniferous cherts, turbidites, and volcanic rocks in the northern Independence Range, Nevada: Geological Society of America Bulletin, v. 73, p. 595–612.

Ferguson, H. G., Roberts, R. J., and Muller, S. W., 1951, Geologic map of the Golconda Quadrangle, Nevada: U.S. Geological Survey Geological Quadrangle Map GQ-15 (reprinted 1976).

Folk, R. L., 1974, Petrology of sedimentary rocks: Austin, Texas, Hemphill's Bookstore, 182 p.

Gilluly, J., and Gates, O., 1965, Tectonic and igneous geology of the northern Shoshone Range, Nevada: U.S. Geological Survey Professional Paper 465, 153 p.

Gordey, S. P., Abbott, J. G., Tempelman-Kluit, D. J., and Gabrielse, H., 1987, "Antler" clastics in the Canadian Cordillera: Geology, v. 15, p. 103–107.

Harbaugh, D. W., 1980, Depositional facies and provenance of the Mississippian Chainman shale and Diamond Peak Formation, central Diamond Mountains, Nevada [M.S. thesis]: Stanford, California, Stanford University, 81 p.

Harwood, D. S., 1983, Stratigraphy of upper Paleozoic volcanic rocks and regional unconformities in part of the northern Sierra terrane, California: Geological Society of America Bulletin, v. 94, p. 413–422.

—— , 1988, Tectonism and metamorphism in the northern Sierra terrane, northern California, *in* Ernst, W. G., ed., Metamorphism and crustal evolution of the western United States; Rubey Volume 7: Englewood Cliffs, New Jersey, Prentice-Hall, p. 764–788.

Hiscott, R. N., 1980, Depositional framework of sandy mid-fan complexes of Tourelle Formation, Ordovician, Quebec: American Association of Petroleum Geologists Bulletin, v. 64, p. 1052–1077.

Hotz, P. E., and Willden, R., 1964, Geology and mineral deposits of the Osgood Mountains Quadrangle, Humboldt County, Nevada: U.S. Geological Survey Professional Paper 431, 128 p.

Ingersoll, R. V., 1978, Submarine fan facies of the Upper Cretaceous Great Valley Sequence, northern and central California: Sedimentary Geology, v. 21, p. 205–230.

Jenkyns, H. C., and Winterer, E. L., 1982, Palaeoceanography of Mesozoic ribbon radiolarites: Earth and Planetary Science Letters, v. 60, p. 351–375.

Jones, A. E., 1988, The Poverty Peak subterrane of the Golconda allochthon in the northern Hot Springs Range, Humboldt County, Nevada: Geological Society of America Abstracts with Programs, v. 20, p. 171.

Ketner, K. B., 1966, Comparison of Ordovician eugeosynclinal and miogeosynclinal quartzites of the Cordilleran geosyncline, *in* Geological Survey Research 1966: U.S. Geological Survey Professional Paper 550-C, p. C54–C60.

—— , 1977, Late Paleozoic orogeny and sedimentation, southern California, Nevada, Idaho, and Montana, *in* Stewart, J. H., Stevens, C. H., and Fritsche, A. E., eds., Paleozoic paleogeography of the western United States: Pacific Section, Society of Economic Paleontologists and Mineralogists, Pacific Coast Paleogeographic Symposium 1, p. 363–369.

Link, M. H., and Nilsen, T. H., 1980, The Rocks Sandstone, an Eocene sand-rich deep-sea fan deposit, northern Santa Lucia Range, California: Journal of Sedimentary Petrology, v. 50, p. 583–602.

Little, T. A., 1987, Stratigraphy and structure of metamorphosed upper Paleozoic rocks near Mountain City, Nevada: Geological Society of America Bulletin, v. 98, p. 1–17.

Madrid, R. J., 1987, Stratigraphy of the Roberts Mountains allochthon in north-central Nevada [Ph.D. thesis]: Stanford, California, Stanford University, 336 p.

Middleton, G. V., and Hampton, M. A., 1973, Sediment gravity flows; Mechanics of flow and deposition, *in* Turbidites and deep water sedimentation: Anaheim, Society of Economic and Petroleum Geologists Pacific Section Short Course, p. 1–38.

Miller, E. L., Bateson, J., Dinter, D., Dyer, J. R., Harbaugh, D. W., and Jones, D. L., 1981, Thrust emplacement of the Schoonover sequence, northern Independence Mountains, Nevada: Geological Society of America Bulletin, v. 92, p. 730–737.

Miller, E. L., Kanter, L. R., Larue, D. K., Turner, R. J., Murchey, B., and Jones, D. L., 1982, Structural fabric of the Paleozoic Golconda allochthon, Antler Peak Quadrangle, Nevada; Progressive deformation of an oceanic sedimentary assemblage: Journal of Geophysical Research, v. 87, p. 3795–3804.

Miller, E. L., Holdworth, B. K., Whiteford, W. B., and Rodgers, D., 1984, Stratigraphy and structure of the Schoonover sequence, northeastern Nevada; Implications for Paleozoic plate margin tectonics: Geological Society of America Bulletin, v. 95, p. 1063–1076.

Miller, M. M., 1987, Dispersed remnants of a northeast Pacific fringing arc; Upper Paleozoic island arc terranes of Permian McCloud faunal affinity, western U.S.: Tectonics, v. 6, p. 807–830.

—— , 1989, Intra-arc sedimentation and tectonism; Late Paleozoic evolution, eastern Klamath terrane, California: Geological Society of America Bulletin, v. 101, p. 170–187.

Murchey, B., 1982, Chert facies in the Havallah sequence near Battle Mountain, Nevada: Geological Society of America Abstracts with Programs, v. 14, p. 219.

Mutti, E., 1977, Distinctive thin-bedded turbidite facies and related depositional environments in the Eocene Hecho Group (south-central Pyrenees, Spain): Sedimentology, v. 24, p. 107–131.

Mutti, E., and Ricci-Lucchi, F., 1978 (English translation of 1972 paper), Turbidites of the northern Appennines; Introduction to facies analysis: International Geological Review, v. 20, no. 2, p. 127–166.

Nelson, C. H., Normark, W. R., Bouma, A. H., and Carlson, P. R., 1978, Thin-bedded turbidites in modern submarine canyons and fans, *in* Stanley,

D. J., and Kelling, G., eds., Sedimentation in submarine canyons, fans, and trenches: Stroudsberg, Pennsylvania, Dowden, Hutchinson and Ross, p. 177–189.

Nilsen, T. H., and Stewart, J. H., conveners, 1980, The Antler orogeny; Mid-Paleozoic tectonism in western North America: Geology, v. 8, p. 1094–1101.

Normark, W. R., 1978, Fan valleys, channels, and depositional lobes on modern submarine fans; Characters for recognition of sandy turbidite environments: American Association of Petroleum Geologists Bulletin, v. 62, p. 912–931.

——, 1980, Modern and ancient submarine fans; Reply: American Association of Petroleum Geologists Bulletin, v. 64, p. 1108–1112.

Normark, W. R., and Hess, H., 1980, Quaternary growth patterns of California submarine fans, *in* Field, M. E., Bouma, A. H., and others, eds., Quaternary depositional environments of the Pacific coast: Pacific Section, Society of Economic Paleontologists and Mineralogists Pacific Coast Paleogeographic Symposium 4, p. 183–208.

Oldow, J. S., 1984, Spatial variability in the structure of the Roberts Mountains allochthon, western Nevada: Geological Society of America Bulletin, v. 95, p. 174–185.

Poole, F. G., 1974, Flysch deposits of the Antler foreland basin, western United States, *in* Dickinson, W. R., ed., Tectonics and sedimentation: Society of Economic and Petroleum Geologists Special Publication 22, p. 58–82.

Poole, F. G., and Sandberg, C. A., 1977, Mississippian paleogeography and tectonics of the western United States, *in* Stewart, J. H., Stevens, C. H., and Fritsche, A. E., eds., Paleozoic paleogeography of the western United States: Pacific Section, Society of Economic Paleontologists and Mineralogists Pacific Coast Paleogeographic Symposium 1, p. 67–85.

Ricci-Lucchi, F., 1975, Depositional cycles in two turbidite formations of the northern Appennines (Italy): Journal of Sedimentary Petrology, v. 45, p. 3–43.

Rich, M., 1977, Pennsylvanian paleogeographic patterns in the western United States, *in* Stewart, J. H., Stevens, C. H., and Fritsche, A. E., eds., Paleozoic paleogeography of the western United States: Pacific Section, Society of Economic Paleontologists and Mineralogists Pacific Coast Paleogeographic Symposium 1, p. 87–111.

Roberts, R. J., 1964, Stratigraphy and structure of the Antler Peak Quadrangle, Humboldt and Lander Counties, Nevada: U.S. Geological Survey Professional Paper 459-A, 93 p.

Roberts, R. J., Hotz, P. E., Gilluly, J., and Ferguson, H. G., 1958, Paleozoic rocks of north-central Nevada: American Association of Petroleum Geologists Bulletin, v. 42, p. 2813–2857.

Saller, A. H., and Dickinson, W. R., 1982, Alluvial to marine facies transition in the Antler overlap sequence, Pennsylvanian and Permian of north-central Nevada: Journal of Sedimentary Petrology, v. 52, p. 925–940.

Schweickert, R. A., and Snyder, W. S., 1981, Paleozoic plate tectonics of the Sierra Nevada and adjacent regions, *in* Ernst, W. G., ed., The geotectonic evolution of California; Rubey Volume 1: Englewood Cliffs, New Jersey, Prentice-Hall, p. 182–202.

Schweickert, R. A., Saleeby, J. B., Tobisch, O. T., and Wright, W. H., 1977, Paleotectonic and paleogeographic significance of the Calaveras complex, western Sierra Nevada, California, *in* Stewart, J. H., Stevens, C. H., and Fritsche, A. E., eds., Paleozoic paleogeography of the western United States: Pacific Section, Society of Economic Paleontologists and Mineralogists Pacific Coast Paleogeographic Symposium 1, p. 381–394.

Silberling, N. J., 1973, Geologic events during Permian-Triassic time along the Pacific margin of the United States, *in* Logan, A., and Hills, L. V., eds., The Permian and Triassic Systems and their mutual boundary: Calgary, Alberta, Canada, Alberta Society of Petroleum Geologists, p. 345–362.

——, 1975, Age relations of the Golconda thrust fault, Sonoma Range, north-central Nevada: Geological Society of America Special Paper 163, 28 p.

Silberling, N. J., and Roberts, R. J., 1962, Pre-Tertiary stratigraphy and structure of northwestern Nevada: Geological Society of America Special Paper 72, 58 p.

Snyder, W. S., and Brueckner, H. K., 1983, Tectonic evolution of the Golconda allochthon, Nevada; Problems and Perspectives, *in* Stevens, C. H., ed., Pre-Jurassic rocks in western North American suspect terranes: Los Angeles, California, Society of Economic and Petroleum Geologists Pacific Section, p. 103–123.

Speed, R. C., 1977, Island-arc and other paleogeographic terranes of late Paleozoic age in the western Great Basin, *in* Stewart, J. H., Stevens, C. H., and Fritsche, A. E., eds., Paleozoic paleogeography of the western United States: Pacific Section, Society of Economic Paleontologists and Mineralogists Pacific Coast Paleogeographic Symposium 1, p. 349–362.

Speed, R. C., and Sleep, N. H., 1982, The Antler orogeny and foreland basin; A model: Geological Society of America Bulletin, v. 93, p. 815–828.

Stewart, J. H., 1980, Geology of Nevada: Nevada Bureau of Mines and Geology Special Publication 4, 136 p.

Stewart, J. H., and Suczek, C. A., 1977, Cambrian and latest Precambrian paleogeography and tectonics in the western United States, *in* Stewart, J. H., Stevens, C. H., and Fritsche, A. E., eds., Paleozoic paleogeography of the western United States: Pacific Section, Society of Economic Paleontologists and Mineralogists Pacific Coast Paleogeographic Symposium, p. 1–17.

Stewart, J. H., MacMillan, J. R., Nichols, K. M., and Stevens, C. H., 1977, Deep-water upper Paleozoic rocks in north-central Nevada; A study of the type area of the Havallah Formation, *in* Stewart, J. H., Stevens, C. H., and Fritsche, A. E., eds., Paleozoic paleogeography of the western United States: Pacific Section, Society of Economic Paleontologists and Mineralogists Pacific Coast Paleogeographic Symposium 1, p. 337–347.

Stewart, J. H., Murchey, B., Jones, D. L., and Wardlaw, B. R., 1986, Paleontologic evidence for complex tectonic interlayering of Mississippian to Permian deep water rocks of the Golconda allochthon in Tobin Range, north-central Nevada: Geological Society of America Bulletin, v. 97, p. 1122–1132.

Suczek, C. A., 1977, Tectonic relations of the Harmony Formation, northern Nevada [Ph.D. thesis]: Stanford, California, Stanford University, 96 p.

Tomlinson, A. J., 1988, Tectonostratigraphic units of the Golconda allochthon, Nevada: Geological Society of America Abstracts with Programs, v. 20, p. 238.

Tomlinson, A. J., and Wright, J. E., 1986, The Klamath-Sierran arc; A possible source region for Permian volcaniclastic and conglomeratic units within the Golconda allochthon: Geological Society of America Abstracts with Programs, v. 18, p. 193.

Turner, R. J., 1982, The geology of east-central Tobin Range, Nevada [M.S. thesis]: Stanford, California, Stanford University, 113 p.

Turner, R. J., Madrid, R. J., and Miller, E. L., 1989, Roberts Mountains allochthon; Stratigraphic comparisons with lower Paleozoic outer continental margin strata of the northern Canadian Cordillera: Geology, v. 17, p. 341–344.

Walker, R. G., and Mutti, E., 1973, Turbidite facies and facies association, *in* Turbidites and deep water sedimentation: Anaheim, Society of Economic and Petroleum Geologists Pacific Section Short Course, p. 79–118.

Whiteford, W. B., 1984, Age relations and provenance of upper Paleozoic sandstones in the Golconda allochthon; The Schoonover sequence, northern Independence Mountains, Elko Co., Nevada [M.S. thesis]: Stanford, California, Stanford University, 126 p.

Williams, H., and McBirney, A. R., 1979, Volcanology: San Francisco, California, Freeman, Cooper, 397 p.

Young, S. W., 1976, Petrographic textures of detrital polycrystalline quartz as in aid to interpreting crystalline source rocks: Journal of Sedimentary Petrology, v. 46, p. 595–603.

MANUSCRIPT ACCEPTED BY THE SOCIETY MARCH 27, 1990

Printed in U.S.A.

Geological Society of America
Special Paper 255
1990

Age and depositional setting of siliceous sediments in the upper Paleozoic Havallah sequence near Battle Mountain, Nevada; Implications for the paleogeography and structural evolution of the western margin of North America

Benita L. Murchey
U.S. Geological Survey, MS 915, 345 Middlefield Road, Menlo Park, California 94025

ABSTRACT

The upper Paleozoic Havallah sequence of central Nevada is a folded and thrust-faulted association of greenstone, siliceous marine sedimentary rocks, and deep-water clastic rocks. Microfossil assemblages (radiolarians, sponge spicules, and conodonts) are used as tools to unravel the stratigraphy and to interpret the paleoenvironments of the siliceous sedimentary rocks. Nine radiolarian assemblages (Osagean to Guadalupian) are described and used for delineation and correlation of fault-bounded lithotectonic units. The biostratigraphic zonation reveals that the oldest rocks in each lithotectonic unit are progressively younger from the structurally highest to the lowest units, suggesting progressive west-to-east upsection stepping of the Golconda sole thrust with accretion of each unit. Analyses of the radiolarian and sponge spicule faunas permit lateral and temporal comparisons of depositional environments. The lower structural units are coarsening-upward sequences of hemipelagic slope deposits overlain by sponge spicule-rich turbidites derived from a shallow source. The uppermost structural unit is a coarsening-upward basinal sequence. Permian sponge spicules in turbidites of the slope sequences and redeposited fusulinids in the basin sequence are similar to those in adjacent autochthonous (North American) regions. Permian radiolarians and sponge spicules in hemipelagic siliceous argillite of the slope sequences are similar to those in the Northern Sierra terrane to the west; the Havallah basin and the Northern Sierra arc terrane were overlain, therefore, by a similar water mass and may have been in proximity during the Permian. Clastic dikes and sills containing volcanic, metamorphic, and sedimentary rock clasts are Leonardian or younger.

INTRODUCTION

A reappraisal of the stratigraphy of the upper Paleozoic Havallah sequence in the Antler Peak area of north-central Nevada is presented herein, with emphasis on the age and environmental significance of microfossil faunas in siliceous rocks (Fig. 1, locality map). The Havallah sequence is a structurally complex association of bedded chert, argillite, greenstone, sandy limestone, sandstone, and conglomerate. As used here, the Havallah sequence includes the Havallah sequence of Silberling and Roberts (1962), as well as other upper Paleozoic rocks in the Golconda terrane such as the Schoonover sequence (Miller and others, 1984), allochthonous upper Paleozoic rocks in the Toquima Range (Laule and others, 1981), and the Pablo Formation (Speed, 1977). Ranges in which the Havallah sequence crops out are shown in black in Figure 1. Deformed rocks of the Havallah

Murchey, B. L., 1990, Age and depositional setting of siliceous sediments in the upper Paleozoic Havallah sequence near Battle Mountain, Nevada; Implications for the paleogeography and structural evolution of the western margin of North America, *in* Harwood, D. S., and Miller, M. M., eds., Paleozoic and early Mesozoic paleogeographic relations; Sierra Nevada, Klamath Mountains, and related terranes: Boulder, Colorado, Geological Society of America Special Paper 255.

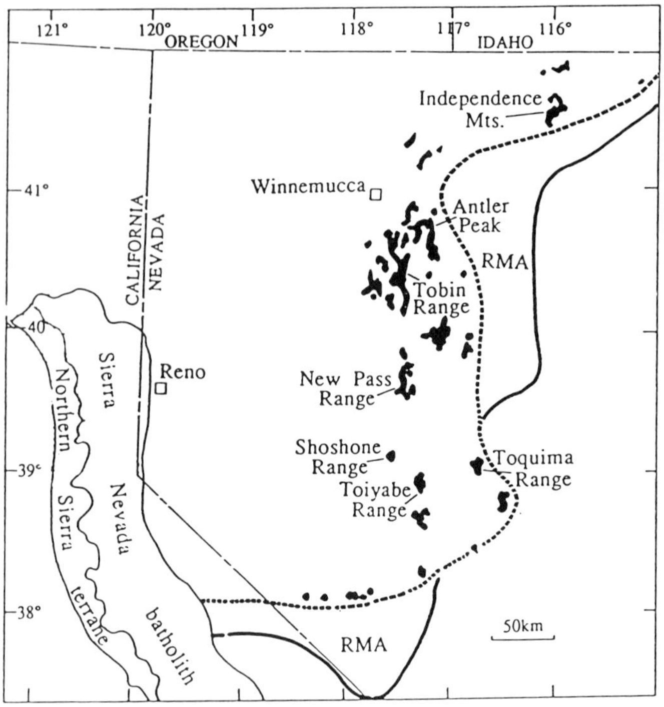

Figure 1. Locality map showing ranges in which the Havallah sequence crops out in central Nevada (black areas) and showing the area of the northern Sierra terrane in California. Dotted line marks the eastern limit of the Golconda allochthon. Solid line marks the eastern limit of the Roberts Mountains allochthon.

sequence are locally unconformably overlain by the Triassic Koipato sequence. The Golconda terrane, or Golconda allochthon, was thrust eastward onto the North American continent sometime after the early Late Permian (early Guadalupian). The Golconda thrust fault, a single or multiple sole thrust, separates the Golconda terrane from underlying Paleozoic rocks of the allochthonous Roberts Mountains terrane and the unconformably overlying Antler sequence. The eastern limit of the Roberts Mountains terrane or allochthon is marked in Figure 1.

In the 1970s the development of new techniques for extracting microfossils from siliceous rocks (Pessagno and Newport, 1972; Dumitrica, 1970) provided a new method for studying the rocks of the Havallah assemblage. Radiolarians, conodonts, and sponge spicules have all been extracted from siliceous rocks in the Havallah sequence by these methods. Radiolarians and conodonts are well suited for dating, while sponge spicules are useful for environmental analyses. The new dating techniques revealed greater structural complexity within the Havallah sequence than was previously recognized, and created the need to reappraise previous stratigraphic interpretations. A series of structure and stratigraphy papers using radiolarian age data soon appeared (Laule and others, 1981; Snyder and Brueckner, 1983; Sosson and others, 1983; Brueckner and Snyder, 1985; Babaie, 1987). Only a few of these studies (Miller and others, 1982, 1984; Stewart and other, 1986) included high-density sampling for fossil data, however. These latter types of studies are required in order to make accurate analyses of the Havallah depositional basin.

In this study, the siliceous rocks of the Antler Peak area are dated with microfossils, and their depositional setting is interpreted by analysis of their microfossil faunas and lithologies. The ages, provenance, and depositional settings of the clastic units in the Havallah assemblage are also discussed briefly. The purpose of this study is to delimit the constraints on the size, duration, and distance from the continental margin of the Havillah depositional basin.

PREVIOUS AGE DATA

In 1964, Roberts published a classic volume on the stratigraphy and structure of the Antler Peak quadrangle and laid the groundwork for all subsequent studies of the area. He (1964) mapped and described the upper Paleozoic Havallah sequence that crops out in the western part of the quadrangle (Fig. 2). Roberts correlated the rock units with the type upper Paleozoic Havallah and Pumpernickel Formations, which lie to the west in the Sonoma and Tobin Ranges (Roberts and others, 1958; Fig. 1). However, subsequent structural and stratigraphic interpretations of the two regions showed the difficulty in making generalized correlations between the Sonoma and Tobin Ranges and the Antler Peak area (Galena Range) without adequate age data (Miller and others, 1982; Stewart and others, 1986).

Fusulinid and conodont data from clastic units led Roberts (1964) to conclude that the Havallah sequence in the Antler Peak area was a homoclinal Pennsylvanian and Permian sequence. Radiolarian collections made between 1977 and 1979 by Roberts and by W. S. Snyder, D. L. Jones, and B. K. Holdsworth included both Mississippian and Permian radiolarians (Holdsworth and Jones, 1980b). The relative structural positions of the fossil localities indicated that older rocks structurally overlie younger rocks in the Havallah sequence of the Antler Peak area. Miller subsequently conducted a Stanford University field camp project in the vicinity of Willow Creek, in the Antler Peak area, in order to examine this problem; radiolarian samples collected by Miller and her students were examined by the author (Miller and others, 1982) and are included in this report. The fossil data enabled Miller and others to map faults within the Havallah sequence in the Willow Creek area.

NEW AGE DATA

New fossil collections in this study were dated primarily with radiolarians, although some were dated with conodonts or fusulinids. The fossils in the more than 90 samples are listed in Murchey (1989), who identified the radiolarians. Bruce Wardlaw of the U.S. Geological Survey identified the conodonts, and Calvin Stevens of San Jose State University identified the fusulinids.

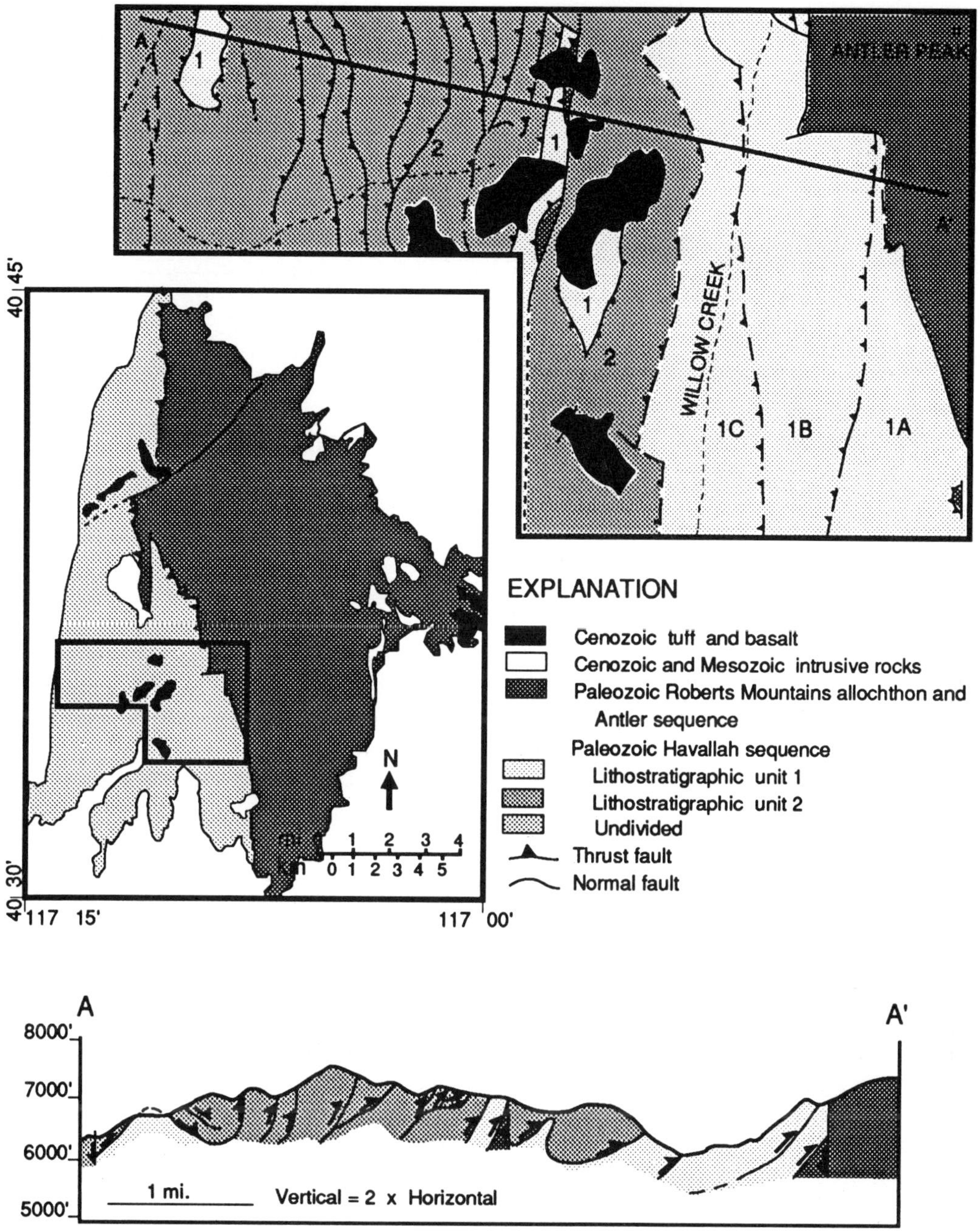

Figure 2. Geologic map and cross section of the Havallah sequence in the Galena Range, Antler Peak area, near Battle Mountain, Nevada. Base after Roberts (1964); geology of Havallah sequence in Willow Creek area after Miller and others (1982).

RADIOLARIAN BIOSTRATIGRAPHY

I have recognized nine radiolarian assemblages in the Havallah sequence of the Antler Peak area and the Tobin Range. The radiolarian faunas range in age from Mississippian to Permian. From oldest to youngest, the radiolarians are assigned to the following nine assemblages: *Albaillella* sp. cf. *A. paradoxa—Albaillella undulata* (assemblage 1), *Scharfenbergia impella* gp.—*A. undulata* (assemblage 2), *Scharfenbergia impella* gp.—*Albaillella* sp. aff. *A. cartalla* (assemblage 4), *Scharfenbergia tailleurense* (assemblage 5), *Pseudoalbaillella* sp. G—*Paronaella? triporosa* (assemblage 6), *Pseudoalbaillella bulbosa/Pseudoalbaillella u-forma—Paronaella?* sp. A (assemblage 7), *Pseudoalbaillella scalprata—Pseudotormentus kamigoriensis* (assemblage 8), and *Follicucullus ventricosus—Pseudoalbaillella fusiformis* (assemblage 9). Thirty-five key taxa in these assemblages are figured in Plate 1, and their ranges are given in Figure 3. The biostratigraphic order of these assemblages was determined by correla-

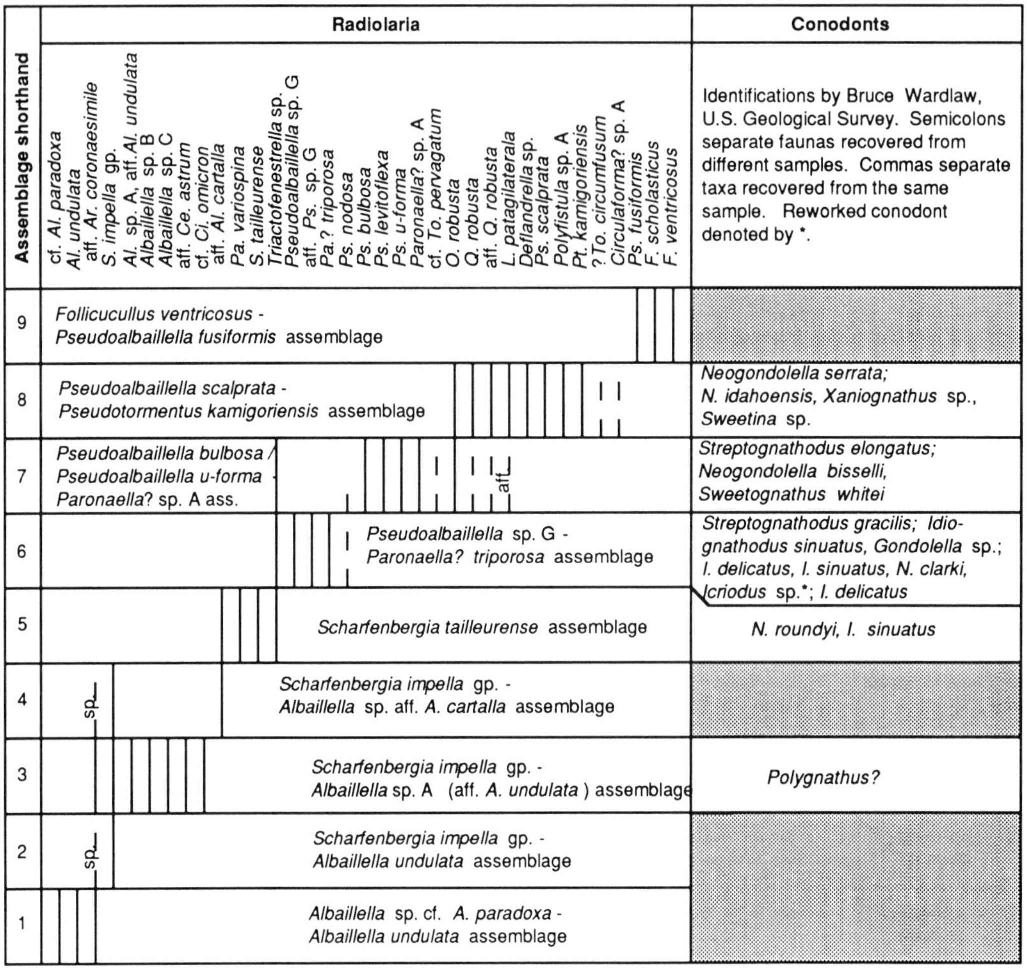

Figure 3. Schematic radiolarian biostratigraphy of the Havallah sequence, Galena and Tobin Ranges, Nevada.

tion with intact sequences in northern Alaska (Holdsworth and Murchey, 1988), in the Northern Sierra terrane of California (Murchey and others, 1986; Harwood and Murchey, this volume) and in Sonora, Mexico (Murchey, 1989), as well as with published data of other workers (Ormiston and Lane, 1976; Ormiston and Babcock, 1979; Holdsworth and Jones, 1980a, b; Sandburg and Gutschick, 1984; Nazarov and Ormiston, 1985a, b, 1986; Cornell and Simpson, 1985; Cheng, 1986; Ishiga, 1986) (Fig. 4). In addition, conodonts, identified by Bruce Wardlaw, have been found with five of the radiolarian assemblages (Fig. 3; Murchey, 1989). A discussion of systematic taxonomy, ranges, and occurrences of radiolarian taxa is given in Murchey (1989); a very brief discussion of some aspects of systematics and correlation is given in the Appendix.

ENVIRONMENTAL INDICATORS

Radiolarians provide information about the water column above the depositional site, while sponges provide information about the basin floor environment. Assuming that the habitats of Paleozoic radiolarians did not differ markedly from those of modern radiolarians, their presence in sediments indicates a relatively stable marine environment, typically off the shelf. Radiolarians now accumulate in settings ranging from borderland basins, arc basins, and fjords to deep oceanic basins. Siliceous sponges (demosponges and hexactinellids) occupy subtidal to abyssal oxygenated environments. Modern marine demosponges inhabit a wide bathymetric range but are most abundant in shallower settings such as reefs (Rigby, 1983). Except in the Antarctic where faunas are anomalously shallow, almost all modern hexactinellid sponges live at depths more than 90 m, with the majority living between 200 and 2,000 m (Reid, 1988, after Schulze, 1899, and Ijima, 1927). Few sponges, mostly hexactinellids, occupy the deep ocean basins.

Based on a study of upper Paleozoic siliceous rocks collected in the Cordillera from Mexico to Alaska, I have developed a paleobathymetric model using siliceous sponge spicule abundance and diversity (Murchey, 1989). In this model, the ratio of spicules to radiolarians decreases with depth and/or distance from a shelf. Discounting monaxon spicules of uncertain affinity,

Period	Age	Ass. shorthand	Havallah sequence Nevada	Rock units with correlative radiolarian faunas on the craton or craton margin of North America*		Holdsworth and Jones, 1980 a,b	Ishiga, 1986	Nazarov and Ormiston, 1985 a,b, 1986	Allochthonous Cordilleran sequences: Nigu Bluff sequence, Alaska Holdsworth and Murchey, 1988	Northern Sierra sequence, California Murchey, 1989	Barita de Sonora sequence, Sonora Murchey, 1989
Permian	Ochoan						N. ornithoformis N. optima				
Permian	Guadalupian	9	F. ventricosus - P. fusiformis	Lamar Ls. Texas	Ormiston and Babcock, 1979 Nazarov and Ormiston, 1985	Follicucullus	F. scholasticus F. sp. A P. globosa P. sp. C A. sinuata	F. ventricosus O. cornelli C.? simulens			
Permian	Leonardian	8	P. scalprata - P. kamigoriensis	Bone Spring Ls. Texas	Cornell and Simpson, 1985 Nazarov and Ormiston, 1985	Pseudoalbaillella	P. scalprata m. rhombothoracata P. lomentaria	Correlation not certain		P. scalprata - P. kamigoriensis	
Permian / Pennsylvanian	Wolfcampian Virgilian Missourian	7	P. bulbosa/P. u-forma - Paronaella? sp. A	Riepetown Fm. Nevada	Author's unpublished data		P. u-forma m. 2 P. u-forma m. I P. bulbosa	L. crux T. protei T. pervagatum		unconformity	P. u-forma - Paronaella? sp. A
Pennsylvanian	Desmoinesian Atokan	6	Pseudoalbaillella sp. G - P.? triporosa	Excello Sh. Oklahoma	Author's unpublished data		P. nodosa	?-Tormentum n. sp.	Pseudoalbaillella sp P.? triporosa	P. sp. G - P.? triporosa	unconformity?
Pennsylvanian / Mississippian	Morrowan Chesterian	5	S. tailleurense	?Jackfork Gp., ?"Caney" Sh., Ok	Cheng, 1986	Albaillella-3		?-Tetratormentum n. sp.	S. tailleurense	S. tailleurense	S. tailleurense
Mississippian	Meramecian	4	S. impella gp. - A. sp. aff. A. cartalla	Sycamore Ls. Oklahoma	Ormiston and Lane, 1976; age revision, Holdsworth and Murchey, 1988	pre-A. pennata-2		A. cartalla; L. impella	S. impella - A. sp. cf. A. cartalla	S. impella gp.	S. impella gp. - A. cartalla
Mississippian	Meramecian / ?	3	S. impella gp. - Albaillella sp. A, aff. A. undulata	Woodman Fm. Utah	Sandburg and Gutschick, 1984	pre-A. pennata-1		Correlation not certain			missing section?
Mississippian	Osagean	2	S. impella gp. - A. undulata					A. paradoxa			S. impella gp. - A. undulata
Mississippian	Osagean	1	A. sp. cf. A. paradoxa - A. undulata			Albaillella-2					

Figure 4. Correlation of radiolarian assemblages in the Havallah sequence with other North American radiolarian occurrences and with published zonation schemes.

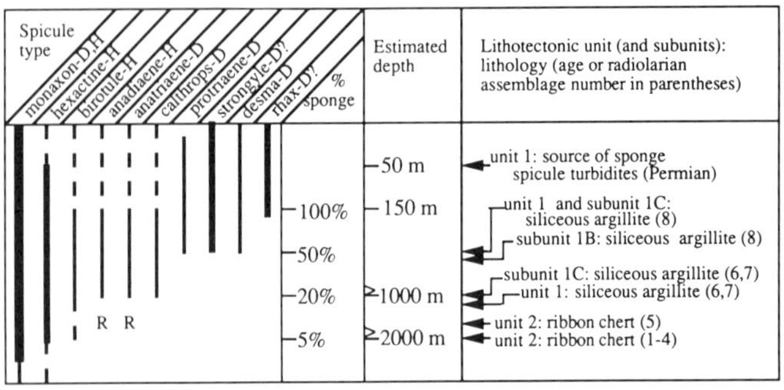

Figure 5. Model of paleobathymetric distribution of late Paleozoic sponge spicules in siliceous basins of the Cordillera of western North America, showing best fit with sponge spicule faunas in lithotectonic units of the Havallah sequence in the Antler Peak area of Nevada. Percent sponge determined from total microfossil population collected on Tyler equivalent 250-mesh sieve. H = Hexactinellida; D = Demospongiae; R = rare occurrence.

shallower water faunas are strongly dominated by demosponge spicules; intermediate-depth faunas contain high-diversity spicule assemblages (including both demosponge and hexactinellid spicules), and deeper water faunas are dominated by hexactinellid spicules (Fig. 5). The 10 most common types of sponge spicules found in the siliceous rocks of the Havallah sequence are shown in Plate 2; their distributions are recorded in Table 1 and Figure 5.

Sponge spicule abundance was recorded as the percentage of spicules in the total microfossil population (radiolarians and spicules) in the 63- to 180-μm-size fraction collected on a Tyler equivalent 250-mesh screen. Table 1 shows a summary of the spicule data (occurrence, abundance, and diversity) for 62 samples from the Havallah sequence organized by lithotectonic units and lithology. Faunal and locality data for individual samples are listed in Murchey (1989).

GEOLOGIC SETTING AND STRUCTURE

In the Antler Peak area, the Golconda thrust fault separates rocks of the lower and middle Paleozoic Roberts Mountains allochthon and upper Paleozoic Antler sequence from the overlying rocks of the Havallah sequence (Fig. 2, geologic map and cross section). The Havallah sequence itself is internally complicated by thrusts and folds related to eastward-directed thrusting of the rocks (for discussion of small-scale structures in the Willow Creek area, see Miller and others, 1982; Snyder and Brueckner, 1983; Brueckner and Snyder, 1985).

Two major lithotectonic units separated by a thrust fault can be recognized in the Havallah sequence. The fault was first mapped, in part, by Miller and others (1982). The fault is tightly folded into a west-dipping synform in the area between Willow Creek and Rocky Canyon, but it is apparently involved in broad, open folds in the western part of the range. These folds are interpreted as late-stage deformation resulting from emplacement of the Havallah assemblage onto the continental margin (Murchey, 1986).

Within the two major lithotectonic units is a hierarchy of bedding-parallel faults distinguished by their different scales of offset. Some divide the major units into smaller mappable bodies that are distinguished from neighboring units by lithology or age. Others separate rock bodies into even smaller packets of similar lithology and age (Snyder and Brueckner, 1983).

Lithotectonic unit 1

The lower lithotectonic unit (LU-1) is dominantly argillite with lesser amounts of sponge spicule chert, rare quartz-rich sandstone, sandy limestone, volcaniclastic sandstone (in clastic dikes and sills), and greenstone. In the Willow Creek area, LU-1 is equivalent to the rocks that Roberts (1964) assigned to the lower, middle, and part(?) of the upper Pumpernickel Formation. In this area, LU-1 is subdivided into three fault-bounded subunits: LU-1A, LU-1B and LU-1C (Figs. 2, 6). West of the Willow Creek canyon the unit is not subdivided. The lithologies of LU-1 and their contained faunas are described successively from what is argued to be the oldest to the youngest: (1) greenstone, (2) argillite, (3) sponge spicule turbidites. Lastly, clastic rocks of uncertain age are described.

Greenstone. Greenstone occurs in at least two localities within LU-1. A small fault-bounded sliver of greenstone lies between LU-1B(?) and overlying LU-1C at the north end of Willow Creek Canyon. In the northern part of the range, greenstone forms the base of LU-1C(?); overlying sedimentary rocks include red jasperoid and maroon siliceous shale with some manganiferous staining, and lithologies suggest a depositional contact between the greenstone and overlying rocks. Because siliceous argillite in LU-1C contains the oldest faunas (Middle Pennsylvanian; see discussion below) in LU-1, the greenstone is interpreted to be the oldest rock type in the unit (Fig. 6).

Argillite and siliceous argillite. Argillite is the principal rock type in LU-1. It contains conodonts but is otherwise unfossiliferous. In LU-1A the argillite is gray and black with thin chert lenses (Roberts, 1964). In LU-1B, LU-1C, and undifferentiated LU-1 to the west, the argillite is red, green, and purple. Thin beds (1 to 2 cm) of siliceous argillite interbedded with the argillite contain radiolarians, sponge spicules, and conodonts. Both the radiolarian and sponge spicule faunas are delicate and do not show indications of current transport.

The structurally highest subunit, LU-1C, contains radiolarian faunas ranging from Middle Pennsylvanian to Early Permian (Leonardian) age. The LU-1C faunas are assigned to the *Pseudoalbaillella* sp. G of Ishiga (1982)—*Paronaella? triporosa* assemblage (assemblage 6), *Pseudoalbaillella bulbosa/Pseudoalbaillella u-forma—Paronaella?* sp. A assemblage (assemblage 7), and *Pseudoalbaillella scalprata—Pseudotormentus kamigoriensis* assemblage (assemblage 8) (Figs. 3, 4, 6). Undifferentiated outcrops of LU-1 lying west of the Willow Creek Canyon contain the same three assemblages present in LU-1C. The mid-

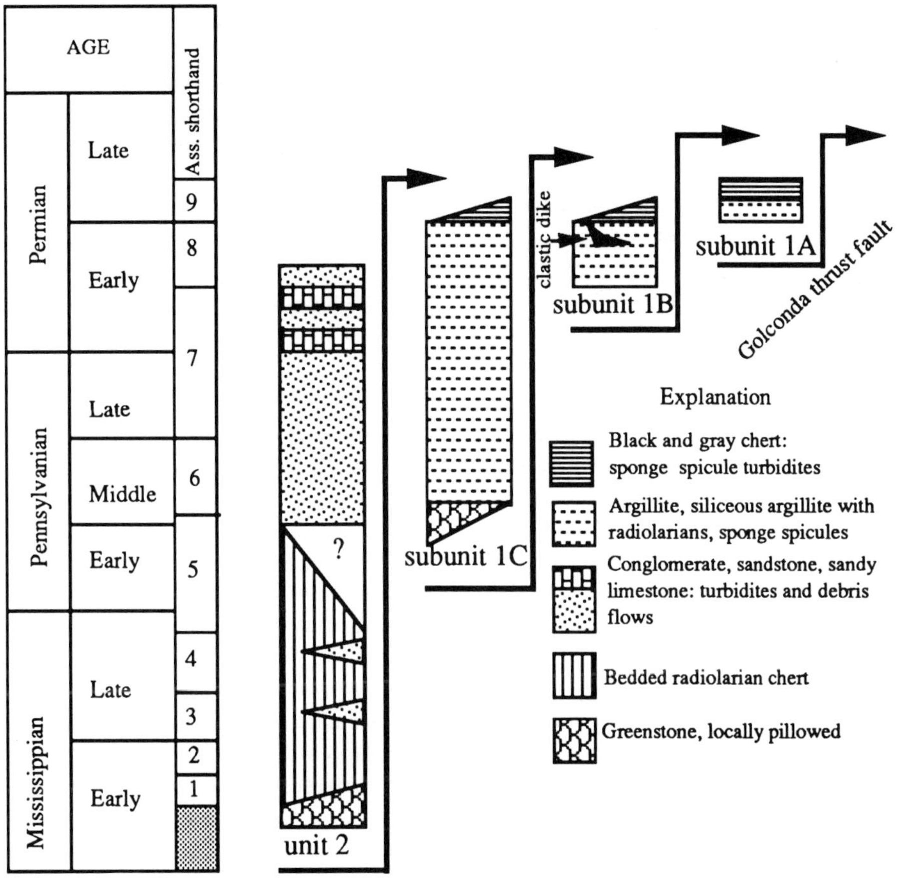

Figure 6. Schematic stratigraphy of lithotectonic units of the Havallah sequence in the Galena Range, Antler Peak area, near Battle Mountain, Nevada. Chert and siliceous argillite dated with radiolarians (assemblage numbers listed next to age) and conodonts. Clastic units dated with fusulinids. Note that the oldest rocks in each lithotectonic unit are progressively younger from the structurally highest to the lowest.

dle subunit, LU-1B, contains the Leonardian *Pseudoalbaillella scalprata—Pseudotormentus kamigoriensis* assemblage (assemblage 8). One sample (USGS MR 0196) from the structurally lowest subunit, LU-1A, contains the Guadalupian *Follicucullus ventricosus—Pseudoalbaillella fusiformis* radiolarian assemblage (assemblage 9) (see Figs. 3, 4, 6). The age of the oldest rocks in each subunit decreases progressively within the structural stack.

The relative abundance and diversity of sponge spicules in siliceous argillite of LU-1 increases through time within structural subunits and increases down structural section between subunits for coeval assemblages (Table 1). In the older and structurally higher rocks, spicule abundance and diversity is lowest, and faunas have a higher ratio of hexactinellids to demosponges. Of LU-1C faunas assigned to Pennsylvanian to early Early Permian, *P. bulbosa/P. u-forma—Paronaella?* sp. A (assemblage 7) and *P.* sp. G—*Paronaella? triporosa* (assemblage 6) assemblages, sponge spicules form 4 to 78 percent (median 19 percent samples counted) of the microfossils (radiolarians and sponge spicules) in the Tyler equivalent 250-mesh sieve residue. Spicules include small monaxons, hexactines, birotules (paraclavules and hemidiscs), anadiaenes (rare), anatriaenes, and calthrops, but lack protriaenes, strongyles (a type of monaxon), and desmas; individual samples contain two to six (median 5) spicule types. In correlative faunas from western undifferentiated LU-1, sponge spicules form 8 to 26 percent (median 17 percent, 5 samples counted) of the microfossil population. Spicules include monaxons, hexactines, anadiaenes (rare), and calthrops (rare), but lack birotules, anatriaenes, protriaenes, strongyles, and desmas; individual samples contain two to four (median 2) spicule types. In LU-1C and western undifferentiated LU-1 faunas assigned to the Leonardian *Pseudoalbaillella scalprata—Pseudotormentus kamigoriensis* assemblage (assemblage 8), sponge spicules form 11 to 60 percent (median 50 percent), four samples counted) of the microfossil faunas; the same spicules have been observed in correlative rocks of LU-1B, but individual samples from the structurally higher units have only two to six spicule types (median 5), and the demosponges are less abundant than in LU-1B. In LU-1B faunas assigned to the *Pseudoalbaillella scalprata—Pseudotormentus kamigoriensis* radiolarian assemblage (assemblage 8), sponge spicules form 20 to 95 percent (median approximately 50 percent, eight samples counted) of the total microfossil faunas. Spicule faunas are diverse and include small monaxons, hexactines, birotules (paraclavules, hemidiscs, and rare amphidiscs), anadiaenes, anatriaenes, calthrops, protriaenes, strongyles, and desmas; sponge spicule faunas typically contain four to eight spicule types (median 7). The single sample from LU-1A is too poorly preserved to compare with others.

Based on the occurrence together of radiolarians and abundant and diverse sponge spicules in hemipelagic sediments showing no strong evidence for transport, the depositional setting for the siliceous argillite and argillite of LU-1 is interpreted to be at depths between 150 and >1,000 m (Fig. 5). Because the sediments do not appear to have been transported, bottom waters were oxygenated and sustained a relatively abundant sponge population. Older and structurally higher faunas contain the fewest sponge spicules, especially demosponge spicules, and were probably deposited in the deepest water.

The diverse spicule types found in the Pennsylvanian and Permian siliceous argillite of the Havallah sequence in the Antler Peak area are also found in Permian siliceous argillite of allochthonous "arc" terranes to the west (Northern Sierra, Grindstone, Bilk Creek terranes). In contrast, Permian sponge faunas in bedded chert of "oceanic" terranes (North Fork, San Juan, Hozameen, Baker terranes) farther west and north contain low-diversity faunas strongly dominated by small monaxons and a few hexactines (Murchey, 1989).

In the Permian *Pseudoalbaillella scalprata—Pseudotormentus kamigoriensis* assemblage (assemblage 8) faunas of the Havallah sequence, stauraxon radiolarians are more abundant than Albaillellaria radiolarians. This relative abundance ratio is characteristic of Leonardian and younger radiolarian faunas in autochthonous rocks to the east and in allochthonous "arc" terranes to the west (Northern Sierra, Grindstone, Bilk Creek, Eastern Klamath Mountains terranes); the reverse is true in "oceanic" terranes (North Fork, San Juan Hozameen, Baker terranes) farther north and west (Murchey and others, 1990). Similarities in faunas from sediments must reflect similarities in the overlying water columns. Siliceous argillite containing radiolarian assemblage 8 faunas in both the Havallah sequence and the Northern Sierra terrane of California also contain nearly identical spicule faunas (Murchey, 1989) suggesting similar basin depths as well.

Sponge spicule chert. Thickly bedded black or gray sponge spicule chert units are scattered throughout LU-1 but are most common in LU-1A and LU-1B. The thick-bedded chert is a ridge and ledge former. On the east side of Willow Creek, laterally persistent bedded black chert packages alternate with argillite units; the two rock types are tightly folded (Roberts, 1964).

The thick-bedded chert contains abundant sponge spicules (grain supported in some cases) and quartz silt but no radiolarians. Bedding is well developed, varying from 4 to as much as 80 cm thick. The thinner beds are typically reddish-weathering,

TABLE 1. LITHOLOGIC AND STRUCTURAL DISTRIBUTION PATTERNS OF TEN TYPES OF SILICEOUS SPONGE SPICULES IN SILICEOUS SEDIMENTARY ROCKS OF THE HAVALLAH SEQUENCE IN THE ANTLER PEAK AREA NEAR BATTLE MOUNTAIN, NEVADA

	Spicule types: D = Demospongiae; H = Hexactinellida											Abundance	Diversity	Age		
	Monaxon-D,H	Hexactine-H	Birotule-H	Anadiaene-H	Anatriaene-D	Calthrops-D	Protriaene-D	Strongyle-D?	Desma-D	Rhax-D	Number of samples examined	Sponge spicules in total fauna collected on Tyler 230 sieve (median in parentheses) (%)	Number spicule types in single sample (median in parentheses)	Radiolarian assemb. (see Fig. 2)	Lithotectonic Unit	Lithology
	100	40R	20R	20	20	10	10	60	50R	50	10	100 (100)	1-8(3-4)	Perm	1	Chert, turbidites
Percent samples in which spicule is present. R = spicule is rare (only 1-2 seen) in at least 50% of samples	100	100	75	88	88	50	50R	38	63	0	8	20-95(50)	4-8(7)	8	1B	Siliceous argillite
	100	75	50	75	50	25	25R	25	25	0	4	11-60(50)	2-6(5)	8	1C, 1undiff. (western)	
	100	88	63	38R	75R	50	0	0	0	0	8	4-78(19)	2-6(5)	6,7	1C	
	100	100	0	20R	0	20R	0	0	0	0	5	8-26(17)	2-4(2)	6,7	1undiff. (western)	
	100	50	0	0	13R	0	0	0	0	0	8	1-76(11)	1-3(2)	5	2	Ribbon chert
	100	73	9	0	9R	0	0	0	0	0	11	1-26(5)	1-3(2)	1-4	2	

black silty sponge spicule chert with partial and complete Bouma T_{a-e} sequences (Bouma, 1962). These beds have rather flat bedding surfaces and break into flagstone-like pieces. Bedding sequences within units of this lithology typically thicken and coarsen upward. Deep-water trace fossils and sole marks are visible on the bedding surfaces (Roberts, 1964). The thickest beds (more than 30 cm) are gray and vitreous and lack apparent sedimentary structures.

Demosponges spicules strongly dominate the sponge spicule faunas (Table 1; Murchey, 1989). Monaxon spicules of probable demosponge origin are abundant in all samples, and the axial canals of large spicules are commonly filled with dark material. Strongyles and rhax spicules are also common to abundant. Hexactine, birotule, and anadiaene spicules are rare. Larger spicules are current-oriented in some samples. The faunas in the sponge turbidites are indistinguishable from faunas in the lower Guadalupian Rex Chert member of the Phosphoria Formation in southwestern Idaho (Murchey, 1989). The rhax spicules found in both the Phosphoria Formation and the Havallah sequence are not known to occur farther west in the "arc" terranes. The thick-bedded chert units are interlayered with Leonardian and lower(?) Guadalupian argillite and siliceous argillite, but they have not been independently dated. It is possible that they are correlative with the Rex Chert. At the least, they represent spicules and silt that accumulated in a Rex Chert–type environment, probably less than 50 m beneath sea level, and that were deposited downslope into the Havallah depositional basin. The close affinity of the sponge faunas to the Rex Chert faunas suggests a tie to North America, and probably even a tie to the Phosphoria Formation region. The known faunas of sponge spiculites in the "arc" terranes to the west (Bilk Creek, Northern Sierra, Eastern Klamath Mountains terranes) are dissimilar; so far as is known, they lack rhax spicules and large monaxons filled with dark material and they contain different varieties of desmas.

Clastic rocks. Rare clastic rocks in LU-1 include calcite-cemented quartzitic sandstone, sandy limestone, and volcaniclastic lithic sandstone. The clastic rocks are not independently dated but are closely associated with Permian (Leonardian) argillite. The stratigraphic relation between the quartz-rich sandstone and limestone turbidites and the volcaniclastic sandstone is unclear because outcrops of the latter appear to be dikes and sills.

Quartz-rich sandstone and limestone turbidites are only minor components of LU-1; they crop out in structurally higher parts of the lithotectonic unit. Lower Permian turbidites of similar composition are common elsewhere in the Havallah sequence. They are volumetrically important in LU-2 (see below) and in the Tobin Range (Stewart and othes, 1986). Similar rocks are intercalated with Leonardian chert and argillite in the Reese River Valley area to the south (A. Tomlinson, written communication, 1989).

Lithic sandstone units with volcanic detritus (average "bed" about 25 cm thick) occur in Leonardian rocks of LU-1B(?) on the west side of the Willow Creek Canyon and in the northern part of the range. They contain volcanic (dacite, rhyolite?) rock fragments, metamorphic rocks fragments, chert, quartz, quartz arenite, and phosphate-coated grains; their composition is compatible with an arc-related provenance (Snyder and Brueckner, 1983). Snyder and Brueckner (1983, p. 115), noted that volcaniclastic strata in the Willow Creek area were injected into surrounding rocks as clastic dikes and sills "during . . . folding and thrusting, indicating high fluid pressures and deformation shortly after deposition" (see also Brueckner and Snyder, 1985). Whether all the volcaniclastic "beds" are actually sills is not certain. One seemingly conformable bed within argillite in the Willow Creek section hs an internal structure of wavy, bedding-parallel laminae, possibly indicating flow during intrusion. Conodonts collected from argillite beds immediately above and below the sill(?) have anomalously high conodont color alteration indices (CAI) (see Epstein and others, 1977) of 5 to 7. Possibly these anomalous values are related to the movement of fluids through the host rocks during the emplacement of the clastic sill(?). Harris and others (1987) noted that moving fluids are able to oxidize the organic matter in corroded conodonts, thereby producing anomalously high CAI values. Because none of the quartz-rich sandstones or sponge spicule turbidites are known to have been remobilized and injected as clastic intrusives, one can argue that they predate the lithic sandstones; no cross-cutting relations were observed, however. The sedimentary source beds for the LU-1 clastic intrusive rocks probably represent some of the last deposits (late Leonardian?, Guadalupian?) within the Havallah basin. Elsewhere in the Havallah sequence, sandstone beds of similar composition have sedimentary structures that indicate a depositional mode of emplacement during Leonardian time (A. Tomlinson, Stanford University, written communication, 1989). During(?) and subsequent to the deposition of the lithic sandstones in LU-1, the rocks of this unit were involved in the eastward-directed folding and thrusting that closed the Havallah basin and that characterize the Sonoma orogeny. In the Tobin Range west of the study area, an angular unconformity separates the deformed Havallah sequence from the Triassic Koipato sequence; the base of the Koipato sequence contains volcaniclastic detritus as well as reworked Havallah sequence material. The similarity between Leonardian sponge and radiolarian faunas in argillite associated with volcaniclastic rocks in LU-1 and in the Northern Sierra arc terrane in California indicates that they were overlain by a similar water mass and is possible evidence that they were in proximity during the Early Permian (Harwood and Murchey, this volume).

Lithotectonic unit 2

Lithotectonic unit 2 (LU-2) is stratigraphically more coherent than LU-1, in part because its lithologies are more competent. Basal basalt is overlain by radiolarian ribbon chert, which is overlain by quartz- and chert-rich clastic rocks (Fig. 6). Polyphase folding in the chert is well studied, but the clastic rocks also locally exhibit more than one phase of folding.

The basal basalt, locally faulted out, incudes pillow basalt.

The basalt is overlain by Mississippian maroon mudstone and chert as well as by thick, red jasperoid sills. The contact between the basalt and overlying sediments is commonly manganiferous. The maroon mudstone and chert grades upward into gray-green ribbon chert.

The ribbon chert is a laterally persistent facies that commonly forms the structural base of LU-2. Roberts (1964) typically mapped the chert as the Trenton Canyon member of the Havallah Formation, but on the ridge west of Willow Creek, he mapped it as the uppermost part of the Pumpernickel Formation. The ribbon chert facies contains rhythmically alternating beds of vitreous radiolarian chert (25 to 50 mm thick) and shale (1 to 20 mm thick). Where they are not folded, the chert beds can be continuous in thickness. Many chert beds have small, mound-shaped diapiric structures (0.5 cm high by 1 cm wide) that give the bedding surfaces a lumpy or knobby appearance (Snyder and Brueckner, 1983; Brueckner and Snyder, 1985). In the Antler Peak region, these diagenetic structures are restricted to the chert beds of LU-2. Similar structures were illustrated and discussed by Fagan (1962, Fig. 6) in chert of the Schoonover sequence. Interestingly, they are the same age (Mississippian to Early Pennsylvanian) in both areas (E. Miller and B. K. Holdsworth, personal communication, 1987).

The LU-2 chert contains abundant radiolarians and few sponge spicules, but the fossils are generally matrix-supported. The radiolarian faunas are Early Mississippian to Early Pennsylvanian in age and include all radiolarian assemblages from assemblage 1 to assemblage 5 (Figs. 3, 4, 6; Murchey, 1989).

The ribbon chert in LU-2 is predominantly a pelagic unit deposited in a deeper environment farther from a sediment source than the hemipelagic argillite and siliceous argillite of LU-2. The sediments were deposited in an open marine environment above pillow basalt. Sponge spicules form only <1 to 26 percent (median 5 percent) of the microfossil population in the Tyler equivalent 250-mesh sieve residue of the older (assemblages 1 through 4) radiolarian faunas (Table 1). In the younger *S. tailleurense* assemblage (assemblage 5) faunas, sponge spicules form 1 to 76 percent (median 11 percent) of the residue.

In addition to small monaxon spicules, small oxyhexactines are present in many samples, but birotules and anatriaenes are rare. No other spicule types were observed in these samples. Water depth may have been more than 1,000 m. Sponge spicules are more abundant in these rocks, however, than in rocks in some purely "oceanic" terranes where spicules commonly form less than 1 percent of the microfossil population. Based on these observations, the water depth may have been greater than 1,000 m, but may not have been as deep as a typical ocean-floor basin (Fig. 5). The upsection increase in sponge spicules within the chert sequence may indicate shallowing or minor redeposition of spicules from a shallower source through time.

Clastic rocks form the upper part of the stratigraphic sequence in LU-2. Roberts (1964) generally mapped these rocks as the Jory and Mill Canyon members of the Havallah Formation. Quartz silt in radiolarian-bearing (assemblage 3) silty chert proves that some clastic material larger than clay size was deposited in the LU-2 basin during Mississippian time. Other clastic rocks interlayered with Mississippian chert may also be Mississippian, but their ages are not proven because the interlayering may be the result of bedding-parallel faults at the base of the chert sequences. Along the ridge west of Willow Creek, the structurally lowest clastic rocks in LU-2 are quartz siltstone and fine- to medium-grained silica- or calcite-cemented quartz sandstone. These are structurally or depositionally interlayered with chert and silty chert having Mississippian *Albaillella* and *S. impella* gp. radiolarian faunas (assemblages 2 through 4). The siltstone to medium-grained sandstone on the ridge is structurally overlain by coarse-grained sandstone and conglomerate in which Roberts reported abraded fusulinids of Middle Pennsylvanian (Atokan) age. In the northern part of the mountain range, coarse-grained sandstone and conglomerate are interlayered with Mississippian bedded chert. Although a Mississippian age for the beginning of large-scale deposition of clastic sediments in the LU-2 basin has not been firmly established, the apparent absence of bedded chert younger than Late Mississippian to Early Pennsylvanian *Scharfenbergia tailleurense* assemblage (assemblage 5) establishes a complete transition from a pelagic to a clastic depositional regime by Early to Middle Pennsylvanian time. Roberts reported abraded fusulinids of Middle Pennsylvanian (Atokan) age from two localities, including the previously mentioned occurrence in the type locality of the Jory member on the ridge west of Willow Creek (fusulinids identified by L. Henbest and R. Douglass, written communication, *in* Roberts, 1964); the fusulinids establish a minimum age of deposition for their host rocks. In the northwestern part of the range, two new fossil localities in coarse-grained sandy bioclastic limestone, mapped by Roberts as the Jory member, contain abundant fusulinids (including *Schwagerina wellsensis*?) of probable middle Wolfcampian age with affinities to autochthonous North American faunas (C. Stevens, written communication, 1985). Roberts also reported Permian, Hueco to Leonardian, fusulinids from many rocks in the "Mill Canyon member." Typically, the bioclastic limestones with abundant Permian fusulinids are poorly sorted with chert and sandstone clasts, well-rounded quartz grains, phosphatic coated grains, and other bioclastic material. These limestone units commonly are interbedded with well-sorted medium-grained quartz sandstone sequences. Whereas the conglomerates and coarse-grained sandstone (locally with abraded Atokan fusulinids) directly overlie ribbon chert, the limestones with Permian fusulinids do not appear to do so. A lateral, interfingering relation between the bioclastic limestones with Permian fusulinids and some of the coarser grained sandstones and conglomerates has not been observed; the former are generally higher in the structural section than the latter. These observations suggest, but do not prove, that the limestone turbidites with Permian fusulinids are generally younger than the conglomerates.

Chert clasts in the conglomerates include large, angular green rip-up clasts as well as smaller black, gray, or green subrounded clasts. Chert rip-up clasts from Atokan or younger conglomerate west of Willow Creek contain *S. impella* gp. faunas of Mississippian age; they apparently were derived from underlying chert. Middle(?) Devonian, Late Devonian, and earliest Early Mississippian radiolarians (*Ceratoikiscum* spp., *Helenifore* sp., *Staurodruppa*? sp., and diverse pylentonemiids) have also been recovered from smaller subrounded chert clasts in the conglomerate. Possibly older radiolarians were not dated with certainty. A provenance tie between the Havallah sequence and the structurally underlying Roberts Mountains terrane has been suggested before on the basis of lithology (Dickinson and others, 1983; Snyder and Brueckner, 1983, Miller and others, 1984). Radiolarian-bearing chert in the Roberts Mountains terrane is predominantly Ordovician to Late Devonian in age and is the probable source for any correlative chert clasts in the Havallah sequence. Latest Late Devonian and earliest Early Mississippian radiolarians faunas have been found in one area of the Roberts Mountains terrane (per B. L. Murchey, *in* Coles and Snyder, 1985) and also in part of the Havallah sequence (per B. K. Holdsworth, *in* Miller and others, 1984); either the Roberts Mountains terrane or eroded Havallah sequence itself may have been the source for correlative clasts in the conglomerates of LU-2. As first suggested by Roberts (1964), the possible Pennsylvanian age of the LU-2 conglomerates permits a stratigraphic tie with autochthonous Pennsylvanian conglomeratic rocks (Battle Formation) of the Antler sequence that unconformably overlie deformed rocks of the Roberts Mountains allochthon in the Antler Peak area.

SUMMARY AND DISCUSSION

Within the four lithostratigraphic units or subunits (LU-1A, -1B, -1C, and LU-2), the oldest rocks in each unit are progressively older from the lowermost unit, LU-1A, to the highest unit, LU-2 (Fig. 6). The oldest rocks in LU-1A are Late Permian in age; in LU-1B they are Early Permian in age, in LU-1C they are Middle Pennsylvanian in age, and, in LU-2 they are Mississippian in age. This pattern suggests successive upsection cutting of the sole thrust from west to east as each new unit was accreted. The interpretation is consistent with the east-directed sense of vergence in the fold geometries and in the eastward direction of motion indicated by analysis of slickensides (Miller and others, 1982). Progressive accretion is a different interpretation from that of Snyder and Brueckner (1983), who documented the imbricated structure of the Havallah sequence but have argued for a chaotic mixture of lithotectonic units.

The interpretation that the Havallah sequence in the Antler Peak area (Galena Range) was progressively stacked from west to east is also consistent with the interpretations of depositional setting based on fossil faunas and lithologies (Fig. 5). The deepest part of the basin is represented by the facies in LU-2. Submarine basalts are overlain by Lower Mississippian to Lower Pennsylvanian pelagic sediments that may have been deposited more than 1,000 m beneath sea level. The basin was not distant from a terrigenous source, because quartz silt appeared within the pelagic sediments as early as Late Mississippian and quartz sands soon followed. From Pennsylvanian(?) to Early Permian time, the sedimentary precursors of bioclastic and chert-quartz conglomerates, silica- and calcite-cemented sandstones, and sandy limestones were deposited in the basin as debris flows and turbidites. The lithologic components in these clastic rocks suggest a North American source similar to the Antler Highlands. The radiolarians in chert clasts also indicate that some clasts may indeed have come from uplifted Roberts Mountains allochthon exposed in the Antler Highlands, but that other clasts are apparent rip-ups derived from the Havallah sequence itself. Permian fusulinids suggest a North American affinity. Final internal folding and thrusting of LU-2 postdates the deposition of middle Wolfcampian and Hueco to Leonardian fusulinid-bearing sediments. Emplacement of LU-2 over LU-1 did nto occur before early Guadalupian time, the age of the youngest material in the LU-1 unit.

The lowermost structural unit, LU-1, is dominated by Middle Pennsylvanian to lower Upper Permian hemipelagic red and green argillite and siliceous argillite deposited at depths estimated between 150 and 1,000 m in an oxygenated open-marine setting (Fig. 5). The sponge faunas in siliceous argillite suggest a lateral shallowing to the east and an upsection shallowing within structural subunits. The deeper and older portions of LU-1 may have been deposited upon basalt, as discussed previously. The absence of easterly derived coarse-grained bioclastic quartz- and chert-bearing turbidites and debris flows in LU-1 suggests that sediments bypassed deposition on the slope setting, presumably through channels outside the areal extent of LU-1, and were instead deposited in deeper parts of the basin (LU-2). Sponge spicule turbidites of Leonardian or Guadalupian age were redeposited into the LU-1 slope setting from a shelf environment similar to, and possibly coeval with, that in which the Rex Chert member of the Phosphoria Formation was deposited; for this reason, a North American or Antler Highlands origin for the sponge spicule turbidites is preferred over a western source. Leonardian or younger clastic sediments containing igneous, metamorphic, and sedimentary rock fragments mark the first appearance in LU-1 of sediments probably derived from a western arc. Sometime between the Leonardian and the Early Triassic, these sands were injected into Leonardian strata during the process of thrust faulting and folding. Because the dikes do not appear to cut major faults, their injection may predate the emplacement of LU-2 over LU-1 and the subsequent emplacement of the Havallah sequence over the Antler sequence.

From the Mississippian to early Late Permian, the slope and basin facies of the Havallah sequence that are represented in the Antler Peak area had faunal and/or sedimentary ties to North America and, arguably, to the Nevada and Idaho region. These parts of the Havallah sequence appear to be parautochthonous. The Lower Mississippian (Osagean) basalt of LU-2 and Pennsylvanian(?) basalt of LU-1, therefore, formed relatively close to the margin of North America and the newly created Antler Highlands in a tectonic setting that included rifting and sea-floor basaltic volcanism. Basement underlying the basalt is not present in the Antler Peak area, but the Schoonover sequence in the Independence Mountains includes older (Upper Devonian and earliest Early Mississippian age) volcanic and volcaniclastic rocks (Miller and others, 1984). The relative abundance (>5 percent) of sponge spicules in sediments overlying basalt in the Antler Peak area is possible evidence that the basin was shallower than typical oceanic rift settings. There is no compelling sedimentologic or faunal evidence in the Havallah sequence of the Antler Peak area suggesting proximity to active arc volcanism before the Leonardian or later, when lithic sandstones with a probable western source were deposited before and/or during basin closure.

The two most popular models for the Havallah depositional basin include a back-arc basin (Burchfiel and Davis, 1972, 1975; Silberling, 1973; Miller and others, 1984) or an oceanic basin in front of an advancing arc (Speed, 1979; Speed and Sleep, 1982; Snyder and Brueckner, 1983). This study does not prove either model to be correct but does place paleogeographic constraints on the proximity of the Havallah basin to North America and to an active arc. For most or all of the 80-m.y. time interval studied herein (Osagean to early Guadalupian), the Havallah depositional basin represented in the Antler Peak area has a history of rifting and basin formation along the continental margin in a region removed from arc-related volcaniclastic sedimentation until the Leonardian or later. This part of the Havallah sequence could not have been incorporated into an accretionary wedge and fore-arc basin depositional system before late Early Permian. It was conceivably part of a back-arc basin and has stratigraphic similarities to correlative rocks in the Northern Sierra terrane of California, which has, in addition, Upper Devonian, Lower Mississippian, and Permian arc-related volcanic and volcaniclastic rocks (Harwood and Murchey, this volume). Direct sedimentologic evidence for an active arc during the Osagean to Wolfcampian interval is lacking in both the Havallah sequence of the Antler Peak area and in the Northern Sierra terrane. Either the arc lay far to the west during this period (as suggested by Harwood and Murchey, this volume), or the Osagean to Wolfcampian plate boundaries were characterized by another tectonic regime such as extension or transform.

ACKNOWLEDGMENTS

Bruce Wardlaw and Calvin Stevens kindly provided me with conodont and fusulinid identifications, respectively. The reviews of Andrew Tomlinson and Allen Ormiston improved the manuscript's clarity. Roy Couzin, Marty Liniecki LaPorte, Andrew Tomlinson, Cynthia Kelley, and Sheraz Khan provided field assistance. Robert Oscarson provided technical expertise with the scanning electron microscope.

APPENDIX

A detailed discussion of radiolarian taxonomy and stratigraphy is not within the scope of this manuscript but is presented in Murchey (1989). Nevertheless, the taxonomic nomenclature used herein requires a few clarifying remarks. Following Ishiga's (1982, 1986) interpretation that the genus *Haplodiacanthus* Nazarov and Rudenko (1981) is a junior synonym of *Pseudoalbaillella* Holdsworth and Jones (1980a), *Haplodiacanthus levitoflexus* Nazarov (*in* Isakova and Nazarov, 1986) was called *Pseudoalbaillella levitoflexa* (Nazarov) n. comb (Plate 1, Fig. 6). Poorly preserved bipennate *Albaillella* specimens (*Albaillella* sp. aff. *A. cartalla* Ormiston and Lane, Plate 1, Fig. 15) were considered more closely related to *Albaillella cartalla* Ormiston and Lane (1976) than to other previously described taxa. The inclusion of *Paronaella impella* Ormiston and Lane (1976) in the genus *Scharfenbergia* Won (1983) rather than the genus *Latentifistula* Nazarov and Ormiston (1983) was based on taxonomic considerations explained in Holdsworth and Murchey (1988); *Scharfenbergia impella* (Nazarov and Ormiston) group includes forms with doubly tapering arms (*S. impella* s.s.) and forms with cylindrical arms that taper slightly only at the termination (Plate 1, Fig. 35).

Holdsworth and Murchey (1988) believed that the Sycamore Limestone fauna of Oklahoma (Ormiston and Lane, 1976) (Fig. 4) was probably Meramecian rather than lower Osagean and that the formation had been misdated because of reworked conodonts. Recent work on the Sycamore Limestone confirms that the base is no older than middle Meramecian and that some intervals near the base of the Sycamore Limestone contain abundant reworked conodonts of older Mississippian age (A. Harris, U.S. Geological Survey, and J. Schwartzapfel, University of Texas at Dallas, written communications, 1989). In Figure 4, the Sycamore Limestone fauna was thus considered younger than the Woodman Formation faunas of Utah described by Sandburg and Gutschick (1984).

Plate 1

Late Paleozoic radiolarians from the Havallah sequence, Nevada, and from the Barita de Sonora sequence, Sonora, Mexico. Photographed with scanning electron microscope (International Scientific Instruments model Super IIIA).

Figure 1. *Archocyrtium* sp. aff. *A. coronaesimile* Won 1983; Havallah sequence, USGS MR 7632-st57; bar = 330 μ.
Figure 2. *Follicucullus ventricosus* Ormiston and Babcock 1979; Havallah sequence, USGS MR 0196-st72C4; bar = 330 μ.
Figure 3. *Follicucullus scholasticus* Ormiston and Babcock 1979; Havallah sequence, USGS MR 0196-st72C6; bar = 330 μ.
Figure 4. *Pseudoalbaillella fusiformis* (Holdsworth and Jones, 1980a); Havallah sequence, USGS MR 0196-st72C5; bar = 330 μ.
Figure 5. *Pseudoalbaillella scalprata* Holdsworth and Jones 1980a; Havallah sequence, USGS MR 1875-st53C19; bar = 330 μ.
Figure 6. *Pseudoalbaillella levitoflexa* (*Haplodiacanthus levitoflexus* Nazarov in Isakova and Nazarov, 1986) n. comb.; Havallah sequence, USGS MR 2261-st50CR3C bar = 460 μ.
Figure 7. *Pseudoalbaillella bulbosa* Ishiga; Havallah sequence, USGS MR 2261-st50C; bar = 500 μ.
Figure 8. *Pseudoalbaillella u-forma* Holdsworth and Jones, morphotype I of Ishiga; Havallah sequence, USGS MR 7633 st72A1; bar = 330 μ.
Figure 9. *Paleoxyphostylus variospina* Won; Barita de Sonora sequence, Mexico, USGS MR 3645-st71D7; bar = 330 μ.
Figure 10. *Albaillella* sp. cf. *A. paradoxa* Deflandre; Havallah sequence, USGS MR 7086-st62C29; bar = 330 μ.
Figure 11. *Albaillella undulata* Deflandre; Havallah sequence, USGS MR 7086-st62C5; bar = 330 μ.
Figure 12. *Albaillella* sp. A (aff. *A. undulata*): narrowly conical *Albaillella* species possessing 10 to 13 undulations of lamellar test; it is distinguished from *A. undulata* by having more undulations of test and smaller length-to-width ratio; commonly possesses one row of pores per undulation. Havallah sequence, USGS MR 7632-st57BB26, bar = 330 μ. Holotype.
Figure 13. *Albaillella* sp. B (= *Albaillella* sp. of Sandberg and Gutschick, 1984, Plate 5, Fig. K); Havallah sequence; USGS MR 7632-st57BC4; bar = 330 μ.
Figure 14. *Albaillella* sp. C (= *Albaillella* sp. of Sandberg and Gutschick, 1984, Plate 5, Figs. U, V, Q); Havallah sequence, USGS MR 7632-st57BC21, bar = 500 μ.
Figure 15. *Albaillella* sp. aff. *A. cartalla* Ormiston and Lane; Havallah sequence, USGS MR 2271; bar = 310 μ.
Figure 16. *Pseudoalbaillella* sp. G of Ishiga, 1982; Havallah sequence, USGS MR 2289-st53B9; bar = 330 μ.
Figure 17. *Pseudoalbaillella* sp. aff. *P.* sp. G of Ishiga; this taxon has a longer pseudoabdomen than *P.* sp. G, differs from *P. elegans* Ishiga in that it lacks an apertural slit and possesses a very straight pseudoabdomen. Havallah sequence, USGS MR 2289-st53B8; bar = 330 μ.
Figure 18. *Pseudoalbaillella nodosa* Ishiga; Havallah sequence, USGS MR 0208; bar = 350 μ.
Figure 19. *Triactofenestrella* Nazarov and Ormiston sp.; Havallah sequence, USGS MR 2261-st53D; bar = 330 μ.

Figure 20. Deflandrella DeWever and Caridroit sp.; Havallah sequence, USGS MR 1875-st53C; bar = 330 μ
Figure 21. *Quinqueremus robusta* Nazarov and Ormiston; Havallah sequence, USGS MR 2261; bar = 330 μ.
Figure 22. *Polyfistula* Nazarov and Ormiston (s.l., six-rayed form) sp.; Havallah sequence, USGS MR 1875-st53C; bar = 460 μ
Figure 23. *Quinqueremus* sp. aff. *Q. robusta*; differs from *Q. robusta* only by the presence of small pores over all parts of the test rather than having pores limited to the distal portion of the arms. Havallah sequence, USGS MR 2261, bar = 330 μ
Figure 24. *Pseudotormentus kamigoriensis* DeWever and Caridroit; Havallah sequence, USGS MR 1875-st53C; bar = 330 μ.
Figure 25. *Ormistonella robusta* DeWever and Caridroit; Havallah sequence, USGS MR 2261-st53D; bar = 300 μ.
Figure 26. *Paronaella*? *triporosa* Holdsworth and Murchey; Havallah sequence, USGS MR 0208-st42A; bar = 500 μ.
Figure 27. *Scharfenbergia tailleurense* Holdworth and Murchey; Havallah sequence, USGS MR 7634; bar = 600 μ.
Figure 28. ?*Circulaforma* Cheng sp. A; Havallah sequence, USGS MR 4233; bar = 330 μ.
Figure 29. *Circulaforma* sp. cf. *C. omicron* (Ormiston and Lane), n. comb.; Havallah sequence, USGS MR 7632-st578BD19; bar = 330 μ.
Figure 30. *Ceratoikiscum* sp. aff. *C. astrum* Cheng; Havallah sequence, USGS MR 7632-st57BD33; bar = 330 μ.
Figure 31. ?*Tormentum circumfusum* Nazarov and Ormiston; Havallah sequence, USGS MR 1875-st53C; bar = 380 μ.
Figure 32. *Paronaella*? sp. A (aff. *P*? *triporosa*); this species differs from *P.*? *triporosa* in that it has a larger test with a circular, rather than triangular, outline; test is formed from lace-like spongy material with larger pores than that of *P.*? *triporosa.* Havallah sequence, USGS MR 2262-st50B; bar = 660 μ.
Figure 33. *Tormentum* sp. cf. *T. pervagatum* Nazarov and Ormiston; Havallah sequence, USGS MR 2076.
Figure 34. *Latentifistula patagilaterala* Nazarov and Ormiston; Havallah sequence, USGS MR 1875-st53C; bar = 660 μ.
Figure 35. *Scharfenbergia impella* (Ormiston and Lane) group; three-rayed scharfenbergids with spongy undifferentiated tests are included in this group. Havallah sequence, USGS MR 2081.

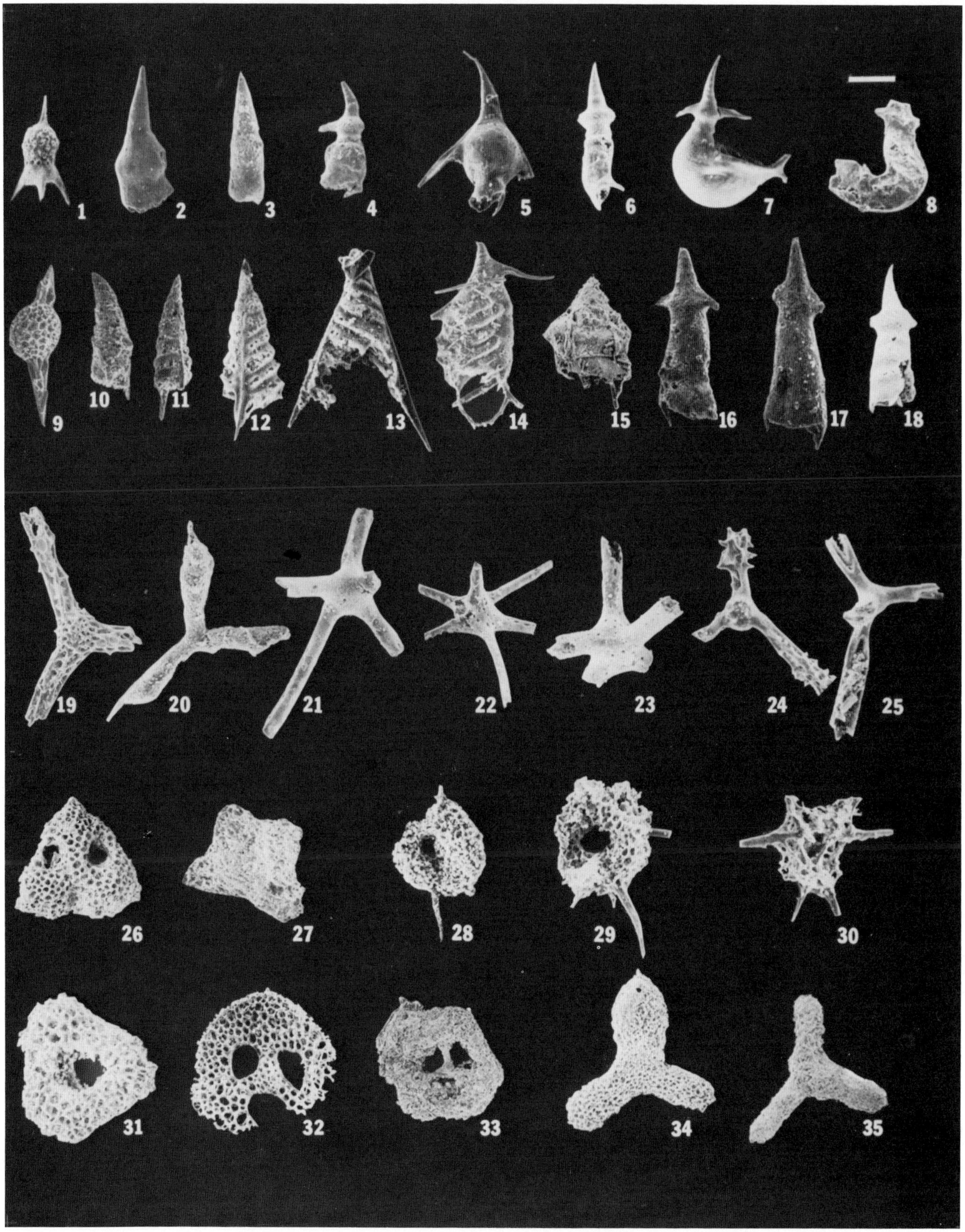
1
2
3
4
5
6
7
8
9
10
11
12
13
14
15
16
17
18
19
20
21
22
23
24
25
26
27
28
29
30
31
32
33
34
35

Plate 2

Late Paleozoic sponge spicules from Nevada and California.

Figure 1. Oxea; Havallah sequence, MY23, 66CA4; bar = 160 μ.
Figure 2. Strongyle; Havallah sequence, MY19, 66C5; bar = 160 μ.
Figure 3. Hexactine; Havallah sequence, MY20, 67C.1; bar = 100 μ.
Figure 4. Hexactine; Havallah sequence, MY20, 67C.2; bar = 70 μ.
Figure 5. Birotule; Havallah sequence, DK-8, 67B.1; bar = 100 μ.
Figure 6. Anadiaene; Havallah sequence, MY 167B, 67A.1; bar = 100 μ.
Figure 7. Anatriaene; Havallah sequence, MYTC3, 66C.D5; bar = 100 μ.
Figure 8. Protriaene; Havallah sequence, MYTC3, 66C.D4; bar = 100 μ.
Figur e9. Desma; Havallah sequence, MY19, 66C.G2; bar = 100 μ.
Figure 10. Desma; Havallah sequence, MY16X, 66C.B3; bar = 160 μ.
Figure 11. Calthrop; Eastern Klamath terrane, USGS MR 3871, 67D.1; bar = 250 μ.
Figure 12. Rhax; Havallah sequence, MY19, 66C.C7; bar = 50 μ.

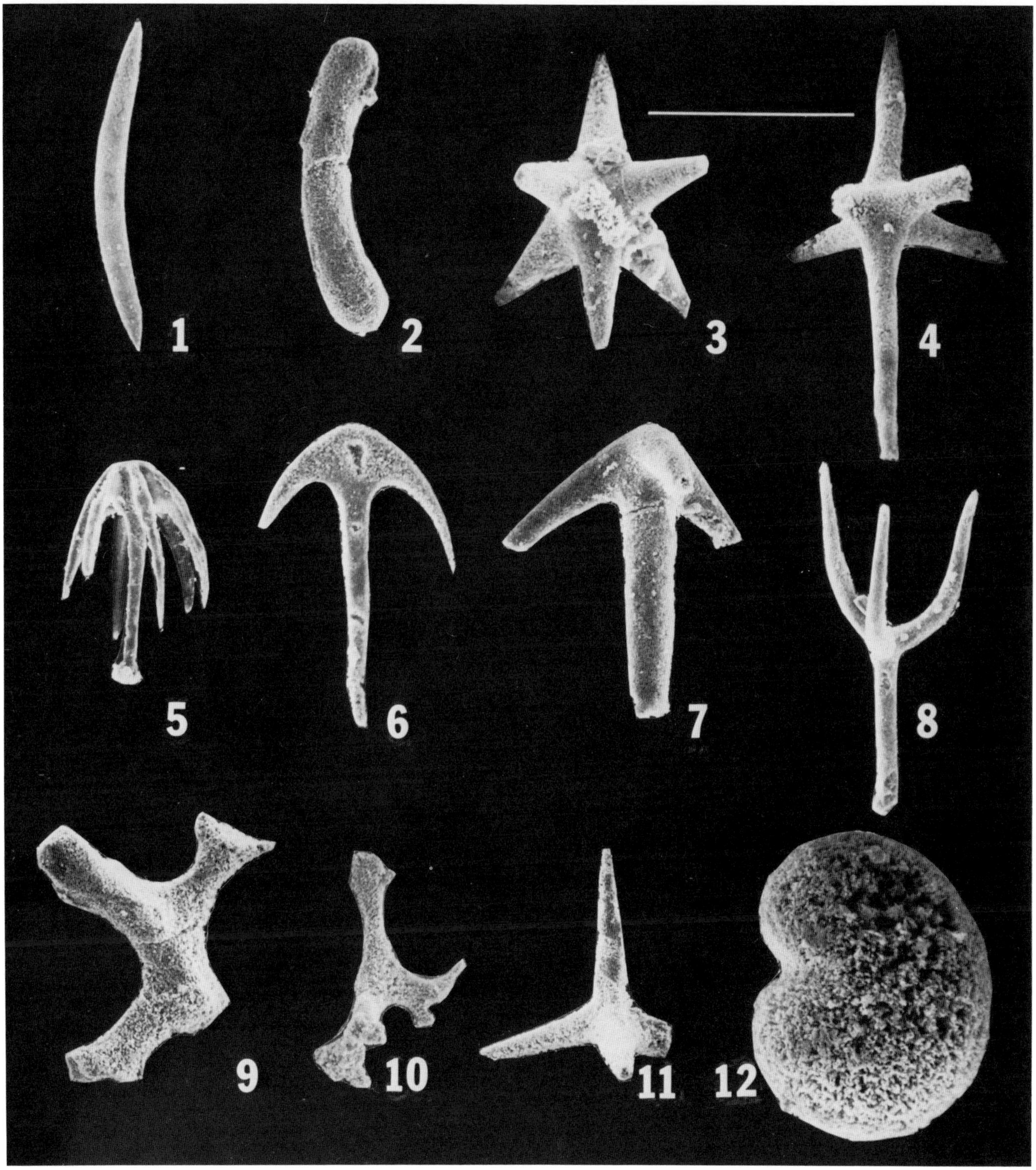
1
2
3
4
5
6
7
8
9
10
11
12

REFERENCES CITED

Babaie, H. A., 1987, Paleogeographic and tectonic implications of the Golconda allochthon, southern Toiyabe Range, Nevada: Geological Society of America Bulletin, v. 99, p. 231–243.

Bouma, A. H., 1962, Sedimentology of some flysch deposits; A graphic approach to facies interpretation: Amsterdam, Elsevier, 168 p.

Brueckner, H. K., and Snyder, W. S., 1985, Structure of the Havallah sequence, Golconda allochthon, Nevada; Evidence for prolonged evolution in an accretionary prism: Geological Society of America Bulletin, v. 96, p. 1113–1130.

Burchfiel, B. C., and Davis, G. A., 1972, Structural framework and evolution of the southern part of the Cordilleran orogen, western United States: American Journal of Science, v. 272, p. 97–118.

—— , 1975, Nature and controls of Cordilleran orogenesis, western United States; Extensions of an earlier synthesis: American Journal of Science, v. 275, p. 363–396.

Cheng, Y.-N., 1986, Taxonomic studies on Upper Paleozoic Radiolaria: Taiwan, Republic of China National Museum of Natural Science Special Publication 1, 311 p.

Coles, K. S., and Snyder, W. S., 1985, Significance of lower and middle Paleozoic phosphatic chert in the Toquima Range, central Nevada: Geology, v. 13, p. 573–576.

Cornell, W. C., and Simpson, R. D., 1985, New Permian albaillellid radiolarians from West Texas: Micropaleontology, v. 31, no. 3, p. 271–279.

Dickinson, W. R., Harbaugh, D. W., Saller, A. H., Heller, P. L., and Snyder, W. S., 1983, Detrital modes of upper Paleozoic sandstones from Antler orogen in Nevada; Implications for the nature of the Antler orogeny: American Journal of Science, v. 283, p. 4811–509.

Dumitrica, P., 1970, Cryptocephalic and cryptothoracic Nassellaria in some Mesozoic deposits of Romania: Revue Roumaine de Géologie, Geophysique et Geographie, Série de Geologie, v. 14, p. 45–124.

Epstein, A. G., Epstein, J. B., and Harris, L. D., 1977, Conodont color alteration; An index to organic metamorphism: U.S. Geological Survey Professional Paper 995, 27 p.

Fagan, J. J., 1962, Carboniferous cherts, turbidites, and volcanic rocks in northern Independence Range, Nevada: Geological Society of America Bulletin, v. 73, p. 595–612.

Harris, A. G., Lane, H. R., Tailleur, I. L., and Ellersieck, I., 1987, Conodont thermal maturation patterns in Paleozoic and Triassic rocks, northern Alaska, geologic and exploration implications, *in* Tailleur, I., and Weimer, P., eds., Alaskan North Slope geology: Pacific Section, Society for Economic Paleontologists and Mineralogists and Alaska Geological Society, p. 181–191.

Holdsworth, B. K., and Jones, D. L., 1980a, Preliminary radiolarian zonation for Late Devonian through Permian time: Geology, v. 8, p. 281–285.

—— , 1980b, A provisional radiolarian biostratigraphy, Late Devonian through Late Permian: U.S. Geological Survey Open-File Report 80-876, 32 p.

Holdsworth, B. K., and Murchey, B. L., 1988, Paleozoic radiolarian biostratigraphy of the northern Brooks Range, Alaska: U.S. Geological Survey Professional Paper 1399, p. 777–792.

Ijima, I., 1927, The Hexactinellida of the Siboga Expedition, v. 6: Leiden, Siboga-Expeditie, 383 p.

Isakova, T. N., and Nazarov, B. B., 1986, Stratigraphy and microfauna of the Late Carboniferous–Early Permian of the Southern Urals: Moscow Nauka, Trudy GIN AN SSSR, v. 402, 213 p. (in Russian).

Ishiga, H., 1982, Late Carboniferous and Early Permian radiolarians from the Tamba Belt, southwest Japan: Earth Science (Chikyu Kagaku), v. 36, p. 333–339.

—— , 1986, Late Carboniferous and Permian radiolarian biostratigraphy of southwest Japan: Osaka City University Journal of Geosciences, v. 29, p. 89–100.

Laule, S. W., Snyder, W. S., and Ormiston, A. R., 1981, Willow Canyon Formation, Nevada; An extension of the Golconda Allochthon: Geological Society of America Abstracts with Programs, v. 13, p. 66.

Miller, E. L., and 5 others, 1982, Structural fabric of the Paleozoic Golconda allochthon, Antler Peak Quadrangle, Nevada; Progressive deformation of an oceanic sedimentary assemblage: Journal of Geophysical Research, v. 87, p. 3795–3804.

Miller, E. L., Holdsworth, B. K., Whiteford, W. B., and Rodgers, D., 1984, Stratigraphy and structure of the Schoonover sequence, northeastern Nevada; Implications for Paleozoic plate-margin tectonics: Geological Society of America Bulletin, v. 95, p. 1063–1076.

Murchey, B. L., 1989, Late Paleozoic siliceous basins of the western Cordillera of North America (Nevada, California, Mexico, and Alaska); Three studies using radiolarians and sponge spicules for biostratigraphic, paleobathymetric, and tectonic analyses [Ph.D. thesis]: Santa Cruz, University of California, 188 p.

Murchey, B. L., Harwood, D. S., and Jones, D. L., 1986, Correlative chert sequences from the northern Sierra Nevada and Havallah sequence, Nevada: Geological Society of America Abstracts with Programs, v. 18, p. 162.

Murchey, B. L., Jones, D. L., and Blome, C. D., 1990, The environmental and tectonic significance of two coeval Permian radiolarian-sponge associations in eastern Oregon: U.S. Geological Survey Professional Paper 1483 (in press).

Nazarov, B. B., and Ormiston, A. R., 1985a, Radiolaria from the late Paleozoic of the southern Urals, USSR and West Texas, USA: Micropaleontology, v. 31, p. 1–54.

—— , 1985b, Evolution of Radiolaria in the Paleozoic and its correlation with the development of other marine fossil groups: Senckenbergiana lethaea, v. 66, p. 203–215.

—— , 1986, Origin and biostratigraphic potential of the stauraxon polycystine Radiolaria: Marine Micropaleontology, v. 11, p. 33–54.

Nazarov, B. B. and Rudenko, V. S., 1981, Some bilaterally symmetrical late Paleozoic radiolarians of the south Urals: Voprosy Micropaleontologii, v. 24, p. 129–139 (in Russian).

Ormiston, A. R., and Babcock, L., 1979, *Follicucullus,* new radiolarian genus from the Guadalupian (Permian) Lamar Limestone of the Delaware Basin: Journal of Paleontology, v. 53, p. 328–334.

Ormiston, A. R., and Lane, H. R., 1976, A unique radiolarian fauna from the Sycamore Limestone (Mississippian) and its biostratigraphic significance: Paleontographica, pt. A, v. 154, no. 4-6, p. 158–180.

Pessagno, E. A., Jr., and Newport, R. L., 1972, A technique for extracting Radiolaria from radiolarian cherts: micropaleontology, v. 18, p. 231–324.

Reid, R.E.H., 1986, Bathymetric distributions of Calcarea and Hexactinellida in the present and the past: Geological Magazine, v. 105, p. 546–559.

Roberts, R. J., 1964, Stratigraphy and structure of the Antler Peak Quadrangle, Humboldt and Lander counties, Nevada: U.S. Geological Survey Professional Paper 459-A, p. A1–A93.

Roberts, R. J., HOtz, P. E., Gilluly, J., and Ferguson, H. G., 1985, Paleozoic rocks of north-central Nevada: American Association of Petroleum Geologists Bulletin, v. 42, p. 2813–2857.

Sandberg, C. A., and Gutschick, R. C., 1984, Distribution, microfauna, and source-rock potential of Mississippian Delle Phosphatic Member of Woodman Formation and equivalents, Utah, and adjacent states, *in* Woodward, J., Messner, F. F., and Clayton, J. L., eds., Hydrocarbon source rocks of the Greater Rocky Mountain region: Denver, Colorado, Rocky Mountain Association of Geologists, p. 135–178.

Schulze, F. E., 1899, Amerikanische Hexactinelliden nach dem Materiale der Albatross-Exepdition: Jena, Gustav Fischer.

Silberling, N. J., and Roberts, R. J., 1962, Pre-Tertiary stratigraphy and structure of northwestern Nevada: Geological Society of America Special Paper 72, 58 p.

Snyder, W. S., and Brueckner, H. K., 1983, Tectonic evolution of the Golconda allochthon, Nevada; Problems and perspectives, *in* Stevens, D. H., ed., Pre-Jurassic rocks in western North America suspect terranes: Pacific Section, Society of Economic Paleontologists and Mineralogists Special Publication, p. 103–123.

Sosson, M., DeWever, P., and Vrieljinck, B., 1983, Datation et analyse structurale de l'unite ophiolitique du Hot Springs Range (NW du Nevada, Etats Unis): consequences sur l'age de mise en place de ce domain oceanique sur le craton nord americain: Comptes Rendus Academic des Sciences, Paris, v. 298, no. 6, p. 235–240.

Speed, R. G., 1977, An appraisal of the Pablo Formation of presumed Late Paleozoic age, central Nevada, *in* Stewart, J. H., Stevens, C. H., and Fritsche, A. E., eds., Paleozoic paleogeography of the western United States: Pacific Section, Society of Economic Paleontologists and Mineralogists Pacific Coast Paleogeography Symposium 1, p. 315–324.

Speed, R. G., 1979, Collided Paleozoic platelet in the western U.S.: Journal of Geology, v. 87, p. 279–292.

Speed, R. G., and Sleep, N. H., 1982, Antler orogeny and foreland basin: A model: Geological Society of America Bulletin, v. 93, p. 815–828.

Stewart, J. H., MacMillan, J. R., Nichols, K. M., and Stevens, C. H., 1977, Deep-water Upper Paleozoic rocks in north-central Nevada; A study of the type area of the Havallah Formation, *in* Stewart, J. H., Stevens, C. H., and Fritsche, A. E., Paleozoic paleogeography of the western United States: Pacific Section, Society of Economic Paleontologists and Mineralogists Pacific Coast Paleogeography Symposium 1, p. 337–348.

Stewart, J. H., Murchey, B., Jones, D. L., and Wardlaw, B. R., 1986, Paleontologic evidence for complex tectonic interlayering of Mississippian to Permian deep-water rocks of the Golconda allochthon in Tobin Range, north-central Nevada: Geological Society of America Bulletin, v. 97, p. 1122–1132.

Won, M.-Z., 1983, Radiolarien aus dem Unterkarbon des Rheinischen Schiefergebirges (Deutschland) (Radiolaria from the Lower Carboniferous of the Rhenish Schiefergebirge [Germany]): Palaeontographica, pt. A, v. 182, p. 116–175.

MANUSCRIPT RECEIVED BY THE SOCIETY MARCH 27, 1990

Printed in U.S.A.

Geological Society of America
Special Paper 255
1990

Biostratigraphic, tectonic, and paleogeographic ties between upper Paleozoic volcanic and basinal rocks in the northern Sierra terrane, California, and the Havallah sequence, Nevada

David S. Harwood and Benita L. Murchey
U.S. Geological Survey, 345 Middlefield Road, Menlo Park, California 94025

ABSTRACT

Biostratigraphic data, based mostly on radiolarian assemblages, establish synchronous deposition in the northern Sierra terrane and the Havallah basin beginning in the Late Devonian and extending into the early Late Permian. Lower Mississippian and mid-Permian arc-derived volcaniclastic debris was deposited in parts of the Havallah basin during episodes of arc volcanism in the northern Sierra terrane. Between these episodes of arc volcanism, from late Early Mississippian to at least Middle Pennsylvanian, the northern Sierra terrane collected siliceous pelagic deposits that correlate with dominantly chert-argillite sections in the Havallah sequence. These intermixed lithic assemblages suggest shared stratigraphic evolution and geographic proximity between the Sierran arc terrane and the Havallah basin during the late Paleozoic.

During Late Devonian and Early Mississippian arc volcanism in the northern Sierra terrane, lower Paleozoic rocks of the Roberts Mountains allochthon were thrust over coeval deposits on the North American shelf. Chert-quartz–rich siliciclastic debris, derived from the Antler orogenic belt, is interbedded with Upper Devonian and Lower Mississippian distal volcanic rocks in the northern Sierra terrane and with Kinderhookian volcaniclastic rocks and chert in the Schoonover sequence. These quartzose-clastic deposits not only provide an independent lithologic link between the Sierran arc terrane and the Havallah basin, they also tie the arc terrane and basin to North America at the time of the Antler orogeny.

Late Devonian and Early Mississippian arc volcanism in the northern Sierra terrane occurred in an extensional regime. Extensional tectonism began locally in the Havallah basin during the Famennian and continued into the early Meramecian. Contemporaneous extension in the arc and basin during emplacement of the Roberts Mountains allochthon is difficult to reconcile with existing arc-continent collision models for the Antler orogeny.

Harwood, D. S., and Murchey, B. L., 1990, Biostratigraphic, tectonic, and paleogeographic ties between upper Paleozoic volcanic and basinal rocks in the northern Sierra terrane, California, and the Havallah sequence, Nevada, *in* Harwood, D. S., and Miller, M. M., eds., Paleozoic and early Mesozoic paleogeographic relations; Sierra Nevada, Klamath Mountains, and related terranes: Boulder, Colorado, Geological Society of America Special Paper 255.

INTRODUCTION

Paleozoic arc-volcanic rocks in the northern Sierra Nevada were deposited during two distinct episodes: the first, in Late Devonian and Early Mississippian time, and the second, in mid-Permian time (D'Allura and others, 1977; Schweickert, 1981; Harwood, 1983, 1988). During the early volcanic episode, lower Paleozoic eugeoclinal rocks of the Roberts Mountains allochthon (Fig. 1) were thrust eastward over coeval shelf deposits in northern Nevada during the Antler orogeny (Speed and Sleep, 1982). Following the Permian volcanic episode, rocks of the Havallah sequence in the Golconda allochthon were thrust eastward over the Roberts Mountains allochthon and its overlap assemblage during the Late Permian to Early Triassic Sonoma orogeny (Silberling and Roberts, 1962; Roberts, 1964; Silberling, 1973).

The similarities in timing between Paleozoic arc volcanism in the northern Sierra Nevada and these periods of thrusting in northern Nevada have prompted numerous models relating emplacement of the allochthons to some form of arc-continent convergence (Moores, 1970; Burchfiel and Davis, 1972, 1975; Silberling, 1973; Dickinson, 1977; Speed, 1979; Schweickert and Snyder, 1981; Speed and Sleep, 1982; Snyder and Brueckner, 1983; Dickinson and others, 1983; Miller and others, 1984; Miller, 1987).

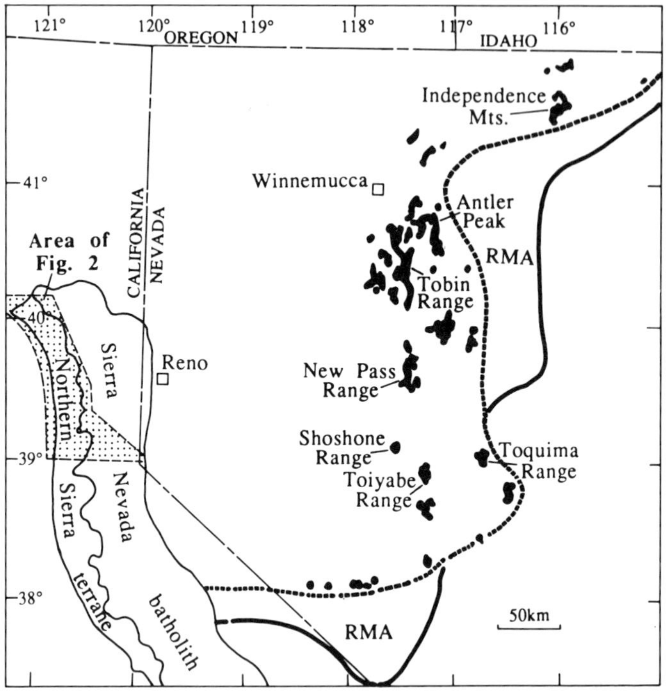

Figure 1. Regional map of eastern California and Nevada showing area of exposure of the Roberts Mountains allochthon (RMA), inferred eastern extent of the Golconda thrust (short dashed line), exposures of the Havallah and Schoonover sequences (black), and areas mentioned in the text.

None of the models has won universal acceptance, however, and the tectonic configuration of the Paleozoic continental margin and the paleogeographic affinity of the Sierran arc-volcanic rocks to North America are subjects of an ongoing, multifaceted debate. One side views the Sierran Paleozoic volcanic rocks as part of an extensive, west-facing, continent-fringing arc system that formed, migrated, and reformed off the active continental margin of North America (Burchfiel and Davis, 1972, 1975; Miller and others, 1984; Miller, 1987). In these models, the Antler and Sonoma orogenies are viewed as back-arc thrusting episodes. This interpretation implies not only temporal but also stratigraphic ties between the Sierran Paleozoic arc-volcanic rocks and deposits of the Roberts Mountains and Golconda allochthons.

A second set of models interprets the Paleozoic volcanic rocks in the northern Sierra Nevada as part of a composite, far-traveled microplate that collided with the passive margin of North America in the Sonoma orogeny, thrusting an accretionary wedge, the Golconda allochthon, over the Roberts Mountains allochthon (Speed, 1979; Speed and Sleep, 1982; Snyder and Brueckner, 1983). This interpretation precludes any stratigraphic and paleogeographic ties between pre–Late Permian rocks in the exotic Sierran arc terrane and allochthonous rocks in the northern Nevada. Furthermore, the location of the Antler arc, which collided with the continent and thrust accreted sediments of the Roberts Mountains allochthon onto the continental margin during the Antler orogeny, is a major problem not adequately explained by the passive-accretionary models.

Schweickert and Snyder (1981) proposed a model involving arc-polarity reversals and repeated collisions of a North American fringing arc. In this model, the Roberts Mountains and Golconda allochthons were emplaced as accretionary wedges above west-dipping subduction zones by arc convergence and the Havallah basin formed in the intervening period by back-arc extension above an east-dipping subduction zone.

Recently, Burchfiel and Royden (1989) suggested that the Antler orogen may have formed in a tectonic setting like that of the late Cenozoic Apennine and East Carpathian orogens. There, foreland thrusting was apparently driven by outward migration of the subduction zone as the underthrust plate sank into the mantle. Simultaneous extension occurred in the hinterland as the overriding plate failed to keep pace with the retreating subduction zone (Malinverno and Ryan, 1986). Although this model meets a number of geological constraints for the Antler orogen (Burchfiel and Royden, 1989), available evidence indicates that extension in the Havallah basin occurred simultaneously with foreland thrusting of the Roberts Mountain allochthon and that extension and compression occurred on the same plate inboard of the northern Sierra arc.

The occurrence of lower and middle Paleozoic phosphatic chert in the northern Sierra terrane and the Roberts Mountains allochthon (Varga, 1982; Coles and Snyder, 1985) suggests paleogeographic affinity of the Sierran arc terrane to North America and lends support to the active North American plate-margin models. Paleontologic and stratigraphic data gathered over the last 5 years have established additional correlations between the Sierran Paleozoic arc-volcanic rocks and the Havallah sequence (Murchey and others, 1986; Harwood and Murchey, 1989; Murchey, this volume). The correlative rocks range in age from the latest Devonian (Famennian) to the early Late Permian) (early Guadalupian) and span the emplacement of the Roberts Mountains allochthon in earliest Mississippian time (Speed and Sleep, 1982; Miller and others, 1984; Stewart and others, 1986). Lithologic and stratigraphic variations along strike in the Sierran Paleozoic arc-volcanic rocks also point to paleogeographic and tectonic connections with the Havallah basin and the Antler orogenic belt. The biostratigraphic data provide additional support to earlier published conclusions that Upper Devonian and Lower Mississippian arc-volcanic rocks in the northern Sierra terrane (Fig. 2) and early deposits in the Havallah basin originated as integral and outboard parts of the Antler orogen and that they maintained paleogeographic ties with North America through the late Paleozoic (Burchfiel and Davis, 1972, 1975; Schweickert and Snyder, 1981; Varga and Moores, 1981; Varga, 1982; Miller and others, 1984; Coles and Snyder, 1985).

STRATIGRAPHY OF THE NORTHERN SIERRA TERRANE

The oldest rocks in the northern Sierra terrane belong to the lower Paleozoic Shoo Fly Complex. North of the 39th parallel, the Shoo Fly Complex is composed of four regionally extensive thrust blocks informally referred to (in ascending order) as: (1) Lang sequence, (2) Duncan Peak allochthon, (3) Culbertson Lake allochthon, and (4) Sierra City melange (Schweickert and others, 1984; Harwood, 1988). Conodonts from a lens of limestone interbedded with the dominant quartzite-pelite turbidites of the Lang sequence are indicative of a late Middle to Late Ordovician age (Table 1, no. 1; Harwood and others, 1988). The Duncan Peak allochthon, which is composed of undated radiolarian chert and siliceous argillite, structurally overlies and locally is structurally interlayered with the Lang sequence. It is structurally overlain by pillow basalt, limestone, chert, quartz arenite, and volcanic-lithic sandstone of the Culbertson Lake allochthon (Girty, 1983; Girty and Wardlaw, 1985; Girty and others, this volume). The Culbertson Lake allochthon and the Lang sequence are structurally overlapped by the Sierra City melange, which contains blocks of serpentinite, gabbro, greenstone, chert, and limestone set in a matrix of disrupted sandstone and pelite. Blocks of limestone in the melange have yielded Late Ordovician conodonts (Table 1, no. 2) (Hannah and Moores, 1986) and megafossils (Potter and others, this volume, Montgomery Limestone chapter) that have North American affinities. Recent U/Pb studies on zircon from a felsic tuff interbedded with the matrix of the Sierra City melange yielded a late Early Silurian age (423 ± 10 Ma; Saleeby and others, 1987). This is the youngest age obtained thus far from the Shoo Fly Complex, and it indicates that at least the uppermost fault blocks in the Shoo Fly were thrust together after the late Early Silurian.

The amalgamated thrust blocks of the Shoo Fly Complex form the basement to middle and upper Paleozoic volcanic and basinal rocks. Discontinuous lenses of Upper Devonian epiclastic rocks of the Grizzly Formation and regionally extensive submarine pyroclastic rocks of the Upper Devonian Sierra Buttes Formation (Hanson and Schweickert, 1986) rest unconformably on the Shoo Fly Complex. Conodonts from the Grizzly and Sierra Buttes Formations are indicative of a mid-Famennian age (Table 1, nos. 3, 5). Ammonoids from the Sierra Buttes Formation are also indicative of a mid-Famennian age (Table 1, no. 4), and the same ammonoid genera were recovered from the type-Woodruff Formation, which, at that locality, is the youngest unit in the Roberts Mountains allochthon (Smith and Ketner, 1968; Johnson and Pendergast, 1981).

The Sierra Buttes Formation is overlain by andesite flows, tuff breccia, lapilli tuff, and volcaniclastic turbidites of the Taylor Formation. No fossils have been found in the Taylor Formation, but it can be no older than Late Devonian (mid-Famennian) and no younger than Early Mississippian (late Kinderhookian), the age of conodonts and megafossils in the overlying lower (volcanic) member of the Peale Formation (Table 1, nos. 7, 8).

The Peale Formation consists of a lower member characterized by alkali-feldspar phyric keratophyre flows, tuff breccia, tuff and tuffaceous siltstone, and sandstone. Lenses of limestone, which have yielded the late Kinderhookian conodonts, are interbedded with fine-grained tuff in the northern part of the terrane and signal the waning stages of the Late Devonian and Early Mississippian episode of arc volcanism. By Osagean time, the Late Devonian and Early Mississippian arc had cooled and subsided sufficiently to collect and preserve, at least locally, a blanket of siliceous pelagic sediment now represented by the chert member of the Peale Formation. Based on radiolarian faunas recovered from the chert member of the Peale, pelagic sedimentation persisted without obvious interruption or influx of coarser detritus until the Middle Pennsylvanian (Desmoinesian?) (Table 1, nos. 9 through 17), a period of about 45 m.y. Pelagic sedimentation may have persisted for an additional 25 m.y., but the depositional record is missing along an erosional unconformity that separates Middle Pennsylvanian chert from Lower Permian (late Wolfcampian? to early Leonardian) volcanic rocks (Fig. 3) (Schweickert, 1981; Harwood, 1983, 1988).

The Permian volcanic sequence was deposited from late Wolfcampian(?) through early Guadalupian time and consists of three major lithofacies. The Arlington Formation was deposited in the northwestern part of the terrane and consists of volcaniclastic sandstone, siltstone, and slate with locally significant accumulations of pebbly mudstone, blocky breccia, and conglomerate. The breccia is composed of blocks of chert and volcanic rocks as large as 50 m, and the pebbly mudstone and conglomerate contain abundant clasts of chert and volcanic rocks, as well as scattered quartzite clasts apparently derived from the Shoo Fly Complex. Conodonts from sparse limestone blocks in the breccia range from early to late Leonardian in age (Table 1, nos. 18, 26). The Arlington Formation is gradationally overlain on Peale

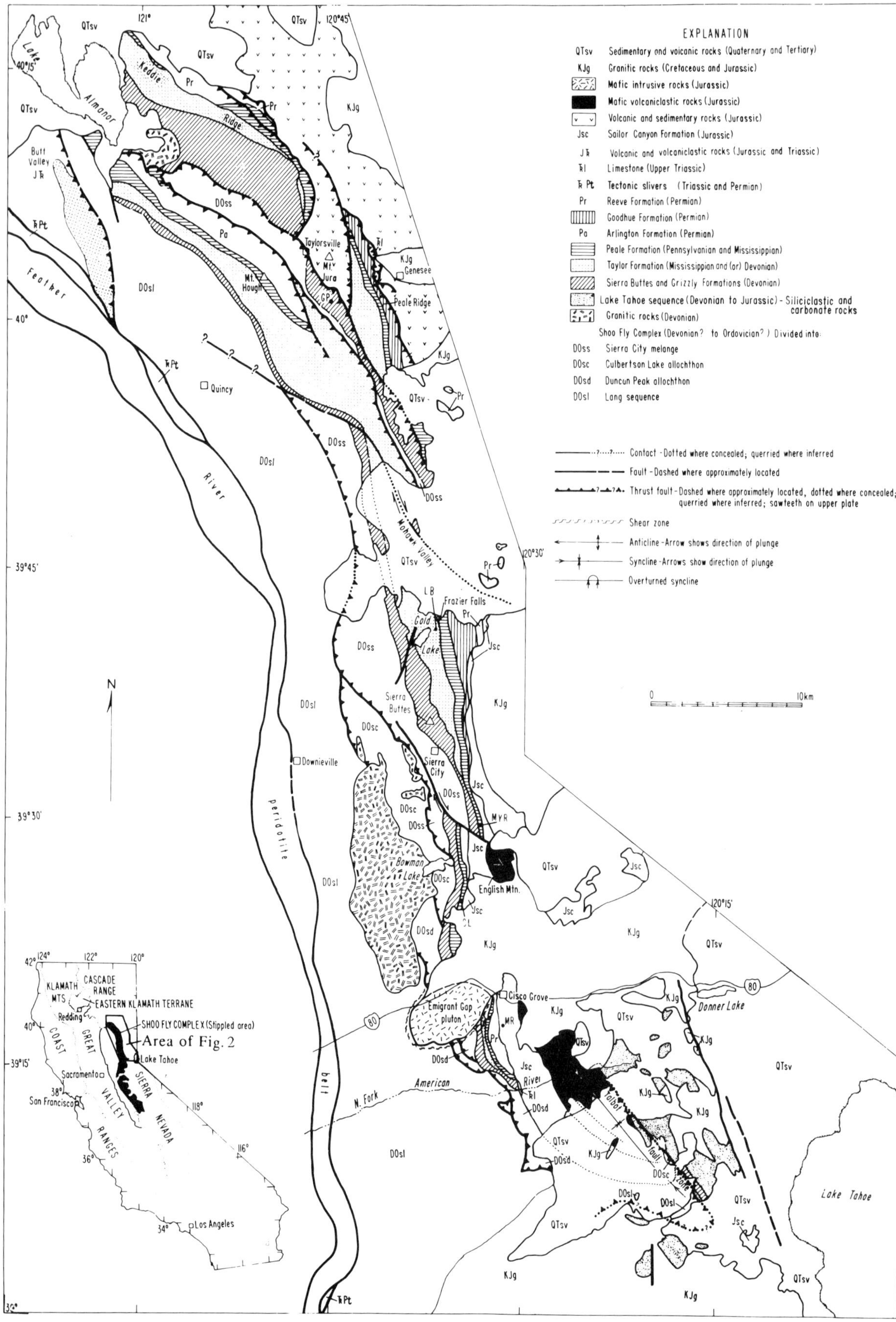

EXPLANATION
QTsv Sedimentary and volcanic rocks (Quaternary and Tertiary)
KJg Granitic rocks (Cretaceous and Jurassic)
Mafic intrusive rocks (Jurassic)
Mafic volcaniclastic rocks (Jurassic)
Volcanic and sedimentary rocks (Jurassic)
Jsc Sailor Canyon Formation (Jurassic)
JTr Volcanic and volcaniclastic rocks (Jurassic and Triassic)
Trl Limestone (Upper Triassic)
TrPt Tectonic slivers (Triassic and Permian)
Pr Reeve Formation (Permian)
Goodhue Formation (Permian)
Pa Arlington Formation (Permian)
Peale Formation (Pennsylvanian and Mississippian)
Taylor Formation (Mississippian and (or) Devonian)
Sierra Buttes and Grizzly Formations (Devonian)
Lake Tahoe sequence (Devonian to Jurassic) - Siliciclastic and carbonate rocks
Granitic rocks (Devonian)
Shoo Fly Complex (Devonian? to Ordovician?) Divided into:
DOss Sierra City melange
DOsc Culbertson Lake allochthon
DOsd Duncan Peak allochthon
DOsl Lang sequence
Contact - Dotted where concealed; querried where inferred
Fault - Dashed where approximately located
Thrust fault - Dashed where approximately located, dotted where concealed; querried where inferred; sawteeth on upper plate
Shear zone
Anticline - Arrow shows direction of plunge
Syncline - Arrows show direction of plunge
Overturned syncline
0 10km
N
121° 120°45' 120°30' 120°15'
40°15' 40° 39°45' 39°30' 39°15'
Lake Almanor
Keddie Ridge
Butt Valley
Feather River
Quincy
Taylorsville
Mt. Jura
Mt. Hough
Genesee
Peale Ridge
Mohawk Valley
Frazier Falls
Gold Lake
Sierra Buttes
Sierra City
Downieville
peridotite belt
Bowman Lake
English Mtn.
Cisco Grove
Emigrant Gap pluton
Donner Lake
N. Fork American River
Valley fault zone
Lake Tahoe
KLAMATH MTS
CASCADE RANGE
EASTERN KLAMATH TERRANE
Redding
SHOO FLY COMPLEX (Stippled area)
Area of Fig. 2
Lake Tahoe
Sacramento
San Francisco
COAST RANGES
GREAT VALLEY
SIERRA NEVADA
Los Angeles
124° 122° 120° 118° 116°
42° 40° 38° 36° 34° 32°

Figure 2. Generalized geologic map of the northern Sierra terrane north of the 39th parallel compiled from detailed mapping by McMath (1958), D'Allura (1977), Durrell and D'Allura (1977), Hannah (1980), Harwood (1983, 1988), Schweickert and others (1984), and Jayko (1988). Cl = Culbertson Lake; GP = Grizzly Peak; LB = Lakes Basin; MYR = Middle Yuba River; MR = Monumental Ridge.

Ridge (Fig. 2) by clinopyroxene–phyric pillow basalt of the Goodhue Formation. The Goodhue Formation is conformably overlain by characteristically plagioclase–phyric andesitic flows, tuff breccia, tuff, and volcaniclastic sandstone of the Reeve Formation. Calcareous tuff in the Reeve Formation has yielded late Early Permian (early Leonardian) fusulinids equivalent to zone H of the McCloud Limestone in the eastern Klamath terrane (Table 1, no. 23). Early Guadalupian conodonts (Table 1, no. 24) occur higher in the section in a lens of limestone interlayered with volcanic breccia mapped as the Robinson Formation by McMath (1958, 1966).

Mesozoic rocks, ranging in age from Middle Triassic (mid-Ladinian) to late Middle Jurassic (Callovian) overlie the Paleozoic arc-volcanic rocks. Relations between the Middle Triassic and Permian rocks in the Taylorsville area are uncertain; locally the contact may be a Late Jurassic fault. To the south, however, Upper Triassic (late Karnian to early Norian) rocks rest unconformably on the underlying strata. As shown in Figure 3, Mesozoic rocks unconformably truncate progressively older units toward the southern part of the terrane, until at the North Fork of the American River (Fig. 2), Upper Triassic limestone and conglomerate rest directly on the Shoo Fly Complex. Late Permian to Early Triassic deformation, uplift, and erosion, related to the Sonoma orogeny, increased in intensity southward through the northern Sierra terrane (Harwood, 1983, 1988; Davis, 1989).

STRATIGRAPHIC VARIATIONS IN THE NORTHERN SIERRA TERRANE

As shown in Figure 2, the Paleozoic volcanic assemblage thins significantly toward the south over its 120-km strike length. However, other important depositional patterns occur within the Devonian to Mississippian and mid-Permian arc sequences that either do not show at this map scale or are obscured by Late Jurassic structural features.

For example, three different parts of the Paleozoic arc assemblage are juxtaposed by Late Jurassic faults. North of Mohawk Valley (Fig. 2), Paleozoic volcanic rocks and their basement of Shoo Fly Complex are structurally repeated and

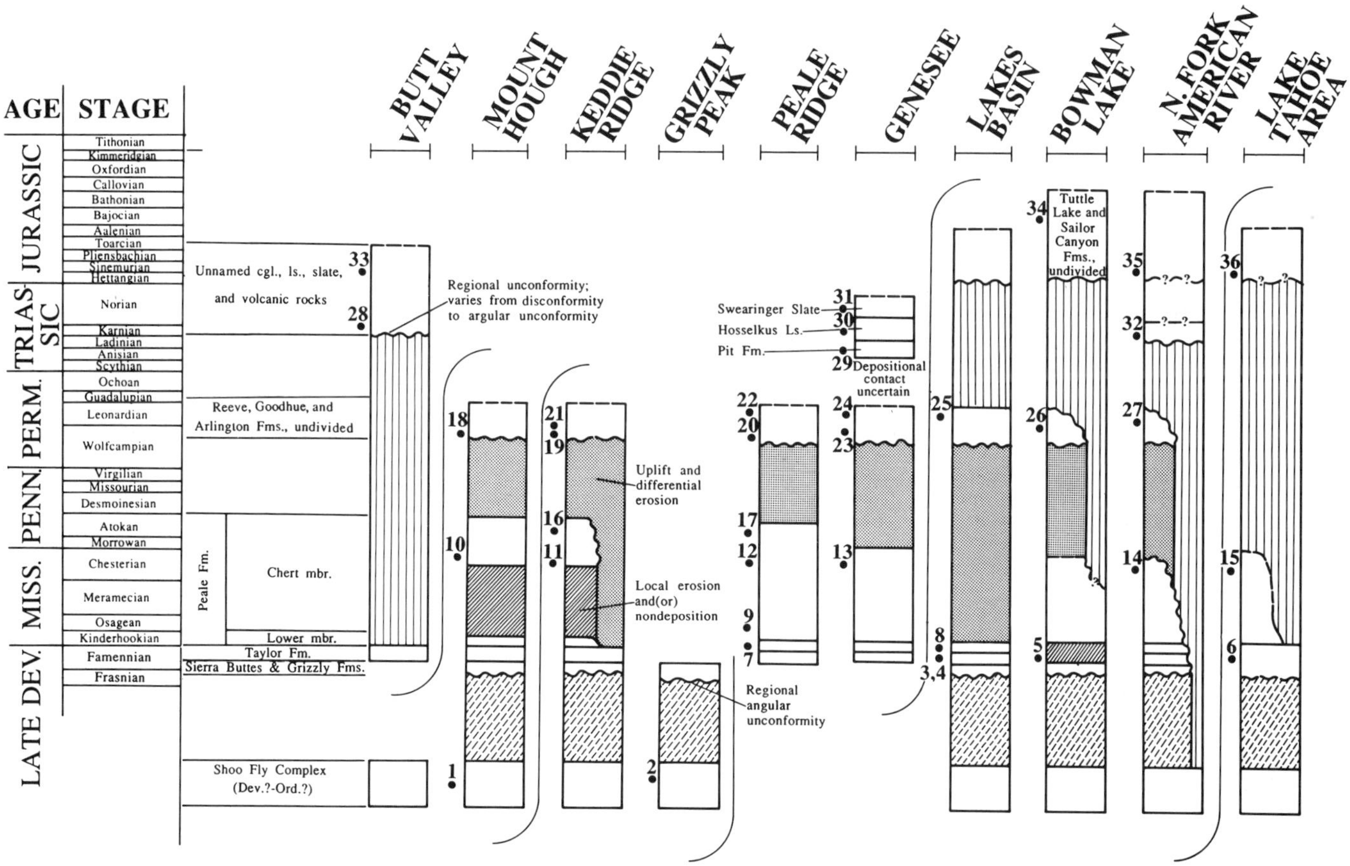

Figure 3. Time-lithology columns of the northern Sierra terrane showing regional and local unconformities in the Sierran arc terrane. Numbered dots refer to fossil localities in Table 1.

TABLE 1. SELECTED FOSSIL DATA FROM PALEOZOIC AND MESOZOIC ROCKS IN NORTHERN SIERRA TERRANE
(Locality numbers correspond to those in Fig. 3)

Locality Number	Quadrangle	Section, Township, Range	Map Unit	Diagnostic Fauna	Fauna Age	Reference
1	Blue Nose Mtn. 7-1/2'	20,22N,11E	Lang sequence, Shoo Fly Complex	*Amorphognathus* indet. sp.	mid-Middle to Late Ordovician	A. G. Harris, 1987, written communication
2	Taylorsville 7-1/2'	11,25N,10E	Limestone block, Sierra City melange, Shoo Fly Complex	*Belodina* sp., *Drepanoistodus suberectus, Panderodus gracilis, Protopanderodus insculptus*	late Late Ordovician	A. G. Harris, 1985, written communication; Hannah and Moores, 1986
3	Sierra City 15'	20,20N,12E	Black chert, Grizzly Formation	*Palmatolepis glabra*	Late Devonian (mid-Famennian)	Hanson and Schweickert, 1986
4	Sierra City 15'	1,20N,11E	Metachert, Sierra Buttes	*Platyclymenia* sp., *Tornoceras* sp.	Late Devonian (mid-Famennian)	Anderson and others, 1974
5	Emigrant Gap 15'	11,19N,12E	Black chert, Grizzly Formation	*Palmatolepis glabra*	Late Devonian (mid-Famennian)	Hanson and Schweikert, 1986
6	Granite Chief 7-1/2'	23,15N,11E	Unnamed limestone lens or block in Picayune Valley Formation	echinoderm debris, *Lampterocrinus(?)*	Silurian-Devonian	J. T. Dutro, Jr., 1988, written communication
7	Genesee Valley 7-1/2'	15,25N,11E	Limestone lens in volcanic member of Peale Formation	*Bispathodus stabilis, Bispathodus* n. sp. *Polygnathus triangulus triangulus, Pseudopolygnathus* sp. indet., *Siphonodella* sp. indet.	Early Mississippian, upper *crenulata-isosticha,* zone of late Kinderhookian	K. Denkler, 1985, written communication
8	Sierra City 15'	5,21N,12E	Limestone lens in volcanic member of Peale Formation	*Gnathodus* sp. indet., *Polygnathus communis communis, Polygnathus* cf *P. symmetricus, Pseudopolygnathus* sp. indet. *Siphonodella*	Early Mississippian, upper *crenulata-isosticha,* zone of late Kinderhookian	A. G. Harris, 1984, written communication
9	Genesee Valley 7-1/2'	15,25N,11E	Chert member of Peale Formation	*Scharfenbergia impella* gp.	Early Mississippian (late Osagean-Meramecian)	B. Murchey, 1985, Written communication
10	Greenville 15'	8,25N,10E	Chert member of Peale Formation	*Scharfenbergia tailleurense*	Late Mississippian to Early Pennsylvanian (Chesterian to Morrowan)	D. L. Jones and B. Murchey, 1984, written communication
11	Greenville 15'	17,27N,10E	Chert member of Peale Formation	*Scharfenbergia tailleurense*	Late Mississippian to Early Pennsylvanian (Chesterian to Morrowan)	D. L. Jones and B. Murchey, 1984, written communication
12	Genesee Valley 7-1/2'	15,25N,11E	Chert member of Peale Formaation	*Scharfenbergia tailleurense*	Late Mississippian to Early Pennsylvanian (Chesterian to Morrowan)	B. Murchey, 1985, written communication
13	Genesee Valley 7-1/2'	9,25N,11E	Chert member of Peale Formation	*Scharfenbergia tailleurense*	Late Mississippian to Early Pennsylvanian (Chesterian to Morrowan)	B. Murchey, 1985, written communication
14	Cisco Grove 7-1/2'	7,17N,13E	Chert member of Peale Formation	*Scharfenbergia tailleurense*	Late Mississippian to Early Pennsylvanian (Chesterian to Morrowan)	D. L. Jones and B. Murchey, 1981, written communication

TABLE 1. SELECTED FOSSIL DATA FROM PALEOZOIC AND MESOZOIC ROCKS IN NORTHERN SIERRA TERRANE (continued)
(Locality numbers correspond to those in Fig. 3)

Locality Number	Quadrangle	Section, Township, Range	Map Unit	Diagnostic Fauna	Fauna Age	Reference
15	Soda Springs 7-1/2'	10,16N,15E	Quartzite, calc-silicate rock, marble	Bivalve fragments	Post-Ordovician; Paleozoic	J. T. Dutro, Jr., 1983, written communication; J. H. Pojeta, 1984, written communication
16	Greenville 15'	17,27N,10E	Chert member of Peale Formation	*Pseudoalbaillella* sp. G (Ishiga), *Pseudoalbaillella* sp., *Latentifistula* or *Latentibifistula*	Middle Pennsylvanian (Desmoinesian?)	B. Murchey, 1985, written communication
17	Genesee Valley 7-1/2'	15,25N,11E	Chert member of Peale Formation	*Pseudoalbaillella* sp. G (Ishiga), *Latentibifistula* sp, *Latentibifistula* or or *Latentifistula* sp.	Middle Pennsylvanian (Desmoinesian?)	B. Murchey, 1985, written communication
18a	Greenville 15'	5,25N,10E	Limestone block in Arlington Formation	*Hindeodus* sp. indet., *Neostreptognathodus* sp. indet. *Xaniognathus* sp.	late Early Permian (early Leonardian)	K. Denkler, 1985, written communication
18b	Greenville 15'	5,25N,10E	Arlington Formation (matrix of conglomerate	*Neospirifer*? sp., *Omphalotrocus*? sp; large orthotetid brachiopod; dictyoclostid brachiopod (perhaps *Penicularis*)	Early Permian (late Wolfcampian through Leonardian)	J. T. Dutro, Jr., 1986, written communication
19	Greenville 15'	31,28N,9E	Reeve Formation (included in chert member of Peale Formation by Hannah and Moores, 1986)	Schwagerinidae including probable *Schwagerina*, and a possible, very primitive *Parafusilina*	Early Permian (late Wolfcampian to middle Leonardian)	C. A. Ross, 1980, written communication (cited in Hannah and Moores, 1986)
20	Genesee Valley 7-1/2'	15,25N,11E	Arlington Formation	*Neospirifer*? sp.	Early Permian(?)	J. T. Dutro, Jr., 1986, written communication
21	Greenville 15'	4,28N,9E	Reeve Formation	*Pseudoalbaillella scalprata*, *Pseudoalbaillella* sp. cf., *sakarensis*	Early Permian (late Wolfcampian or early Leonardian)	B. Murchey, 1985, written communication
22	Blairsden 15'	12,24N,11E	Reeve Formation	*Timaniella* sp. (abundant), *Chonetinella*? sp.	middle Permian (Wordian)	J. T. Dutro, Jr., 1986, written communication
23	Genesee Valley 7-1/2'	4,25N,11E	Reeve Formation	*Parafusulina mccloudensis*, *Parafusulina* sp. aff. *Parafusulina nosonensis*, *Schwagerina*? sp.	late Early Permian (late Wolf-campian to early Leonardian)	C. H. Stevens, 1989, written communication
24	Genesee Valley 7-1/2'	3,25N,11E	Robinson Formation of McMath (1966)	*Neogondolella* cf. *N. rosenkrantzi*, Neogondolella sp. indet.	early Late Permian (Wordian-Captanean)	K. Denkler and A. Harris, 1986, written communication
25	Sierra City 15'	14,22N,12E	Reeve Formation	*Megousia* sp., *Neospirifer* sp.	late Early to early Late Permian (Leonardian or Wordian)	Durrell and D'Allura, 1977
26	English Mtn. 7-1/2'	35,19N,12E	Limestone block in conglomerate, lower part of Reeve Formation	*Hindeodus excavatus, Neostrep-tognathodus* sp., *Sweetognathus* sp. n. gen. et n. sp.	late Early Permian (middle late Leonardian)	K. Denkler and A. Harris, 1986, written communication
27	Cisco Grove 7-1/2'	7,16N,13E	Reeve Formation	*Chonetinetes* sp., *Megousia* sp. *Neospirifer* sp., *Aviculopecten* sp., *Cyrtorostra*? sp. indet.	mid-Permian (Roadian or Wordian)	M. Gordon, Jr., 1980, written communication; Harwood, 1983

TABLE 1. SELECTED FOSSIL DATA FROM PALEOZOIC AND MESOZOIC ROCKS IN NORTHERN SIERRA TERRANE (continued)
(Locality numbers correspond to those in Fig. 3)

Locality Number	Quadrangle	Section Township, Range	Map Unit	Diagnostic Fauna	Fauna age	Reference
28	Almanor 15'	8,26N, 8E	Hosselkus Limestone	*Epigondolella primitia, Neogondolella navicula*	early Late Triassic (late Karnian through early Norian)	B. R. Wardlaw, 1986, written communication to A. S. Jayko (Jayko, 1988)
29	Genesee Valley 7-1/2'	3,25N,11E	Pit Formation	*Daonella* cf. *D. degeeri, Meginoceras meginae, Protrachyceras*	late Middle Triassic (mid-Ladinian)	N. J. Silberling, 1987, written communication
30	Genesee Valley 7-1/2'	3,25N,11E	Hosselkus Limestone	*Neogondolella polygnathiformis Neogondolella* sp. indet.	Late Triassic (Karnian-early Norian)	A. G. Harris, 1987, written communication
31	Genesee Valley 7-1/2'	3,25N,11E	Swearinger Slate	*Monotis subcircularis*, Gabb	Late Triassic (lower upper Norian)	N. J. Silberling, 1987, oral communication
32	Duncan Peak 7-1/2'	28,16N,13E	Unnamed limestone correlative with Hosselkus Limestone	*Epigondolella* cf. *E. abneptis*	Late Triassic (late Karnian through Norian)	A. G. Harris, 1981, written communication; Harwood, 1983
33	Almanor 15'	20,27N,7E	Unnamed slate, graywacke, and conglomerate	ammonites, *Posidonia*	Possibly Early Jurassic	N. J. Silberling, 1989, written communication to A. S. Jayko
34	English Mtn. 7-1/2'	6,18N,13E	Unnamed volcanic tuff and breccia	*Posidonia* cf. *P. ornati*	late Early through Middle Jurassic	N. J. Silberling, 1983, written communication to D. Stuart Alexander
35	Royal Gorge 7-1/3	34,16N,13E	Sailor Canyon Formation	*Harpoceras* sp., *Orthildaites* cf. *O. orthus, Hildaites* cf. *H. serpentinum, Dactylioceras* cf. *D. tenuicostatum, Protogrammoceras*? sp.	Early Jurassic (late Pleinsbachian to early Toarcian)	R. W. Imlay, 1961, written communication to L. D. Clark, (this is one of several fossil localities in the lower part of the Sailor Canyon Formation, see Imlay, 1968)
36	Homewood 7-1/2'	9,14N,16E	Blackwood Creek Formation (Harwood, 1991), which is probably correlative with lower part of the Sailor Canyon Formation	poorly preserved molds of ammonites	probable Early Jurassic	N. J. Silberling, 1987, oral communication

locally interlayered with Mesozoic rocks in a series of southwest-dipping, east-vergent imbricate thrust slices. Rocks in the thrust slices are overturned toward the northeast, dip moderately to steeply southwest, and rest structurally on a thick, southwest-dipping, upright section of Jurassic volcanic rocks that contains Callovian ammonites in its upper part (McMath, 1958; Imlay, 1961; Christe, 1987). From Mohawk Valley south to the North Fork of the American River (Fig. 2), the Shoo Fly Complex and overlying Paleozoic and Mesozoic rocks form an east-facing, near-vertical, homoclinal section. To the north, the homoclinal section is overlain by imbricate thrust slices of Paleozoic and Mesozoic rocks, and in its southern part the homoclinal section is juxtaposed against rocks of the Lake Tahoe sequence along the Talbot fault zone (Fig. 2). Cross sections through the northern, central, and southern parts of the terrane (Fig. 4) show depositional variations in the arc sequences that have important bearing on their paleogeographic setting.

A restored section across the northern part of the terrane, which accommodates an estimated 30 percent shortening due to Late Jurassic thrusting, is shown in Figure 4A. Upper Devonian volcanic rocks of the Sierra Buttes Formation are about 2.5 km thick on Grizzly Mountain (McMath, 1958) and Keddie Ridge (Hannah, 1980) but they thin dramatically to less than 0.5 km at Mt. Hough (McMath, 1958). A thin fault sliver composed of Sierra Buttes Formation also occurs in the Butt Valley section (Jayko, 1988; this volume), but its depositional thickness is indeterminate.

Because the overlying Taylor Formation maintains a relatively constant thickness across the northern part of the terrane, it appears that the western extent of the Sierra Buttes Formation was restricted by an outboard ridge of the Shoo Fly Complex in Famennian time. Whether the ridge of Shoo Fly was a fault-bounded horst or a simple promontory is unknown at this time, and available evidence fits either interpretation. McMath (1958) and Hanson and Schweickert (1986) concluded that the Upper Devonian Grizzly Formation accumulated in relatively deep water, certainly below storm-wave base, as submarine mass-flow deposits. Wedges of coarse breccia composed of Shoo Fly debris, which might have collected in a trough along a fault-bounded ridge have not been reported in the Grizzly or Sierra Buttes Formation near Grizzly Peak (Fig. 2). However, Hanson and Schweickert (1986) reported a thick wedge of coarse-grained polymict conglomerate derived from the Shoo Fly at Sierra Buttes (Fig. 2). They also found a large slump block of shallow-water sandstone and bioclastic limestone, which contains shallow-water gastropod and coral faunas of Frasnian age, in the Famennian deep-water deposits of the Grizzly above the polymict conglomerate. These relations suggest that the outboard ridge of Shoo Fly extended for at least 40 km in a roughly north-south direction, and also that it may have been more prominent in its southern extent during Famennian time, shedding coarse polymict conglomerate and slump blocks toward the arc. The strike of this outboard ridge of Shoo Fly cannot be determined from existing data.

The three lithofacies of the Permian volcanic sequence also show pronounced east-west variations in the northern part of the terrane (Fig. 4A). Chert-rich pebbly mudstone, conglomerate, and coarse breccia dominate the lower part of the Arlington Formation at Mt. Hough. The upper part of the Arlington at this locality contains volcaniclastic sandstone, siltstone, and a lens of feldspar–phyric tuff breccia correlated with the Reeve Formation by McMath (1958). Significantly thinner accumulations of chert-rich conglomerate, assigned here to the Arlington Formation, also occur on Keddie Ridge below volcanic rocks of the Reeve Formation (see Hannah and Moores, 1986) and on Peale Ridge below an eastward and southward thickening wedge of pillow basalt of the Goodhue Formation (Harwood, 1991). On Keddie Ridge, Lower Permian epiclastic rocks of the Arlington Formation rest locally on the Upper Devonian and/or Lower Mississippian Taylor Formation (Harwood, 1991), indicating that this area experienced locally prominent and differential uplift prior to the late Early Permian. The pattern of deep erosion on Keddie Ridge, coupled with thick epiclastic deposits on Mt. Hough, and the southeastward-thickening wedge of Goodhue pillow basalt, indicates the region was differentially uplifted and eroded prior to the late Early Permian (late Wolfcampian and/or early Leonardian). Harwood (1988) concluded that this erosion may have occurred on uplifted blocks during local thermal expansion and extension of the crust related to initial phases of late Early Permian volcanism; however, uplift may have occurred earlier during the Late Pennsylvanian.

Stratigraphic variations along strike in the east-facing homoclinal section of the Paleozoic arc-volcanic rocks are shown in Figure 4B. Depositional patterns here were profoundly influenced by the Bowman Lake batholith, which has been interpreted as a local magma chamber for Upper Devonian submarine extrusive and hypabyssal rocks of the Sierra Buttes Formation (Schweickert, 1981; Hanson and others, 1988). Asymmetric uplift occurred over the batholith and was accompanied by north-down normal faulting at its north end (Hanson and Schweickert, 1986). Uplift and normal faulting began after the mid-Famennian because fossils of that age occur in the Grizzly and Sierra Buttes Formations in the Culbertson Lake, Middle Yuba, and Sierra Buttes sections (Fig. 4B) (Table 1, no. 3, 4, 5), but the Taylor Formation is missing over the batholith. The lower member of the Peale Formation, which contains late Kinderhookian conodonts (Table 1, no. 8) at Frazier Falls (Fig. 2), is also missing over the batholith, indicating either that uplift continued or the area was highstanding into the Early Mississippian. The area over the batholith is overlapped by the chert member of the Peale Formation, but unfortunately no fossils have been recovered from the chert member in that area. To the north, the lower part of the chert member contains radiolarians indicative of an Early Mississippian (Osagean to early Meramecian) age, whereas radiolarians of latest Late Mississippian to Early Pennsylvanian age occur near the top of the chert member at Monumental Ridge (Fig. 4B; Table 1, no. 14) and at scattered localities north of Frazier Falls (Table 1, nos. 9 through 13). In sections where the base and top

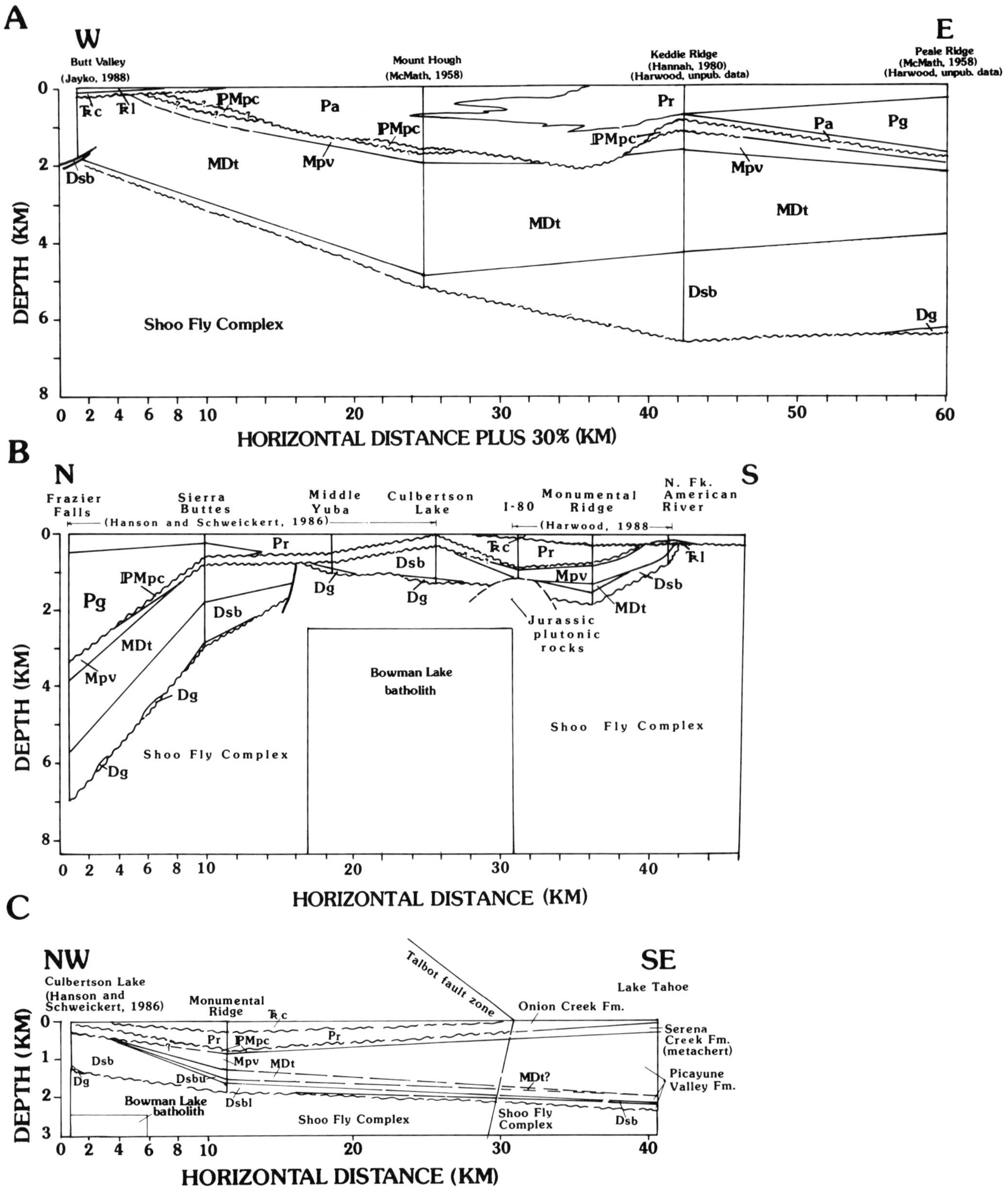
A
W
E
Butt Valley
(Jayko, 1988)
Mount Hough
(McMath, 1958)
Keddie Ridge
(Hannah, 1980)
(Harwood, unpub. data)
Peale Ridge
(McMath, 1958)
(Harwood, unpub. data)
TRc
TRl
PMpc
Pa
Pr
Pg
Mpv
MDt
Dsb
Dg
Shoo Fly Complex
DEPTH (KM)
HORIZONTAL DISTANCE PLUS 30% (KM)
B
N
S
Frazier Falls
Sierra Buttes
Middle Yuba
Culbertson Lake
I-80
Monumental Ridge
N. Fk. American River
(Hanson and Schweickert, 1986)
(Harwood, 1988)
Pr
PMpc
Pg
MDt
Mpv
Dsb
Dg
TRc
TRl
Jurassic plutonic rocks
Bowman Lake batholith
Shoo Fly Complex
DEPTH (KM)
HORIZONTAL DISTANCE (KM)
C
NW
SE
Culbertson Lake
(Hanson and Schweickert, 1986)
Monumental Ridge
Talbot fault zone
Onion Creek Fm.
Lake Tahoe
Serena Creek Fm. (metachert)
Picayune Valley Fm.
TRc
Pr
PMpc
Mpv
MDt
MDt?
Dsb
Dsbu
Dsbl
Dg
Bowman Lake batholith
Shoo Fly Complex
DEPTH (KM)
HORIZONTAL DISTANCE (KM)

of the chert member are dated, pelagic sedimentation rates, uncorrected for compaction, are about 1 m/m.y. This rate, combined with the thickness of the chert member in the Culbertson Lake and Middle Yuba sections, suggests that the Bowman Lake batholith had cooled and the overlying region had subsided by late Osagean time.

Prior to overlap by the chert member, however, the uplifted area over the batholith profoundly influenced depositional patterns in the Upper Devonian and Lower Mississippian rocks. The intra-arc graben or half-graben (Fig. 4B) that formed north of the batholith collected a northward-thickening wedge of typically coarse-grained submarine pyroclastic debris, predominantly felsic in the Sierra Buttes Formation, but andesitic in the Taylor Formation. Relatively fine-grained tuffaceous debris-flow deposits, locally intercalated with phosphate-streaked black chert, occur in the Sierra Buttes Formation associated with blocky tuff breccia, intrusive hyaloclastic breccia, hypabyssal intrusive rocks, air-fall tuff, and tuffaceous turbidites (Hanson and Schweickert, 1986). Andesitic tuff breccia with blocks 1.5 m in size, pyroxene-phyric pillowed lava flows, and tuff with a maximum thickness of about 2.5 km, make up the Taylor Formation north of the batholith (McMath, 1958; Durrell and D'Allura, 1977). Keratophyric tuff, lapilli tuff, and sparse flows, all characterized by alkali-feldspar phenocrysts, compose the northern exposures of the lower member of the Peale Formation (McMath, 1958). Lenses of late Kinderhookian bioclastic limestone (Table 1, nos. 7, 8) occur near the top of the lower member of the Peale and mark the end of early volcanism and a brief moment of shallow-water carbonate deposition before the Late Devonian to early Mississippian arc subsided to basinal depths.

South of Bowman Lake batholith, Upper Devonian and Lower Mississippian rocks are significantly thinner and finer grained than their northern counterparts (Harwood, 1983, 1988). Furthermore, volcaniclastic units in the south, except for a thin wedge of Taylor Formation, contain interbedded siliciclastic rocks not found in deposits of the intra-arc graben north of the batholith. The lower part of the Sierra Buttes Formation on Monumental Ridge (Fig. 4B) consists of 500 m of coarse felsic tuff breccia with abundant intraclasts of phosphate-streaked black chert. The tuff breccia debris-flow deposits are gradationally overlain by about 500 m of thin-bedded, locally cross-laminated tuffaceous slate and siltstone that contain scattered lenses of chert-quartzite–granule conglomerate. This unit of mixed volcaniclastic and siliciclastic debris has no counterpart north of the Bowman Lake batholith. A discontinuous unit of andesitic crystal–lithic tuffaceous turbidites, correlated with the Taylor Formation, overlies the Sierra Buttes Formation and apparently represents distal deposits of the blocky andesitic tuff breccia and flows found in the Taylor to the north. The lower member of the Peale Formation in the southern part of the terrane contains some alkali-feldspar phyric lapilli tuff in its lower part that grades up into interbedded tuffaceous siltstone and black slate that contains abundant lenses of chert-rich granule- and pebble-conglomerate, including well-rounded quartzite clasts derived from the Shoo Fly Complex. The conglomerate lenses in the lower member of the Peale increase in abundance, and the clast size increases to the southeast. These variations in the conglomerate lenses, coupled with the absence of chert-quartz–rich debris in the lower member of the Peale north of the Bowman Lake batholith, indicate a southeastern source for the siliciclastic debris.

It appears that the uplifted area over the Bowman Lake batholith formed a major, but not complete, divide in Late Devonian and Early Mississippian depositional patterns. Coarse-grained volcanic debris, abundant flows, vent-proximal hyaloclastite breccias and hypabyssal intrusive rocks filled the intra-arc graben to the north, while fine-grained volcaniclastic and quartzose-clastic debris collected in an arc-marginal basin to the south and southeast.

Permian volcanic rocks in the homoclinal section (Fig. 4B) show depositional patterns similar to those in the older volcanic rocks. The most striking similarity is the northward-thickening wedge of Goodhue Formation that coincides with the underlying half graben north of the Bowman Lake batholith. It appears that tectonic heredity influenced the location of the "Goodhue trough," but the controlling mechanism is not obvious. The normal fault bounding the underlying half-graben apparently was not reactivated in the Permian, because the chert member of the Peale Formation is not offset and the Goodhue appears to interfinger with the Reeve Formation at its southern limit (Fig. 4B) (Hanson and Schweickert, 1986; Schweickert and Hanson, unpublished map, Sierra City Quadrangle). The Reeve Formation, like the underlying Upper Devonian and Lower Mississippian volcanic rocks, thins and is significantly finer grained south of Interstate Highway 80, but the transition is not exposed. Coarse-grained, chert-rich conglomerate, typical of the Arlington Formation on Mt. Hough, occurs in the lower part of the Reeve east of Bowman Lake (Fig. 3) (Schweickert, 1981). However, *Nereites* assemblage trace fossils, typical of deep-water environments, occur in fine-grained, thin-bedded tuffaceous slate and siltstone of the Reeve south of Interstate Highway 80, suggesting relatively deep-water deposition prevailed in the southern part of the terrane in the mid-Permian.

The Lake Tahoe sequence provides additional data on distal

Figure 4. Generalized restored sections through the northern Sierra terrane. A, Restored section through Late Jurassic imbricate thrust slices north of Mohawk Valley. B, North-south section from Frazier Falls to North Fork American River. C, Schematic, unrestored section across Talbot fault zone showing probable correlations between arc-volcanic rocks and siliciclastic rocks of Lake Tahoe sequence. Datum is base of Jurassic rocks, except in the eastern part of 3A where datum is the top of Paleozoic rocks. Unit symbols: Dg = Grizzly Formation, Dsb = Sierra Buttes Formation; Dsbl = lower member, Sierra Buttes Formation; Dsbu = upper member, Sierra Buttes Formation; MDt = Taylor Formation; Mpv = volcanic member, Peale Formation; PMpc = chert member, Peale Formation; Pa = Arlington Formation; Pg = Goodhue Formation; Pr = Reeve Formation; TRc = conglomerate and tuff; TRl = limestone.

deposits of the Paleozoic volcanic rocks. East of the Talbot fault zone (Fig. 2), black slate, quartzite, and chert-rich conglomerate of the Picayune Valley Formation (Harwood, 1991) rest unconformably on the Shoo Fly Complex. The western and stratigraphically lowest part of the formation contains scattered interbeds and mappable accumulations of felsic volcaniclastic rocks that are lithologically identical to the Sierra Buttes Formation (Fig. 4C). Most of the volcaniclastic rocks are fine-grained, quartz-bearing debris-flow deposits, as much as 1.5 m thick, that are interlayered with very fine-grained, aquagene, felsic tuff. Intraclasts of phosphate-streaked black chert, like those in the lower member of the Sierra Buttes Formation on Monumental Ridge, are widely scattered in the fine-grained felsic debris-flow deposits. Sparse lenses of andesitic tuff breccia and andesitic tuffaceous turbidites are interbedded with the felsic debris-flow deposits and locally occur as interbeds in the black pelite and siliciclastic rocks. The andesitic volcanic rocks may be correlative with more proximal deposits in either the Sierra Buttes or Taylor Formation.

No volcanic rocks have been found in the upper part of the Picayune Valley Formation, which contains abundant lenses of chert-rich conglomerate in a matrix of black slate. White, gray, and black chert clasts compose most of the conglomerate and are associated with lesser amounts of quartzite, micaceous quartzite, and black pelite clasts. The clasts are commonly rounded and increase in size from granules and pebbles in the west to cobbles in the eastern exposures, implying an eastern or southeastern source for the siliciclastic debris. Lenses or very large blocks of coarsely recrystallized limestone occur in the upper part of the formation. Silicified echinoderm fragments, extracted from the limestone in a search for conodonts, indicate a Silurian or Devonian age (Table 1, no. 6) and suggest, at best, a middle Paleozoic age for the Picayune Valley Formation. A sample of felsic volcaniclastic rocks from the lower part of the formation was crushed and separated for zircons, but none were found.

The Picayune Valley Formation is conformably overlain by metachert of the Serena Creek Formation (Harwood, 1991), which is tentatively correlated lithologically with the chert member of the Peale Formation (Fig. 4C). No fossils have been recovered from the thoroughly recrystallized chert, and the correlation is based on lithologic similarity and its stratigraphic position above the Picayune Valley Formation, which contains interbeds of felsic volcanic rocks correlative with the Sierra Buttes in its lower part and probable middle Paleozoic fossil scraps in its upper part.

The tentative lithologic correlation of the Serena Creek Formation with the chert member of the Peale is strengthened by the occurrence of Paleozoic bivalve fragments in the overlying Onion Creek Formation (Harwood, 1991) (Table 1, no. 15). The Onion Creek Formation consists of quartzite, calc-silicate rock, and limestone, all thoroughly recrystallized by Mesozoic plutonic rocks. Coarse-ribbed bivalve fragments, which occur a few tens of meters above the base of the formation, are indicative of a post-Ordovician Paleozoic age; their exact age is unknown, though, and the lithologic assemblage has no obvious counterpart in the Paleozoic volcanic sequence west of the Talbot fault zone. At present, the Onion Creek Formation is tentatively interpreted as an eastward-thickening unit of upper middle to upper Paleozoic quartzose-clastic and carbonate rocks that formed during the period marked by the unconformity at the base of the Reeve Formation in the arc terrane (Fig. 4C).

CORRELATIONS WITH THE HAVALLAH SEQUENCE

The Havallah sequence and its correlative, the Schoonover sequence, range in age from Late Devonian through early Late Permian (Miller and others, 1984; Stewart and others, 1986; Tomlinson, 1988; Murchey, this volume; Whiteford, this volume), and thus span the episodes of Paleozoic volcanism in the northern Sierra terrane. Thrust faults repeat Devonian to lower Upper Permian rocks of the Havallah sequence and interleave different lithofacies into complex structural stacks, some more than 5,000 m thick (Stewart and others, 1984). In spite of the structural complexity, original stratigraphic sequences have been reconstructed for local parts of the Golconda allochthon that provide important information on the evolution of the Havallah basin and its possible paleogeographic relations to the northern Sierra terrane.

Reconstructed sections of the Havallah sequence are compared to a composite section of the northern Sierra terrane in Figure 5. These generalized sections show lithologic variations through time and ignore the thickness of units. For example, thick accumulations of Paleozoic volcanic rocks, which dominate the map pattern of the northern Sierra terrane (Fig. 2), appear in their proper perspective on the time-lithology plot as relatively short-lived volcanic episodes widely separated in time. During much of the late Paleozoic, in fact, the so-called "northern Sierra arc" was actually a basin that collected siliceous pelagic deposits now represented by the thin but widespread chert member of the Peale Formation. Radiolarian assemblages in the chert member correlate directly with Mississippian and Pennsylvanian chert and argillite in the Havallah sequence (Fig. 6). In addition, radiolarians from chert interbedded in the Permian volcanic rocks in the northern Sierra terrane correlate directly with Early Permian radiolarians in the Havallah sequence. In spite of fewer radiolarian assemblages and lower diversity in the northern Sierra Nevada, which undoubtedly reflect poorer radiolarian preservation in Sierra cherts, there can be no doubt that the Havallah sequence and the Sierran Paleozoic arc sequences are synchronous.

Reconstructed sections of the Havallah sequence (Fig. 5) show considerable lithologic variation at any given time, reflecting different depositional patterns in the basin. All of these Havallah sections, however, show a similar lithologic progression through time—from early basalt to dominant chert-argillite, with increasing limestone turbidites—that reflects the influence of regional tectonic patterns on sedimentation. The lower part of the Havallah sequence, at least locally, contains pillow basalt and basaltic greenstone associated with manganiferous chert, jasper,

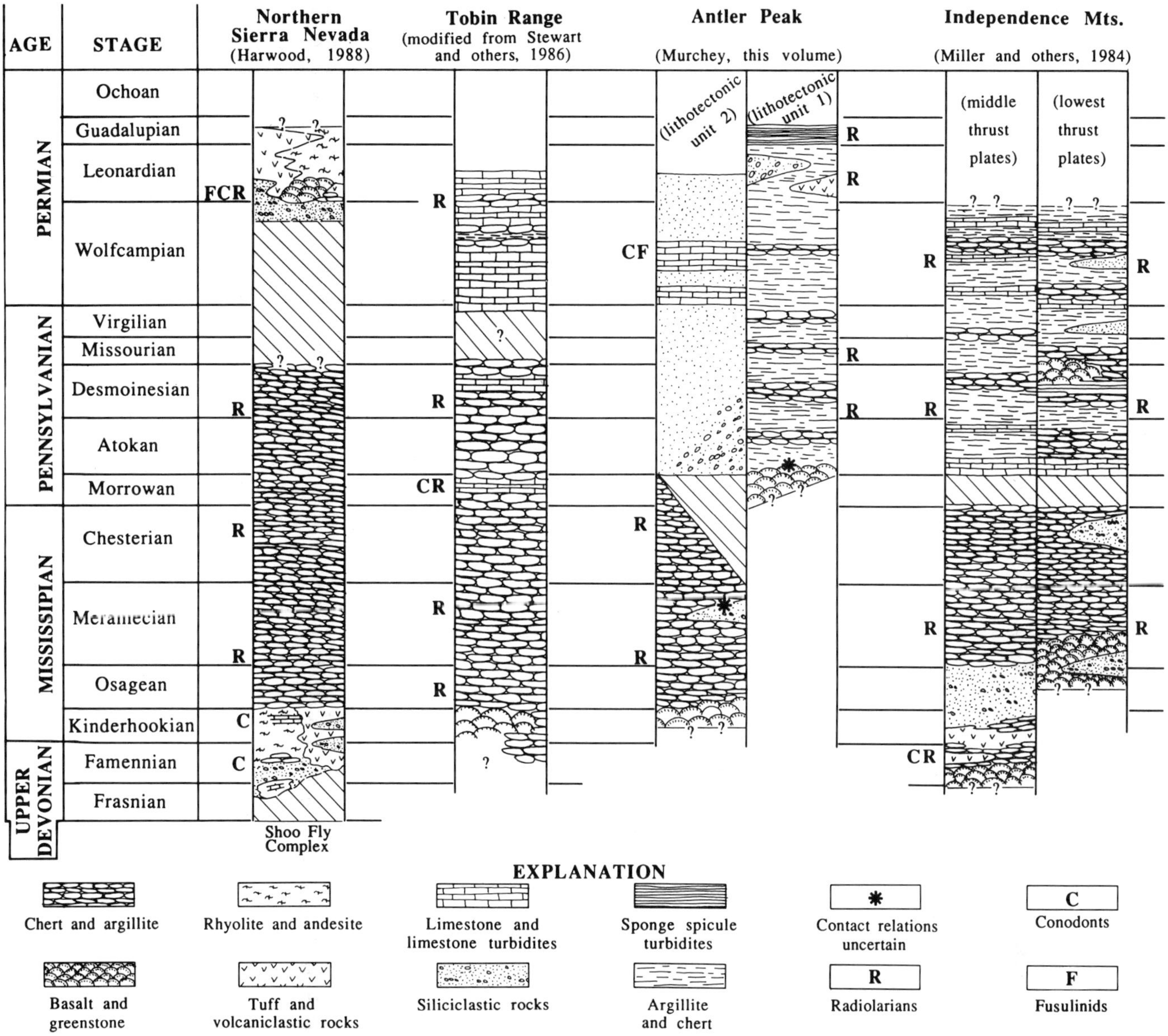

Figure 5. Generalized composite sections comparing time-lithologic data from the northern Sierra terrane and selected areas of the Havallah sequence.

and local concentrations of massive sulfides (Snyder, 1977; Snyder and Brueckner, 1983; Miller and others, 1984). In the Tobin Range, basalt underlies Lower Mississippian (Osagean) chert, whereas at Antler Peak the basalt is undated but is shown to be older than Osagean or Meramecian, based on radiolarians in overlying rocks (Murchey, this volume). In the Independence Mountains, basaltic greenstone at the base of the middle thrust plates underlies Famennian chert that grades up into tuff and volcaniclastic rocks of Kinderhookian age. The lower thrust plates in the Independence Mountains contain a relatively thick section of basaltic greenstone overlain by Meramecian chert. Boulder beds in this greenstone contain clasts of chert, limestone, and quartzite derived from the Valmy Formation in the underlying Roberts Mountains allochthon (Miller and others, 1984).

Pillow basalt, basaltic greenstone, manganiferous chert, and exhalative massive sulfides have been interpreted as rift deposits produced at an oceanic spreading center(s) during extension of the Havallah basin (Snyder, 1977; Snyder and Brueckner, 1983; Miller and others, 1984). If this interpretation is correct, extension occurred in the basin during Late Devonian and Early Mississippian time (Famennian through early Meramecian), and coincided, in part, with early arc volcanism and extension in the northern Sierra Nevada and with emplacement of the Roberts Mountains allochthon.

Other than the Kinderhookian volcaniclastic rocks in the Independence Mountains (Miller and others, 1984), no arc-volcanic debris has been reported in the Havallah sequence that might be related to the early episode of volcanism in the northern

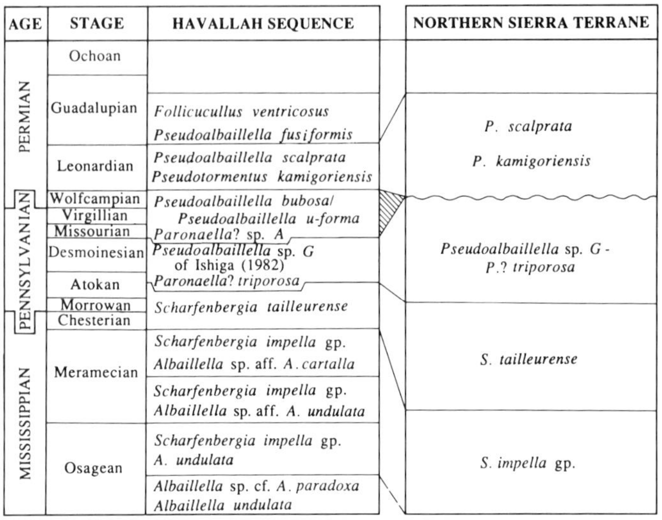

Figure 6. Correlative radiolarian assemblages in the Havallah sequence and the northern Sierra terrane.

Sierra Nevada. No siliciclastic rocks occur with the Famennian chert in the Independence Mountains, but locally, coarse-grained quartzose-clastic turbidites interfinger with and overlie the Kinderhookian volcaniclastic rocks (Miller and others, 1984). In the older Sierran arc sequence, on the other hand, sparse lenses of sedimentary-lithic rock debris appear first as arc-basement–derived deposits in the Grizzly Formation and then as eastward-thickening deposits in distal parts of the mid-Famennian Sierra Buttes Formation. Relatively abundant lenses of chert- and quartzite-pebble conglomerate are interbedded with tuff and volcaniclastic rocks in the lower member of the Peale Formation of Kinderhookian age, and black chert-quartz–rich rocks with only minor volcaniclastic rocks dominate the presumed Upper Devonian and Lower Mississippian Picayune Valley Formation in the Lake Tahoe sequence. From the essentially two-dimensional picture we have of the northern Sierra terrane, there is a well-defined and progressive south and southeastward decrease in volcaniclastic rocks and a parallel increase in quartzose-clastic debris, away from the Late Devonian and Early Mississippian volcanic centers. The original spatial distribution of middle Paleozoic lithofacies in the Havallah sequence is unknown.

From the Osagean and Meramecian through the mid-Permian, large segments of the Havallah sequence are dominated by interbedded chert and argillite, which reflect background sedimentation in the basin. Pennsylvanian and Lower Permian quartzite and conglomerate are interbedded with limestone turbidites in the Antler Peak area (LU-2, Murchey, this volume), whereas chert-quartzite-lithic sandstones occur sporadically above the Kinderhookian quartzose-lithic turbidites in the Independence Mountains (Miller and others, 1984). Limestone turbidites, which appear first in the late Meramecian–earliest Morrowan(?) chert-argillite sequence in the Independence Mountains and in Lower to Middle Pennsylvanian rocks in the Tobin Range (Stewart and others, 1986), make up a progressively larger part of the Havallah sequence throughout the Pennsylvanian and Early Permian. In contrast, from Osagean through Desmoinesian(?) time at least, the raw materials of chert, undiluted by coarse-grained quartzose or carbonate debris, were deposited over the submerged Late Devonian and Early Mississippian arc in the northern Sierra terrane.

Sparse Lower Permian volcaniclastic rocks occur with chert and argillite in the Antler Peak area (Murchey, this volume), but none have been reported from the Independence Mountains where thin, widely scattered lenses of greenstone and/or mafic sills occur in Middle Pennsylvanian to Lower Permian chert-argillite (Miller and others, 1984). Thrust slices of the Golconda allochthon in the New Pass and Shoshone Ranges, the southern part of the Toiyabe Range, and small areas to the south contain significant concentrations of Pennsylvanian(?) and Permian volcaniclastic debris (Speed, 1977a, b; Tomlinson and Wright, 1986; Tomlinson, 1988). In the Toiyabe Range, sparse boulder beds in volcaniclastic rocks (Tomlinson and Wright, 1986) contain granitoid clasts that give a late Early Permian (267 Ma) crystallization age with an inherited xenocrystic zircon component of about 2.2 Ga, which is the upper-intercept age of detrital zircon from the Shoo Fly Complex (Girty and Wardlaw, 1985). In areas of the Golconda allochthon where Permian volcaniclastic rocks are reasonably well dated, they range in age from Leonardian to early Guadalupian, and thus are coeval with the Permian volcanic sequence in the northern Sierra terrane.

An erosional unconformity spans the Late Pennsylvanian and early Permian (Desmoinesian(?) through early Wolfcampian) and separates Permian volcanic rocks from older strata in the northern Sierra terrane (Fig. 3). Ketner (1977) documented an erosional unconformity of equivalent time in rocks of the miogeocline and Antler overlap sequence, which he suggested was related to regional uplift along the shelf edge. Stone and Stevens (1988) also inferred Early or Middle Pennsylvanian to late Early Permian tectonism, including southeast-trending sinistral transform faulting along the continental margin, from deposition patterns along the shelf-slope break. Uplift and erosion are recorded locally in the Schoonover sequence where uppermost Pennsylvanian chert-argillite and interbedded limestone turbidites rest on rocks as old as Osagean (Miller and others, 1984). These relations suggest that regional tectonism, uplift, and erosion broadly affected the western continental margin from the outboard arc through at least local parts of the Havallah basin and onto the miogeocline from Early or Middle Pennsylvanian to late Early Permian time.

PALEOGEOGRAPHY

Several lines of evidence suggest that the northern Sierra terrane evolved within depositional proximity of the Havallah basin. The paleontologic data discussed above establish synchronous deposition in the northern Sierra terrane and the Havallah basin from the Late Devonian through the early Late Permian.

Although this is a necessary condition for correlation, evidence for paleogeographic proximity comes from stratigraphic and tectonic data.

Intermixed volcaniclastic and basinal deposits in the Sierran arc terrane and the Havallah basin indicate periodic shifts in depositional patterns in both terranes. The influx of Kinderhookian volcaniclastic debris in the Schoonover sequence (Miller and others, 1984) occurred during the waning stages of Sierran Late Devonian and Early Mississippian arc volcanism when at least part of the arc over the Bowman Lake batholith was uplifted and being eroded (Fig. 4B). Similarly, the influx of mid-Permian volcaniclastic debris in the Havallah sequence was coeval with late Wolfcampian or early Leonardian to early Guadalupian volcanism in the northern Sierra terrane. Between these episodes of arc volcanism, the Sierran arc terrane received a blanket of Lower Mississippian to Middle Pennsylvanian (Desmoinesian?) deep-water ratiolarian ooze that is directly correlative with the dominant chert-argillite of the Havallah sequence.

Siliciclastic debris, apparently derived from the Antler orogenic belt, establishes an independent lithologic tie between the Sierran arc volcanic rocks and the Havallah sequence. Eastward-thickening and -coarsening lenses of conglomerate are interbedded with Famennian to Kinderhookian distal volcanic rocks west of the Talbot fault zone. An eastward-thickening wedge of chert-quartz–rich rocks, with relatively thin lenses of volcanic rocks that are lithologically correlative with the Sierra Buttes Formation and possibly the Taylor Formation, characterize the middle Paleozoic Picayune Valley Formation east of the Talbot fault zone and suggest deposition in an arc-marginal basin outboard of the Antler orogenic belt. Kinderhookian volcaniclastic rocks in the Schoonover sequence are interbedded with and overlain by chert-rich turbidites derived from the Antler highland (Miller and others, 1984). Chert of Osagean and Meramecian age disconformably overlies the chert-rich turbidites. In the Antler Peak area, Upper Mississippian chert-lithic fragment-rich turbidites are interlayered with chert and argillite, which is unconformably overlain by Pennsylvanian and Lower Permian quartzose rocks and limestone turbidites (Murchey, this volume). The quartz–rich rocks, which were not derived from either the Sierran arc terrane or the Havallah basin, yet occur in both, establish important lithologic ties between the northern Sierra terrane and the Havallah sequence and connect both the arc and basin to the Antler orogen. Similar conclusions, based on other lithologic evidence, were reached by Varga (1982), Coles and Snyder (1985), and Hanson and Schweickert (1986).

Late Devonian and Early Mississippian arc-volcanism in the northern Sierra terrane occurred in an extensional regime marked by local uplift over a magmatic center (Bowman Lake batholith) and rapid deposition in an intra-arc basin to the north. Extensional tectonism also controlled early deposition in the Havallah basin where pillow basalt, manganiferous chert, and local massive sulfide deposits form the lower parts of various thrust slices of the Golconda allochthon (Snyder and Bruechner, 1983; Miller and others, 1984). Earliest rift deposits in the basin are Famennian in age and coincide with the onset of arc volcanism in the northern Sierra terrane, but this period of rifting and basaltic volcanism lasted into the Meramecian in the basin and apparently extended to the western margin of the Antler orogenic belt during that time (Miller and others, 1984; Little, 1987). The evidence for Late Devonian to early Late Mississippian extension in the Sierran arc terrane and the Havallah basin seems compelling, and it appears to conflict with the hypothesis of an exotic-arc collision for the Antler orogeny.

Although sparse Meramecian limestone turbidites occur in the Havallah sequence, indicating local transport of shelf carbonate debris across the Antler orogen at that time, carbonate sediment did not constitute a significant volume of the basin deposits until the Pennsylvanian, and it increased in abundance into the Early Permian. No limestone turbidites have been found in the northern Sierra terrane, suggesting that the submerged arc terrane either lay beyond the limit of carbonate transport or below the carbonate lysocline. Local uplift in the basin or along the continental margin may have blocked the westward transport of carbonate sediment, or possibly the width of the basin precluded deposition of carbonate sediment in the Sierran arc terrane.

The brief but intense episode of Permian volcanism in the northern Sierra terrane apparently occurred in an extensional or transtensional environment. Locally thick deposits of breccia, pebbly mudstone, conglomerate, and volcaniclastic sandstone characterize the Arlington Formation, which is the outboard, western facies of the Permian volcanic sequence (Fig. 4A). Contemporaneous pillow basalt of the Goodhue Formation was deposited in a restricted basin in the central part of the terrane. The Reeve Formation at Keddie Ridge locally rests disconformably on rocks as old as the Taylor Formation. These depositional patterns indicate pronounced local relief in the northern Sierra terrane during mid-Permian time, but fault-bounded horsts and grabens have not been mapped.

CONCLUSIONS

Biostratigraphic data establish a direct correlation between Upper Devonian (Famennian) through lower Upper Permian (early Guadalupian) rocks in the northern Sierra terrane and the Havallah sequence. Volcaniclastic rocks were deposited in the Havallah basin during synchronous episodes of Late Devonian to Early Mississippian and mid-Permian volcanism in the Sierran arc terrane. Between these episodes of volcanism, the Sierran arc terrane was part of the Havallah basin and collected correlative siliceous pelagic sediment for a period of at least 45 m.y., between the Early Mississippian and Middle Pennsylvanian (Desmoinesian?). These synchronous deposits establish important lithologic ties between the Sierran arc terrane and the Havallah basin and indicate that they shared parallel tectonic histories for about 110 m.y.

During the Late Devonian and Early Mississippian, lower Paleozoic rocks of the Roberts Mountains allochthon were thrust over coeval shelf deposits along the continental margin. Chert-quartz-rich debris provides an independent lithologic tie between early deposits of the Sierran arc terrane and the Havallah basin, and it also links those tectonic provinces spatially and tectonically to the Antler orogen.

There is little doubt that the Upper Devonian and Lower Mississippian volcanic arc in the northern Sierra terrane was active during emplacement of the Roberts Mountains allochthon and that this arc received chert-quartz-rich debris from an eastern source, probably the Antler highland. It appears that the Upper Devonian and Lower Mississippian Sierran volcanic rocks represent the elusive "Antler arc." However, that arc volcanism occurred in an extensional regime that was accompanied by Famennian through early Meramecian extension in parts of the Havallah basin. This widespread region of extensional tectonism, which stretched progressively from the Sierran arc through the Havallah basin to the western margin of the Antler orogenic belt, is difficult to explain in terms of arc-collision models for the Antler orogeny and suggests that this arc formed along the active North American margin.

ACKNOWLEDGMENTS

The efforts of K. Denkler, J. T. Dutro, Jr., M. Gordon, Jr., A. G. Harris, D. L. Jones, J. H. Pojeta, N. J. Silberling, C. H. Stevens, and B. R. Wardlaw to identify commonly deformed and fragmentary fossil debris are gratefully acknowledged. The manuscript was significantly improved by comments from E. L. Miller, M. M. Miller, R. A. Schweickert, J. H. Stewart, and A. H. Tomlinson.

REFERENCES CITED

Burchfiel, B. C., and Davis, G. A., 1972, Structural framework and evolution of the southern part of the Cordilleran orogen, western United States: American Journal of Science, v. 272, p. 97–118.

—— , 1975, Nature and controls of Cordilleran orogenesis, western United States; Extensions of an earlier synthesis: American Journal of Science, v. 275A, p. 363–396.

Burchfiel, B. C., and Royden, L. H., 1989, The Antler orogen; An Apennine-East Carpathian type orogen: Geological Society of America Abstracts with Programs, v. 21, p. 61.

Christe, G., 1987, The geology and petrology of the eastern Mesozoic belt, northern Sierra, California [M.S. thesis]: Burlington, University of Vermont, 350 p.

Coles, K. S., and Snyder, W. S., 1985, Significance of lower and middle Paleozoic phosphatic chert in the Toquima Range, central Nevada: Geology, v. 13, p. 573–576.

D'Allura, J. A., 1977, Stratigraphy, structure, petrology, and regional correlations of metamorphosed upper Paleozoic volcanic rocks in portions of Plumas, Sierra, and Nevada Counties, California [Ph.D. thesis]: Davis, University of California, 338 p.

Davis, D. A., 1989, The Paleozoic-Mesozoic unconformity of the northern Sierra Nevada (NSN), California, and its significance: Geological Society of America Abstracts with Programs, v. 21, p. 71–72.

Dickinson, W. R., 1977, Paleozoic plate tectonics and the evolution of the Cordilleran continental margin, *in* Stewart, J. H., and others, eds., Paleozoic paleogeography of the western United States: Pacific Section, Society of Economic Paleontologists and Mineralogists Pacific Coast Paleogeography Symposium 1, p. 137–156.

Dickinson, W. R., Harbaugh, D., Saller, A. H., Heller, P. L. and Snyder, W. S., 1983, Detrital modes of upper Paleozoic sandstones derived from Antler Orogen in Nevada; Implications for nature of Antler orogeny: American Journal of Science, v. 283, p. 481–509.

Durrell, C., and D'Allura, J. A., 1977, Upper Paleozoic section in eastern Plumas and Sierra Counties, Sierra Nevada, California: Geological Society of America Bulletin, v. 88, p. 844–852.

Girty, G. H., 1983, The Culbertson Lake allochthon, a newly identified structural unit within the Shoo Fly Complex; Sedimentologic, stratigraphic, and structural evidence for extension of the Antler orogenic belt to the northern Sierra Nevada, California [Ph.D. thesis]: New York, Columbia University, 155 p.

Girty, G. H., and Wardlaw, M. S., 1985, Petrology and provenance of pre–Late Devonian sandstones, Shoo Fly Complex, northern Sierra Nevada, California: Geological Society of America Bulletin, v. 97, p. 516–521.

Hannah, J. L., 1980, Stratigraphy, petrology, paleomagnetism, and tectonics of Paleozoic arc complexes, northern Sierra Nevada, California [Ph.D. thesis]: Davis, University of California, 323 p.

Hannah, J. L., and Moores, E. M., 1986, Age relationships and depositional environments of Paleozoic strata, northern Sierra Nevada, California: Geological Society of America Bulletin, v. 97, p. 787–797.

Hanson, R. E., and Schweickert, R. A., 1986, Stratigraphy of mid-Paleozoic island-arc rocks in part of the northern Sierra Nevada, Sierra and Nevada Counties, California: Geological Society of America Bulletin, v. 97, p. 986–998.

Harwood, D. S., 1983, Stratigraphy of upper Paleozoic volcanic rocks and regional unconformities in part of the northern Sierra terrane: Geological Society of America Bulletin, v. 94, p. 413–422.

—— , 1988, Tectonism and metamorphism in the northern Sierra terrane, northern California, *in* Ernst, W. G., ed., Metamorphism and crustal evolution, western conterminous United States; Rubey Volume 7: Englewood Cliffs, New Jersey, Prentice-Hall, p. 764–788.

—— , 1991, Stratigraphy of Paleozoic and lower Mesozoic rocks in the northern Sierra terrane: U.S. Geological Survey Bulletin 1957 (in press).

Harwood, D. S., Jayko, A. S., Harris, A. G., Silberling, N. J., and Stevens, C. H., 1988, Permian-Triassic rocks slivered between the Shoo Fly Complex and the Feather River periodite belt, northern Sierra Nevada, California: Geological Society of America Abstracts with Programs, v. 20, p. 167–168.

Imlay, R. W., 1961, Late Jurassic ammonites from the western Sierra Nevada, California: U.S. Geological Survey Professional Paper 374-D, p. D-1-D-30.

—— , 1968, Lower Jurassic (Pliensbachian and Toarcian) ammonites from eastern Oregon and California: U.S. Geological Survey Professional Paper 593-C, p. C-1-C-51.

Jayko, A. S., 1988, Paleozoic and Mesozoic rocks of the Almanor 15′ Quadrangle, Plumas County, California: U.S. Geological Survey Open-File Report 88-757, 12 p.

Johnson, J. G., and Pendergast, A., 1981, Timing and mode of emplacement of the Roberts Mountains allochthon, Antler orogeny: Geological Society of America Bulletin, part 1, v. 92, p. 648–658.

Ketner, K. B., 1977, Late Paleozoic orogeny and sedimentation, southern California, Nevada, Idaho, and Montana, *in* Stewart, J. H. Stevens, C. H., and Fritsche, A. E., eds., Paleozoic paleogeography of the western United States: Pacific Section, Society of Economic Paleontologists and Mineralogists Pacific Coast Paleogeography Symposium 1, p. 363–369.

Little, T. A., 1987, Stratigraphy and structure of metamorphosed upper Paleozoic rocks near Mountain City, Nevada: Geological Society of America Bulletin, v. 98, p. 1–17.

Malinverno, A., and Ryan, W.B.F., 1986, Extension in the Tyrrehenian Sea and shortening in the Apennines as result of arc migration driven by sinking of the lithosphere: Tectonics, v. 5, p. 227–245.

McMath, V. E., 1958, The geology of the Taylorsville area, Plumas County, California [Ph.D. thesis]: Los Angeles, University of California, 199 p.

—— , 1966, Geology of the Taylorsville area, northern Sierra Nevada, *in* Bailey, E. H., ed., Geology of northern California: California Division of Mines and Geology Bulletin 190, p. 173–183.

Miller, E. L., Holdsworth, B. K., Whiteford, W. B., and Rogers, D., 1984, Stratigraphy and structure of the Schoonover Sequence, northeastern Nevada; Implications for Paleozoic plate margin tectonics: Geological Society of America Bulletin, v. 95, p. 1063–1076.

Miller, M. M., 1987, Dispersed remnants of a northeast Pacific fringing arc; Upper Paleozoic terranes of Permian McCloud faunal affinity, western U.S.: Tectonics, v. 6, p. 807–830.

Moores, E. M., 1970, Ultramafics and orogeny, with models of the U.S. Cordillera and the Tethys: Nature, v. 228, p. 837–842.

Murchey, B. L., Harwood, D. S., and Jones, D. L., 1986, Correlative chert sequences from the northern Sierra Nevada and Havallah sequence, Nevada: Geological Society of America Abstracts with Programs, v. 18, p. 162.

Roberts, R. J., 1964, Stratigraphy and structure of the Antler Peak Quadrangle, Humboldt and Lander Counties, Nevada: U.S. Geological Survey Professional Paper 495A, 93 p,

Saleeby, J., Hannah, J. L., and Varga, R. J., 1987, Isotopic age constraints on Middle Paleozoic deformation in the northern Sierra Nevada, California: Geology, v. 15, p. 757–760.

Schweickert, R. A., 1981, Tectonic evolution of the Sierra Nevada range, *in* Ernst, W. G., ed., The geotectonic development of California; Rubey Volume 1: Englewood Cliffs, New Jersey, Prentice-Hall, p. 87–131.

Schweickert, R. A., and Snyder, W. S., 1981, Paleozoic plate tectonics of the Sierra Nevada and adjacent regions, *in* Ernst, W. G., ed., The geotectonic development of California; Rubey Volume 1: Englewood Cliffs, New Jersey, Prentice-Hall, p. 182–201.

Schweickert, R. A., Harwood, D. S., Girty, G. H., and Hanson, R. E., 1984, Tectonic development of the Northern Sierra terrane; An accreted Late Paleozoic island arc and its basement, *in* Lintz, J., Jr., ed., Western geological exclusions; Geological Society of America Guidebook: Reno, University of Nevada Department of Geology, v. 4, p. 1–65.

Silberling, N. J., 1973, Geologic events during Permian-Triassic time along the Pacific margin of the United States, *in* Logan, S., and Hills, L. V., eds., The Permian and Triassic systems and their mutual boundary: Calgary, Canada, Alberta Society of Petroleum Geology, p. 345–362.

Silberling, N. J., and Roberts, R. T., 1962, Pre-Tertiary stratigraphy and structure of northwestern Nevada: Geological Society of America Special Paper 72, 58 p.

Smith, J. F., and Ketner, K. B., 1968, Devonian and Mississippian rocks and the date of the Roberts Mountains thrust in the Carlin-Pinon range area, Nevada: U.S. Geological Survey Bulletin 1251-1, p. 11–118.

Snyder, W. S., 1977, Origin and exploration for ore deposits in upper Paleozoic chert-greenstone complexes of northern Nevada [Ph.D. thesis]: Stanford, California, Stanford University, 159 p.

Snyder, W. S., and Brueckner, H. K., 1983, Tectonic evolution of the Golconda allochthon, Nevada; Problems and perspectives, *in* Stevens, C. H., ed., Pre-Jurassic rocks in western North America suspect terranes: Pacific Section, Society of Economic Paleontologists and Mineralogists Special Publication, p. 103–123.

Speed, R. C., 1977a, Excelsior Formation, west-central Nevada; Stratigraphic appraisal, new divisions, and paleogeographic interpretations, *in* Stewart, J. H., and others, eds., Paleozoic paleogeography of the western United States: Pacific Section, Society of Economic Paleontologists and Mineralogists Pacific Coast Paleogeography Symposium 1, p. 349–362.

—— , 1977b, An appraisal of the Pablo Formation of presumed late Paleozoic age, central Nevada, *in* Stewart, J. H., and others, eds., Paleozoic paleogeography of the western United States: Pacific Section, Society of Economic Paleontologists and Mineralogists Pacific Coast Paleogeography Symposium 1, p. 315–324.

—— , 1979, Collided Paleozoic platelet in the western U.S.: Journal of Geology, v. 87, p. 279–292.

Speed, R. C., and Sleep, N. H., 1982, Antler orogeny and foreland basin; A model: Geological Society of America Bulletin, v. 93, p. 815–828.

Stewart, J. H., Murchey, B., Jones, D. L., and Wardlaw, B. R., 1986, Paleontologic evidence for complex tectonic interlayering of Mississippian to Permian deep-water rocks of the Golconda allochthon in Tobin Range, north-central Nevada: Geological Society of America Bulletin, v. 97, p. 1122–1132.

Stone, P., and Stevens, C. H., 1988, Pennsylvanian and Early Permian paleogeography of east-central California; Implications for the shape of the continental margin and the timing of continental truncation: Geology, v. 16, p. 330–333.

Tomlinson, A. J., 1988, Tectonostratigraphic units of the Golconda allochthon, Nevada: Geological Society of America Abstracts with Programs, v. 20, p. 238.

Tomlinson, A. J., and Wright, J. E., 1986, The Klamath-Sierran arc; A possible source for Permian volcaniclastic and conglomeratic units within the Golconda allochthon: Geological Society of America Abstracts with Programs, v. 18, p. 193.

Varga, R. J., 1982, Implications of Paleozoic phosphates in the northern Sierra Nevada range: Nature, v. 297, p. 217–220.

Varga, R. J., and Moores, E. M., 1981, Age, origin, and significance of an unconformity that predates island-arc volcanism in the northern Sierra Nevada: Geology, v. 9, p. 512–518.

Manuscript Accepted by the Society March 27, 1990

Printed in U.S.A.

Geological Society of America
Special Paper 255
1990

Paleogeographic setting of upper Paleozoic rocks in the northern Sierra and eastern Klamath terranes, northern California

M. Meghan Miller
Jet Propulsion Laboratory, California Institute of Technology, 183-501, 4800 Oak Grove Drive, Pasadena, California 91109
David S. Harwood
U.S. Geological Survey, 345 Middlefield Road, Menlo Park, California 94025

ABSTRACT

Upper Paleozoic rocks of the northern Sierra and eastern Klamath terranes provide detailed stratigraphic records of ensimatic arc-related sedimentation and magmatism. Comparison of Paleozoic stratigraphic relations between the two terranes, however, suggests certain contrasts in depositional environments and the nature, volume, and timing of volcanism for given time intervals. Some lithologic and provenance ties indicate a paleogeographic relation. Variations in stratigraphy between the two terranes and within terranes imply differences in geodynamic setting. These and regional geologic relations indicate an early and persistent paleogeographic tie between the two areas, and they further suggest that the eastern Klamath terrane may have lain outboard and trenchward of the northern Sierra terrane during much of their late Paleozoic evolution. Stratigraphic ties for mid-Permian and lower Mesozoic rocks imply a closer relation and more similar geodynamic setting during subsequent evolution.

INTRODUCTION

Stratigraphic similarities, provenance ties, biogeographic relations, and shared histories of structural, metamorphic, or magmatic evolution are used to constrain paleogeographic relations between differing ancient rock sequences. The usefulness of some of these criteria in establishing original proximity between any two terranes is compromised by global events such as fluctuations in ocean basin volume resulting from variation in ocean ridge spreading rates, sea-level and climate changes, worldwide patterns in rates and loci of subduction, widespread evolutionary changes, and global changes in rates of carbonate or other precipitate production. These factors may result in similarities between contemporaneous sequences based on their synchronous evolution, rather than on their original spatial distribution. Similarly, difficulties arise when comparing assemblages formed in differing tectonic settings that may have been paleogeographically related, such as a volcanic arc and related subduction complex. Thus, no routine approach is possible in evaluating paleogeographic relations between ancient rock sequences. Realistic inferences are based on large bodies of data that characterize depositional history and source regions, biostratigraphy and biogeography, geochronology and geochemistry of magmatic rocks, and other shared characteristics. Patterns that emerge from such syntheses can then be evaluated in light of variations expected from similar tectonic regimes in the modern setting, and known global distribution patterns in rocks of a given age.

Pre-Cretaceous rocks of the Klamath Mountains and Sierra Nevada record the evolution and accretion of oceanic and volcanic-arc assemblages. Both regions contain an eastern belt that consists of rocks formed in a volcanic arc setting from Devonian to Middle Jurassic time. Regional structural and lithologic similarities between rocks of the Klamath Mountains and Sierra Nevada have long been recognized (e.g., Davis, 1969). Together

Miller, M. M., and Harwood, D. S., 1990, Paleogeographic setting of upper Paleozoic rocks in the northern Sierra and eastern Klamath terranes, northern California, *in* Harwood, D. S., and Miller, M. M., eds., Paleozoic and early Mesozoic paleogeographic relations; Sierra Nevada, Klamath Mountains, and related terranes: Boulder, Colorado, Geological Society of America Special Paper 255.

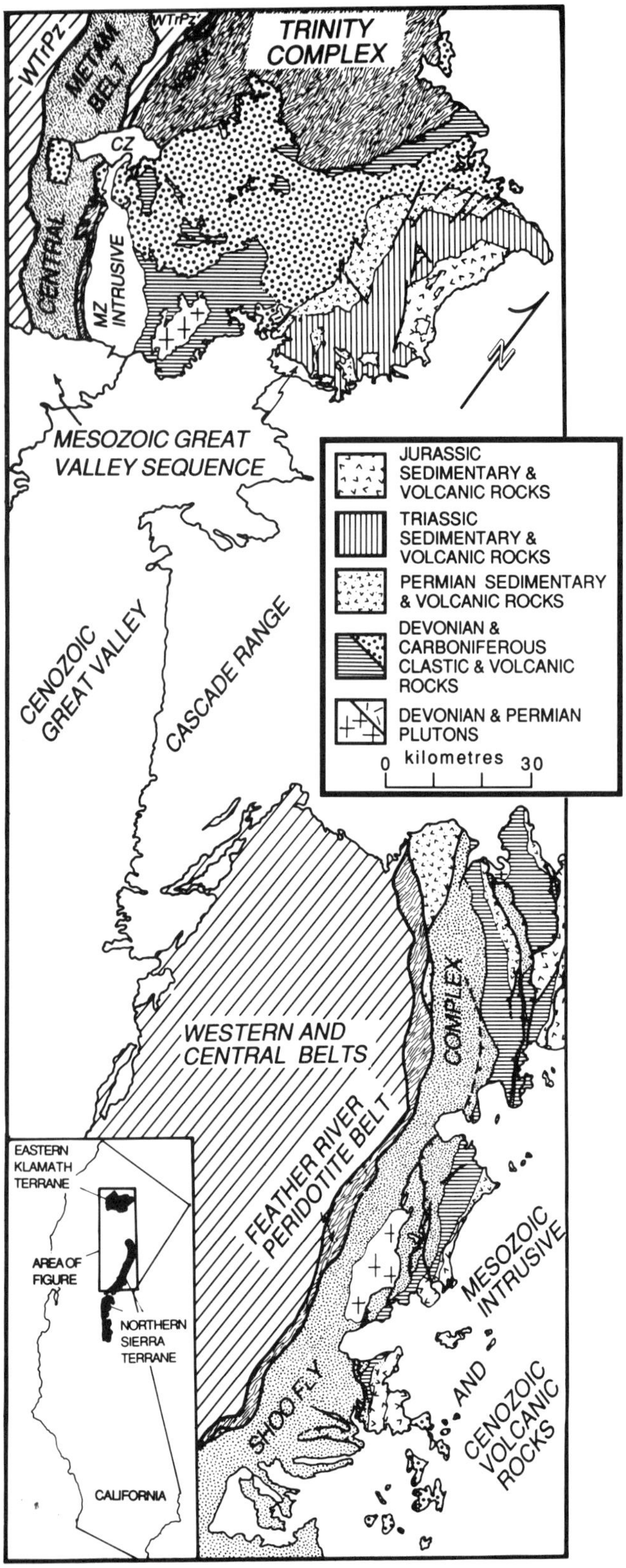

Figure 1. Generalized geologic map of part of northern California showing the eastern Klamath terrane and northern part of the northern Sierra terrane (patterned). Cz = Cenozoic, TrPz = Western Paleozoic and Triassic belt. Simplified from Harwood (1988) and Miller (1989, and unpublished compilation).

with other lines of evidence, these similarities have led to the inference that arc-related sequences in the eastern parts of both ranges once belonged to a single northeast Pacific fringing arc (Burchfiel and Davis, 1981; Miller, 1987). Although stratigraphic similarities exist in parts of the sections, many important differences are present between rocks of the two areas, and the significance of these differences has not previously been explored in detail. We summarize and update our stratigraphic studies and recent syntheses (Harwood, 1983, 1988; Miller, 1987, 1989), and use these data to evaluate paleogeographic relations in light of stratigraphic differences and common provenance of specific upper Paleozoic units. In particular, we focus on the depositional setting, magmatic history, provenance, and age relations of Devonian and Carboniferous rocks, where pronounced and well-documented differences exist. Basement assemblages that underpin the Devonian to Jurassic arc rocks are also discussed, as important similarities and differences are now better known between the two areas, and the basement complexes sporadically provided detritus to basinal deposits in the overlying successions.

GEOLOGIC SETTING

The Klamath Mountains and Sierra Nevada have been subdivided into numerous terranes (e.g., Irwin, 1972; Silberling and others, 1987) that evolved in oceanic and volcanic arc environments during the late Paleozoic and early Mesozoic, and that underwent subsequent addition to western North America in an accretionary orogen (Saleeby, 1983). In both areas, the easternmost and structurally highest regionally extensive terranes (eastern Klamath and northern Sierra terranes; Fig. 1), comprise Devonian to Middle Jurassic volcanic arc–related sequences that record sporadic magmatism and sedimentation. Shared histories of voluminous mid-Permian intermediate volcanism form the oldest strong lithostratigraphic tie between the two terranes, and lithologic similarities exist between younger units in both terranes. Thus our discussion focuses on pre–Late Permian events and sequences that record differing histories but may reflect a common paleogeographic origin.

BASEMENT TERRANES AND STRATIGRAPHIC RELATIONS

Northern Sierra terrane

The northern Sierra terrane contains three principal lithotectonic units: (1) lower Paleozoic supracrustal rocks of the Shoo Fly Complex (Fig. 2) that were imbricated and eroded prior to deposition of the overlying (2) Upper Devonian to Upper Permian volcanic and sedimentary rocks of the Taylorsville and laterally equivalent Tahoe sequences and their Mesozoic cover (Fig. 3; Schweickert and others, 1984; Harwood, 1988), and (3) Permian and Triassic sedimentary rocks in tectonic slivers that lie between the Shoo Fly Complex and the Feather River peridotite belt to the west (Harwood and others, 1988; Jayko, 1988).

Shoo Fly Complex. The Shoo Fly Complex consists of four principal allochthons that were tectonically imbricated prior to construction of the Late Devonian and Early Mississippian volcano-plutonic complex (Fig. 2). Structurally upward, these consist of: (1) the areally extensive Lang sequence, composed of quartzite-pelite turbidites, with lesser quartz-microcline grit, chert, and limestone; (2) radiolarian chert and black siliceous argillite of the Duncan Peak allochthon; (3) the Culbertson Lake allochthon, comprising a lower pillow basalt and basalt breccia with intercalated and overlying chert, limestone, and argillite, and an upper subarkose and quartz arenite; and (4) the structurally highest Sierra City melange, containing blocks of serpentinite, gabbro, pillowed and massive basalt, chert, sandstone, limestone, and dolostone in a disrupted slate, sandstone, and chert matrix (Schweickert and others, 1984; Harwood, 1988).

Limited U-Pb data for detrital zircon populations from clastic units within the Shoo Fly Complex are available. Multifraction analyses of detrital zircon from mature quartzose sandstone in the Culbertson Lake allochthon yield an upper intercept age of ~2.09 Ga (Figs. 2, 3), indicating an Early to Middle Proterozoic average age, and a continental source terrane for these rocks (Girty and Wardlaw, 1985). Monocyclic detrital zircon was obtained from volcaniclastic sandstone in the Sierra City melange, and yields an upper intercept age of 506 ± 22 Ma; the Mesozoic lower intercept age indicated younger thermal disturbance (Fig. 2; Girty and Wardlaw, 1984). The early Paleozoic upper intercept suggests derivation from a Cambrian to Ordovician volcano-plutonic source terrane. Other zircon data from the Shoo Fly Complex include ages of 423 ± 8 Ma on a tuff within the Sierra City melange matrix (Saleeby and others, 1987), and 600 ± 10 Ma from a plagiogranite block in melange within the same unit (J. B. Saleeby and R. A. Schweickert, unpublished data).

Taylorsville and Tahoe sequences. Two upper Paleozoic sequences overlie the Shoo Fly Complex in the northern Sierra terrane: the Taylorsville and Tahoe sequences. The Taylorsville sequence comprises two volcanic successions, Late Devonian to early Early Carboniferous and mid-Permian in age (Fig. 3), and occurs in the central and northern parts of the northern Sierra terrane. These successions are separated by a widespread middle Carboniferous radiolarian chert and overlying unconformity (Harwood, 1988). The Tahoe sequence lies to the south, west of Lake Tahoe, and contains a succession of metasedimentary rocks that are laterally equivalent to rocks of the Taylorsville sequence. Both sequences are associated with lower Mesozoic rocks that have common lithostratigraphic elements.

The oldest rocks within the Taylorsville sequence form a Late Devonian volcanic-plutonic complex that marks the initiation of arc magmatism after imbrication of the Shoo Fly Complex. The basal Grizzly Formation consists of areally restricted chert, quartzite, and limestone conglomerate, turbiditic shale, and quartzo-feldspathic and chert-rich sandstone (Fig. 3; Schweickert and others, 1984; Hanson and Schweickert, 1986). The Grizzly Formation contains middle Famennian conodonts in interbedded chert (B. R. Wardlaw, *in* Hanson, 1983) and is interpreted to record epiclastic sedimentation that succeeded and perhaps accompanied imbrication of the Shoo Fly Complex (Hanson and Schweickert, 1986). The Grizzly Formation, where present, and

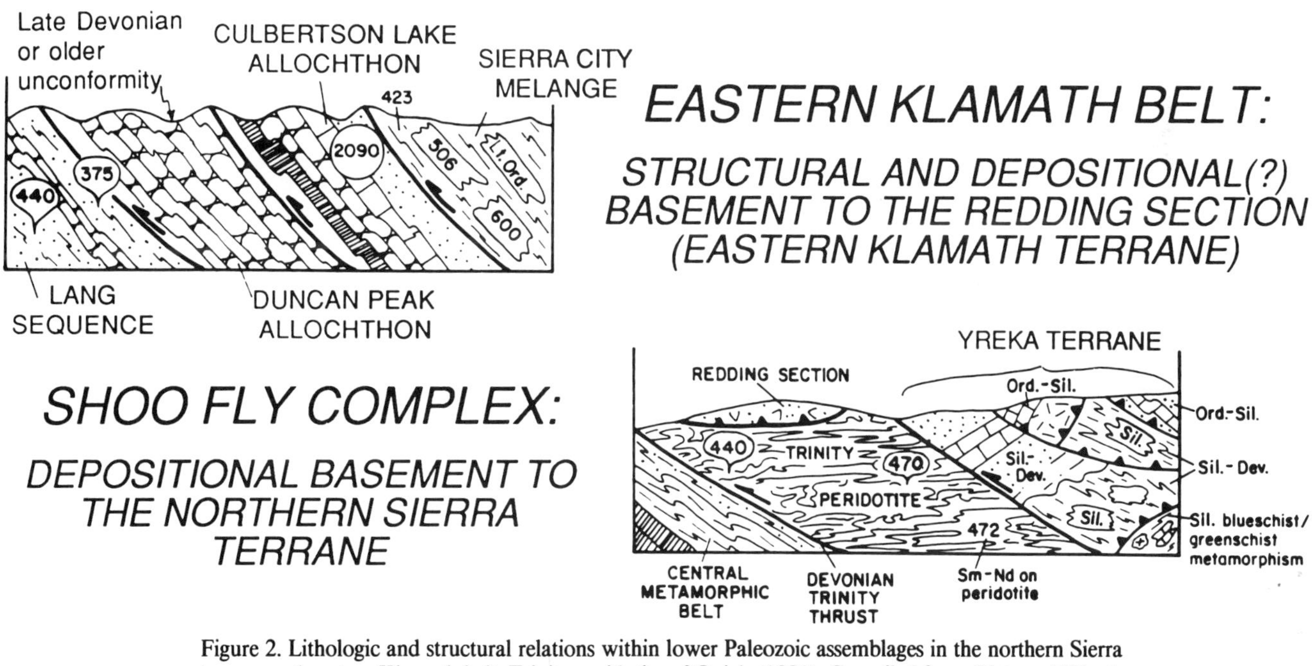

Figure 2. Lithologic and structural relations within lower Paleozoic assemblages in the northern Sierra terrane and eastern Klamath belt. Trinity peridotite of Quick (1981). Compiled from Girty and Wardlaw (1984, 1985), Hanson and others (1988), Harwood (1988), Hotz (1977), Jacobsen and others (1984), Lanphere and others (1968), Peacock (1987), Potter and others (1977), J. B. Saleeby and R. A. Schweickert (unpublished data), Schweickert and others (1984), and Wallin and others (1988).

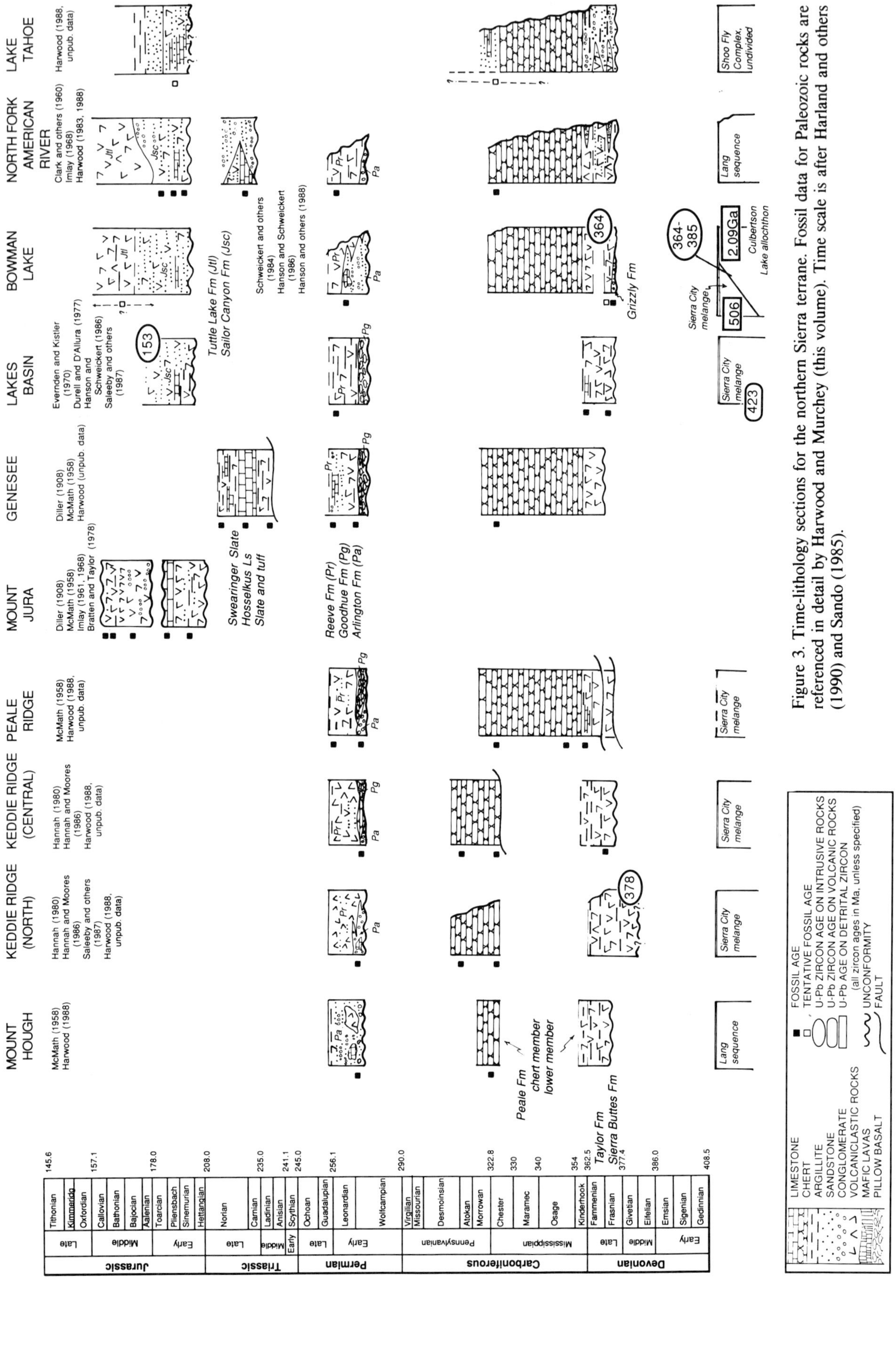

Figure 3. Time-lithology sections for the northern Sierra terrane. Fossil data for Paleozoic rocks are referenced in detail by Harwood and Murchey (this volume). Time scale is after Harland and others (1990) and Sando (1985).

elsewhere the Shoo Fly Complex are overlain by water-laid pyroclastic debris of the Sierra Buttes Formation, which is predominantly rhyolitic to dacitic at the base and increasingly andesitic upward (Hanson and Schweickert, 1986). Interbedded phosphatic chert indicates deposition adjacent to the eastern edge of an open ocean basin, where upwelling occurred along a continental margin (Varga, 1982). The Sierra Buttes Formation is middle Famennian in age based on radiolarians and ammonoids (Hanson and Schweickert, 1986). Geochemical data imply a continental crustal contribution to the arc magmas that fed the Sierra Buttes Formation (Rouer and others, 1989); however, less altered intrusive rocks yield intermediate initial $^{87}Sr/^{86}Sr$ ratios (~0.704) and only slightly negative ϵ_{Nd}, implying transitional or perhaps more primitive basement. These data are consistent with assimilation of continent-derived sedimentary rocks of the Shoo Fly Complex and a substrate of transitional or oceanic crust.

The Bowman Lake batholith intrudes the Shoo Fly Complex, and associated dikes and sills form feeders to the Sierra Buttes Formation (Hanson and others, 1988). Discordant zircon from the Bowman Lake batholith suggest a range of ages from 385 to 364 Ma; a crystallization age of 364 Ma is favored based on field relations and calibration of the radiometric time scale (Hanson and others, 1988). Sills of approximately the same age (368 Ma) intrude the overlying Sierra Buttes Formation (Fig. 3; Hanson and others, 1988). In the northern part of the northern Sierra terrane, the Wolf Creek stock intrudes, and may be cogenetic with, the Sierra Buttes Formation (Saleeby and others, 1987). The Wolf Creek stock yields a discordant U-Pb zircon age of 378 ± 5 Ma (Saleeby and others, 1987), implying a latest Middle Devonian age (based on the time scale of Harland and others, 1990), which is in conflict with field data indicating an intrusive contact with Upper Devonian (Famennian) rocks of the Sierra Buttes Formation. These relations perhaps imply a discrepancy in calibration of the radiometric and biostratigraphic time scales for the Middle and Late Devonian. Alternatively, the Wolf Creek stock is slightly younger or the Sierra Buttes is slightly older than existing evidence suggests. Because of the conflicting data, the lower range of the Sierra Buttes Formation at Keddie Ridge is queried in Figure 2.

The Sierra Buttes Formation is conformably overlain by andesitic augite-bearing flows, tuff breccia, and tuffaceous turbidites of the Upper Devonian and/or Lower Carboniferous Taylor Formation (Fig. 3; Harwood, 1988), which is in turn overlain by the volcanogenic Lower Carboniferous lower member of the Peale Formation (Harwood, 1988; Keddie Ridge Formation of Hannah and Moores, 1986). The lower part of the Peale Formation contains trachyte and quartz latite flows, felsic tuff, tuff breccia, volcanogenic turbidites, and minor chert-granule conglomerate. The lower Peale is dated by Early Mississippian brachiopods and trilobites (McMath, 1966) and Early Mississippian (late Kinderhookian) conodonts (A. Harris, *in* Harwood, 1988). Volcanogenic rocks of the Taylor and lower member of the Peale Formations extend the record of arc-related magmatism into early Early Carboniferous time, although the presence of chert-granule conglomerate, particularly common in the southern part of the terrane, indicates epiclastic sedimentation was active as well. All three Devonian and Mississippian volcanic units show northward thickening and contain vent-proximal facies northward. Sparse epiclastic debris is more widespread to the south (Harwood, 1988; Harwood and Murchey, this volume).

The (upper) chert member of the Peale Formation (restricted Peale Formation of Hannah and Moores, 1986) comprises radiolarian-bearing ribbon chert that overlies volcanogenic beds of the lower part of the Peale Formation. At least the base of the chert member may be in part volcanogenic, as suggested by bipyramidal quartz and volcanic debris in the oldest beds at Peale Ridge (B. Murchey, *in* Harwood, 1988). The chert member has been dated in numerous locations and is primarily middle Carboniferous (Chesterian to Desmoinesian(?) in age; at one locality, however, it is as old as Early Carboniferous (Osagean) (Fig. 3; Harwood and Murchey, this volume). The regionally widespread chert member of the Peale marks an important hiatus in magmatism throughout the Taylorsville sequence; volcanic and tectonic quiescence resulted in the dominance of pelagic sedimentation.

Two distinct stratigraphic successions, of Paleozoic and Mesozoic ages respectively, are now recognized in the Tahoe sequence (Fig. 3; Harwood and Murchey, this volume; Fisher, this volume), which is restricted to the southern part of the northern Sierra terrane. The Paleozoic part of the Tahoe sequence depositionally overlies the Shoo Fly Complex and is laterally equivalent to the lower part of the Taylorsville sequence (Fig. 3). The Shoo Fly Complex is there overlain by a succession, several kilometers thick, composed of chert-rich turbidites, associated channel-fill sequences, and intercalated quartzo-feldspathic volcanogenic beds that represent a lateral variation of the Sierra Buttes Formation (D. S. Harwood, unpublished data; Fig. 3). Limestone in the upper part of the clastic succession contains echinoderm debris of middle Paleozoic, possibly Silurian or Devonian, age (J. T. Dutro, Jr., written communication, 1988). The strata are younger than the unconformity over the imbricated Shoo Fly Complex and older than the overlying metachert, which is probably correlative with the mid-Carboniferous chert member of the Peale Formation (Fig. 3). They are thus considered broadly Late Devonian(?) and Early Carboniferous(?) in age.

The chert-rich turbidites and conglomerates are overlain by thin-bedded, sugary, medium- to coarse-grained quartzite (D. S. Harwood, unpublished data). Structureless beds are characteristic, and together with primary $\delta^{18}O$ values of +23.7 and +24.0 (J. R. O'Neil, written communication, 1987), indicate that this unit represents metachert. Broad age constraints based on bracketing of the unit between the Shoo Fly Complex and poorly dated but probable upper Paleozoic overlying beds, together with a unique regional lithologic correlation, suggest that this unit is correlative with the chert member of the Peale Formation (Fig. 3).

Overlying the metachert is a poorly understood unit consisting of quartzite, calc-silicate, and marble (Fig. 3), which probably represents shallow-water deposits of continental affinity. Scattered shelly fragments include a bivalve of likely post-Ordovician Paleozoic age (J. T. Dutro, Jr., written communication, 1988).

In the Taylorsville sequence, the chert member of the Peale is unconformably overlain by a upper Lower Permian to lower Upper Permian (Leonardian to Guadalupian) volcanic succession that comprises the overlapping and laterally equivalent Arlington, Goodhue, and Reeve Formations (Fig. 3; Harwood, 1988). Locally thick deposits of the Arlington Formation contain epiclastic Shoo Fly–derived lithologies and rest above an unconformity, which together with contemporaneous basaltic volcanism in the Goodhue Formation, may record extensional tectonism (Harwood, 1988). Early Leonardian fusulinids that occur as solitary fossils within a volcaniclastic horizon of the Reeve indicate a biogeographic relation to the McCloud Limestone of the eastern Klamath terrane (discussed below; Stevens, 1987), and date the Reeve as Early Permian (Leonardian). Elsewhere the chert member of the Peale Formation is directly overlain by intermediate to mafic volcanogenic rocks of the Lower Permian (Leonardian) Goodhue Formation, or shallow-marine to subaerial andesitic tuff, breccia, and lavas of the Reeve Formation (Fig. 3; Harwood, 1988). These rocks collectively record a pulse in magmatism during mid-Permian time. No similar rocks are known in the Tahoe sequence.

The Permian volcanic sequence is locally overlain by Upper Triassic limestone and conglomerate, or by younger, Jurassic volcanogenic strata (Harwood, 1983; Fig. 3). At Genesee (Fig. 3), however, Middle Triassic (middle Ladinian) slate and tuffaceous rocks are present that are lithologically and biostratigraphically identical to the Pit Formation (discussed below) in the eastern Klamath terrane (N. J. Silberling, personal communication, 1987). The tuffaceous rocks are overlain by the Upper Triassic (late Carnian and early Norian) Hosselkus Limestone, which, in turn, is overlain by the Upper Triassic (late Norian) Swearinger Slate (Diller, 1908; also see Genesee, Fig. 3).

Correlative Jurassic rocks occur above both the Taylorsville and Tahoe sequences (Fisher, this volume), although stratigraphic relations with the Tahoe sequence are obscured by intervening, late Mesozoic intrusive rocks (Fig. 1). Metamorphosed argillite, feldspathic sandstone turbidites, tuff, minor limestone, and pebbly mudstone characterize the Lower and Middle Jurassic Sailor Canyon Formation (Harwood, 1988; Fisher, this volume) and are locally overlain by coarse clastic units and intercalated volcanic rocks of the Middle and Upper(?) Jurassic Tuttle Lake Formation (Fisher, this volume). Quartz-rich sandstone occurs in the matrix of Tuttle Lake conglomerates that are thought to record active extensional faulting (Fisher, this volume). Thick successions of quartz-rich eolian sands that were reworked in a marine setting also occur west of Lake Tahoe (Fig. 3) and must have been derived from the craton (Harwood, unpublished data). These beds record a direct tie to a continental source during Early or Middle Jurassic time.

Tectonic slivers of Permian and Triassic rocks. At the northern end of the northern Sierra terrane, a tectonic sliver of Permian and Triassic sedimentary rocks, referred to as the Soda Ravine block (Jayko, 1988, this volume), occurs between the Feather River peridotite belt and rocks presumably belonging to the Taylorsville sequence. The lithologies and age relations in the Soda Ravine block are relatively well documented (Jayko, this volume), and similar blocks occur sporadically far to the south (Fig. 1; Harwood and others, 1988). Fossiliferous Lower Permian and Middle and Upper Triassic limestone occur in a matrix of argillite and tuffaceous slate (Jayko, 1988). Fusulinids from the Permian limestone are earliest Permian (early Wolfcampian) and have biogeographic affinity to the McCloud Limestone (C. H. Stevens, *in* Jayko, 1988). Fossil ages from younger limestone are Middle and Late Triassic (Ladinian, and Carnian-Norian; N. J. Silberling, *in* Jayko, 1988). Associated tuffaceous argillite is thought to be Middle Triassic, and intermediate in age with respect to the limestone blocks (Jayko, 1988). Some crystal- and lithic-rich clastic rocks are associated with the Permian limestone. The relation between the Taylorsville sequence and Soda Ravine and lithologically and biostratigraphically related rocks to the south is not well established. Upper Triassic conglomerate that occurs in the Taylorsville sequence to the east may record the juxtaposition of the Permian and Triassic tectonic slivers and the Taylorsville sequence (Jayko, this volume).

Eastern Klamath belt

The eastern Klamath belt forms the easternmost and structurally highest lithotectonic belt in the Klamath Mountains, and structurally overlies the Devonian central metamorphic belt (Irwin, 1966). It consists of three terranes: (1) the structurally lowest, mostly Ordovician Trinity complex (Figs. 2, 4); (2) deformed lower Paleozoic sedimentary and metamorphic rocks of the Yreka terrane that overlie the Trinity complex to the north (Fig. 2); and (3) Devonian to Jurassic arc-related rocks of the eastern Klamath terrane (Redding section) that overlie the Trinity complex to the south (Fig. 5), and are separated from the Yreka terrane by intervening exposures of peridotite and gabbro of the Trinity complex. The central metamorphic belt, which lies to the west, was metamorphosed to upper greenschist and amphibolite facies during Late Devonian time. The fault between the Trinity complex and the central metamorphic belt represents an ancient subduction zone of probable Late Devonian age (Peacock, 1987; Peacock and Norris, 1989). Lower Paleozoic rocks of the eastern Klamath and central metamorphic belts were imbricated during middle Paleozoic thrust faulting related to emplacement of the Trinity complex, probably during late Middle or Late Devonian time (Boucot and others, 1974). This deformation has recently been recognized in Devonian rocks of the eastern Klamath terrane (Charvet and others, 1989).

Trinity complex. The polygenetic Trinity complex (see Boudier and others, 1989) contains several generations of mafic

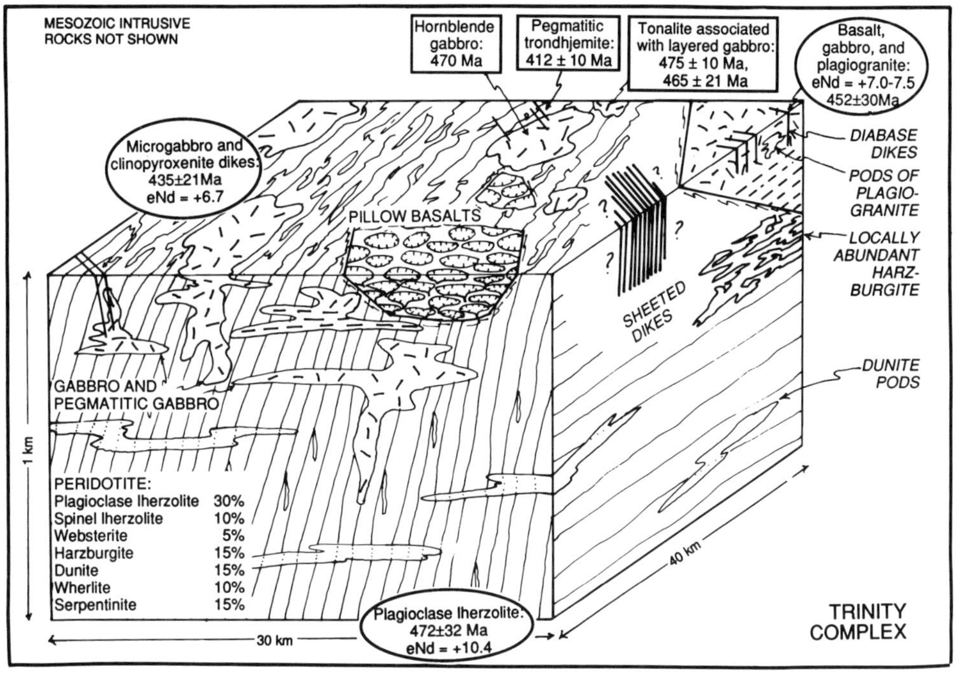

Figure 4. Lithologic and structural relations between polygenetic elements of the Trinity complex. Isotopic data, petrology, and field relations are summarized from Quick (1981), Jacobsen and others (1984), Brouxel and Lapierre (1988), Wallin and others (1988), and Boudier and others (1989). Mafic rocks everywhere cross-cut or are in fault contact with the Ordovician peridotite. Some mafic suites are internally cogenetic (as in upper right, Brouxel and Lapierre, 1988), but show strong isotopic and geochemical contrasts with peridotite. Zircon crystallization ages (in boxes) range over 63 m.y., indicating a protracted evolution for many mafic components. Sm-Nd ages (in ovals) are also long-ranging, although associated errors prevent exact resolution of magmatic events.

and ultramafic intrusive rocks (Fig. 4; simplified in Fig. 2). Although some workers have interpreted disparate elements of the Trinity complex as members of a cogenetic ophiolite (Brouxel and Lapierre, 1988; Boudier and others, 1989), differences in isotope geochemistry and radiometric ages indicate different origins for the ultramafic and mafic rocks, and multiple generations of mafic rocks (Quick, 1981; Jacobsen and others, 1984; Wallin and others, 1988; Fig. 4). An Ordovician or older age for the peridotite is indicated by a Sm-Nd two-mineral isochron age of 472 ± 21 Ma (Jacobsen and others, 1984), and by Ordovician zircon ages on cross-cutting gabbro (Wallin and others, 1988, Fig. 4). A six-point whole-rock Sm-Nd isochron on basalt, gabbro, and plagiogranite (no peridotite is included) yields an age of 452 ± 40 Ma (Brouxel and Lapierre, 1988). Although the error ranges for the age determinations overlap, the data do not strongly suggest a similar age, and contrasting ϵ_{Nd} values (Fig. 4) imply differing origins. Dates from gabbro and tonalite that intrude the peridotite indicate a range of ages from 475 ± 10 to 412 ± 10 (Wallin and others, 1988; Fig. 4). Based on the Sm-Nd mineral isochron and on U-Pb and Sm-Nd ages of cross-cutting mafic rocks, the main part of the ultramafic sheet is Ordovician or older. Mafic rocks that intrude the Trinity complex yield U-Pb zircon ages that span 63 m.y. suggesting a protracted and polygenetic origin. Such a history of mafic magmatism corroborates an origin as an arc-related mantle diapir, where repeated episodes of magmatism that were imposed on the upper mantle are preserved as cross-cutting mafic sequences that range over tens of millions of years in age (Quick, 1981; Jacobsen and others, 1984; Saleeby, this volume; also see Boudier and others, 1988, and Brouxel and Lapierre, 1989, for differing views).

Even older ages (565 to 571 Ma, U-Pb upper intercept, zircon) have also been reported from the area (Wallin and others, 1988). These ages come from the plagiogranite of Gregg Ranch, which structurally overlies the main mafic and ultramafic units of the Trinity complex (Wallin and others, 1988) and may not have a direct bearing on the genesis of the Trinity complex. Similar ages were obtained from tonalite melange blocks in the schist of Skookum Gulch, an imbricate fault slice between the Trinity complex and overlying rocks of the Yreka terrane (Wallin and others, 1988).

Yreka terrane. The Yreka terrane is primarily composed of a structurally and stratigraphically complex assemblage of sparsely fossiliferous lower Paleozoic sedimentary and metasedimentary rocks (Potter and others, 1977, this volume). Structurally disrupted assemblages, such as the plagiogranite of Gregg Ranch (discussed above) and the melange of the schist of Skookum Gulch, which was metamorphosed to blueschist facies in early Paleozoic time (Cotkin and Armstrong, 1987), lie at the base of

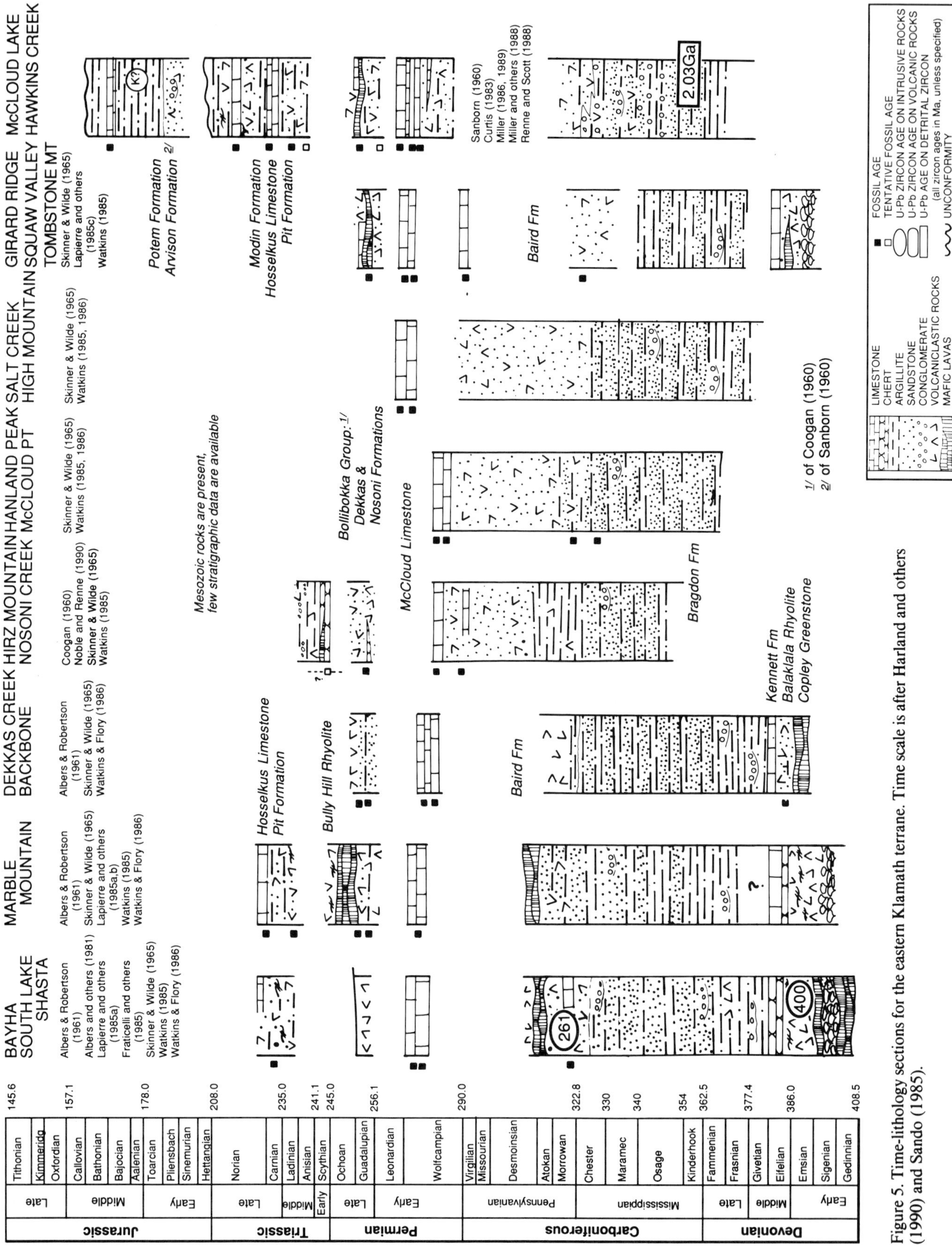

Figure 5. Time-lithology sections for the eastern Klamath terrane. Time scale is after Harland and others (1990) and Sando (1985).

the Yreka terrane or between the Yreka terrane and Trinity complex. The Yreka terrane contains at least three distinct lithostratigraphic successions that are regionally widespread; each of these records the influence of craton-derived sedimentation by the presence of early Proterozoic and, in places, Archean detrital zircon (Wallin, 1989) and multi-cycle quartz arenite (Bond and Devay, 1980). In two of the successions, contributions from an Ordovician and/or Silurian volcanic-plutonic source are also recorded by compositions of volcaniclastic rocks and monocyclic, relatively homogeneous detrital zircon suites (Wallin, 1989). These data imply an important stratigraphic relation between rocks derived from isotopically and paleogeographically distinct source terranes.

Eastern Klamath terrane. The eastern Klamath terrane is composed of a Devonian to Middle Jurassic succession of volcaniclastic, volcanic, and basinal rocks with interstratified shallow-water carbonate rocks that record evolution predominantly in an oceanic arc setting. The oldest rocks in the eastern Klamath terrane belong to the Middle(?) Devonian and older(?) Copley Greenstone, which consists of a lower unit of massive lava and mafic volcaniclastic rocks and an upper unit of pillow basalt (Fig. 5; Lapierre and others, 1985a). In the southern part of the terrane, the Copley Greenstone interfingers with and is overlain by areally restricted keratophyric water-laid tuff and lavas of the Middle Devonian and older(?) Balaklala Rhyolite (Albers and Robertson, 1961; Albers and Bain, 1985; Lapierre and others, 1985b; this report). Aquagene tuff of the Balaklala Rhyolite contains an Eifelian (early Middle Devonian) fossil fish plate (Boucot and others, 1974). The fossil age is contradicted by a single fraction of concordant U-Pb zircon age of 400 Ma from the Mule Mountain stock (Fig. 5; Albers and others, 1981), a trondjhemite that is thought to be cogenetic with and intrusive into the Balaklala Rhyolite (Barker and others, 1979). The 400-Ma age suggests an early Early Devonian origin (after the time scale of Harland and others, 1990), but again, the discrepancy may be attributable to errors in calibration of the radiometric and biostratigraphic time scales for the Devonian Period. Alternatively, the single-fraction zircon age may be in error. It is unlikely that the relatively thin, geochemically coherent Copley-Balaklala rocks were extruded over a protracted time interval. These rocks record bimodal magmatism in a primitive, extensional island-arc setting (Lapierre and others, 1985a; Brouxel and others, 1988) where sea-floor graben formation accompanied silicic magmatism and Kuroko-type massive sulfide mineralization (Lindberg, 1985).

Some workers suggest that the Middle Devonian and older(?) mafic volcanic sequence, together with the Trinity complex, belonged to the same island-arc–back-arc basin system (Brouxel and others, 1988). In this context, the Copley-Balaklala rocks are interpreted to be Silurian in age, based on the cross-cutting 400-Ma (apparently Early Devonian, possibly younger, as discussed above) Mule Mountain stock; a Silurian age is also inferred for the peridotite, based on the large error limits for the Sm-Nd ages. The implied overlap in age is ruled out by Ordovician U-Pb ages from mafic rocks that intrude Trinity peridotite, and Devonian U-Pb ages from the Mule Mountain stock, and by the Middle Devonian fossil constraint from the Balaklala Rhyolite, which interfingers with and is geochemically related to the Copley Greenstone. The inference that the mostly Ordovician Trinity complex and Middle Devonian and older(?) volcanic units are cogenetic is thus in conflict with existing constraints that indicate a $\geqslant$65-m.y. difference between the age of the Copley mafic volcanic rocks and the ages of Ordovician mafic and ultramafic components of the Trinity complex.

Tuffaceous rocks and limestone of the Middle Devonian Kennett Formation interfinger with the underlying Balaklala Rhyolite. Basal, shallow-marine volcaniclastic rocks are overlain by more basinal slope turbidites, siliceous mudstone, and limestone debris aprons (Fig. 5) that were deposited in multiple small, extension-related basins constructed on the older arc edifice (Watkins and Flory, 1986). An early Middle Devonian (Eifelian) age is indicated by conodonts and brachiopod faunas (Savage, 1976). These relations indicate rapid subsidence during the Middle Devonian accompanied by persistent volcanism.

The Upper Devonian(?) and lower Carboniferous Bragdon Formation overlies all the older Devonian units, but the nature of the stratigraphic contact is in dispute. Early workers argued that an unconformity separates the Bragdon Formation from underlying rocks, based on apparent incision of the contact into the Devonian volcanic succession, and on the presence of limestone debris (presumably Kennett-derived) in the Bragdon Formation (Diller, 1905; Kinkel and others, 1956). This view is perhaps corroborated by the suggestion that outliers of the Bragdon Formation rest depositionally on the Trinity complex (e.g., Brouxel and others, 1988; Fig. 2); one such outlier contains peridotite debris in its basal beds (Charvet and others, 1989). The outliers of supposed Bragdon Formation are undated, however, and their correlation to the Bragdon Formation where it occurs in stratigraphic succession with other units of the eastern Klamath terrane has not been confirmed by age or petrographic data. Furthermore, the limestone clasts that were thought to contain Kennett fossils may have been derived from an older source (R. Watkins, personal communication, 1988). Based on conformable relations at Backbone Creek (Fig. 5) and French Gulch, and on the bathymetric relief implied by the vent-proximal (Balaklala) and basinal (Kennett) facies of the underlying rocks, Watkins (1986) suggested that the basal contact is conformable and onlapped the older, subsided volcanic topography. This view implies that deposition of the Bragdon Formation spanned Late Devonian to late Early Carboniferous time. The only fossils that directly date the Bragdon Formation are the late Early Carboniferous in age, and come from the uppermost part of the formation near Hanland Peak (Fig. 5; Watkins, 1973).

The possible unconformity between the Bragdon Formation and the underlying Trinity complex and Middle Devonian (and older?) volcanic strata of the eastern Klamath terrane has been taken as evidence for the imbrication of the older eastern Klamath rocks prior to deposition of the Bragdon Formation (Boucot and others, 1974; Charvet and others, 1989). Direct constraints on the age of deformation come from the youngest fossils in the deformed rocks (Eifelian) and the age of the Bragdon Formation, which extended through much of the Early Carboniferous and possibly part of the Late Devonian, as discussed above. These relations may indicate a late Middle or Late Devonian deformation within the eastern Klamath Mountains (Boucot and others, 1974), a view that is perhaps corroborated by similar isotopic ages for dynamothermal metamorphism of the central metamorphic belt (Fig. 2; Lanphere and others, 1968; Cashman, 1980; Hacker and Peacock, this volume). In this context, deformation of lower Middle Devonian and older(?) rocks in the eastern Klamath and central metamorphic belts above an eastward-dipping subduction zone resulted in metamorphism, thrusting, and uplift that provided chert and sedimentary-lithic–rich clastic debris derived from the Yreka terrane or its lithologic equivalents. These postorogenic deposits are now represented by epiclastic turbiditic strata of the Bragdon Formation.

The Bragdon Formation consists of a crudely progradational succession of sand-rich turbidites, and intercalated channel-fill conglomerates and debris-flow deposits (Miller and Cui, 1989; Watkins, 1986). Sparse, fine-grained facies and apparently high sedimentation rates imply deposition within a rapidly subsiding, ponded basin. Bragdon sandstones contain three principal components: (1) a quartz-, chert-, and sedimentary-lithic–rich component; (2) reworked volcanic-lithic– and feldspar-rich component; and (3) a juvenile crystal-rich and tuffaceous component that was reworked downslope in a submarine fan setting (Miller and Cui, 1989). These compositions respectively reflect basement uplift, arc dissection, and the persistence of volcanism. Epiclastic sandstone composition records a source terrane like the lower Paleozoic Shoo Fly Complex of the northern Sierra terrane, or lithologically similar, coeval rocks of the Yreka terrane within the Klamath Mountains, and is interpreted to indicate erosion of locally uplifted basement (Miller and Cui, 1989; Miller and Saleeby, 1990). Recent collections of rare fossiliferous limestone clasts in the Bragdon Formation (by M. L. Sorenson, 1985) contain Middle Devonian and Carboniferous corals, bryozoans, and stromatoporoids (W. J. Sando, written communication to M. L. Sorenson, 1986, 1987), as well as conodonts that suggest a Middle Devonian (probably Givetian) age (A. Harris, written communication to M. L. Sorensen, 1987). The clasts were likely derived from intrabasinal sources. If this is so, the clasts date deposition and/or date the sparse, reworked volcaniclastic source, rather than restricting the age of the predominant sedimentary-lithic–rich source terrane for other clastic components of the Bragdon Formation.

U-Pb isotopic study of detrital zircon populations from sandstone of the Bragdon Formation indicates a complex and heterogeneous source (Miller and others, 1988; Miller and Saleeby, 1990). Proterozoic average ages for the older zircon component (upper intercept ages ~2.03 to ~2.07 Ga, $^{207}Pb/^{206}Pb$ ages over 1.9 Ga) indicate an ultimate Precambriam continental source, and together with sandstone clast composition, corroborate an immediate source that is lithologically and isotopically similar to parts of the Shoo Fly Complex or the Antelope Mountain Quartzite in the Yreka terrane. Other isotopic components in the complicated Bragdon zircon populations, such as a poorly defined early Paleozoic lower intercept, are consistent with other, younger sources in the northern Sierra or Yreka terranes, as also implied by compositions of lithic fragments (Miller and Cui, 1989; Miller and Saleeby, 1990). Although other source terranes are permissible using only the zircon data, taken together with the sandstone framework compositions (Miller and Cui, 1989), the data substantiate a northern Sierra and Yreka terrane source for Bragdon detritus.

Little attention has been paid to juvenile tuffaceous and crystal-rich debris that forms a volumetrically important component of the Bragdon Formation. Coarse, crystal-rich turbidites and channel-fill sequences, and finer grained tuffaceous turbidites form as much as 30 percent of the section in the northern part of the terrane, where studied in detail (Miller and Cui, 1989). The tuffaceous beds are intimately interfingered with, but not mixed with, epiclastic turbidites, and were distributed from localized sources within the same depositional basin. Crystal populations in these units correspond to phenocryst assemblages for rhyolitic to andesitic volcanic rocks; silicic units contain pumice fragments and chloritized glassy matrix, recording incorporation of juvenile volcanic debris. These strata record the important but little noted influence of persistent felsic volcanism during the Late Devonian(?) and early Carboniferous.

The Bragdon Formation is gradationally overlain by water-laid pyroclastic beds of the middle and upper Carboniferous Baird Formation. The varied Baird Formation generally consists of two parts; a lower argillaceous shelf facies, and an upper part made up of intermediate volcanogenic debris flows, water-laid tuff, and mafic lavas (Fig. 5; volcaniclastic dispersal aprons of Watkins, 1985; see also Miller, 1989). Although the Baird Formation is regionally widespread, either one of the parts is commonly missing. Shallow-marine strand and shelf facies are present in the northern part of the terrane (Miller, 1989). Mafic lavas are present in the Baird Formation in the southern part of the terrane, and water-laid volcanogenic facies dominate elsewhere (Fig. 5; Watkins, 1973, 1985). The Baird Formation records penecontemporaneous volcanism and shoaling during variable outbuilding of volcanic debris aprons. Lateral variations in the age of the Baird Formation are relatively well established, and range from late early Carboniferous to late late Carboniferous or Early Permian(?), according to Watkins (1973, 1985; see Fig. 5). Localized lacunas within the formation are also well documented (Watkins, 1985).

The Baird Formation locally interfingers with the conformably overlying Lower Permian (Wolfcampian to early Leonardian) McCloud Limestone, although in places the contact is disconformable (Fig. 5; Watkins, 1985; Miller, 1989). The McCloud Limestone comprises areally restricted, fossiliferous, shallow-water carbonate platform facies. Relative ages are well documented by detailed biostratigraphic zonation of biogeographically distinctive fusulinids (Skinner and Wilde, 1965) and indicate rapid lateral variations in limestone facies and ages. Carbonate platform development took place in two principal stages (Watkins, 1985). In both stages, shallow-water facies occur to the south, and more distal facies to the north (Watkins, 1985; Miller, 1989).

The McCloud Limestone is locally unconformably overlain by rocks of the Upper Permian Bollibokka Group of Coogan (1960; Nosoni and Dekkas Formations), although in many places the contact is intruded by mafic quartz diorite dikes (Albers and Robertson, 1961). The Bollibokka Group rocks collectively represent more than 5 km of volcanogenic debris and lavas that were deposited in a submarine setting, although original stratigraphic thickness may have been as little as half that (Miller, 1989). Proximal lavas are abundant in the southern part of the terrane (Albers and Robertson, 1961). More distal volcaniclastic dispersal facies are widespread in the north, characterized by volcanogenic debris flows, interstratified turbidites, and other volcanogenic sediment gravity flows. Late Permian (Guadalupian) fusulinids (Coogan, 1960) and corals (Stevens and others, 1987) occur in limestone lenses and as clasts in tuffaceous debris flows throughout the Bollibokka Group (Fig. 5). Based on fusulinid assemblages (C. H. Stevens, *in* Miller, 1989) from the basal part of the more basinal northerly section near McCloud Lake, the Nosoni Formation there may be slightly older, spanning some part of Leonardian time (Fig. 5; Miller, 1989). Brachiopods from limestone lenses near the top of the Dekkas Formation indicate a Wordian (early Guadalupian or earliest Late Permian) age (J. T. Dutro, Jr., personal communication, 1988), implying that the very thick succession of volcanogenic rocks was deposited in a short time interval (about 4 m.y.) during the earliest Late Permian. This is confirmed by recent studies of radiolarian biostratigraphy from shallow-water (spiculitic) chert in the upper part of the Dekkas Formation in the central part of the terrane, where lower to middle Guadalupian radiolarians date the uppermost part of the formation (Fig. 5; Noble and Renne, 1990).

In the south, the Dekkas Formation is overlain by keratophyric tuff and lava of the Bully Hill Rhyolite (Albers and Robertson, 1961). The trondjhemitic Pit River stock is believed to be cogenetic with the Dekkas Formation, and Bully Hill Rhyolite and has yielded a Late Permian radiometric age of approximately 261 Ma (Fig. 5; discordant U-Pb, zircon, Fraticelli and others, 1985). Although this age is slightly older than the depositional age of the underlying rocks (after the time scale of Harland and others, 1989), it supports the inference of short-lived Permian magmatism in the eastern Klamath terrane indicated by the new biostratigraphic age constraints discussed above.

The Lower(?) and Middle Triassic Pit Formation overlies the Dekkas Formation (Coogan, 1960) and, in the south, overlies the Bully Hill Rhyolite (Albers and Robertson, 1961). The Pit Formation contains siliceous argillite, silicic tuffaceous and crystal-rich turbidites, and radiolarian-bearing hemipelagic sedimentary rocks in its basal part, recording basinal and volcanogenic sedimentation (Sanborn, 1960; Curtis, 1983; Noble and Renne, 1990). These strata are very sparsely fossiliferous, and thus are generally poorly dated. In the northern part of the terrane, east of McCloud Lake, Middle Triassic radiolarians are preserved in the overlying Brock Shale, and Early Triassic radiolarians have been tentatively identified from the basal part of the Pit Formation (Fig. 5; B. Murchey, *in* Curtis, 1983). The Pit Formation may be Lower(?) and Middle Triassic in this area. At Nosoni Creek (Fig. 5), Late Permian radiolarians (no stage given) were reported from medium- and thin-bedded white and gray chert beds assigned to the Pit Formation (Silberling and Jones, 1982). The Permian age determination was not confirmed, however, by a more detailed biostratigraphic study in the same area by Noble and Renne (1990). These workers tentatively suggest an Early Triassic age for the base of the Pit Formation in this area based on poorly preserved radiolarians and long-ranging (Late Permian and Early Triassic) conodont faunas. Near Lake Shasta, late Middle Triassic fossils occur tens of meters above the base of the Pit Formation, and both Middle and Late Triassic fossils are recognized within the formation (Albers and Robertson, 1961). Thus the base of the Pit Formation appears to be time transgressive. It may be Early Triassic in the northern and central parts of the terrane, and Middle Triassic in age in the south (Fig. 5).

Most previous work from the eastern Klamath terrane has suggested that continuous deposition occurred during the Late Permian and Early Triassic, based on broad age constraints and apparent conformity of the generally poorly exposed Dekkas-Pit contact (e.g., Albers and Robertson, 1961; Miller, 1987; Renne and Scott, 1988). Better biostratigraphic resolution for ages of units, summarized above, strongly suggests that no post–middle Guadalupian Permian rocks are present in the eastern Klamath terrane (Fig. 5). This is consistent with an isotopic age of 261 Ma for the Pit River stock that is inferred to be cogenetic with the Bully Hill Rhyolite (Fraticelli and others, 1985), limiting the Bollibokka volcanic section to the lowermost Upper Permian. Lower Triassic strata are only tentatively documented. Thus, existing age and stratigraphic data indicate that most of the Upper Permian and possibly the Lower Triassic (post–middle Guadalupian to pre–Early or Middle Triassic) rocks are missing within the eastern Klamath terrane. This hiatus supports the inference that deformation of the Paleozoic sequence occurred prior to deposition of the Pit Formation, and that a structural break occurs at the Permian-Triassic stratigraphic horizon (Miller, 1988, 1990).

The Pit Formation is overlain by the Upper Triassic (Carnian) Hosselkus Limestone and Upper Triassic Brock Shale, which sequentially record shallow-water and basinal sedimentation (Sanborn, 1960). The Hosselkus Limestone of the eastern Klamath terrane was originally correlated with the Hosselkus Limestone of the northern Sierra terrane by Diller (1906). Calcareous volcaniclastic rocks of the Upper Triassic Modin Formation record deposition of volcanic debris. An unconformity separates the Triassic succession from overlying Lower and Middle Jurassic volcanic and sedimentary rocks of the Arvison Formation (of Sanborn, 1960) and Potem Formation (Fig. 5; Sanborn, 1960). Much of what was previously mapped by Sanborn (1960) as his Middle Jurassic Bagley Andesite, based on intercalation with the Potem Formation, is now thought to represent hypabyssal intrusive rocks of Jurassic and Cretaceous age (Renne and Scott, 1988; Fig. 5).

DISCUSSION

Although many broad parallels exist between the eastern Klamath and northern Sierra terranes, important differences in stratigraphy, structural development, and relations within related basement complexes are also present. Differences in basement terranes have long obscured the possible paleogeographic relation between the terranes. Rocks within the Trinity complex and central metamorphic belt (Klamath Mountains), however, are now thought to have correlatives in the Sierra Nevada (Feather River peridotite belt and associated rocks; Hacker and Peacock, this volume), but the exact relation between the Feather River peridotite belt and rocks of the northern Sierra terrane is poorly understood. Similarities between lower Paleozoic volcaniclastic, quartzose, and sedimentary-lithic–rich strata of the Yreka terrane and Shoo Fly Complex have long been recognized (Bond and Devay, 1980), and have recently been expanded upon (Wallin and others, 1988; Potter and others, this volume). A new data set suggesting that outliers of the Bragdon Formation (eastern Klamath terrane) may overlie the Trinity complex and contain peridotite detritus along a basal unconformity (Charvet and others, 1989) provides a critical middle Paleozoic depositional tie between upper Paleozoic arc strata in the eastern Klamath terrane and the regionally widespread lower Paleozoic rocks. A similar tie has long been recognized in the northern Sierra terrane, where upper Paleozoic arc rocks rest depositionally on the imbricated Shoo Fly Complex. Detritus from lower Paleozoic supracrustal rocks in both areas was sporadically reworked into the overlying arc-related successions.

The tectonic development of upper Paleozoic sequences in the eastern Klamath terrane and the Taylorsville and Tahoe sequences in the northern Sierra terrane are interpreted from the stratigraphy detailed above (Fig. 6). In the eastern Klamath terrane, the Copley Greenstone, Balaklala Rhyolite, and cogenetic intrusive rocks record primitive, ensimatic arc volcanism in an extensional setting during the early Middle Devonian and possibly earlier. The sporadic influence of continent-derived sedimentation is present, however, in older rocks of the Yreka terrane and recycled into younger rocks of the Bragdon Formation.

The oldest Devonian volcanic rocks in the eastern Klamath and northern Sierra terranes are of different ages and record differing tectonic and magmatic evolution. Comparison of either the radiometric or biostratigraphic data for Devonian magmatism in both the eastern Klamath and northern Sierra terranes indicates that Middle Devonian and older(?) Copley-Balaklala volcanism and Mule Mountain plutonism preceded the initiation of arc magmatism in the northern Sierra terrane. Zircon ages on intrusive rocks that are cogenetic with volcanic sections from each terrane differ by 15 to 22 m.y.; absolute age differences of 13 to 24 m.y. are implied by differences in fossil ages from the volcanic sections and calibration of the biostratigraphic time scale (Figs. 3, 5; Harland and others, 1990). An overlap in the age of magmatism in both areas can be construed from discrepancies between the U-Pb zircon and biostratigraphic data, if the oldest zircon ages for intrusive rocks in the northern Sierra terrane (378 Ma and 364 to 385 Ma) are compared to the fossil ages for the youngest volcanic rocks in the eastern Klamath terrane (Eifelian, 386 to 380.8 Ma; after Harland and others, 1990). Such overlap appears unlikely, however, based on internal discrepancies for age data from each terrane. Furthermore, in contrast to Late Devonian and early Carboniferous magmatism in both the eastern Klamath and northern Sierra terranes, Middle Devonian and older(?) magmatism in the eastern Klamath terrane preceded inferred regionally widespread compressional tectonism. We consider the possible paleotectonic relation between the oldest volcanic rocks in both areas unlikely in light of the minimal implied age overlap, the differing nature of the age data sets that allow such overlap, differences in the nature of magmatism, and differing temporal relations to late Middle or Late Devonian deformation. These differences may account for differences in geochemistry previously ascribed to paleogeographic separation (e.g., Rouer and others, 1989). These interpretations are subject to modification as new and better age and field constraints are acquired.

Late Devonian and earliest Mississippian localized, voluminous volcanism is recorded by rocks of the Taylorsville sequence, although sparse, basement-derived epiclastic deposits are also known (Fig. 6). The Taylorsville and Tahoe sequences were both deposited in the Shoo Fly Complex. Interfingering of the Taylorsville pyroclastic rocks with chert- and metasedimentary-lithic–rich epiclastic rocks of the Tahoe sequence (Fig. 3, 6), together with shared depositional basement, establishes lateral equivalence between the two sequences (Harwood and Murchey, this volume). Broadly coeval strata of the Bragdon Formation in the eastern Klamath terrane also contain chert- and metasedimentary-lithic–rich rocks that are interfingered with water-laid silicic pyroclastic debris. Lithologic equivalents to crystal-rich beds in the Sierra Buttes Formation may indicate a lithostratigraphic tie to the Taylorsville sequence. Isotopic data from detrital zircon in the Bragdon Formation suggest a source isotopically similar to older elements of the eastern Klamath and northern Sierra terranes. Although this interpretation is merely permissible from the isotopic data, it is strongly supported by regional geologic relations and by the provenance of framework constituents in sandstone of the Bragdon Formation. The provenance tie implies a middle Paleozoic paleogeographic relation between the two terranes (Fig. 6).

Some workers interpret the lower contact of the lower Carboniferous and older(?) Bragdon Formation as an angular unconformity that records imbrication of lower Paleozoic rocks in the Klamath Mountains (Fig. 6; Boucot and others, 1974; Charvet and others, 1989). In this context, deformation occurred prior to deposition of the Bragdon Formation and, based on available fossil ages, occurred during post–early Middle Devonian (post-

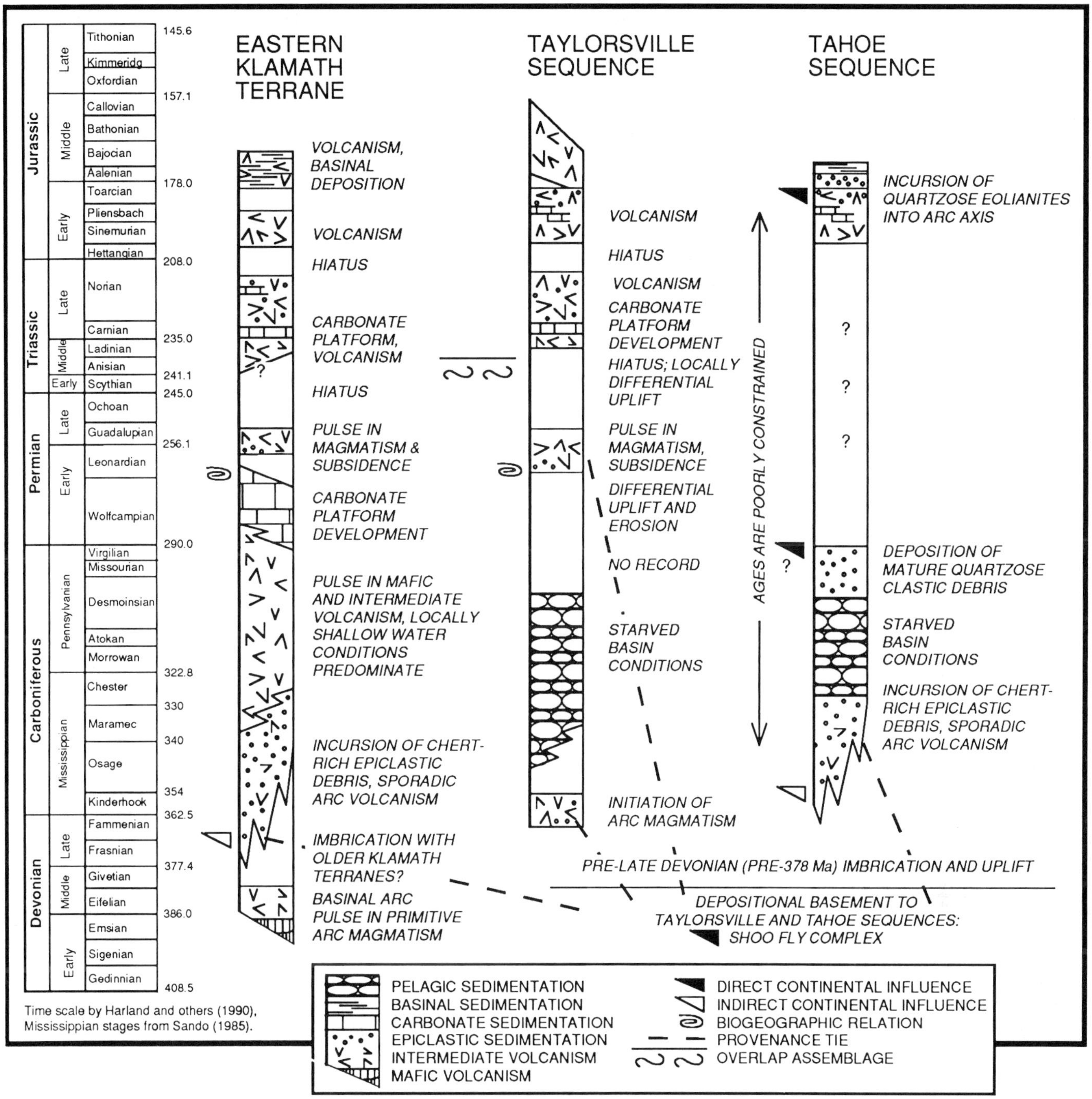

Figure 6. Summary of inferred lithostratigraphic and paleogeographic relations between the eastern Klamath and northern Sierra terranes.

Eifelian) and pre–early Carboniferous time, and was thus broadly contemporaneous with deformation and metamorphism of the central metamorphic belt. The age constraints, although broad, suggest that deformation may have been contemporaneous with late Middle or early Late Devonian imbrication of the Shoo Fly Complex in the northern Sierra terrane. If this is true, a new parallel is established between the structural evolution of the eastern Klamath and northern Sierra terranes (Fig. 6).

The Tahoe and Taylorsville sequences record a widespread hiatus in volcanism, during which pelagic sedimentation was active at least until the Middle(?) Pennsylvanian (Fig. 6). In the eastern Klamath terrane, sporadic volcanism and volcaniclastic sedimentation were accompanied by localized uplift during the Late Mississippian and Pennsylvanian. Rocks of this age in both terranes differ significantly, and a bathymetric or geographic barrier is implied.

In the Taylorsville sequence, subaerial(?) uplift either initiated during or persisted until the early Early Permian (Wolfcampian), whereas in the eastern Klamath terrane, differential uplift was punctuated by deposition of laterally variable and time transgressive shallow-water carbonate platform deposits that record slightly deeper water conditions to the north (Fig. 6). Mid-Permian (Leonardian to early Guadalupian) volcanogenic rocks are widespread and volumetrically important in both terranes; they form the base of a mid-Permian to Middle Jurassic lithostratigraphic succession that is common to both the eastern Klamath terrane and the Taylorsville sequence. Both terranes record a break in deposition during the late Late Permian and Early(?) Triassic. In the northern Sierra terrane, the hiatus is marked in one area by a pronounced angular unconformity, but elsewhere in both terranes Triassic or Jurassic strata rest disconformably on Permian rocks. The Triassic and Jurassic successions, although laterally variable, record parallel stratigraphic and tectonic development between the two terranes (Fig. 6).

Much of the post-Carboniferous tectonic development of the Tahoe sequence is not known due to sparse preservation in roof pendants and as scattered exposures of wall rocks in Mesozoic plutons. The relatively well preserved Jurassic section of the Tahoe sequence, however, has correlatives in the Sailor Canyon Formation that occur in stratigraphic succession with the Taylorsville sequence (Fisher, this volume). Jurassic rocks in the Tahoe sequence also uniquely record the incursion of quartz-rich, wind-sorted sand from the continent into the axis of the Middle Jurassic arc (Busby-Spera, 1988; Busby-Spera and others, this volume).

Proximity between the eastern Klamath and northern Sierra terranes throughout their late Paleozoic and early Mesozoic development is supported by provenance ties, common stratigraphic elements, structural evolution, and regional geologic relations. Differences in stratigraphy indicate, however, that each terrane played a different role in the evolution of a convergent plate boundary during given time intervals. For instance, eastward-dipping subduction is recorded by late Middle or Late Devonian deformation, metamorphism, and synorogenic to postorogenic sedimentation in the Klamath Mountains, and more specifically by the relation of the central metamorphic belt subduction zone complex to the eastward, upper plate rocks of the eastern Klamath belt. In contrast, coeval deformation in the northern Sierra terrane appears to have occurred in a fore-arc or intra-arc setting, and evidence for subduction zone metamorphism is not well established. During the middle and late Carboniferous, rocks of the northern Sierra terrane record a common stratigraphic evolution with marginal basin rocks of the Golconda allochthon to the east (Harwood and Murchey, this volume), whereas the eastern Klamath terrane records active arc volcanism, indicating relative proximity to a subduction zone. Again, subduction was eastward-dipping, as implied by paleogeographic ties between the arc (eastern Klamath terrane), marginal basin (northern Sierra terrane and Havallah and Schoonover sequences; see Whiteford, this volume; Murchey, this volume), and the western margin of the North American continent.

In the Early Permian, both areas were relatively high-standing, as recorded by an unconformity in the northern Sierra terrane, and the deposition of shallow-water carbonate platforms in the eastern Klamath terrane. During the mid-Permian, common histories of volcanism in both the eastern Klamath and northern Sierra terranes imply that they probably lay approximately along strike above an oceanic convergent plate boundary. During this time interval, continued deposition of the Havallah sequence indicates persistence of a back-arc basin to the east of the arc. Biogeographic relations may imply a substantial separation of the arc terranes from the continent during this time interval (Stevens and others, this volume). During the latest Permian, however, the polarity of subduction is poorly constrained. Closing of the Havallah-Schoonover basin may have occurred during back-arc contraction or during a short-lived interval of subduction polarity reversal. Terranes of Cache Creek affinity throughout the Cordillera represent fragments of a subduction complex that developed west of the volcanic arc rocks, and was active during the Late Triassic, based on metamorphic ages of blueschist blocks and the youngest rocks contained in the complexes, and for some unknown interval prior to the Late Triassic (see discussion in Miller and Wright, 1987; Miller, 1987). Thus, by Early or Middle Triassic time, eastward-dipping subduction was active or reestablished, and the volcanic arc was hinged to the continent in the Mojave Desert region (Walker, 1988). Progressively more oceanic environments are recorded by the northern Sierra and eastern Klamath terranes.

The significance of the tectonic slivers, such as the Soda Ravine block, that lie between the Feather River peridotite belt and other rocks of the northern Sierra terrane is enigmatic. Based on lithologic, biogeographic, and age similarities, Permian and Triassic rocks within the slivers are thought to be correlative to coeval rocks of the eastern Klamath terrane (Jayko, 1988, this volume; Harwood and others, 1988). If this is true, the tectonic slivers potentially hold the key to our understanding of the relation between the northern Sierra and eastern Klamath terranes. Lower Permian, Middle Triassic, and Upper Triassic rocks occur in a largely undated matrix. In the Soda Ravine block, however, matrix tuffaceous argillite is considered to be lithologically equivalent to rocks in the Pit Formation of the eastern Klamath terrane, and is associated with similar faunal assemblages in fossiliferous limestone (N. J. Silberling, *in* Jayko, 1988); associated lithic-rich clastic rocks may correlate to the Baird Formation (Jayko, this volume). These relations imply that the matrix and limestone are the same age and form parts of an originally coherent stratigraphic succession. Upper Triassic rocks are the youngest known from the Soda Ravine block and provide a maximum age constraint for final tectonic juxtaposition to other rocks in the northern Sierra terrane. Rocks to the east, in the northern Sierra terrane, were imbricated after deposition of upper Middle Jurassic (Callovian) strata; involvement of rocks with eastern Klamath affinity (at Genesee, Fig. 3), indicate their proximity prior to late Middle or Late Jurassic deformation. Furthermore, limestone clasts thought to have been derived from the McCloud Limestone occur in Middle Jurassic conglomerate in the western belt of the Sierra Nevada (see summary in Miller, 1987), indicating that the slivers were probably emplaced by that time. More specific age constraints come from the inference that Upper Triassic conglomerate lying above the Taylorsville sequence in the Lake Almanor area records the juxtaposition of the tectonic slivers (Jayko, this volume). This suggestion is corroborated by the occurrence of Late Triassic(?) blueschist to the west in both the Klamath Mountains and Sierra Nevada (Schweickert and others, 1980; Hacker and Goodge, this volume), and perhaps by the absence of post-Triassic rocks within the slivers.

The nature of the faulting that juxtaposed the tectonic slivers to the Shoo Fly Complex is also ambiguous, but is interpreted to have occurred along a strike-slip fault (Jayko, this volume), an intra-arc transform (e.g., Saleeby, 1990), or perhaps even within a thrust belt setting, during the Late Triassic or somewhat later. Future study of the kinematic relations of this event may provide important constraints on the palinspastic restoration of the eastern Klamath and northern Sierra terranes to a pre–Late Triassic configuration, and substantially increase our understanding of the details of their paleogeographic relations.

SUMMARY AND CONCLUSIONS

In summary, Middle Devonian and older(?) primitive arc magmatism recorded in the eastern Klamath terrane has no known analog in the northern Sierra terrane. Late Devonian deformation imbricated older lithologic units in the northern Sierra; contemporaneous deformation in the eastern Klamath belt has been suggested (Boucot and others, 1974; Charvet and others, 1989), and occurred above an eastward-dipping subduction zone (Peacock, 1987). Deformation was succeeded in both areas by deposition of volcanogenic rocks and chert- and sedimentary-lithic–rich clastic rocks in varying proportions. Strong lithologic similarities exist between the Bragdon Formation in the eastern Klamath terrane and chert-rich clastic rocks in the Tahoe sequence (southern part of the northern Sierra terrane), which may indicate that during Late Devonian(?) and Early Carboniferous time, the eastern Klamath terrane lay farther south (in present coordinates) than is suggested by its present location, and perhaps outboard of the northern Sierra terrane. The two terranes record different histories for the middle Carboniferous; starved basin conditions were established in the northern Sierra terrane, which was geographically or bathymetrically isolated from an influx of volcaniclastic debris into shallow-marine settings like those recorded in the eastern Klamath terrane (Fig. 6). Uplift of the Taylorsville sequence during the early Early Permian and perhaps somewhat earlier (Late Pennsylvanian) was paralleled by deposition of Pennsylvanian volcaniclastic rocks in a shallow-marine and strand setting and development of Early Permian shallow-water carbonate platforms on blocks that were uplifted differentially in the eastern Klamath terrane. Shared histories of mid-Permian volcanism and subsequent similar stratigraphic development form a lithostratigraphic tie between the two terranes, overlapping the stratigraphically distinct but paleogeographically related Paleozoic substrates (Fig. 6).

A growing body of stratigraphic, paleontologic, isotopic, geochemical, and structural data allow detailed comparison of the tectonic evolution of the eastern Klamath and northern Sierra terranes. Apparent problems in calibration of the radiometric and biostratigraphic time scales, conflicting age and field relations, and poorly understood structural histories leave unanswered questions concerning the details of the possible relation between the terranes. Discrepancies in paleogeographic constraints on the relation between the arc terranes and western North America yielded by biogeographic (Stevens and others, this volume) and geologic data (Miller, 1987, 1989; Harwood and Murchey, this volume) also warrant further attention. However, new data have substantially increased our understanding of the late Paleozoic relations and paleotectonic setting of both terranes. Although upper Paleozoic rocks in both terranes vary significantly in detail, sporadic but persistent evidence for a paleogeographic relation is present (Fig. 6), and arc-related sequences in both terranes are inferred to be linked to a common, dynamic convergent margin or sequential convergent margins that evolved through time. The differing paleotectonic role of each terrane during any given time interval is suggested by differences in stratigraphy; these changes can therefore be charted as data increase constraining depositional environments, the nature, timing, and tectonic setting of magmatism, deformation, and the paleogeographic relations between various terranes.

ACKNOWLEDGMENTS

The National Research Council Resident Research Associate program (NASA) at the Jet Propulsion Laboratory and the National Academy of Sciences are gratefully acknowledged for supporting M. M. Miller during writing of this manuscript. George Gehrels, John Goodge, and Rodney Watkins provided helpful reviews. This work was carried out by the Jet Propulsion Laboratory, California Institute of Technology, under contract with the National Aeronautics and Space Administration.

REFERENCES CITED

Albers, J. P., and Bain, J.H.C., 1985, Regional setting and new information of some critical geologic features of the West Shasta District, California: Economic Geology, v. 80, p. 2072–2091.

Albers, J. P., and Robertson, J. F., 1961, Geology and ore deposits of East Shasta copper-zinc district, Shasta County, California: U.S. Geological Survey Professional Paper 338, 107 p.

Albers, J. P., Kistler, R. W., and Kwak, L., 1981, The Mule Mountain stock; An early Middle Devonian pluton in northern California: Isochron/West, v. 31, p. 17.

Barker, F., Millard, H. T., Jr., and Knight, R. J., 1979, Reconnaissance geochemistry of Devonian island-arc volcanic and intrusive rocks, West Shasta District, California, *in* Barker, F., ed., Trondhjemites, dacites, and related rocks: Amsterdam, Elsevier, p. 531–545.

Bond, G. C., and Devay, J. C., 1980, Pre–Upper Devonian quartzose sandstones in the Shoo Fly Formation, northern California; Petrology, provenance, and implications for regional tectonics: Journal of Geology, v. 88, p. 285–288.

Boucot, A. J., Dunkle, D. H., Potter, A., Savage, N. M., and Rohr, C., 1974, Middle Devonian orogeny in western North America?; A fish and other fossils: Journal of Geology, v. 82, p. 691–708.

Boudier, F., LeSueur, E., and Nicolas, A., 1989, Structure of an atypical ophiolite; The Trinity complex, eastern Klamath Mountains, California: Geological Society of America Bulletin, v. 101, p. 820–833.

Bratten, B. L., and Taylor, D. A., 1978, Reassessment of the age of the Thompson Limestone, Lower Jurassic, Plumas County, California: Journal of Paleontology, v. 52, p. 208–209.

Brouxel, M., and Lapierre, H., 1988, Geochemical study of an early Paleozoic island-arc–back-arc basin system; Part 1, The Trinity Ophiolite (northern California): Geological Society of America Bulletin, v. 100, p. 1111–1119.

Brouxel, M., Lapierre, H., Michard, A., and Albarede, F., 1988, Geochemical study of an early Paleozoic island-arc–back-arc basin system; Part 2, Eastern Klamath, early to middle Paleozoic island-arc rocks (northern California): Geological Society of America Bulletin, v. 100, p. 1120–1130.

Burchfiel, B. C., and Davis, G. A., 1981, Triassic and Jurassic tectonic evolution of the Klamath Mountains–Sierra Nevada geologic terrane, *in* Ernst, W. G., ed., The geotectonic development of California; Rubey Volume 1: Englewood Cliffs, New Jersey, Prentice-Hall, p. 50–70.

Busby-Spera, C. J., 1988, Speculative tectonic model for the early Mesozoic arc of the southwest Cordilleran United States: Geology, v. 16, p. 1121–1125.

Cashman, S. M., 1980, Devonian metamorphic event in the northeastern Klamath Mouintains, California: Geological Society of America Bulletin, v. 91, p. 453–459.

Charvet, J., Lapierre, H., and Campos, C., 1989, Les effets de la phase Antler (Devonien superieur-Carbonifere inferieur) dans les Klamath orientales (Californie, U.S.A.): Implications geodynamiques: Comptes Rendu Academie Science Paris, Serie II, p. 1629–1635.

Clark, L. D., Imlay, R. W., McMath, V. E., and Silberling, N. J., 1962, Angular unconformity between Mesozoic and Paleozoic rocks in the northern Sierra Nevada, California: U.S. Geological Survey Professional Paper 450-B, p. B15–B19.

Coogan, A. H., 1960, Stratigraphy and paleontology of the Permian Nosoni and Dekkas Formations (Bolibokka Group): University of California Publications in Geological Sciences, v. 5, p. 243–316.

Cotkin, S. J., and Armstrong, R. L., 1987, Rb/Sr age, geochemistry, and tectonic significance of blueschist from the schist of Skookum Gulch, eastern Klamath Mountains; Introducing the Callahan event: Geological Society of America Abstracts with Programs, v. 19, p. 367–368.

Curtis, D. R., 1983, Stratigraphy and origin of the Pit Formation, northern California [M.Sc. thesis]: Berkeley, University of California, 56 p.

Davis, G. A., 1969, Tectonic correlations, Klamath Mountains and western Sierra Nevada, California: Geological Society of America Bulletin, v. 80, p. 1095–1108.

Diller, J. S., 1905, The Bragdom Formation: American Journal of Science, 4th Series, v. 19, p. 379–387.

——, 1906, Description of the Redding Quadrangle: U.S. Geological Survey Geologic Atlas, Folio 138.

——, 1908, Geology of the Taylorsville region, California: U.S. Geological Survey Bulletin 353, 128 p.

Durrell, C., and D'Allura, J. A., 1977, Upper Paleozoic section in eastern Plumas and Sierra Counties, Sierra Nevada, California: Geological Society of America Bulletin, v. 88, p. 844–852.

Fraticelli, L. A., Albers, J. P., and Zartman, R. E., 1985, The Permian Pit River stock of the McCloud plutonic belt, eastern Klamath terrane, northern California: Isochron/west, no. 44, p. 6–8.

Girty, G. H., and Wardlaw, M. S., 1984, Was the Alexander terrane a source of feldspathic sandstones in the Shoo Fly Complex, Sierra Nevada, California?: Geology, v. 12, p. 339–342.

——, 1985, Petrology and provenance of pre–Late Devonian sandstones, Shoo Fly Complex, northern Sierra Nevada, California: Geological Society of America Bulletin, v. 96, p. 516–521.

Hannah, J. L., 1980, Stratigraphy, petrology, paleomagnetism, and tectonics of Paleozoic arc complexes, northern Sierra Nevada, California [Ph.D. thesis]: Davis, University of California, 323 p.

Hannah, J. L., and Moores, E. M., 1986, Age relationships and depositional environments of Paleozoic strata, northern Sierra Nevada: Geological Society of America Bulletin, v. 97, p. 787–797.

Hanson, R. E., 1983, Volcanism, plutonism, and sedimentation in a Late Devonian submarine island-arc setting, northern Sierra Nevada, California [Ph.D. thesis]: New York, Columbia University, 201 p.

Hanson, R. E., and Schweickert, R. A., 1986, Stratigraphy of mid-Paleozoic island-arc rocks in part of the northern Sierra Nevada, Sierra and Nevada Counties, California: Geological Society of America Bulletin, v. 97, p. 986–998.

Hanson, R. E., Saleeby, J. B., and Schweickert, R. A., 1988, Composite Devonian island-arc batholith in the northern Sierra Nevada, California: Geological Society of America Bulletin, v. 100, p. 446–457.

Harland, W. B., and 5 others, 1990, A Geologic time scale 1989: Cambridge, England, Cambridge University Press, 263 p.

Harwood, D. S., 1983, Stratigraphy of upper Paleozoic volcanic rocks and regional unconformities in part of the northern Sierra terrane, California: Geological Society of America Bulletin, v. 94, p. 413–422.

——, 1988, Tectonism and metamorphism in the northern Sierra terrane, northern California, *in* Ernst, W. G., ed., Metamorphism and crustal evolution of the western United States; Rubey Volume 7: Englewood Cliffs, New Jersey, Prentice-Hall, p. 764–788.

Harwood, D. S., Jayko, A. S., Harris, A. G., Silberling, N. J., and Stevens, C. H., 1988, Permian-Triassic rocks silvered between the Shoo Fly Complex and the Feather River peridotite belt, northern Sierra Nevada, California: Geological Society of America Abstracts with Programs, v. 20, p. 167–168.

Hotz, P. E., 1977, Geology of the Yreka Quadrangle, Siskiyou County, California: U.S. Geological Survey Bulletin 1476, 72 p.

Imlay, R. W., 1961, Late Jurassic ammonites from the western Sierra Nevada, California: U.S. Geological Survey Professional Paper 374-D, 30 p.

——, 1968, Lower Jurassic (Pleinsbachian and Toarcian) ammonites from eastern Oregon and California: U.S. Geological Survey Professional Paper 593, 51 p.

Irwin, W. P., 1966, Geology of the Klamath Mountains province: California Division of Mines and Geology, Bulletin 190, p. 19–38.

——, 1972, Terranes of the western Paleozoic and Triassic belt in the southern Klamath Mountains, California: U.S. Geological Survey Professional Paper 800-C, C103-C111.

Jacobsen, S. B., Quick, J. E., and Wasserburg, G. J., 1984, A Nd and Sr isotopic study of the Trinity peridotite; Implications for mantle evolution: Earth and

Planetary Science Letters, v. 68, p. 361–378.

Jayko, A. S., 1988, Paleozoic and Mesozoic rocks of the Almanor 15′ Quadrangle, Plumas County, California: U.S. Geological Survey Open-File Report 88-757, 12 p.

Kinkel, A. R., Jr., Hall, W. E., and Albers, J. P., 1956, Geology and base-metal deposits of West Shasta Copper-Zinc district, Shasta County, California: U.S. Geological Survey Professional Paper 285, 156 p.

Lanphere, M. A., Irwin, W. P., and Hotz, P. E., 1968, Isotopic age of the Nevadan orogeny and older plutonic and metamorphic events in the Klamath Mountains, California: Geological Society of America Bulletin, v. 79, p. 1027–1052.

Lapierre, H., Albarede, F., Albers, J., Coulon, C., and Cabanis, B., 1985a, Early Devonian volcanism in the eastern Klamath Mountains, California; Evidence for an immature island arc: Canadian Journal of Earth Sciences, v. 22, p. 214–227.

Lapierre, H., Cabanis, B., Coulon, C., Brouxel, M., and Albarede, F., 1985b, Geodynamic setting of Early Devonian Kuroko-type sulfide deposits in the eastern Klamath Mountains (northern California) inferred by the petrological and geochemical characteristics of the associated island-arc volcanic rocks: Economic Geology, v. 80, p. 2100–2113.

Lindberg, P. A., 1985, A volcanogenic interpretation for massive sulfide origin, West Shasta district, California: Economic Geology, v. 80, p. 2240–2254.

McMath, V. E., 1958, The geology of the Taylorsville area, Plumas County, northern Sierra Nevada [Ph.D. thesis]: Los Angeles, University of California, 199 p.

——, 1966, Geology of the Taylorsville area, northern Sierra Nevada: California Division of Mines and Geology Bulletin 190, p. 173–183.

Miller, M. M., 1986, Tectonic evolution of late Paleozoic island arc sequences in the western U.S. Cordillera; with detailed studies from the Klamath Mountains, northern California [Ph.D. thesis]: Stanford, California, Stanford University, 201 p.

——, 1987, Dispersed remnants of a northeast Pacific fringing arc; Upper Paleozoic island arc terranes of Permian McCloud faunal affinity, western U.S.: Tectonics, v. 6, p. 807–830.

——, 1988, Permo-Triassic(?) deformation within the eastern Klamath terrane, northern California: Geological Society of America Abstracts with Programs, v. 20, p. 216.

——, 1989, Intra-arc sedimentation and tectonism; Late Paleozoic evolution, eastern Klamath terrane, California: Geological Society of America Bulletin, v. 101, p. 170–187.

——, 1990, Episodic compressional tectonism in an intra-arc setting, eastern Klamath terrane, northern California, *in* Wiley, T. J., Howell, D. G., and Wang, F., eds., Terrane analysis of China and the Pacific Rim: Houston, Texas, Circum Pacific Council for Energy and Mineral Resources, Earth Science Series, no. 13, in press.

Miller, M. M., and Cui, B., 1989, Carboniferous arc-related sedimentation, Klamath Mountains, California; Hybrid fan characteristics and dual sediment provenance: Canadian Journal of Earth Sciences, v. 26, p. 927–940.

Miller, M. M., and Saleeby, J. B., 1990, Continental detrital zircon in Carboniferous ensimatic arc rocks, Bragdon Formation, eastern Klamath terrane, northern California: Geological Society of America Bulletin, in press.

Miller, M. M., and Wright, J. E., 1987, Paleogeographic implications of Permian Tethyan corals from the Klamath Mountains, California: Geology, v. 16, p. 266–269.

Miller, M. M., Saleeby, J. B., and Cui, B., 1988, Provenance of the Carboniferous Bragdon Formation, eastern Klamath terrane, California; Proterozoic detrital zircon and quartzose detritus as continental tracers in oceanic convergent margin systems, *in* Howell, D. G., and Wiley, T. J., eds., Provisional Proceedings, 4th International Tectonostratigraphic Terrane Conference, Nanjing University, September, 1988, Nanjing, P.R.C., U.S. Geological Survey, p. 67–69.

Noble, P., and Renne, P. R., 1990, Paleoenvironmental and biostratigraphic significance of siliceous microfossils of the Permian-Triassic Redding section, eastern Klamath Mountains, California: Marine Micropaleontology, v. 15, no. 3/4, p. 379–391.

Peacock, S. M., 1987, Serpentinization and infiltration metasomatism in the Trinity peridotite, Klamath province, northern California: implications for subduction zones: Contributions to Mineralogy and Petrology, v. 95, p. 55–70.

Peacock, S. M., and Norris, P. J., 1989, Metamorphic evolution of the CMB, Klamath province, California; An inverted metamorphic gradient beneath the Trinity peridotite: Journal of Metamorphic Geology, v. 7, p. 191–209.

Potter, A. W., Hotz, P. E., and Rohr, D. M., 1977, Stratigraphy and inferred tectonic framework of lower Paleozoic rocks in the eastern Klamath Mountains, northern California, *in* Stewart, J. H., Stevens, C. H., and Fritsche, A. E., eds., Paleozoic paleogeography of the western United States: Pacific Section, Society of Economic Paleontologists and Mineralogists Pacific Coast Paleogeographic Symposium 1, p. 421–440.

Quick, J. E., 1981, Petrology and petrogenesis of the Trinity peridotite, an upper mantle diapir in the eastern Klamath Mountains, northern California: Journal of Geophysical Research, v. 86, p. 11837–11863.

Renne, P. R., and Scott, G. R., 1988, Structural chronology, oroclinal deformation, and tectonic evolution of the southeastern Klamath Mountains, California: Tectonics, v. 7, p. 1223–1242.

Rouer, O., Lapierre, H., Mascle, G., Coulon, C., and Albers, J., 1989, Geodynamic implications of Devonian silicic arc magmatism in the Sierra Nevada and Klamath Mountains, California: Geology, v. 17, p. 177–180.

Saleeby, J. B., 1983, Accretionary tectonics of the North American Cordillera: Annual Reviews of Earth and Planetary Sciences, v. 15, p. 45–73.

Saleeby, J. B., 1990, Tectonic and petrogenetic settings of U.S. Cordillera ophiolites, *in* Burchfiel, B. C., Lipman, P., and Zoback, M. L., eds., The Cordilleran Orogen; Conterminous U.S.: Boulder, Colorado, Geological Society of America, The Geology of North America, v. G-3 (in press).

Saleeby, J. B., Hannah, J. L., and Varga, R. J., 1987, Isotopic age constraints on middle Paleozoic deformation in the northern Sierra Nevada, California: Geology, v. 15, p. 757–760.

Sanborn, A. F., 1960, Geology and paleontology of the southwest quarter of the Big Bend Quadrangle Shasta County, California: California Division of Mines and Geology Special Report 63, 26 p.

Sando, W. J., 1985, Revised Mississippian time scale, western interior region, conterminous United States: U.S. Geological Survey Bulletin 1605-A, p. A-15–A-26.

Savage, N. M., 1976, Middle Devonian species of *Polygnathus* from the Kennett Formation, northern California: Journal of Paleontology, v. 50, p. 374–379.

Schweickert, R. A., Armstrong, R. L., and Harakal, J. E., 1980, Lawsonite blueschist in the northern Sierra Nevada, California: Geology, v. 8, p. 27–31.

Schweickert, R. A., Harwood, D. S., Girty, G. H., and Hanson, R. E., 1984, Tectonic development of the northern Sierra terrane; An accreted late Paleozoic island arc and its basement, *in* Lintz, J., Jr., ed., Western geological excursions: Reno, University of Nevada Department of Geological Sciences, p. 1–65.

Silberling, N. J., and Jones, D. L., 1982, Tectonic significance of Permian-Triassic strata in northwestern Nevada and northern California: Geological Society of America Abstracts with Programs, v. 14, p. 234.

Silberling, N. J., Jones, D. L., Blake, M. C., Jr., and Howell, D. G., 1987, Lithotectonic terrane map of the western conterminous United States, *in* Lithotectonic terrane maps of the North American Cordillera: U.S. Geological Survey Map MR—1874-C, scale 1:2,500,000.

Skinner, J. W., and Wilde, G. L., 1965, Permian biostratigraphy and fusulinid faunas of the Lake Shasta area, northern California: University of Kansas Paleontological Contributions, v. Protozoa, article 6, 98 p.

Stevens, C. H., 1987, Affinities of Permian fusulinid faunas in the Golconda allochthon and northern Sierra Nevada: Geological Society of America Abstracts with Programs, v. 19, p. 455.

Stevens, C. H., Miller, M. M., and Nestell, M., 1987, A new Permian waagenophyllid coral from the Klamath Mountains, California: Journal of Paleontology, v. 61, p. 690–699.

Varga, R. J., 1982, Implications of Paleozoic phosphorites in the northern Sierra Nevada Range: Nature, v. 297, p. 217–220.

Wallin, E. T., 1989, Provenance of lower Paleozoic sandstones in the eastern Klamath Mountains and the Roberts Mountains allochthon, California and Nevada [Ph.D. thesis]: Lawrence, University of Kansas, 152 p.

Wallin, E. T., Mattinson, J. M., and Potter, A. W., 1988, Early Paleozoic magmatic events in the eastern Klamath Mountains, northern California: Geology, v. 16, p. 144–148.

Walker, J. D., 1988, Permian and Triassic rocks of the Mojave Desert and their implications for timing and mechanisms of continental truncation: Tectonics, v. 7, p. 685–709.

Watkins, R., 1973, Carboniferous faunal associations and stratigraphy, Shasta County, northern California: American Association of Petroleum Geologists Bulletin, v. 57, p. 1743–1764.

—— , 1985, Volcaniclastic and carbonate sedimentation in late Paleozoic island-arc deposits, eastern Klamath Mountains, California: Geology, v. 13, p. 709–713.

—— , 1986, Late Devonian to Early Carboniferous turbidite facies and basinal development of the eastern Klamath Mountains, California: Sedimentary Geology, v. 49, p. 51–71.

Watkins, R., and Flory, R. A., 1986, Island arc sedimentation in the Middle Devonian Kennett Formation, eastern Klamath Mountains, California: Journal of Geology, v. 94, p. 753–761.

Manuscript Accepted by the Society March 27, 1990

Printed in U.S.A.

Geological Society of America
Special Paper 255
1990

Carboniferous and Permian island-arc deposits of the eastern Klamath terrane, California

Rodney Watkins
Geology Department, Milwaukee Public Museum, 800 West Wells Street, Milwaukee, Wisconsin 53233

ABSTRACT

Carboniferous and Permian strata of the eastern Klamath terrane, California, include sedimentary and volcanic facies deposited in an island-arc setting. Carboniferous arc volcanism is first recorded in Upper Visean to Namurian portions of the Baird Formation, in which volcaniclastic sediments of marine deltaic origin grade southward to fine-grained, more offshore shelf deposits. During the Westphalian and Stephanian, a volcanic center developed at the southern end of the terrane, and volcaniclastic dispersal aprons of delta-front origin were deposited to the north. Pauses in deltaic progradation are represented by carbonate bank deposits. The Wolfcampian to Lower Leonardian McCloud Limestone represents a period of volcanic quiescence. The McCloud formed as a series of diachronous carbonate platforms that developed on topographic highs of the Carboniferous arc; platforms were bordered by deeper water ramps and escarpments. Renewed arc volcanism in Late Permian time is represented by Lower Guadalupian volcaniclastic sediments of the Nosoni Formation. More proximal volcanic deposition is represented by the overlying Dekkas and Bully Hill formations, with volcanic centers located at the southern end of the terrane. The Carboniferous and Permian sequence of the eastern Klamath terrane resembles Cenozoic deposits of topographically complex intra-arc settings and represents only a small portion of a larger Late Paleozoic island arc.

INTRODUCTION

The eastern Klamath terrane (EKT) of northern California includes an essentially continuous Lower Devonian to Middle Jurassic section deposited during several phases of island-arc construction and related basin development. Tectonic setting of the EKT was discussed by Miller (1987, 1989) and Renne and Scott (1988). Figure 1 summarizes the Paleozoic stratigraphy and depositional history of the terrane. This chapter is a study of Carboniferous and Permian paleogeography of the EKT as recorded by the distribution of depositional facies. Broad similarities to sedimentation in Cenozoic island arcs are apparent and suggest an intra-arc setting. The Carboniferous and Permian formations discussed here are exposed for a distance of 50 km along the terrane (Fig. 1) and represent only a limited areal portion of an ancient island arc. Although the formations are deflected from a northward to northeastward trend over this distance, there is no significant structural repetition of strata, and paleogeographic reconstructions are restricted to an essentially north-south axis.

DEPOSITIONAL FACIES

Seven major depositional facies are present within Upper Paleozoic strata of the EKT. (1) The *volcanic facies* consists of basaltic to andesitic flow-rocks and breccias with associated shallow intrusive and pyroclastic rocks. This facies is present within Upper Carboniferous and Permian formations and represents centers of arc volcanism. (2) The *rhyolitic volcanic facies,* consisting of the Upper Permian Bully Hill Rhyolite, contains felsic flow-rocks, pyroclastic rocks, shallow intrusions and massive sulfide bodies. It represents a local volcanic center formed in conjunction with rifting. (3) Marine sedimentary sequences domi-

Watkins, R., 1990, Carboniferous and Permian island-arc deposits of the eastern Klamath terrane, California *in* Harwood, D. S., and Miller, M. M., eds., Paleozoic and early Mesozoic paleogeographic relations; Sierra Nevada, Klamath Mountains, and related terranes: Boulder, Colorado, Geological Society of America Special Paper 255.

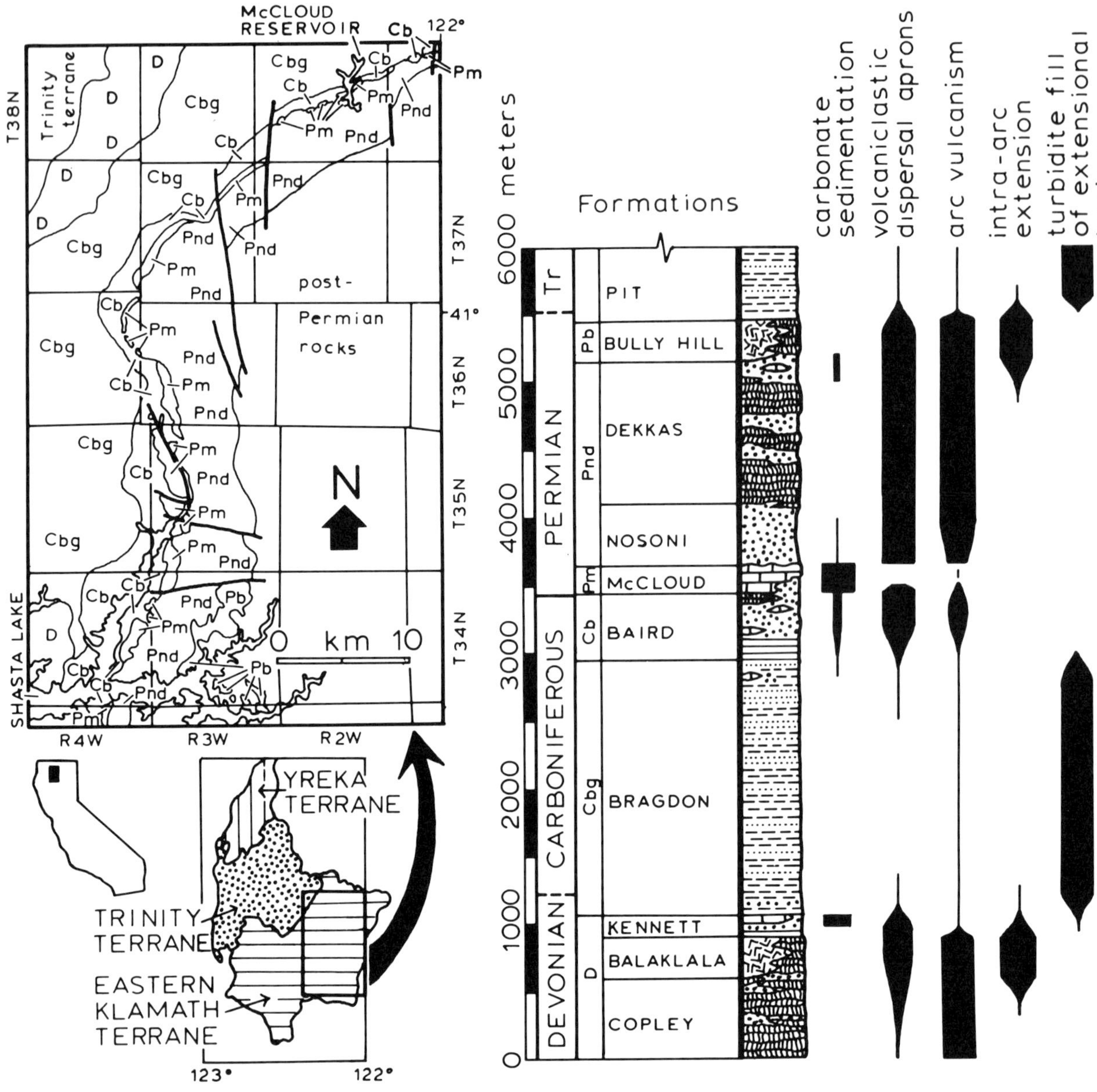

Figure 1. Location, general geology, and Paleozoic stratigraphy of study area within the eastern Klamath terrane. Thicknesses shown for Paleozoic formations are generalized and do not remain constant along strike. The Permian Pit River stock and other intrusives are not shown on the geologic map.

nated by coarse-grained volcanic debris are grouped as the *volcaniclastic dispersal facies,* which occurs through most of the Upper Paleozoic section. This facies, containing both epiclastic and pyroclastic sediment, formed in proximity to volcanic centers and indicates both active volcanism and erosion of arc volcanoes. Much of the facies is of marine deltaic origin. (4) The *fine clastic shelf facies* consists of mud-grade epiclastic sediment. Shallow-marine faunas are locally common, and the facies represents shelf areas beyond the limits of coarse-grained volcaniclastic influx. (5) *Carbonate banks* are represented by shallow-water limestone lenses within Carboniferous and Upper Permian volcaniclastic strata. These lenses are 2 km or less in length. (6) Larger *carbonate platforms* are represented by the Lower Permian McCloud Limestone. The platforms ranged up to 25 km in length and generated limestone sections hundreds of meters in thickness. (7) An additional major facies, not shown on the paleogeographic maps, is represented by turbidite sequences of Early Carboniferous age, which are discussed below.

EARLY CARBONIFEROUS PALEOGEOGRAPHY

Lower Carboniferous strata of the EKT occur within the Bragdon Formation, which contains Upper Visean faunas in its uppermost portion, probably spans Early Visean and Tournaisian

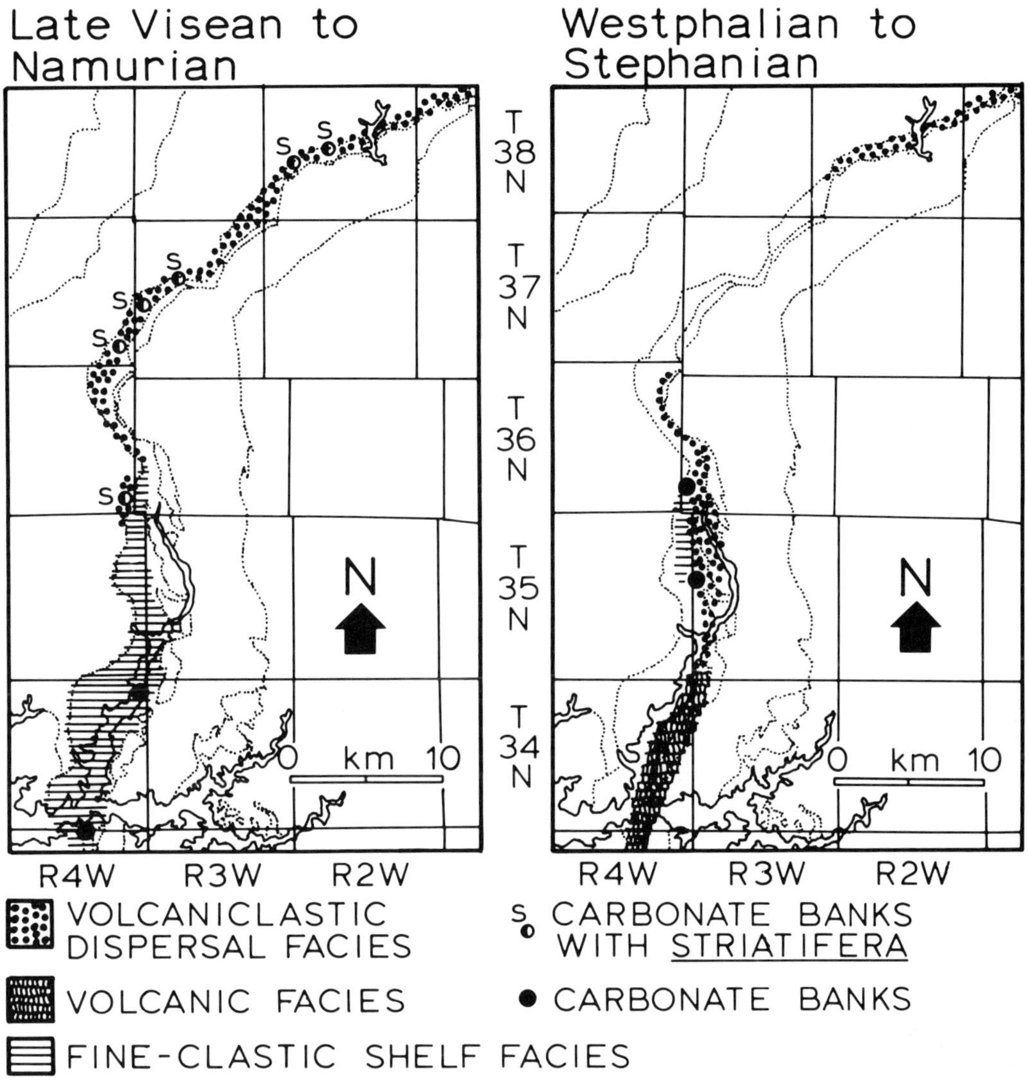

Figure 2. Middle and Late Carboniferous paleogeography shown as depositional facies of the Baird Formation.

times, and may be of Devonian age at its base (Watkins, 1986). The Bragdon consists of turbidite sequences deposited in an arc-related basin developed by Devonian extension (LaPierre and others, 1985; Albers and Bain, 1985; Watkins and Flory, 1986). Origin of the formation as submarine fan complexes was discussed by Watkins (1986), Miller (1989), and Miller and Cui (1989). Miller and others (1988) and Miller and Cui (1989) documented a multicomponent source for Bragdon sediments, which include recycled Proterozoic zircons potentially derived from the Lower Paleozoic Yreka terrane (Fig. 1) or the Shoo Fly Complex of the northern Sierra Nevada region. Other sediment components include material eroded from Devonian or older arc deposits and rare tuff, which indicates distant volcanism (Miller and Cui, 1989). The Bragdon Formation represents an Early Carboniferous basin encompassing at least two-thirds of the present-day area of the EKT. Terrestrial conditions in uplifted basement source areas are indicated by land plant debris within Bragdon turbidites.

MIDDLE TO LATE CARBONIFEROUS PALEOGEOGRAPHY

By late Visean (Late Mississippian) time, upper fan turbidites of the Bragdon were succeeded by slope and outer shelf deposits (Watkins, 1986). Overlying Upper Visean to Namurian facies of the Baird Formation are shown in Figure 2. A volcaniclastic dispersal facies, extending across the northern part of the study area, records the initiation of Carboniferous arc volcanism in the immediate area of the EKT. This facies consists mainly of volcaniclastic sandstone, pebbly volcaniclastic sandstone and tuffaceous mudstone. Nearshore deposits are present in the McCloud Reservoir area (Miller, 1989). To the southwest, between

T37N,R3W and T36N, R4W, the volcaniclastic dispersal facies is predominately of deltaic origin and consists of upward-thickening and coarsening sequences that contain sparse debris of land plants. Sequences begin as interbedded volcaniclastic turbidites and tuffaceous mudstone of delta-front origin; they shoal upward to thick-bedded, cross-stratified sandstone and conglomerate that represent proximal deposits of prograding deltas. Tops of deltaic sequences are capped by carbonate bank deposits, with thick packstone beds containing the large brachiopod *Striatifera*. Complete delta-front to carbonate bank sequences reach 160 m in thickness and are repeated up to three times in local sections.

In the southern part of the terrane, deltaic volcaniclastic sequences grade to a fine clastic shelf facies (Fig. 2). The lower stratigraphic part of this facies consists of unfossiliferous, parallel-laminated mudstone with thin tuff interbeds; it possibly represents a dysaerobic, deep shelf environment. The upper part of the facies consists of bioturbated mudstone with several subtidal marine communities dominated by pelmatozoans, brachiopods, bryozoans, molluscs, and corals (Watkins, 1973). Interbeds of fine-grained, hummocky cross-stratified sandstone indicate deposition above storm wave base.

More proximal arc facies occur within the Upper Carboniferous part of the Baird Formation. In the southern part of the EKT, a volcanic facies is represented in the Baird by a thick greenstone unit described by Albers and Robertson (1961). This greenstone extends 12 km along strike (Fig. 2) and is tentatively assigned a Westphalian-Stephanian (Pennsylvanian) age, based on its stratigraphic position above Namurian beds and below the Lower Permian McCloud Limestone. A volcaniclastic dispersal facies of Westphalian-Stephanian age occurs to the north of the greenstone unit (Fig. 2). This facies includes volcaniclastic sandstone, pebbly volcaniclastic sandstone, tuffaceous mudstone and thick, lenticular units of tuff and tuff breccia. Upward-thickening and coarsening sequences of volcaniclastic turbidites represent delta-front deposits. Two occurrences of carbonate bank deposits are present (Fig. 2). Bank-top deposits consist of cross-stratified skeletal packstone and osagid-skeletal grainstone, grading laterally to bank-margin deposits with convolute lamination and soft-sediment folding. The bank deposits are bordered by aprons of carbonate turbidites and interbedded mudstone. Some of the volcaniclastic strata of the Baird Formation near McCloud Reservoir are also potentially of Westphalian-Stephanian age (Miller, 1989). An apparent absence of Upper Carboniferous sediments between this area and the northern reaches of Shasta Lake (Fig. 2) suggests an uplifted area of the arc.

EARLY PERMIAN PALEOGEOGRAPHY

The Early Permian was a time of volcanic quiescence in the EKT and is represented by carbonate platform sequences of the McCloud Limestone. The McCloud contains a diverse fauna and a variety of shallow-marine facies (Demirmen and Harbaugh, 1965; Wilson, 1982; Watkins, 1985). Platform subsidence, combined with carbonate deposition, generated limestone sections several hundreds of meters thick. Skinner and Wilde (1965) defined eight fusulinid zones within the McCloud, ranging from zone A (Basal Wolfcampian) to zone H (Lower Leonardian). The distribution of these zones is used here to plot locations of several separate carbonate platforms. The Lower Wolfcampian part of the McCloud, represented by zones A, B and C, defines a platform at least 18 km in north-south extent in the central part of the terrane (Fig. 3). Much of this area of the McCloud overlies parts of the Baird in which Upper Carboniferous strata are thin or absent, suggesting carbonate platform development in the Early Permian on a topographic high of Middle Carboniferous volcaniclastic rocks. In the southern part of this platform sequence, near McCloud Bridge, the lower part of zone A includes interbedded limestone, volcaniclastic turbidites, mudstone, and mixed carbonate-volcaniclastic conglomerates deposited in a ramp setting (Watkins, 1985). The ramp sequence is overlain by a thick section of pure limestone that includes bioturbated skeletal wackestone and mudstone, cross-stratified crinoidal packstone, channelized beds of fusulinid packstone, and *Palaeoaplysina* and algal biostromes (Watkins and Wilson, 1989). These relations indicate platform growth by progradation of shallow-water carbonates over ramp deposits. South of this platform is an area of Baird volcaniclastic strata dated as Lower Permian because of their conformable position above a Stephanian limestone lens and below zone D strata of the McCloud (Watkins, 1973). This occurrence probably represents redeposition of relict volcaniclastic sediment rather than persistent volcanism. Volcaniclastic strata of the Baird around McCloud Reservoir, which conformably underlie and interfinger with the McCloud Limestone (Miller, 1989), may also be of Early Permian age.

Zones D and E of the McCloud Limestone, of Middle Wolfcampian age, are restricted to the southern part of the EKT, where they represent a carbonate platform at least 16 km in north-south extent (Fig. 3). This platform was developed on Upper Carboniferous volcanic rocks of the Baird Formation that represent another topographic high within the arc. Carbonate deposition continued on this platform during zone F (Late Wolfcampian) time and is represented by a shallow-water facies association of cross-stratified skeletal packstones, bioturbated skeletal wackestones, and beds rich in algae and colonial corals. A separate, northern carbonate platform was also established during zone F time (Fig. 4). Miller (1989) described a progradational sequence from turbidites to shallow-water facies, which indicates ramp-type margins of this northern platform.

Zones G and H of the McCloud, Late Wolfcampian and Early Leonardian in age, are restricted to the northern part of the EKT; they indicate growth of the carbonate platform in this area to a length of 25 km or more (Fig. 4). At Tombstone Mountain, some G consists of thick-bedded limestone conglomerates of debris-flow origin with clasts containing fusulinids from older McCloud zones (Skinner and Wilde, 1965). Zone G is overlain by shallow-water skeletal packstones and wackestones of zone H. These relations indicate a platform margin of escarpment type, with platform growth in zone H time occurring as progradation of shallow-water facies over escarpment-related conglomerates of zone G.

LATE PERMIAN PALEOGEOGRAPHY

An Upper Permian arc sequence is represented in the EKT by the Nosoni, Dekkas, and Bully Hill Formations (Fig. 1). The Nosoni Formation, of Early Guadalupian age, represents the spread of a volcaniclastic dispersal facies across the north-south extent of the EKT (Fig. 5). In the southern part of the terrane, Albers and Robertson (1965) described a disconformity between the Nosoni and the underlying McCloud. A locally developed basal conglomerate of the Nosoni, containing clasts of McCloud Limestone, indicates erosion of Early Permian carbonate platforms. Sedimentation from Early to Late Permian times may have been continuous in the northern part of the EKT, where siliceous siltstones of the basal Nosoni conformably overlie the McCloud and may be of Leonardian age (Miller, 1989). Most of the Nosoni Formation consists of volcaniclastic sandstone, pebbly volcaniclastic sandstone, tuffaceous mudstone, and pyroclastic rocks (Coogan, 1960; Albers and Robertson, 1961; Miller, 1989). Lenses of bioturbated mudstone containing shallow-marine communities of pelmatozoans, brachiopods, fusulinids, bryozoans, and molluscs are also locally present. Minor volcanic flow-rocks are present, as well as densely welded ash-flow tuff of possible subaerial origin (Renne and others, 1988). Sediments of the Nosoni were derived from volcanic source areas in the immediate vicinity of the EKT.

The volcanic Dekkas Formation, which contains a Guadalupian marine fauna, overlies and partly intergrades with the Nosoni. Basaltic to andesitic flow-rock is the main constituent of the Dekkas in the southern part of the EKT (Albers and Robertson, 1961), indicating volcanic centers in this area (Fig. 5). The magmatic source of arc volcanism is represented in the same vicinity by the Permian Pit River stock, which intrudes the Dekkas and underlying formations (Albers and Robertson, 1961; Miller, 1987). To the north, the Dekkas consists of interbedded flow-rocks, pyroclastics, and volcaniclastic sediments that repre-

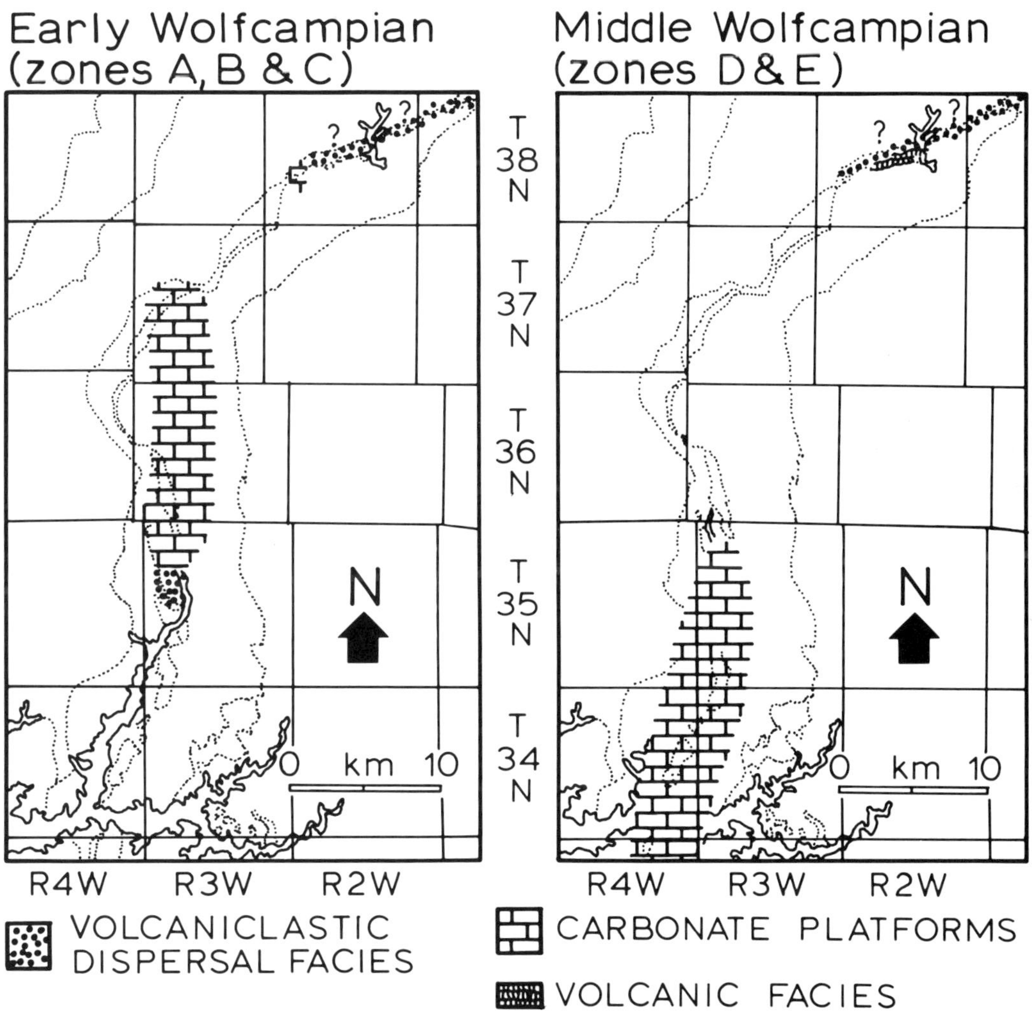

Figure 3. Early Permian paleogeography shown as distribution of zones A through E of the McCloud Limestone and possibly correlative portions of the Baird Formation. East-west dimensions of carbonate platforms are speculative and were probably larger than those shown.

sent deeper water deposition on the flanks of arc volcanoes (Miller, 1989). Shallow-marine carbonate bank deposits are also present in the upper part of the Dekkas in the northern part of the EKT (Stevens and others, 1987).

In the southern part of the EKT, the Dekkas is overlain by the Bully Hill Rhyolite, which consists of flow-rocks, pyroclastic rocks, and shallow intrusions (Albers and Robertson, 1961). The Bully Hill, of Late Permian age (Miller, 1987), represents development of a local rhyolitic volcanic center (Fig. 5). Extension and subsidence of arc deposits near the end of Permian time (Eastoe and others, 1987; Renne and Scott, 1988) were followed by deposition of the Pit Formation (Fig. 1). The Pit ranges from Late Permian to Triassic in age and includes mudstone, chert, gravity-displaced volcaniclastic sediments, and a thick olistostrome in the Shasta Lake area (Eastoe and others, 1987).

CONCLUSIONS

The Upper Paleozoic succession in the eastern Klamath terrane is broadly similar to deposits of Cenozoic island arcs. Within the Baird, Nosoni, and Dekkas Formations, the presence of interbedded volcanic, pyroclastic, epiclastic, and carbonate rocks, as well as rapid lateral changes in facies and thicknesses, suggests deposition in a topographically complex, intra-arc setting. Comparable Cenozoic sequences of intra-arc origin include those of Guam in the Marianas arc (Tracey and others, 1964) and Malekula Island in the New Hebrides arc (Mitchell, 1970). Deposits of the EKT formed in relatively shallow water in close proximity to arc volcanic centers, as indicated by shallow-marine faunas, the presence of volcanic and carbonate bank deposits, and the lack of deep-marine pelagic sediments. Stratigraphic and sedimentologic data for the McCloud Limestone indicate carbonate deposition on a shifting complex of shallow-marine platforms that developed on topographic highs of the Carboniferous arc. These relations also suggest a temporally variable topography characteristic of intra-arc settings.

Middle Carboniferous to Upper Permian formations in the EKT record two phases of island-arc volcanism, with intervening quiescence and carbonate deposition in the Early Permian. An analogue to this sequence is present in the Cenozoic sections of

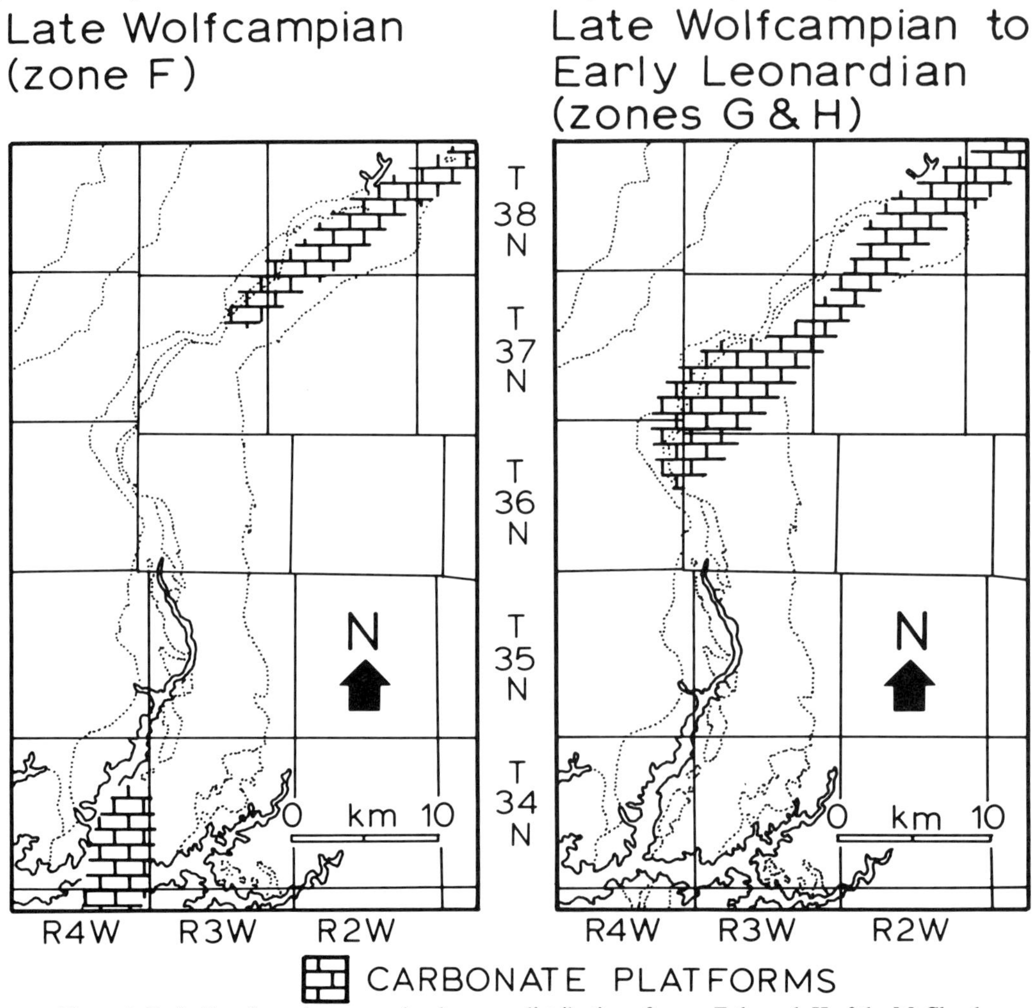

Figure 4. Early Permian paleogeography shown as distribution of zones F through H of the McCloud Limestone. East-west dimensions of carbonate platforms are speculative and were probably larger than those shown.

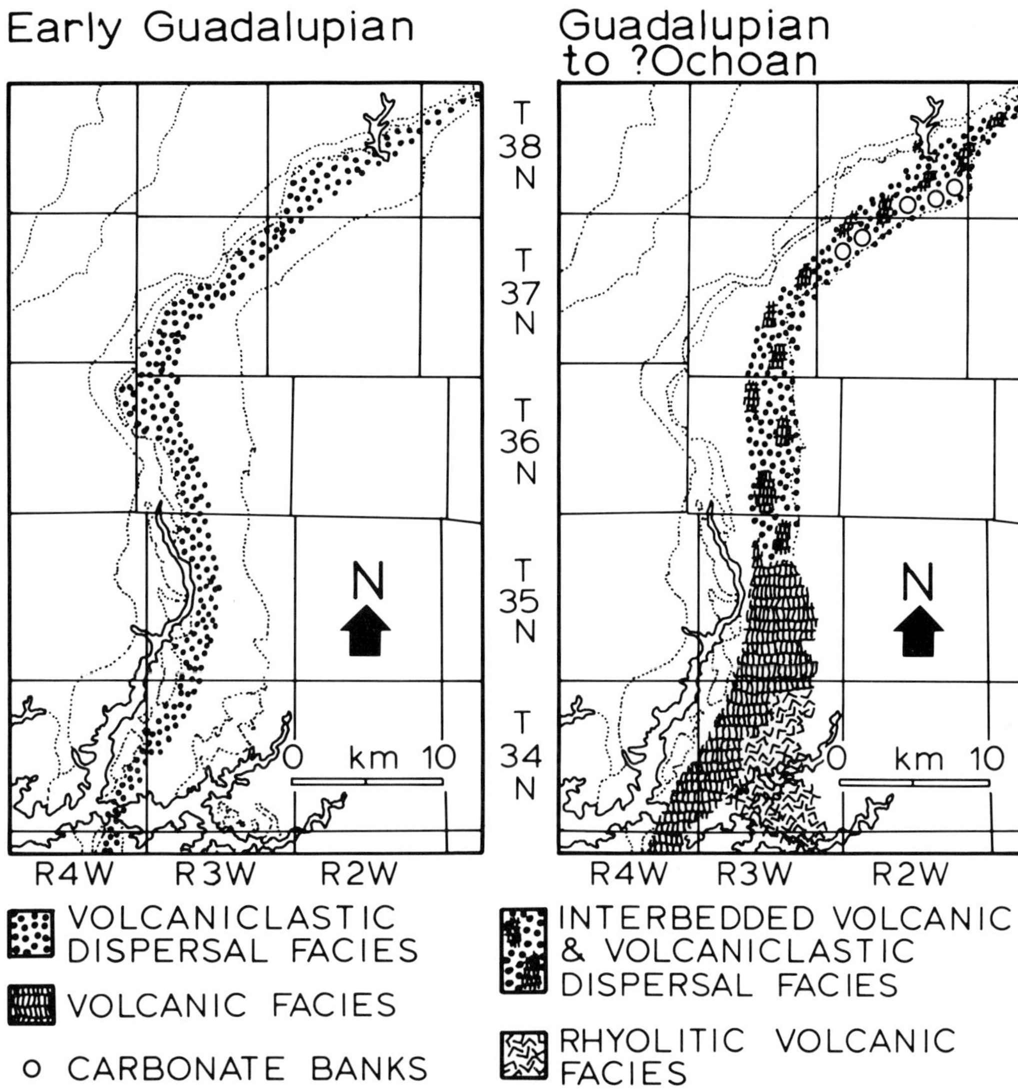

Figure 5. Late Permian paleogeography. Map for the early Guadalupian represents depositional facies of the Nosoni Formation, and map for the Guadalupian to Ochoan(?) represents depositional facies of the Dekkas and Bully Hill Formations.

the Bismark, New Britain, and Soloman island arcs of the South Pacific, as summarized by Hamilton (1979). In these arcs, Paleogene volcanism is represented by interbedded andesitic volcanic and volcaniclastic rocks. Volcanism ceased by middle Miocene time, and many separate islands preserve Miocene sections of shallow-marine limestones. Arc volcanism in all three chains resumed during the late Neogene, and Miocene limestones were covered by a younger sequence of volcanic and volcaniclastic deposits.

Whereas modern island arcs have lengths of several hundred kilometers to more than 2,000 km, Upper Paleozoic formations of the EKT cover a north-south distance of only about 50 km. In the context of the scale of modern island arcs, virtually all of the facies trends in a single EKT formation could be generated in proximity to a single volcanic edifice. Detailed differences between the Upper Paleozoic succession in the EKT and those of other Cordilleran terranes do not preclude their origin within a single large-scale island arc. Miller (1987) proposed that the EKT formed part of a single continent-fringing arc in the Late Paleozoic, which also included the northern Sierra Nevada, the Grindstone terrane of Oregon, the Bilk Creek terrane of Nevada, the Chilliwack terrane of Washington, and the Stikine terrane of British Columbia. This proposal is entirely tenable with respect to the local paleogeography and depositional facies discussed here.

ACKNOWLEDGMENTS

Acknowledgment is made to the Donors of the Petroleum Research Fund, administered by the American Chemical Society, for partial support of this research.

REFERENCES CITED

Albers, J. P., and Bain, J.H.C., 1985, Regional setting and new information on some critical geologic features of the West Shasta district: Economic Geology, v. 80, p. 2072–2091.

Albers, J. P., and Robertson, J. F., 1961, Geology and ore deposits of the East Shasta Copper-Zinc District, Shasta County, California: U.S. Geological Survey Professional Paper 338, 107 p.

Coogan, A. H., 1960, Stratigraphy and paleontology of the Permian Nosoni and Dekkas formations (Bollibokka Group): University of California Publications in Geological Sciences, v. 36, p. 243–316.

Demirmen, F., and Harbaugh, J. W., 1965, Petrology and origin of Permian McCloud Limestone of northern California: Journal of Sedimentary Petrology, v. 35, p. 136–154.

Eastoe, C. J., Gustin, M. M., and Nelson, S. E., 1987, Problems of recognition of olistostromes; An example from the lower Pit Formation, eastern Klamath Mountains, California: Geology, v. 15, p. 541–544.

Hamilton, W., 1979, Tectonics of the Indonesian region: U.S. Geological Survey Professional Paper 1078, 345 p.

LaPierre, H., Albarede, F., Albers, J., Cabanis, B., and Coulon, C., 1985, Early Devonian volcanism in the eastern Klamath Mountains, California; Evidence for an immature island arc: Canadian Journal of Earth Sciences, v. 22, p. 214–226.

Miller, M. M., 1987, Dispersed remnants of a northeast Pacific fringing arc; Upper Paleozoic terranes of Permian McCloud faunal affinity, western U.S.: Tectonics, v. 6, p. 807–830.

——, 1989, Intra-arc sedimentation and tectonism; Late Paleozoic evolution of the eastern Klamath terrane, California: Geological Society of America Bulletin, v. 101, p. 170–187.

Miller, M. M., and Cui, B., 1989, Submarine fan characteristics and dual sediment provenance, Lower Carboniferous Bragdon Formation, eastern Klamath terrane, California: Canadian Journal of Earth Sciences, v. 26, p. 927–940.

Miller, M. M., Saleeby, J. B., and Cui, B., 1988, Provenance of the Carboniferous Bragdon Formation, eastern Klamath terrane, California; Proterozoic detrital zircon and quartzose detritus as continental tracers in oceanic convergent margin systems, *in* Howell, D. G., and Wiley, T. J., eds., Proceedings, 4th International Conference on Circum-Pacific Terranes, Nanjing, P.R.C.: U.S. Geological Survey, p. 67–69.

Mitchell, A.H.G., 1970, Facies of an Early Miocene volcanic arc, Malekula Island, New Hebrides: Sedimentology, v. 14, p. 201–243.

Renne, P. R., and Scott, G. R., 1988, Structural chronology, oroclinal deformation, and tectonic evolution of the southeastern Klamath Mountains, California: Tectonics, v. 7, p. 1223–1242.

Renne, P. R., Scott, G. R., and Bazard, D. R., 1988, Multicomponent paleomagnetic data from the Nosoni Formation, eastern Klamath Mountains, California; Cratonic Permian primary directions with Jurassic overprints: Journal of Geophysical Research, v. 93, p. 3387–3400.

Skinner, J. W., and Wilde, G. L., 1965, Permian biostratigraphy and fusulinid faunas of the Shasta Lake area, northern California: University of Kansas Paleontological Contributions, Protozoa, Article 6, 98 p.

Stevens, C. H., Miller, M. M., and Nestell, M., 1987, A new Permian waagenophyllid coral from the Klamath Mountains, California: Journal of Paleontology, v. 61, p. 690–699.

Tracey, J. I., Schlanger, S. O., Stark, J. T., Doan, D. B., and May, H. G., 1964, General geology of Guam: U.S. Geological Survey Professional Paper 403-A, 104 p.

Watkins, R., 1973, Carboniferous faunal associations and stratigraphy, Shasta County, northern California: American Association of Petroleum Geologists Bulletin, v. 57, p. 1743–1764.

——, 1985, Volcaniclastic and carbonate sedimentation in late Paleozoic island-arc deposits, eastern Klamath Mountains, California: Geology, v. 13, p. 709–713.

——, 1986, Late Devonian to Early Carboniferous turbidite facies and basinal development of the eastern Klamath Mountains, California: Sedimentary Geology, v. 49, p. 51–71.

Watkins, R., and Flory, R. A., 1986, Island arc sedimentation in the Middle Devonian Kennett Formation, eastern Klamath Mountains, California: Journal of Geology, v. 94, p. 753–761.

Watkins, R., and Wilson, E. C., 1989, Paleoecologic and biogeographic significance of the biostromal organism *Palaeoaplysina* in the Lower Permian McCloud Limestone, eastern Klamath Mountains, California: Palaios, v. 4, p. 181–192.

Wilson, E. C., 1982, Wolfcampian rugose and tabulate corals (Coelenterata: Anthozoa) from the Lower Permian McCloud Limestone of northern California: Natural History Museum of Los Angeles County Contributions in Science 337, 90 p.

Manuscript Accepted by the Society March 27, 1990

Printed in U.S.A.

Geological Society of America
Special Paper 255
1990

Significance of the provincial signature of Early Permian faunas of the eastern Klamath terrane

Calvin H. Stevens
Department of Geology, San Jose State University, San Jose, California 95192
Thomas E. Yancey
Texas A&M University, College Station, Texas 77843
Rex A. Hanger
Department of Geology, University of California, Berkeley, California 94720

ABSTRACT

The abundant and well-studied Early Permian faunas of the eastern Klamath terrane, which contain a high percentage of endemic species and genera, constitute part of a biotic province (the McCloud province) that is distinct from the better known central Cordilleran and Tethyan provinces. The McCloud province evidently fringed an island arc that was located within the tropical climatic zone of the Paleopacific Ocean during the Early Permian. The development of an endemic biota in this province required considerable isolation for long periods of time, and because the shallow-water biotic provinces of the eastern Klamath terrane, Tethys, and the central Cordillera lay along the same general latitudinal band and were open to waters of the same ocean basin, their separate development must have been the result of geographic barriers. We conclude that geographic isolation was produced by large distances of deep water that lay between these shallow-marine provinces in the Early Permian Paleopacific Ocean. By analogy with biotic distributions in modern ocean basins, these deep-water barriers must have been several thousand kilometers wide.

INTRODUCTION

Interpretations of the origin and history of the eastern Klamath tectonostratigraphic terrane (herein referred to as the eastern Klamath terrane) in north-central California and its geographic position during pre-accretion times have been changing as new evidence has become available about the geologic character of the terrane. Especially important data are coming from the petrologic and geochemical characteristics of the rocks and their contained fossil biotas. Initial recognition of the exotic character of the eastern Klamath terrane came from the realization that the biotas, contained primarily in limestones of Early Permian age, are quite different (e.g., Ross, 1967) from those of the central part of continental western North America (herein referred to as the central Cordillera). More recently, faunal elements of the eastern Klamath terrane (mostly fusulinids) have been discovered in a series of terranes stretching from California to Alaska (summarized by Ross and Ross, 1983). These terranes, with rocks containing the distinctive fossils characteristic of the eastern Klamath faunal province, were considered by Miller (1987) to constitute a related group of terranes she called the McCloud belt.

The igneous rocks of the McCloud belt suggest an origin as an island arc (Miller, 1987) that developed somewhere in the Paleopacific Ocean. Closing of the oceanic basin that originally separated the eastern Klamath terrane and the central Cordillera with accretion of the eastern Klamath Mountains to continental North America was complete by the Middle Jurassic at the latest (Dilles and Wright, 1988), but substantial disagreement still exists on the width of the ocean that separated this island-arc complex from the central Cordillera during the Permian. Paleomagnetic data of Mankinen and others (1989) show a similarity between Permian and Triassic paleolatitudes of the central Cordillera and

Stevens, C. H., Yancey, T. E., and Hanger, R. A., 1990, Significance of the provincial signature of Early Permian faunas of the eastern Klamath terrane, *in* Harwood, D. S., and Miller, M. M., eds., Paleozoic and early Mesozoic paleogeographic relations; Sierra Nevada, Klamath Mountains, and related terranes: Boulder, Colorado, Geological Society of America Special Paper 255.

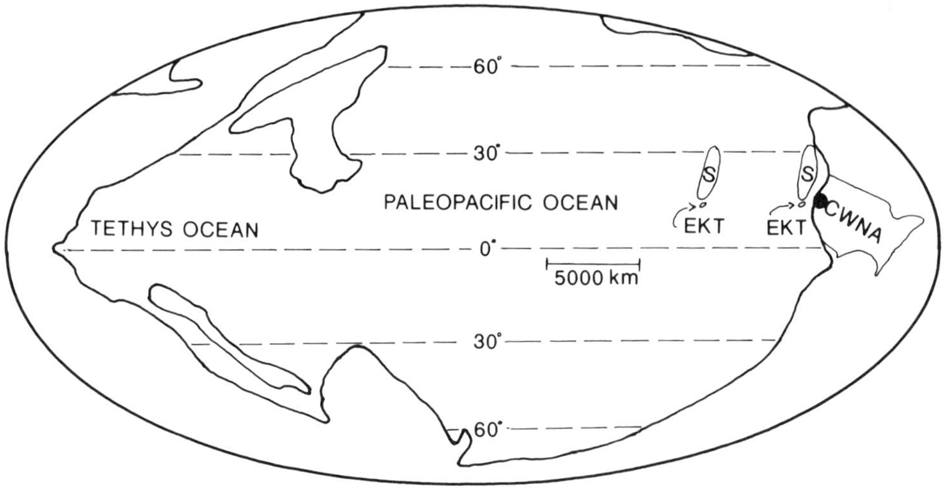

Figure 1. Paleogeographic map of the Early Permian world showing the limits of interpretation of the position of the McCloud belt island arc consisting of the eastern Klamath terrane (EKT) plus Stikinia (S) plus other smaller terranes. We favor the more distant position from cratonal western North America (CWNA).

the eastern Klamath terrane, but the eastern Klamath terrane apparently has been greatly rotated compared to the central Cordillera. Thus, little can be deduced from these data in terms of the relative paleolongitudinal placement of the two regions.

Paleontologists have found the Permian tropical shallow-marine biotas around the Pacific rim to be especially interesting because of the great dissimilarity among faunas. Two biotic provinces, the central Cordilleran and the Tethyan, have long been recognized. Work in the eastern Klamath terrane, especially on the fusulinids (Skinner and Wilde, 1965) and corals (Wilson, 1982, 1985), has shown that these faunas also are distinct, and that the endemism—at least among the corals—is on a level similar to, or more often greater than, that of corals of various biotic provinces of other ages (e.g., Silurian and Devonian [Oliver and Pedder, 1976] and Mississippian [Sando and others, 1975]). A number of workers (e.g., Stevens and Rycerski, 1983; Yancey and Hanger, 1986) have suggested that such differences reflect very large original distances between the shallow-water faunas involved, whereas others (e.g., Ross and Ross, 1987) apparently would place the eastern Klamath terrane somewhat closer to North America during the Early Permian.

Newton (1988) has proposed an explanation for the presence of characteristic Permian Tethyan marine fossils in Texas (Ross, 1967) and in various North American terranes, including the eastern Klamath terrane, reasoning that dispersion of modern marine species across the tropical Indo-Pacific region has been so common in the recent past that there should be no problem explaining the presence of Tethyan fossils in terranes (e.g., eastern Klamath terrane) that may have been adjacent to North America. Although we disagree with her basic conclusions concerning Permian faunas, her observations on modern faunas actually reinforce our answer to the most important question we are exploring here: if a terrane (specifically the eastern Klamath terrane) was close to North America during the Early Permian, why are the Permian faunas so much more different from one another than are modern faunas on opposite sides of the Pacific Ocean, where open-ocean distances of greater than 5,000 km have been traversed by many shallow-water organisms? In this chapter we attempt to answer this question, present the paleontologic data on the three diverse Permian biotic provinces, interpret the biogeographic relations, and evaluate the evidence concerning the position of the eastern Klamath terrane relative to the central Cordillera during the latest Paleozoic. At present, the range of interpretations of the distance between the eastern Klamath terrane and North America during the Permian is from greater than 5,000 km to almost nothing (Fig. 1).

PALEONTOLOGICAL OVERVIEW

The brachiopods, corals, fusulinids, and gastropods of the three great, diverse Early Permian shallow-marine provinces in the Paleopacific Ocean basin (the eastern Klamath, central Cordilleran, and Tethyan provinces) have been studied in moderate to great detail, and are known well enough for paleobiogeographic analysis. These groups will be compared using the Otsuka similarity index, which is defined as $C\sqrt{N1N2} \times 100$, where C is the number of genera or subgenera common to the two biogeographic provinces being compared, and $N1$ and $N2$ are the total number of these taxa in each of the two provinces (Sando and others, 1975). Scores of 0 to 27 indicate low similarity; 28 to 54, moderate similarity; and above 54, high similarity. This index was used because it attempts to reduce the effects of sample size disparities (Savage and others, 1976). The similarity indices involving these fossils are shown in Table 1, and the genera used in all computations presented here are listed in Appendices 1 through 4. The percentages of endemic genera and subgenera, which have been very conservatively estimated, range from 7 percent in the fusulinids to 33 percent in the corals (Table 2).

TABLE 1. DATA ON THE EARLY PERMIAN TROPICAL ARCHAEOGASTROPOD, BRACHIOPOD, CORAL, AND FUSULINID FAUNAS OF THE PALEOPACIFIC OCEAN BASIN*

	Permian Fossils						
	EKT Genera	CWNA Genera	Tethyan Genera	Genera of EKT and CWNA	EKT-CWNA Otsuka Index	Genera of EKT and Tethys	EKT-Tethys Otsuka Index
Gastropods	31	54	29	19	46	15	50
Brachiopods	33	60	73	15	34	18	37
Corals	28	35	45	11	35	3	8
Fusulinids	14	20	36	12	72	12	53

*Otsuka similarity indices between the eastern Klamath terrane (EKT) and crational western North America (CWNA), and the eastern Klamath Mountains and the Tethyan province also are shown.

Percentages of 0 to 7 are considered low; 8 to 14, moderate; and above 14, high.

TABLE 2. ENDEMISM IN THE McCLOUD BIOTIC PROVINCE

	Total Genera	Endemic Genera	Endemic Genera (%)
Brachiopods	33	3	9
Corals	27	9	33
Fusulinids	14	1	7

Fusulinids

The fusulinids of the eastern Klamath terrane have been documented in detail by Skinner and Wilde (1965), who recognized eight zones in the McCloud Formation ranging in age from earliest Wolfcampian into the Leonardian. Genera of the eastern Klamath Mountains are listed in Appendix 1. Considerably more advanced fusulinids of early Guadalupian age are present in the overlying Nosoni and Dekkas Formations (Coogan, 1960; Skinner and Wilde, 1965).

Works on Early Permian fusulinids in the central Cordillera are numerous; all genera listed in Appendix 1, however, are represented in Douglass (1974). Douglass and Nestell (1974), Hoare (1963), Magginetti and others (1988), Verville and Laule (1979), and Verville and others (1986). Our data on Tethyan fusulinid faunas probably are less complete than on the eastern Klamath terrane and the central Cordillera, but are adequate for their use here. All Tethyan genera listed in Appendix 1 are represented in the works of Kahler and Kahler (1980), Kochansky-Devide (1970), and Toriyama (1982, 1984).

Fourteen genera and subgenera of fusulinids have been recognized from the Wolfcampian to Leonardian strata of the eastern Klamath terrane, compared to 20 in coeval rocks from the central Cordillera. The Otsuka similarity index of these two provinces is 72 (high), considerably higher than that between the eastern Klamath and Tethyan provinces, which is about 53 (Table 1). Only one genus, *Klamathina,* apparently is endemic. Several central Cordilleran genera, including the relatively common *Leptotriticites* and *Stewartina,* are not present in the eastern Klamath faunas, and *Mccloudia* is not known from the central Cordillera. Furthermore, several of the very common eastern Klamath genera, especially *Cuniculinella* and *Pseudofusulinella,* are rare in the central Cordillera. Major differences also appear in the geologic ranges of some genera common to both the central Cordillera and the eastern Klamath Mountains. For instance, species of *Pseudofusulinella* are numerous in the eastern Klamaths in rocks that extend well into the Leonardian (zone H, Skinner and Wilde, 1965), whereas in the central Cordillera, species are few and the genus apparently disappears in the middle Wolfcampian.

On the species level, most eastern Klamath fusulinids are endemic, as shown by the fact that, of the 167 species described from there, Skinner and Wilde (1965) assigned only one to a species described from the central Cordillera. Similarly, in a study of 60 species from cratonal eastern California, Magginetti and others (1988) considered only four species to be closely similar to eastern Klamath forms, and none to be conspecific. The only Leonardian species of *Parafusulina* from the eastern Klamath Mountains, *P. mccloudensis,* is strikingly dissimilar to central Cordilleran species. Several eastern Klamath species, however, have been reported from the Wolfcampian of Ellesmere Island (Nassichuk and Wilde, 1977), which was part of Pangaea during the Permian, indicating some tie to North America. Thus, although on the generic level the similarity index of the Wolfcampian to Leonardian fusulinid faunas of the eastern Klamath faunas and those of the Cordillera is high, on the specific level there is virtually no similarity except in the modern Arctic.

Early Guadalupian fusulinids in the eastern Klamath terrane are anomalous; no Guadalupian fusulinids of any kind are known from the Cordilleran miogeocline.

Corals

Data on eastern Klamath coral faunas (Appendix 2) have been taken from Wilson (1982, 1985), as modified by Stevens and Rycerski (1989), with additional unpublished data. Works on Early Permian coral faunas in the central Cordillera are numerous; all genera tabulated in Appendix 2, however, are represented in Hoare (1964), Laule (1978), Sando (1984), Stevens (1967), Stevens and Rycerski (1983), and Wilson and Langenheim (1962), supplemented by some unpublished data. The Tethyan coral data are taken from Douglas (1950), Federowski (1986), Fontaine (1986), Minato (1975), Minato and Kato (1965a, b), Yoh and Huang (1932), Zhao (1976), and Zhao and Yang (1985).

On the family and genus level, the Wolfcampian and Leonardian corals of the eastern Klamath terrane are much less similar to those of the Tethyan province than to those of the central Cordillera, but even so, on the generic level the corals of the eastern Klamath terrane are very different from those of the central Cordillera. At least nine genera are endemic to the eastern Klamaths, and the overall generic similarity index between the eastern Klamaths and the central Cordillera is only about 35 (moderate). Genera common to both the eastern Klamaths and the central Cordillera also differ significantly in having strikingly different patterns of occurrence. For instance, the many species of the genus *Heritschioides* occur in the eastern Klamath terrane from earliest to latest Wolfcampian. In eastern Nevada there are five known Leonardian species of *Heritschioides* (Stevens, 1967), but only a single specimen from Wolfcampian strata is known; in southeastern California, one Wolfcampian species is present (C. H. Stevens, unpublished data).

On the species level, the faunas of the eastern Klamath terrane are almost totally different from those of the eastern Cordillera. Of the 52 species described by Wilson (1982, 1985) from the eastern Klamaths, only 2 species of the tabulate coral *Neosyringopora,* a broadly conceived coralline genus, are known from the adjacent craton.

The only definitely Tethyan coral genus in the eastern Klamaths, *Waagenophyllum* (Stevens and others, 1987), is present in the Guadalupian Dekkas Formation.

Brachiopods

The Permian brachiopod fauna of the eastern Klamaths is largely undescribed, but the Early Permian species have been tabulated and studied as part of an ongoing project to document the brachiopods and molluscs of the McCloud Formation (T. E. Yancey and R. A. Hanger, unpublished data). Coogan (1960) reported on the brachiopods of the Late Permian Nosoni Formation. The McCloud Limestone contains 40 species distributed in 33 genera (Appendix 3); an additional 8 genera occur in the Nosoni Formation (Coogan, 1960). There are many works on early Permian brachiopod faunas in the central Cordillera; genera tabulated in Appendix 3 are reported in Gilluly and others (1954), Kirkland (1963), Mills and Langenheim (1987), Pederson (1954), Winters (1963), Yancey (1978), and Yochelson (1968), supplemented by unpublished data for southeastern California by M. E. Kaasa. Data on Tethyan brachiopod genera are taken from Grant (1976).

The early Permian brachiopods of the eastern Klamaths are as distinct as the fusulinids and corals, with a large majority (90 percent) of the species and 3 of the 33 known genera (9 percent) being endemic. This brachiopod fauna is more similar to Tethyan brachiopod faunas than to those of the central Cordillera. The Otsuka generic similarity index between the eastern Klamath and Tethyan brachiopods of Wolfcampian and Leonardian age is 37 (moderate similarity); the similarity index of the eastern Klamaths and the central Cordillera also is moderate (34). There are three eastern Klamath genera (*Marginifera* s.s., *Crurícella,* and *Orbocoelia*) that are shared with Tethyan biotas but that do not occur in the central Cordillera, and two genera (*Teguliferina* and *Glossothyropsis*) that are present in the eastern Klamaths and the central Cordillera but that do not occur in Tethyan biotas.

Gastropods

The Wolfcampian to Leonardian gastropod fauna of the eastern Klamaths also has been tabulated (Appendix 4) and studied in Yancey's ongoing project on the McCloud Limestone. Gastropod genera in the central Cordillera (Appendix 4) are taken from several monographic works on early Permian gastropod faunas (Batten, 1958; Chronic, 1952; Erwin, 1988; Winters, 1963; Yochelson, 1956, 1960). Tethyan gastropods are documented in reports by Batten (1972, 1979, 1985).

The Wolfcampian to Leonardian gastropod faunas of the eastern Klamaths contain 75 species in 53 genera (T. E. Yancey, unpublished data). There is a higher degree of cosmopolitan distribution among Permian gastropod genera than among the corals and brachiopods, so the provincial character is less apparent. However, the eastern Klamath gastropod fauna is still quite different from other gastropod faunas, and contains several undescribed endemic genera. Comparisons of gastropods between the eastern Klamath, central Cordilleran, and Tethyan provinces are based on a total of 70 genera from the families of archaeogastropods and subulitaceans that have been fully documented in all three provinces. The eastern Klamath genera, of which 31 belong to these groups, are slightly more similar to those in the Tethyan province than to those of the central Cordillera. The Otsuka index between the eastern Klamath and Tethyan gastropods is 50 (moderate similarity); the Otsuka index between the eastern Klamath and central Cordillera gastropods is 46 (moderate similarity). Some common central Cordilleran genera, such as *Platyworthenia* and *Goniasma,* are not present in the eastern Klamath fauna.

POSSIBLE CAUSES OF DIFFERENCES BETWEEN PERMIAN BIOTAS

Pronounced differences between biotas of the eastern Klamath, central Cordilleran, and Tethyan areas conceivably could have arisen from either environmental or geographic factors. Environmental factors include differences in (1) water depths, (2) substrates and energy regimes, (3) water temperatures, (4) salinities, and (5) degrees of turbidity, or other, possibly undetectable differences among biotas within a single province, but they should not mask the identity of a province when comparisons are made between biotas from similar environmental settings. Geographic factors include separation by (6) land barriers, (7) oceanographic barriers, and (8) large distances across deep ocean basins. Because our major interest here lies in the comparison of the faunas of the eastern Klamaths and the central Cordilleran region, most of our discussion is limited to these two regions.

Geologic data are available to evaluate the importance of most factors that might have affected the shallow-marine faunas of the eastern Klamath Mountains and the central Cordillera, and we believe that most factors can be ruled out as being responsible for the observed biotic differences. For the convenience of the reader, different elements in the following discussion are keyed by number to the factors listed above.

1. A similar spectrum of water depths in the two regions is indicated by the presence of depositional environments that range from high-energy shoals to deep-water turbidite basins in both areas.

2. There are no apparent differences in the range of wave energy or substrates in the two areas. Demirmen and Harbaugh (1965) and Watkins (1985) documented the occurrence of carbonates in the McCloud Formation in the eastern Klamaths ranging from relatively deep-water micrites interbedded with submarine debris-flow beds to shallow-water, subtidal shelly muds and well-washed oncoid-bearing grainstones. Many of the Permian faunas in eastern California occur in conglomeratic deposits of gravity-flow origin, but clasts in these deposits, consisting of micritic limestone, skeletal wackestones and packstones, and rare oolitic grainstones, suggest original deposition in as wide a range of environments and substrates as in the McCloud Formation. The Permian section in the central Cordillera is much richer in quartzose sand than that in the eastern Klamaths, but many areas on the craton also contain great thicknesses of carbonate rocks; it seems unlikely that the amount of quartzose sand was a meaningful factor affecting the biota.

3. Differences in temperatures resulting from differing paleolatitudes can be ruled out for this study on the basis of paleomagnetic studies that place the eastern Klamaths at $17.8° \pm 8.4°$, and the central Cordillera at about $18.5° \pm 4.5°$ (Mankinen and others, 1989). Latitudinal diversity gradients for brachiopods computed by Hanger (1986) for these areas, using methods developed by Stehli (1970, 1973) and Waterhouse and Bonham-Carter (1975), suggest similar latitudinal placements; both the paleomagnetic and diversity data show that both biotic provinces were tropical.

4. There is no indication that the salinity of the water masses surrounding the eastern Klamaths and covering the central Cordillera was substantially different. Faunas in both provinces include large numbers of stenohaline marine organisms, such as corals and crinoids, that suggest the waters were of normal marine salinity.

5. The lack of shale and mudstone in both areas suggests that turbidity was low, unless lime mud was suspended; even so, the presence of abundant filter feeders in both areas suggests that turbidity was not a limiting factor.

6. Because the eastern Klamath terrane was located within the Paleopacific Ocean and was exposed to open-ocean conditions, whereas much of the Cordilleran miogeocline was separated from the open ocean by uplifts along the old Antler orogenic belt, the free interchange of organisms between the provinces may have been hindered. However, displaced fusulinid faunas present in the Havallah sequence of north-central Nevada west of the uplifted belt suggest that this was not the reason for the biotic differences. These fossils, which we interpret to have been derived from shallow-water environments on the open ocean (west) side of the uplifts at the western edge of North America, are typically central Cordilleran rather than eastern Klamath in aspect (Stevens, 1987). In addition, eastern California, which has a typical central Cordilleran fauna, is interpreted to have faced the same ocean as the eastern Klamaths in Early Permian time, any fringing barriers having been removed during a slightly earlier continental truncation (Stone and Stevens, 1988).

7. No oceanographic barriers between the eastern Klamaths and central Cordillera are suspected. Modeled Permian circulation patterns are similar to those in the modern Pacific Ocean, and the northern subtropical convergence zone presumably was north of both the eastern Klamaths and the central Cordillera (P. Wilde, written communication, 1989); thus, the same water mass presumably bathed both provinces. Evidence of unusually strong upwelling, except later in the Permian in the Idaho region, also generally is lacking.

8. Biotic isolation across a wide ocean basin (wide paleolongitudinal separation) is the remaining factor to account for the differences in Early Permian faunas between the eastern Klamaths and the central Cordillera. As this factor has not been discounted, it will be further investigated.

PALEOLONGITUDINAL SEPARATIONS

Longitudinal position is not measurable in rocks, and in the open ocean must be inferred by using methods for estimating distance of separation within latitudinal bands. Paleontologic data provide the only practical means of estimation of such longitudinal separation, because dispersal and evolution of biotas are strongly related to isolation (in this setting, large distances of separation across deep water). As there are excellent Permian fossil records of the tropical biotas of the eastern Klamath terrane, the central Cordillera, and the Tethys, good comparisons are possible.

The early Permian was a time of strong biotic provincialism worldwide (Yancey, 1979; Ziegler and others, 1981), due to the presence of strong climatic gradients and land barriers oriented

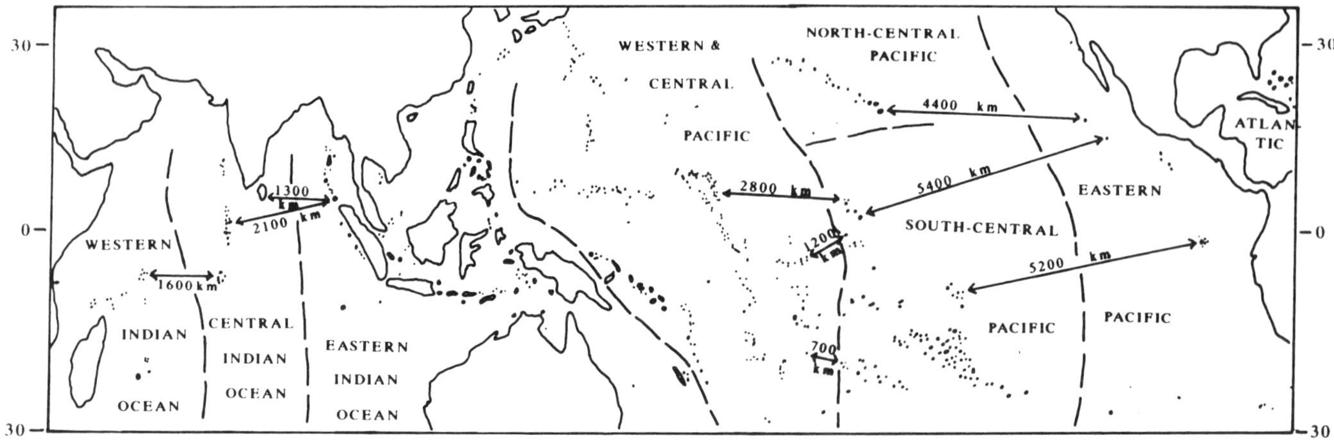

Figure 2. Map of the Indo-Pacific Ocean showing geographic subprovinces. Some of the greater distances between suitable habitats for shallow-water marine organisms also are indicated.

along longitudinal lines. Permian marine provincialism apparently was somewhat more pronounced than that present on the Earth today, probably because of the greater width of the Pacific Ocean basin during the Permian, but we still believe it is feasible to use modern biotic provincialism and degrees of endemism to estimate the amount of geographic separation between Permian biotas. Calibration can be achieved by reference to modern examples where the distance of separation and the degree of biotic similarity can be related. We will utilize information on modern reef corals, supplemented by some on modern archaeogastropods, to illustrate this.

DISTRIBUTION OF MODERN CORALS AND ARCHAEOGASTROPODS: A STANDARD FOR CALIBRATING A SCALE OF LONGITUDINAL SEPARATION FOR PERMIAN ANALYSES

Monographic study combining uniformly reliable taxonomic and distribution data is available for modern reef corals, and to a lesser extent, for modern archaeogastropods. A correlation between distance of separation and degree of dissimilarity is apparent in the distribution of members of both groups.

Modern reef corals range from approximately 30° north to 30° south latitude, and related corals in early Miocene coral-algal reefs ranged from Japan to New Zealand (about 40°N to 40°S), according to Fulthorpe and Schlanger (1989). This is similar to the distribution of Permian shelf-inhabiting massive rugose corals that ranged from about 40°N to 40°S (Stevens, 1985). Both assemblages of corals inhabited warm, shallow-marine water, and thus seem to be quite comparable ecologically (Stevens, 1985). Wood (1983) has compiled maps showing the geographic distribution of 100 living reef corals and gives geographic information on four others. This information, revised on the basis of data from Veron (1986), provides a satisfactory basis for analyzing the biogeography of reef corals worldwide and provides a standard by which the distribution of Permian corals and other invertebrates can be measured.

In the modern oceans, there are two rich coral provinces: the western Atlantic and the Indo-Pacific (Wood, 1983), the latter of which, for the purposes of this analysis, has been divided into geographic subprovinces based on the largest distances between shallow-water environments (Fig. 2). Distribution of genera in these provinces and subprovinces (from Wood, 1983, and Veron, 1986) is indicated in Appendix 5, and the Otsuka similarity indices between these various geographic elements are shown in Table 3.

The coral faunas of the western Atlantic and Indo-Pacific provinces are substantially different. The Otsuka similarity index for genera is about 22 (low). Within the Indo-Pacific province the number of genera decreases both eastward and westward from a high diversity in the western Pacific. Although similarity indices remain relatively high throughout the province, there is a rough correlation between similarity index values and the distance of separation between shallow-water environments within this province, a generality pointed out by Newton (1988). The faunas of the Western Indian Ocean and the Central Indian Ocean subprovinces, separated by about 1,600 km of water, have a similarity index of 95. The faunas of the Central Indian Ocean subprovince and the Nicobar Islands in the eastern Indian Ocean have a similarity index of about 84. They are separated by a minimum of 1,300 m of deep water, if Sri Lanka provides a stopping point for corals, or about 2,200 km if it does not. As about 25 percent of the widely distributed corals in the Indian Ocean do not occur in Sri Lanka (Veron, 1986), this does not appear to have been a very good stopping place for corals. The South-central Pacific and Central and Western Pacific subprovinces are separated by distances ranging from 2,800 km near the equator, where the equatorial countercurrent disperses corals eastward most effectively, to 700 km south of the equator; the similarity index is about 76. The westernmost islands in the Eastern Pacific subprovince are separated from the South-central Pacific subprovince by more than 5,000 km of open ocean, and there is a reduced similarity index of about 49. This moderate similarity is due to a large decrease in the number of genera in the eastern Pacific caused by the "eastern Pacific barrier," the vast expanse of deep water (Veron, 1986).

Thus, within the equatorial zone of the Indo-Pacific province, coral similarity indices generally are inversely proportional to the distance between shallow-water habitats. In the modern Indo-Pacific Ocean, however, there are no areas in the equatorial zone where similarity values are as low as those between most of the Permian brachiopod, coral, or gastropod faunas of the eastern Klamath, central Cordilleran, and Tethyan provinces, even though there are distances in the modern ocean greater than 5,000 km. In order to equalize the similarity index between Permian coral faunas of the eastern Klamaths and the central Cordillera, and that between the modern South-central and Eastern Pacific faunas, 11 non–Indo-Pacific coral genera would have to be added to the modern Eastern Pacific subprovince. Thus, it is clear that the differences between the pairs of coral faunas of the two different ages are not trivial and cannot be due to factors such as degree of study, nature of preservation, or degree of generic subdivision.

The fossil corals of the Middle American region suggest that the history of coral colonization and extinction in the region between the western Atlantic and the eastern Pacific was complex. From the Cretaceous through Oligocene, when there were good marine connections between the western Atlantic and the eastern Pacific Oceans, the faunas were highly similar (Glynn and Wellington, 1983); however, few corals were dispersed eastward across the Pacific from the south-central Pacific region during this time, presumably because of even greater distances across deep water then than now (Dana, 1975). During formation of the Central American landbridge (25 to 3 Ma), the similarities between the western Atlantic and eastern Pacific faunas decreased somewhat, but they were still high. The faunas between these provinces evidently did not diverge radically until after completion of the land bridge in the Pliocene, extinction of the eastern Pacific coral fauna during the Pleistocene glacial epochs, and dispersal of Indo-West Pacific corals to replace them in relatively recent times (Glynn and Wellington, 1983). This history suggests that even though marine connections may be poor (as during construction of the Middle American land bridge), communication between faunas can be maintained, and they tend to retain their similarities if the distance is not too great. Thus, it appears that large differences in shallow-water marine faunas in similar latitudinal bands may arise only in cases of complete separation by land bridges or separation by large distances across deep water.

Archaeogastropods show patterns somewhat similar to those of the corals. The high diversity, great range in size (a few millimeters to a couple of decimeters), and variety of environments inhabited by modern archaeogastropods has inhibited tabulation of taxonomic and distributional data to the level presently available for reef corals, but monographic reports are available on archaeogastropods of the Panamic (eastern Pacific) province and parts of the Indo-Pacific province. The Otsuka similarity index of modern archaeogastropods in the north-central Pacific (Kay, 1979) and the central-eastern Pacific Ocean (Keen, 1971) is moderate (33), somewhat lower than for the corals. This lower level of generic similarity recorded for modern archaeogastropods compared to modern corals may be a function of the lower level of sampling and study for the group, or the greater extent of planktotrophic larval dispersal employed by many of the modern corals, enabling them to disperse more readily than gastropods. It is not apparent which factor is more important; both may apply.

TABLE 3. DATA ON MODERN REEF CORALS*

Faunal Group	Areas Compared	Smiilarity Index	Barrier
Modern Corals	Indo-Pacific and west Atlantic provinces	22	Land
Modern Corals	Western and central Indian Ocean	95	1,600 km water
Modern Corals	Central and eastern Indian Ocean	84	2,100 or 1,300 km water
Modern Corals	Central and western and south-central Pacific	76	700–2,800 km water
Modern Corals	South-central and eastern Pacific	49	>5,000 km water

*Otsuka similarity indices between adjacent provinces and subprovinces separated by barriers of differing significance also are shown.

DISTANCE OF SEPARATION OF THE EASTERN KLAMATH TERRANE FROM THE CENTRAL CORDILLERA DURING THE PERMIAN

Corals seem to be the best group for comparison of modern and Permian biogeography because the compositions of both faunas are relatively well determined and the Permian coral data set is comparable to that of the modern corals. Otsuka similarity indices in the equatorial zone in the modern open ocean are generally inversely proportional to the distance between shallow-water habitats. In the modern ocean, more than 5,000 km of deep-ocean separation within a single ocean basin (between the South-central and Eastern Pacific subprovinces) has resulted in an Otsuka similarity index of about 49 for modern corals. This modern scale can be used for Permian coral faunas if modern and ancient corals dispersed similarly, if the circulation pattern and current velocities in the Permian Paleopacific Ocean were similar to that at present, and if part of the Klamath arc lay in the path of the Permian equatorial countercurrent. Assuming these similarities, most of which already have been shown to be reasonable, an Otsuka generic similarly index of 35 (for eastern Klamath–central Cordilleran Permian corals) suggests that the two terranes were much farther apart than the maximum of 1,000 km given by Miller (1987) and probably were separated by a distance greater than 5,000 km, the limit of calibration in the modern data set. Even if some of these conditions do not apply, several thousand kilometers of distance between the two areas is indicated by data from the modern Indian Ocean.

SUMMARY

The fossil evidence for placement of the eastern Klamath terrane relative to North America is good for the Early Permian, at which time the deposits in both areas contained a rich and varied biota of shallow-marine invertebrates. The eastern Klamath biota is part of a biogeographic realm that was distinct from that of the North American realm, which contains the eastern Cordilleran province, suggesting a great distance of separation of the two areas. Comparisons of fusulinids, corals, brachiopods, and gastropods between the eastern Klamath, central Cordilleran, and Tethyan provinces all show relatively low similarities; comparison with modern coral distributions allow calculation of at least 5,000 km of separation of the eastern Klamaths relative to North America during the Early Permian. Our data and interpretations are in conflict with the inferences drawn from some current plate tectonic models, pointing out the need to reassess some of the assumptions inherent in those models.

ACKNOWLEDGMENTS

Work on the eastern Klamath Terrane biotas was supported by National Science Foundation Grant EAR-8217222. In addition, we are grateful to the various reviewers who made suggestions that resulted in improvement of the manuscript: David Andersen, Gary Gray, David Harwood, Meghan Miller, Charles Ross, Norm Silberling, George Stanley, Jr., and Edward Wilson.

APPENDIX 1. DISTRIBUTION OF EARLY PERMIAN FUSULINIDS

	Eastern Klamath Mountains	Central Cordillera	Tethys
Alpinoschwagerina			x
Bartramella		x	
Biwaella			x
Boultonia			x
Chalaroschwagerina	x	x	x
Chusenella		x	x
Cuniculinella	x	x	x
Darvasites			x
Dutkevitchia			x
Eoparafusulina	x	x	x
Hayasakaina			x
Klamathina	x		
Leptotriticites		x	
Mccloudia	x		x
Minojapanella			x
Misellina			x
Nagatoella			x
Nankinella			x
Occidentoschwagerina			x
Oketaella		x	
Ozawainella			x
Pamirina			x
Parafusulina	x	x	x
Paraschwagerina	x	x	x
Paratriticites			x
Pravitoschwagerina			x
Pseudofusulina	x	x	x
Pseudofusulinella	x	x	x
Pseudofusulinoides			x
Pseudoreichelina		x	x
Pseudoschwagerina	x	x	x
Quasifusulina			x
Reticulosepta		x	
Robustoschwagerina			x
Rugosochusenella			x
Schubertella			
Schwagerina	x	x	x
Sphaeroschwagerina			x
Staffella			x
Stewartina		x	
Thompsonella	x	x	
Triticites	x	x	x
Zellia			x

All Early Permian fusulinid genera and subgenera from the eastern Klamath Mountains are recorded by Skinner and Wilde (1965).

Early Permian fusulinid genera and subgenera from the central Cordillera. Authors' names indicate one reference to the presence of the taxon; = means that Stevens has reassigned the genus for the purpose of this chapter..
Bartramella (Douglass, 1974)
Boultonia (Verville and others, 1986)
Chalaroschwagerina (Magginetti and others, 1988)
Chusenella (Magginetti and others, 1988)
Cuniculinella (Magginetti and others, 1988)
Eoparafusulina (Magginetti and others 1988)
Leptotriticites (Magginetti and others 1988)
Oketaella (Douglass, 1974)
Parafusulina (Magginetti and others 1988)
Paraschwagerina (Magginetti and others 1988)
Pseudofusulina (Hoare, 1963)
Pseudofusulinella (Verville and Laule, 1979)
Pseudoreichelina (Douglass and Nestell, 1974)
Pseudoschwagerina (Magginetti and others, 1988)
Reticulosepta (Magginetti and others, 1988)
Schubertella (Verville and Laule, 1979)
Schwagerina (Magginetti and others, 1988)
Stewartina (Magginetti and others, 1988)
Thompsonella (= *Pseudofusulinella* of Hoare, 1963)
Triticites (Magginetti and others, 1988)

Early Permian fusulinid genera and subgenera from the Tethyan province. Authors' names indicate one reference to the presence of the taxon; = means that Stevens has reassigned the genus for the purpose of this chapter.
Alpinoschwagerina (Kahler and Kahler, 1980)
Biwaella (Kochansky-Devide, 1970)
Boultonia (Kahler and Kahler, 1980)
Chalaroschwagerina (Toriyama, 1984)
Chusenella (Kahler and Kahler, 1980)
Cuniculinella (Toriyama, 1984)
Darvasites (Kahler and Kahler, 1980)
Dutkevitchia (Toriyama, 1984)
Eoparafusulina (Toriyama,. 1984)
Hayasakaina (Kahler and Kahler, 1980)
Mccloudia (Kahler and Kahler, 1980)
Minojapanella (Kahler and Kahler, 1980)
Misellina (Toriyama, 1984)
Nagatoella (Kahler and Kahler, 1980)
Nankinella (Kochansky-Devide, 1970)
Occidentoschwagerina (Kahler and Kahler, 1980)
Ozawainella (Toriyama, 1982)
Pamirina (Toriyama, 1982)
Parafusulina (= *Praeparafusulina* of Kahler and Kahler, 1980)
Paraschwagerina (Kahler and Kahler, 1980)
Paratriticites(Kahler and Kahler, 1980)
Pravitoschwagerina (Toriyama, 1982)
Pseudofusulina (Toriyama)
Pseudofusulinella (= *Neofusulinella* of Kochansky-Devide, 1970)
Pseudofusulinoides (Kahler and Kahler, 1980)
Pseudoreichelina (= *Reichelina* of Kochansky-Devide, 1970)
Pseudoschwagerina (Kahler and Kahler, 1980)
Quasifusulina (Kahler and Kahler, 1980)
Robustoschwagerina (Kahler and Kahler, 1980)
Rugosochusenella (Kochansky-Devide, 1970)
Schubertella (Kahler and Kahler, 1980)
Schwagerina (Kahler and Kahler, 1980)
Sphaeroschwagerina (= *Schwagerina* of Kochansky-Devide, 1970)
Staffella (Kochansky-Devide, 1970)
Triticites (Kahler and Kahler, 1980)
Zellia (Kahler and Kahler, 1980)

APPENDIX 2. DISTRIBUTION OF EARLY PERMIAN CORALS

	Eastern Klamath Mountains	Central Cordillera	Tethys
Acaciapora		x	
Akagophyllum			x
Allotropiophyllum			x
?Amydalophylloides		x	
Amandophyllum			x
Amplexocarinia			x
?Aulophyllum	x		
Bassius	x		
Bayhaium	x	x	
Bradyphyllum		x	
Caninia		x	
Carinthiaphyllum			x
Chaoiphyllum			x
Chihsiaphyllum			x
Chusenophyllum			x
Cladochonus		x	x
Clisiophyllum	x		
Cornwallatia		x	
?cyathopsid coral (new?)	x		
?Darwasophyllum	x		
?Diffingia		x	
Dillerium	x		
Diphyphyllum		x	
"Dobrolyubovia"		x	
"Duplophyllum"			x
Durhamina		x	
Enigmalites	x		
Eoheritschioides		x	
Fomichevella		x	
Geyerophyllum			x
Gshelia	x		
Hayasakaia			x
Heintzella		x	
Heritschioides	x	x	?
Heterocaninia	x		
Ipciphyllum			x
Iranophyllum			x
Kabakovitchiella			x
Kepingophyllum			x
Kleopatrina	x	x	
Langenheimia	x		
Laophyllum			x
?Lithostrotionella		x	
Lonsdaleiastraea			x
Lonsdaleoides			x
Lophocarinophyllum			x
Lophophyllidium			x
Lytvophyllum	x		
Mccloudius	x		
Michelinia	x	x	x
Multimurinus			x
Multithecopora		x	
Neomultithecopora	x	x	
Neosyringopora	x	x	
Orionastraea		x	
Paracaninia			x
Paraduplophyllum			x
Paraheritschioides	x	x	
Pararachnastraea	x		
Parawentzelella			x
Pavastehphyllum			x
Petalaxis brokawi group	x	x	
Petalaxis occidentalis group	x	x	
Polythecalis			x
Praewentzelella			x
Protomichelinia			x
Protowentzelella		x	
Pseudocarniaphyllum			x
Pseudohuangia			x
Sakamotosawaenella			x
Sciophyllum		x	
Sochkineophyllum			x
Stereostylus		x	
Stikineastraea	x		
Stylastraea		x	
Szechuanophyllum			x
Tachylasma			x
Thomasiphyllum			x
Thysanophyllum		x	
Timania		x	
Traskina	x		
Uralnevadaphyllum	x	x	
Waagenophyllum			x
Wentzelella			x
Wentzelophyllum			x
Wexolina	x		
Wilsonastraea	x		
Yabeiphyllum			x
Yatsengia			x
Yokoyamaella			x
New cerioid genus (aff. *Bassius*)		x	
New phacelloid genus C	?	x	?
New phacelloid genus (aff. *Petalaxis*)		x	

Early Permian coral genera and subgenera from the eastern Klamath Mountains. Authors' names indicate one reference to the presence of the taxon; = means that Stevens has reassigned the genus for the purpose of this chapter.

Aulophyllum? (Wilson, 1982)
Bassius (Wilson, 1982)
Bayhaium (Wilson, 1982)
Clisiophyllum ((Wilson, 1982)
?cyathopsid coral (Wilson, 1982)
?Darwasophyllum (C. H. Stevens, unpublished data)
Dillerium (Wilson, 1982)
Enigmalites (Wilson, 1982)
Gshelia (Wilson, 1982)
Heritschioides (Wilson, 1982)
Heterocaninia (Wilson, 1982)
Kleopatrina (Wilson, 1982)
Langenheimia (Wilson, 1982)
Lytvophyllum (= *Siphonodendron* of Wilson, 1982)
Mccloudius (Wilson, 1982)
Michelinia (Wilson, 1982)
Neomultithecopora (Wilson, 1982)
Neosyringopora (= *Syringopora* Wilson, 1982)
Paraheritschioides (= part of *Heritschioides* of Wilson, 1982)
Pararachnastraea (= part of Arachnastraea of Wilson, 1982)
Petalaxis brokawi group (Wilson, 1982)
Petalaxis ocidentalis group (Wilson, 1985)
Stikineastraea (= part of *Arachnastraea* of Wilson, 1982)
Traskina (Wilson, 1982)
Uralnevadaphyllum (= *Porfirievella* of Wilson, 1982)
Wexolina (Wilson, 1985)
Wilsonastraea (= part of *Yatsengia* of Wilson, 1982)
?New phacelloid genus C (= *Durhamina* of Wilson, 1982)

Early Permian coral genera and subgenera from the central Cordillera. Authors' names indicate one reference to the presence of the taxon; = means that Stevens has reassigned the genus for the purpose of this chapter.
Acaciapora (Hoare, 1964)
?Amydalophylloides (C. H. Stevens, unpublished data)
Bayhaium (Sando, 1984)
Bradyphyllum (Hoare, 1964)
Caninia (Wilson and Langenheim, 1962)
Cladochonus (Wilson and Langenheim, 1962)
Cornwallatia (Sando, 1984)
?Diffingia (C. H. Stevens, unpublished data)
Diphyphyllum (Wilson and Langenheim, 1962)
"Dobrolyubovia" (Stevens and Rycerski, 1983)
Durhamina (Wilson and Langenheim, 1962)
Eoheritschioides (Stevens and Rycerski, 1983)
Fomichevella (Stevens and Rycerski, 1983)
Heintzella (Stevens and Rycerski, 1983)
Heritschioides (Wilson and Langenheim, 1962)
Kleopatrina (= *Ptolemaia* of Wilson and Langenheim, 1962)
?Lithostrotionella (C. H. Stevens, unpublished data)
Michelinia (C. H. Stevens, unpublished data)
Multithecopora (Sando, 1984)
Neomultithecopora (Sando, 1984)
Neosyringopora (Sando, 1984)
Orionastraea (Wilson and Langenheim, 1962)
Paraheritschioides (C. H. Stevens, unpublished data)
Patalaxis brokawi group (= *Lithostrotionella* and tentatively *Eastonoides* of Wilson and Langenheim, 1962)
Petalaxis occidentalis group (C. H. Stevens, unpublished data)
Protowentzelella (Stevens and Rycerski, 1983)
Sciophyllum (Wilson and Langenheim, 1962)
Stereostylus (Wilson and Langenheim, 1962)
Stylastraea (Stevens and Rycerski, 1983)
Thysanophyllum (Wilson and Langenheim, 1962)
Timania (Laule, 1978)
Uralnevadaphyllum (Stevens and Rycerski, 1983)
New cerioid genus, much more simple than *Bassius* (C. H. Stevens, unpublished data)
New phacelloid genus C (Stevens and Rycerski, 1983)
new phacelloid genus like *Petalaxis* (C. H. Stevens, unpublished data)

Early Permian coral genera and subgenera from the Tethyan Province. Authors' names indicate one reference to the presence of the taxon; = means that Stevens has reassigned the genus for the purpose of this chapter.
Akagophyllum (Minato and Kato, 1965a)
Allotropiophyllum (Yoh and Huang, 1932)
Almandophyllum (Minato and Kato, 1965b)
Amplexocarinia (Zhao and Yang, 1985)
Carinthiaphyllum (Fedorowski, 1986)
Chaoiphyllum (Minato and Kato, 1965a)
Chihsiaphyllum (Minato and Kato, 1965a)
Chusenophyllum (Minato and Kato, 1965a)
Cladochonus (= *Monilopora* of Yoh and Huang, 1932)
"Duplophyllum" (Fedorowski, 1986)
Geyerophyllum (Minato, 1975)
Hayasakaia (= *Tetrapora* of Yoh and Huang, 1932)
?Heritschioides (= *Heritschioides* of Minato and Kato, 1965b)
Ipciphyllum (Minato and Kato, 1965a)
Iranophylkum (Minato and Kato 1965a)
Kabakovitchiella (Fedorowski, 1986)
Kepingophyllum (Fontaine, 1986)
Laophyllum (Minato and Kato, 1965a)
Lonsdaleiastraea (Minato and Kato, 1965a)
Lonsdaleoides (Fedorowski, 1986)
Lophocarinophyllum (Fedorowski, 1986)
Lophophyllidium (= *Sinophyllum* of Yoh and Huang, 1932)
Michelinia (Yoh and Huang, 1932)
Multimurinus (Zhao and Yang, 1985)
Paracaninia (Zhao, 1976)
Paraduplophyllum (Federowski, 1986)
Parawentzelella (Minato and Kato, 1965a)
Pavastehphyllum (Minato and Kato, 1965a)
Polythecalis (Minato and Kato, 1965a)
Praewentzelella (Minato and Kato, 1965a)
Protomichelinia (Fontaine, 1986)
Pseudocarniaphyllum (Minato and Kato, 1965a)
Pseudohuangia (Minato and Kato, 1965a)
Sakamotosawaenella (Minato and Kato, 1965a)
Sochkineophyllum (Minato, 1975)
Szechuanophyllum (Minato and Kato, 1965a)
Tachylasma (Douglas, 1950)
Thomasiphyllum (Minato and Kato, 1965a)
Waagenophyllum (Minato and Kato, 1965a)
Wentzelella (Minato and Kato, 1965a)
Wentzelophyllum (Minato and Kato, 1965a)
Yabeiphyllum (Minato and Kato, 1965b)
Yatsengia (Minato, 1975)
Yokoyamaella (Minato and Kato, 1965a)
?New phacelloid genus C (= *Durhamina* of Minato and Kato, 1965b)

APPENDIX 3. DISTRIBUTION OF EARLY PERMIAN BRACHIOPODS

	Eastern Klamath Mountains	Central Cordillera	Tethys
Acritosia			x
Adriana			x
Anidanthus		x	
Anomaloria			x
anopliinid (new genus)	x		
Antiquatonia		x	
Astegosia			x
Bathymyonia		x	
Bibatiola			x
Bilotina			x
brachythyridid (new genus?)	x		
Bryorhynchus		x	
Buxtonia		x	
Calliprotonia		x	
Callispirina			x
Camerisma (Callaiapsida)	x		
Cancrinella		x	x
Caricula			x
Celebetes			x
Cenorhynchia		x	
Ceocypea			x
Chonetina			x
Chonetinella	x	x	x
Cleiothyridina	x	x	x
Composita		x	x
Comuquia			x
Cooperina			x
Costellarina		x	
Crenispirifer		x	
Cruricella	x		x
Crurithyris		x	
Demonedys			x
Derbyia	x	x	x
dictyoclostid (new genus?)	x	?	
Dictyoclostus		x	
Dielasma	x	x	x
Dielasmina			x
Dyschrestia			x
Echinaria	x		
Echinauris		x	x
Elliotella		x	
Eolyttonia			x
Gefonia			x
Girtyella		x	
Glossothyropsis	x		
Goleomixa			x
Gratiosina			x
Gypospirifer		x	
Haydenella			x
Heterelasma		x	
Hemiptychina			x
Heteralosia			x
Hustedia	x	x	x
Incisius			x
Juresania		x	
Keyserlingina			x
Kochiproductus	x	x	
Kozlowskia		x	x
Kutorginella		x	x
Leiorhynchus		x	
Leptodus			x
Leurosina		x	
Linoproductus	x	x	x
Liosotella		x	
Litocothia			x
Lissochonetes		x	
Lowenstamia		x	
Marginifera	x		x
Martiniopsis	x		x
Meekella	x	x	x
Muirwoodia		x	
Neochonetes		x	
Neophricodothyris	x		
Neospirifer	x	x	x
Notothyris			x
Orbicoelia	x		x
Orthotichia	x	x	x
overtoniid (new genus?)	x		
Paralyttonia			x
Paraspiriferina			x
Peniculauris		x	
Perigeyerella			x
Permophricodothyris			x
Phricodothyris		x	
Pontisia		x	x
Pseudodielasma		x	
Pseudoleptodus			x
Psilocamera		x	
Punctospirifer	x	x	
Ramavectus			x
Reticulatia		x	
Reticulariina		x	
Retimarginifera			x
Rhipidomella	x	x	x
Rhynchopora	x	x	
Rigbyella			x
Rugaria			x
Rugatia		x	
rugosochonetid (new genus)	x		
Schuchertella	x		x
Septacamera	x		
Sphenalosia		x	
Sphenosteges		x	
Spiriferella			x
Spiriferellina	x	x	x
Spiriferina		x	
spiriferid (new genus)	x		
Squamaria		x	
Squamularia		x	
Stenoscisma	x		x
Stereochia			x
Stictozoster			x
Streptorhynchus	x		x
Striochonetes			x
Strophalosia		x	
Teguliferina	x		
Terebratuloidea			x
Timorina			x
Tipispirifer			x
Tornquistia			x

	Eastern Klamath Mountains	Central Cordillera	Tethys
Uncinella			x
Uncinunellina			x
Urushtenia			x
Waagenites			x
Waagenoconcha		x	
Wellerella		x	
Wilberrya		x	

All Early Permian brachiopod genera from the eastern Klamath Mountains are from unpublished data of T. E. Yancey and R. A. Hanger.

Early Permian brachiopod genera from the central Cordillera. Authors' names indicate one reference to the presence of the genus.

Anidanthus (Yochelson, 1968)
Antiquatonia (Yochelson, 1968)
Bathymyonia (Yancey, 1978)
Bryorhynchus (Yancey, 1978)
Buxtonia (Gilluly and others, 1954)
Calliprotonia (Mills and Langenheim, 1987)
Cancrinella (Yochelson, 1968)
Cenorhynchia (MIlls and Langenheim, 1987)
Chonetinella (M. E. Kaasa, unpublished data)
Cleiothyridina (Yancey, 1978)
Composita (Yancey,1978)
Costellarina (Yancey, 1978)
Crenispirifer (Mills and Langenheim, 1987)
Crurithyris (Mills and Langenheim, 1987)
Derbyia (Yancey, 1978)
Dictyoclostus (Pederson, 1954)
Dielasma (Yancey, 1978)
Echinaurus (Yancey, 1978)
Elliotella (Mills and Langheim, 1987)
Girtyella (Kirkland, 1963)
Gypospirifer (Mills and Langenheim, 1987)
Heterelasma (Kirkland, 1963)
Hustedia (Mills and Langenheim, 1987)
Juresania (Pederson, 1954)
Kochiproductus (Yochelson, 1968)
Kozlowskia (Mills and Langenheim, 1987)
Kutorginella (Mills and Langenheim, 1987)
Leiorhynchus (Yochelson, 1968)
Leurosina (Mills and Langenheim, 1987)
Linoproductus (Gilluly and others, 1954)
Liosotella (Yochelson, 1968)
Lissochonetes (Kirkland, 1963)
Lowenstamia (M. E. Kaasa, unpublished data)
Meekella (Yancey, 1978)
Muirwoodia (Yochelson, 1968)
Neochonetes (M. E., Kaasa, unpublished data)
Neospirifer (Pederson, 1954)
Orthotichia (M. E. Kaasa, unpublished data)
Peniculauris (Kirkland, 1963)
Phricodothyris (Yochelson, 1968)
Pontisia (Yancey, 1978)
Pseudodielasma (Winters, 1963)
Psilocamera (M. E. Kaasa, unpublished data)
Punctospirifer (Gilluly and others, 1954)
Reticulatia (M. E. Kaasa, unpublished data)
Reticulariina (Yancey, 1978)
Rhipidomella (Yancey, 1978)
Rhynchopora (Yochelson, 1968)
Rugatia (Yochelson, 1968)
Sphenalosia (Yochelson, 1968)
Sphenosteges (Yochelsohn, 1968)
Spiriferellina (Kirkland, 1963)
Spiriferina (Yochelson, 1968)
Squamaria (Yancey, 1978)
Squamularia (Kirkland, 1963)
Strophalosia (Gilluly and others, 1954)
Waagenoconcha (Kirkland, 1963)
Wellerella (Pederson, 1954)
Wilberrya (Yancey, 1978)

All Early Permian brachiopod genera from the Tethyan province are from Grant (1976).

APPENDIX 4. DISTRIBUTION OF EARLY PERMIAN GASTROPODS

	Eastern Klamath Mountains	Central Cordillera	Tethys
Ambozone			x
Amphiscapha		x	
Ananias		x	
Anomphalus	x	x	x
anomphalid (new genus)	x		
Apachella	x	x	x
Babylonites		x	
Baylea	x		
Bellerophon	x	x	x
Borestus	x		x
Brochidium		x	
Callitomaria		x	
Ceraunocochlis		x	x
Cyclites		x	
Cylicioscapha		x	
Cylindritopsis	x	x	x
Dichostasia		x	x
Discotomaria		x	
Discotropis		x	
Diploconula		x	
Eirlysia		x	
eotomariid #1 (new genus)	x		
eotomariid #2 (new genus)	x		
Euconospira	x	x	x
Euomphalus	x	x	x
Euphemites		x	
Euphemitopsis		x	
Girtyspira	x		
Glabrocingulum	x	x	x
Glyptomaria		x	
Gosseletina	x		
Hesperiella	x		
Ianthinopsis	x		
Ischoptygma		x	
Knightites		x	
Labridens		x	
Lacunospira	x	x	x
Lamellispira	x	x	x
lamellispirid (new genus)	x		
Lepetopsis		x	x
Leptomphalus		x	
Lucinellina			x
Meekospira	x	x	x
Metoptoma		x	
Mourlonia		x	x
Omphalotrochus	x	x	
Paragoniozona		x	x
Pernotrochus		x	
Peruvispira	x	x	
Pharkidonotus	x	x	
Phymatopleura	x	x	x
Planotectus		x	
Platyceras	x	x	
Platyworthenia		x	
Platyzona			x
Retispira	x	x	x
Sallya		x	x
Schwedogonia		x	x
Shansiella		x	
Sinuitina		x	
Soleniscus	x	x	x
Spiroscala		x	
Stenozone	x		x
Strobeus		x	x
Tapinotomaria	x	x	
Trepospira	x		
Turbonilopsis			x
Warthia		x	
Worthenia	x	x	x
Yunnania		x	x

All Early Permian archaeogastropod and subulitid genera from the eastern Klamath Mountains are from unpublished data of T. E. Yancey.

Early Permian archaeogastropod and subulitid genera from the central Cordillera. Authors' names indicate one reference to the presence of the genus.

Amphiscapha (Yochelson, 1956)
Ananias (Chronic, 1952)
Anomphalus (Yochelson, 1956)
Apachella (Winters, 1963)
Babylonites (Yochelson, 1956)
Bellerophon (Yochelson, 1960)
Brochidium (Yochelson, 1956)
Callitomaria (Batten, 1958)
Ceraunochohlis (Erwin, 1988)
Cyclites (Yochelson, 1956)
Cylicioscapha (Yochelson, 1956)
Cylindritopsis (Erwin, 1988)
Dichostasia (Yochelson, 1956)
Discotomaria (Batten, 1958)
Discotropis (Yochelson, 1956)
Diploconula (Yochelson, 1956)
Eirlysia (Batten, 1958)
Euconospira (Batten, 1958)
Euomphalus (Yochelson, 1956)
Euphemites (Yochelson, 1960)
Euphemitopsis (Yochelson, 1960)
Glabrocingulum (Chronic, 1952)
Glyptomaria (Batten, 1958)
Ischoptygma (Erwin, 1988)
Knightites (Yochelson, 1960)
Labridens (Erwin, 1988)
Lacunospira (Batten, 1958)
Lamellispira (Batten, 1958)
Lepetopsis (Yochelson, 1960)
Leptomphalus (Yochelson, 1956)
Meekospira (Erwin, 1988)
Metoptoma (Yochelson, 1960)
Mourlonia (Chronic, 1952)
Omphalotrochus (Yochelson, 1956)
Paragoniozona (Batten, 1958)
Pernotrochus (T. E. Yancey, unpublished data)
Peruvispira (T. E. Yancey, unpublished data)
Pharkidonotus (Yochelson, 1960)
Phymatopleura (Batten, 1958)
Planotectus (Yochelson, 1956)
Platyceras (Yochelson, 1956)
Platyworthenia (Chronic, 1952)
Retispira (Yochelson, 1960)
Sallya (Yochelson, 1956)
Schwedogonia (Batten, 1958)
Shansiella (Batten, 1958)
Sinuitina (Yochelson, 1960)
Soleniscus (Erwin, 1988)
Spiroscala (Batten, 1958)
Strobeus (Erwin, 1988)
Tapinotomaria (Batten, 1958)
Warthia (Yochelson, 1960)
Worthenia (Chronic, 1952)
Yunnania (T. E. Yancey, unpublished data)

All Early Permian archaeogastropod and subulitid genera from the Tethyan province are from Batten (1972, 1979, 1985).

APPENDIX 5. DISTRIBUTION OF MODERN REEF CORALS (WOOD, 1983) WITH REVISIONS FROM VERON (1986)

	Indian Ocean			Pacific Ocean		
	West	Central	East Indian–West Pacific Ocean	South-central	East	West Atlantic
Acanthostrea	x	x	x	x		
Acrhelia			x			
Acropora	x	x	x	x	x	x
Agaricia						x
Alveopora	x	x	x	x		
Anacropora	x	x	x			
Anomastrea	x					
Astreopora	x	x	x	x		
Astreosmilia	x					
Australogyra			x			
Australomussa			x			
Balanophyllia	x	x	x	x	x	x
Barabattoia			x			
Blastomussa	x	x	x			
Boninastrea			x			
Catalaphyllia	x	x	x			
Caulastrea	x	x	x			
Cladocora						x
Coeloseris			x			
Colpophyllia						x
Coscinaraea	x	x	x	x		
Ctenella		x				
Cycloseris	x	x	x	x	x	
Cynarina	x	x	x			
Cyphastrea	x	x	x	x		
Dendrogyra						x
Dendrophyllia	x	x	x	x	x	x
Diaseris	x	x	x	x	x	
Dichocoenia						x
Diploastrea	x	x	x			
Diploria						x
Distichopora	x		x	x		
Duncanopsammia			x			
Echinophyllia	x	x	x	x		
Echinopora	x	x	x	x		
Euphyllia	x	x	x			
Eusmilia						x
Favia	x	x	x	x		x
Favites	x	x	x	x		
Fungia	x	x	x	x		
Galaxea	x	x	x	x		
Gardineroseris	x	x	x	x	x	
Goniastrea	x	x	x	x		
Goniopora	x	x	x	x		
Goreaugyra						x
Halomitra	x	x	x	x		
Heliofungia			x			
Heliopora	x	x	x			
Herpolitha	x	x	x	x		
Heteropsammia	x	x	x			
Horastrea	x	x				
Hydnophora	x	x	x	x		
Isophyllastrea						x
Isophyllia						x
Leptastrea	x	x	x	x		
Leptoria	x	x	x	x		
Leptoseris	x	x	x	x	x	x
Lithophyllom			x			
Lobophyllia	x	x	x	x		
Madracis			x			x

APPENDIX 5 (CONTINUED)

	Indian Ocean			Pacific Ocean		
	West	Central	East Indian–West Pacific Ocean	South-central	East	West Atlantic
Manicina						x
Meandrina						x
Merulina	x	x	x	x		
Millepora	x	x	x	x		x
Montastraea	x	x	x	x		x
Montigyra			x			
Montipora	x	x	x	x		
Moseleya			x			
Mussa						x
Mussismilia						x
Mycedium	x	x	x	x		
Mycetophyllia						x
Nemenzophyllia	x					
Oculina						x
Oulastrea			x			
Oulphyllia	x	x	x			
Oxypora	x	x	x	x		
Pachyseris	x	x	x	x		
Palauastrea			x			
Paraclavarina			x			
Pavona	x	x	x	x	x	
Pectinia	x	x	x	x		
Physogyra	x	x	x			
Physophyllia	x	x	x			
Platygyra	x	x	x	x		
Plerogyra	x	x	x			
Plesiastraea	x	x	x	x		
Pocillopora	x	x	x	x	x	
Podabacia	x	x	x	x		
Polyphyllia	x	x	x			
Porites	x	x	x	x	x	x
Psammocora	x	x	x	x	x	
Pseudosiderastrea			x			
Sandalolitha			x	x		
Scapophyllia			x			
Scolymia		x	x	x		
Seriatopora	x	x	x			
Siderastrea	x	x				x
Solenastrea						x
Stephanocoenia						x
Stylaraea	x		x	x		
Stylaster	x	x	x	x		x
Stylocoeniella	x	x	x	x		
Stylophora	x	x	x	x		
Symphyllia	x	x	x			
Trachyphyllia	x	x	x			
Tubastraea	x	x	x	x	x	x
Turbinaria	x	x	x	x		
Tubipora	x	x	x			
Wellsophyllia			x			
Zoopilus			x			

REFERENCES CITED

Batten, R. L., 1958, Permian gastropoda of the southwestern United States; Part 2, Pleurotomariacea: Portlockiellidae, Phymatopleuridae, and Eotomariidae: American Museum of Natural History Bulletin, v. 114, Article 2, p. 159–246.

——, 1972, Permian gastropods and chitons from Perak, Malaysia; Part 1, Chitons, bellerophontids, euomphlids, and pleurotomarians: American Museum of Natural History Bulletin, v. 147, Article 1, 44 p.

——, 1979, Permian gastropods from Perak, Malaysia; Part 2, The trochids, patellids, and neritids: American Museum of Natural History Novitates, no. 2685, 26 p.

——, 1985, Permian gastropods from Perak, Malaysia; Part 3, The murchisoniids, cerithids, loxonematids, and subulitids: American Museum of Natural History Novitates, no. 2829, 40 p.

Chronic, H., 1952, Molluscan fauna from the Permian Kaibab Formation, Walnut Canyon, Arizona: Geological Society of America Bulletin, v. 63, p. 95–165.

Coogan, A. H., 1960, Stratigraphy and paleontology of the Permian Nosoni and Dekkas Formations: University of California Publications in Geological Sciences, v. 36, no. 5, p. 243–316.

Dana, T. F., 1975, Development of contemporary eastern Pacific coral reefs: Marine Biology, v. 33, no. 4, p. 355–375.

Demirmen, F., and Harbaugh, J. W., 1965, Petrography and origin of Permian McCloud Limestone of northern California: Journal of Sedimentary Petrology, v. 35, p. 136–154.

Dilles, J. H., and Wright, J. E., 1988, The chronology of early Mesozoic arc magmatism in the Yerington district of western Nevada and its regional implications: Geological Society of America Bulletin, v. 100, p. 644–652.

Douglas, J. A., 1950, The Carboniferous and Permian faunas of south Iran and Iranian Baluchistan: Palaeontologia Indica, n.s., v. 22, no. 7, 27 p.

Douglass, R. C., 1974, Fusulinids in the Basin and Range province in California, Nevada, and Utah: Journal of Paleontology, v. 48, p. 846–853.

Douglass, R. C., and Nestell, M. K., 1974, The fusulinid genus *Pseudoreichelina* from the Pequop Formation, Spruce Mountains, Elko County, Nevada: Journal of Paleontology, v. 38, p. 1170–1173.

Erwin, D. H., 1988, Permian gastropoda of the southwestern United States; Subulitacea: Journal of Paleontology, v. 62, p. 56–69.

Fedorowski, J., 1986, The rugose coral faunas of the Carboniferous/Permian boundary interval: Acta Palaeontologica Polonica, v. 31, no. 3-4, p. 253–275.

Fontaine, H., 1986, Discovery of Lower Permian corals in Sumatra; Geosea V Proceedings, v. 1: Geological Society of Malaysia Bulletin 19, p. 183–191.

Fulthorpe, C. S., and Schlanger, S. O., 1989, Paleo-oceanographic and tectonic settings of early Miocene reefs and associated carbonates of offshore southeast Asia: American Association of Petroleum Geologists Bulletin, v. 73, p. 729–756.

Gilluly, J., Cooper, J. R., and Williams, J. S., 1954, Late Paleozoic stratigraphy of central Cochise County, Arizona: U.S. Geological Survey Professional Paper 266, 49 p.

Glynn, P. W., and Wellington, G. M., 1983, Corals and coral reefs of the Galapagos Islands: Berkeley, University of California Press, 330 p.

Grant, R. E., 1976, Permian brachiopods from southern Thailand: Paleontological Society Memoir 9, 269 p.

Hanger, R. A., 1986, Biotic communities and brachiopod paleoecology of the Early Permian McCloud Formation, northern California [M.S. thesis]: College Station, Texas A&M University, 127 p.

Hoare, R. D., 1963, Permian fusulinids from the Sunflower Reservoir area of northern Nevada: Journal of Paleontology, v. 37, p. 1143–1149.

——, 1964, Permian corals from northern Nevada: Journal of Paleontology, v. 38, p. 496–504.

Kahler, F., and Kahler, G., 1980, Fusuliniden aus den Kalken der Trogkofel-Schichten der Karnischen Alpen: Sonderdruck aus Sonderheft 36 der Carinthia II: Die Trogkofelstufe im Unterperm der Karnischen Alpen, p. 183–254.

Kay, E. A., 1979, Hawaiian marine shells: Bernice P. Bishop Museum Special Publication 64, section 4, 653 p.

Keen, A. M., 1971, Sea shells of tropical west America: Stanford, California, Stanford University Press, 1064 p.

Kirkland, R. L., 1963, Permian stratigraphy and stratigraphic paleontology of a part of the Colorado Plateau: Four Corners Geological Society Guidebook for 1963, p. 80–100.

Kochansky-Devide, V., 1970, Permski mikrofosili zahodnih Karavank: Geologija, Razprave in Porocila, 13. Ljubljana, p. 175–256.

Laule, S. W., 1978, A reinterpretation of the Permian Wildcat Peak Formation, central Nevada [M.S. thesis]: Reno, University of Nevada, 90 p.

Magginetti, R., Stevens, C. H., and Stone, P., 1988, Early Permian fusulinids from the Owens Valley Group: Geological Society of America Special Paper 217, 61 p.

Mankinen, E. A., Irwin, W. P., and Gromme, C. S., 1989, Paleomagnetic study of the Eastern Klamath Terrane, California, and implications for the tectonic history of the Klamath Mountains Province: Journal of Geophysical Research, v. 94, p. 10444–10472.

Miller, M. M., 1987, Dispersed remnants of a northeast Pacific fringing arc; Upper Paleozoic terranes of Permian McCloud faunal affinity, western U.S.: Tectonics, v. 6, p. 807–830.

Mills, P. C., and Langenheim, R. L., Jr., 1987, Wolfcampian brachiopods from the Bird Spring Group, Warp Spring area, Las Vegas Range, Clark County, Nevada: Journal of Paleontology, v. 61, p. 32–55.

Minato, M., 1975, Japanese Palaeozoic corals: Journal of the Geological Society of Japan, v. 81, p. 103–126.

Minato, M., and Kato, M., 1965a, Waagenophyllidae: Journal of the Faculty of Science, Hokkaido University, ser. 4, Geology and Mineralogy, v. 12, no. 3-4, 241 p.

——, 1965b, Durhaminidae (Tetracoral): Journal of the Faculty of Science, Hokkaido University, ser. 4, Geology and Mineralogy, v. 13, no. 1, p. 11–86.

Nassichuk, W. W., and Wilde, G. L., 1977, Permian fusulinaceans and stratigraphy at Blind Fiord, southwestern Ellesmere Island: Geological Survey of Canada Bulletin 268, 59 p.

Newton, C. R., 1988, Significance of "Tethyan" fossils in the American Cordillera: Science, v. 242, p. 385–391.

Oliver, W. A., Jr., and Pedder, A.E.H., 1976, Biogeography of Late Silurian and Devonian rugose corals in North America, *in* Gray, J., and Boucot, A. J., eds., Historical biogeography, plate tectonics, and the changing environment: Corvallis, Oregon State University Press, p. 131–145.

Pederson, S. L., 1954, A Permian (Wolfcampian) fauna of the Casper Formation of southeastern Wyoming: Journal of Paleontology, v. 28, p. 17–21.

Ross, C. A., 1967, Development of fusulinid (Foraminiferida) faunal realms: Journal of Paleontology, v. 41, p. 1341–1354.

Ross, C. A., and Ross, J.R.P., 1983, Late Paleozoic accreted terranes of western North America, *in* Stevens, C. H., ed., Pre-Jurassic rocks in western North America suspect terranes: Pacific Section, Society of Economic Paleontologists and Mineralogists, p. 23–35.

——, 1987, Biostratigraphic zonation of Late Paleozoic depositional sequences: Cushman Foundation for Foraminiferal Research Special Publication 24, p. 151–168.

Sando, W. J., 1984, Biostratigraphic utility of Upper Paleozoic syringoporoid corals, western interior region, conterminous U.S.A.: Palaeontographica Americana, no. 54, p. 453–458.

Sando, W. J., Bamber, E. W., and Armstrong, A. K., 1975, The zoogeography of North American Mississippian corals; Second International Symposium on corals and fossil coral reefs: Paris, Bureau du Recherche Geologique et Mineres, no. 89, p. 175–184.

Savage, N. M., Perry, D. G., and Boucot, A. J., 1976, A quantitative analysis of Lower Devonian brachiopod distribution, *in* Gray, J., and Boucot, A. J.,

Historical biogeography, plate tectonics, and the changing environment: Corvallis, Oregon State University Press, p. 169–200.

Skinner, J. W., and Wilde, G. L., 1965, Permian biostratigraphy and fusulinid faunas of the Shasta Lake area, northern California: University of Kansas Paleontological Contributions, Protozoa, article 6, 98 p.

Stehli, F. G., 1970, A test of the Earth's magnetic field during Permian time: Journal of Geophysical Research, v. 75, p. 3325–3342.

——, 1973, Permian brachiopods, *in* Hallam, A., ed., Atlas of paleobiogeography: New York, Elsevier, p. 143–149.

Stevens, C. H., 1967, Leonardian (Permian) compound corals of Nevada: Journal of Paleontology, v. 41, p. 423–431.

——, 1985, Reconstruction of Permian paleogeography based upon distribution of Tethyan faunal elements: IX-ICC Compte Rendu, v. 5, p. 383–393.

——, 1987, Affinities of Permian fusulinid faunas in the Golconda allochthon and northern Sierra Nevada: Geological Society of America Abstracts with Programs, v. 19, p. 455.

Stevens, C. H., Miller, M. M., and Nestell, M., 1987, A new Permian waagenophyllid coral from the Klamath Mountains, California: Journal of Paleontology, v. 61, p. 690–699.

Stevens, C. H., and Rycerski, B., 1983, Permian colonial rugose corals in the western Americas; Aids in positioning of suspect terranes, *in* Stevens, C. H., ed., Pre-Jurassic rocks in western North American suspect terranes: Pacific Section, Society of Economic Paleontologists and Mineralogists, p. 23–36.

——, 1989, Early Permian rugose corals from the Stikine River area, British Columbia, Canada: Journal of Paleontology, v. 63, p. 158–181.

Stone, P., and Stevens, C. H., 1988, Pennsylvanian and Early Permian paleogeography of east-central California; Implications for the shape of the continental margin and the timing of continental truncation: Geology, v. 16, p. 330–333.

Toriyama, R., 1982, Fusuline fossils from Thailand; Part 15, Peculiar spirothecal structure of schwagerinid from east of Wang Saphung, Changwat Loei, central north Thailand, *in*, Kobayashi, T., Toriyama, R., and Hashimoto, W., eds., Geology and palaeontology of Southeast Asia: University of Tokyo Press, v. 23, p. 1–7.

——, 1984, Summary of the fusuline faunas in Thailand and Malaysia, *in* Kobayashi, T., Toriyama, R., and Hashimoto, W., eds., Geology and palaeontology of Southeast Asia: University of Tokyo Press, v. 25, p. 137–146.

Veron, J.E.N., 1986, Corals of Australia and the Indo-Pacific: North Ryde, NSW, Australia, Angus and Robertson, 644 p.

Verville, G. J., and Laule, S. W., 1979, Early Permian fusulinids from the Wildcat Peak Formation, central Toquima Range, Nye County, Nevada, *in* Newman, G. W., and Goode, H. D., eds., Basin and Range Symposium: Rocky Mountain Association of Geologists and Utah Geological Association, p. 273–276.

Verville, G. J., Sanderson, G. A., and Drowley, D. D., 1986, Wolfcampian fusulinids from the Antler Peak Limestone, Battle Mountain, Lander County, Nevada: Journal of Foraminiferal Research, v. 16, p. 353–362.

Waterhouse, J. B., and Bonham-Carter, G. F., 1975, Global distribution and character of Permian biomes based on brachiopod assemblages: Canadian Journal of Earth Sciences, v. 12, p. 1085–1146.

Watkins, R., 1985, Volcaniclastic and carbonate sedimentation in late Paleozoic island-arc deposits, eastern Klamath Mountains, California: Geology, v. 13, p. 709–713.

Wilson, E. C., 1982, Wolfcampian rugose and tabulate corals (Coelenterata: Anthozoa) from the Lower Permian McCloud Limestone of northern California: Natural History Museum of Los Angeles County Contributions in Science, no. 337, 90 p.

——, 1985, Rugose corals (Coelenterata, Anthozoa) from the Lower Permian McCloud Limestone at Tombstone Mountain, northern California: Natural History Museum of Los Angeles County Contributions in Science, no. 366, 11 p.

Wilson, E. C., and Langenheim, R. L., Jr., 1962, Rugose and tabulate corals from Permian rocks in the Ely Quadrangle, White Pine County, Nevada: Journal of Paleontology, v. 36, p. 495–520.

Winters, S. S., 1963, Supai Formation (Permian) of eastern Arizona: Geological Society of America Memoir 9, 269 p.

Wood, E. M., 1983, Reef corals of the world: Hong Kong, T.F.H. Publications, 256 p.

Yancey, T. E., 1978, Brachiopods and molluscs of the lower Permian Arcturus Group, Nevada and Utah; Part 1, Brachiopods, scaphopods, rostroconchs, and bivalves: Bulletins of American Paleontology, v. 74, no. 303, p. 257–367.

——, 1979, Permian positions of the northern hemisphere continents as determined from marine biotic provinces, *in* Gray, J., and Boucot, A. J., eds., Historical biogeography, plate tectonics, and the changing environment: Corvallis, Oregon State University Press, p. 239–247.

Yancey, T. E., and Hanger, R. A., 1986, Endemism among Permian invertebrates, with implications for paleogeographic position of the Eastern Klamath Terrane: Geological Society of America Abstracts with Programs, v. 18, p. 797.

Yochelson, E. L., 1956, Permian Gastropoda of the southwestern United States; Part 1, Euomphalacea, Trochonematacea, Pseudophoracea, Anomphalacea, Craspedostomatacea, and Platyceratacea: American Museum of Natural History Bulletin, v. 110, Art. 3, p. 177–275.

——, 1960, Permian Gastropoda of the southwestern United States; Part 3, Bellerophontacea and Patellacea: American Museum of Natural History Bulletin, v. 119, Art. 4, p. 209–293.

——, 1968, Biostratigraphy of the Phosphoria, Park City, and Shedhorn Formations: U.S. Geological Survey Professional Paper 313-D, p. 571–660.

Yoh, S. S., and Huang, T. K., 1932, The coral fauna of the Chihsia Limestone of the lower Yangtze Valley: Palaeontologia Sinica, Ser. B, v. 8, fasc. 1, 72 p.

Zhao, J., 1976, Late Permian rugose corals from Anshun, Luzhi, and Qinglong, Guizhou Province: Acta Palaeontologica Sinica, v. 15, no. 2, p. 213–222.

Zhao, J., and Yang, D.-R., 1985, Discovery of the *Multimurinus* coral fauna from Xiuzhumuqinqi of Nei Mongol: Acta Palaeontologica Sinica, v. 24, no. 4, p. 440–448.

Ziegler, A. M., and 7 others, 1981, Paleozoic biogeography and climatology, *in* Niklas, K. H., ed., Paleobotony, paleoecology, and evolution, v. 2: New York, Praeger Publishers, p. 231–266.

MANUSCRIPT ACCEPTED BY THE SOCIETY MARCH 27, 1990

Printed in U.S.A.

Geological Society of America
Special Paper 255
1990

Paleozoic and Mesozoic rocks of the Pine Forest Range, northwest Nevada, and their relation to volcanic arc assemblages of the western U.S. Cordillera

Sandra J. Wyld*
Department of Geology, Stanford University, Stanford, California 94305

ABSTRACT

New geologic mapping and fossil data from the Pine Forest Range, Black Rock Desert, northwest Nevada, indicates that the range contains a structurally intact sequence of variably metamorphosed middle (and early?) Paleozoic through latest Triassic strata. The oldest rocks in the range include metamorphosed quartzo-feldspathic sedimentary rocks and mafic volcanic and volcaniclastic rocks of Mississippian and/or older age. Overlying fan facies chert/argillite/quartz-rich clastic rocks are of post–Late Devonian(?) and pre–Late Mississippian age, and are succeeded by shallower marine Upper Mississippian to Lower Pennsylvanian(?) volcanic rocks, volcanic-lithic–rich clastic rocks, and limestone. The remainder of Paleozoic time is characterized mostly by shallow marine conditions and the development of several unconformities. A thin sequence of shallow marine carbonates and clastic sediments, yielding early Late Permian fossils at the top, overlies Pennsylvanian(?) strata across an unconformity that may span early Pennsylvanian through Early Permian time. Upper Permian(?) chert and shale unconformably overlie older rocks and reflect some subsidence in Late Permian(?) time. A third unconformity separates Paleozoic and Triassic rocks and spans latest Permian(?) through Middle or Late Triassic time. Triassic strata in the Pine Forest Range record two distinct periods of deposition: (1) fan facies sedimentary-lithic–rich sediments and basinal carbonates were deposited from Ladinian or Carnian (late Middle or early Late Triassic) through early Norian (late late Triassic) time, and (2) mafic to intermediate composition lavas and associated volcanic-lithic– and crystal-rich fan facies sediments were deposited during most of the remainder of Norian time. Lavas exhibit the trace-element characteristics of volcanic arc magmas. Relatively deep marine conditions of deposition occurred throughout Middle(?) to Late Triassic time.

The Paleozoic stratigraphic record in the Pine Forest Range shows important similarities to that of other Paleozoic arc sequences in the western U.S. Cordillera, including those in the northern Sierra Nevada and eastern Klamath Mountains (California), Blue Mountain province (Oregon), and Chilliwack terrane (Washington). These similarities support an interpretation of paleogeographic and tectonic ties between the Black Rock Desert and these other arc sequences in Mississippian (and early Paleozoic?) through Permian time. In addition, the presence of a Permo-Triassic unconformity in the Pine Forest Range represents new evidence that these Paleozoic arc sequences were

*Present address, Geoscience Department, University of Houston, 4800 Calhoun Road, Houston, Texas 77204.

Wyld, S. J., 1990, Paleozoic and Mesozoic rocks of the Pine Forest Range, northwest Nevada, and their relation to volcanic arc assemblages of the western U.S. Cordillera, *in* Harwood, D. S., and Miller, M. M., eds., Paleozoic and early Mesozoic paleogeographic relations; Sierra Nevada, Klamath Mountains, and related terranes: Boulder, Colorado, Geological Society of America Special Paper 255.

characterized by uplift and erosion during the time of the Sonoma orogeny. Early Mesozoic strata in the Pine Forest Range provide a record of volcanism and sedimentation that is similar to that in other early Mesozoic volcanic arc sequences from the southwestern United States through northern California. These similarities support an interpretation that early Mesozoic arc sequences in northwest Nevada, as well as northern California, form the northern continuation of the west-facing early Mesozoic arc documented in the southwestern United States. In addition, the Triassic record in the Pine Forest Range suggests that extension-related intra-arc subsidence, inferred to have characterized the southwestern United States during early Mesozoic time, may also have affected early Mesozoic rocks of the Black Rock Desert.

INTRODUCTION

Most studies of Paleozoic and Mesozoic arc and related rocks of the western United States Cordillera have focused on the Sierra Nevada and Klamath Mountain provinces of northern California, and on the Blue Mountain province of eastern Oregon (Fig. 1). Considerably less is known about the stratigraphy and history of Paleozoic to Mesozoic volcanic and sedimentary rocks in the Black Rock Desert region of northwest Nevada (Fig. 1), although these strata are widely interpreted to have developed within the same arc system as that of rocks farther to the west (Burchfiel and Davis, 1972, 1975; Silberling, 1973; Davis and others, 1978; Speed, 1979; Russell, 1984; Miller, 1987; Speed and others, 1988). Previous detailed studies in the Black Rock Desert have centered mostly on a small area of exposure of Permian (and older?) to Middle Triassic rocks in the Bilk Creek Mountains, and Triassic to Cretaceous rocks in the Jackson Mountains (Fig. 1; Ketner and Wardlaw, 1981; Russell, 1981, 1984; Maher and Saleeby, 1986, 1988; Jones, this volume). These studies, however, have been able to provide only limited information on the pre-Permian history of the region or the development of the area from Late Permian into Triassic time.

This chapter focuses on new work in the Pine Forest Range (Fig. 1) where an extensive and structurally intact section of Mississippian (and older?) through latest Triassic volcanic and sedimentary rocks is exposed. Detailed stratigraphic, lithologic, and age data from this sequence of rocks are presented and used to better constrain (1) the Paleozoic to early Mesozoic development of the Black Rock Desert region; (2) the tectonic and paleogeographic relation between Paleozoic and early Mesozoic rocks in the Black Rock Desert and those in adjacent arc sequences; and (3) tectonic models, involving the Black Rock Desert region, for the Paleozoic to early Mesozoic development of the western U.S. Cordillera.

PREVIOUS WORK AND GEOLOGIC FRAMEWORK

The Pine Forest Range was first mapped by Willden (1964) and Smith (1966, 1973), who distinguished "Permian or older" volcanic rocks in the south, "Triassic" limestone and siliciclastic rocks, and "Triassic or Jurassic" sedimentary and metasedimentary rocks in the north. Very few age data were available at the time of these studies, however, and the ages of most rocks in the range were not well constrained. Silberling and others (1987) were the first to establish that volcanic rocks in the south are early Mesozoic and are underlain progressively to the north by Triassic sedimentary rocks and sedimentary rocks that are, at least in part, of Permian age. Geologic mapping for the present study builds on the mapping of Smith (1973), and confirms the general facing direction of strata in the range as determined by Silberling and others (1987). Facing direction in this study is indicated by fossil and radiometric age control as well as by graded beds within many of the units.

The eastern side of the Pine Forest Range is bounded by a major east-dipping normal fault, and several smaller displacement normal faults have also been mapped near the range front (Fig. 2). The principal fault that is not related to normal faulting is the Buckaroo fault, between Bishop and Buckaroo Canyons (Fig. 2), which repeats the stratigraphy and is of relatively minor displacement. Two other faults, of unspecified displacement and style but apparently broadly of bedding-parallel orientation, were also reported by Silberling and others (1987) as occurring within the pre-Tertiary section; one of them was considered to be a terrane-bounding fault (boundary between the "Jackson" and "Black Rock" terranes). The present study did not confirm the existence of these faults; thus the terrane terminology proposed by Silberling and others (1987) for the Black Rock Desert is difficult to apply and is not used in this chapter. Data indicating a stratigraphic, rather than structural, contact across the 'terrane boundary' of Silberling and others (1987) are discussed in a later section.

Paleozoic and Mesozoic rocks in the east-central Pine Forest Range have been regionally metamorphosed and deformed, and the metamorphic grade and intensity of deformation increase with stratigraphic depth from subgreenschist facies (and weak foliation) in the youngest Mesozoic rocks through amphibolite grade (and moderate to strong schistosity) in the oldest Paleozoic rocks (S. J. Wyld, unpublished data). This variation in metamorphic grade is progressive and not associated with any structural juxtaposition of high-grade against low-grade rocks. Because the metamorphic grade increases with stratigraphic depth, older Paleozoic rocks must be described in terms of metamorphic rocks, whereas younger Paleozoic and Mesozoic rocks can be described in terms of protoliths.

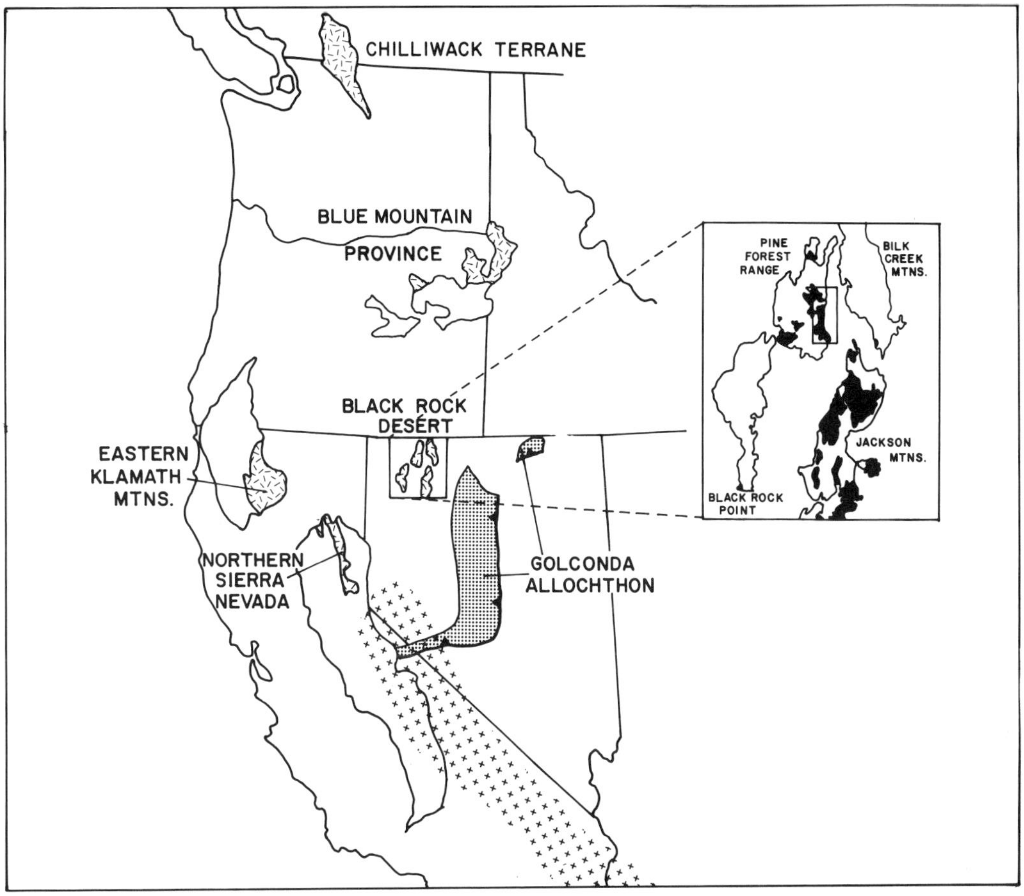

Figure 1. Index map of the western United States. Dash patterns = areas of exposure of Paleozoic and lower Mesozoic volcanic arc and related rocks discussed in text; dot pattern = Paleozoic basinal deposits of the Golconda allochthon; X pattern = general area of exposure of lower Mesozoic arc sequences inferred to represent a west-facing arc constructed on or adjacent to the North American continent (Burchfiel and Davis, 1972; Schweickert, 1978; Speed, 1978; Busby-Spera, 1988; Dilles and Wright, 1988; Tosdal and others, 1989). Inset of Black Rock Desert region from Stewart and Carlson (1978). Shaded areas in inset are Paleozoic and Mesozoic rocks. Boxed area in Pine Forest Range is the area shown in Figure 2A.

PALEOZOIC STRATIGRAPHY

Paleozoic(?) quartzo-feldspathic schist (unit Pz(?)qs)

Unit Pz(?)qs is the northernmost unit in the mapped area and consists mostly of quartzo-feldspathic schist (Fig. 2A). Other lithologies include minor amphibolite and rare quartzite. All of the rocks within this unit are moderately to strongly schistose and at amphibolite grade.

Quartzo-feldspathic schist typically contains 30 to 60 percent quartz, 30 to 55 percent plagioclase and potassium feldspar, and 1 to 20 percent biotite ± hornblende ± epidote. Millimeter- to centimeter-scale compositional layering, consisting of more and less biotite-rich layers, is common in the quartzo-feldspathic schist, but relict bedding features are only rarely discernible. Quartzite consists of quartz plus rare muscovite, chlorite, biotite, and epidote, and occurs as massive layers 1 to 3 m thick. Amphibolite is variable in composition and typically compositionally layered on a scale of millimeters to centimeters; layers vary from hornblende-rich to hornblende-poor and plagioclase-, epidote-, and/or quartz-rich, but overall, hornblende-rich schists are the most abundant rock type.

Metamorphic mineral assemblages in the quartzo-feldspathic schist suggest that these rocks were originally variably muddy quartzo-feldspathic sediments. Quartz, plagioclase, and potassium feldspar clearly formed a significant portion of the clasts in the protolith sediments, but the metamorphic grade makes it impossible to assess the possible contribution of such clast types as chert, or sedimentary, metamorphic or silicic volcanic lithics. The paucity of epidote or amphibole in these schists, however, suggests that mafic volcanic clasts were probably not an important component of the original sediments. Quartzites could have been derived from chert or quartzose sandstones. Amphibolites were probably derived mostly from bedded volcaniclastic deposits, as

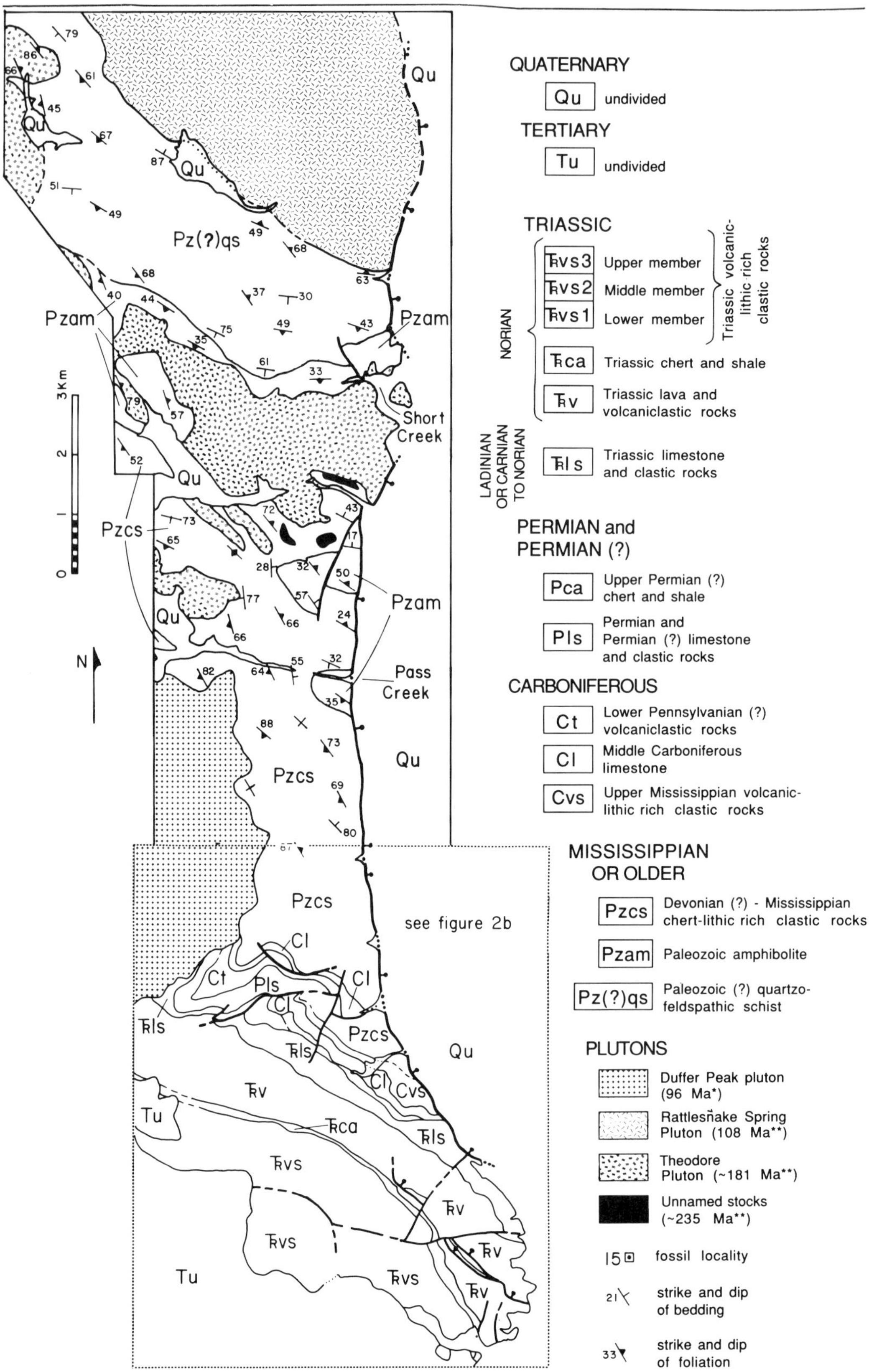

Figure 2A. Geologic map of the east-central Pine Forest Range (boxed area shown in inset in Fig. 1). *K/Ar age data (Smith and others, 1971). **Preliminary U/Pb zircon age data (J. E. Wright, personal communication, 1989).

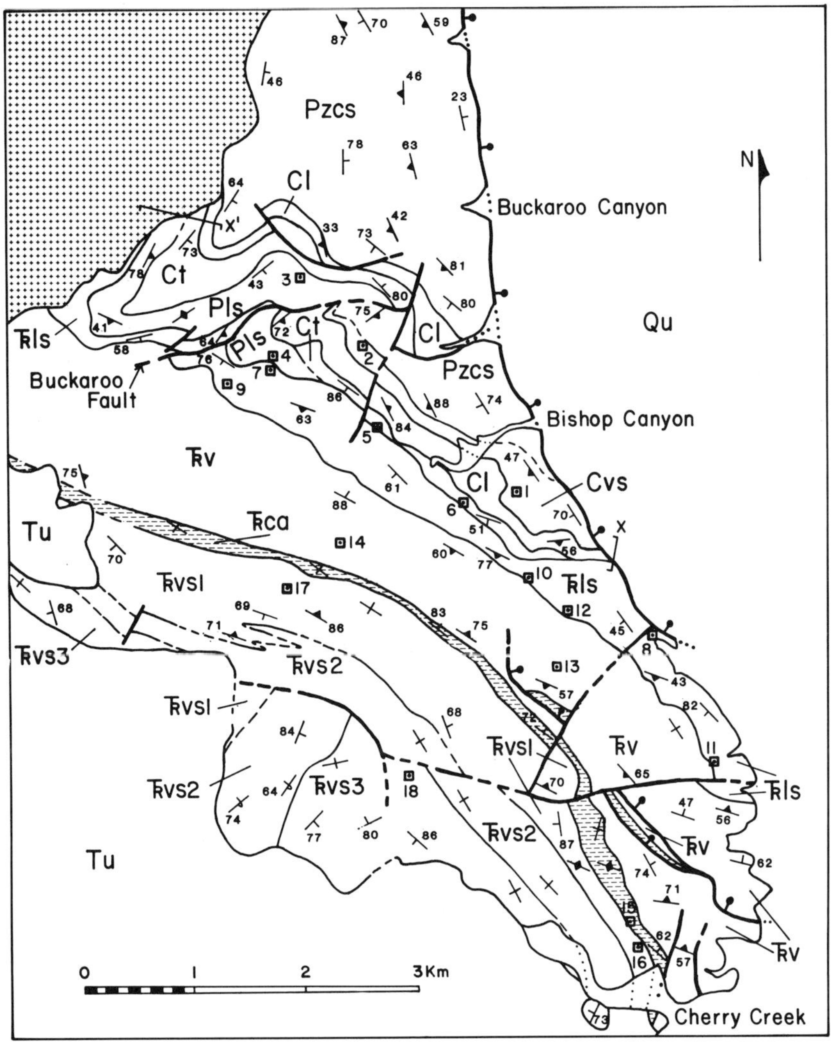

Figure 2B. Enlarged map of the southern part of Figure 2A (area enclosed by dotted-line box). Symbols same as in Figure 2A.

compositional layering is common and there is a wide variation in the composition of the amphibolite layers.

Contact relations between unit Pz(?)qs and the adjacent amphibolite unit are somewhat ambiguous, making interpretation of the age of the undated unit Pz(?)qs more difficult. Foliation and sparse bedding measurements in both the amphibolite unit and unit Pz(?)qs near the contact are parallel to subparallel (Fig. 2A), and there is no evidence of fault-related rocks or fabrics along the contact; these features suggest that the contact is not a fault. The contact is not lithologically gradational, however, and if stratigraphic, could be either conformable or unconformable. At the present time, unit Pz(?)qs is tentatively considered to be older than the adjacent amphibolite unit because there is no evidence of a fault between the two units, and because unit Pz(?)qs is lithologically unlike rocks interpreted to stratigraphically overlie the amphibolite unit to the south (see below). As the amphibolite unit is constrained only to be Mississippian or older, unit Pz(?)qs can also only be interpreted to be Mississippian or older.

Paleozoic Amphibolite (Unit Pzam)

Unit Pzam consists of metamafic schist, less common micaceous schist, minor felsic metavolcanic rock, and rare marble. Rocks in this unit are at amphibolite grade and moderately to strongly schistose.

Metamafic rocks consist mostly of amphibolite that contains 40 to 90 percent hornblende, 5 to 55 percent plagioclase, 0 to 20 percent epidote, and minor quartz. Compositional layering, ranging from <1 mm to 3 cm in thickness, is locally common in the amphibolite and juxtaposes hornblende-rich layers with epidote- or plagioclase-rich layers. As with amphibolite of the Paleozoic(?) quartzo-feldspathic schist unit, this common compositional layering is inferred to reflect derivation of the layered amphibolite

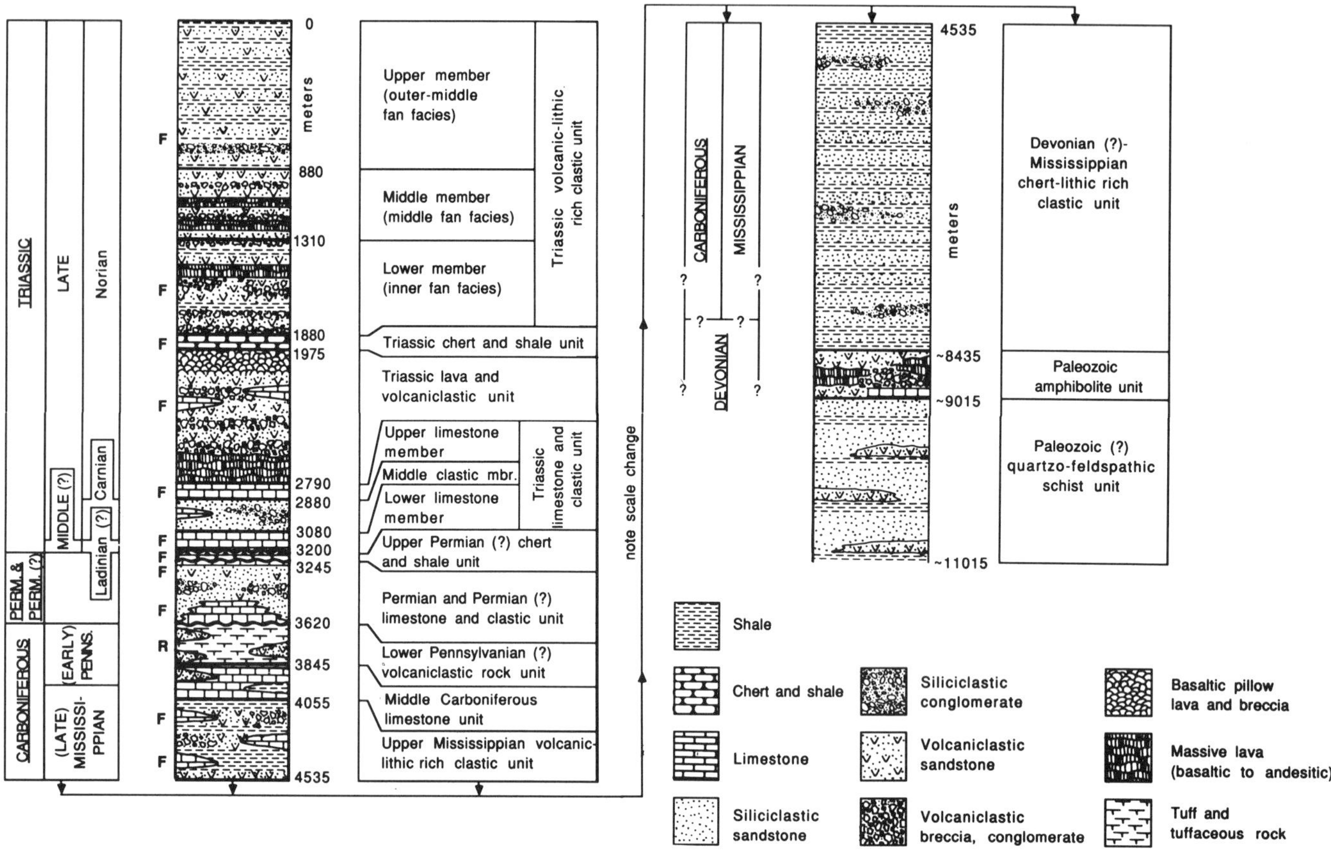

Figure 3. Generalized stratigraphic column of Paleozoic and Mesozoic rocks in the east-central Pine Forest Range. F denotes approximate stratigraphic positions of fossil localities (see text and Table 1). R denotes radiometric age data (see text). All unit thicknesses are structural. Thicknesses for Devonian(?)-Mississippian chert-lithic–rich clastic and Paleozoic(?) quartzo-feldspathic schist units are approximate. Thickness of Paleozoic amphibolite unit is a minimum. All lithologic data depicted in terms of unit protoliths.

from a bedded volcaniclastic protolith. Large-scale bedding between coarser and finer grained amphibolite is also locally present, and supports an interpretation of volcaniclastic protoliths. Other metamafic lithologies include massive, homogeneous amphibolite with variably abundant amygdules or phenocrysts, interpreted to have been derived from mafic lava, and volcanic breccia that contains porphyritic or amygdaloidal volcanic clasts. Felsic metavolcanic rocks are very fine grained, siliceous, and contain relict feldspar ± quartz phenocrysts. Micaceous schists are fine grained and consist of abundant quartz, less common biotite, and minor muscovite and feldspar.

In summary, protoliths for unit Pzam consisted of common mafic lava and volcaniclastic rock, less common fine-grained sedimentary rock, minor felsic lava, and rare limestone (Fig. 3). Unit Pzam is undated, but in the Pass Creek area (Fig. 2A), the contact between this unit and the adjacent chert-lithic-rich clastic unit is clearly gradational. As discussed below, the chert-lithic–rich clastic unit is inferred to be gradationally overlain to the south by progressively younger rocks, none of which are of appropriate lithology to be the protolith for unit Pzam. Unit Pzam is therefore considered to be older than the chert–lithic–rich clastic unit, which is Mississippian or older.

Devonian(?)-Mississippian chert-lithic–rich clastic rocks (unit Pzcs)

Unit Pzcs is the most widespread unit in the mapped area and consists of variably metamorphosed and deformed sedimentary rocks (Figs. 2, 3). Metapelitic rocks in the unit vary from amphibolite-grade, fine-grained muscovite + biotite schist in the north, through finer grained muscovite schist and phyllite, to greenschist-grade slate (foliation defined by very fine grained phyllosilicates) in the south. Metamorphic grade is indicated by the metamorphic grade of the underlying amphibolite unit in the north and metamorphic mineral assemblages in metamafic dikes that intrude the unit in the south. In the lithologic description below, rocks are discussed in terms of protoliths, as protolith lithologies and textures are still reasonably identifiable within the unit.

Unit Pzcs consists mostly of shale, siltstone, and fine- to medium-grained sandstone. Conglomerate and coarser grained sandstone are considerably less abundant except near Bishop Canyon (Fig. 2B), where sandstone and conglomerate predominate. North of the Bishop Canyon area, shale and sandstone are typically thinly bedded, although sandstone and conglomerate beds are locally as thick as 1 m. Sedimentary structures observed in sandstone and conglomerate beds include graded bedding and partial Bouma sequences. In the Bishop Canyon area, sandstone and conglomerate beds are up to 2 to 3 m and 3 to 4 m thick, respectively, and are generally graded. Facies associations in unit Pzcs, coupled with sedimentary structures that indicate common deposition by sediment gravity flows including turbidity currents, are consistent with deposition of the unit in a submarine fan environment (e.g., Mutti and Ricci-Lucchi, 1978). Sediments in the Bishop Canyon area are interpreted to have been deposited in a fan channel or channels.

Original sandstone compositions are impossible to determine in much of unit Pzcs due to the effects of metamorphism. Some sandstones, however, were originally sufficiently coarse-grained that as much as 20 percent of the original clasts are still recognizable. In these sandstones, relict grains consist of variable amounts of monocrystalline quartz and plagioclase with less common chert clasts and polycrystalline quartz. Sandstone compositions in the Bishop Canyon area are more easily identified due to a coarser average grain size and a lower metamorphic grade; these sandstones consist of abundant chert and argillite clasts, variably abundant monocrystalline quartz, and minor polycrystalline quartz. Pebble-sized clasts in conglomerates are generally identifiable and consist of abundant chert and argillite, minor polycrystalline quartz, and rare monocrystalline quartz, quartz-rich sandstone, foliated meta-quartzite, and volcanic clasts.

The only direct age constraint for unit Pzcs comes from the age of deformed stocks that intrude the lower part of the unit (Fig. 2A). Preliminary U/Pb data from two zircon size fractions from one of these stocks indicate an age of ~235 Ma (J. E. Wright, 1989, personal communication), which restricts unit Pzcs to an age of Middle Triassic or older. Indirect age constraints for unit Pzcs are discussed in the following section.

Upper Mississippian volcanic-lithic–rich clastic rocks (unit Cvs)

Unit Cvs consists mostly of shale, siltstone, relatively fine-grained sandstone, and minor limestone, while conglomerate and coarser grained sandstone are also present south of Bishop Canyon. Finer grained sediments are rarely calcareous and locally contain common fossil fragments, including crinoid stems as large as 15 mm in diameter. Limestone typically forms thin, discontinuous lenses that may contain as much as 40 percent fossil fragments, including crinoid stems as large as 20 mm in diameter. Conglomerate and coarser sandstone contain abundant plagioclase and porphyritic volcanic clasts, and rare chert clasts and mono- and polycrystalline quartz. Clastic hornblende and possible pyroxene (both mostly replaced by metamorphic minerals) are present in minor amounts in some beds. The depositional environment of unit Cvs is not well constrained, but a comparatively shallow marine environment may be suggested by the abundance of large, relatively intact crinoid stems in some limestone and clastic beds. Limestone lenses from the lower part of unit Cvs have yielded Late Mississippian and Late Mississippian(?) conodonts (Table 1, nos. 1–2).

Unit Cvs occurs only in the southern part of the mapped area; to the southeast, it is as thick as 480 m, but the unit thins and pinches out to the northwest beneath the overlying limestone (Fig. 2B). Detailed structural studies (S. J. Wyld, unpublished data) indicate that there is no fault between unit Cvs and the overlying limestone, and the decreasing thickness of unit Cvs is therefore related either to facies changes or the presence of an unconformity at the top of the unit. Age data are insufficient to demonstrate the presence or absence of an unconformity (Fig. 3), but because the decreasing thickness of the unit to the northwest is associated with a marked decrease in the abundance of clastic rocks coarser than siltstone, the variation in unit thickness is considered to be most likely related to facies changes. A conformable, gradational upper contact is also suggested by the occurrence of sandstone of similar composition and limestone containing large crinoid stems in both units Cvs and the overlying limestone unit.

Contact relations between unit Cvs and the adjacent chert-lithic–rich clastic unit (unit Pzcs; Fig. 2B) provide important constraints on the age of the latter. These two units are very similar in that they contain an abundance of shale and siltstone, and are difficult to distinguish from one another in many places—relations that, in the field, strongly suggest a gradational contact between the units. The principal differences between the two units are the presence of limestone in Cvs, the possibly more shallow marine environment of deposition for Cvs, and the volcanic-lithic–rich character of clastic rocks in Cvs. As unit Cvs is in contact with deep marine clastics to the north (unit Pzcs) and overlain by a shallow marine limestone to the south, the first two differences can be explained as the result of an environment that was gradually shoaling and becoming increasingly influenced by carbonate deposition over time. The change in composition of clastic rocks is rather abrupt, but the abundance of volcanic clasts and subhedral to euhedral plagioclase crystals, coupled with the occurrence of some unstable mafic crystal fragments, in clastic rocks of unit Cvs suggests that the change in composition of clastic rocks may be primarily a function of inception of nearby volcanism. This interpretation is supported by a continued influx of volcanic-lithic–rich clastic as well as primary volcanic deposits in strata above unit Cvs. In addition, chert and quartz clasts continue to occur in minor amounts in clastic rocks of unit Cvs, suggesting that the chert/quartz-rich source supplying sediment to unit Pzcs continued to supply detritus to the basin in Late Mississippian time but was overwhelmed by an influx of material related to inception of nearby volcanism. In summary, the field-based conclusion that the contact is clearly gradational is supported by detailed analysis of depositional environments and sediment provenance in these two units.

TABLE 1. FOSSIL DATA FROM PALEOZOIC AND MESOZOIC ROCKS OF THE EAST-CENTRAL PINE FOREST RANGE

Locality Number*	Map Unit	Diagnostic Fauna	Age	Reference
1 (PF-1862)	Upper Mississippian volcanic-lithic–rich clastic rocks (unit Cvs)	*Cavusgnathus?* sp. *Idioprioniodus*? sp.	Tentatively Late Mississippian	M. J. Orchard, 1989, written communication
2 (PF-902)	Upper Mississippian volcanic-lithic–rich clastic rocks (unit Cvs)	*Cavusgnathus* sp. *Gnathodus texanus* Roundy *Idioprioniodus*	Late Mississippian	M. J. Orchard, 1989, written communication
3 (PF-1666 and NJS locality 80-S-222)	Lower member of Permian and Permian(?) limestone and clastic rocks unit (Unit Pls)	Corals	Carboniferous or Permian	C. Stevens, 1989, oral communication, and 1982, written communication to N. J. Silberling
4 (NJS localities 80-S-241 and 81-5-157)	Upper member of Permian and Permian(?) limestone and clastic rocks unit (unit Pls)	*Thamnosia depressa* (Cooper)	Wordian (early Guadalupian)	B. R. Wardlaw, 1981, written communication to N. J. Silberling
5 (PF-850)	Upper Permian(?) chert and shale (unit Pca)	*Neogondolella* sp. indet.	Probably Permian, Leonardian or younger	M. J. Orchard, 1989, written communication
6 (PF-J-39)	Upper Permian(?) chert and shale (unit Pca)	*Neogondolella* sp. indet.	Probably Permian, Leonardian or younger	M. J. Orchard, 1989, written communication
7 (NJS locality 81-S-159)	Lower limestone member of Triassic limestone and clastic rocks unit (unit TRls)	*Neogondolella navicula*	Middle or Late Triassic	B. R. Wardlaw, 1981, written communication to N. J. Silberling
8 (PF-913)	Lower limestone member of Triassic limestone and clastic rocks unit (unit TRls)	*'Paragondolella'* sp.	Ladinian or Carnian	M. J. Orchard, 1989, written communication
9 (NJS locality 80-S-231)	Middle clastic or upper limestone member of Triassic limestone and clastic rocks unit (unit TRls)	*Septocardia*?	Probably late Carnian or Norian	N. J. Silbering, 1989, written communication
10 (SM 1981 locality PF-3)	Upper limestone member of Triassic limestone and clastic rocks unit (unit TRls)	*Epigondolella* cf. *E. multidentata*	Norian	B. R. Wardlaw, 1981, written communication to N. J. Silberling
11 (Sm 1981 "Arcestes" locality	Upper limestone member of Triassic limestone and clastic rocks unit (unit TRls)	*Epigondolella multidentata*	Norian	B. R. Wardlaw, 1981, written communication to N. J. Silberling
12a (NJS locality 79-S-58)	Upper limestone member of Triassic limestone and clastic rocks unit (unit TRls)	*Epigondolella* cf. *E. primitia*	early Norian	B. R. Wardlaw, 1981, written communication to N. J. Silberling
12b (NJS locality 79-S-58)	Upper limestone member of Triassic limestone and clastic rocks unit (unit TRls)	*Septocardia*? sp.	Late Triassic	N. J. Silberling, 1989, written communication

TABLE 1. FOSSIL DATA FROM PALEOZOIC AND MESOZOIC ROCKS OF THE EAST-CENTRAL PINE FOREST RANGE (continued)

Locality Number*	Map Unit	Diagnostic Fauna	Age	Reference
13 (PF-911)	Triassic lava and volcaniclastic rocks (unit TRv)	*Epigondolella* sp. indet	Norian	M. J. Orchard, 1989, written communication
14 (PF-916)	Triassic lava and volcaniclastic rocks (unit TRv)	*Epigondolella* sp. indet.	Norian	M. J. Orchard, 1989, written communication
15 (NJS locality 80-S-211)	Triassic chert and shale (unit TRca)	*Rhabdoceras* sp.	early late Norian	N. J. Silberling, 1989, written communication
16a (NJS locality 79-S-52)	Limestone clasts in lower member of Triassic volcanic-lithic–rich clastic rocks unit (unit TRvs1)	*Neogondolella polygnathi-Formis angusta, Carinella diebeli*	late Carnian	B. R. Wardlaw, 1981, written communication to N. J. Silberling
16b (KBK 1978 locality 8602)	Limestone clasts in lower member of Triassic volcanic-lithic–rich clastic rocks unit (unit TRvs1)	*Epigondolella* sp.	Norian	B. R. Wardlaw, 1981, written communication to N. J. Silberling
16c (KBK 1978 locality 8603)	Limestone clasts in lower member of Triassic volcanic-lithic–rich clastic rocks unit (unit TRvs1)	*Neogondolella*	Permian or Triassic	B. R. Wardlaw, 1981, written communication to N. J. Silberling
17 (KBK 1978 locality 8596)	Limestone clasts in lower member of Triassic volcanic-lithic–rich clastic rocks unit (unit TRvs1)	*Epigondolella abneptis*	early Norian	B. R. Wardlaw, 1981, written communication to N. J. Silberling
18 (NJS locality 80-S-202)	Upper member of Triassic volcanic-lithic–rich clastic rocks unit (unit TRvs3)	*Choristoceras* sp.	late late Norian	N. J. Silberling, 1989, written communication

*Locality numbers without parentheses refer to those in Figure 2B. Locality numbers in parentheses refer to actual locality numbers. NJS, SM, and KBK refer to the fossil localities of N. J. Silberling, S. McDaniel, and K. B. Ketner, respectively. Sample 15 (NJS locality 80-S-211) was discovered by B. J. Russell. All other fossil localities are those of the author.

This interpretation of a gradational contact forms an important basis for constraining the age of unit Pzcs. Numerous graded beds occur within unit Pzcs near its southern contact, and all uniformly indicate that stratigraphic "up" is to the south. As the contact between units Pzcs and Cvs is interpreted to be gradational, and as unit Cvs is overlain to the south by progressively younger strata that bear no resemblance to unit Pzcs (Fig. 2B), unit Pzcs is thus considered to be older than the Upper Mississippian unit Cvs and Mississippian in age at the top. Unit Pzcs is also similar, both lithologically and in terms of sedimentary structures, to chert/quartz–rich, graded sandstone and conglomerate in the Bilk Creek Mountains (Fig. 1; "Block 2" rocks of Ketner and Wardlaw [1981], and the chert-quartzite arenite of Jones [this volume]), and is herein interpreted to be correlative with these rocks. Chert clasts in these Bilk Creek rocks have yielded Late Devonian radiolaria, and the rocks are therefore Late Devonian in age or younger (Jones, this volume). Available age constraints thus suggest that unit Pzcs was deposited sometime between Late Devonian and Late Mississippian time.

Middle Carboniferous limestone (unit Cl)

Unit Cl is dominated by limestone, but variably calcareous interbedded siliciclastic sediments are important constituents in some areas. A characteristic feature throughout the unit is an abundance of large crinoid stems (2 to 30 mm in diameter and 2 to 15 cm in length), many of which are largely unbroken. A moderately strong foliation is developed in finer grained rocks,

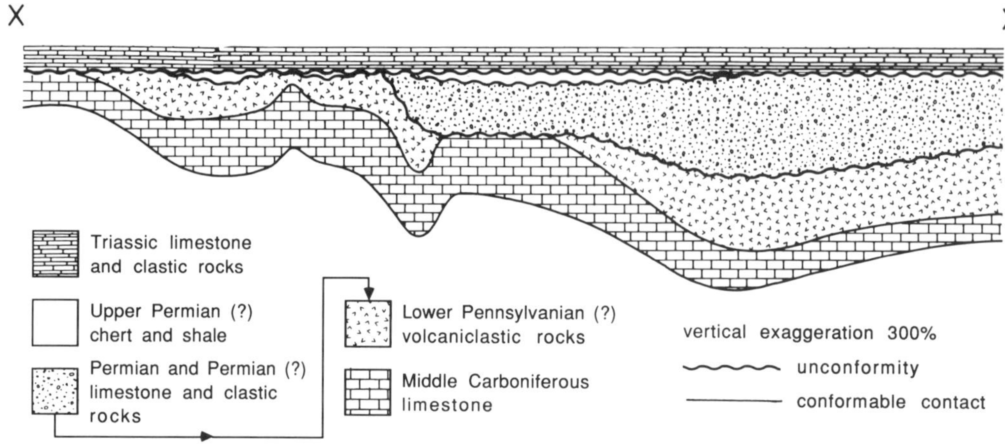

Figure 4. Approximate down-dip view showing unconformities and stratigraphic rleations between middle Carboniferous through Middle or Upper Triassic units in the east-central Pine Forest Range. Figure shows relations between locations X (south of Bishop Canyon) and X′ (southeast side of Duffer Peak pluton) (see Fig. 2B for locations of X and X′). Figure was constructed by: (1) restoring movement (approximate) along the Buckaroo fault, (2) calculating the approximate thicknesses of Paleozoic units beneath the Triassic limestone and clastic unit (unit ₮ ls) along the base of unit ₮ ls from X to X′, and (3) plotting this data while holding the base of unit ₮ ls horizontal. Figure is intended only as a diagrammatic means of showing unconformities and stratigraphic relations that are otherwise difficult to discern in Figure 2B. Note 300 percent vertical exaggeration.

and greenschist-facies mineral assemblages in metamafic dikes that intrude the unit near Buckaroo Canyon indicate that unit Cl is at greenschist grade.

South of the Buckaroo fault area (Fig. 2B), unit Cl can be generally divided into lower and upper parts. The lower part of the unit consists of variably calcareous sandstone, sandy fossiliferous limestone, limestone, and shale interbedded on a scale of a few centimeters to a few meters. The upper part of the unit in this area consists mostly of massive, light-gray to white fossiliferous limestone in which possible bioturbation features are common. Thin interbeds of shale are uncommon, and sandstone beds are rare. North of the Buckaroo fault area, the entire unit consists predominantly of gray to white fossiliferous limestone with less common thin interbeds of shale and sandstone. Most sandstones in unit Cl contain abundant plagioclase and porphyritic volcanic clasts, and minor amounts of chert clasts, and mono- and polycrystalline quartz.

Unit Cl is inferred to have been deposited in a shallow marine setting, based on several factors: the abundance of light colored limestone and possible bioturbation features that indicate a well-oxygenated environment; the massive character of most of the limestone and a scarcity of fine-scale laminations; the common occurrence of shale and sandstone interbeds; and the abundance of largely unbroken fossils (e.g., Wilson and Jordan, 1983). Unit Cl has yielded no datable fossils but is constrained to be younger than Late Mississippian, the age of the underlying volcanic-lithic clastic unit, and older than the overlying volcaniclastic unit that is interpreted to be Early Pennsylvanian in age. Unit Cl is therefore considered to be of Late Mississippian and/or Early Pennsylvanian age.

Pennsylvanian(?) and Permian rocks

Strata overlying the middle Carboniferous limestone can be divided into three different units separated by two slightly angular unconformities, which are in turn overlain unconformably by Middle or Upper Triassic rocks (Figs. 2–4). Age data, discussed below, indicate that the upper unconformity within the Pennsylvanian to Permian section probably spans some part of late Permian time, while the time missing across the lower unconformity is not well constrained but may include Late Pennsylvanian, and possibly Early Permian, time.

Lower Pennsylvanian(?) volcaniclastic rocks (unit Pt). Unit Pt concordantly overlies the Middle Carboniferous limestone and consists predominantly of crystal tuff, tuffaceous(?) rock, and volcanic lithic-rich sandstone (Figs. 2-4). Crystal tuff contains 45 to 55 percent crystals set in a fine-grained matrix of quartz, muscovite, and minor chlorite and biotite. Crystals include common broken and angular plagioclase, and less common water-clear quartz crystals that frequently have delicate embayed edges and are clearly of volcanic origin. The matrix-supported character of these rocks, coupled with the abundance of broken crystals and volcanic quartz, suggests that they were originally crystal-vitric tuffs. Tuffaceous(?) rock contains scattered angular quartz and feldspar grains in a matrix of very fine-grained quartz and feldspar, with less common muscovite and chlorite. The pale green color of these rocks, coupled with the abundance of metamorphic quartz and feldspar, the occurrence of sparse angular crystal fragments, and the similarity in composition of these rocks to the matrix of the crystal tuffs, collectively argue for a tuffaceous, rather than fine-grained epiclastic, protolith. Volcanic-lithic-rich sandstone contains intermediate to silicic composition volcanic clasts, minor quartz and plagioclase, and rare chert clasts.

Although the depositional environment of these rocks is not clear, they are interpreted to reflect simultaneous dacitic volcanism and erosion of an intermediate to silicic volcanic terrane. Preliminary U/Pb age data on two fractions of zircon from a crystal tuff bed suggest an age between ~310 and ~320 Ma (J. E. Wright, personal communication, 1989), and this age (Early Pennsylvanian) is considered to be the approximate age of deposition of this unit.

Permian and Permian(?) limestone and clastic rocks (unit Pls). This unit, which sits unconformably on both the Lower Pennsylvanian(?) volcaniclastic unit and the middle Carboniferous limestone (Figs. 2-4), can generally be divided into three members. The lower member consists of gray limestone with thin interbeds of calcareous shale, and often contains unbroken rugose and/or tabulate corals. Gradationally overlying this member is a clastic member that consists of sandstone, conglomerate, shale, and minor limestone. Sandstone and conglomerate contain abundant sedimentary and volcanic clasts, less common feldspar, minor quartz, and rare plutonic clasts. Interbedded limestones are thin and sometimes contain up to 90 percent fossils, including abundant crinoid stems. The upper member is a limestone that consists of up to 90 percent unbroken fossils (mostly brachiopods) in a muddy carbonate matrix. Unit Pls is inferred to have been deposited in a shallow marine environment based on the occurrence, throughout the unit, of limestones that contain abundant unbroken fossils, including corals. Corals in the lower member are Carboniferous or Permian (Table 1, no. 3), while brachiopods in the upper member are early Guadalupian (early Late Permian; Table 1, no. 4). Although the age of the base of this unit is not well constrained, the unit is no more than 375 m thick, and it is considered unlikely that it spans Middle Pennsylvanian through Late Permian time. The entire unit is therefore tentatively considered to be Permian, and the underlying unconformity is tentatively considered to span some part of Pennsylvanian and possibly Early Permian time.

Upper Permian(?) chert and shale (unit Pca). This unit, consisting of thinly interbedded chert and shale, is too thin to be shown on the map, but its outcrop distribution and unconformable relation with underlying rocks is diagramatically illustrated in Figure 4. Conodonts from chert beds indicate a probable Permian age of deposition (Table 1, nos 5–6); the unit is also constrained to be younger than early Guadalupian and is therefore considered to be Late Permian(?) in age.

TRIASSIC STRATIGRAPHY

Triassic strata in the east-central Pine Forest Range sit unconformably on older Paleozoic rocks and have been divided into four main units (in ascending order; Figs. 2, 3): (1) a limestone and clastic rock unit, (2) a lava and volcaniclastic rock unit, (3) a chert and shale unit, and (4) a volcanic lithic-rich clastic unit. Metamorphic grade varies within these strata from greenschist facies in the older rocks to subgreenschist facies in the younger rocks, while older rocks typically exhibit a moderately strong foliation and younger rocks exhibit only a weak cleavage or no cleavage at all.

Triassic limestone and clastic rocks (unit Ŧ ls)

This unit generally consists of a lower limestone, a middle clastic, and an upper limestone member (Fig. 3), although both the middle and upper members pinch out stratigraphically to the northwest. Conodonts of Ladinian or Carnian (late Middle or early Late Triassic) age and of Middle or Late Triassic age have been found in the lower limestone (Table 1, nos. 7–8), while the upper limestone has yielded Norian and early Norian (late Late Triassic) conodonts (Table 1, nos. 10–12a) and Late Triassic bivalves (Table 1, no 12b).

The lower limestone consists predominantly of medium to light gray, fine- to medium-grained limestone with common fine-scale laminations. Fossil fragments are locally common but nowhere exceed sand size. Conglomeratic to sandy limestone turbidites are common near the base of the unit, as are thin interbeds of shale or chert. The middle clastic member consists of conglomerate, sandstone, siltstone, and limestone turbidites. Conglomerate beds are thick (as much as 4.5 m), massive, normally, or inverse to normally graded, and often lenticular. Sandstone beds are massive or graded, commonly amalgamated, and range in thickness from 0.5 to 2.5 m. Conglomerates and sandstones contain abundant sedimentary lithic clasts (mostly siliceous argillite and chert), less abundant volcanic clasts and plagioclase, and minor mono- and polycrystalline quartz. The upper limestone member is characteristically fine- to very fine-grained, and medium to dark gray, with abundant fine-scale laminations. Conglomeratic to sandy limestone turbidites are also locally present, and fossils are extremely rare. Both limestone members are inferred to have been deposited in relatively deep water in a basinal or base-of-slope setting based on their fine grain size, the abundance of fine-scale laminations, the paucity of fossils, and the occurrence of limestone turbidites (e.g., Cook and Mullins, 1983). Facies in the middle clastic member are consistent with a proximal (inner) fan setting of deposition (Walker, 1975; Mutti and Ricci-Lucchi, 1978).

The contact between the lower limestone and middle clastic members of unit Ŧls (as defined here) was interpreted by Silberling and others (1987) to be a terrane-bounding fault that separates rocks of differing structural styles. Detailed structural studies (S. J. Wyld, unpublished data), however, indicate a strong parallelism in bedding and foliation orientations, and an identical intensity and style of foliation in equivalent rock types across this contact. These relations do not support an interpretation of differing structural styles. In addition, the contact is exposed in one area and there is no evidence of structural displacement at this location, while elsewhere, no fault-related structures or rock types are associated with the contact. Finally, there are no lithologic or

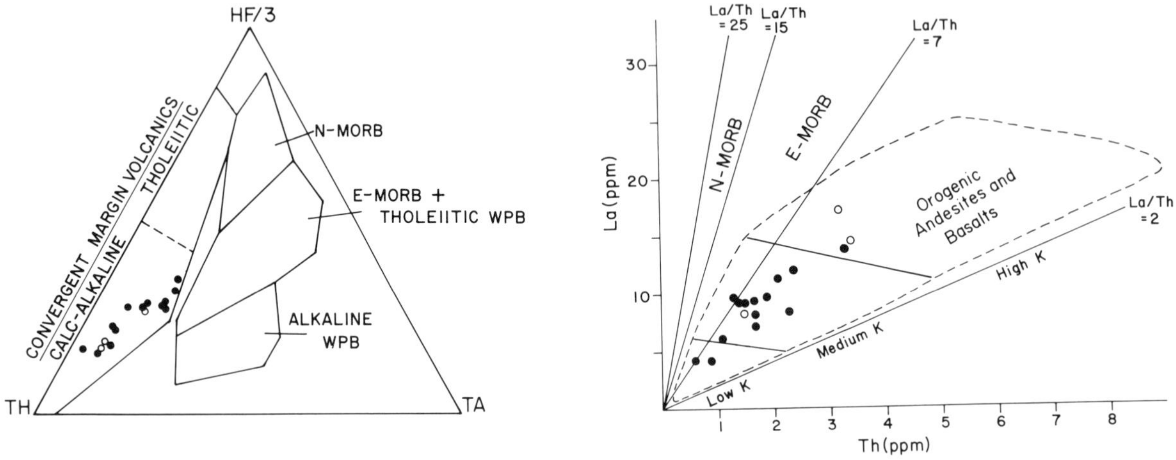

Figure 5. Trace-element characteristics of Triassic lavas and dikes from the east-central Pine Forest Range. Solid circles = lava with plagioclase + clinopyroxene ± olivine phenocrysts from the Triassic lava and volcaniclastic unit. Open circles = lava and dike with plagioclase + hornblende + clinopyroxene phenocrysts from the Triassic volcanic-lithic–rich clastic unit. A, Th-Hf/3-Ta discrimination plot of Wood (1980). N-MORB = normal mid-ocean ridge basalts and their differentiates; E-MORB = enriched mid-ocean ridge basalts and their differentiates; WPB = within-plate basalts and their differentiates. B, La-Th discrimination plot of Gill (1981). N-MORB = normal mid-ocean ridge basalts. E-MORB = enriched mid-ocean ridge basalts. Within-plate basalts typically plot in E-MORB field (Gill, 1981). K = potassium content.

age data to support any interpretation of stratigraphic repetition or omission along this contact (Fig. 3; Table 1). In summary, the present study indicates that this contact is stratigraphic rather than structural, and does not support division of the east-central Pine Forest Range into two tectonostratigraphic terranes as proposed by Silberling and others (1987).

Triassic lava and volcaniclastic rocks (unit Ŧ v)

This unit consists mostly of massive basaltic lava, basaltic pillow lava and pillow breccia, and volcanic breccia and sandstone. Rare limestone beds occur within the unit, and two of these have yielded Norian conodonts (Table 1, nos. 13–14). In addition, the unit is overlain by lower upper Norian rocks, and can therefore be considered to be early Norian or early late Norian (late Late Triassic) in age.

Massive lava generally occurs only in the lower part of the unit (Fig. 3), is highly porphyritic, and contains the phenocryst assemblage plagioclase + clinopyroxene (pseudomorphed) ± olivine (pseudomorphed). Pillow lava and breccia, in contrast, occurs only near or at the top of the unit (Fig. 3); these volcanic rocks range from aphyric to moderately porphyritic and contain phenocrysts of plagioclase ± clinopyroxene (pseudomorphed) only. The middle part of the unit is a thick sequence of volcanic breccia and sandstone that contains clasts of volcanic rock similar in composition to associated lavas, as well as fragments of plagioclase, clinopyroxene (pseudomorphed), and minor pumice. Bedding within this middle part of the unit is generally obscure, but where visible, beds are arranged into sequences of thick (as much as 5 m) graded breccias and relatively coarse-grained sandstones, and thinner (10 to 75 cm) graded sandstones and rare shales.

Trace-element characteristics of pillowed and massive lavas from this unit indicate a calc-alkaline volcanic arc affinity (Fig. 5; S. J. Wyld, unpub. data). The abundance of lava, as well as the occurrence of some pumice-bearing sediments, indicates depositional proximity to volcanic vents. Volcanism occurred in a marine environment, as indicated by interbedded limestone and calcareous interpillow matrix, while sedimentary facies in volcaniclastic rocks are consistent with deposition of at least part of the unit in a proximal submarine fan environment.

Triassic chert and shale (unit Ŧca)

This unit, which has yielded early late Norian (late Late Triassic) ammonites (Table 1, no. 15), consists mostly of chert, shale, and calcareous shale, while siltstone, sandstone, and limestone are uncommon. The unit reflects a waning of arc volcanism in the area during early late Norian time, but indicates continued deposition in a relatively deep marine setting.

Triassic volcanic-lithic–rich clastic rocks (unit Ŧvs)

The uppermost Triassic unit consists mostly of volcanic-lithic–rich sedimentary rocks, which can be divided into three main members (Figs. 2B, 3). The lower member depositionally overlies lower upper Norian rocks, and also contains limestone clasts that have yielded late Carnian to early Norian conodonts (Table 1, nos. 16–17), while the upper member contains late late Norian age ammonites near the base (Table 1, no. 18). The entire unit is therefore considered to be late Norian (late Late Triassic) in age.

The lower member contains a variety of facies that are consistent with an inner submarine fan environment of deposition (e.g., Walker, 1975; Mutti and Ricci-Lucchi, 1978). These facies include thick beds of massive to graded breccia, sandstone, and debris flow deposits, as well as uncommon slide or slump blocks (as much as 11 × 14 m in size) of thin-bedded shale, chert and shale, or sandstone and shale. Associated sequences of thin-bedded shale and sandstone are also locally common and interpreted to represent overbank or interchannel deposits. The middle member differs in containing no chaotic beds, and consists instead mostly of thick-bedded breccia and sandstone. Graded beds are very common and typically represent top-absent turbidites that grade from breccia or coarse sandstone at the base through plane-laminated and locally rippled finer grained sandstone at the top. Facies in the middle member suggest a channelized middle or proximal fan environment (e.g., Walker, 1975; Mutti and Ricci-Lucchi, 1978). Sparse basaltic to basaltic andesite lavas and dikes are present within both the lower and middle members and contain phenocrysts of plagioclase, hornblende, and clinopyroxene. Trace-element analyses from one of these lavas and one dike indicate chemical signatures typical of calc-alkaline volcanic arc lavas (Fig. 5; S. J. Wyld, unpublished data).

The upper member consists of thin- to medium-bedded shale and fine-grained sandstone, with much less common thicker bedded, coarser grained sandstone. Partial to complete Bouma sequences are common and include base-present as well as top-present turbidites. This member is inferred to have been deposited in a middle- to outer-fan environment (e.g., Mutti and Ricci-Lucchi, 1978).

Breccias and debris flow deposits in the lower two members contain abundant porphyritic volcanic and rare sedimentary clasts, while sandstones contain abundant plagioclase and porphyritic volcanic clasts, less common hornblende and clinopyroxene, and very rare sedimentary clasts and quartz. Sandstones in the upper member are similarly rich in plagioclase and volcanic clasts, but they also contain rare metamorphic clasts, more common sedimentary and quartzose clasts, and no mafic crystal fragments.

In summary, the uppermost Triassic unit reflects sedimentation in an apparently retrogradational submarine fan setting during late late Norian time. Sediments within all three members contain the same basic clast types, but the lower two members are much richer than the upper member in volcanic clasts and crystals believed to be derived from the same volcanic source (i.e., plagioclase, hornblende, and clinopyroxene). Volcanic clasts within the lower two members are all very similar in composition and texture, suggesting derivation from a relatively monolithologic volcanic terrane. In addition, these volcanic clasts are identical in texture and phenocryst assemblage to lavas and dikes within the lower two members. Collectively, these features suggest that at least the lower two members reflect clastic deposition associated with renewed arc volcanism in the region in late Norian time, rather than erosion of a pre-existing volcanic terrane.

SUMMARY OF PALEOZOIC AND TRIASSIC STRATIGRAPHY

Middle (and lower?) Paleozoic strata in the Pine Forest Range record siliciclastic sedimentation punctuated by periodic influxes of generally mafic volcanic and volcaniclastic deposits (Figs. 3, 6). Siliciclastic metasedimentary rocks in the Paleozoic(?) quartzo-feldspathic schist unit appear to record an influx of quartz- and feldspar-rich detritus, while later Devonian(?)-Mississippian clastic rocks record submarine fan deposition of sediments that are rich in chert, argillite, and quartz. In Late Mississippian to Early Pennsylvanian(?) time, the source of sediments changed and the region shoaled (at least during part of this time interval). Although chert and quartz continued to be a component in clastic sediments, these clast types were overwhelmed in middle Carboniferous time by plagioclase and largely intermediate composition volcanic detritus. Volcanism was clearly active nearby by Early Pennsylvanian(?) time, as manifested by deposition of Lower Pennsyvlanian(?) crystal tuffs, although the influx of volcanic-lithic– and crystal-rich sediments in underlying units may indicate that volcanism commenced near this area as early as the Late Mississippian.

The remainder of the Pennsylvanian and most of the Permian appear to have been a time of uplift and/or minimal subsidence in the region, as indicated by the presence of several unconformities and the shallow marine character of strata of these ages (Figs. 3, 6). Lower Pennsylvanian(?) rocks are overlain unconformably by a relatively thin sequence of rocks dated as early Late Permian at the top. Much of the Pennsylvanian, and possibly Early Permian, may be missing across this unconformity, although this cannot be documented at present. Permian and Permian(?) strata above this unconformity record carbonate deposition and erosion of a mixed sedimentary and igneous source terrane. An unconformity, apparently spanning some part of the Late Permian, separates these shallow marine deposits from Upper Permian(?) chert and shale that may indicate some subsidence of the region in Late Permian time. An unconformity between Upper Permian(?) and upper Middle or lower Upper Triassic rocks is inferred to reflect renewed uplift of the region in Permo-Triassic time.

Triassic rocks record deposition of more than 3 km of sedimentary and volcanic arc-related strata from Ladinian or Carnian (late Middle or early Late Triassic) through latest Norian (late Late Triassic) time (Figs. 3, 6). All of these strata were deposited in a relatively deep marine environment, indicating subsidence of the area and a return to deep marine conditions of deposition in Ladinian or Carnian time. Carbonates and siliciclastic sediments characterize Ladinian or Carnian to early Norian time, and there is no record of active volcanism during this period. During early(?) Norian time, arc volcanism commenced, and mafic volcanic flows and associated volcaniclastic sediments were deposited. Volcanism waned in the early late Norian, but was renewed again later in Norian time. Volcanic centers were

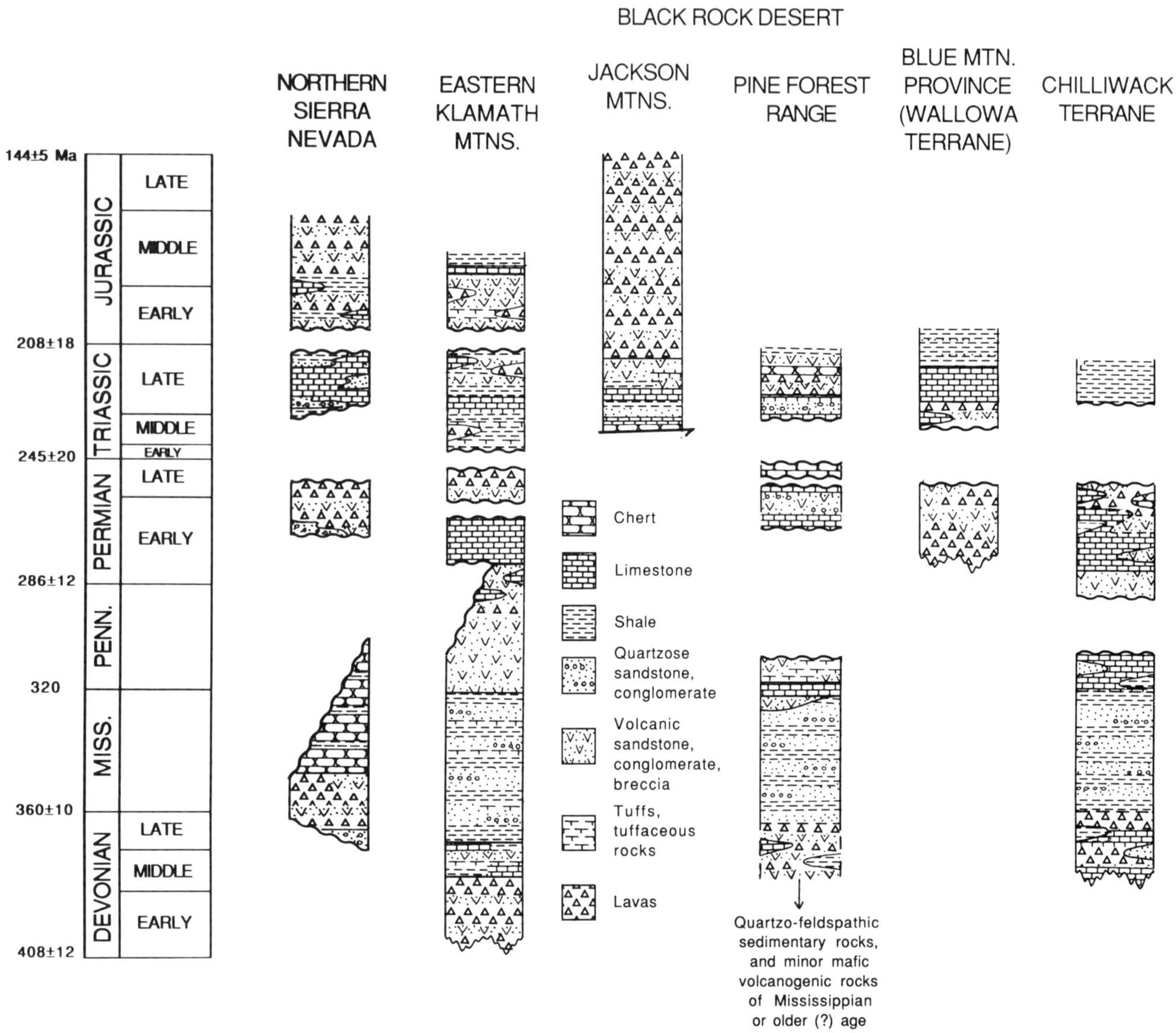

Figure 6. Generalized stratigraphy for Paleozoic through Jurassic-age volcanic assemblages of the western U.S. Cordillera. The oldest rocks in the Pine Forest Range (written description below column) form depositional basement for the younger rocks, but their age is too poorly constrained to depict them in the stratigraphic column. Depositional basement for Middle Paleozoic rocks in the northern Sierra Nevada is the early Paleozoic Shoo Fly complex (Harwood, 1988). Early Paleozoic rocks of the Yreka terrane are present in the eastern Klamath Mountains but are not in stratigraphic continuity with Paleozoic rocks shown in this diagram (Potter and others, 1977; Miller, 1987). Data from Sanborn (1960), Albers and Robertson (1961), Danner (1977), Monger (1977), Vallier (1977), Brooks and Vallier (1978), Schweickert (1978), Russell (1981, 1984), Watkins (1985), Hannah and Moores (1986), Maher and Saleeby (1986, 1988), Harwood (1988), Miller (1989), Miller and Harwood (this volume). Time scale from Palmer (1983).

more distal during this second period of arc volcanism, since volcanic-lithic and crystal-rich sediments predominate while lavas are uncommon.

REGIONAL IMPLICATIONS

Paleozoic time

The northern Chilliwack terrane, Blue Mountain province, eastern Klamath Mountains, and northern Sierra Nevada (Fig. 1) are interpreted to represent Paleozoic volcanic arc sequences, based on the presence of Paleozoic volcanic and volcaniclastic deposits in these areas, and numerous authors have proposed that these sequences probably represent related parts of what was once a single arc system (Hamilton, 1969; Burchfiel and Davis, 1972, 1975; Danner, 1977; Durrel and D'Allura, 1977; Vallier, 1977; Davis and others, 1978; Speed, 1979; Hannah and Moores, 1986; Miller, 1987, 1989; Miller and Harwood, this volume). Paleozoic rocks in the Black Rock Desert have also been interpreted to represent part of this Paleozoic arc system (Burchfiel and Davis, 1972, 1975; Davis and others, 1978; Speed, 1979; Miller, 1987; Speed and others, 1988). This interpretation was based, in older studies, on the assumption that most volcanic rocks in the area were Permian or older, an assumption now known to be incorrect. Current interpretations of an arc setting are based largely on the presence of sparse Permian volcanic rocks in the Bilk Creek Mountains and on strong similarities in Early Permian fauna of the Bilk Creek Mountains and other arc sequences in California, Oregon, and Washington (Skinner and Wilde, 1966; Ketner and Wardlaw, 1981; Wardlaw and others, 1982; Miller, 1987; Jones, this volume). These relations, however, offer no insight into the pre-Permian history of the Black Rock Desert or how this history might be related to that documented in adjacent arc sequences. The Paleozoic stratigraphy of the Pine Forest Range provides new constraints on these questions.

The oldest rocks in the Pine Forest Range are inferred to be Mississippian in age or older and were derived from clastic sedimentary rocks rich in quartz and feldspar, and less common mafic volcanogenic rocks (quartzo-feldspathic schist unit; Figs. 3, 6). Similar strata occur in the Shoo Fly complex (northern Sierra Nevada) and Yreka terrane (eastern Klamath Mountains) (Fig. 6) where abundant Ordovician- to Devonian-age quartzose and quartzo-feldspathic sedimentary rocks are exposed in association with less common to rare volcanic lithic and feldspathic sedimentary rocks, basalt, chert, limestone, and melange (Potter and others, 1977; Harwood, 1988). These similarities in lithology and possibly in age suggest a tentative interpretation that the quartzo-feldspathic schist unit of the Pine Forest Range is broadly equivalent to early Paleozoic rocks of the northern Sierra Nevada and eastern Klamath Mountains. Although pre–Late Devonian compressive deformation is recorded in rocks of the Yreka terrane and Shoo Fly complex (Lindsley-Griffin and Griffin, 1983; Saleeby and others, 1987) while a pre-Mesozoic deformation has not been recognized in the quartzo-feldspathic schist unit of the Pine Forest Range, this could be related to obscuring of an older fabric by Mesozoic amphibolite-grade metamorphism and deformation in the Pine Forest Range.

Comparisons between the Pine Forest Range and other arc sequences for Devonian-Mississippian and younger time are more reliable due to better age and lithologic control for strata of these ages. The amphibolite unit in the Pine Forest Range, derived mostly from mafic (to silicic) volcanic and volcaniclastic deposits, is only constrained to be Mississippian or older, but is likely Devonian or Mississippian based on its gradational contact with overlying Devonian(?)- to Mississippian-age rocks. This unit is similar in both age and lithology to Devonian and locally Lower Missisippian mafic to silicic volcanic and volcaniclastic rocks of the Chilliwack terrane, eastern Klamath Mountains, and northern Sierra Nevada (Fig. 6; Danner, 1977; Miller, 1987; Harwood, 1988). Middle Paleozoic strata in the Pine Forest Range record a post–Late Devonian to Late Mississippian period of deep marine deposition of sediments rich in chert, argillite, and quartz, and a Late Mississippian to Early Pennsylvanian(?) period of shallower marine volcanism and deposition of carbonate and sediments rich in volcanic detritus (Figs. 3, 6). This history is remarkably similar to that of the eastern Klamath Mountains where sediments rich in chert, argillite, and quartz were deposited in a submarine fan environment in Late Devonian to Late Mississippian time, and succeeded by shallower marine volcanic and volcanic-lithic–rich clastic deposits in Late Mississippian through Pennsylvanian time (Fig. 6; Watkins, 1986; Miller, 1989). Turbiditic siliciclastic sediments of Mississippian(?) age are also present in the Chilliwack terrane and are overlain by shallow marine Lower Pennsylvanian limestone and sediments of unspecified provenance (Fig. 6; Danner, 1977). Sediments rich in chert, quartz, and argillite are also locally present in Late Devonian and Early Mississippian strata of the northern Sierra Nevada (Harwood and Murchey, this volume). During the remainder of Mississippian and Early Pennsylvanian time, however, the stratigraphic development of the northern Sierra Nevada differed markedly from that of the Pine Forest Range, as well as that of other arc terranes farther north (Fig. 6; Miller, 1987; Miller and Harwood, this volume).

Early Late Paleozoic rocks are missing in the Pine Forest Range across an unconformity that spans some part of Pennsylvanian, and possibly Early Permian, time (Figs. 3, 6). Late Paleozoic uplift and erosion also characterize the northern Sierra Nevada, where strata of at least middle Pennsylvanian through Early Permian age are absent (Fig. 6; Harwood, 1988; Harwood and Murchey, this volume). In the eastern Klamath Mountains, shallow marine facies are common in Pennsylvanian and Early Permian rocks, and unconformities locally span as much as Early Pennsylvanian to Early Permian time (Fig. 6; Watkins, 1985; Miller, 1989). A Pennsylvanian unconformity may also be present in the Chilliwack terrane (Fig. 6; Danner, 1977; Monger, 1977).

Comparison of the Permian stratigraphy of the Pine Forest Range with that of adjacent arc sequences is hampered by the fact that only a very thin sequence of Permian and Permian(?) strata are exposed in the Pine Forest Range (Fig. 2). These rocks, however, include shallow marine carbonates and clastic sediments derived from a mixed sedimentary and igneous source (Fig. 3, 6), while Permian rocks in the nearby Bilk Creek Mountains include shallow to deeper marine carbonates and volcaniclastic deposits (Ketner and Wardlaw, 1981; Jones, this volume). This Permian record in the Black Rock Desert is similar to the history of deposition of shallow to deep marine carbonates, volcanic sequences, and clastic sediments containing volcanic detritus during Permian time in the northern Sierra Nevada, eastern Klamath Mountains, Blue Mountain province, and Chilliwack terrane (Fig. 6; Danner, 1977; Vallier, 1977; Wardlaw and others, 1982; Harwood, 1988; Miller, 1989).

In summary, a comparison of the Paleozoic stratigraphic record in the Pine Forest Range with that in other Paleozoic arc sequences of the western U.S. Cordillera indicates many similarities in the histories of these regions from middle (and early?) to late Paleozoic time. These similarities include: (1) Mississippian or older quartzo-feldspathic siliciclastic sedimentation and mafic to silicic volcanism; (2) Mississippian deposition of deep marine siliciclastic sediments; (3) Pennsylvanian deposition of shallow marine carbonate and/or volcanogenic strata; (4) Pennsylvanian or Pennsylvanian to Early Permian unconformities; and (5) deposition of variably shallow to deep marine carbonate, volcanic, and clastic deposits in Permian time (Fig. 6). These similarities support previous interpretations of a paleogeographic tie between Permian rocks in the Black Rock Desert and those in adjacent arc sequences, but in addition, suggest that ties can be established in Carboniferous (and possibly earlier?) time as well. Further, these data support the conclusion of Miller (1987) that Paleozoic arc sequences of the western U.S. Cordillera all shared a generally similar Paleozoic history.

Permo-Triassic time

The Paleozoic arc represented by strata in western Nevada, northern California, eastern Oregon, and northern Washington is generally interpreted to have accreted to the continental margin of North America sometime in the Late Permian to Early Triassic, during the Sonoma Orogeny, either via back-arc basin closure or arc-continent collision (Silberling and Roberts, 1962; Burchfiel and Davis, 1972, 1975; Silberling, 1973; Davis and others, 1978; Speed, 1979; Miller and others, 1984; Brueckner and Snyder, 1985; Speed and others, 1988; Tomlinson and Wardlaw, 1988). It has been suggested, however, that a lack of evidence for any Permo-Triassic age deformation in the Bilk Creek or eastern Klamath Mountains casts doubt on their tectonic involvement in the Sonoma orogeny (Silberling and Jones, 1982; Jones and others, 1988; Jones, this volume). In the Black Rock Desert, this interpretation has been based on the presence of a conformable or nearly conformable section of Permian and Triassic rocks in the Bilk Creek Mountains, which argues for a lack of any Permo-Triassic age tectonism in the area (Jones and others, 1988; Jones, this volume). In the Pine Forest Range, however, Carboniferous through Upper Permian strata are overlain unconformably by upper Middle or lower Upper Triassic rocks (Figs. 3, 4, 6). The Black Rock Desert was therefore at least locally affected by tectonism sometime in latest Permian(?) to Middle or Late Triassic time. Recent work also suggests that a Late Permian to Early Triassic unconformity may be present in the eastern Klamath Mountains and associated with some Permo-Triassic compressive deformation (Fig. 6; Miller, 1988, 1990; Miller and Harwood, this volume). Late Permian to Middle or Late Triassic unconformities are also present in the northern Sierra Nevada, Blue Mountain province, and Chilliwack terrane (Fig. 6; Monger, 1977; Brooks and Vallier, 1978; Harwood, 1988). Together these relations suggest that arc terranes of the western U.S. Cordillera experienced widespread uplift at approximately the same time as compressive deformation, related to the Sonoma orogeny, affected basinal strata farther east. Although the structural style of Permo-Triassic tectonism differs between the Golconda allochthon and the Paleozoic arc sequences, the similarity in timing of tectonism in these two groups of rocks is consistent with an interpretation that uplift in the arc sequences was tectonically related to Sonoma-age compressive deformation in the Golconda allochthon. The existence of a conformable or nearly conformable Permo-Triassic section in the Bilk Creek Mountains may indicate that a history of tilting and uplift, as well as local downwarping, affected the Paleozoic arc sequences during Permo-Triassic time.

Triassic time

Postdating the Sonoma orogeny, an early Mesozoic west-facing arc was constructed on or adjacent to the North American craton, as documented in Arizona, southern California, the eastern and southeastern Sierra Nevada, and western Nevada (Fig. 1; e.g., Burchfiel and Davis, 1972). Lower Mesozoic volcanic rocks in northwest Nevada, as well as the eastern Klamath Mountains and northern Sierra Nevada (Fig. 1), are generally inferred to have been constructed on the North American plate as well, based on the interpretation that underlying Paleozoic rocks in these areas were accreted to the continent by Early Triassic time (Burchfiel and Davis, 1972, 1975; Dilles and Wright, 1988; Speed and others, 1988). It can therefore be concluded that these northern volcanic rocks represent a continuation of the west-facing arc documented farther south, as has been suggested by Burchfiel and Davis (1972), Schweickert (1978), Dilles and Wright (1988), and Speed and others (1988). Few studies, however, have attempted to demonstrate any stratigraphic or paleogeographic ties between these two regions. This can be done by looking at the stratigraphy and subsidence/uplift history of early Mesozoic rocks in the Black Rock Desert.

As discussed earlier, Triassic rocks in the Pine Forest Range record a history of Middle(?) to early Late Triassic carbonate and clastic sedimentation succeeded by late Late Triassic arc volcanism (Figs. 3, 6). An identical history is recorded in the Jackson

Mountains where studies also indicate copious arc volcanism continuing into the Jurassic (Fig. 6; Russell, 1984; Maher and Saleeby, 1986, 1988). This record is similar to that in the northern Sierra Nevada and eastern Klamath Mountains where Middle and Late Triassic time is represented by deposition of carbonates, clastic sediments, and variably abundant volcanogenic deposits, and where a major pulse of arc volcanism occurred in the Early and Middle Jurassic (Fig. 6; Sanborn, 1960; Schweickert, 1978; Harwood, 1988). The early Mesozoic arc of the southwest United States also experienced a very similar history: Triassic time, where represented in the rock record, is characterized by carbonate and clastic sedimentation coupled with sporadic and localized arc volcanism, while Early and Middle Jurassic time is characterized by voluminous arc magmatism (Fiske and Tobisch, 1978; Speed, 1978; Karish and others, 1987; Busby-Spera, 1988; Dilles and Wright, 1988; Tosdal and others, 1989; Fisher, 1989). These stratigraphic similarities support an interpretation that lower Mesozoic volcanic rocks in northwest Nevada, as well as the eastern Klamath Mountains and northern Sierra Nevada, represent part of the same west-facing arc documented in the southwestern United States.

As discussed by Karish and others (1987), Busby-Spera (1988), Tosdal and others (1989), and Wright and Miller (1988), the early Mesozoic arc of the southwestern United States appears to have been characterized by extensional tectonism and associated subsidence. Extension-related subsidence may also have occurred in the Black Rock Desert region, at least during Triassic time. This inference is based on stratigraphic evidence from the Pine Forst Range. Specifically, deep marine Triassic strata in the Pine Forest Range were deposited on Permian rocks across an unconformity that spans latest Permian(?) through Ladinian or Carnian time (Fig. 3). This indicates subsidence of, or transgression in, the region beginning in late Middle or early Late Triassic time. As there is no evidence for any significant sea-level rise during Triassic time (Vail and others, 1977), the inception of deep marine conditions of deposition is interpreted to be related to tectonically driven subsidence rather than transgression. Tectonic subsidence appears to be better explained by crustal extension than by thermal contraction as, regionally, there is no evidence to support the occurrence of any significant Early or Middle Triassic magmatic event whose demise could account for thermal contraction associated with major subsidence in late Middle or early Late Triassic time (Sanborn, 1960; Albers and Robertson, 1961; Ketner and Wardlaw, 1981; Jones, this volume; Miller and Harwood, this volume). Continued subsidence in the Black Rock Desert throughout the remainder of the Triassic is suggested by the deposition of more than 3,000 m of deep marine strata in the Pine Forest Range during Norian time, as well as the absence of any progradational facies within these rocks. Subsidence during this period of time is also better explained by extension than by thermal cooling and contraction, as there is abundant evidence of thermal input, in the form of arc magmatism, into the region during Norian time. In summary, subsidence associated with extensional tectonism appears to have characterized Triassic rocks of the Black Rock Desert region as well as the early Mesozoic arc of the southwestern United States. While a similar tectonic record does not prove that the two arc sequences represent part of a single arc system, it suggests, in conjunction with the stratigraphic similarities, that such an interpretation is correct.

CONCLUSIONS

This study indicates that the Pine Forest Range contains an extensive and structurally intact section of pre-Tertiary rocks whose history provides important constraints on the Paleozoic and Mesozoic development of northwest Nevada, and on the relation between Paleozoic and Mesozoic rocks in this region and those in adjacent arc sequences of the western U.S. Cordillera. The main conclusions of this chapter can be summarized as follows.

1. Stratigraphic and age relations in the Pine Forest Range indicate that the pre-Tertiary section spans at least Mississippian through latest Triassic time, while regional correlations further suggest the possibility that some of the oldest rocks in the range may be early Paleozoic and broadly equivalent to early Paleozoic strata in the eastern Klamath Mountains and northern Sierra Nevada. These relations indicate that the Pine Forest Range contains a stratigraphic record extending to much older ages than have previously been documented in this part of the Cordillera.

2. Ties between Paleozoic rocks of the Black Rock Desert and those in adjacent arc sequences of the western U.S. Cordillera have previously been based largely on lithologic and faunal similarities between Permian rocks. The Paleozoic section in the Pine Forest Range indicates strong stratigraphic similarities in Carboniferous, and possibly early Paleozoic, time as well. These similarities suggest a tectonic and/or paleogeographic tie between the Black Rock Desert and these other arc sequences during middle (and early?) to late Paleozoic time, and add further support to interpretations that all of these arc sequences were originally related parts of a single Paleozoic arc system.

3. The presence of a Late Permian(?) to Middle or Late Triassic unconformity in the Pine Forest Range suggests at least localized uplift in the Black Rock Desert during this period of time and is consistent with the record of widespread Late Permian to Middle or Late Triassic unconformities in other arc sequences of the western U.S. Cordillera. These relations support an interpretation that these Paleozoic arc sequences were affected by tectonism associated with the Late Permian to Early Triassic Sonoma orogeny.

4. Lower Mesozoic strata in the Black Rock Desert record a history of arc volcanism and sedimentation similar to that documented in the early Mesozoic arc of the southwestern United States, as well as the eastern Klamath Mountains and northern Sierra Nevada. This supports an interpretation that all of these arc sequences are related parts of a single west-facing arc system. Intra-arc extensional tectonism and related subsidence, which characterized the early Mesozoic arc of the southwestern United

States, may also have affected early Mesozoic arc development in the Black Rock Desert, suggesting a further tectonic link between these volcanic arc sequences.

ACKNOWLEDGMENTS

This project was supported by Geological Society of America Penrose Grants 3905-87 and 4121-88; by Shell, Chevron, and Fowler Fund research grants from Stanford University; and by American Chemical Society Petroleum Research Grant 21861-AC8. I gratefully acknowledge N. J. Silberling, B. R. Wardlaw, and C. H. Stevens for allowing me to make use of and publish the details of their paleontological studies in the Pine Forest Range. New fossil identifications by M. R. Orchard and C. H. Stevens were invaluable, as were U/Pb zircon analyses by J. E. Wright. U/Pb analyses were done by J. E. Wright at the U.S. Geological Survey, Menlo Park, California, in the laboratory of J. Wooden, and were supported in part by National Science Foundation Grant EAR-88-04814. Rare earth element analyses were done by the Oregon State University Radiation Center as part of its reactor sharing program. Field work benefited greatly from the generosity and assistance of the Nuffer and Woodward families of Denio, Nevada. Discussions with A. J. Tomlinson, N. J. Silberling, J. E. Wright, and E. L. Miller have been important to the progress of this study, and reviews by these people, as well as by R. C. Speed and D. S. Harwood, improved earlier versions of the manuscript.

REFERENCES CITED

Albers, J. P., and Robertson, J. F., 1961, Geology and ore deposits of east Shasta copper-zinc district, Shasta County, California: U.S. Geological Survey Professional Paper 338, 107 p.

Brooks, H. C., and Vallier, T. L., 1978, Mesozoic rocks and tectonic evolution of eastern Oregon and western Idaho, *in* Howell, D. G., and McDougall, K. A., eds., Mesozoic paleogeography of the western U.S.: Pacific Section, Society of Economic Paleontologists and Mineralogists Pacific Coast Paleogeography Symposium 2, p. 133–146.

Brueckner, H. K., and Snyder, W. S., 1985, Structures of the Havallah sequence, Golconda allochthon, Nevada; Evidence for prolonged evolution in an accretionary prism: Geological Society of America Bulletin, v. 96, p. 1113.

Burchfiel, B. C., and Davis, G. A., 1972, Structural framework and evolution of the southern part of the Cordilleran orogen, western U.S.: American Journal of Science, v. 272, p. 97.

—— , 1975, Nature and controls of Cordilleran orogenesis, western United States; Extensions of an earlier synthesis: American Journal of Science, v. 275-A, p. 363.

Busby-Spera, C. J., 1988, Speculative tectonic model for the early Mesozoic arc of the southwest Cordilleran United States: Geology, v. 16, p. 1121–1125.

Cook, H. E., and Mullins, H. T., 1983, Basin margin environment, *in* Scholle, P. A., Bebout, D. G., and Moore, C. H., eds., Carbonate depositional environments: American Association of Petroleum Geologists Memoir 33, p. 539–618.

Danner, W. R., 1977, Paleozoic rocks of northwest Washington and adjacent parts of British Columbia, *in* Stewart, J. H., and others, eds., Paleozoic paleogeography of the western U.S.: Pacific Section, Society of Economic Paleontologists and Mineralogists Pacific Coast Paleogeography Symposium 1, p. 395–408.

Davis, G. A., Monger, J.W.H., and Burchfiel, B. C., 1978, Mesozoic construction of the Cordilleran "collage," "central British Columbia to central California, *in* Howell, D. G., and McDougall, K. A., eds., Mesozoic paleogeography of the western U.S.: Pacific Section, Society of Economic Paleontologists and Mineralogists Pacific Coast Paleogeography Symposium 2, p. 1–32.

Dilles, J. H., and Wright, J. E., 1988, The chronology of Early Mesozoic arc magmatism in the Yerington district of western Nevada, and its regional implications: Geological Society of America Bulletin,v . 100, p. 644–652.

Durrel, C., and D'Allura, J., 1977, Upper Paleozoic section in eastern Plumas and Sierra Counties, northern Sierra Nevada, California: Geological Society of America Bulletin, v. 88, p. 844–852.

Fisher, G. R., 1989, The Mt. Tallac roof pendant, California; Mesozoic stratigraphic link between the northern Sierra terrane and west-central Nevada: Geological Society of America Abstracts with Programs, v. 21, p. 78.

Fiske, R. S., and Tobisch, O. T., 1978, Paleogeographic significance of volcanic rocks of the Ritter Range pendant, central Sierra Nevada, California, *in* Howell, D. G., and McDougall, K. A., eds., Mesozoic paleogeography of the western U.S.: Pacific Section, Society of Economic Paleontologists and Mineralogists Pacific Coast Paleogeography Symposium 2, p. 209–222.

Gill, J. B., 1981, Orogenic andesites and plate tectonics: Berlin, Springer-Verlag, 390 p.

Hamilton, W., 1969, Mesozoic California and the underflow of Pacific mantle: Geological Society of America Bulletin, v. 80, p. 2409–2430.

Hannah, J. L., and Moores, E. M., 1986, Age relationships and depositional environments of Paleozoic strata, northern Sierra Nevada, California: Geological Society of America Bulletin, v. 97, p. 787.

Harwood, D. S., 1988, Tectonism and metamorphism in the northern Sierra terrane, northern California, *in* Ernst, W. G., ed., Metamorphism and crustal evolution of the western United States; Rubey Volume 7: Englewood Cliffs, New Jersey, Prentice-Hall, p. 764–788.

Jones, A. E., Langenheim, V. E., Jones, D. L., and Renne, P., 1988, New age dasta from Quinn River Crossing, Humboldt County, Nevada; Implications for Paleozoic and Mesozoic tectonics: Geological Society of America Abstracts with Programs, v. 20, p. A109.

Karish, C. R., Miller, E. L., and Sutter, J. F., 1987, Mesozoic tectonic and magmatic history of the central Mojave Desert, *in* Dickinson, W., and Klute, M., eds., Mesozoic rocks of southern Arizona and adjacent regions: Arizona Geological Society Digest, v. 18, p. 15–32.

Ketner, K. B., and Wardlaw, B. R., 1981, Permian and Triassic rocks near Quinn River Crossing, Humboldt County, Nevada: Geology, v. 9, p. 123.

Lindsley-Griffin, N., and Griffin, J. R., 1983, The Trinity terrane; An early Paleozoic microplate assemblage, *in* Stevens, C. H., ed., Pre-Jurassic rocks in western North American suspect terranes: Pacific Section, Society of Economic Paleontologists and Mineralogists, p. 63–75.

Maher, K. A., and Saleeby, J., 1986, Geology of the Jackson Mountains, NW Nevada: Geological Society of America Abstracts with Programs, v. 18, p. 679.

—— , 1988, Age constraints on the geologic evolution of the Jackson Mountains, NW Nevada: Geological Society of America Abstracts with Programs, v. 20, p. 177.

Miller, E. L., Holdsworth, B. K., Whiteford, W. B., and Rogers, D., 1984, Stratigraphy and structure of the Schoonover sequence, northern Nevada; Implications for Paleozoic plate margin tectonics: Geological Society of America Bulletin, v. 95, p. 1063.

Miller, M. M., 1987, Dispersed remnants of a northeast Pacific fringing arc; Upper Paleozoic terranes of Permian McCloud faunal affinity, western U.S.: Tectonics, v. 6, p. 807.

—— , 1988, Permo-Triassic(?) deformation within the eastern Klamath terrane, northern California: Geological Society of America Abstracts with Programs, v. 20, p. 216.

REFERENCES CITED

Albers, J. P., and Stewart, J. H., 1972, Geology and mineral deposits of Esmeralda County: Nevada Bureau of Mines and Geology Bulletin 78, 80 p.

Aleinikoff, J. N., Dusel-Bacon, C., and Foster, H. L., 1986, Geochronology of augen gneiss and related rocks, Yukon-Tanana terrane, east-central Alaska: Geological Society of America Bulletin, v. 97, p. 626–637.

Armstrong, R. L., 1975, Precambrian (1500 m.y. old) rocks of central Idaho; The Salmon River Arch and its role in Cordilleran sedimentation and tectonics: American Journal of Science, v. 275-A, p. 437–467.

Bennett, V. C., and DePaolo, D. J., 1987, Proterozoic crustal history of the western United States as determined by neodymium isotopic mapping: Geological Society of America Bulletin, v. 99, p. 674–685.

Black, L. P., Williams, I. S., and Compston, W., 1986, Four zircon ages from the rock; The history of a 3930 Ma old granulite from Mount Sones, Enderby Land, Antarctica: Contributions to Mineralogy and Petrology, v. 94, p. 427–437.

Bowring, S. A., and Podosek, F. A., 1988, Nd isotopic data from Wopmay Orogen; Implications for models of crustal growth: EOS Transactions of the American Geophysical Union, v. 69, no. 16, p. 517.

Burchfiel, B. C., and Davis, G. A., 1972, Structural framework and evolution of the southern part of the Cordilleran orogen, western United States: American Journal of Science, v. 272, p. 97–118.

——, 1975, Nature and controls of Cordilleran orogenesis, western United States; Extensions of an earlier synthesis: American Journal of Science, v. 275-A, p. 363–396.

Burchfiel, B. C., and Royden, L. H., 1989, The Antler orogen; An Apenine—East Carpathian type orogen: Geological Society of America Abstracts with Programs, v. 21, p. 61.

Churkin, M., Jr., and Kay, M., 1967, Graptolite-bearing Ordovician siliceous and volcanic rocks, northern Independence Range, Nevada: Geological Society of America Bulletin, v. 78, p. 651–668.

Condie, K. C., 1982, Plate tectonics model for Proterozoic continental accretion in the southwestern United States: Geology, v. 10, p. 37–42.

Dickinson, W. R., 1977, Paleozoic plate tectonics and the evolution of the Cordilleran continental margin, *in* Stewart, J. H., Stevens, C. H., and Fritsche, A. E., eds., Paleozoic paleogeography of the western United States: Pacific section, Society of Economic Paleontologists and Mineralogists Pacific Coast Paleogeography Symposium Volume 1, p. 137–156.

Dickinson, W. R., and Suczek, C. A., 1979, Plate tectonics and sandstone compositions: American Association of Petroleum Geologists Bulletin, v. 63, p. 2164–2182.

Dover, J. H., 1980, Status of the Antler Orogeny in Idaho; Clarification and constraints, *in* Fouch, T. D., and Magathan, E. R., Paleozoic paleogeography of the west-central United States: Rocky Mountain Section, Society of Economic Paleontologists and Mineralogists Rocky Mountain Paleogeography Symposium, v. 1, p. 371–386.

Erickson, R. L., and Marsh, S. P., 1974, Paleozoic tectonics in the Edna Mountain Quadrangle, Nevada: U.S. Geological Survey Journal of Research, v. 2, no. 3, p. 331–337.

Evans, K. V., and Fischer, L. B., 1986, U-Pb geochronology of two augen gneiss terranes, Idaho; New data and tectonic implications: Canadian Journal of Earth Sciences, v. 23, p. 1919–1927.

Evans, K. V., and Zartman, 1981, U-Th-Pb zircon geochronology of Proterozoic Y granitic intrusions in the Salmon area, east-central Idaho: Geological Society of America Abstracts with Programs, v. 13, p. 195.

Farmer, G. L., and DePaolo, D. J., 1983, Origin of Mesozoic and Tertiary granite in the western United States and implications for pre-Mesozoic crustal structure; 1, Nd and Sr isotopic studies in the geocline of the northern Great Basin: Journal of Geophysical Research, v. 88, p. 3779–3401.

——, 1984, Origin of Mesozoic and Tertiary granite in the western United States and implications for pre-Mesozoic crustal structure; 2, Nd and Sr isotopic studies of unmineralized and Cu- and Mo-mineralized granite in the Precambrian craton: Journal of Geophysical Research, v. 89, no. 10, p. 10141–10160.

Gilluly, J., and Gates, O., 1965, Tectonic and igneous geology of the northern Shoshone Range, Nevada: U.S. Geological Survey Professional Paper 465, 153 p.

Gilluly, J., and Masursky, H., 1965, Geology of the Cortez Quadrangle: U.S. Geological Survey Bulletin 1175, 117 p.

Girty, G. H., and Wardlaw, M. S., 1984, Was the Alexander terrane a source of feldspathic sandstones in the Shoo Fly Complex, Sierra Nevada, California?: Geology, v. 12, p. 339–342.

——, 1985, Petrology and provenance of pre-Late Devonian sandstones, Shoo Fly Complex, northern Sierra Nevada, California: Geological Society of America Bulletin, v. 96, p. 516–521.

Girty, G. H., Reiland, D. N., and Wardlaw, M. S., 1985, Provenance of the Silurian Elder Sandstone, north-central Nevada: Geological Society of America Bulletin, v. 96, p. 925–930.

Hanson, R. E., Saleeby, J. B., and Schweickert, R. A., 1988, Composite Devonian island-arc batholith in the northern Sierra Nevada, California: Geological Society of America Bulletin, v. 100, p. 446–457.

Hoffman, P. F., 1988, United Plates of America, the birth of a craton; Early Proterozoic growth and assembly of Laurentia: Annual Reviews of Earth and Planetary Sciences, v. 16, p. 543–603.

Hotz, P. E., and Willden, R., 1964, Geology and mineral deposits of the Osgood Mountains Quadrangle, Humboldt County, Nevada: U.S. Geological Survey Professional Paper 431, 128 p.

Jaffey, A. H., Flynn, K. F., Glendenin, L. E., Bentley, W. C., and Essling, A. M., 1971, Precision measurements of the half lives and specific activities of U-235 and U-2;38: Physics Review, v. C4, p. 1889–1906.

Johnson, J. G., and Pendergast, A., 1981, Timing and mode of emplacement of the Roberts Mountains allochthon, Antler orogeny: Geological Society of America Bulletin, part 1, v. 92, p. 648–658.

Kay, M., and Crawford, J. P., 1964, Paleozoic facies from the miogeosynclinal to the eugeosynclinal belt in thrust slices, central Nevada: Geological Society of America Bulletin, v. 75, p. 425–454.

Kerr, J. W., 1962, Paleozoic sequences and thrust slices of the Seetoya Mountains, Independence Range, Elko County, Nevada: Geological Society of America Bulletin, v. 73, p. 439–460.

Ketner, K. B., 1966, Comparison of Ordovician eugeosynclinal and miogeosynclinal quartzites of the Cordilleran geosyncline: U.S. Geological Survey Professional Paper 550-C, p. C54–C60.

——, 1977, Deposition and deformation of lower Paleozoic western facies rocks, northern Nevada, *in* Stewart, J. H., Stevens, C. H., and Fritsche, A. E., Paleozoic paleogeography of the western United States: Pacific section, Society of Economic Paleontologists and Mineralogists, Pacific Coast Paleogeography Symposium Volume 1, p. 251–258.

Kirk, E., 1983, The Eureka Quartzite of the Great Basin region: American Journal of Science, v. 26, no. 151, p. 27–44.

Kistler, R. W., and Peterman, Z. E., 1978, Reconstruction of crustal blocks of California on the basis of initial strontium isotopic compositions of Mesozoic granitic rocks: U.S. Geological Survey Professional Paper 1071, 17 p.

Madrid, R. J., 1987, Stratigraphy of the Roberts Mountains allochthon in north-central Nevada [Ph.D. thesis]: Stanford, California, Stanford University, 342 p.

Malinverno, A., and Ryan, W.B.F., 1986, Extension in the Tyrrhenian Sea and shortening in the Appenines as result of arc migration driven by sinking of the lithosphere: Tectonics, v. 2, p. 227–246.

Merriam, C. W., and Anderson, C. A., 1942, Reconnaissance survey of the Roberts Mountains, Nevada: Geological Society of America Bulletin, v. 53, p. 1675–1728.

Miller, E. L., and Larue, D. K., 1983, Ordovician quartzite in the Roberts Mountains allochthon, Nevada; Deep sea fan deposits derived from cratonal North

America, *in* Stevens, C. H., ed., Pre-Jurassic rocks in western North America suspect terranes: Pacific Section, Society of Economic Paleontologists and Mineralogists, p. 91–102.

Nilsen, T. H., and Stewart, J. H., 1980, The Antler orogeny; Mid-Paleozoic tectonism in western North America: Geology, v. 8, p. 298–302.

Nolan, T. B., Merriam, C. W., and Williams, J. S., 1956, The stratigraphic section in the vicinity of Eureka, Nevada: U.S. Geological Survey Professional Paper 276, 77 p.

Oldow, J. S., 1984, Spatial variability in the structure of the Roberts Mountains allochthon, western Nevada: Geological Society of America Bulletin, v. 95, p. 174–185.

Palmer, A. R., 1971, The Cambrian of the Great Basin and adjoining areas, western United States, *in* Holland, C. H., ed., Cambrian of the New World, New York, Wiley-Interscience, p. 1–78.

Parrish, R. R., and Wheeler, J. O., 1983, A U-Pb zircon age for the Kuskanax batholith, southeastern British Columbia: Canadian Journal of Earth Sciences, v. 20, p. 1751–1756.

Poole, F. G., 1974, Flysch deposits of the Antler foreland basin, western United States, *in* Dickinson, W. R., ed., Tectonics and sedimentation: Society of Economic Paleontologists and Mineralogists Special Publication 22, p. 28–57.

Poole, F. G., and Sandberg, C. A., 1977, Mississippian paleogeography and tectonics of the western United States, *in* Stewart, J. H., Stevens, C. H., and Fritsche, A. E., eds., Paleozoic paleogeography of the western United States: Pacific Section, Society of Economic Paleontologists and Mineralogists Pacific Coast Paleogeography Symposium Volume 1, p. 137–156.

Poole, F. G., Sandberg, C. A., and Boucot, A. J., 1977, Silurian and Devonian paleogeography of the western United States, *in* Stewart, J. H., Stevens, C. H., and Fritsche, A. E., eds., Pacific Section, Society of Economic Paleontologists and Mineralogists Pacific Coast Paleogeography Symposium Volume 1, p. 137–156.

Roberts, R. J., 1964, Stratigraphy and structure of the Antler Peak Quadrangle, Humboldt and Lander Counties, Nevada: U.S. Geological Survey Professional Paper 459-A, 93 p.

Roberts, R. J., Hotz, P. E., Gilluly, J., and Ferguson, H. G., 1958, Paleozoic rocks of north-central Nevada: American Association of Petroleum Geologists Bulletin, v. 42, no. 12, p. 2813–2857.

Ross, G. M., Parrish, R. R., Bowring, S. A., and Tankard, A. J., 1988, Tectonics of the Canadian Shield in the Alberta subsurface: Geological Society of Canada Programs with Abstracts, v. 13, p. A106.

Ross, R. J., 1961, Distribution of Ordovician graptolites and eugeosynclinal facies in western North America and its paleogeographic implications: American Association of Petroleum Geologists Bulletin, v. 45, p. 330–344.

——, 1964, Relations of Middle Ordovician time and rock units in Basin Ranges, western United States: American Association of Petroleum Geologists Bulletin, v. 49, no. 9, p. 1526–1554.

——, 1977, Ordovician paleogeography of the western United States, *in* Stewart, J. H., Stevens, C. H., and Fritsche, A. E., eds., Paleozoic paleogeography of the western United States: Pacific section, Society of Economic Paleontologists and Mineralogists Pacific Coast Paleogeography Symposium Volume 1, p. 19–38.

Ross, R. J., and Berry, W.B.N., 1963, Ordovician graptolites of the Basin Ranges in California, Nevada, Utah, and Idaho: U.S. Geological Survey Bulletin 1134, 177 p.

Rowell, A. J., Rees, M. N., and Suczek, C. A., 1979, Margin of the North American continent in Nevada during Late Cambrian time: American Journal of Science, v. 279, p. 1–18.

Saleeby, J. B., Hannah, J. L., and Varga, R. J., 1987, Isotopic age constraints on middle Paleozoic deformation in the northern Sierra Nevada, California: Geology, v. 15, p. 757–760.

Scharer, U., and Allegre, C. J., 1982, Uranium-lead system in fragments of a single zircon grain: Nature, v. 295, p. 585–587.

Schweickert, R. A., 1981, Tectonic evolution of the Sierra Nevada Range, *in* Ernst, W. G., ed., The geotectonic development of California; Rubey Volume 1: Englewood Cliffs, New Jersey, p. 182–202.

Schweickert, R. A., and Lahren, M. M., 1987, Continuation of Antler and Sonoma orogenic belts to the eastern Sierra Nevada, California, and Late Triassic thrusting in a compressional arc: Geology, v. 15, p. 270–273.

Schweickert, R. A., and Snyder, W. S., 1981, Paleozoic plate tectonics of the Sierra Nevada and adjacent regions, *in* Ernst, W. G., ed., The geotectonic development of California; Rubey Volume 1: Englewood Cliffs, New Jersey, p. 182–202.

Solomon, G. C., and Taylor, H. P., Jr., 1989, Isotopic evidence for the origin of Mesozoic and Cenozoic granitic plutons in the northern Great Basin: Geology, v. 17, p. 591–594.

Speed, R. C., 1979, Collided Paleozoic microplate in the western United States: Journal of Geology, v. 87, p. 279–292.

——, 1983, Evolution of the sialic margin in the central western United States: American Association of Petroleum Geologists Memoir 34, p. 457–470.

Speed, R. C., and Sleep, N. H., 1982, Antler orogeny and foreland basin; A model: Geological Society of America Bulletin, v. 93, p. 815–828.

Stacey, J. S., and Kramers, J. D., 1975, Approximation of terrestrial lead isotope evolution by a two-stage model: Earth and Planetary Science Letters, v. 26, p. 207–221.

Stewart, J. H., 1980, Geology of Nevada: Nevada Bureau of Mines and Geology Special Publication 4, 136 p.

Stewart, J. H., and Poole, F. G., 1974, Lower Paleozoic and uppermost Precambrian Cordilleran miogeocline, Great Basin, western United States, *in* Dickinson, W. R., Tectonic and sedimentation: Society of Economic Paleontologists and Mineralogists Special Publication 22, p. 28–57.

Stewart, J. H., and Suczek, C. A., 1977, Cambrian and latest Precambrian paleogeography and tectonics in the western United States, *in* Stewart, J. H., Stevens, C. H., and Fritsche, A. E., Paleozoic paleogeography of the western United States: Pacific Section, Society of Economic Paleontologists and Mineralogists Pacific Coast Paleogeography Symposium Volume 1, p. 1–18.

Suczek, C. A., 1977, Tectonic relations of the Harmony Formation, northern Nevada [Ph.D. thesis]: Stanford, California, Stanford University, 96 p.

Thomas, J. J., Schuster, R. D., and Bickford, M. E., 1984, A terrane of 1,350- to 1,450 m.y. old silicic volcanic and plutonic rocks in the buried Proterozoic of the mid-continent and in the Wet Mountains, Colorado: Geological Society of America Bulletin, v. 95, p. 1150–1157.

Turner, R.J.W., Madrid, R. J., and Miller, E. L., 1989, Roberts Mountains allochthon; Stratigraphic comparison with lower Paleozoic outer continental margin strata of the northern Canadian Cordillera: Geology, v. 17, p. 341–344.

Van Schmus, W. R., Bickford, M. E., and Zietz, I., 1987, Early and Middle Proterozoic provinces in the central United States, *in* Kroner, A., Proterozoic lithospheric evolution: American Geophysical Union Geodynamics Series, v. 17, p. 43–68.

Wallin, E. T., 1988, Evolution of lower Paleozoic terrigenous clastic sedimentation in the Yreka-Callahan area, eastern Klamath Mountains, California: Geological Society of America Abstracts with Programs, v. 20, p. A231.

——, 1989, Provenance of lower Paleozoic sandstones in the eastern Klamath Mountains and the Roberts Mountains allochthon, California and Nevada [Ph. D. thesis]: Lawrence, University of Kansas, 152 p.

Wallin, E. T., Mattinson, J. M., and Potter, A. W., 1988, Early Paleozoic magmatic events in the eastern Klamath Mountains, California: Geology, v. 16, p. 144–148.

Manuscript Accepted by the Society March 27, 1990

Printed in U.S.A.

——, 1989, Intra-arc sedimentation and tectonism; Late Paleozoic evolution of the eastern Klamath terrane, California: Geological Society of America Bulletin, v. 101, p. 170–187.

——, 1990, Episodic compressional tectonism in an intra-arc setting, eastern Klamath terrane, northern California, *in* Howell, D. G., and Wiley, T. J., eds., Tectonostratigraphic terranes of the Circum-Pacific region, *in* Proceedings, 4th International Circum-Pacific Terranes Conference, Nanjing, China: Circum-Pacific Council for Energy and Mineral Resources, Earth Science Series 3 (in press).

Monger, J. W.H., 1977, Upper Paleozoic rocks of the western Canadian Cordillera and their bearing on Cordilleran evolution: Canadian Journal of Earth Sciences, v. 14, p. 1832–1859.

Mutti, E., and Ricci-Lucchi, F., 1978, Turbidites of the northern Apennines; Introduction to facies analysis: International Geology Review, v. 20, p. 125–166.

Palmer, A. R., 1983, The Decade of North American Geology 1983 geologic time scale: Geology, v. 11, p. 503–504.

Potter, A. W., Hotz, P. E., and Rohr, D. M., 1977, Stratigraphy and inferred tectonic framework of lower Paleozoic rocks in the eastern Klamath Mountains, northern California, *in* Stewart, J. H., and others, eds., Paleozoic paleogeography of the western U.S.: Pacific Section, Society of Economic Paleontologists and Mineralogists Pacific Coast Paleogeography Symposium 1, p. 421–440.

Russell, B. J., 1984, Mesozoic geology of the Jackson Mountains, northwestern Nevada: Geological Society of America Bulletin, v. 95, p. 313–323.

Russell, B. R., 1981, Pre-Tertiary paleogeography and tectonic history of the Jackson Mountains, northwestern Nevada [Ph.D. thesis]: Evanston, Illinois, Northwestern University, 206 p.

Saleeby, J., Hannah, J. L., and Varga, R. J., 1987, Isotopic age constraints on middle Paleozoic deformation in the northern Sierra Nevada, California: Geology, v. 15, p. 757–760.

Sanborn, A. F., 1960, Geology and paleontology of the southwest quarter of the Big Bend Quadrangle, Shasta County, California: California Division of Mines and Geology Special Report 6, 26 p.

Schweickert, R. A., 1978, Triassic and Jurassic paleogeography of the Sierra Nevada and adjacent region, California and western Nevada, *in* Howell, D. G., and McDougall, K. A., eds., Mesozoic paleogeography of the western U.S.: Pacific Section, Society of Economic Paleontologists and Mineralogists Pacific Coast Paleogeography Symposium 2, p. 361–384.

Siberling, N. J., 1973, Geologic events during Permo-Triassic time along the Pacific margin of the United States, *in* Logan, S., and Hills, L. V., eds., The Permian and Triassic systems and their mutual boundary: Calgary, Canada, Alberta Society of Petroleum Geology, p. 345–362.

Siberling, N. J., and Jones, D. L., 1982, Tectonic significance of Permian-Triassic strata in northwestern Nevada and northern California: Geological Society of America Abstracts with Programs, v. 14, p. 234.

Siberling, N. J., and Roberts, R. J., 1962, Pre-Tertiary stratigraphy and structure of northwestern Nevada: Geological Society of America Special Paper 72, p. 58.

Siberling, N. J., Jones, D. L., Blake, M. C., Jr., and Howell, D. G., 1987, Lithotectonic terrane map of the western conterminous United States; Lithotectonic terrane maps of the North American Cordillera: U.S. Geological Survey Map MF-1874-C, scale 1:2,500,000.

Skinner, J. W., and Wilde, G. L., 1965, Permian biostratigraphy and fusulinid faunas of the Lake Shasta area, northern California: Lawrence, University of Kansas Paleontology Contributions, Protozoa, article 6, p. 1–98.

Smith, J. G., 1966, Petrology of the southern Pine Forest Range, Humboldt County, Nevada [Ph.D. thesis]: Stanford, California, Stanford University, 136 p.

——, 1973, Geologic map of the Duffer Peak Quadrangle, Humboldt County, Nevada: U.S. Geological Survey Miscellaneous Field Investigations Map I-606, scale 1:48,000.

Smith, J. G., McKee, E. H., Tatlock, D. B., and Marvin, R. F., 1971, Mesozoic granitic rocks in northwest Nevada; A link between the Sierra Nevada and Idaho batholiths: Geological Society of America Bulletin, v. 82, p. 2933–2944.

Speed, R. C., 1978, Paleogeographic and plate tectonic evolution of the early Mesozoic marine province of the western Great Basin, *in* Howell, D. G., and McDouglass, K. A., eds., Mesozoic paleogeography of the western U.S.: Pacific Section, Society of Economic Paleontologists and Mineralogists Pacific Coast Paleogeography Symposium 2, p. 253–270.

——, 1979, Collided Paleozoic microplate in the western U.S.: Journal of Geology, v. 87, p. 279.

Speed, R. C., Elison, M. W., and Heck, F. R., 1988, Phanerozoic tectonic evolution of the Great Basin, *in* Ernst, W. G., ed., Metamorphism and crustal evolution of the western United States; Rubey Volume 7: Englewood Cliffs, New Jersey, Prentice-Hall, p. 572–606.

Stewart, J. H., and Carlson, J. G., 1978, Geologic map of Nevada: U.S. Geological Survey, scale 1:500,000.

Tomlinson, A. J., and Wardlaw, B. R., 1988, Permian-Triassic Sonoma orogeny, western U.S., and late Cenozoic orogenesis in Japan; Examples of back-arc basin closure: Geological Society of America Abstracts with Programs, v. 20, p. A109.

Tosdal, R. M., Haxel, G. B., and Wright, J. E., 1989, Jurassic geology of the Sonoran Desert region, southern Arizona, southeastern California, and northernmost Sonora; Construction of a continental-margin magmatic arc: Arizona Geological Society Digest, v. 17, p. 397–434.

Vail, P. R., Mitchum, R. M., Jr., and Thompson, S., III, 1977, Seismic stratigraphy and global changes of sea level; Part 4, Global cycles of relative changes of sea level, *in* Payton, C. E., ed., Seismic stratigraphy; Applications to hydrocarbon exploration: American Association of Petroleum Geologists Memoir 26, p. 83–98.

Vallier, T. L., 1977, The Permian and Triassic Seven Devils Group, western Idaho and northeastern Oregon: U.S. Geological Survey Bulletin 1437, 59 p.

Walker, R. G., 1975, Generalized facies models for resedimented conglomerates of turbidite association: Geological Society of America Bulletin, v. 86, p. 737–748.

Wardlaw, B. R., Nestell, M. K., and Dutro, J. T., Jr., 1982, Biostratigraphy and structural setting of the Permian Coyote Butte Formation of central Oregon: Geology, v. 10, p. 13–16.

Watkins, R., 1985, Volcaniclastic and carbonate sedimentation in late Paleozoic island-arc deposits, eastern Klamath Mountains, California: Geology, v. 13, p. 709–713.

——, 1986, Late Devonian to early Carboniferous turbidite facies and basinal development of the eastern Klamath Mountains, California: Sedimentary Geology, v. 49, p. 51–71.

Willden, R., 1964, Geology and mineral deposits of Humboldt County, Nevada: Nevada Bureau of Mines Bulletin, v. 59, 154 p.

Wilson, J. L., and Jordan, C., 1983, Middle shelf environment, *in* Scholle, P. A., Bebout, D. G., and Moore, C. H., eds., Carbonate depositional environments: American Association of Petroleum Geologists Memoir 33, p. 297–344.

Wood, D. A., 1980, The application of a Th-Hf-Ta diagram to problems of tectonomagmatic classification and to establishing the nature of crustal contamination of basaltic lavas of the British Tertiary volcanic province: Earth and Science Letters, v. 50, p. 11.

Wright, J. E., and Miller, E. L., 1988, The early Mesozoic APW path of North America and its correspondence to western U.S. Cordilleran continental margin tectonics: Geological Society of America Abstracts with Programs, v. 20, p. 350–351.

Manuscript Accepted by the Society March 27, 1990

Printed in U.S.A.

Geological Society of America
Special Paper 255
1990

Geology and tectonic significance of terranes near Quinn River Crossing, Nevada

A. Elizabeth Jones
Department of Geology and Geophysics, University of California Berkeley, Berkeley, California 94720

ABSTRACT

Paleozoic and Mesozoic rocks exposed near Quinn River Crossing, Humboldt County, Nevada, constitute two terranes that have played important roles in the tectonic history of this region. Data collected from these terranes provide insight into the tectonic events that shaped the margin of western North America during the late Paleozoic and the early Mesozoic. The Jackson terrane at Quinn River is composed of a chert-quartzite arenite containing Late Devonian and pre–Late Devonian chert pebbles. This unit sits below a Jurassic(?) ash-flow tuff that has embedded limestone blocks, mafic and felsic volcanic clasts, and hornblende porphyry blocks K-Ar dated at 191 ± 9 Ma. Thrust over the Jurassic(?) ash-flow tuff is the Black Rock terrane, which includes the Lower Permian Bilk Creek limestone, Permian clastic and arc-derived volcaniclastic rocks, and the Upper Permian to Middle Triassic Quinn River Formation, composed of chert, siltstone, shale, sandstone, and tuffaceous and volcaniclastic rocks.

The Jackson terrane has an enigmatic paleogeographic relation with the Black Rock terrane. Lithologic, age, structural, and biogeographic data indicate that the Black Rock terrane at Quinn River is correlative with the eastern Klamath terrane in northern California. In contrast, the Black Rock terrane at Quinn River is lithologically, biogeographically, and structurally distinct from coeval rocks of the Golconda allochthon in the nearby Osgood Mountains and Hot Springs Range. The lack of Sonoman deformation in the Black Rock terrane, the lack of Permian arc-derived volcaniclastic rocks in the Golconda allochthon in the Osgood Mountains and Hot Springs Range, the absence of any linking facies between these two areas, and the presence of post–Early Jurassic deformation in rocks between these areas all do not support models that imply relative proximity of the Black Rock arc terrane to the northwesternmost part of the Golconda allochthon during Permian-Triassic time. The evidence implies instead that the present relative positions of the Black Rock terrane and the northwesternmost exposures of the Golconda allochthon are the result of younger Mesozoic tectonic events unrelated to the original paleogeographic settings of these rocks.

INTRODUCTION

The unusual structural, stratigraphic, and paleogeographic information revealed by the pre-Tertiary rocks at the southern end of the Bilk Creek Mountains near Quinn River Crossing, Nevada, has led to contradicting interpretations of the important role this small area has played in the geologic history of the western Cordillera. The structural significance of this small exposure (less than 3 km^2) was first recognized by Willden (1961), when he identified the post–Middle Triassic Quinn River thrust, which places Permian to Triassic carbonate, chert, and volcaniclastic rocks above a Mesozoic volcanic complex. The deformational event that was responsible for this emplacement remains inadequately understood.

Jones, A. E., 1990, Geology and tectonic significance of terranes near Quinn River Crossing, Nevada, *in* Harwood, D. S., and Miller, M. M., eds., Paleozoic and early Mesozoic paleogeographic relations; Sierra Nevada, Klamath Mountains, and related terranes: Boulder, Colorado, Geological Society of America Special Paper 255.

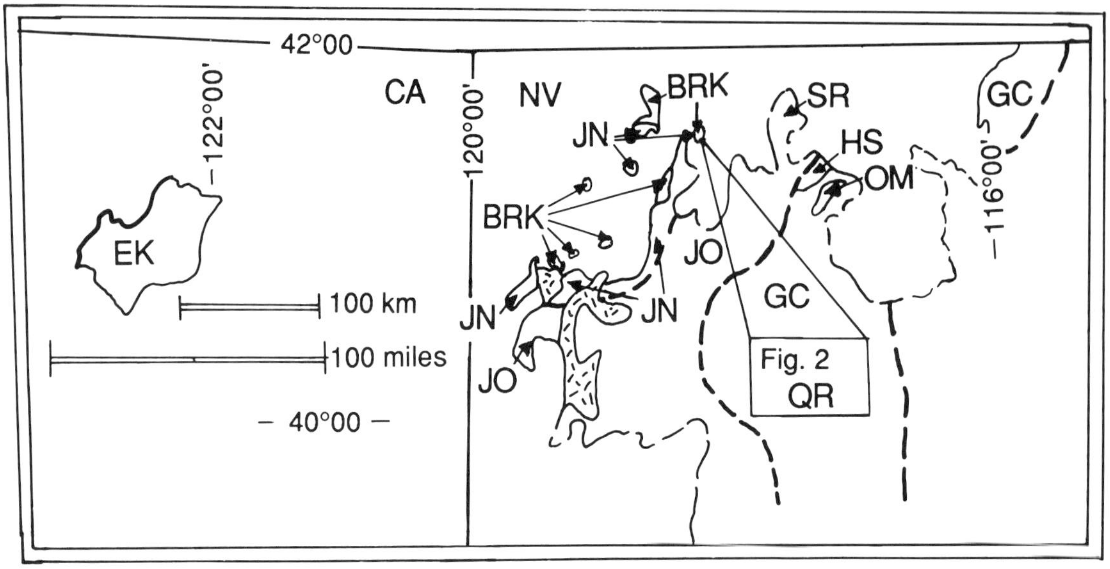

Figure 1. Regional location map showing eastern Klamath (EK), Jackson Mountains (JN), Black Rock (BRK), Jungo (JO), and Golconda (GC) terranes, Santa Rosa Range (SR), Hot Springs Range (HS), Osgood Mountains (OM), study area at Quinn River (QR) shown in Figure 2. Modified from Silberling and others, 1987.

Regional stratigraphic relations of these rocks were discussed by Skinner and Wilde (1966), Ketner and Wardlaw (1981), and Silberling and Jones (1982). Skinner and Wilde (1966) reported on the early Permian fusulinid species from the limestone at Quinn River, and recognized that many of these species are identical to those in the McCloud limestone in the eastern Klamath Mountains more than 300 km to the west. Silberling and Jones (1982), who reported on Permian and Triassic radiolarians and Triassic molluscs from the Quinn River Formation, also concluded that the Quinn River Formation had both age and stratigraphic ties with coeval rocks of the eastern Klamath terrane to the west (Fig. 1). They also recognized the importance of the absence at Quinn River Crossing of Late Permian–Early Triassic Sonoman deformation that strongly affected rocks of the Golconda allochthon to the east in north-central Nevada. In contrast to Skinner and Wilde (1966) and Silberling and Jones (1982), Ketner and Wardlaw (1981) made lithologic correlations between each of the units at Quinn River and allochthonous and parautochthonous upper Paleozoic rocks of the Farrel Canyon, Edna Mountain, Adam Peak, and Antler Peak Formations in the Osgood Mountains area, 120 km to the east (Fig. 1).

The regional tectonic interpretation of Speed (1979) included the rocks at Quinn River as part of the far-traveled, east-facing, arc-microplate "Sonomia," which according to his model, accreted to North America with the closure of the Golconda basin during the Late Permian–Early Triassic Sonoma orogeny. Stratigraphic correlations between the eastern Klamath terrane and the units at Quinn River, however, also make them part of the west-facing Klamath-Sierran arc of Burchfiel and Davis (1981), which is also interpreted to have been emplaced on the North American margin during the Sonoma orogeny. Subsequent tectonic models for the evolution of this area have been based primarily on one or the other of these two contrasting models (Miller, 1989; Miller and others, 1984, and references therein; Russell, 1984).

This chapter aims to more clearly present the structural, lithologic, and biostratigraphic data that have been gathered from Quinn River both in this work and by earlier workers, in an effort to address contradicting interpretations of the tectonic history of these rocks. The chapter demonstrates that the lithologic, structural, and biostratigraphic characteristics of the terranes at Quinn River do not support models that imply either relative proximity of the Klamath arc to the northwesternmost part of the Golconda allochthon during Permian-Triassic time, or involvement of an arc terrane in the Sonoman deformation. Instead, the evidence implies that the present relative positions of the Black Rock terrane and the northwesternmost exposures of the Golconda allochthon are the result of younger Mesozoic tectonic events unrelated to the original paleogeographic settings of these rocks.

STRATIGRAPHY OF TERRANES AT QUINN RIVER

The strata at Quinn River can be divided into two major structural units separated by the Quinn River thrust of Willden (1961) (Fig. 2). The lower plate to the north is a post-Mid-Devonian chert-quartzite arenite and conglomerate overlain by a Jurassic(?) ash-flow tuff containing large blocks of volcanic and intrusive(?) rocks, and limestone. On the upper plate to the south of the Quinn River thrust is a Lower Permian limestone, here referred to as the Bilk Creek limestone, Permian volcaniclastic rocks and mudstone, and the Permian-Triassic Quinn River

Formation (Willden, 1961) composed of chert, shale, siltstone, sandstone, tuff, and volcaniclastic rocks.

Silberling and others (1987) have included all the pre-Tertiary strata at Quinn River as part of the Black Rock terrane. The northern part of the exposure below the Quinn River thrust that includes both the ash-flow tuff and the chert-quartzite arenite is considered here, instead, as part of the Jackson terrane, for regional stratigraphic reasons discussed later. The Black Rock terrane is restricted to units above (south of) the Quinn River thrust. Other exposures of these two terranes are scattered across the northwest corner of Nevada (Fig. 1). This chapter only discusses exposures near Quinn River Crossing.

Jackson terrane at Quinn River

Chert-quartzite arenite (Pzcqa). The strongly silicified post–Mid-Devonian chert-quartzite arenite (Figs. 2, 3) is composed of interbeds of chert, argillite, and quartz pebbles in a quartzite matrix, coarse-grained quartz and chert sandstone, quartzite, siltstone, and bedded chert. Bed thickness ranges from 10 cm to 1 m. The quartzite and coarse-grained quartz and chert sandstone beds are generally well sorted and graded and have subrounded to rounded grains. In contrast, the pebble beds are poorly sorted, reverse graded, clast-supported beds of subrounded to rounded pebbles in a quartzite matrix (Fig. 3). Dikes from the pebble beds are injected into the underlying spiculitic chert beds. Some of the quartzite pebbles in the pebble beds show a strong metamorphic fabric, and the gray and cream colored chert pebbles contain well-preserved radiolarians. The chert beds that are interlayered with the pebble beds lack characteristics of replacement cherts such as silicified calcite and dolomite rhombs, silicified fossils, or dissolution boundaries with interbedded siliceous layers, and are therefore considered primary.

The interbeds of reverse graded chert, argillite, and quartz

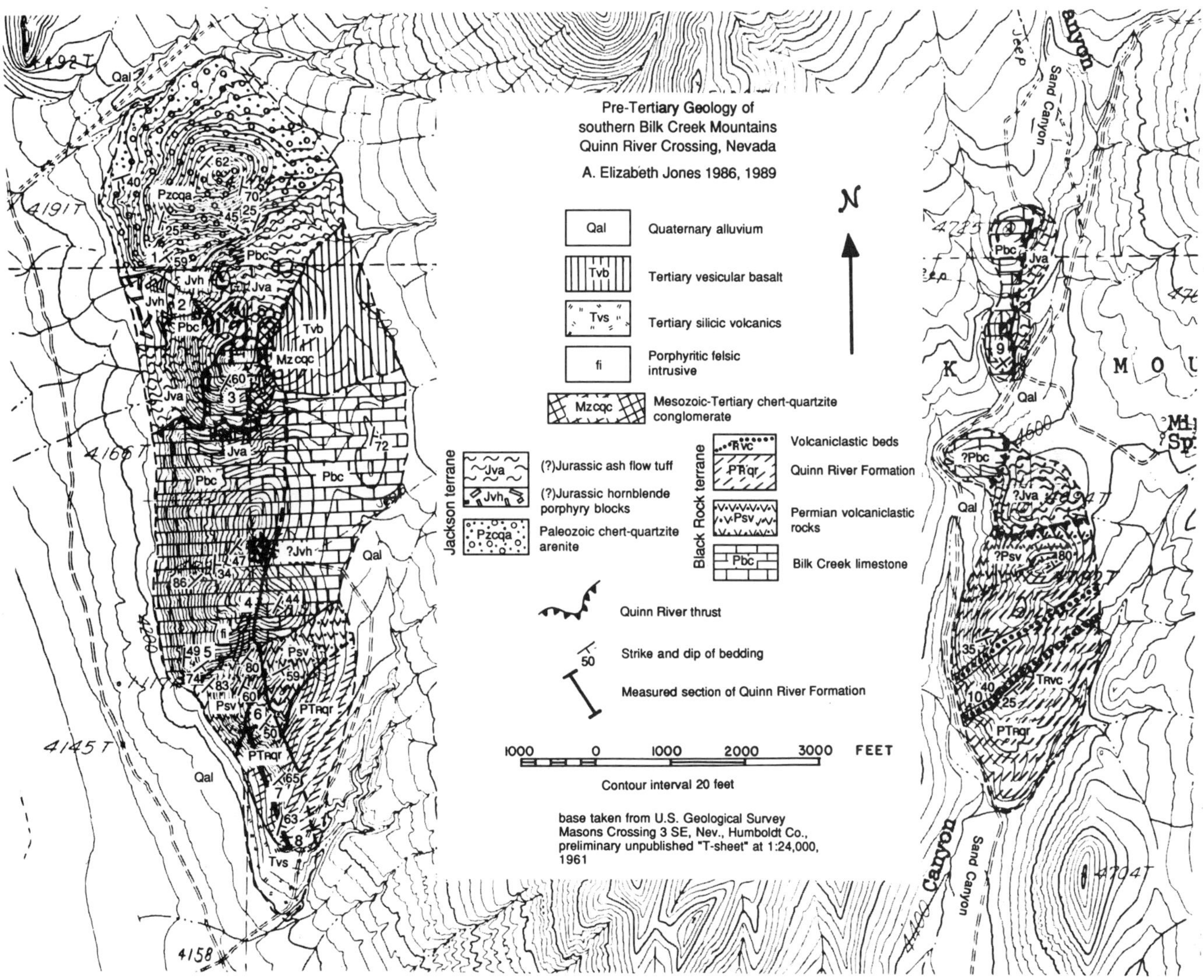

Figure 2. Geologic map of southern Bilk Creek Mountains, Quinn River Crossing, Nevada. Numbers refer to fossil/age localities listed in Tables 1 and 2. Section measured perpendicular to bedding, not as drawn.

Figure 3. Paleozoic chert-quartzite arenite. Interbedded chert, argillite, and quartzite pebbles and black spiculitic chert. Sample measures 8 by 12.5 cm.

pebbles with black chert beds are unusual (Fig. 3). The poor sorting and grading, abrupt grain-size change from pebbles to chert beds, and injection structures of the pebble beds into the chert beds, suggest that the pebble beds were deposited by grain flows or high-density turbidity currents associated with part of a deep-sea fan.

No radiolarians were found in the black chert beds interbedded with the pebble beds. The only age data for the chert-quartzite arenite comes from radiolarians extracted from pebbles in the chert, argillite, and quartz pebble beds. A very small population of Spumellarians, which have large, bladed spines typical of Late Devonian forms, was found together with a much larger population of Spumellarians that have multiple simple spines characteristic of Ordovician, Silurian, and Devonian forms (Nazarov, 1975). Many of the radiolarian fragments have characteristics in common with both Late Devonian and pre-Middle Devonian forms. If all of the radiolarians were Late Devonian or younger, the population would be expected to have a much larger percentage of bladed spine-bearing Spumellarians than was actually found. The sparsity of these forms suggests that they are mixed with an older, pre-Middle Devonian population of simple spined forms. No radiolarians that are indicative of post-Late Devonian age were found. The good preservation of many of the specimens suggests that this absence is a result of age and not sample preservation, but lack of positively identified indicative species prevents a more specific age identification. The location and age of the radiolarian sample are given in Table 1 and Figure 2, and identifications of these radiolarians are given in Appendix 1.

Beds of the chert-quartzite arenite are gently folded, but strike generally northeast and dip moderately to steeply southeast underneath the Jurassic(?) ash-flow tuff to the south. The contact between them is nowhere clearly exposed, but they crop out within 1 m of one another without showing evidence of a major structure separating them. The nature of the contact between them is therefore unknown.

Jurassic(?) ash-flow tuff (Jva). A silicic ash-flow tuff matrix dominates the eastern half of the Jurassic(?) ash-flow tuff exposure, while the western half of the exposure is strongly clast supported. The clasts consist of volcanic breccia, andesitic and basaltic volcanic rocks, large blocks (1 to 100 m across) of a distinctive hornblende porphyry (Fig. 4b), and limestone blocks (0.5 to 5 m across), commonly silicified. The massive silicic ash-flow matrix has embayed quartz and fractured feldspar phenocrysts, devitrified glass shards, and relict pumice (Fig. 4a). There is no evidence of reworking of the ash-flow tuff into epiclastic sedimentary rocks.

The andesitic clasts in the ash-flow tuff have a strongly chloritized matrix with altered feldspars, and the abundant basaltic clasts are composed of microlaths of feldspar and larger altered mafic and plagioclase phenocrysts (Fig. 4a). A minimum K-Ar age of 191 ± 9 Ma (Jones and others, 1988), obtained from exceptionally large (0.5 to 2 cm) hornblende phenocrysts in the hornblende porphyry blocks, suggests that the blocks are no younger than Lower Jurassic (Table 2; Fig. 2). Ketner and Wardlaw (1981) have reported faunas from the limestone blocks in the ash-flow tuff that have Permian eastern Klamath affinities similar to the Bilk Creek limestone described in the next section (Table 1, location F1; Appendix 1).

Black Rock terrane at Quinn River

Bilk Creek limestone (Pbc). More than 600 m of fossiliferous massively bedded biosparite, referred to here as the Bilk Creek limestone, makes up the lower two-thirds of the Black Rock terrane at Quinn River. It is truncated at the base by the conspicuous Quinn River thrust (Figs. 2, 5) and overlapped at the northeast edge by Tertiary vesicular basalt. The limestone beds strike north-northeast and dip consistently 40° to 70° to the southeast, with beds rarely overturned steeply to the northwest. The limestone becomes siltier and sandier toward the top of the

Figure 4. A, Photomicrograph of Jurassic ash-flow tuff with broken lithic clasts, plagioclase, and embayed quartz phenocrysts. Note relict devitrified glass shards between broken quartz and plagioclase phenocrysts. Outline shows mafic volcanic clast. Field of view, approximately 5 mm by 2 mm. B, Hornblende porphyry with large hornblende phenocrysts used for K-Ar dating. Scale in centimeters.

TABLE 1. FOSSIL LOCALITIES COLLECTED AT QUINN RIVER

Location	Lat. & Long.	Formation	Fossil	Age	Reference
1	41°33.15' 118°20.00'	cqa	Ra	Pre–Mid-Devonian and post–Mid-Devonian	this chapter
3	41°32.70' 118°19.80'	Bcl	Cr	Wolfcampian	C. Stevens, 1988, personal communication
4	41°32.25' 118°19.80'	Bcl	Br	Wolfcampian–early Leonardian	R. Hanger, 1989, personal communication
5	41°32.20' 118°19.90'	Bcl	Fu	Early Leonardian	C. Stevens, 1988, personal communication
6	41°31.95' 118°19.80'	Qrf	Ra	Early Guadalupian	this chapter
6	41°31.95' 118°19.80'	Qrf	Cn	Early Guadalupian	A. Harris and B. Wardlaw, 1988, personal communication
7	41°31.90' 118°19.75'	Qrf	Am	Early Anisian	N. Silberling, 1988, personal communication
8	41°31.85' 118°19.70'	Qrf	Cn	?Ladinian	A. Harris and B. Wardlaw, 1988, personal communication
9	41°32.80' 118°17.70'	Bcl	Cr	Wolfcampian	C. Stevens, 1989, personal communication
10	41°32.05' 118°17.70'	Qrf	Am	Mid Ladinian	N. Silberling and T. Tozer, 1988, personal communication
F1*	41°32.85' 118°19.90'	Jaft	Fu	Wolfcampian–early Leonardian	Ketner and Wardlaw, 1981
A*	----------	Bcl	Br	Late Leonardian	Ketner and Wardlaw, 1981
F2*	41°32.15' 118°19.85'	Psv	Fu	Early Leonardian	Ketner and Wardlaw, 1981
B*	41°32.00' 118°19.80'	Qrf	Br	Late Permian	Ketner and Wardlaw, 1981
C*	----------	Bcl	Fu	Mid–late Wolfcampian	Skinner and Wilde, 1966

Note: Locations correspond to circled numbers shown on geologic map (Fig. 2). cqa = chert-quartzite arenite; Bcl = Bilk Creek Limestone; Qrf = Quinn River Formation; Jaft = Jurassic(?) ash-flow tuff; Psv = Permian volcaniclastic rocks; Ra = radiolarian; Cr = coral; Br = brachiopod; Fu = fusulinid; Cn = conodont; Am = ammonite.

*See reference for exact location. A = no specific location given; B = no specific location given, but exposure is very small; C = fossils collected throughout the section, see reference for specific locations. See Appendix 1 for listing of faunas collected at each locality.

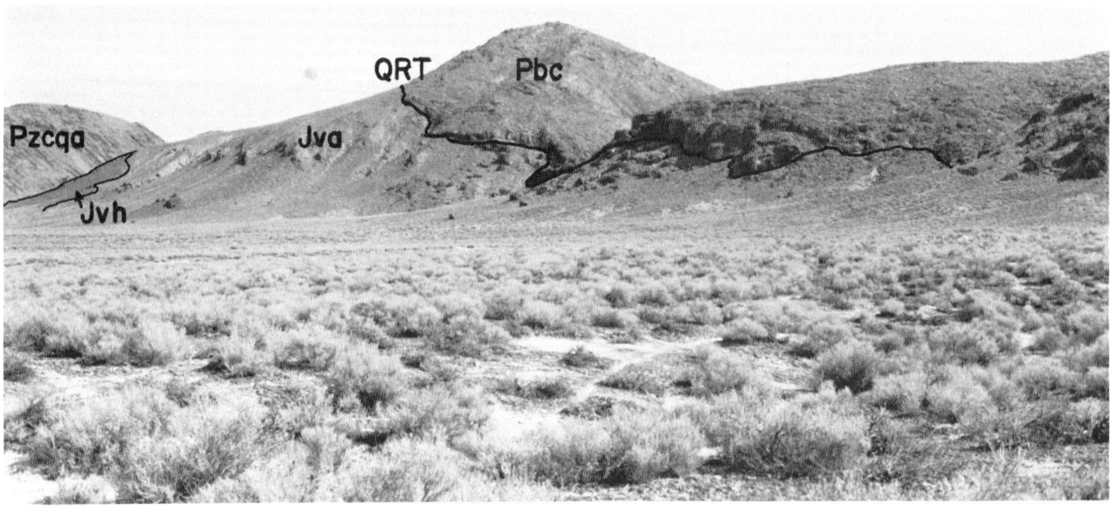

Figure 5. Quinn River thrust (QRT), which places the Permian Bilk Creek limestone (Pbc) over the Jurassic(?) ash-flow tuff (Jva).

section, marking a depositional transition to the overlying volcaniclastic sedimentary rocks. Ketner and Wardlaw (1981) interpreted this contact as a fault, but other than minor folding and shearing between beds, there is no evidence of a major structure or unconformity at the contact that disrupts the section, and the succession of beds is clear.

A second exposure of the Bilk Creek limestone crops out on the east side of Sand Canyon approximately 2.5 km due east of the principal exposure (Fig. 2). Here, the Quinn River thrust places the Bilk Creek limestone over a red volcanic breccia that is inferred to correlate with the Jurassic(?) ash-flow tuff.

Abundant Early Permian fusulinids, corals, crinoids, and brachiopods are well preserved at various locations throughout the section in spite of complete silicification in many areas. The fusulinids from the Bilk Creek limestone were first reported by Skinner and Wilde (1966), who recognized their similarity to the McCloud species in the eastern Klamaths (Skinner and Wilde, 1965) and assigned to them a middle to late Wolfcampian age. Abundant brachiopods present near the top of the section were dated by Ketner and Wardlaw (1981) as late Leonardian, but fusulinids collected a few meters above this are early Leonardian, suggesting the brachiopods could be older than originally reported. Wolfcampian and early Leonardian corals, brachiopods, and fusulinids were collected in this study. The locations and ages of the fossils collected are shown in Figure 2 and listed in Table 1, locations 3, 4, 5, 9, A, C, and the faunas are listed in Appendix 1. A detailed study of the paleoenvironmental significance of this unit is presently underway by Hanger (1989).

Permian volcaniclastic rocks (Psv). The base of the Permian volcaniclastic unit is defined as the first appearance of predominantly silty material. This section, approximately 125 m thick, includes thin interbeds of spiculitic dark gray argillite, bright green tuffaceous argillite, and greenish brown and yellow turbiditic volcaniclastic sandstones, siltstones, and mudstones. Volcaniclastic crystal-rich sandstone contains broken phenocrysts of K-feldspar, plagioclase, quartz, and chloritized biotite, and mafic volcanic clasts that have fine feldspar microlaths. Secondary micaceous alteration commonly defines grain boundaries within the volcaniclastic sandstone. The volcaniclastic beds become thicker and coarser grained toward the top of the section.

Abundant intermediate to mafic volcaniclastic rocks in this section indicate its proximity to an active arc source terrane during submarine deposition by turbidity currents. The lithologic trend from the carbonate of the Bilk Creek limestone to the overlying interbedded spiculitic argillite and volcaniclastic rocks suggests that deposition occurred in a subsiding basin receiving an increasing influx of volcaniclastic sediment.

Ketner and Wardlaw (1981) collected early Leonardian fusulinids from the lower part of this section (Table 1, location F2; Appendix 1). Although no age data have been collected from the upper 100 m of the Permian volcaniclastic rocks, no major structures have been found, and the continuous succession of repetitive volcaniclastic lithologies does not suggest a sudden gap in time.

TABLE 2. K-Ar ANALYSIS FROM HORNBLENDE PORPHYRY BLOCK IN JURASSIC(?) ASH-FLOW TUFF*

	Mass (g)	^{*40}Ar (x 10^{-10}mol/g)	Atmos. ^{40}Ar (%)	Date (Ma)
Run 1	0.21070	2.052	16.2	184 ± 7
Run 2	0.19965	2.184	15.0	191 ± 9

Note: K analysis yielded 0.626 ± 0.002 percent K. Analysis by P. Renne, 1988.
*Location 2, Figure 2.

Figure 6. Quinn River Formation (PTRqr) overlying Permian volcaniclastic rocks (Psv). Prominent southeast-dipping bed of brown bioclastic calcirudite marks the base of the measured section (arrow). Contact between Permian volcaniclastic rocks and Quinn River Formation is underneath the cobble tuff immediately below the calcirudite. On the far right in the shadow; an open box fold in the Quinn River Formation shows gentle folding.

The gently folded beds of the Permian volcaniclastic rocks strike slightly obliquely into a cobble-bearing tuff at the base of the Quinn River Formation, indicating either a fault or an angular unconformity at this contact. Minor shearing along parts of the contact suggests limited movement.

Quinn River Formation (PTrqr). The base of the Quinn River Formation is a pale green, fine-grained siliceous tuff that has well-preserved devitrified bubble wall glass shards, rare altered volcanic clasts, and well-rounded, matrix-supported cobbles of cherty tuff. A prominent, brown, silicified, bioclastic calcirudite 2 m thick, just above the tuff (Fig. 6), forms the base of the 170-m measured section of the Quinn River Formation (Fig. 7). A 15-m-thick section of light brown dolomite beds sits above the calcirudite. These beds are overlain by 25 m of interbedded gray radiolarian chert and argillite. Above the gray chert and argillite is a 30-m-thick section of gray to black carbonaceous siltstones. Interbedded siltstones, laminated siliceous shales, radiolarian cherts, and crystal-rich volcaniclastic rocks mark the upper hundred meters of the exposed section.

Late Wordian (early Guadalupian) brachiopods were reported by Ketner and Wardlaw (1981) from the brown dolomite beds near the base of the section. The overlying interbedded gray chert and argillite beds yielded well-preserved Guadalupian (Late Permian) radiolarians and conodonts (Silberling and Jones, 1982; this study). Primitive Early Triassic Spumellarian radiolarians were recovered near the top of the gray to black carbonaceous siltstones above the Permian cherts. The position of the Permian-Triassic boundary within this section is uncertain, but it may be very near the base of the section, just above the Upper Permian cherts and argillites. No angular unconformity is present within the Quinn River Formation, although a disconformity may exist between the Guadalupian and Lower Triassic strata. A late Spathian to early Anisian (late Early to early Middle Triassic) ammonite and primitive undescribed Triassic Nasselarian radiolarians were collected from a sandy layer 30 m from the top of the exposed section. A few meters from the top, well-developed Nasselarians were recovered, along with a Middle Triassic (Ladinian?) conodont.

A second exposure of the Quinn River Formation crops out on the east side of Sand Canyon approximately 2.5 km due east of the principal exposure (Figure 2). Mid-Ladinian ammonites occur adjacent to volcaniclastic beds near the top of the exposed section at Sand Canyon. The fossil locations from the Quinn River Formation are shown in Figures 2 and 8; the ages are listed in Table 1, locations 6, 7, 8, 10, B, and the faunas are listed in Appendix 1.

Although the contact between the Permian volcaniclastic rocks and the Quinn River Formation is clearly some kind of fault or unconformity, the age data across it do not require either a major gap or a repetition of section. The general similarity of depositional and lithologic characteristics across the contact suggests that these two units were paleogeographically linked to one another. Regional stratigraphic considerations discussed later imply that very little movement has occurred at this contact.

UPPER MESOZOIC TO LOWER TERTIARY SEDIMENTARY ROCKS (Mzcqc)

A Mesozoic to Tertiary chert-quartzite conglomerate that has large boulders (1 to 5 m) of the chert-quartzite arenite of the Jackson terrane in a red mudstone matrix forms an irregular north-south–trending strip on the east side of the hill that exposes

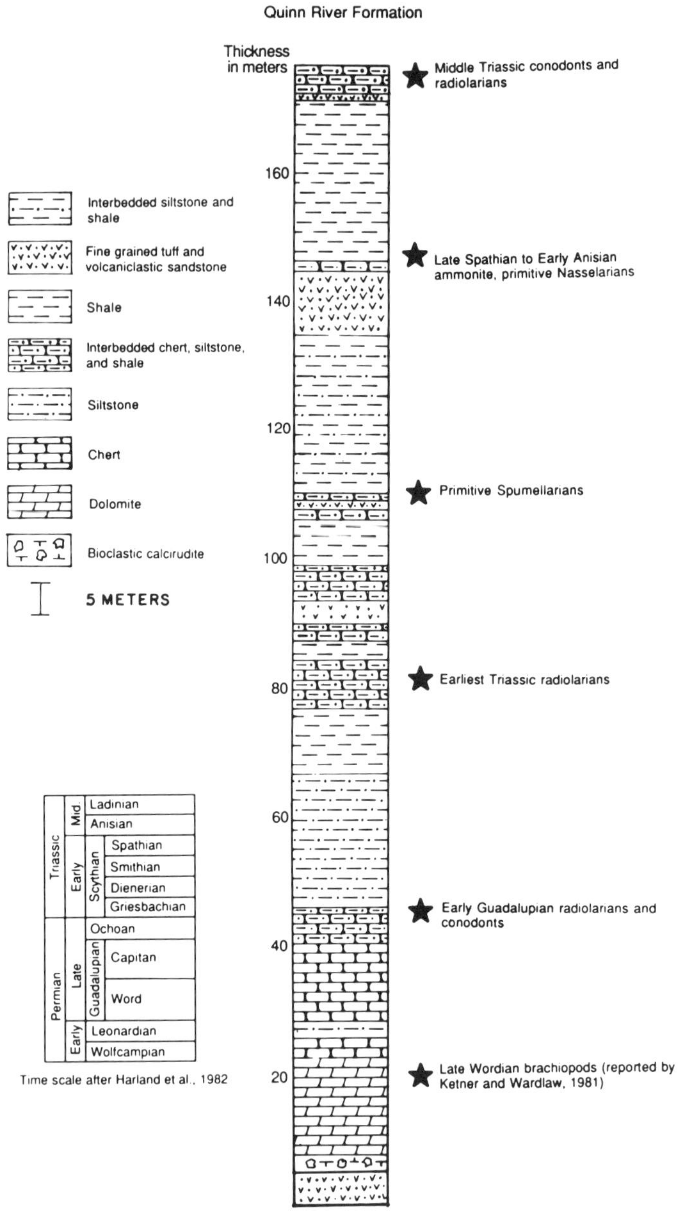

Figure 7. Measured section of the Quinn River Formation. Locality plotted in Figure 2.

the Quinn River thrust (Fig. 2). This strip is covered by Tertiary basalt on the east and overlies silicified Bilk Creek limestone and the Jurassic(?) ash-flow tuff to the west and southwest. Several small boulders (<1 m), containing cobbles of the chert-quartzite arenite in a quartz arenite matrix, are scattered in gullies at the north part of this strip, suggesting erosion and redeposition of the chert-quartzite arenite. Since structural relations indicate that the main exposure of the chert-quartzite arenite is dipping to the southeast below the Jurassic(?) ash-flow tuff, the presence of boulders of the chert-quartzite arenite above the ash-flow tuff indicates that an uplift event took place that eroded a significant quantity of the chert-quartzite arenite onto the Bilk Creek limestone and the ash-flow tuff, prior to the arrival of the Tertiary basalt. This unit forms an overlap between the Jackson terrane and the Black Rock terrane, as it overlaps the Quinn River thrust.

REGIONAL STRATIGRAPHIC RELATIONS

The rocks at Quinn River are divided into the Jackson terrane and the Black Rock terrane based on stratigraphic and structural similarities with rocks in the nearby Jackson Mountains. The rocks to the north of the Quinn River thrust more closely resemble those of the Jackson terrane in the Jackson Mountains (Silberling and others, 1987) than any unit of the Black Rock terrane reported from this region. The same structural relation that places the Black Rock terrane over the Jackson terrane on the Quinn River thrust is found in the Jackson Mountains where the McGill Canyon Unit of the Black Rock terrane, an upper Paleozoic assemblage of volcanogenic and siliceous turbidites and hemipelagic to pelagic rocks (Russell, 1984), is thrust over the Boulder Creek Beds and the Happy Creek Igneous Complex, described below, of the Jackson terrane (see Fig. 4, Russell, 1984).

Jackson terrane

Chert-quartzite arenite. Ketner and Wardlaw (1981) correlated the chert-quartzite arenite of the Jackson terrane with parts of the Edna Mountain and Adam Peak Formations in the Osgood Mountains based largely on the presence of chert and quartzite. The Pennsylvanian-Permian Adam Peak Formation and Permian Edna Mountain Formation are fossiliferous and contain common calcareous beds. In contrast, there are very few, if any, calcareous beds in the chert-quartzite arenite at Quinn River, and only pre–Middle Devonian and post–Mid-Devonian radiolarians in chert pebbles have been recovered. Although the chert pebbles in conglomeratic horizons of the chert-quartz arenite appear to have a bimodal chert source based on the ages of the radiolarians, the age of the unit itself is poorly constrained. It is, however, lithologically similar to and falls within the same broad age constraints as Devonian(?)-Mississippian chert-lithic–rich clastic rocks described from the Pine Forest Range (Wyld, this volume).

Jurassic(?) ash-flow tuff. The volcanic assemblage at Quinn River was originally reported by Willden (1961) as part of the Happy Creek Volcanic Series of Permian or older age. The Happy Creek Volcanic Series in the Jackson Mountains was renamed the Happy Creek Igneous Complex by Russell (1984), who determined a Late Triassic to mid-Jurassic age for the complex based on Late Triassic faunas collected from the conformably underlying Boulder Creek Beds, and on a whole-rock Rb/Sr isochron age of 160 ± 35 Ma. Maher and Saleeby (1988) determined U/Pb zircon dates of approximately 160 Ma from diorite sills within the Happy Creek Igneous Complex. The Happy Creek Igneous Complex is included in the Jackson terrane of Silberling and others (1987) in the Jackson Mountains.

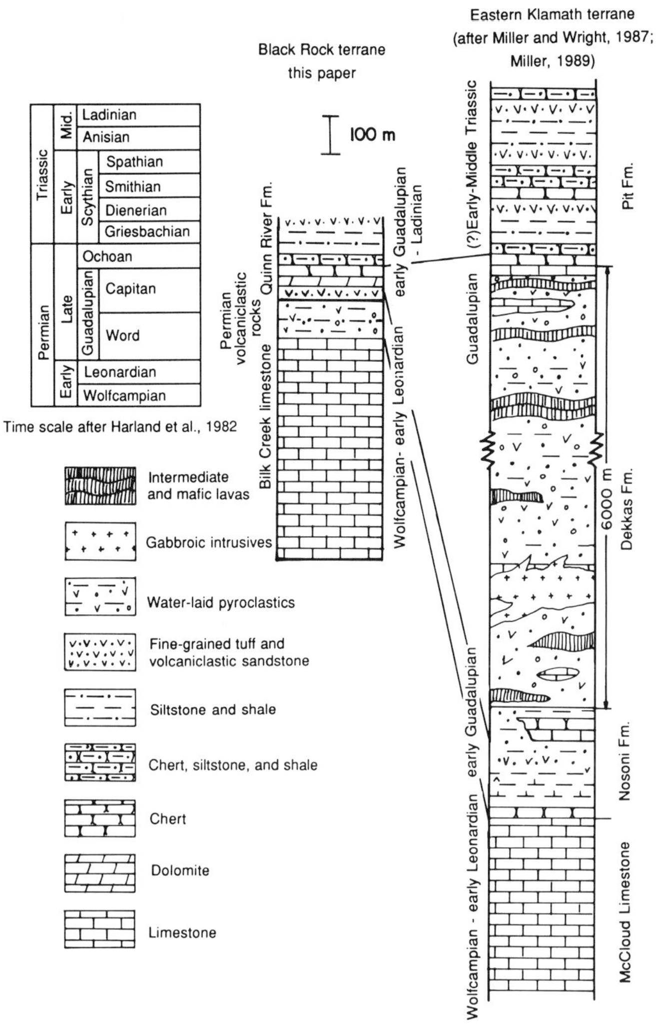

Figure 8. Comparative stratigraphic columns of the eastern Klamath terrane and the Black Rock terrane at Quinn River. Eastern Klamath column compiled from Miller and Wright (1987) and Miller (1989).

Speed (1977, 1979) included the ash-flow tuff at Quinn River in his arc terrane of "Sonomia" under the assumption that it was Permian or older. Similar lithologies and the presence of blocks in the ash-flow tuff that could be as young as 191 ± 9 Ma support its correlation with the Late Triassic to Mid-Jurassic Happy Creek Igneous Complex in the Jackson Mountains. This suggests that the ash-flow tuff would be time correlative with younger Mesozoic cover rocks in that area (Speed, 1979, p. 283), and not therefore uniquely associated with the older allochthonous postulated Sonomia.

The presence of "Bilk Creek" limestone blocks embedded in the Jurassic(?) ash-flow tuff, as reported by Ketner and Wardlaw (1981), suggests an interesting enigmatic paleogeographic relation between the Jackson terrane and the Black Rock terrane at the time the ash-flow tuff formed. Their original paleogeographic relations, however, remain obscured by younger deformation.

Black Rock terrane

Bilk Creek limestone. The Bilk Creek limestone has distinctive faunas that closely resemble the faunas of the McCloud Limestone in the eastern Klamath mountains (Skinner and Wilde, 1966). These faunas are distinct from North American faunas and are indicative of a different paleobiogeographic province as explained by Stevens and Rycerski (1988), Yancey and others, 1988) and Stevens and others (this volume), although contrary views are presented by Watkins and Wilson (1989), see also Newton (1988). The presence of these faunas at Quinn River demonstrates a closer relation between the Bilk Creek limestone and the McCloud Limestone in the Eastern Klamath terrane than with the Antler Peak Limestone in north-central Nevada, as suggested by Ketner and Wardlaw (1981). Only North American faunas have been collected from the Antler Peak Limestone.

Permian volcaniclastic rocks. Ketner and Wardlaw (1981) combined the Permian volcaniclastic rocks and the Jurassic(?) ash-flow tuff at Quinn River, and noted similarities between them and the Farrel Canyon Formation in the Osgood Mountains. The lithologic distinction between the two units at Quinn River is clear: one is a massive Jurassic(?) ash-flow tuff containing Permian and possibly Lower Jurassic blocks, and the other is a section of volcaniclastic sedimentary rocks that conformably overlies Lower Permian limestone and contains early Leonardian fossils at its base. The Farrel Canyon Formation in the Osgood Mountains is a Mississippian and Pennsylvanian(?) assemblage of bedded chert, lithic sandstone, and minor basaltic tuff (Jones and Jones, 1990; McCollum and McCollum, 1989; Hotz and Willden, 1964) strongly folded by the Permian-Triassic Sonoman deformation that is notably absent in the Black Rock terrane. The Farrel Canyon Formation does not contain any volcaniclastic material similar to the Permian volcaniclastic rocks at Quinn River.

Overlying the McCloud Limestone in the eastern Klamath terrane is the volcanogenic Upper Permian Bollibokka Group composed of the Leonardian and early Guadalupian Nosoni Formation and the Guadalupian Dekkas Formation (see Miller, 1989, for references). Remarkable stratigraphic correlation can be made between the Permian volcaniclastic rocks at Quinn River and the lower 200 m of the Nosoni Formation as described by Miller (1989). Calcareous siltstones at the base of the Nosoni grade into more siliceous siltstone and sandstone, interstratified with apple green volcanogenic sandstone and tuffaceous argillite (Miller, 1989, Fig. 5). This same succession of lithologies is recorded in the Permian volcaniclastic rocks at Quinn River (Fig. 8). The new Leonardian fossil collections from near the base of the Nosoni reported by Miller (1989) agree with the Leonardian age reported for the base of the Permian volcaniclastic rocks at Quinn River by Ketner and Wardlaw (1981).

Quinn River Formation. The oldest rocks in the Quinn River Formation are early Guadalupian (Upper Permian) and hence younger than the upper Pennsylvanian to Lower Permian Adam Peak Formation in the Osgood Mountains (Hotz and

Willden, 1964) to which Ketner and Wardlaw (1981) compared them.

Speed (1978) referred to the Permian-Triassic Quinn River Formation as the base of his "basinal terrane," which includes deformed terrigenous clastic rocks of the Norian to Lower(?) Jurassic Auld Lang Syne Group in the Jungo terrane (Fig. 1). Lupe and Silberling (1985, p. 267) discussed reservations about this assumption on the basis of both age and lithology. As they pointed out, the youngest rocks exposed at Quinn River are Ladinian (late Middle Triassic); rocks older than Norian (mid-Late Triassic) have not been recognized in the Auld Lang Syne Group, although the base of the deformed rocks of the Auld Lang Syne Group in the Jungo terrane is nowhere exposed. Lupe and Silberling (1985) also noted the presence of thick beds of Ladinian volcaniclastic and tuffaceous rocks in the Quinn River Formation, which contrasts with the terrigenous detritus in the Norian and younger Auld Lang Syne Group of the Jungo terrane. Sharply contrasting structural styles in each terrane also suggest a lack of continuity between them and do not support the hypothesis that the Quinn River Formation forms the base of the Auld Lang Syne Group (see Compton, 1960, for structural descriptions of the Auld Lang Syne Group).

The two lowest beds of the Quinn River Formation are lithologically very similar to beds in the upper half of the Nosoni Formation of the Bollibokka Group in the eastern Klamath terrane as shown by Miller (1989, Fig. 5). The apple green, cobble-bearing, matrix-supported tuffaceous mudstone just above the middle of the section of the Nosoni is very similar to the pale green, fine-grained, cobble-bearing, matrix-supported tuff that forms the base of the Quinn River Formation. The prominent, brown, silicified bioclastic calcirudite that sits above the cobble-bearing tuff at Quinn River is similar to the massive calcirudite above the cobble-bearing tuff in the Nosoni Formation described by Miller (1989). The stratigraphic correlation between the Nosoni Formation and both the Permian volcaniclastic rocks and the Quinn River Formation suggests only minor movement along the contact between the two units at Quinn River.

The Dekkas Formation is a thick sequence (greater than 5 km) of volcaniclastic rocks and lava flows that overlie the Leonardian and early Guadalupian Nosoni Formation in the eastern Klamath terrane. The upper part of the Dekkas Formation contains early Guadalupian brachiopods and radiolarians and is the same age as the upper part of the Nosoni Formation, indicating that it was deposited over a very short period of time (see Miller, 1989, for references). However, Guadalupian brachiopods and radiolarians in the Quinn River Formation lie only 5 to 40 m above rocks correlated with the Nosoni Formation based on lithology. The thick pile of volcaniclastic rocks and lava flows of the Dekkas Formation is not present at Quinn River. Silberling and Jones (1982) noted lithologic and age similarities between the lower part of the Quinn River Formation and the Pit Formation of the Eastern Klamath terrane; however, based on age, stratigraphic position, and lithology, the Upper Permian chert in the Eastern Klamaths terrane to which they correlated the Quinn River chert may actually represent a lateral equivalent of the Dekkas Formation (Miller and Harwood, this volume). The presence of Guadalupian radiolarians in similar facies at the top of the Dekkas has since been confirmed (Noble and Renne, 1990).

The Pit Formation in the eastern Klamath terrane is a Lower(?) and Middle Triassic section of siliceous argillite, silicic tuffaceous, and crystal-rich turbidites, and radiolarian-bearing hemipelagic sedimentary rocks (Miller and Harwood, this volume). The lithologic and age similarities of this unit to the upper 130 m of the Quinn River Formation strongly suggest that they formed in the same Triassic basin under similar paleoenvironmental conditions.

At least 70 m of Early Triassic (Spathian) rocks are preserved in the Quinn River Formation, evidenced by very primitive Triassic radiolarians found 70 m below the late Spathian to early Anisian ammonite. This indicates that much of the Early Triassic record thought to be missing in correlative sections in the eastern Klamath terrane could be preserved at Quinn River, and any disconformity present here is probably restricted to the latest Permian and/or the earliest Triassic.

Upper Mesozoic to lower Tertiary sedimentary rocks

The Upper Mesozoic to Tertiary chert-quartzite conglomerate that overlaps the Black Rock terrane and the Jackson terrane could be equivalent to either the Cretaceous King Lear Formation or the Cretaceous or Tertiary Pansy Lee Conglomerate. The King Lear Formation consists of locally derived clastic rocks, including pebble to boulder conglomerate interbedded with siltstone, graywacke, and lenses of limestone (Willden, 1963). The Cretaceous or Tertiary Pansy Lee Conglomerate, in the Jackson Mountains, is a poorly sorted pebble-to-boulder conglomerate of partially locally derived material that contains exotic chert and quartzite clasts (Willden, 1963).

Summary of regional stratigraphic relations

The chert-quartzite arenite in the Jackson terrane at Quinn River is lithologically and sedimentologically similar to the Devonian(?)-Mississippian chert-lithic–rich clastic rocks (unit Pzcs) described from the Pine Forest Range (Wyld, this volume). The Jurassic(?) ash-flow tuff in the Jackson terrane is lithologically similar and possibly age-correlative to parts of the Upper Triassic–Mid-Jurassic Happy Creek Volcanic Complex of the Jackson Mountains. The presence of McCloud faunas in blocks of the ash-flow tuff, as reported by Ketner and Wardlaw (1981), suggests a puzzling paleogeographic relation between the Jackson terrane and the Black Rock terrane.

The Black Rock terrane at Quinn River has many lithologic and age similarities with the eastern Klamath terrane in the eastern Klamath Mountains of northern California (Fig. 8). The Bilk Creek limestone has long been recognized as having affinities with the McCloud Limestone of the eastern Klamath terrane

(Skinner and Wilde, 1966). The Permian volcaniclastic sedimentary rocks above the Bilk Creek limestone have striking lithologic similarities to the lower part of the Nosoni Formation. The base of the Quinn River Formation matches closely with lithologies in the upper part of the Nosoni Formation. Beds depositionally above this are the same age as, and lithologically similar to, the upper part of the Dekkas Formation in the eastern Klamath terrane. The absence of thick beds of volcaniclastic rocks and lava flows representing the Dekkas Formation at Quinn River suggests that the Black Rock terrane at Quinn River represents a more distal volcanogenic facies than does the eastern Klamath terrane. The upper part of the Quinn River Formation has strong lithologic and age ties with the Pit Formation in the eastern Klamath terrane. In summary, the abundant faunal, age, lithologic, and structural similarities between the Black Rock terrane at Quinn River and the eastern Klamath terrane imply that the Black Rock terrane formed and evolved in the same basin and was part of the same tectonic evolution that affected the eastern Klamath terrane during Permian through Middle Triassic time.

REGIONAL TECTONIC INTERPRETATION

A number of tectonic models have been put forth to explain the late Paleozoic and early Mesozoic tectonic evolution of this part of western North America. Burchfiel and Davis (1972, 1975, 1981) were among the first to propose the closing of a back-arc basin of unknown size, represented by rocks of the Golconda allochthon between a west-facing Klamath-Sierran arc and the passive margin of western North America, as the culminating tectonic event of Permian-Triassic time (the Sonoma orogeny). Speed (1979, 1983) proposed another model, suggesting instead that the arc was east-facing and far-traveled, originally part of a plate separate from North America (his microplate Sonomia), accreting and subsiding during the final closure of an indeterminately large basin, leaving little evidence of its existence. Both of these models assume that a volcanic arc played a primary role in the deformation and closure of the Golconda basin, the remains of which are now preserved as deformed rocks of the Golconda allochthon in north-central Nevada.

The location of the Black Rock terrane at Quinn River provides important insights into the paleogeographic and paleotectonic relations of the Permian-Triassic rocks of the eastern Klamath terrane and the northwestern exposures of the Golconda allochthon. The Black Rock terrane is located more than 300 km east of the eastern Klamath terrane, and although it is missing the extensive volcaniclastic rocks and lava flows present in the Dekkas Formation in the eastern Klamath terrane, the Black Rock terrane at Quinn River has obvious stratigraphic and age ties with the Eastern Klamath terrane. In contrast, the Quinn River section lies only about 120 km from the northwesternmost exposure of rocks belonging to the Golconda allochthon in the northern Hot Springs Range (Stewart, 1980; Jones, 1988), to which it has no recognized structural or stratigraphic ties. The Black Rock terrane at Quinn River is separated from the Golconda rocks by the extensively deformed Upper Triassic to Lower Jurassic Jungo terrane in the Santa Rosa Range (Fig. 1), which is composed of fine-grained terrigenous clastic rocks (Silberling and others, 1987; Willden, 1964; Compton, 1960).

The Poverty Peak subterranc of the Golconda allochthon at the northern end of the Hot Springs Range (Jones, 1988) is composed dominantly of mid-Mississippian to Lower Permian red chert, argillite, pillow basalts, and basaltic tuff that is tightly folded and imbricated with a thick sequence of interbedded siltstone, calcareous siltstone, and quartz-rich sandstone (Sosson, 1984; Jones, 1988). The Poverty Peak subterrane clearly shows the east-vergent, tight, isoclinal folding, with north- to northeast-trending axial cleavages characteristic of Sonoman deformation recognized elsewhere throughout the Golconda allochthon (Brueckner and Snyder, 1985; Miller and others, 1982, 1984). It also shows evidence for a second partially coaxial compressive event overprinting the older deformation (A. E. Jones, unpublished data).

Several lines of evidence suggest that the present positions of the oceanic and arc-derived terranes of northern Nevada are the result of major post–mid-Mesozoic tectonic events and are not related to the original paleogeographic settings of these terranes. The lack of a characteristic Sonoman-age deformation within the Black Rock terrane precludes the assumption of a tectonic relation between it and the Golconda allochthon based on structural considerations. Permian arc-derived volcaniclastic rocks are absent in the Poverty Peak subterrane of the Golconda allochthon, but have been reported from the Shoshone, New Pass, and Toiyabe Ranges 200 km to the south in central Nevada (Tomlinson and Wright, 1986) and from Battle Mountain (Murchey, this volume). Mississippian arc-derived volcaniclastic rocks have been recognized in the Schoonover sequence in the Independence Mountains of northeastern Nevada (Miller and others, 1984), but no Permian arc-derived volcaniclastic rocks have been reported from that area. Mississippian(?) volcaniclastic rocks are also present to the south of the Poverty Peak subterrane in the Hot Springs Range, but the nature of their provenance is still unknown (A. E. Jones, unpublished data). There clearly is evidence that links parts of the Golconda basin to arc terranes at different times during its evolution, as would be expected of a basin as complex and long-lived as the Golconda. However, the isolated distribution of Permian arc-derived volcaniclastic rocks in the Golconda allochthon makes stratigraphic ties with the volcaniclastic-rich Black Rock terrane, and hence the eastern Klamath terrane, difficult. The arc-derived Black Rock terrane and the contrasting deep-marine, rift-related facies in parts of the Poverty Peak subterrane of the Golconda allochthon have no intermediate linking facies in northwestern Nevada. This dissimilarity of lithologies in the two terranes, and the absence of linking facies between them, provides no evidence to suggest that they were juxtaposed near one another during Permian to Early Triassic time, as has been suggested by earlier tectonic models.

The presence of the Upper Triassic to Lower Jurassic rocks of the Jungo terrane, characterized by complex southeast-verging isoclinal folds and sandwiched between the Golconda allochthon and the Black Rock terrane, indicates post–Early Jurassic relative movement between the Poverty Peak subterrane and the Black Rock terrane. The absence of even this younger deformation within the Black Rock terrane provides no evidence to structurally link it to terranes to the east until at least the mid-Mesozoic or later, when movement along the Quinn River thrust must have occurred. In summary, the Black Rock terrane at Quinn River does not provide structural or stratigraphic evidence that links the Permian eastern Klamath arc to the Permian Golconda basin, implying that the present juxtaposition of these terranes must be the result of post-Permian tectonic events.

CONCLUSION

The pre-Tertiary rocks exposed at Quinn River belong to the Jackson terrane and the Black Rock terrane. Volcanogenic rocks in the Jackson terrane have lithologic and age similarities with the early to mid-Mesozoic Happy Creek Igneous Complex of the Jackson terrane in the Jackson Mountains. Blocks of limestone with reported faunal affinities to the Black Rock terrane embedded in the volcanic unit of the Jackson terrane suggest an enigmatic paleogeographic relation between the two terranes.

The structural, lithologic, and biogeographic similarities between the Black Rock terrane at Quinn River and the eastern Klamath terrane suggest that these two terranes formed in the same basin as part of the same volcanic arc complex throughout the Permian and Early Triassic. The lack of extensive volcaniclastic rocks and flows in the Black Rock terrane similar to those preserved in the eastern Klamath terrane suggests that the Black Rock terrane was farther from the volcanic source than the eastern Klamath terrane.

In contrast to this, the structural, lithologic, and biogeographic dissimilarity between the Black Rock terrane and nearby rocks of the Golconda allochthon indicate that they were tectonically isolated from one another during this same period. The tectonic models for the history of these rocks imply proximity and involvement of an arc terrane in the tectonic events that emplaced the Golconda basin onto the continental margin. Although there is evidence for volcanic arc detritus in parts of the Golconda allochthon at different times throughout its history, there is no specific evidence in northwestern Nevada that ties the eastern Klamath terrane to the Golconda allochthon during the Permian, or that suggests that an arc was involved in the Permian-Triassic deformation of the allochthon. This lack of supporting evidence to link these terranes, together with evidence demonstrating post-Triassic displacement of these terranes relative to one another, suggests that the present juxtaposition of allochthonous terranes in this area is the result of multiple independent tectonic events and cannot readily be explained by a single Permian-Triassic event.

ACKNOWLEDGMENTS

This research was supported by funds from the Department of Geology and Geophysics at the University of California at Berkeley, Geological Society of America Research Grant 3802-87, Sigma Xi Grants-in-Aid of Research, and the Chancellor's Patent Fund of the University of California, Berkeley. Discussions with David L. Jones and Norm Silberling were helpful in providing focus for this manuscript. Constructive reviews by Cathy Busby-Spera, David Jones, Meghan Miller, Norm Silberling, Walt Snyder, and Bruce Wardlaw greatly improved the original manuscript. Thanks to Rex Hanger, Anita Harris, Norm Silberling, Cal Stevens, Tim Tozer, and Bruce Wardlaw for fossil identification, and to Paul Renne for K-Ar analysis. Field assistance from Victoria E. Langenheim and Daniel J. Stein is gratefully acknowledged. Thanks also to James Kennett and Cathy Busby-Spera for the generous use of their laboratory equipment at the University of California, Santa Barbara.

APPENDIX 1. FAUNAS COLLECTED AT QUINN RIVER

Location	Fauna
1	Ceratoikiscidae similar to Goodbody 1982, Plate 2, Figs. 5-8
	Ceratoikiscum sp. (Nazarov and Ormiston 1983, Plate 2, Fig. 4)
	Ceratoikiscum bujugum Foreman 1963
	Ceratoikiscum spinosiarcuatum Foreman 1963
	Entactinia? sychnacanthina Foreman 1963
	Entactiniid forms similar to Goodbody 1982, Plate 2, Figs. 9-12
	Entactinosphaera? erebenna Foreman 1963
	Entactinosphaera riedeli Foreman 1963
	Haplentactinia rhinophyusa Foreman 1963
	Palaeoscenidium cladophorum Deflandre 1953
3	*Heterocaninia* sp.
	Lytvophyllum sp.
4	*Peniculauris* cf. *ivesi*
5	*Parafusulina klamathensis* Skinner and Wilde 1965
6	*Follicuculus scholasticus* Ormiston and Babcock 1979
	Follicuculus ventricosis Ormiston and Babcock 1979
	Foreman cf. *F. helena triangula* DeWever and Caridroit 1984
	Hegleria mammifera Nazarov and Ormiston 1985
	Ishigaum obesum DeWever and Caridroit 1984
	Ishigaum cf. *I. trifustis* DeWever and Caridroit 1984
	Latentifistula cf. L. similicutus Caridroit and DeWever 1986
	Latentifistula sp. cf. *L. texana* Nazarov and Ormiston 1985
	Nabispecha leonardia Cornell and Simpson 1986
	Pseudotormentus cf. *P. kamigoriensis* DeWever and Caridroit 1984
	Pseudoalbaillella lanceolata Ishiga and Imoto 1982
	Scharfenbergia sp.
6	*Neogondolella* sp.
	Neogondolella phosphoriensis Youngquist, Hawley, and Miller
	Neogondolella aff. *N. prolongata* Wardlaw and Collinson
7	acrochordiceratid
8	*Neogondolella* cf. *N. constricta* Mosher and Clark
9	*Heterocaninia langenheimi* Wilson
10	*Protrachyceras* sp.
F1	*Pseudofusulina* sp.
	Pseudofusulinella sp.
	Pseudofusulinella sp. aff. *P. fragilis* Skinner and Wilde, 1965
	Pseudofusulinella aff. *P. spississepta*
	Schwagerina sp. of the general group of *S. longissimoidea* (Beede) aff *Schwagerina wheeleri* Skinner and Wilde, 1965
A	*Acolosia* n. sp.
	Acosarina mesoplatys (R. E. King)
	Anidanthus sp.
	Antiquatonia sp.
	Beecheria sp.
	Bothrostegium sp.
	Crurithyris sp.
	Derbyia sp.
	Dielasma sp.
	Liosotella sp.
	Meekella sp.
	Neophricadothyris crassibecca (Cooper and Grant)
	Paraspiriferina sp.
	Rhynchopora cf. *R. magna* (Cooper)
	Spiriferella cf. *drashei* (Toula)
	Stenocisma oregonense (Cooper)
	Yakovlevia transversa (Cooper)
F2	aff. *Schwagerina jeffordsi* Skinner and Wilde, 1966, but more advanced with cuniculi *(Parafusulina)*
B	Ctenalosia fixata Cooper and Stehli
	Stenocisma sp.
C	*Chalaroschwagerina ampla* Skinner and Wilde 1966
	Chalaroschwagerina decora, Skinner and Wilde 1966
	Chalaroschwagerina eximia Skinner and Wilde 1966
	Chalaroschwagerina formosa Skinner and Wilde 1966
	Chalaroschwagerina globularis Skinner and Wilde 1966
	Chalaroschwagerina solita Skinner and Wilde 1966
	Eoparafusulina bellula Skinner and Wilde 1966
	Eoparafusulina concisa Skinner and Wilde 1966
	Pseudofusulinella jeffordsi Skinner and Wilde 1966
	Pseudofusulinella pusilla Skinner and Wilde 1966
	Pseudofusulinella rotunda Skinner and Wilde 1965
	Schwagerina jeffordsi Skinner and Wilde 1966
	Schwagerina munda Skinner and Wilde 1966

Note: See Figure 2 for locations and Table 1 for references.

REFERENCES CITED

Brueckner, H. K., and Snyder, W. S., 1985, Structure of the Havallah sequence, Golconda allochthon, Nevada; Evidence for prolonged evolution in an accretionary prism: Geological Society of America Bulletin, v. 96, p. 1113–1130.

Burchfiel, B. C., and Davis, G. A., 1972, Structural framework and evolution of the southern part of the Cordilleran orogen, western United States: American Journal of Science, v. 272, p. 97–118.

——, 1975, Nature and controls of Cordilleran orogenesis, western United States; Extensions of an earlier synthesis: American Journal of Science, v. 275A, p. 363–396.

——, 1981, Triassic and Jurassic tectonic evolution of the Klamath Mountains-Sierra Nevada geologic terrane, *in* Ernst, W. G., ed., The geotectonic development of California; Rubey Volume 1: Englewood Cliffs, New Jersey, Prentice-Hall, p. 50–70.

Caridroit, M., and DeWever, P., 1986, Some Late Permian radiolarians from pelitic rocks of the Tatsuno Formation (Hyogo Prefecture), southwest Japan: Marine Micropaleontology, v. 11, p. 55–90.

Compton, R. R., 1960, Contact metamorphism in Santa Rosa Range, Nevada: Geological Society of America Bulletin, v. 71, p. 1383–1416.

Cornell, W. C., and Simpson, R. D., 1986, *Nabespecha leonardia*, n. gen., n. sp.; An unusual radiolarian from the Permian of West Texas: Micropaleontology, v. 32, no. 3, p. 286–288.

Deflandre, G., 1953, Radiolaires fossiles, *in* Grasse, P. P., ed., Traite de Zoologie: Paris, Masson et Cie, v. 1, no. 2, p. 389–436.

DeWever, P., and Caridroit, M., 1984, Description de quelques nouveaux Latentifistulidea (Radiolaires Polycystines) Paleozoiques du Japon: Revue de Micropaleontologie, v. 27, p. 98–106.

Foreman, H. P., 1963, Upper Devonian Radiolaria from the Huron member of the Ohio shale: Micropaleontology, v. 9, no. 3, p. 267–304.

Furutani, H., 1983, Middle Palaeozoic Palaeoscenidiidae (Radiolaria) from Mt. Yokokura, Shikoku, Japan, Part 1: Transactions of the Proceedings of the Palaeontological Society of Japan, no. 130, p. 96–116.

Goodbody, Q. H., 1982, Silurian Radiolaria from the Cape Phillips Formation, Canadian Arctic Archipelago: Proceedings, 3rd North American Paleontological Convention, v. 1, p. 211–216.

Hanger, R. A., 1989, Ecostratigraphy of Permian Strata of the Black Rock terrane near Quinn River Crossing, Nevada: Geological Society of America Abstracts with Programs, v. 21, p. 89.

Harlan, W. B., Cox, A. V., Llewellyn, P. G., Acton, C.A.G., Smith, A. G., and Walters, R., 1982, A geologic time scale: Cambridge, England, Cambridge University Press, 131 p.

Hotz, P. E., and Willden, R., 1964, Geology and mineral deposits of the Osgood Mountains Quadrangle, Humboldt County, Nevada: U.S. Geological Survey Professional Paper 431, 128 p.

Ishiga, H., Kito, T., and Imoto, N., 1982, Middle Permian radiolarian assemblages in the Tamba District and an adjacent area, southwest Japan: Earth Science; Journal of the Association for the Geological Collaboration in Japan, v. 36, no. 5, p. 272–281.

Jones, A. E., 1988, The Poverty Peak subterrane of the Golconda Allochthon in the northern Hot Springs Range, Humboldt County, Nevada: Geological Society of America Abstracts with Programs, v. 20, p. 171.

Jones, A. E., and Jones, D. L., 1990, Paleogeographic significance of the subterranes of the Golconda allochthon; Geology and ore deposits of the Great Basin: Geological Society of Nevada Great Basin Symposium, Abstracts with Programs, p. 46.

Jones, A. E., Langenheim, V. E., Jones, D. L., and Renne, P., 1988, New age data from Quinn River Crossing, Humboldt County Nevada; Implications for Paleozoic and Mesozoic tectonics: Geological Society of America Abstracts with Programs, v. 20, p. A109–A110.

Ketner, K. B., and Wardlaw, B. R., 1981, Permian and Triassic rocks near Quinn River Crossing, Humboldt County, Nevada: Geology, v. 10, p. 298–303.

Lupe, R., and Silberling, N. J., 1985, Genetic relationship between Lower Mesozoic continental strata of the Colorado Plateau and marine strata of the western Great Basin; Significance for accretionary history of Cordilleran lithotectonic terranes, *in* Howell, D. G., ed., Tectonostratigraphic terranes of the Circum-Pacific region: Circum-Pacific Council for Energy and Mineral Resources, Earth Science Series, no. 1, p. 263–271.

Maher, K. A., and Saleeby, J., 1988, Age constraints on the geologic evolution of the Jackson Mountains, NW Nevada: Geological Society of America Abstracts with Programs, v. 20, p. 177.

McCollum, L. B., and McCollum, M. B., 1989, The Antler orogeny; A transpressional orogen within a transcurrent fault system: Geological Society of America Abstracts with Programs, v. 21, p. 114.

Miller, E. L., and 5 others, 1982, Structural fabric of the Paleozoic Golconda allochthon, Antler Peak Quadrangle, Nevada; Progressive deformation of an oceanic sedimentary assemblage: Journal of Geophysical Research, v. 87, no. B5, p. 3795–3804.

Miller, E. L., Holdsworth, B. K., Whiteford, W. B., and Rodgers, D., 1984, Stratigraphy and structure of the Schoonover sequence, northeastern Nevada; Implications for Paleozoic plate-margin tectonics: Geological Society of America Bulletin, v. 95, p. 1063–1076.

Miller, M. M., 1989, Intra-arc sedimentation and tectonism; Late Paleozoic evolution of the eastern Klamath terrane, California: Geological Society of America Bulletin, v. 101, p. 170–187.

Miller, M. M., and Wright, J. E., 1987, Paleogeographic implications of Permian Tethyan corals from the Klamath Mountains, California: Geology, v. 15, p. 266–269.

Nazarov, B. B., 1975, Lower and Middle Paleozoic Radiolarians of Kazakhstan: Akademiya Nauk SSSR, Ordena Trudovogo Krasnogo znameni Geologicheski Institut, Trudy, v. 275, 202 p.

Nazarow, B. B., and Ormiston, A. R., 1985, Radiolaria from the Late Paleozoic of the southern Urals, USSR, and West Texas, USA: Micropaleontology, v. 31, no. 1, p. 1–54.

Newton, C. R., 1988, Meaning of "Tethyan" fossils in the American cordillera: Geological Society of America Abstracts with Programs, v. 20, p. A309.

Noble, P., and Renne, P., 1990, Paleoenvironmental and biostratigraphic significance of siliceous microfossils from Permo-Triassic rocks of the Redding section, eastern Klamath Mountains, California: Marine Micropaleontology, v. 15, p. 379–391.

Ormiston A., and Babcock, L., 1979, *Follicucullus,* New Radiolarian genus from the Guadalupian (Permian) Lamar limestone of the Delaware Basin: Journal of Paleontology, v. 53, no. 2, p. 328–334.

Russell, B. J., 1984, Mesozoic geology of the Jackson Mountains, northwestern Nevada: Geological Society of America Bulletin, v. 95, p. 313–323.

Silberling, N. J., and Jones, D. L., 1982, Tectonic significance of Permian-Triassic strata in northwestern Nevada and northern California: Geological Society of America Abstracts with Programs, v. 14, p. 234.

Silberling, N. J., Jones, D. L., Blake, M. C., Jr., and Howell, D. G., 1987, Lithotectonic terrane map of the western conterminous United States: U.S. Geological Survey Miscellaneous Field Studies Map MF-1874-C, scale 1:2,500,000.

Skinner, J. W., and Wilde, G. L., 1965, Permian biostratigraphy and fusulinid faunas of the Shasta Lake area, northern California: Lawrence, University of Kansas Paleontological Contributions 392, Protozoa, article 6, 98 p.

——, 1966, Permian fusulinids from northwestern Nevada: Lawrence, University of Kansas Paleontological Contributions Paper 4, part 1, p. 1–10.

Sosson, M., 1984, Contribution à L'Etude Géologique des Cordillères Nord-Américaines. Une coupe de part et d'autre du Chevauchement de Golconda dans le Nord-ouest du Nevada, Etas-Unis: Thèse présentée pour l'obtention du Diplome de Docteur de 3 ème cycle à L'Université Pierre et Marie Curie, Paris 6, 202 p.

Speed, R. C., 1977, Island arc and other paleogeographic terranes of late Paleozoic age in the western Great Basin, *in* Stewart, J. H., Stevens, C. H., and Fritsche, A. E., eds., Paleozoic paleogeography of the western United States: Pacific Section, Society of Economic Paleontologists and Mineralogists Pacific Coast Paleogeography Symposium 1, p. 315–324.

——, 1978, Paleogeographic and plate tectonic evolution of the early Mesozoic marine province of the western Great Basin, *in* Howell, D. G., and McDougall, K. A., eds., Mesozoic paleogeography of the western United States: Pacific Section, Society of Economic Paleontologists and Mineralogists Pacific Coast Paleogeography Symposium 2, p. 253–270.

——, 1979, Collided Paleozoic microplate in the western United States: Journal of Geology, v. 87, p. 279–292.

——, 1983, Evolution of the Sialic Margin in the central western United States, *in* Watkins, J. S., and Drake, C. L., eds., Continental margin geology: American Association of Petroleum Geologists Memoir 34, p. 457–468.

Stevens, C. H., and Rycerski, B. A., 1988, The Eastern Klamath Mountains Permian coral province, western North America: Geological Society of America, Abstracts with Programs, v. 20, p. 235.

Stewart, J. H., 1980, Geology of Nevada: Nevada Bureau of Mines and Geology Special Publication 4, 136 p.

Tomlinson, A. J., and Wright, J. E., 1986, The Klamath-Sierran arc; A possible source region for Permian volcaniclastic and conglomeratic units within the Golconda allochthon: Geological Society of America Abstracts with Programs, v. 18, p. 193.

Watkins, R., and Wilson, E. C., 1989, Paleoecologic and biogeographic significance of the biostromal organism *Palaeoaplysina* in the Lower Permian McCloud Limestone, Eastern Klamath Mountains, California: Palaios, v. 4, p. 181–192.

Willden, R., 1961, Major westward thrusting of Post-Middle Triassic age in northwestern Nevada, *in* Short papers in the geologic and hydrologic sciences, Articles 147-292, Geological Survey Research: U.S. Geological Survey Professional Paper 424-C, p. C116-C120.

——, 1963, General geology of the Jackson Mountains, Humboldt County, Nevada: U.S. Geological Survey Bulletin 1141-D, 65 p.

——, 1964, Geology and mineral deposits of Humboldt County, Nevada: Nevada Bureau of Mines Bulletin 59, 154 p.

Yancey, T. E., Stevens, C. H., and Hanger, R. A., 1988, Geographic placement of the Eastern Klamath terrane in the Pacific Ocean basin during the Early Permian: Geological Society of America Abstracts with Programs, v. 20, p. A308.

MANUSCRIPT ACCEPTED BY THE SOCIETY MARCH 27, 1990

Printed in U.S.A.

Geological Society of America
Special Paper 255
1990

Tectono-magmatic evolution of Paleozoic and early Mesozoic rocks in the eastern Klamath Mountains, California, and the Blue Mountains, eastern Oregon–western Idaho

J. Charvet, H. Lapierre, O. Rouer
Laboratoire de Géologie structurale, URA/CNRS No. 1366, Université d' Orléans, BP 6759, 45067 Orléans Cedex 02, France
C. Coulon
Laboratoire de Pétrologie Magmatique, URA/CNRS No. 1277, Faculté des Sciences de St Jérome, BP 441, 13397 Marseille Cedex 13, France
C. Campos
Laboratoire de Géologie structurale, URA/CNRS No. 1366, Université d' Orléans, BP 6759, 45067 Orléans Cedex 02, France
P. Martin
Laboratoire de Pétrologie, Université de Nancy 1, BP 239, 54506 Vandoeuvre/Nancy Cedex, France
C. Lecuyer
Laboratoire de Géochimie, Université de Rennes, CAESS/CNRS, 35042 Rennes, France

ABSTRACT

Rocks in the eastern Klamath Mountains were affected by polyphase tectonism. The older Callahan tectonic event, of Late Ordovician–Early Silurian age, produced high-pressure and epizonal metamorphism and is expressed in the Yreka-Callahan nappes by intrafolial folds. The main tectonic event occurred during Late Devonian and Early Mississippian time and is interpreted here to be coeval with the Antler orogeny, defined in the Roberts Mountain allochthon to the east in Nevada. It resulted in the amalgamation of lower Paleozoic Trinity ophiolite and the pre-Devonian to Early Devonian intra-oceanic island arc with flysch-type sediments deposited in a continent margin (Yreka-Callahan assemblage), and a metamorphic sequence of amphibolite, marble, and quartz-mica-schist of the Central Metamorphic Belt.

The magmatic evolution of Paleozoic and lower Mesozoic rocks in the eastern Klamath Mountains is characterized by the predominance of convergent plate boundary magmatism. Nevertheless, lower Paleozoic pillow basalts associated with coralline limestone occur as olistoliths in the Yreka-Callahan assemblage. They display either T mid-ocean ridge basalt (MORB) affinities with flat rare earth element (REE) patterns or mild alkalic affinities marked by higher contents in TiO_2 and HFS elements and LREE-enriched patterns.

The Ashgillian calc-alkaline differentiated suite, characterized by LREE-enriched patterns and low ϵ_{Nd} (T) + 3.5, was erupted through continental crust in an Andean margin setting. The pre-Devonian to Early Devonian volcanic rocks that show tholeiitic affinities with very depleted to somewhat enriched LREE patterns and high ϵ_{Nd} (T) values represent the remnants of an intra-oceanic immature island arc. After the Late Devonian(?)–Early Mississippian tectonic event, island-arc magmatism resumed with the eruption of depleted low-K tholeiitic basalt and andesite, characterized by depleted

Charvet, J., Lapierre, H., Rouer, O., Coulon, C., Campos, C., Martin, P., and Lecuyer, C., 1990, Tectono-magmatic evolution of Paleozoic and early Mesozoic rocks in the eastern Klamath Mountains, California, and the Blue Mountains, eastern Oregon–western Idaho, *in* Harwood, D. S., and Miller, M. M., eds., Paleozoic and early Mesozoic paleogeographic relations; Sierra Nevada, Klamath Mountains, and related terranes: Boulder, Colorado, Geological Society of America Special Paper 255.

LREE patterns during Late Devonian(?) or Carboniferous-Permian(?) time (Gregg Ranch and Copper Peak sections).

A major pulse in volcanism occurred during Middle to Late Permian and is now represented by the voluminous sequence of volcanogenic debris and flows of the Nosoni, Dekkas, and Bully Hill Formations. This volcanism, formed of predominantly low-K tholeiitic basalt and andesite, characterized by flat REE patterns and high ϵ_{Nd} (T) + 4 to 8, with minor calc-alkaline, alkaline (enriched LREE patterns), and boninitic (Mg-Cr-rich) basalts, was erupted through a thickened crust. At the end of the Permian island-arc volcanism, uplift and faulting occurred in the arc. The magmatic affinities of the middle to late Paleozoic volcanic rocks in the eastern Klamath Mountains suggest an evolution through time from an immature to a mature island-arc setting.

The Permian-Triassic history of the Blue Mountains (Seven Devils Group; Huntington Formation) presents many similarities with that of the eastern Klamath terrane. In both areas, evidence for Permian volcanism is similar with flat REE patterns and high ϵ_{Nd} (T) + 6.7 to 8.4. Triassic volcanic rocks display low-K tholeiitic affinities (flat REE patterns) with minor calc-alkaline flows in the uppermost levels of the volcanic pile. Apparently, compressive structures related to the Sonoma orogeny are absent in both areas. Thus the eastern Klamath and Blue Mountains could have formed part of the same arc in Permian-Triassic time.

INTRODUCTION

The western Cordillera of North America is interpreted by most geologists (e.g., Hellwig, 1974; Monger, 1977; Davis and others, 1978) as a collage of oceanic and continental fragments that were juxtaposed against each other and to North America mostly in Mesozoic time. According to Roure and Sosson (1986), the building of the Cordillera resulted in the accretion of the volcanogenic American-Mexican block to North America in Late Jurassic. The American-Mexican block was formed during Late Triassic–Early Jurassic by the collision of various Paleozoic to Early Triassic exotic island arcs. These island-arc terranes are presently fault-bounded, but geologic and paleomagnetic data indicate ties between them and cratonal North America prior to their disruption by strike-slip faults. These inferred ties are based on paleontologic provinciality (Miller, 1987), occurrence and distribution of orogenic disturbances (Roberts, 1964; Davis, 1969; Burchfiel and Davis, 1972; Silberling, 1973; Davis, 1977; Ketner, 1977; Wright, 1982; Cotkin and Armstrong, 1987; Schweickert and Lahren, 1987; Charvet and others, 1989), sedimentation type (Watkins, 1973), petrology and geochemistry of the island-arc magmas (Churkin, 1974; Rogers and others, 1974; Brooks and Coles, 1980; Lapierre and others, 1987; Vallier, 1989; Rouer and Lapierre, 1989), and metamorphic histories (Schweickert and others, 1980; Cashman, 1980; Cotkin, 1986).

Unfortunately, paleomagnetic data are not consistent enough for periods prior to the Cretaceous because of the destruction of primary magnetization in older rocks by large Mesozoic plutonic intrusions (Irwin, 1985; Mankinen and others, 1989). The eastern Klamath Mountains, northern Sierra terrane, and Blue Mountains have in common island-arc series, Permian and Upper Triassic limestones, and Triassic blueschists, but they display differences in the distribution and age of tectonic phases, and the type and timing of the island-arc activity. This study provides a tectonic and magmatic history of Paleozoic and lower Mesozoic rocks in the eastern Klamath Mountains and a comparison with the pre-Cenozoic formations of the Blue Mountains.

All analyses of major, trace, and rare earth elements reported here were carried out by inductively coupled plasma atomic emissions spectroscopy at the Centre de Recherches Pétrographiques et Géochimiques (CRPG) (Nancy), with analytical errors of 0.5 ppm for contents <10 ppm and 5 percent for contents >10 ppm, and by x-ray fluorescence with analytical accuracy of 1 to 5 percent. Sr and Nd contents were analyzed at the CRPG, Nancy, by isotopic dilution using techniques described by Alibert and others (1983) and Michard and others (1985).

GEOLOGIC SETTING AND LITHOSTRATIGRAPHIC SUCCESSIONS IN THE EASTERN KLAMATH MOUNTAINS

The eastern Klamath Mountains (Figs. 1, 2; Tables 1–4) are composed of four main geologic units: (1) in the north, the lower Paleozoic Yreka-Callahan assemblage; (2) in the center, the Trinity complex; (3) in the south, the pre-Devonian to Lower Devonian island arc and its sedimentary cover; and (4) finally to the east, the Upper Devonian(?)–lower Carboniferous to Lower Jurassic sedimentary and volcanic rocks.

The lower Paleozoic Yreka-Callahan assemblage (Figs. 2, 3) consists of a stack of nappes thrust upon the northwestern margin of the Trinity ophiolitic suite (Potter and others, 1977; Hotz, 1978; Lindsley-Griffin and Griffin, 1983). The lowermost unit is the Gazelle Formation of Early Devonian age (Emsian) (Boucot and others, 1973; Potter and others, 1977). It includes an olistostromal horizon (the Gregg Ranch Melange of Lindsley-Griffin and Griffin, 1983), which consists of reworked calc-

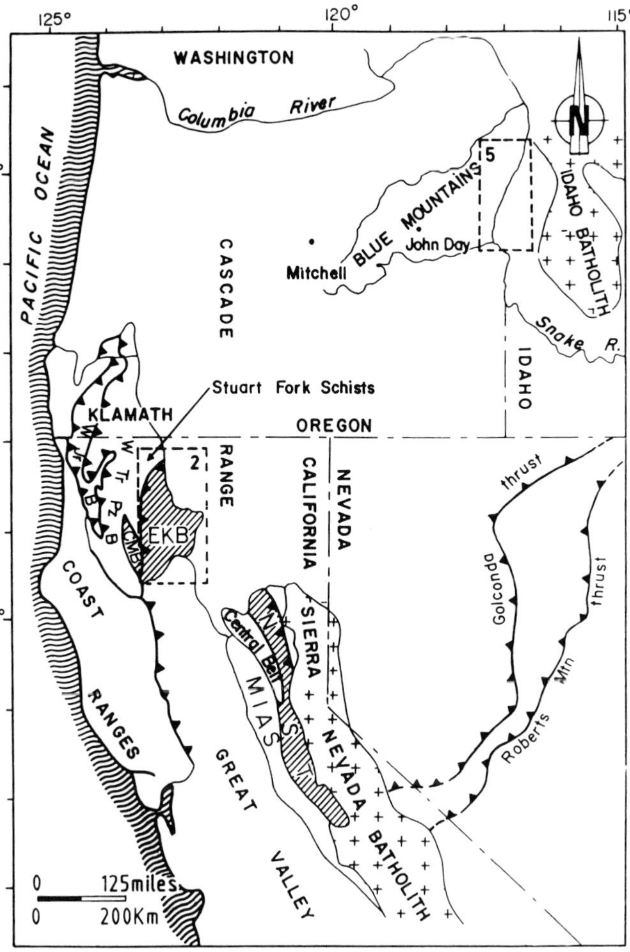

Figure 1. Schematic map of part of the western Cordillera showing the location of the studied areas; modified after Schweickert and others (1980); Irwin, (1981); Roure, (1984); Day and others, (1985); Stewart and others, (1986). W Jr B = Western Jurassic Belt; W Tr Pz B = Western Triassic and Paleozoic Belt; CMB = Central Metamorphic Belt; EKB = Eastern Klamath Belt; MIAS = Mesozoic island-arc sequence; NST = Northern Sierra Terrane. Hatched areas show the main belts comprising Lower to Middle Paleozoic rocks.

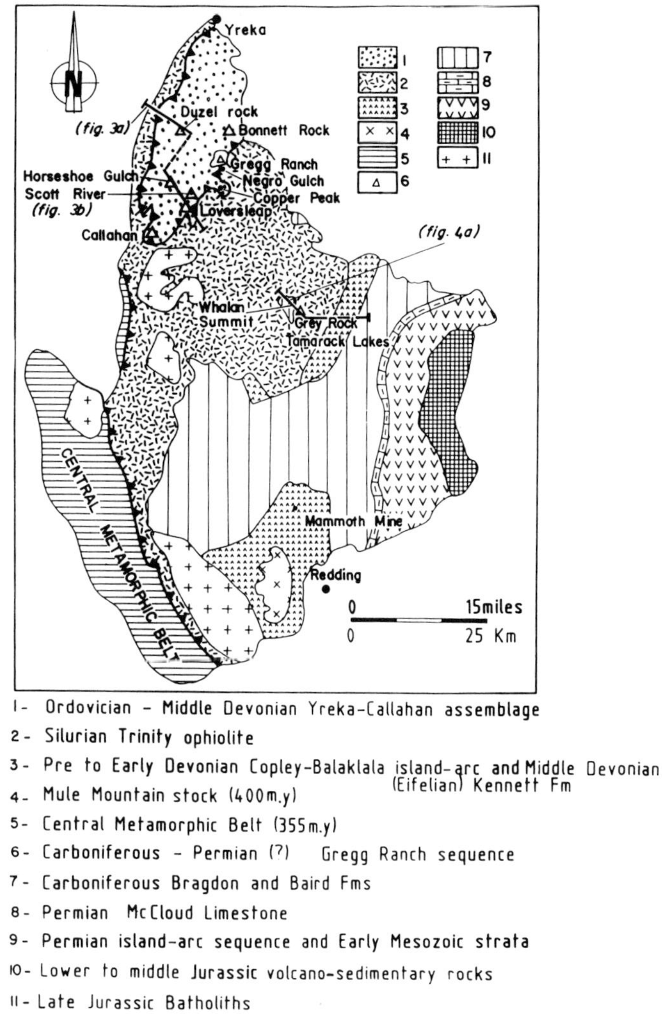

Figure 2. Simplified geologic map of the eastern Klamath Mountains; after Irwin, (1981).

alkaline andesite and dacite (Brouxel and others, 1987; Brouxel and others, 1989b) associated with Ashgillian coralline limestone (Lovers Leap section of Potter and others, 1977; Potter and others, this volume; Figs. 2, 3B). The Silurian(?) Moffett Creek Formation is a nappe (Potter and others, 1977; Hotz, 1978) that structurally overlies the Gazelle Formation. The Moffett Creek Formation consists of pebbly mudstone, sandstone, and minor slate that include exotic blocks (Skookum Gulch and Horseshoe Gulch sections of Potter and others, 1977) of (1) blueschists initially formed in Late Ordovician–Early Silurian (Cotkin, 1987) but reset in the Devonian (Cotkin and Armstrong, 1987), (2) volcanic rocks (Figs. 2, 3C, D; discussed in a later section), and (3) Cambrian quartz-diorite and tonalite (Wallin, 1986; Wallin and others, 1988). The uppermost sheet (Duzel Phyllite–Antelope Quartzite unit) contains a greenschist unit dated at 440 Ma on metamorphic amphibole and mica (Cashman, 1980) in an assemblage of argillite and quartzose sandstone (Potter and others, 1977).

The Trinity complex consists of an ultramafic metamorphic sheet composed mainly of lherzolite; it is referred to as the Trinity peridotite, ultramafic-mafic cumulate rocks, isotropic gabbros, plagiogranites, sheeted dikes, and pillow-basalts. The association of metamorphic harzburgite, plutonic rocks, diabase dikes, and pillow basalts (Brouxel and others, 1988a) are referred to here as the Trinity ophiolitic suite. The plutonic rocks intrude the ultramafic sheet, whereas the pillow basalts and sheeted dikes are thrust over the ultramafic sheet. The Trinity ophiolitic suite was defined by Lindsley-Griffin (1977) near Lovers Leap, but recently Brouxel and others (1989a) have documented that most of the dikes, considered as belonging to the ophiolite, were of Jurassic age and displayed calc-alkaline affinities.

TABLE 1. MAJOR, TRACE, AND RARE EARTH ELEMENTS OF THE EASTERN KLAMATH LOWER PALEOZOIC LAVAS

	Seamount							Tholeiite	
Sample No.	H12	BO13	BO15	DR3A	DR4	DR5B	DR7	H15	H17
SiO_2	49.84	49.47	46.40	47.05	42.78	45.36	45.32	80.4	77.27
TiO_2	1.44	2.16	2.14	2.49	3.08	2.81	3.17	0.23	0.11
Al_2O_3	16.05	13.88	13.34	14.21	13.09	12.79	13.82	9.70	11.10
Fe_2O_3	12.46	11.47	13.60	12.67	13.42	13.49	16.67	2.00	2.94
MgO	5.78	5.86	5.54	4.23	5.35	4.4	5.96	0.20	1.03
MnO	0.20	0.2	0.21	0.18	0.19	0.2	0.19	0.04	0.05
CaO	4.24	8.09	10.38	10.66	10.73	10.24	5.78	0.13	0.07
Na_2O	5.45	4.89	4.23	3.42	3.94	4.00	2.71	5.65	5.10
K_2O	0.17	0.56	0.14	0.57	0.53	0.54	0.53	0.09	0.35
LOI	4.20	2.86	3.84	4.41	6.66	5.45	5.57	0.96	1.57
Total	99.83	99.44	99.82	99.89	99.77	99.28	99.72	99.40	99.59
Cr	149	258	129	157	93	64	119	10	12
Ni	95	112	77	80	86	81	100	16	14
V	247	320	338	288	330	318	375	10	0
Nb	4	9	8	19	26	25	20	3	4
Zr	103	151	137	164	221	218	187	60	88
Y	25	37	38	32	36	36	40	19	40
La	7.58	7.43	5.25	17.15	19.46	19.16	18.69	3.08	7.27
Ca	22.80	22.36	20.53	45.15	48.08	48.69	48.32	8.12	20.54
Nd	13.63	15.79	14.32	26.42	27.78	28.42	25.55	5.29	11.86
Sm	4.23	5.24	5.22	7.39	7.41	7.71	7.62	1.89	4.53
Eu	1.29	1.64	1.76	2.49	2.54	2.53	2.52	0.53	0.94
Gd	4.31	5.49	5.90	7.12	6.75	6.95	7.25	2.08	4.93
Dy	4.92	6.18	7.02	6.04	6.15	6.42	7.44	2.91	6.61
Er	2.82	3.47	4.04	2.79	2.92	3.10	3.87	2.13	4.42
Yb	3.01	3.75	4.39	2.47	2.80	2.87	4.08	2.88	5.47
Lu	0.43	0.51	0.61	0.43	0.41	0.41	0.54	0.14	0.85
(La/Yb)N	1.69	1.45	0.80	4.68	4.69	4.50	3.09	0.72	0.89

The age of the Trinity ophiolitic suite, which is considered to be a remnant of a back-arc basin (Quick, 1981; Brouxel and Lapierre, 1988) formed at a slow-spreading ridge (Boudier and others, 1989), is still very controversial. It has been considered to be Ordovician on the basis of an Sm/Nd isochron (472 ± 32 Ma) on a metamorphic clinopyroxene in a plagioclase lherzolite (Jacobsen and others, 1984); a U/Pb age on zircons (469 ± 21 to 475 ± 10 Ma) on plagiogranites structurally overlying the northwestern margin of the Trinity ophiolitic suite (Wallin and others, 1988); and a U/Pb age on zircon (440 Ma) on a tonalite boulder (Mattinson and Hopson, 1972) considered to be derived from the eroded Trinity ophiolitic suite (Lindsley-Griffin and Griffin, 1983) (discussed below). Ages straddling the Ordovician–Silurian boundary were obtained by an Sm/Nd isochron (435 ± 21 Ma) on a microgabbro dike that intrudes the metamorphic peridotitic sheet (Jacobsen and others, 1984) and by a $^{143}Nd/^{144}Nd$ isochron (452 ± 40 Ma) on basalts, sheeted dikes, and gabbros (Brouxel and Lapierre, 1988). Finally, a Silurian age is given by a U/Pb age on zircons (412 ± 10 to 419 ± 15 Ma) on pegmatite trondjhemite dikes that cut gabbroic rocks (Wallin and others, 1988).

The ϵ_{Nd} values of the gabbros, dikes, and basalts belonging to the Trinity ophiolitic suite, calculated at T = 450 Ma, range between 6.6 and 7.5, suggesting that these rocks derived from a similar depleted mantle source (Brouxel and Lapierre, 1988). These values are in the range of those obtained by Jacobsen and others (1984) on two microgabbros (ϵ_{Nd} = 7.3) intruding the metamorphic peridotitic sheet. These ϵ_{Nd} values are, however, very different from the one obtained by Jacobsen and others (1984) on the metamorphic plagioclase lherzolite. Because isotopic data show that the metamorphic plagioclase lherzolite belonging to the ultramafic sheet is not cogenetic with the gabbros, dikes, and basalts of the Trinity ophiolitic suite, we assume that the age of the magmatic sequence of the Trinity ophiolitic suite lies between 452 ± 40 and 412 ± 10 Ma (Late Ordovician to Late Silurian).

The pre-Devonian to Lower Devonian intra-oceanic island-arc sequence of the Copley Greenstone–Balakala Rhyolite

TABLE 2. MAJOR, TRACE, AND RARE EARTH ELEMENTS OF THE EASTERN KLAMATH LATE ORDOVICIAN CALC-ALKALINE ROCKS

	Q-Diorite		Lovers Leap		Negro Gulch	
Sample No.	K20	K21	Tr 454	Tr 456	NG 13	NG 3
SiO_2	63.50	66.57	51.69	57.85	51.04	66.18
TiO_2	0.67	0.7	0.81	0.74	0.73	0.26
Al_2O_3	15.04	14.73	17.46	19.27	17.25	17.02
Fe_2O_3	5.70	4.4	9.43	7.17	8.79	3.11
MgO	2.3	1.79	5.75	3.11	7.51	1.63
MnO	0.1	0.09	0.12	0.18	0.18	0.05
CaO	2.92	1.87	4.34	1.18	3.15	2.1
Na_2O	4.68	5.43	4.69	5.54	0.97	5.61
K_2O	1.15	1.2	0.8	1.03	2.33	1.57
LOI	2.55	2.38	4.85	3.74	7.91	2.25
Total	98.78	99.44	99.94	99.81	99.86	99.78
Cr	28	10	10	14	144	10
Ni	14	12	102	49	46	16
V	146	140	361	122	173	100
Nb			3	3	4	3
Zr			74	114	99	102
Y	30	23	20	16	13	8
La	12.91	19.74	7.35	6.65		
Ce	31.27	45.93		15.82		
Nd	14.73	22.35	10.01	9.06		
Sm	3.95	5.01	2.64	2.38		
Eu	1.07	1.17	0.87	0.76		
Gd	3.75	3.97	3.00	2.7		
Dy	4.16	3.42	3.17	3.01		
Er	2.54	1.9	1.98	1.88		
Yb	3.14	2.25	2.17	1.98		
Lu	0.43	0.31				
(La/Yb)N	2.77	5.92	13.14	2.26		

(Kinkel and others, 1956; Barker and others, 1979; Lapierre and others, 1985a,b; Brouxel and others, 1987, 1988) is conformably capped by a Middle Devonian (Eifelian) sedimentary sequence of the Kennett Formation (Boucot, 1973; Boucot and others, 1973, 1974; Flory and Watkins, 1983) that represents deep-sea sedimentation around and over a foundering volcanic arc.

The Upper Devonian(?)–Carboniferous Bragdon Formation (Watkins, 1985; Miller, 1989) unconformably overlies both the Trinity ophiolite (Charvet and others, 1989) (Fig. 4) and the Kennett Formation (Kinkel and others, 1956). Volcanogenic shelf, and volcaniclastic and turbiditic rocks containing continental detritus (Miller, 1989) were deposited continuously during all of the Carboniferous and are represented by the Baird Formation (Watkins, 1985; Miller, 1989). The upper part of the Baird locally interfingers with the overlying Lower Permian McCloud Limestone.

An undated sequence (D?v: Devonian? volcanic rocks of Lindsley-Griffin and Griffin, 1983), which rests unconformably on the northwest margin of the Trinity ophiolitic suite, includes pillow basalt and pillow breccia (Gregg Ranch section), as well as basaltic and andesitic massive flows interbedded with breccias (Copper Peak sections; Figs. 1, 2) (Charvet and others, 1989). This sequence may represent an Upper Devonian (post Gazelle Formation) and/or Carboniferous-Permian cover deposited unconformably on the Trinity ophiolitic suite.

Volcanism resumed in the Early Permian with the eruption of the Dekkas, Nosoni, and Bully Hill island-arc volcanic rocks (Fig. 6) (Albers and Robertson, 1961). In the overlying Pit Formation, Lower Triassic rocks are missing (Silberling and Jones, 1982). The Pit Formation includes significant proportions of both pyroclastic and fine-grained clastic sedimentary rocks, and rests disconformably on the older rocks (Albers and Robertson, 1961). An olistostromal horizon with huge blocks of Bully Hill Rhyolite occurs at the base of the Pit Formation (Eastoe and others, 1987), whereas crystal-rich tuffs, with plagioclase and quartz, are present in the uppermost levels of the Pit Formation (Miller, 1989). The Carnian Hosselkus Limestone forms a discontinuous, broken horizon (Eastoe and Nelson, 1988). The Modin

TABLE 3. MAJOR, TRACE, AND RARE EARTH ELEMENTS OF THE LAVAS OF THE GREGG RANCH AND COPPER PEAK SECTIONS (DEVONIAN?) IN THE EASTERN KLAMATH MOUNTAINS

Sample No.	GR10	GR12	GR13	GR16	GR19	GR21	C1	C4	C5
SiO_2	57.51	55.08	56.90	51.85	49.02	53.73	54.38	65.07	48.73
TiO_2	0.55	0.61	0.65	0.49	0.55	0.50	0.50	0.75	0.68
Al_2O_3	15.79	15.76	16.11	15.79	16.80	15.84	16.30	14.68	18.56
Fe_2O_3	6.68	7.53	6.02	6.64	7.71	6.06	6.78	5.61	8.70
MgO	5.24	5.77	5.07	6.86	7.61	6.05	7.25	2.21	7.24
MnO	0.16	0.15	0.15	0.16	0.15	0.15	0.10	0.08	0.18
CaO	5.00	5.97	4.61	9.80	10.04	7.54	4.74	4.83	8.17
Na_2O	5.72	5.79	7.16	4.34	3.94	6.12	6.59	4.00	3.34
K_2O	0.26	0.08	0.09	0.07	0.06	0.23	0.05	0.57	0.11
LOI	2.91	3.17	2.75	3.68	4.24	3.66	2.89	1.98	3.99
Total	99.82	99.91	99.51	99.68	100.12	99.88	99.58	99.78	99.70
Cr	111	123	219	276	267	214	100	10	78
Ni	62	70	69	92	83	82	58	29	67
V	210	236	235	222	267	267	206	274	262
Nb	0	1	1	1	1	1	1	1	1
Zr	28	28	34	25	29	25	26	41	35
Y	13	15	15	9	18	11	16	22	21
La	0.90	0.94	1.51	1.17	1.97	0.16	1.14	1.54	1.13
Ca	6.11	4.77	6.14	5.24	6.72	3.16			
Nd	3.25	3.02	4.28	2.48	4.33	2.48	4.69	5.54	5.12
Sm	1.54	1.55	1.85	1.09	1.80	1.14	2.23	2.77	2.65
Eu	0.53	0.55	0.73	0.45	0.62	0.33	0.75	0.78	3.13
Gd	1.76	1.83	2.21	1.22	2.09	1.40	2.50	3.09	0.82
Dy	2.29	2.34	2.75	1.55	2.92	1.91	2.42	3.24	3.13
Er	1.44	1.48	1.68	0.96	1.84	1.10	1.77	2.25	3.08
Yb	1.55	1.72	1.81	1.16	2.11	1.23	1.74	2.77	2.32
Lu	0.26	0.26	0.24	0.13	0.26	0.15	0.32	0.39	0.41
(La/Yb)N	0.39	0.36	1.05	0.68	0.63	0.08	0.44	0.37	0.32

Formation, composed of detritic deposits associated with minor tuffs and lavas, represents the end of the Triassic sedimentation (Renne and others, 1988). Miller (1989) suggested that a Late Permian or Early Triassic deformational event (Sonoman orogeny of Silberling, 1973) separated Paleozoic volcanic rocks from Triassic and Jurassic rocks in the Redding section. However, Renne and Scott (1988) reported apparent stratigraphic concordance between the Permian and Triassic rocks, which seems to preclude the existence of a Permian-Triassic unconformity corresponding to the Sonoman orogeny.

The Jurassic strata (Arvison and Potem Formations) consist of detritic deposits with few limestones and tuffs. According to Sanborn (1960) and Roure and Sosson (1986), an unconformity occurs between the uppermost Triassic Modin Formation and the Lower Jurassic Arvison Formation and the pre–Early Jurassic strata are affected by deformations that are absent in the overlying rocks. This unconformity is denied by Renne and Scott (1988). Moreover, Renne and Scott (1988) consider the main tectonic event affecting the Redding section to be of early Middle or late Early Jurassic age coincident with the age of major deformation, metamorphism and plutonism in the western Klamath (Wright and Fahan, 1988). Our recent survey in the Redding section suggests the absence of any compressive (folding) tectonic event at the Late Permian–Early Triassic boundary (C. Campos, unpublished data).

The eastern Klamath terrane is thrust over the Central Metamorphic Belt, which consists of garnet-amphibolite, garnet-muscovite quartz-schist, and marble with a metamorphic age of 370 ± 20 Ma (Cashman, 1980). Rocks in the Central Metamorphic Belt are retrograded locally to greenschist facies.

Rocks in the eastern Klamath Mountains were affected by at least three tectonic events: (1) the Late Ordovician–Early Silurian Callahan orogeny of Cotkin and Armstrong (1987), (2) Late Devonian–early Carboniferous deformation, and (3) Early to mid-Jurassic deformation. The two first tectonic disturbances are developed in a later section. The age of the younger tectonic events is still a matter of debate. The Late Triassic deformation and high-pressure metamorphism of the Stuart Fork and Fort Jones schists (Hotz and others, 1977; Mortimer, 1984), as well as the thrusting of the Central Metamorphic Belt onto the Western

TABLE 4. MAJOR, TRACE, AND RARE EARTH ELEMENTS OF THE EASTERN KLAMATH (NOSONI, DEKKAS, AND BULLY HILL FORMATIONS) AND BLUE MOUNTAINS (WINDY RIDGE AND HUNSAKER FORMATIONS) PERMIAN ISLAND-ARC LAVAS

Sample No.	Nosoni Basalts ES248	Nosoni Basalts ES245	Nosoni Dacite ES147	Nosoni Welded Tuff ES143b	Dekkas Basalts ES 223	Dekkas Basalts ES 106	Dekkas Dacite ES116	Dekkas Saddle ES359	Bully Hill Basalt ES73*	Bully Hill Rhyolite ES 74	Windy Ridge BM 75	Hunsaker BM96	Hunsaker BM92	Hunsaker
SiO_2	49.78	47.91	66.48	76.47	49.37	49.60	69.21	51.50	65.85	72.49	78.84	48.28	47.64	
TiO_2	0.45	1.1	0.62	0.37	0.60	0.88	0.63	0.39	0.32	0.32	0.38	0.93	1.67	
Al_2O_3	18.06	15.67	14.32	12.65	12.02	18.72	13.55	11.47	10.95	13.77	12.93	17.62	16.59	
Fe_2O_3	9.75	10.92	7.35	2.34	12.07	11.03	5.28	7.73	4.93	2.55	2.34	10.41	11.43	
MgO	6.35	5.70	1.73	0.06	0.22	4.93	0.93	9.31	5.38	0.66	1.03	8.38	5.79	
MnO	0.18	0.21	0.13	0.03	13.20	0.18	0.15		0.09	0.04	0.08	0.21	0.23	
C_2O	9.67	9.62	0.56	0.46	7.59	4.77	2.02	13.45	5.69	1.24	0.29	3.50	4.03	
Na_2O	1.73	3.60	5.25	6.68	2.76	1.78	4.78	1.62	3.46	5.88	4.96	5.01	3.83	
K_2O	0.14	0.89	0.31	0.06	0.51	0.55	1.08	1.38	1.00	0.67	0.80	0.50	0.90	
P_2O_5							0.04							
LOI	3.17	3.6	2.56	0.39	1.66	8.94	3.54	2.81	2.92	1.38	1.19	4.89	8.41	
Total	99.28	99.36	99.31	99.51	100	99.28	99.41	99.67	100.01	99.00	98.87	99.81	99.92	
Cr	43	87	20	10	1249	26	11	883	398	10	13	151	45	
Ni	41	46	12	16	385	49	10	191	107	47	15	37	32	
V	255	297		10	184	229	41	33		72	66	298	313	
Zr	21	88			37	37			3		172	29	117	
Nb	2	4			3	3			14		5	1	3	
Y	10	22	29.60	54.17	9.60	14	44.75	11.96	22.64	80.09	41	14	28	
La	2.13	12.10	4.83	11.48	2.87	3.29	8.94	78.20	6.84	10.36	8.97	2.44	10.85	
Ca	4.89	25	16.53	33.16	6.31	8.39	25.88	144	13.60	19.86	24.79	9.91	31.73	
Nd	3.43	15.4	11.79	24.47	4.35	4.45	16.32	58	8.44	18.23	15.84	4.49	19.55	
Sm	1.02	3.75	3.68	7.08	1.06	1.90	4.83	7.47	2.12	5.26	4.87	1.78	5.81	
Eu	0.436	1.25	1.11	1.61	0.39	0.73	1.52	2.27	0.551	1.43	0.99	0.71	1.53	
Cd	1.31	4.02	3.80	6.79	1.20	2.16	5.25	4.40	2.59	6.60	4.57	1.92	5.60	
Dy	1.48	3.95	4.47	8.34	1.29	2.71	5.79	2.45	2.91	8.59	5.83	1.95	5.53	
Er	0.965	2.30	2.73	4.96	0.89	1.68	3.41	1.21	1.96	6.19	3.79	1.18	3.07	
Yb	0.948	2.23	3.14	5.27	0.92	1.94	2.66	1.03	2.05	4.77	4.85	1.33	3.53	
Lv			0.49	0.83		0.32	0.37			0.73	0.75	0.22	0.53	
(La/Yb)N	1.52	3.66	0.99	1.47	2.03	0.79	1.65	51.25	2.25	1.47	8.07	1.23	2.07	
ENd(T)	6.60	6.80			8	4.90		3.40	4.20	5.30		6.70		

*Q- xenocrysts.

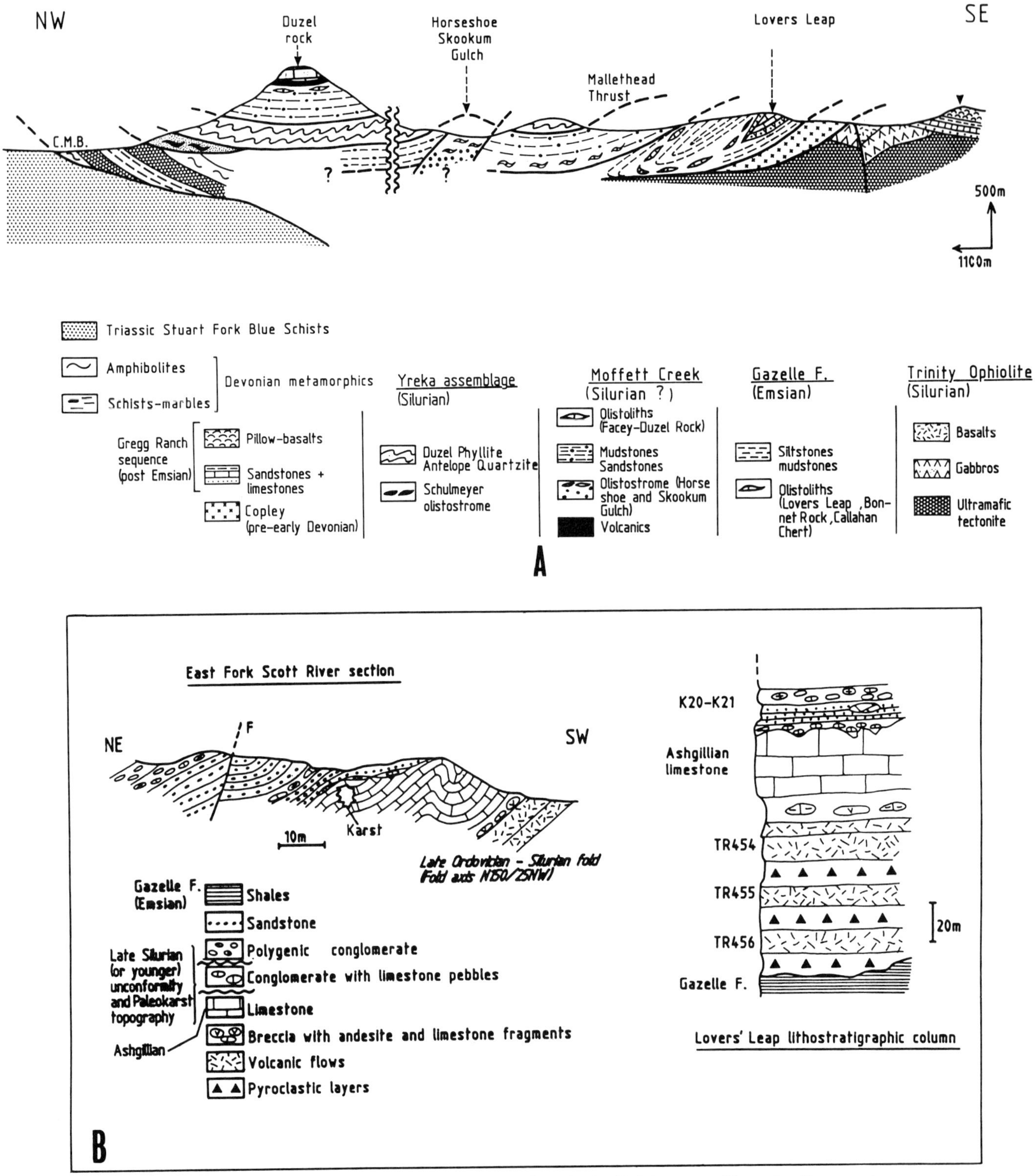

Figure 3A and B. A, Diagrammatic cross section of the Yreka-Callahan assemblage after Potter and others (1977), Lindsley-Griffin and Griffin (1983), and Charvet and others (1989). B, East Fork Scott River cross section showing the stratigraphic relationships between the Ashgillian limestone and the Upper Silurian conglomerate; lithostratigraphic column of the Lovers Leap section showing the location of the analyzed samples.

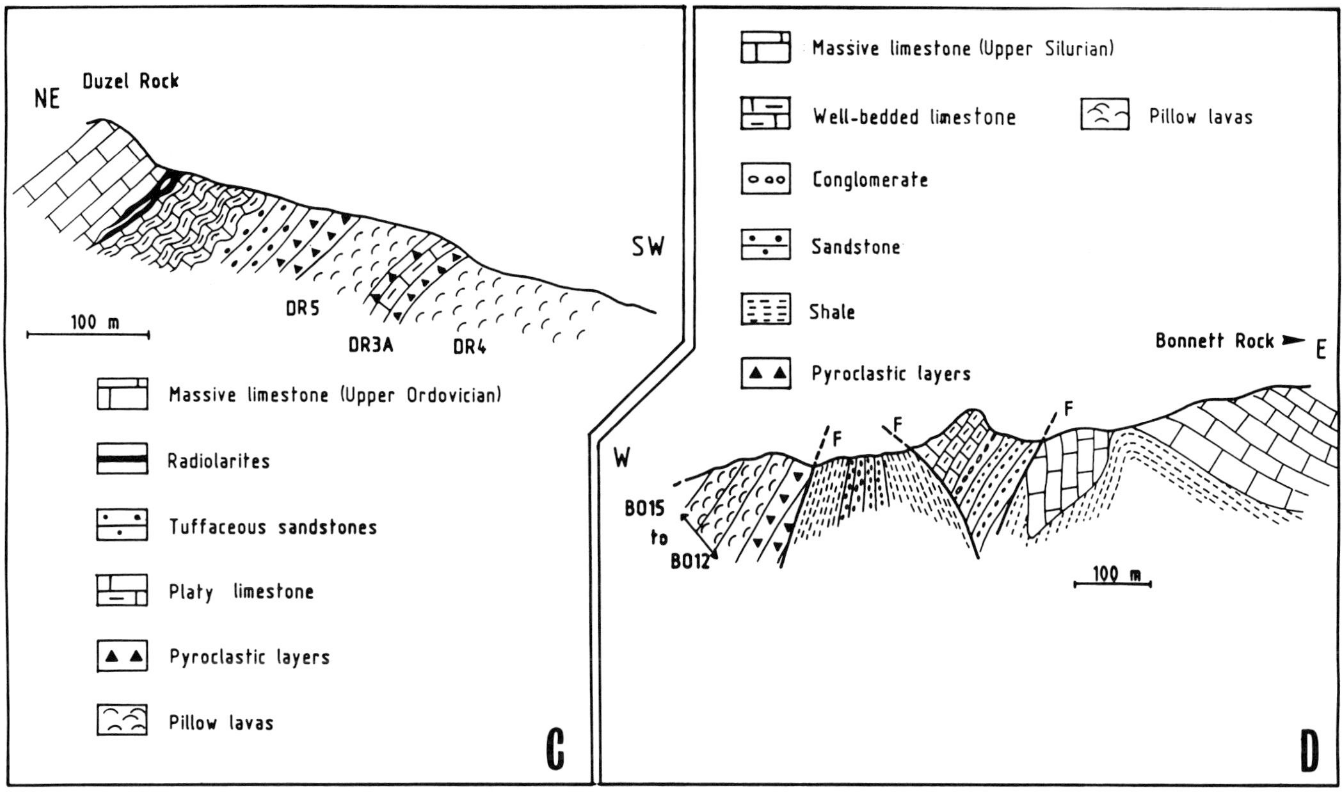

Figure 3C and D. C, Duzel Rock cross section showing the stratigraphic relationships between the pillow basalts and the Late Ordovician coralline limestones. D, Bonnett Rock cross section showing the stratigraphic relationships between the volcanics and the Upper Silurian coralline limestones.

Triassic and Paleozoic Belt, could be related to the debated Late Triassic–Early Jurassic tectonic event. Moreover, late folds associated with greenschist retrometamorphism that are present locally (Coffee Creek section) in the Central Metamorphic Belt (Cannat, 1983) may have formed during the Late Triassic–Early Jurassic event. A Middle to Late Jurassic disturbance is generally acknowledged by most workers.

GEOLOGIC SETTING AND LITHOSTRATIGRAPHIC SUCCESSIONS IN THE BLUE MOUNTAINS

The pre-Cenozoic formations in the Blue Mountains (Figs. 1, 5) are characterized by: (1) a thick sequence of volcanic and sedimentary rocks of Permian and Triassic age (Seven Devils Group of Vallier, 1977, 1989), and Huntington Formation (Brooks, *in* Vallier, 1989); (2) an Upper Triassic clastic sequence (Hurwal Formation, several formations in the Suplee-Izee area; Dickinson, 1979), which includes a melange (Elkhorn Ridge Formation, Miller Mountain Melange, Frenchy Butte Melange, Grindstone–Twelvemile Creek Melange, and Baker Melange); and (3) an Upper Sinemurian to Oxfordian flysch (Coon Hollow and Weatherby Formations in the Wallowa, Snowshoe Formation near Supplee; Figs. 1, 5). The Jurassic sequence unconformably overlies the Triassic flysch and melange, which includes Late Triassic blueschists (Hotz and others, 1977), as well as Permian and Triassic ophiolitic blocs, radiolarian chert, and Paleozoic limestones blocks. The melange fabric may represent both tectonic disruption and gravity sliding. This melange is interpreted as formed either in a forearc environment (Vallier, 1989) or in a subduction complex (Davis and others, 1978), or during a collisional event (Cimmerian orogeny of Roure and Sosson, 1986).

The lithostratigraphic succession of the Permian and Triassic island-arc series (Fig. 5) (Wallowa terrane of Silberling and others, 1984) was established by Vallier (1977) and Brooks and Vallier (1978). According to Vallier (1977), the Middle and Upper Triassic rocks rest unconformably on the Permian sequence. We never recognized this unconformity along the Snake River sections (Lapierre and others, 1988). The deformation of the amphibolites and quartz-diorites, considered to be the basement of the arc by Vallier and others (1977), appears to be associated with a NNE-SSW ductile strike-slip fault system that also affects the Permian to Late Triassic rocks (Kirkwood Bar and Suicide point sections) along the Snake River. If such is the case, the amphibolite and quartz-diorite probably do not represent a basement older than the Permian-Triassic arc (Seven Devils Group of Vallier, 1977), but may instead be a deep part of this arc, synchronous with the volcanic rocks, and later affected by mylonitization.

The Permian arc, dated by Leonardian and Guadalupian

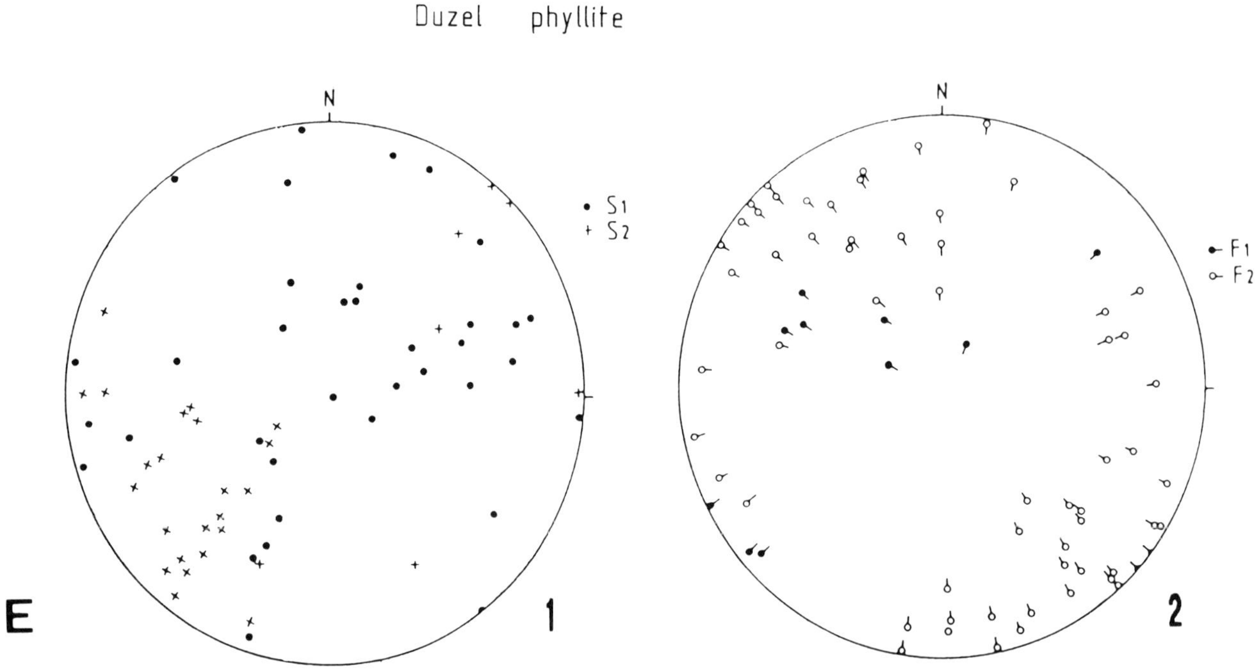

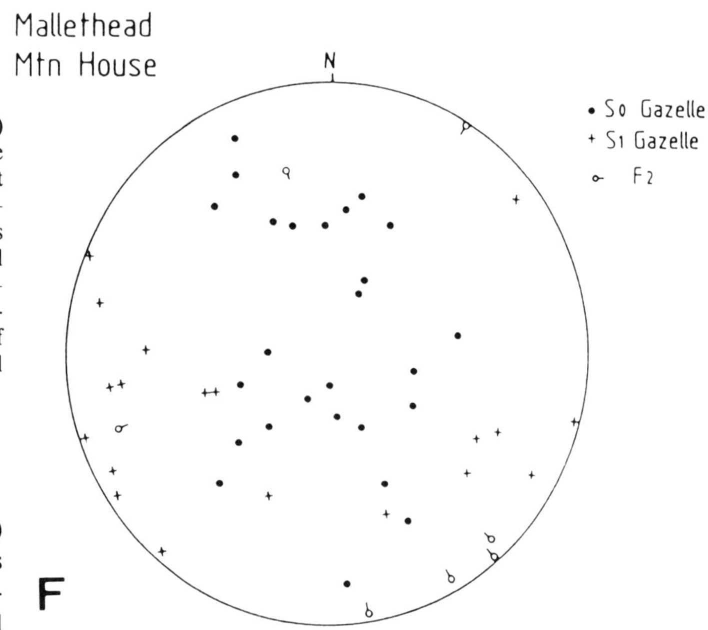

Figure 3E and F. E, Stereograms (equal area net, lower hemisphere) showing the two main generations of folds and related schistosities in the Duzel phyllite (Yreka-Callahan assemblage: 1: S1—poles to the first schistosity planes; 2: S2—poles to the second schistosity planes; 2: F1—axes of first-phase isoclinal intrafolial folds (Callahan event); F2—axes of second-phase folds. F, stereogram (equal area net) of some structural features in the Gazelle Formation (Yreka-Callahan assemblage): SO—pole to bedding plane; S1—pole to the first schistosity of Gazelle Formation (correlative with the second one in the Duzel phyllite). F2—axes of post-Emsian folds, correlated with the second-phase folds of Duzel phyllite.

brachiopods (Vallier, 1977), consists of (in ascending order): (1) rhyolitic flows and acidic pyroclastic rocks intruded by numerous basic dikes (Windy Ridge Formation) (Vallier, 1977); (2) massive basaltic, andesitic, and rhyolitic flows (Table 5) interlayered with graded pyroclastic, volcaniclastic, and clastic beds (base of the Hunsaker Formation (Vallier, 1977); and (3) coarse-grained andesitic breccias overlain by ignimbrites and waterlain tuffs interbedded with graded sandstones (top of the Hunsaker Formation (Vallier, 1977; Lapierre and others, 1988).

The Triassic rocks rest on Permian strata and consist of basaltic and andesitic flows interbedded with pyroclastic rocks, graywackes, and Ladinian-Carnian limestones (Vallier, 1977).

In Hells Canyon (Seven Devils Group of Vallier, 1977; Wallowa terrane of Silberling and others, 1984, and Vallier, 1989), the uppermost levels of the volcanic sequence are characterized by: (1) pillowed flows of basaltic and andesitic composition, (2) fluidal rhyolitic domes, and (3) conglomerate including predominantly cobbles and pebbles of highly porphyritic andesite and fluidal rhyolite. This conglomerate suggests that erosion was important at the end of the volcanic activity.

In the north, near Pittsburg Landing (Fig. 5), the pillowed flows are thicker and are formed of olivine-bearing basalts. They are interbedded with pillow-breccia, hyaloclastite rich in glass and palagonite, and thin limey beds rich in Ladinian fauna (ammonites and *Daonella*) (Vallier, 1977, 1989).

To the south in the vicinity of Huntington (Fig. 5; Olds Ferry terrane of Silberling and others, 1984; Huntington Formation of Brooks, 1979), basaltic and andesitic flows predominate and are interbedded with tuff, volcanic breccia, and conglomer-

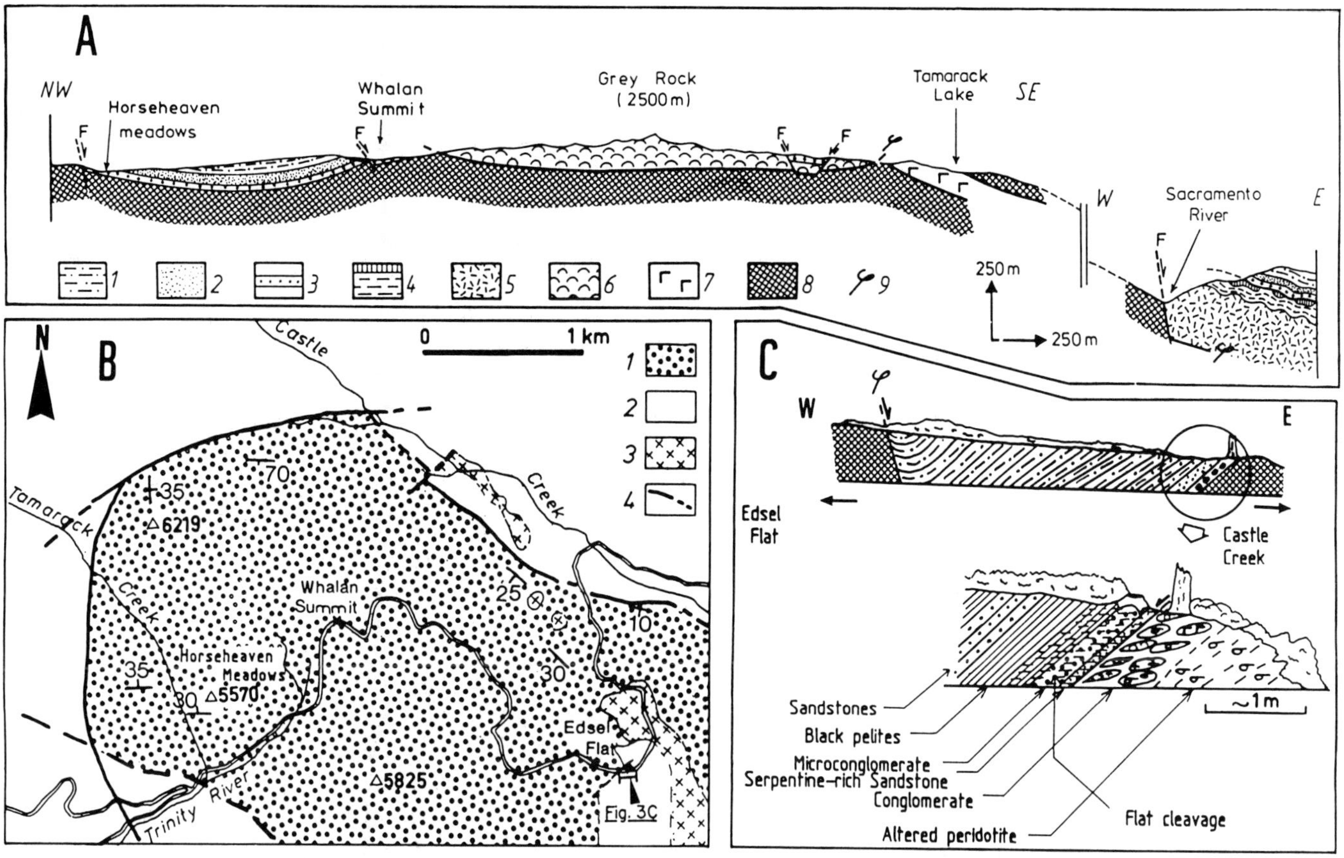

Figure 4. The unconformity of the Bragdon Formation onto the metamorphic peridotite and the Trinity ophiolitic pillow basalts in the Whalan Summit–Grey Rock area. A, Schematic cross section from Grey Rock klippe to Girard Ridge; after Charvet and others (1989). Upper Devonian–lower Mississippian Bragdon Formation: (1) mudstones and fine-grained sandstones; (2) green sandstones with serpentine clasts; (3) basal conglomerate. Eifelian Kennett Formation: (4) volcaniclastic rocks, siliceous shales, and limestones; (5) Pre-Devonian to Lower Devonian Copley-Balaklala island-arc volcanic rocks; Lower Paleozoic Trinity ophiolite: (6) pillow basalts; (7) gabbros; (8) metamorphic peridotite; (9) thrust contact; (10) faults. B, Schematic geologic map of the Whalan Summit area: (1) Upper Devonian(?)–Lower Carboniferous Bragdon Formation; (2) Lower Paleozoic Trinity ophiolite and metamorphic peridotite; (3) Upper Jurassic plutonic rocks; (4) faults. C, Detailed drawing of the outcrop showing the Bragdon basal conglomerate in contact with the metamorphic peridotite.

ate, as well as volcaniclastic deposits. Some conglomerates contain granitoid pebbles suggesting that deeper levels of the arc were exposed. Sandstone, graywacke, and siliceous tuffs predominate, whereas *Halobia*-bearing limestone of late Carnian–early Norian age (Vallier, 1989) are uncommon. Pillowed flows occur also, but, compared to those exposed in Hells Canyon, they are more felsic in composition and their matrix is tuffaceous. Everywhere, the island-arc sequence is capped by Late Triassic limestone (Martin Bridge Formation of Early Norian age, Lime Formation in the sequence in the vicinity of Huntington, and Nelson Marble).

TECTONIC EVOLUTION OF THE EASTERN KLAMATH MOUNTAINS

Rocks in the eastern Klamath Mountains were deformed by two Paleozoic orogenic events. The oldest, Callahan event (Cotkin and Armstrong, 1987) occurred at the boundary of the Ordovician and Silurian. It is evidenced by the isoclinal synmetamorphic and intrafolial folds present in the Duzel Phyllite (Fig. 3E) and in the marbles and blueschist blocks of the Moffett Creek Formation, and by the presence of an Upper Silurian conglomerate that includes limestone and calc-alkaline quartz-diorite and tonalite pebbles and rests unconformably on folded Ashgillian limestone. This conglomerate is exposed in the Lovers Leap section (Figs. 2, 3A, B), and along the East Fork Scott River section (Figs. 2, 3B).

The second phase of deformation occurred during the Late Devonian and/or Early Mississippian. This tectonic disturbance is recorded by the unconformity of the Bragdon Formation on the ultramafic metamorphic sheet, the Trinity ophiolitic pillow basalts exposed at Grey Rock, and the Eifelian Kennett Formation along the shores of Shasta Lake (Diller, 1906) (refer to the Red-

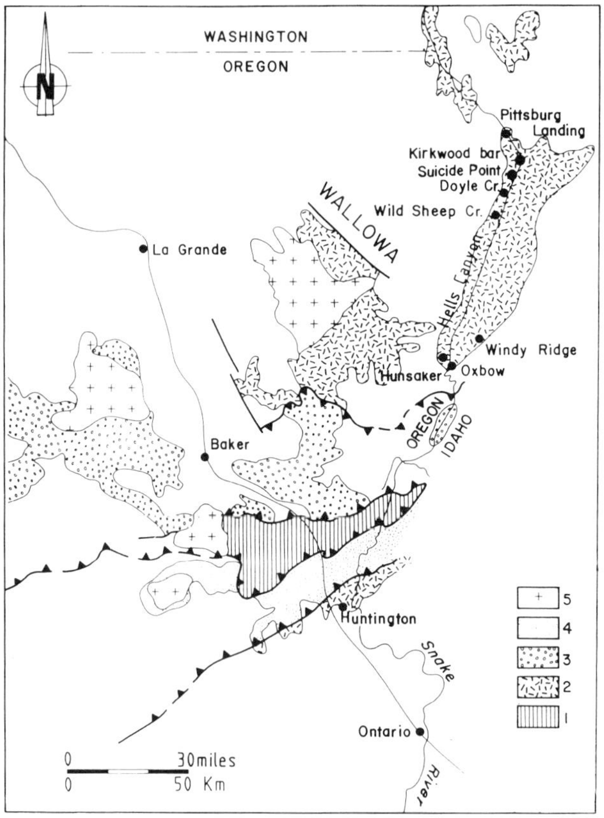

1 Triassic Burnt River Unit (metavolcanics, Nelson marble)
2 Permian-Triassic island-arc rocks (Seven Devils and Huntington)
3 Lower Jurassic olistostromes (Elkorn, Miller Mountain, Frenchy Butte)
4 Jurassic volcaniclastic deposits (Weatherby Flysh)
5 Late Mesozoic or Jura-Cretaceous batholiths

Figure 5. Simplified geologic map of the pre-Cenozoic formations of the Blue Mountains; after Mullen (1985).

ding map of Kinkel and others, 1956, and Albers and Bain, 1985). The Bragdon basal conglomerate is exposed near Whalan Summit, and on the south side of the Grey Rock summit (Fig. 4A, B). Where it rests on the ultramafic metamorphic sheet, it consists of pebbles of peridotite, gabbro, basalt, plagiogranite, and altered pyroxenite. It grades upward into a green sandstone rich in serpentine and chloritized debris, overlain by black shale and silt (Fig. 4B, C). Where it rests on the ophiolitic pillow basalts as at Grey Rock (Fig. 4A), it consists mainly of basaltic and gabbroic pebbles. Locally, it rests unconformably either on both the Copley and Balaklala Formations or only on the Copley Greenstone (Brouxel and others, 1988).

In the vicinity of Redding (Fig. 2), the Bragdon basal conglomerate contains limestones, medium-grained radiolarian cherts, and tuffaceous clasts derived from the underlying Kennett Formation, as well as quartzose sandstone debris most likely derived from the erosion of the Emsian Gazelle Formation. Locally, as in Girard Ridge (Fig. 4A), the contact is disconformable (c. Campos, unpublished data) or has been reported as conformable (Watkins, 1986). But, the basal conglomerate is often present and contains pebbles of the Kennett slates and limestones; the presence of Kennett limestone clasts in the Bragdon conglomerate was previously noticed by Diller (1906). The small folds in the argillite beds interbedded in the uppermost Copley flows (near Redding; Fig. 2) (Albers and Bain, 1985) and in the Kennett Formation (Mammoth Mine; Fig. 2) (Charvet and others, 1989) are generally recumbent to the west. These folds are likely linked to this orogenic event because the Bragdon Formation that overlies the Kennett Formation in the Redding vicinity is only affected, at the same scale, by open folds slightly overturned to the east.

Local west-verging reverse faults restricted to the uppermost flows of the Copley Greenstone and the Lower Balaklala Rhyolite (exposed along the road of Mammoth Mine, Fig. 2) could be related to this event.

In the Yreka-Callahan assemblage, this disturbance is less constrained in time because of the absence of the Bragdon Formation. The Duzel Phyllite and the Emsian Gazelle Formation (Fig. 3A) show also westward-overturned folds (trend of the fold axes: N140° to 150°E, Fig. 3E, F). These post-Emsian structures could be related to the late Devonian–early Carboniferous event (refer to Charvet and others, 1989, for discussion).

The ultramafic metamorphic sheet and the Trinity ophiolite are thrust over the Central Metamorphic Belt. The transport direction during thrusting deduced from lineations observed in the ultramafic sheet and in the metamorphic rocks indicates a westward displacement of the ultramafic sheet over the Central Metamorphic Belt (Cannat, 1983). The structural concordance of the lineations observed in the ultramafic sheet with the foliations and lineations in the amphibolites of the Central Metamorphic Belt added to the coincidence of the temperatures of deformation indicated by the mineral assemblages in the ultramafic sheet and by the almandine-amphibolite facies in the Central Metamorphic Belt (Cannat and Boudier, 1985) show that these units have been deformed together. Thus, this thrusting occurred together with the amphibolite facies metamorphism of the Central Metamorphic Belt during Devonian or Early Mississippian times (Cannat and Boudier, 1985). It appears that the obduction of the Trinity ophiolite onto the Central Metamorphic Belt may have been contemporaneous with the folding and the thrusting in the Copley-Balaklala island arc and the Kennett Formation.

In summary, the eastern Klamath Mountains have been affected during Paleozoic time by two orogenic events. The Late Ordovician–Early Silurian Callahan event is observed only in the Yreka-Callahan assemblage (Duzel Phyllite unit, blocks in the olistostromal horizons in the Moffett Creek and Gazelle Formations). It formed isoclinal folds present in the high-pressure and epizonal metamorphosed rocks. The Late Devonian–Early Mississippian event is well documented in the western, central, and southern area of the eastern Klamath Mountains. It was responsible for the obduction of the Trinity ophiolite onto the Central Metamorphic Belt and the deformation of the Copley-Balaklala island arc and its Eifelian sedimentary cover. It appears reasonable to assume that the emplacement of the Yreka-Callahan

TABLE 5. MAJOR, TRACE, AND RARE EARTH ELEMENTS OF THE BLUE MOUNTAINS TRIASSIC LAVAS (SEVEN DEVILS GROUP AND HUNTINGTON FORMATION)

	Seven Devils							Hunington Dyke	Hunington Pillow	Hunington Flow	Hunington Flow	Pittsburg Landing Pillow	Pillow
Sample No.	BM 115	BM 116	BM 48	BM 26	BM 109	BM 122	BM 18	BM 315	BM 335	BM 304	BM 321	PL 23	PL 2
SiO_2	47.98	53.64	60.36	48.68	54.52	74.70	47.53	49.45	46.23	49.57	51.44	47.17	52.19
TiO_2	1.32	1.81	0.84	0.80	0.84	0.07	1.37	0.87	0.72	0.66	0.88	0.74	1.29
Al_2O_3	17.15	15.75	14.39	15.92	18.61	15.10	14.88	17.27	15.35	18.27	17.21	16.41	17.36
Fe_2O_3	12.40	12.19	8.08	9.90	8.07	0.72	11.15	8.36	8.63	9.13	10.52	8.28	10.18
MgO	4.66	2.60	3.50	6.66	5.24	0.13	6.40	6.87	3.49	5.51	4.12	6.74	4.33
MnO	0.19	0.16	0.15	0.18	0.18	0.06	0.20	0.16	0.28	0.15	0.26	0.15	0.21
CaO	7.76	4.37	5.13	9.79	2.74		9.20	9.24	8.47	4.49	2.53	13.81	8.46
Na_2O	4.58	6.15	4.12	3.20	4.42	4.95	2.18	3.31	5.07	4.09	6.14	0.16	4.02
K_2O	0.13	1.54	0.38	0.81	1.55	2.44	0.41	1.83	1.43	1.50	1.90		
P_2O_5	0.16	0.28	0.11	0.21	0.16	0.07	0.24	0.18	0.17	0.12	0.23	0.06	0.19
LOI	3.80	1.79	2.93	4.24	4.01	1.27	4.73	2.97	10.05	6.34	5.16	5.56	1.86
Total	100.25	100.28	99.99	100.39	100.34	99.28	99.29	100.51	99.89	99.93	100.39	99.08	100.09
Cr	86	20	49	246	52	10	303	300	298	33	5	427	55
Ni	49	25	39	101	55	15	201	76	99	30	13	169	63
V	238	301	200	288	184	20	203	250	226	283	200	285	363
Nb								57	39	27	54		
Zr								5	30	30	5		
Y	35	52	24	19	18	10	24	20	14	16	24	13	39
La	4.97	7.89	7.04	7.62	6.37	6.37	7.13	6.28	4.21	1.89	8.23	2.32	6.50
Cb	11.76	18.64	9.20	9.31	10.76	6.97	10.99	12.50	6.72	4.54	13.78	4.81	10.64
Nd	3.81	6.14	2.96	2.67	2.86	2.15	3.52	4.68	2.03	1.72	4.57	1.75	4.05
Sm	1.27	1.93	1.77	1.33	0.96	0.64	2.09	1.39	0.64	0.66	1.49	0.77	1.15
Eu	4.34	6.82	3.30	2.75	2.95	1.52	3.76	5.65	2.04	2.21	5.35	1.92	4.54
Cd	5.22	7.96	3.45	2.72	2.59	1.34	3.54	4.08	2.16	2.43	4.47	1.95	5.07
Dy	2.99	4.47	1.97	1.49	1.37	0.69	1.89	3.25	1.32	1.45	3.00	1.17	3.09
Er	3.30	4.97	2.31	1.63	1.46	0.79	2.08	1.95	1.37	1.54	2.47	1.24	3.80
Yb	0.48	0.76	0.34	0.22	0.24	0.09	0.37	0.40	0.20	0.23	0.41	0.21	0.57
$(La/Yb)_N$	1.01	1.06	2.06	3.15	2.94	5.47	2.32	2.17	2.07	0.82	2.24	1.26	1.15

nappes was linked to this orogenic event because of (1) the retrometamorphism of the Late Ordovician–Early Silurian blueschist during Devonian time, and (2) the quartzose detritus supplied to the Upper Devonian(?)–Carboniferous Bragdon Formation. The time span of this orogenic event is compatible with that of the Antler orogeny, defined in the Roberts Mountain Allochthon in Nevada by Roberts and others (1958), where Famennian rocks are thrust over Lower Mississippian strata (Murphy and others, 1984). This statement was also done by Johnson (1971), Boucot and others (1974, p. 694) and Boucot and Potter (1977), although the latter concluded in assuming an Eifelian orogeny. Thus, this Late Middle Devonian to Late Devonian(?)–Early Mississippian orogeny (post-Kennett Formation, i.e., Late Eifelian and pre-Bragdon Formation, i.e., Late Devonian(?)–Early Mississippian) could be an early expression of the classic Antler orogeny in the eastern Klamath Mountains, synchronous with the first phase of the Antler orogeny described in the Roberts Mountain Allochthon (Murphy and others, 1984).

MAGMATIC EVOLUTION OF THE EASTERN KLAMATH

Volcanic sequences of different types characterize the Paleozoic to early Mesozoic history of the eastern Klamath Mountains. Magmas formed at convergent plate boundaries are the most common and are dated Late Ordovician, pre-Devonian to Early Devonian, post–Middle-Devonian, and middle to Late Permian. Dated relics of seamounts are restricted in age to Late Ordovician and Late Silurian. The pre-Devonian to lower Devonian and Permian island-arc rocks are well characterized, whereas the petrology and geochemistry of the lower Paleozoic and post–Middle Devonian rocks are less well known. For that reason, we develop here the petrology and magmatic affinities of the Paleozoic rocks exposed in the Yreka terrane and summarize relation in pre-Devonian to lower Devonian and Permian rocks to the south.

In contrast with the Copley-Balaklala volcanic rocks, the

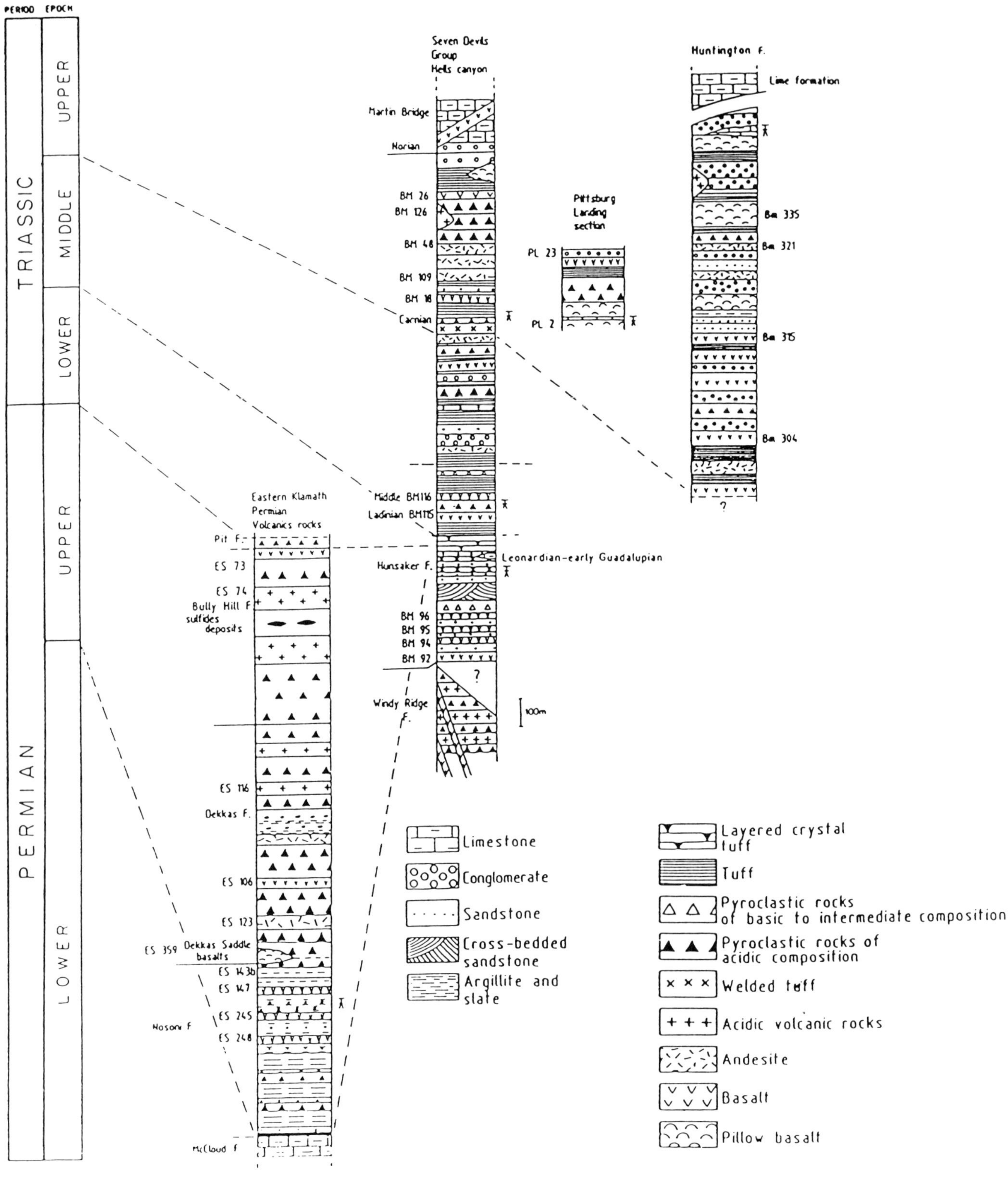

Figure 6. Schematic lithostratigraphic columns of the eastern Klamath and Blue Mountains Permian and Triassic formations showing the location of the analyzed samples. The stratigraphic thicknesses, lithologies, and fossiliferous levels were compiled from Coogan (1960), Miller (1989), Vallier (1974, 1977), Brooks and Vallier (1978), Roure (1984), and our own data.

lower Paleozoic volcanic rocks in the Yreka terrane are found in olistoliths in the Moffett Creek and Gazelle Formations, but they are generally well dated because coralline limestones are interbedded with them.

The Upper Ordovician (Duzel Rock, Horseshoe Gulch sections; Figs. 2, 3C) and Upper Silurian (Bonnett Rock; Figs. 2, 3D) volcanic rocks consist of pillowed basaltic flows, pillow breccia and tuff, interbedded with limey thin-bedded strata and capped by coralline limestones. They are interpreted as seamounts because the flows are pillowed and the limestone layers are rich in corals (Potter and others, 1977). The basalts are generally very altered with albitized plagioclase and olivine pseudomorphs, but often preserved clinopyroxene crystals. They always display quench textures. They are either porphyritic with olivine pseudomorphs and diopside phenocrysts in a fluidal, highly vesicular groundmass (Duzel Rock: DR5b, Table 1), or intersertal with abundant plagioclase (Bonnett Rock: BO12, Table 1), or even slightly porphyritic with few plagioclase microphenocrysts (Horseshoe Gulch: H12, Table 1). Bonnett Rock samples display generally a T MORB affinity with flat REE patterns and locally depleted in LREE (Fig. 7A). The Upper Ordovician basalts (Duzel Rock and Horseshoe Gulch) differ by mild alkalic affinities marked by higher contents in TiO_2, Zr, Nb, and Y (Table 1) and LREE-enriched patterns (Fig. 7A).

The Ashgillian calc-alkaline basalt-andesite-dacite differentiated suite of Lovers Leap and Negro Gulch sections (Figs. 2, 3B) (Lapierre and others, 1987; Brouxel and others, 1988, 1989) exhibits highly porphyritic textures with abundant plagioclase and magmatic hornblende phenocrysts, low TiO_2 content and high levels in incompatible elements (Table 2), LREE-enriched patterns (Fig. 7B), and low ϵ_{Nd} (T) (+3.9). All these features suggest volcanic rocks erupted in an active margin setting. The 440-Ma quartz-diorite to tonalite pebbles included in the Upper Silurian or Lower Devonian(?) conglomerate that rests unconformably on the folded Ashgillian limestone (Fig. 3B) display the same calc-alkaline affinities (LREE-enriched patterns; Fig. 7C, Table 2). These geochemical and age similarities between plutonic pebbles and volcanic rocks suggest a common magmatic suite. The plutonic rocks probably represent the roots of the calc-alkaline flows.

The post–Middle Devonian volcanic rocks of the Gregg Ranch and Copper Peak sections (Fig. 2) consist of diopside-bearing pillow basalts interbedded with pillow breccia and hyaloclastite (Gregg Ranch section) or andesitic massive flows and coarse-grained breccias (Copper Peak section). These volcanic rocks are quartz-normative, low-K tholeiites poor in TiO_2 (Table 3). They are Cr- and Ni-rich, characterized by low incompatible elements (Zr = 39 ppm, Nb = 0.5 ppm and La = 1.28 ppm; Table 3), low REE concentrations (10 times the chondritic abundance), with a marked LREE depletion (Fig. 7D, 7E) (Brouxel and others, 1989).

The pre-Devonian to Lower Devonian tholeiitic island-arc volcanic rocks of the Copley and Balaklala are the remnants of an intra-oceanic island-arc back-arc basin system (Lapierre and others, 1985a, b; Brouxel and Lapierre, 1988; Brouxel and others, 1987, 1988).

The Lower to middle Permian basalt, andesite, dacite, and rhyolite of the Nosoni, Dekkas, and Bully Hill (Table 4), exposed in the eastern part of the eastern Klamath Mountains, exhibit flat REE patterns (Fig. 8A) of island-arc tholeiitic affinities, with the exception of the Dekkas Saddle Cr-rich diopside-bearing basalts that can be compared to calc-alkaline ankaramites or picritic basalts with shoshonitic affinities, as described in Solomon islands and eastern Sunda arc (Martin, 1989). The presence of Cr-rich diopside suggests that these basalts of the lower part of the Dekkas Formation (ϵ_{Nd} (T) = + 3.4; Table 4) represent deep magmas ascended rapidly along transversal fractures (Maury and Westercamp, 1985). Some basalts of the Nosoni Formation show an LREE enrichment, suggesting calc-alkaline affinities (Fig. 8B). In contrast, the Mg-Cr–rich Bully Hill basalts display similarities with boninites. Finally, the range of the ϵ_{Nd} (T) values (ϵ_{Nd} (T) = + 3.7 to + 6.8; Table 4) that are similar to present-day island arcs built on a continental crust (Japan) or a thickened crust (Lesser Antilles) (Gill, 1981; Faure, 1986), suggest to us that the Permian island-arc volcanic rocks formed from melting of a MORB-type mantle source affected by crustal contamination. This crustal contamination could be linked to the arc-crust thickness. Indeed, the ϵ_{Nd} (T) values decrease with time; i.e., the basalts of the Nosoni Formation exhibit higher ϵ_{Nd} (T) values than those in the Dekkas (with the exception of one lowermost Dekkas basaltic flow; Fig. 6; Table 4). The lowermost Nosoni basaltic flows display higher ϵ_{Nd} (T) (+6.8) values than the uppermost ones ϵ_{Nd} (T) (+ 4.9) (in Martin, 1989). Moreover, acidic flows increase with time. Thus, the Permian island-arc volcanic rocks represent the remnant of a mature island arc, built on a thickened crust in sub-aerial conditions (Martin, 1989).

In conclusion, the volcanic rocks of the eastern Klamath Mountains are dominated by tholeiitic island-arc magmas. If we exclude the Lower Paleozoic volcanic rocks of the Yreka-Callahan terrane, which possibly were amalgamated during a post-Emsian and pre–lower Carboniferous tectonic event, the petrology and magmatic affinities of the Paleozoic island-arc volcanic rocks suggest an evolution from a pre-Devonian to Early Devonian island-arc–back-arc system to a Late Permian mature island arc. The basement of these island arcs evolved from a normal oceanic crust to a thick crust, possibly thickened during the post-Eifelian and pre–Late Devonian to Early Mississippian orogeny by the emplacement of the Yreka-Callahan nappes and the underthrusting of the Central Metamorphic Belt.

PERMIAN AND TRIASSIC EVOLUTION OF THE BLUE MOUNTAINS

The petrology and geochemistry of the Permian-Triassic volcanic rocks in the Blue Mountains (Figs. 5, 6) have been less studied than those in the eastern Klamath Mountains (Vallier, 1977, 1989; Vallier and others, 1977; Sarewitz, 1983; Lapierre and others, 1988). All the volcanic rocks are affected by low

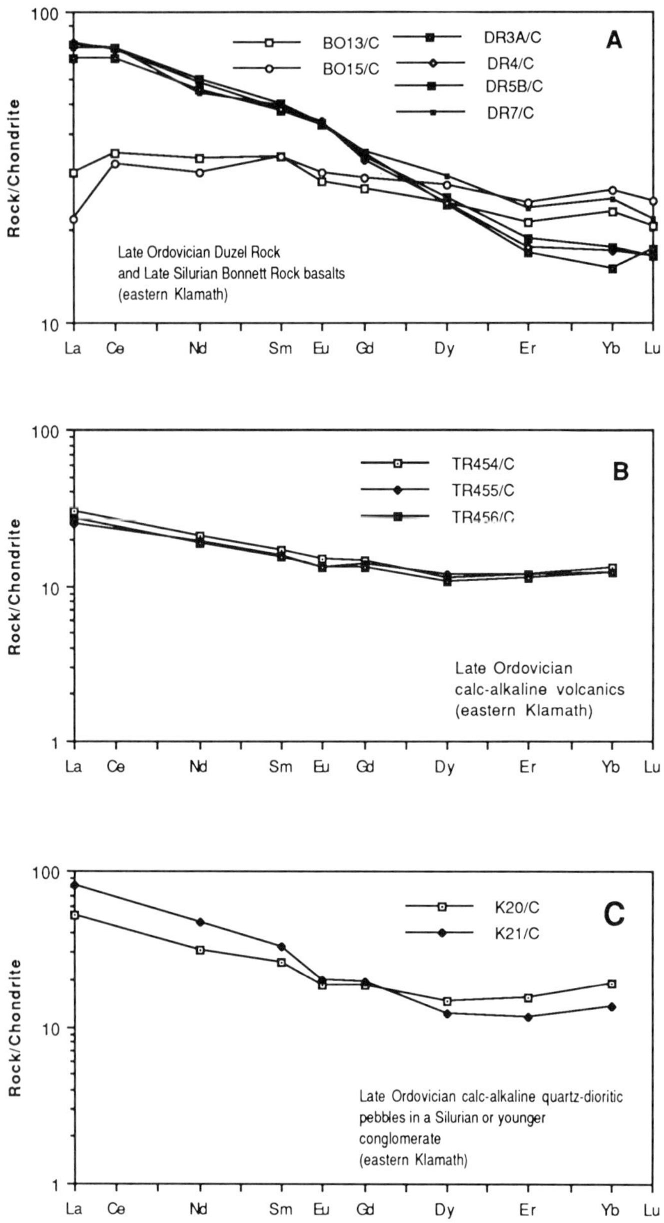

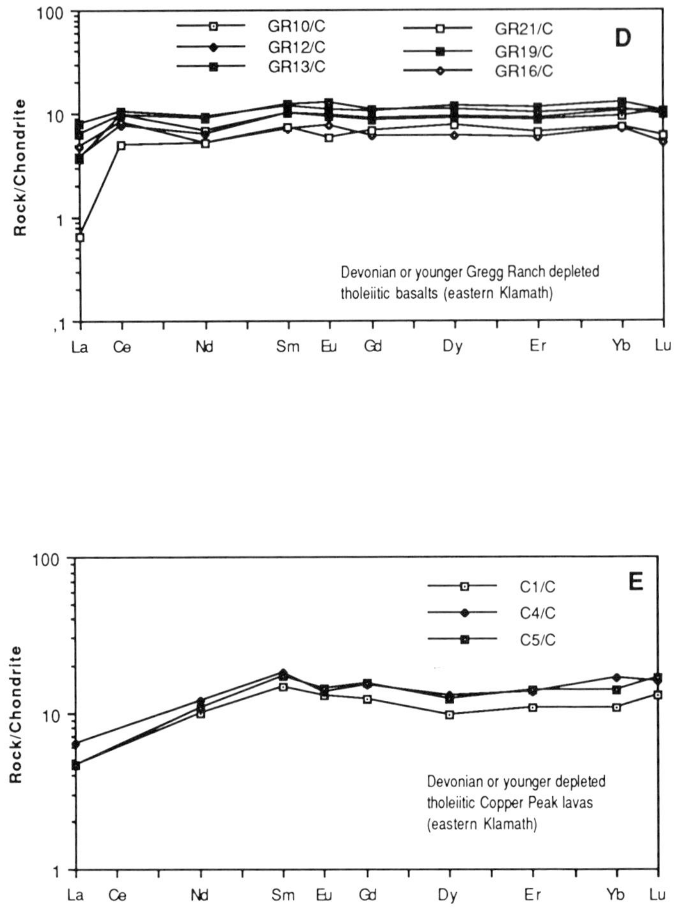

Figure 7. Rare earth patterns (chondrite normalization values, after Evensen and others, 1978) of the eastern Klamath Yreka-Callahan assemblage Paleozoic volcanic rocks. A, Late Ordovician Duzel Rock and Late Silurian Bonnett Rock basalts. B, Late Ordovician Lovers Leap calc-alkaline volcanics. C, calc-alkaline Late Ordovician quartz-dioritic pebbles in a Late Silurian or Devonian conglomerate. D, Post-Devonian Gregg Ranch–depleted tholeiitic basalts. E, Post-Devonian Copper Peak–depleted tholeiitic lavas.

greenschist facies metamorphism. The common assemblage is albite + chlorite + epidote + calcite (Vallier and others, 1977). Clinopyroxene is the only magmatic mineral preserved except in the Triassic lavas where primary olivine and Cr-spinels occur.

The lowermost acidic lavas of the Permian Windy Ridge Formation (Table 4) include quartz and plagioclase phenocrysts in a crystallized groundmass. They display tholeiitic or mild calc-alkaline affinities with flat REE patterns or slightly enriched in LREE ($1.24 < (La/Yb)_N < 2.05$; Fig. 8C). The observed Eu anomaly suggests plagioclase fractionation.

The dacitic to rhyolitic rocks are overlain by a differentiated basalt-andesite-dacite suite (Hunsaker Formation). The basalts (Vallier, 1974, 1989) differ from each other by their textures, mineral composition and geochemistry (Table 4). Olivine basalts (BM 96; Fig. 8D) are medium-grained porphyritic lavas with plagioclase and 1- to 5-mm size olivine pseudomorphs. Their groundmass is rich in flow-aligned feldspar microlites and includes elongated vesicles filled with chlorite and calcite. Their low TiO_2 content (0.93 percent; Table 4) and flat REE patterns ($(La/Yb)_N = 1.23$; 10 times the chondritic abundance; Fig. 8D) are features of island-arc tholeiites.

Clinopyroxene basalts (BM 94-95; Table 4) are highly porphyritic (50 to 60 percent modal composition) with plagioclase phenocrysts in glomeroporphyritic aggregates. Their groundmass displays intersertal texture with plagioclase laths molded by preserved or altered clinopyroxene. Their rather high TiO_2 levels

(TiO_2 = 1.30 percent; Table 4) are linked to the accumulation of late crystallizing Ti-Fe oxides. Their REE patterns differ from the olivine basalts by higher REE concentrations (20 to 30 times the chondritic abundance; Fig. 8D).

Finally, the last type of Permian basalt (BM 86-92, Table 4) is plagioclase-rich and clinopyroxene free with high TiO_2 (TiO_2 = 1.92 percent). It consists of plagioclase either in phenocrysts or in laths in a mesostasis very rich in Fe-Ti oxides and interstitial chlorite. It differs also by a minor negative Eu anomaly, higher Zr content (Zr = 117 ppm) and an REE pattern that shows a slight enrichment in LREE ($(La/Yb)_N$ = 2.08; Fig. 8D).

The Permian lavas belong to tholeiitic suites erupted in an island-arc environment. The plagioclase-rich volcanic rocks show an enrichment in Zr and LREE contents, suggesting that they display mild calc-alkaline affinities. These results are in agreement with data presented by Vallier (1989), except that Vallier (1989) did not report LREE-enriched basalts. The isotopic data differ greatly from one type basalt to another (ϵ_{Nd} (T) = + 6.7 to + 8.4; Table 4) (C. Alibert, unpublished data), and suggest that the basalts were derived from a mantle source of MORB type and that the tholeiitic magmas were more or less contaminated by crust (the plagioclase-rich basalts that display the higher ϵ_{Nd} (T) = + 8.4 are the less contaminated).

The petrology and geochemistry of the Triassic lavas show differences depending on their geographic distribution and their location in the lithostratigraphic succession. In the Seven Devils Group (Wallowa terrane of Silberling and others (1984); Hells Canyon area; Figs. 5, 6) the rocks of basic to intermediate composition can be subdivided into three groups on the basis of their mineralogy and geochemistry. The first group (BM 115-116, Table 5) is characterized by the presence of olivine pseudomorphs and flat REE patterns (Fig. 9A). It occurs generally at the base of the sequence. These olivine basalts are highly vesicular with abundant plagioclase phenocrysts clustered in glomeroporphyritic aggregates. The vesicles are filled with quartz and calcite. Their high Fe_2O_3 and TiO_2 contents (Table 5) are linked to the presence of hematite and Fe-Ti oxides. The second group (BM 18, 48, 109; Table 5) differs by lower REE content, LREE-enriched patterns, and positive Eu anomalies. The partly preserved olivine (Fo-90) locally includes Cr spinels (BM 18; Cr = 303 ppm; Table 5; Figs. 5, 6, 9A). Plagioclase phenocrysts and laths are locally very abundant (BM 109; 60 percent of the modal composition). This abundance in feldspars is marked by an enrichment in Al_2O_3 (Al_2O_3 = 18.61 percent; Table 5) and a positive Europium anomaly suggesting plagioclase accumulation. In the more differentiated lavas (BM 48; Table 5), the texture is porphyritic but the clinopyroxene microphenocrysts are enclosed in the groundmass. Finally, the third group is formed of clinopyroxene-plagioclase phyric basalts that are present at the top of the Triassic sequence. The clinopyroxene (augitic to salitic compositions) forms centimeter-size zoned crystals with plagioclase in glomeroporphyritic aggregates. Secondary epidote is widespread in the mesostasis. The lavas display calc-alkaline affinities marked by an enrichment in LREE (Table 5; Figs. 5, 6,

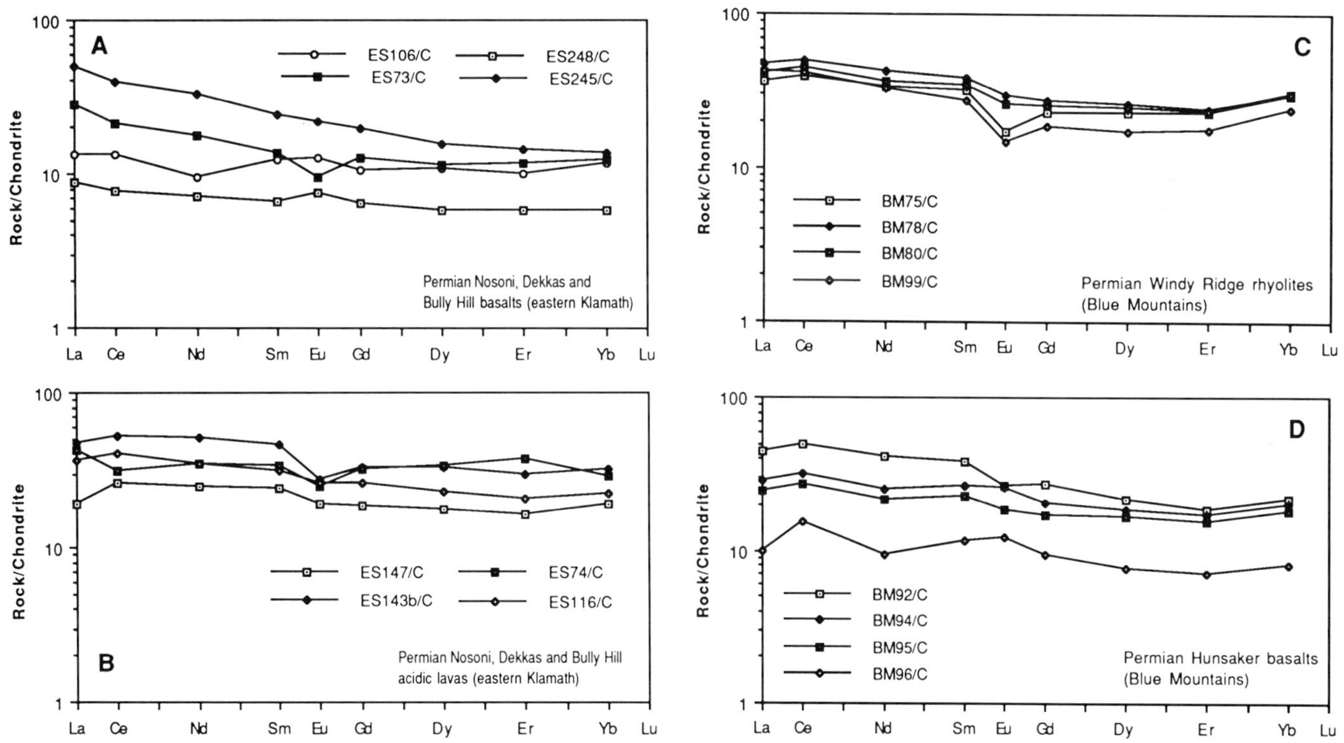

Figure 8. Rare earth patterns (chondrite normalization values, after Evensen and others, 1978) of the eastern Klamath and Blue Mountains Permian lavas. A, Nosoni, Dekkas, and Bully Hill basalts. B, Nosoni, Dekkas, and Bully Hill acidic lavas. C, Windy Ridge rhyolites. D, Hunsaker basalts.

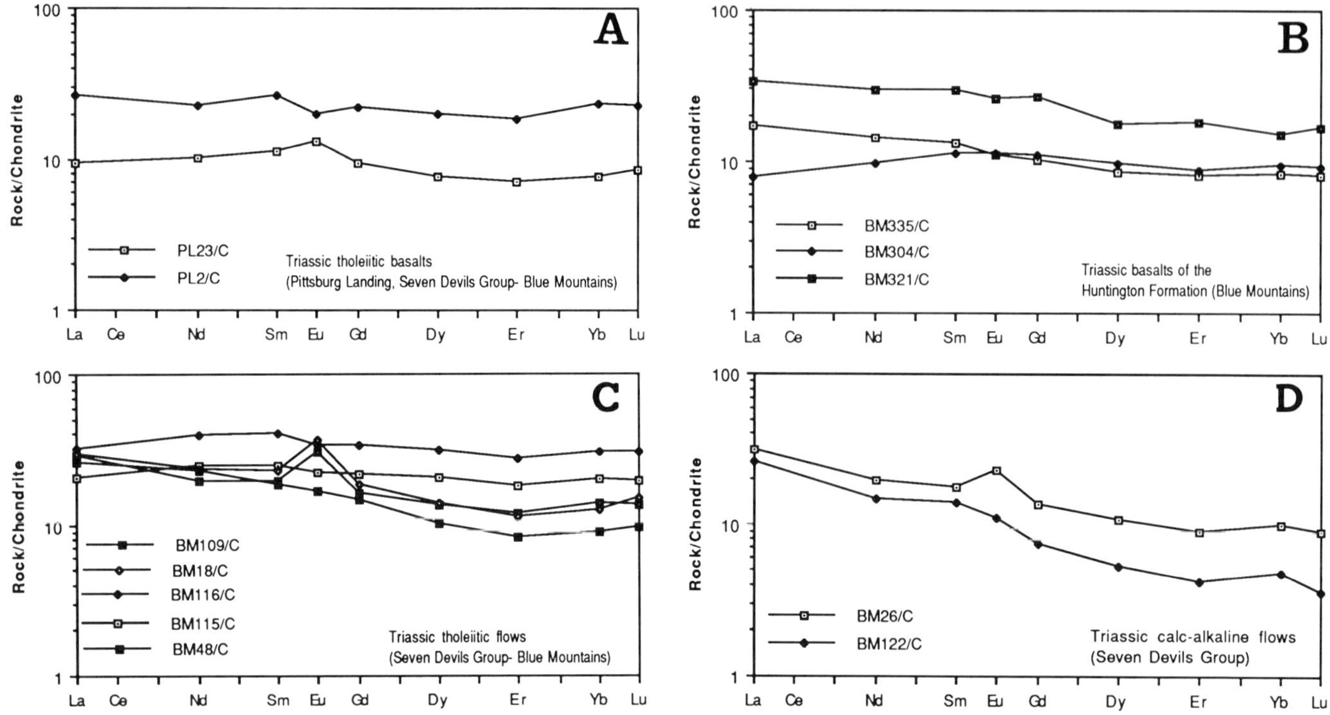

Figure 9. Rare earth patterns (chondrite normalization values, after Evensen and others, 1978) of the Blue Mountains Triassic lavas. A, Triassic tholeiitic flows of the Seven Devils Group. B, Calc-alkaline flows of the Seven Devils Group. C, Triassic tholeiitic basalts of the Pittsburg Landing section. D, Triassic tholeiitic basalts of the Huntington Formation.

9B). The uppermost acidic intrusions (BM 122; Table 5; Fig. 9B) and flows, which contain quartz and plagioclase phenocrysts, are also calc-alkaline with steeper REE patterns and lesser HREE.

The basaltic pillowed and massive flows, north of Pittsburg Landing, are typical low-K tholeiites (flat REE patterns, Table 5; Figs. 5, 6, 9C) with millimeter-size olivine phenocrysts locally including Cr spinel.

In the Huntington Formation, most of the basalts exhibit tholeiitic affinities with low TiO_2 and flat REE patterns (Table 5; Figs. 5, 6, 9D). They are generally plagioclase–clinopyroxene–phyric basalts with the exception of the pillowed flows that are aphyric with flow-aligned feldspar microlites in an abundant groundmass of secondary microcrystalline fibrous chlorite and granular epidote and quartz. The slightly LREE-depleted rocks (BM 304; Fig. 9D), located at the base of the sequence, differ by the presence of orthopyroxene pseudomorphs and the absence of olivine.

In conclusion, the Triassic volcanic rocks of the Seven Devils Group (Vallier, 1974, 1977; Vallier and others, 1977) show an evolution with time from typical island-arc tholeiites to calc-alkaline suites. In the Huntington Formation, the depleted tholeiites are restricted to the lowermost flows and the calc-alkaline rocks are lacking; but most of the lavas display the same features as the Seven Devils Group contemporaneous flows. These data suggest that the Huntington Formation and the Seven Devils Group belong to the same arc. The northernmost olivine low-K tholeiites, sampled north of Pittsburg Landing, are of Late Triassic age, and thus represent lateral facies variation of the LREE-enriched tholeiites and/or calc-alkaline uppermost flows of the Seven Devils Group.

SUMMARY AND CONCLUSIONS

Rocks in the eastern Klamath Mountains were affected by polyphase tectonism. The initial phase of tectonism occurred during the late Middle Devonian to Early Mississippian, and possibly was equivalent to the Antler orogeny, defined in Nevada. This orogenic event resulted in the amalgamation of oceanic remnants, Trinity ophiolite, and Copley-Balaklala island arc, with detritic sedimentary units of the Yreka-Callahan assemblage and metamorphosed basic rocks, limestone, and micaceous quartzitic schist of the Central Metamorphic Belt. On this amalgamated basement, a thick turbiditic and carbonate sequence with episodic flows and tuffs of the Bragdon, Baird, and McCloud Formations was deposited during Mississippian to Lower Permian time (Watkins, 1985; Miller, 1989). Mid-Permian island-arc volcanism resumed with the eruption of differentiated tholeiitic suites and uncommon calc-alkaline flows (Martin, 1989). At the end of the volcanic episode, rifting occurred in the arc, as suggested by the presence of the Bully Hill bimodal volcanism (Martin, 1989) and olistostromal deposits at the base of the Late Permian–Upper Triassic Pit Formation (Eastoe and others, 1987). The presence of a Late Permian–Early Triassic compressive event (Sonoma orogeny) is unlikely here, because no unconformity has been reported between the Pit Formation and the underlying rocks; in contrast with the northern Sierra terrane (Harwood, 1983, 1988) and the Tobin Range (Nevada) (Silberling, 1973), where the Middle and Upper Triassic rocks rest unconformably on deformed Paleozoic island-arc and oceanic rocks. In the Redding section, according to our data, the style and attitude of folds and associated cleavage are identical in Permian and Triassic rocks, which suggests they were all deformed in a post-Triassic event.

The Permian-Triassic history recorded in the eastern Klamath Mountains presents some similarities with that in the Blue Mountains. In both areas, Permian island-arc rocks present and exhibit some common features: (1) predominance of low-K tholeiitic suites with abundant acidic volcanics, (2) minor LREE-enriched basalts, and (3) similar $\epsilon_{(Nd)}$ (T) + 4 to + 8. These features suggest that these island-arc tholeiites were derived from a similar mantle source. In both areas, the Sonoma orogeny is probably lacking. One difference between the eastern Klamath and Blue Mountains lies probably in the fauna provinciality, but this problem remains rather unclear in general (Renne and Scott, 1988). In particular, we do not know if the Leonardian and Guadalupian brachiopods found in the Hunsaker limestone present faunal affinities with the McCloud Belt.

Faunal correlation between the eastern Klamath and Blue Mountains has been suggested for post-Permian rocks (Hosselkus and Martin Bridge limestones; Newton, 1986). Gravity and paleomagnetism data (Wilson and Cox, 1980; Hillhouse and others, 1982; Fagin and Goose, 1983; Mankinen and others, 1984; Riddihough and others, 1986; Renne and Scott, 1988; Renne and others, 1988; Mankinen and others, 1989) are consistent with the assumption that the eastern Klamath and Blue Mountain rocks were part of a single block during Permian-Triassic times. Likely, this block could also include two other correlatives. The first one, as proposed by Mankinen and others (1989), includes all or part of the Stikine terrane of British Columbia, as suggested by paleontologic and paleomagnetic data (Ross and Monger, 1978; Irving and Monger, 1987), but a precise comparison of the magmatic affinities of the volcanic rocks is yet to be done. The second one is represented by some units now cropping out at Quinn River Crossing, in northwestern Nevada; as they show striking similarities in faunal assemblages and lithology with the eastern Klamath (Skinner and Wilde, 1965; Ketner and Wardlaw, 1981; Silberling and Jones, 1982; E. A. Jones, oral communication, 1989; Jones, this volume); however, no paleomagnetic evidence is available, due to Cretaceous to Tertiary remagnetization (Russel and others, 1982). Therefore, the late Paleozoic tectonic evolution of rocks in the eastern Klamath and Blue Mountains can be tentatively summarized as follows: during middle to Late Permian time, rocks in the eastern Klamath and Blue Mountains were part of the same arc, maybe together with some other elements cited above. Late Permian intra-arc rifting occurred, followed by the splitting of the arc. Volcanism continued during the Triassic in the Blue Mountains, the active arc, while the eastern Klamath represented the remnant arc. A back-arc basin developed in which the Pit Formation of the eastern Klamath was deposited. At the end of the island-arc activity, a carbonate platform was deposited. Upper Triassic carbonate rocks, the Hosselkus Limestone, is now found either as sedimentary cover to the arc or as olistoliths in the back-arc basin sedimentary sequence (Eastoe and Nelson, 1988).

ACKNOWLEDGMENTS

Field studies were supported by the ASP Cordillères américaines and the CNRS (action initiative). We thank the U.S. Geological Survey, Menlo Park, for providing a field vehicle, and T. Vallier for his assistance and discussions in the field and at Menlo Park. The geochemical data were financed by the UA 736, Ophiolites et Paléosutures, and by the URA 1366, Modèles géotectoniques et Métallogénie. Constructive reviews of the manuscript by J. Dilles, D. S. Harwood, M. M. Miller, and T. L. Vallier are appreciated.

REFERENCES CITED

Albers, J. P., and Bain, J. H., 1985, Regional setting and new information on some critical geologic features of the west Shasta District, California: Economic Geology, v. 80, p. 2072–2091.

Albers, J. P., and Robertson, J. F., 1961, Geology and ore deposits of east Shasta copper-zinc district, Shasta County, California: U.S. Geological Survey Professional Paper 338, 107 p.

Alibert, C., Michard, A., and Albarède, F., 1983, The transition from alkali basalts to kimberlites; Isotope and trace element evidence from melilite: Contributions to Mineralogy and Petrology, v. 82, p. 176–186.

Barker, F., Millard, H. T., Jr., and Knight, R. J., 1979, Reconnaissance geochemistry of Devonian island-arc volcanic and intrusive rocks, West Shasta district, California, *in* Barker, F., ed., Trondjhemites, dacites, and related rocks: Berlin, Elsevier, p. 531–545.

Boucot, A. J., 1973, Pre-late Middle Devonian biostratigraphy of the eastern Klamath Belt, northern California: Geological Society of America Abstracts with Programs, v. 5, p. 15.

Boucot, A. J., and Potter, A. W., 1977, Middle Devonian orogeny and biogeographical relations in areas along the North American Pacific rim, *in* Murphy, W. A., Berry, N. A., and Sandberg, C. A., eds., Western North America; Devonian: Riverside, University of California Campus Museum Contribution 4, p. 210–219.

Boucot, A. J., and 7 others, 1973, Biogeographic relations of the pre-late Middle Devonian of the eastern Klamath Belt, northern California: Geological Society of America Abstracts with Programs, v. 5, p. 14.

Boucot, A. J., Dunkle, D. H., Potter, A. W., Savage, N. M., and Rohr, D. M., 1974, Middle Devonian orogeny in western North America?; A fish and other fossils: Journal of Geology, v. 82, p. 691–708.

Boudier, F., Le Sueur, E., and Nicolas, A., 1989, Structure of an atypical ophiolite; The Trinity complex, eastern Klamath Mountains, California: Geological Society of America Bulletin, v. 101, p. 820–833.

Brooks, E. R., and Coles, D. G., 1980, Use of immobile trace elements to determine original tectonic setting of eruption of metabasalts, northern Sierra

Nevada, California: Geological Society of America Bulletin, part I, v. 91, p. 665–671.

Brooks, H. C., 1979, Geologic map of the Huntington and part of the Olds Ferry Quadrangles, Baand Malheur Counties, Oregon: Oregon Department of Geology and Mineral Industries Geological Map Series GMS 35, scale 1:62,500.

Brooks, H. C., and Vallier, T. L., 1978, Mesozoic rocks and tectonic evolution of eastern Oregon and western Idaho, *in* Howell, D. G., and McDougall, K. A., eds., Mesozoic paleogeography of the western United States: Pacific Section, Society of Economic Paleontologists and Mineralogists Pacific Coast Paleogeography Symposium 2, p. 147–161.

Brouxel, M., and Lapierre, H., 1988, Geochemical study of an Early Paleozoic island-arc-back-arc basin system; Part 1, The Trinity ophiolite (northern California): Geological Society of America Bulletin, v. 100, p. 1111–1119.

Brouxel, M., Lapierre, H., Michard, A., and Albarède, F., 1987, The deep layers of a Paleozoic; Geochemistry of the Copley-Balaklala series, northern California: Earth and Planetary Science Letters, v. 86, p. 386–400.

Brouxel, M., Lapierre, H., Michard, A., and Albarede, F., 1988, Geochemical study of an early Paleozoic island-arc–back-arc basin system; Part 2, Eastern Klamath, early to middle Paleozoic island arc volcanic rocks (northern California): Geological Society of America Bulletin, v. 100, p. 1120–1130.

Brouxel, M., Lapierre, H., and Zimmermann, J. L., 1989a, Late Jurassic mafic magmas of the eastern Klamath Mountains (northern California); Remnant of a volcanic arc built on a young continental crust: Geology, v. 17, p. 273–276.

Brouxel, M., Lecuyer, C., and Lapierre, H., 1989b, Diversity of magma types in a lower Paleozoic island arc-marginal basin system (eastern Klamath Mountains, California, U.S.A.): Chemical Geology, v. 77, p. 251–264.

Burchfiel, B. C., and Davis, G. A., 1972, Structural framework and evolution of the southern part of the Cordilleran orogen, western United States: American Journal of Science, v. 272, p. 97–118.

Cannat, M., 1983, Cinématique de charriages ophiolitiques (Klamath, Semail, Groix) et Convergence océanique (Thèse de 3ème Cycle): Nantes, France, Université de Nantes, 146 p.

Cannat, M., and Boudier, F., 1985, Structural study of intra-oceanic thrusting in the Klamath Mountains, northern California; Implications on accretion geometry: Tectonics, v. 4, p. 435–459.

Cashman, S. M., 1980, Devonian metamorphic event in the northeastern Klamath Mountain, California: Geological Society of America Bulletin, v. 91, p. 453–459.

Charvet, J., Lapierre, H., and Campos, C., 1989, Mise en évidence de la phase Antler dans les Klamath orientales (N. Californie): Comptes Rendus de l'Académie des Sciences de Paris, v. 308, p. 1629–1635.

Churkin, M., Jr., 1974, Paleozoic marginal ocean basin–volcanic arc systems in the Cordilleran foldbelt, *in* Dott, R. J., and Shaver, R. H., eds., Modern and ancient geosynclinal sedimentation: Society of Economic Paleontologists and Mineralogists, p. 174–192.

Coogan, A. H., 1960, Stratigraphy and paleontology of the Permian Nosoni and Dekkas Formations (Bollibokka Group): Berkeley, University of California Publications in the Geological Sciences, v. 5, p. 243–316.

Cotkin, S. J., 1986, Mineralogy and petrology of blueschist from the schist of Skookum Gulch, eastern Klamath Mountain, California: Geological Society of America Abstracts with Programs, v. 18, p. 96.

——, 1987, Conditions of metamorphism in an Early Paleozoic blueschist; Schist of Skookum Gulch, northern California: Contributions to Mineralogy and Petrology, v. 96, p. 192–200.

Cotkin, S. J., and Armstrong, R. L., 1987, Geochemistry and tectonic significance of blueschists from the schist of Skookum Gulch, eastern Klamath Mountains, California; Introducing the Callahan event: Geological Society of America Abstracts with Programs, v. 19, p. 367–368.

Davis, G. A., 1969, Westward thrust faulting in the south-central Klamath Mountains: Geological Society of America Bulletin, v. 79, p. 911–934.

——, 1977, Tectonic evolution of the Pacific Northwest; Precambrian to present: Washington Public Power Supply System Preliminary Safety Analysis Report, Subappendix 2R.C., Amendment 23, Nuclear Project 1, p. i-2R-c. 46.

Davis, G. A., Monger, J.W.H., and Burchfiel, B. C., 1978, Mesozoic construction of the Cordilleran "collage," central British Columbia to central California, *in* Howell, D. G., and McDougall, K. A., eds., Mesozoic paleogeography of the western United States: Pacific Section, Society of Economic Paleontologists and Mineralogists Pacific Coast Paleogeography Symposium 2, p. 1–32.

Day, H. W., Moores, E. M., and Tuminas, A. C., 1985, Structure and tectonics of the northern Sierra Nevada: Geological Society of America Bulletin, v. 96, p. 436–450.

Dickinson, W. R., 1979, Mesozoic forearc in central Oregon: Geology, v. 7, p. 166–170.

Diller, J. S., 1906, Description of the Redding Quadrangle, California: U.S. Geological Survey Geological Atlas Folio 138, 14 p.

Eastoe, C. J., and Nelson, S. E., 1988, A Permian Kuroko-type hydrothermal system, Afterthought-Ingot area, and geochemical evolution: Economic Geology, v. 83, p. 588–605.

Eastoe, C. J., Gustin, M. M., and Nelson, S. E., 1987, Problems of recognition of olistostromes; An example from the Lower Pit Formation, eastern Klamath, California: Geology, v. 15, p. 541–544.

Evensen, N. M., Hamilton, P. J., and O'Nions, R. K., 1978, Rare earth abundances in chondritic meteorites: Geochimica et Cosmochimica Acta, v. 42, p. 1199–1212.

Fagin, S. W., and Goose, W. A., 1983, Paleomagnetic data from the Redding section in eastern Klamath belt, northern California: Geology, v. 1, p. 505–508.

Faure, G., Principles of isotope geology, 2nd ed.: New York, Wiley, 589 p.

Flory, B. A., and Watkins, R., 1983, Biogeography of Lower Paleozoic tabulate corals from northern California: Geological Society of America Abstracts with Programs, v. 15, p. 409.

Gill, J. B., 1981, Orogenic andesites and plate tectonics: Berlin, Springer-Verlag, 390 p.

Harwood, D. S., 1983, Stratigraphy of upper Paleozoic volcanic rocks and regional unconformities in part of the northern Sierra terrane, California: Geological Society of America Bulletin, v. 94, p. 413–422.

——, 1988, Tectonism and metamorphism in the northern Sierra terrane, northern California, *in* Ernst, W. G., ed., Tectonic and metamorphic evolution of the U.S. Cordillera; Rubey Volume 7: Englewood Cliffs, New Jersey, Prentice-Hall, p. 764–788.

Hellwig, J., 1974, Eugeosynclinal basement and a collage concept for orogenic belts: Society of Economic Paleontologists and Mineralogists Special Publication 19, p. 359–376.

Hillhouse, J. W., Grommé, C. S., and Vallier, T. L., 1982, Paleomagnetism and Mesozoic tectonics of the Seven Devils volcanic arc in northeastern Oregon: Journal of Geophysical Research, v. 87, p. 3777–3794.

Hotz, P. E., 1978, Geology of the Yreka Quadrangle, Siskyou County, California: U.S. Geological Survey Bulletin 1436, 72 p.

Hotz, P. E., Lanphere, M. A., and Swanson, D. A., 1977, Triassic blueschist from northern California and north-central Oregon: Geology, v. 5, p. 659–663.

Irving, E., and Monger, J.W.H., 1987, Preliminary paleomagnetic results from the Permian Asitka group, British Columbia: Canadian Journal of Earth Sciences, v. 24, p. 1490–1497.

Irwin, W. P., 1981, Tectonic accretion of the Klamath Mountains, *in* Ernst, W. G., ed., The geotectonic development of California; Rubey Volume 6: Englewood Cliffs, New Jersey, Prentice-Hall, p. 29–49.

——, 1985, Age and tectonics of plutonic belts in accreted terranes of the Klamath Mountains, California and Oregon, *in* Howell, D. G., ed., Tectonostratigraphic terranes of the circum-Pacific region: American Association of Petroleum Geologists Earth Sciences Series 1, p. 187–199.

Jacobsen, S. B., Quick, J. E., and Wasserburg, G. J., 1984, An Nd and Sr isotopic study of the Trinity Peridotite; Implications for mantle evolution: Earth and Planetary Science Letters, v. 68, p. 361–378.

Johnson, J. G., 1971, Timing and coordination of orogenic, epiorogenic, and eustatic events: Geological Society of America Bulletin, v. 82, p. 3263–3298.

Ketner, K. B., 1977, Deposition and deformation of Lower Paleozoic western facies rocks, northern Nevada, *in* Stewart, J. H., Stevens, C. J., and Fritsche, A. E., eds., Paleozoic paleogeography of western United States: Pacific section, Society of Economic Paleontologists and Mineralogists Pacific Coast Paleogeographic Symposium 1, p. 251–258.

Ketner, K. B., and Wardlaw, B. P., 1981, Permian and Triassic rocks near Quinn River Crossing, Humboldt County, Nevada: Geology, v. 9, p. 123–126.

Kinkel, A. R., Hall, W. E., Jr., and Albers, J. P., 1956, Geology and base-metal deposits of west Shasta copper-zinc district, Shasta County, California: U.S. Geological Survey Professional Paper 285, 156 p.

Lapierre H., Albarède, F., Albers, J., Cabanis, B., and Coulon, C., 1985a, The early Devonian volcanism in the eastern Klamath Mountains, California; Evidence for an immature island arc: Canadian Journal of Earth Sciences, v. 22, p. 214–227.

Lapierre, H., Cabanis, B., Coulon, C., Brouxel, M., and Albarède, F., 1985b, The geodynamic setting of the Early Devonian Kuroko-type sulfide deposits in the eastern Klamath Mountains (northern California) inferred by the petrological and geochemical characteristics of associated island-arc volcanic rocks: Economic Geology, v. 80, p. 2100–2113.

Lapierre, H., and 7 others, 1987, Paleozoic to Lower Mesozoic magmas from the eastern Klamath Mountains (Northern California) and the geodynamic evolution of northwestern America: Tectonophysics, v. 140, p. 155–177.

Lapierre, H., and 5 others, 1988, Les magmatismes d'arc permo-triasiques des Blue Mountains (Oregon oriental): corrélations possibles avec les unités voisines de Californie: Comptes Rendus de l'Académie des Sciences de Paris, v. 306, p. 1103–1108.

Lindsley-Griffin, N., 1977, The Trinity ophiolite, Klamath mountains, California, *in* Coleman, R. G., and Irwin, W. P., eds., North American ophiolites: Oregon Departement of Geology and Mineral Industries Bulletin 95, p. 107–120.

Lindsley-Griffin, N., and Griffin, J. R., 1983, The Trinity terrane; An early Paleozoic microplate assemblage, *in* Stevens, C. H., ed., Pre-Jurassic rocks in western North American suspect terranes; Pacific section, Society of Economic Paleontologists and Mineralogists, p. 63–75.

Mankinen, E. W., Irwin, W. P., and Gromme, C. S., 1989, Paleomagnetic study of eastern Klamath terrane, California, and implications for the tectonic history of the Klamath Mountains: Journal of Geophysical Research, v. 94, p. 10444–10472.

Martin, P., 1989, Le volcanisme permien d'arc insulaire des Klamath orientales (Californie, USA); Pétrogenèse et implications géodynamiques (Thèse d'Université): Nancy, France, Université de Nancy I, 172 p.

Mattinson, J. M., and Hopson, C. A., 1972, Paleozoic ages of rocks from ophiolitic complexes in Washington and northern California: Carnegie Institution of Washington Year Book, v. 71, p. 578–583.

Maury, R. C., and Westercamp, D., 1985, Variations chronologiques et spatiales des basaltes néogènes des Petites Antilles, Implications sur l'evolution de l'arc: Géodynamique des Caraïbes, Symposium Paris, Février 1985: Paris, Technip, p. 77–89.

Michard, A., Gurriet, P., Soudant, M., and Albarède, F., 1985, Nd isotopes in French Phanerozoic shales; External vs. internal aspects of crustal evolution: Geochimica et Cosmochimica Acta, v. 49, p. 601–610.

Miller, M. M., 1987, Dispersed remnants of a northeast Pacific fringing arc; Upper Paleozoic terranes of Permian McCloud faunal affinity, western U.S.: Tectonics, v. 6, p. 807–830.

——, 1989, Intra-arc sedimentation and tectonism; Late Paleozoic evolution of the eastern Klamath terrane, California: Geological Society of America Bulletin, v. 101, p. 170–189.

Miller, M. M., and Wright, J. E., 1987, Paleogeographic implications of Permian Tethyan corals from the Klamath Mountains, California: Geology, v. 15, p. 266–269.

Monger, J.W.H., 1977, Upper Paleozoic rocks of western Cordillera and their bearing on Cordilleran evolution: Canadian Journal of Earth Sciences, v. 14, p. 1832–1859.

Mortimer, N., 1984, Petrology and structure of Permian to Jurassic rocks near Yreka, Klamath Mountains, California [Ph.D. thesis]: Stanford, California, Stanford University, 83 p.

Mullen, E. D., 1985, Petrologic character of Permian and Triassic greenstones from Melange terranes of eastern Oregon and their implications for terrane origin: Geology, v. 13, p. 131–134.

Murphy, M. A., Power, J. D., and Johnson, J. G., 1984, Evidence for Late Devonian movement within the Roberts Mountains allochthon, Roberts Mountains, Nevada: Geology, v. 12, p. 20–23.

Newton, C. R., 1986, Late Triassic bivalves of the Martin Bridge Limestone, Hells Canyon, Oregon, U.S. Geological Survey Professional Paper 1435, p. 7–21.

Potter, A. W., Hotz, P. E., and Rohr, D. M., 1977, Stratigraphy and inferred tectonic framework of lower Paleozoic rocks in the eastern Klamath Mountains, northern California, *in* Stewart, J. H., Stevens, C. H., and Fritsche, A. E., eds., Paleozoic paleogeography of the western United States: Pacific section: Society of Economic Paleontologists and Mineralogists Pacific Coast Paleogeographic Symposium 1, p. 421–441.

Quick, J. E., 1981, Petrology and petrogenesis of the Trinity peridotite; An upper diapir in the eastern Klamath Mountain, northern California: Journal of Geophysical Research, v. 87, p. 3831–3848.

Renne, P. R., and Scott, G. R., 1988, Structural chronology, oroclinal deformation, and tectonic evolution of the southeastern Klamath Mountains, California: Tectonics, v. 7, p. 1223–1242.

Renne, P. R., Scott, G. R., and Bazard, D. R., 1988, Multicomponent paleomagnetic data from the Nosoni Formation; Cratonic Permian primary directions from Jurassic overprints: Journal of Geophysical Research, v. 93, p. 3387–3400.

Riddihough R., Finn, C., and Couch, R., 1986, Klamath–Blue Mountain lineament, Oregon: Geology, v. 14, p. 528–531.

Roberts, R. J., 1964, Stratigraphy and structure of the Antler Peak Quadrangle: U.S. Geological Survey Professional Paper 459, 93 p.

Roberts, R. J., Hotz, P. E., Gilluly, J., and Ferguson, H. G., 1958, Paleozoic rocks of north-central Nevada: American Association of Petroleum Geologists Bulletin, v. 42, p. 2813–1857.

Rogers, J.J.W., and 7 others, 1974, Paleozoic and Lower Mesozoic volcanism and continental growth in the western United States: Geological Society of America Bulletin, v. 85, p. 1913–1924.

Ross, C. A., and Monger, J.W.H., 1978, Carboniferous and Permian fusulinaceans from the Omineca Mountains, British Columbia: Geological Survey of Canada Bulletin, v. 267, p. 43–64.

Rouer, O., and Lapierre, H., 1989, Comparison between two Paleozoic island-arc terranes in northern California (eastern Klamath and northern Sierra Nevada); Geodynamic constraints: Tectonophysics, v. 169, p. 341–349.

Roure, F., 1984, Une coupe géologique de Golconda au Pacifique (Oregon, nord-ouest du Nevada, nord de la Californie); Évolution mésozoïque et cénozoïque de la marge ouest-américaine [thèse de Doctorat d'Etat]: Paris, France, Université Pierre et Marie Curie, 250 p.

Roure, F., and Sosson, M., 1986, Late Jurassic collision between a composite exotic block and the North American continent; A model for the Cordillera building: Bulletin de la Société Geologique de France, 8, II, v. 6, p. 945–959.

Russel, B. J., Beck, M. E., and Speed, R. C., 1982, Cretaceous magnetizations in northwestern Nevada and tectonic implications: Geology, v. 10, p. 423–428.

Sanborn, A. F., 1960, Geology and paleontology of the southwest quarter of the Big Bend Quadrangle, Shasta County, California: California Division of Mines and Geology Special Report 63, 26 p.

Sarewitz, D., 1983, Seven Devils terrane; Is it really a piece of Wrangellia? Geology, v. 2, p. 634–637.

Schweickert, R. A., and Lahren, M. M., 1987, Continuation of Antler and Sonoma orogenic belts to the eastern Nevada, California, and late thrusting in a compressional arc: Geology, v. 15, p. 270–273.

Schweickert, R. A., Armstrong, R. L., and Harakal, J. E., 1980, Lawsonite blueschist in the northern Sierra Nevada, California: Geology, v. 8, p. 27–31.

Silberling, N. J., 1973, Geologic events during Permian-Triassic time along Pacific margin of the United States, *in* Logan, A., and Hills, L. V., eds., The Permian and Triassic system and their mutual boundary: Canadian Society of Petroleum Geologists, v. 2, p. 345–362.

Silberling, N. J., and Jones, D. L., 1982, Tectonic significance of Permian-Triassic strata in northwestern Nevada and northern California: Geological Society of America Abstracts with Programs, v. 14, p. 234.

Silberling, N. J., Jones, D. L., Blake, M. C., Jr., and Howell, D. G., 1984, Lithotectonic terrane map of the western conterminous United States, *in* Silberling, N. J., and Jones, D. L., eds., Lithotectonic maps of the North American Cordillera, U.S. Geological Survey Open-File Report 84-523, p. C1-C43.

Skinner, J. W., and Wilde, G. L., 1965, Permian biostratigraphy and fusulinid faunas of the Shasta Lake area, northern California: Lawrence University Kansas Paleontology Contribution, Protozoa, art. 6, 98 p.

Stewart, J. H., Murchey, B., Jones, D. L., and Wardlaw, B. R., 1986, Paleontologic evidence for complex tectonic interlayering of Mississippian to Permian deep-water rocks of the Golconda allochthon in Tobin Range, north-central Nevada: Geological Society of America Bulletin, v. 97, p. 1122–1132.

Vallier, T. L., 1974, Preliminary report on the geology of parts of the Snake River Canyon: Oregon Department of Geology and Mineral Industries Geological Map Series GMS-6, 28 p., scale 1:125,000.

—— , 1977, The Permian and Triassic Seven Devils Group, western Idaho and northeastern Oregon: U.S. Geological Survey Professional Paper 1437, 58 p.

—— , 1989, Geology of pre-Tertiary igneous rocks in the Blue Mountains province of Oregon and Idaho; Implications for the geologic evolution of a complex island arc: U.S. Geological Survey Professional Paper 1438 (in press).

Vallier, T. L., Brooks, H. C., and Thayer, T. P., 1977, Paleozoic rocks of eastern Oregon and western Idaho, *in* Stewart, J. H., Stevens, C. H., and Fritsche, A. E., eds., Paleozoic paleogeography of the western United States, Pacific Section, Society of Economic Paleontologists and Mineralogists Pacific Coast Symposium, p. 455–466.

Wallin, E. T., 1986, Eocambrian–Early Cambrian plutonic rocks in the eastern Klamath province, northern California: Geological Society of America Abstracts with Programs, v. 18, p. 195.

Wallin, E. T., Mattinson, J. M., and Potter, A. W., 1988, Early Paleozoic magmatic events in the eastern Klamath Mountains, northern California: Geology, v. 16, p. 144–148.

Watkins, R., 1973, Carboniferous faunal associations and stratigraphy, Shasta County, northern California: American Association of Petroleum Geologists Bulletin, v. 57, p. 1743–1764.

—— , 1985, Volcaniclastic and carbonate sedimentation in late Paleozoic island-arc deposits, eastern Klamath Mountains, California: Geology, v. 13, p. 709–713.

—— , 1986, Late Devonian to Early Carboniferous turbidite facies and basinal development of the eastern Klamath Mountains, California: Sedimentary Geology, v. 49, p. 51–71.

Wilson, D., and Cox, A., 1980, Paleomagnetic evidence for tectonic rotation of Jurassic plutons in Blue Mountains, eastern Oregon: Journal of Geophysical Research, v. 85, p. 3681–3689.

Wright, J. E., 1982, Permo-Triassic accretionary subduction complex, southwestern Klamath Mountains, northern California: Journal of Geophysical Research, v. 87, p. 3805–3818.

Wright, J. E., and Fahan, M. R., 1988, An expanded view of Jurassic orogenesis on the western United States Cordillera; Middle Jurassic (pre-Nevadan) regional metamorphism and thrust faulting within an active arc environment, Klamath Mountains, California: Geological Society of America Bulletin, v. 100, p. 859–876.

MANUSCRIPT ACCEPTED BY THE SOCIETY MARCH 27, 1990

Printed in U.S.A.

Geological Society of America
Special Paper 255
1990

Comparison of early Mesozoic high-pressure rocks in the Klamath Mountains and Sierra Nevada

Bradley R. Hacker*
Department of Earth and Space Sciences, University of California, Los Angeles, California 90024
John W. Goodge
Department of Geological Sciences, Southern Methodist University, Dallas, Texas 75275

ABSTRACT

Early Mesozoic blueschist terranes in the Klamath Mountains (the Stuart Fork) and Sierra Nevada (the Red Ant), are lithologically, petrologically and structurally similar. Both contain common crossite-epidote assemblages formed during subduction of oceanic sedimentary and volcanic protoliths that had previously been metamorphosed at an ocean ridge. Both are remnants of a high-P/T metamorphic belt paired with a more easterly magmatic arc along the western margin of North America. These similarities imply that the Mesozoic blueschists in both mountain belts shared a common early history. Important differences, however, indicate that their later postsubduction histories were different: blueschists in the Klamath Mountains were preserved without overprinting by higher T assemblages, whereas blueschists in the Sierra Nevada were overprinted by pumpellyite-actinolite facies metamorphism. From these relations we infer that the Stuart Fork rocks may have been lifted upward more rapidly than the Red Ant terrane, or that the Red Ant terrane was incompletely subducted and not subject to the cooling effect of continuing subduction. The Stuart Fork blueschists were also affected by a contact metamorphic event during widespread Middle to Late Jurassic plutonism, which is not manifest in the northern Sierra Nevada blueschist terrane.

INTRODUCTION

The Klamath Mountains and the Sierra Nevada of the North American Cordillera contain a variety of rocks interpreted as magmatic arcs, oceanic crust, and subduction complexes formed during different phases and styles of ocean-continent plate interactions. Magmatic arc rocks in the eastern Klamath Mountains and eastern Sierra Nevada attest to convergence between continental North America and westward oceanic provinces from at least Devonian to Cretaceous time (e.g., Burchfiel and Davis, 1981). The Stuart Fork terrane in the Klamath Mountains and the Red Ant terrane in the Sierra Nevada are subduction complexes that developed oceanward of the more easterly early Mesozoic magmatic arcs. Both are coherent high-pressure belts of greater extent than previously appreciated, and through their multiple-generation metamorphic parageneses and structural fabrics, they provide detailed records of convergent-margin processes operative during early Mesozoic growth of the Cordillera.

This chapter compares and contrasts the structural features, phase assemblages, and mineral chemistry of the two terranes. Further details may be found in Goodge (1989a, b, 1990) and B. R. Hacker (unpublished data). Textural relations determined by back-scattered electron microscopy constrain the sequence of deformational and metamorphic events, and phase relations and mineral compositions determined by electron-probe microanalysis constrain the P-T conditions of metamorphism. The metamorphic evolution of these terranes provides the petrogenetic framework for a tectonic model linking the late Paleozoic–early Mesozoic history of the Klamath Mountains and the Sierra Nevada.

*Present address: Department of Geology, Stanford University, Stanford, California 94305-2115.

Hacker, B. R., and Goodge, J. W., 1990, Comparison of early Mesozoic high-pressure rocks in the Klamath Mountains and Sierra Nevada, *in* Harwood, D. S., and Miller, M. M., eds., Paleozoic and early Mesozoic paleogeographic relations; Sierra Nevada, Klamath Mountains, and related terranes: Boulder, Colorado, Geological Society of America Special Paper 255.

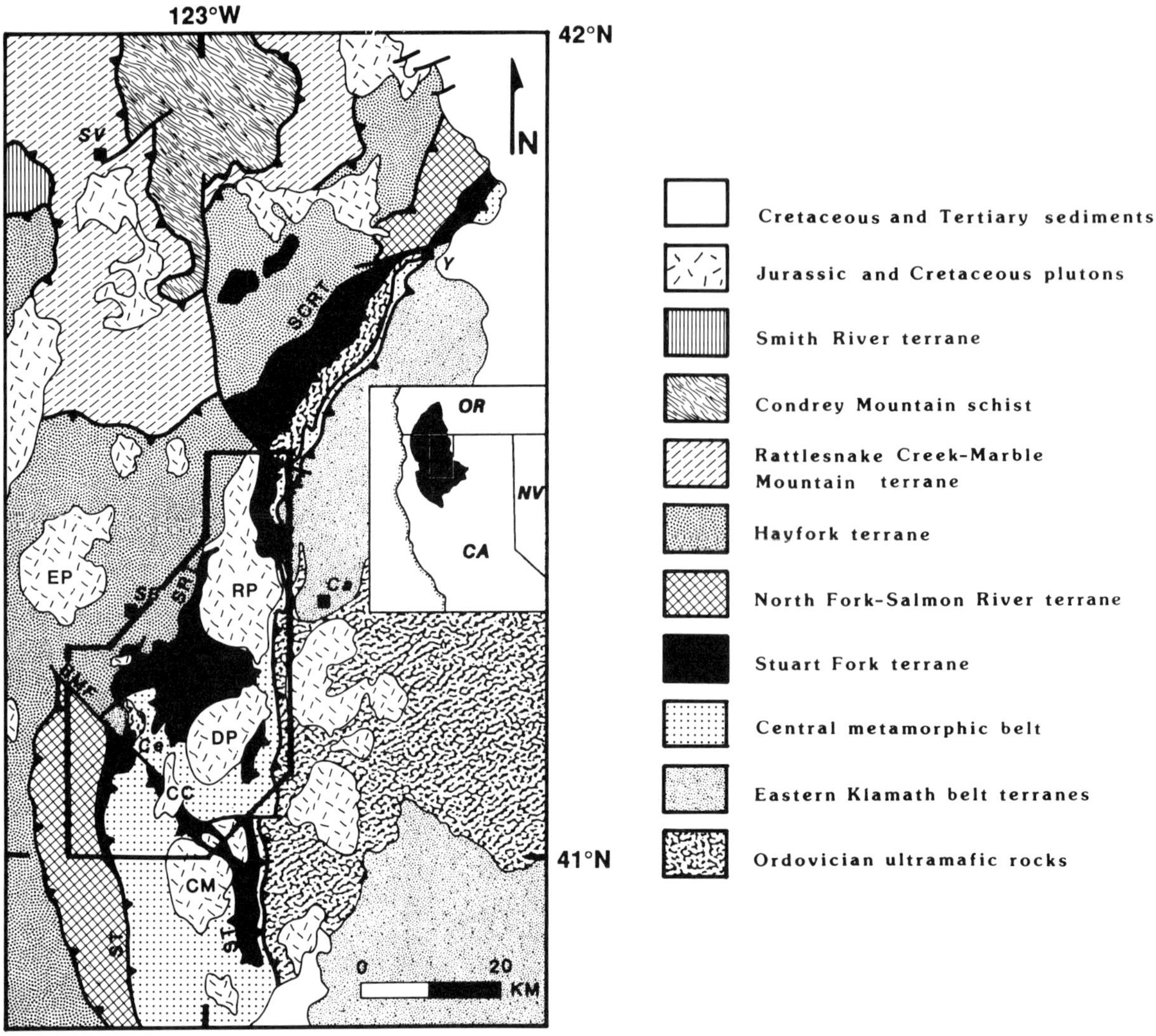

Figure 1. Regional tectonic map of the Klamath Mountains (after Strand, 1962; Hotz, 1971; Wright, 1982; Mortimer, 1984; Wagner and Saucedo, 1988). Regional thrust faults are shown by barbed lines. Eastern Klamath belt includes the Eastern Klamath and Yreka terranes and the Trinity peridotite. Western Paleozoic and Triassic belt includes the Stuart Fork, North Fork–Salmon River, Hayfork, Rattlesnake Creek–Marble Mountain terranes. Western Jurassic belt includes the Condrey Mountain and Smith River terranes. Plutons include Russian Peak (RP), Deadman Peak (DP), China Creek (CC), English Peak (EP), and Caribou Mountain (CM). Major faults include Siskiyou thrust (ST), Salmon River thrust (SRT), Soap Creek Ridge thrust (SCRT), and Brown Meadow fault (BMF). SV = Seiad Valley; Y = Yreka; SB = Sawyers Bar; Ce = Cecilville; Ca = Callahan; E = Etna. Heavy lines delineate the area of the present study.

REGIONAL GEOLOGY

Field and analytical data presented in this chapter come from study of the central Stuart Fork terrane in the vicinity of the South Fork of the Salmon River in the Klamath Mountains (Goodge, 1989a, b; Fig. 1), and the Red Ant terrane in the Yuba River area of the Sierra Nevada (B. R. Hacker, unpublished data; Fig. 2). Previous studies of these units include those of Holdaway (1963), Davis and others (1965), Hotz (1977a, b), Borns (1980, 1984), Schweickert and others (1980), Hietanen (1981), and Edelman and others (1989).

Lithotectonic belts in the Klamath Mountains form a generally westward-younging structural sequence of units typically separated by east-dipping, west-directed regional thrust faults (Fig. 1; Irwin, 1966, 1981; Burchfiel and Davis, 1981). Irwin (1966) identified four Klamath belts, from east to west, as: (1) the eastern Klamath belt, a Paleozoic to Mesozoic volcanic arc and continental margin sequence (Irwin, 1981; Miller, 1989; Brouxel and others, 1988); (2) the central metamorphic belt, a remnant of subducted Devonian oceanic crust (Davis, 1968; Cashman, 1980; Peacock and Norris, 1989; Hacker and Peacock, this volume); (3) the western Paleozoic and Triassic belt, composed of various

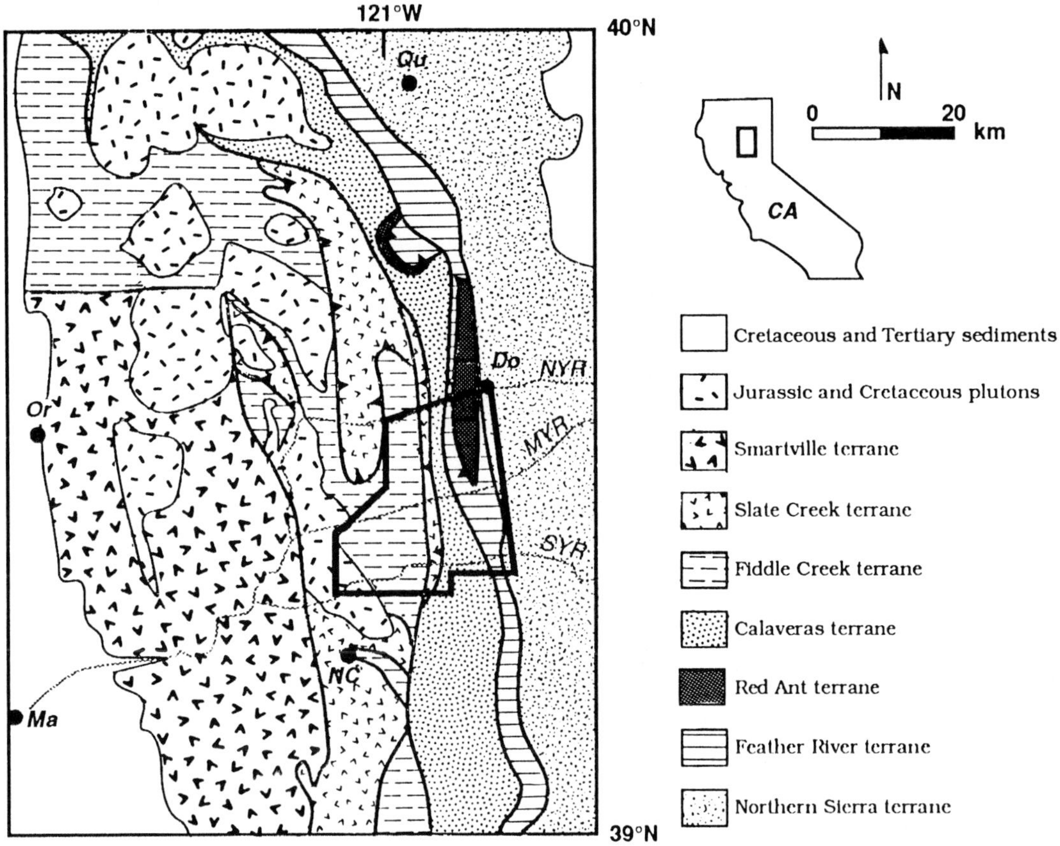

Figure 2. Regional tectonic map of the northern Sierra Nevada (after Day and others, 1988; Edelman and others, 1989). Regional thrust faults are shown by barbed lines. Post-Mesozoic rocks that overlie the lithotectonic belts are not shown. Qu = Quincy; Do = Downieville; Or = Oroville; Ma = Marysville; NC = Nevada City; NYR = North Yuba River = MYR = Middle Yuba River; SYR = South Yuba River. Heavy lines delineate the area of the present study.

accreted oceanic assemblages (Davis and others, 1978; Wright, 1982; Ando and others, 1983; Mortimer, 1984; Donato, 1987; Coleman and others, 1988); and (4) the western Jurassic belt, containing ophiolitic rocks (Saleeby and others, 1982; Harper and others, 1988), moderate- to high-P metamorphic rocks (Donato and others, 1980; Helper, 1986), and arc-derived clastic sedimentary rocks (Garcia, 1979).

In the east-central Klamath Mountains, the western Paleozoic and Triassic belt is further subdivided (from east to west) into the Stuart Fork, North Fork–Salmon River, eastern Hayfork, western Hayfork, and Rattlesnake Creek–Marble Mountain terranes (Fig. 1) (Wright, 1982; Donato and others, 1982; Ando and others, 1983; Mortimer, 1984; Wright and Wyld, 1985; Donato, 1987; Goodge, 1989b; Wright and Fahan, 1988). These western Paleozoic and Triassic belt terranes represent segments of late Paleozoic to mid-Mesozoic oceanic or island-arc complexes that have been deformed to varying degrees as melange or subduction complex during convergence with the western margin of North America (Davis and others, 1978; Irwin, 1981; Burchfiel and Davis, 1981; Coleman and others, 1988).

The Stuart Fork terrane, also referred to as the Fort Jones terrane (Blake and others, 1982; Mortimer, 1984), consists of metabasaltic and siliceous metasedimentary rocks that originated as late Paleozoic(?) oceanic crust (Goodge, 1990). It lies structurally beneath Devonian amphibolite-facies rocks of the central metamorphic belt. The two terranes are separated by the Siskiyou thrust fault that was active between Devonian and Middle Jurassic time, as constrained by the ages of metamorphism of the central metamorphic belt (Lanphere and others, 1968; Hotz, 1977b; Cashman, 1980) and plutons that cut across the thrust fault (Harper and Wright, 1984; Wright and Fahan, 1988). The Stuart Fork terrane overlies the low-grade, ophiolitic North Fork–Salmon River terrane along the Salmon River thrust, which was active between Early and Middle Jurassic time (Goodge, 1989b).

As in the Klamath Mountains, metamorphic rocks in the northern Sierra Nevada are divisible into a number of terranes with different structural, metamorphic, or depositional histories (Fig. 2). The Northern Sierra terrane is composed of lower Paleozoic metasedimentary rocks of the Shoo Fly Complex that are overlain by three successive magmatic arc complexes of Devonian-Mississippian, mid-Permian, and Jurassic age (e.g.,

D'Allura and others, 1977; Schweickert, 1981; Girty and Wardlaw, 1985; Hannah and Moores, 1986; Hanson and Schweickert, 1986; Harwood, 1988). The Feather River terrane has been interpreted as a Paleozoic ophiolite (Ehrenberg, 1975; Weisenberg and Ave Lallement, 1977; Standlee, 1978; Edelman and others, 1989). The Fiddle Creek terrane includes broken formation and melange formed in an east-dipping subduction zone during Triassic to Jurassic time (Hietanen, 1981), and the Calaveras and Red Ant terranes are interpreted to have formed in similar environments at earlier times (Hietanen, 1981; Sharp, 1988). The Slate Creek terrane is an Early Jurassic (Hietanen, 1981; Saleeby and others, 1989), immature intraoceanic arc (Day and others, 1965; Edelman and others, 1989). The Smartville terrane is a Late Jurassic rifted volcanic arc (Beard and Day, 1987) formed in Early Jurassic ophiolitic basement (Saleeby and others, 1989). A more complete description of these rock units is given by Hietanen (1981) and Edelman and others (1989).

The Red Ant terrane lies structurally below Devonian amphibolite-facies rocks of the Feather River terrane (Fig. 2). These terranes are separated by a thrust fault that was active between Devonian and Jurassic time, as constrained by ages of Feather River terrane metamorphism and the age of high-angle faults that cut the thrust fault (Edelman and others, 1989). The Red Ant terrane overlies the Calaveras terrane along a contact also interpreted as an east-dipping thrust fault that was active between late Paleozoic and Jurassic time, as indicated by the age of the Calaveras terrane and the age of high-angle faults that cut the thrust fault (Edelman and others, 1989).

ROCK TYPES

The Stuart Fork and Red Ant terranes consist primarily of metamorphosed, interlayered siliceous sedimentary and mafic volcanic rocks (Fig. 3). Metasedimentary rocks include interlayered quartzite, phyllitic quartzite, phyllite, and argillite. Phyllitic quartzite, which is characterized by alternating millimeter- to centimeter-thick layers of micaceous phyllite and quartzite, is the most common metasedimentary lithology (Fig. 3A, B). Relict bedding, rare microfossils, Mn-rich bulk compositions (as indicated by spessartine-rich garnet), and the high silica content indicate that the protolith was probably deep-sea chert. Feldspars of variable composition are the most common detrital grains in argillaceous layers, implying deposition proximal to a magmatic or metamorphic source.

The protoliths of mafic metavolcanic rocks include massive and pillowed flows in the Stuart Fork terrane, and volcanic breccia, flows, and tuffs in the Red Ant terrane (Fig. 3C, D). Pillowed metavolcanic rocks have massive cores and commonly amygdaloidal rinds; inter-pillow carbonate occurs locally. Concordant depositional contacts are exposed locally between pillowed volcanic and overlying metasedimentary rocks of the Stuart Fork terrane (Goodge, 1989b). This relation documents that, at least locally, the metavolcanic rocks are not tectonic or olistostromal blocks within a metasedimentary matrix. Contacts between metavolcanic rocks and *underlying* metasedimentary rocks are not well exposed and may be tectonic or depositional. Major-, trace-, and rare earth element abundances show that mafic metavolcanic rocks in the Stuart Fork terrane are slightly enriched mid-ocean tholeiites with compositions similar to modern mid-ocean ridge or back-arc basin basalts (Goodge, 1990), whereas relict pyroxenes from the Red Ant terrane share affinities with volcanic-arc basalts. The compositions of the metasedimentary and metavolcanic rocks thus suggest that deposition occurred in an arc-related or larger ocean basin near a source of igneous or metamorphic detritus.

STRUCTURAL RELATIONS

Structural relations, documented more completely by Holdaway (1963), Cebull and Russell (1979), Schweickert and others (1980), Hietanen (1981), Goodge (1989b), and Edelman and others (1989), provide evidence of two major deformation events: D_1, characterized by the formation of complex folds and possible southwest-directed thrust faults; and D_2, which formed simple, open folds related to the emplacement of regional thrust sheets. D_1 structures are interpreted to have formed within a Late Triassic subduction zone, and D_2 structures represent a later mid-Jurassic regional contraction.

D_1 structures

Both terranes are characterized by relatively coherent packets of complexly folded rocks bounded by east- or northeast-dipping reverse faults. Layered metasedimentary rocks contain a well-developed phyllitic to schistose foliation parallel to the compositional layering. The limbs and axial planes of early tight to isoclinal, rootless intrafolial folds are parallel to the foliation. These folds, lineations, and the foliation are folded about mesoscopic tight, similar, and parallel folds with dominantly northwest-striking axial planes and variably plunging axes.

Micaceous metasedimentary rocks containing D_1 mesoscopic structures display well-formed microscopic fabrics expressed by compositional layering and platy mineral preferred orientation. The foliation is commonly folded tightly to isoclinally. Lawsonite commonly occurs as elongate ellipsoids sheathed by white mica, and shows progressive deformation from euhedral, unbroken blocks, to segmented, curved grains. Recrystallized quartz occurs between lawsonite microboudins. Microscopic D_1 fabrics in metabasaltic rocks include a moderate to strong foliation. Although foliation is most commonly defined by amphibole and mica grain-shape preferred orientation, this fabric is in places transected by zones containing recrystallized quartz and feldspar. Segments of boudinaged phases such as epidotes are separated by strain shadows filled with recrystallized phengite and other minerals. Amphiboles and micas commonly are bent around fold hinges; in rare examples they are recrystallized to form an axial planar foliation.

There are clear relations between D_1 structures and minerals that grew during high-pressure metamorphism. Several petro-

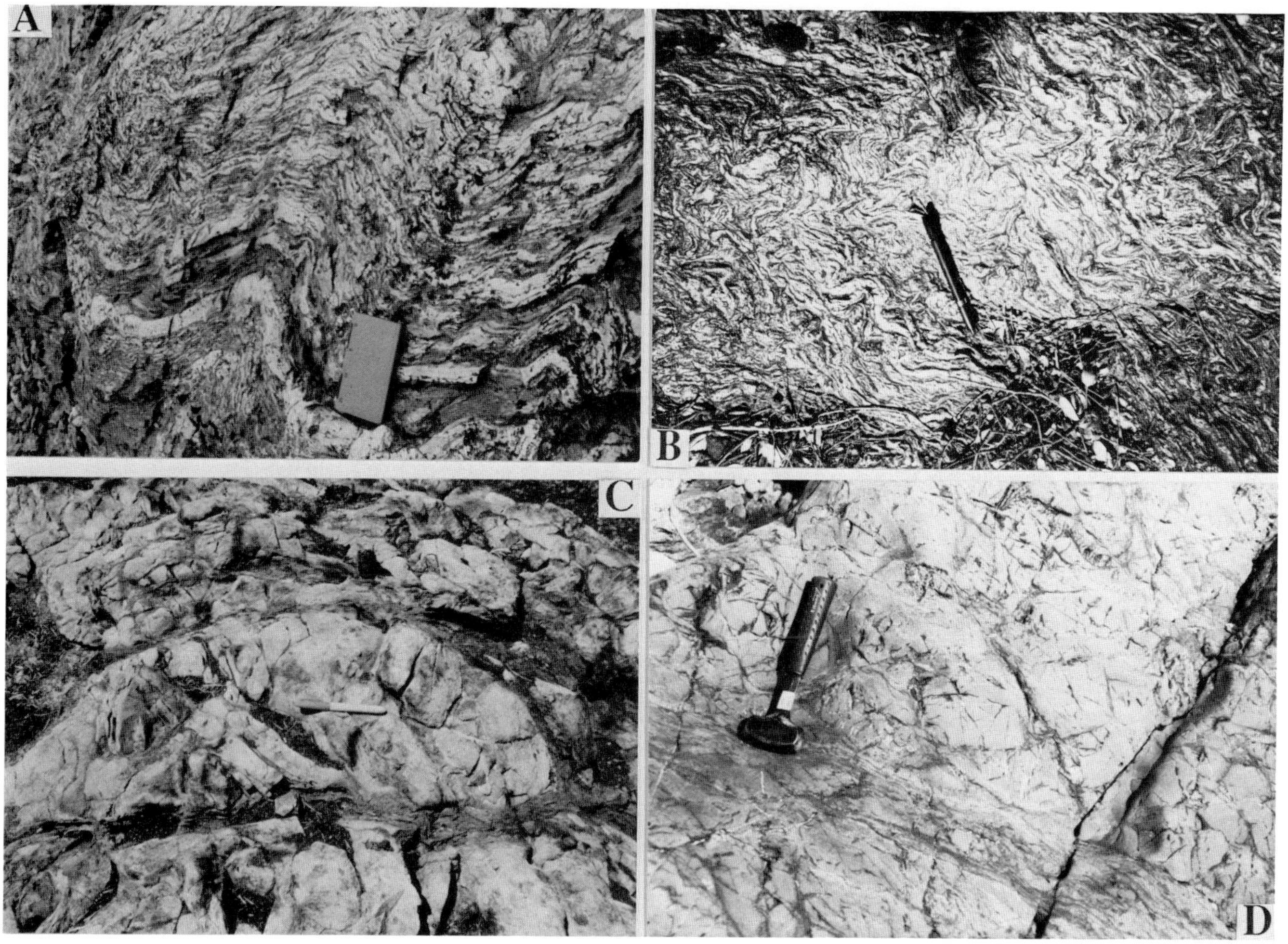

Figure 3. Complexly folded interlayered phyllitic quartzite and phyllite in the Stuart Fork (A) and Red Ant (B) terranes. Metavolcanic rocks include pillowed metabasalt in the Stuart Fork terrane (C) and tuff or massive flows in the Red Ant terrane (D).

graphic observations indicate that recrystallization took place synkinematically, and that deformation outlasted recrystallization: (1) folia formed in mica, sodic amphibole, and lawsonite crystals are folded, and metamorphic phases in fold hinges are partially recrystallized; (2) epidote and lawsonite crystals are boudinaged; (3) a complete spectrum of early, strongly deformed veins to late, weakly deformed veins containing quartz, albite, chlorite, crossite, epidote, lawsonite, and phengite, is present; (4) fibrous vein fillings, particularly of sodic amphibole, phengite, and sphene, nucleated on wall minerals and recorded movements of the vein walls in their crystal shape. There is no indication that recrystallization at high pressures occurred after deformation. Recrystallization of phases such as sodic amphibole, lawsonite, and phengite in the deformation indicates that the synmetamorphic D_1 occurred under blueschist-facies conditions, and the deformation in general is interpreted to have taken place in a subduction-zone environment. D_1 structures are not observed in units adjacent to either the Stuart Fork or Red Ant terranes.

D_2 structures

Both terranes contain decameter- to kilometer-scale open folds with east-dipping axial surfaces. These open folds are morphologically and geometrically distinct from the complex D_1 structures, and they are interpreted to represent a younger period of deformation—D_2. D_2 structures occur in the Stuart Fork terrane mostly within 1 to 2 km of the Salmon River thrust (Fig. 1), and within the underlying North Fork–Salmon River terrane (Trexler, 1968; Ando, 1979; Goodge, 1989b). The Salmon River thrust itself is considered to be a D_2 structure (Goodge, 1989b). The kilometer-scale antiformal structure within the Red Ant terrane (Fig. 2) is interpreted as a D_2 structure. Both the central metamorphic belt overlying the Stuart Fork terrane and the Feather River terrane overlying the Red Ant terrane contain west-vergent folds that may have formed coeval with D_2 structures. D_2 structures are not associated with widespread metamorphic recrytallization.

Summary

Early deformation, partly synchronous with high-pressure metamorphism, produced mesoscopic and microscopic tight to isoclinal folds and possible reverse faults in both terranes. Younger decameter- to kilometer-scale open folds of regional extent formed after high-pressure metamorphism. Constraints on timing of these events, and whether the events were coeval in the Klamath Mountains and Sierra Nevada, are discussed below.

METAMORPHISM

The fine grain size of low-grade rocks makes analysis by conventional thin-section petrography problematic; consequently, emphasis in this study was placed on determining coexisting metamorphic minerals and the sequences of metamorphic events by back-scattered electron microscopy and mineral chemistry by electron-probe microanalysis. All phase analyses were conducted at the University of California, Los Angeles, with a four-spectrometer Cameca Camebax electron-probe microanalyzer and a single set of well-characterized natural and synthetic mineral standards; this is important because rocks from two different mountain ranges are compared. Spot selection and textural analysis were performed in back-scattered electron mode. The electron beam diameter was 2 μm, at 15 kV and 10 nA. One to five counts of 20 seconds of peak intensities and 10 seconds of background intensities were made for each of 10 elements. Detection limits at these conditions are a function of mineral composition, but for this study, conservative limits are approximately 0.02 wt % for SiO_2, Al_2O_3, TiO_2, Cr_2O_3, and MnO; 0.03 wt % for FeO^*, MgO, and CaO; 0.04 wt % for K_2O; and 0.05 wt % for Na_2O. Analyses of inhomogeneous grains were not averaged, but are presented individually. All mineral formulae were calculated following the method of Laird and Albee (1981a, b), except for pumpellyite, which was normalized according to Brown and Ghent (1983). Amphibole names follow the convention of Rock and Leake (1984). "FeO^*" refers to total iron oxide, FeO + Fe_2O_3.

Equilibrium mineral assemblages can be difficult to identify in low-grade rocks, and a brief discussion of techniques used is warranted. Widespread metastable igneous and sedimentary phases indicate that complete textural and chemical equilibrium was not often attained, except in tuffaceous units where recrystallization was pervasive. A combination of techniques was used to identify equilibrium mineral assemblages, including (1) recognizing phases whose mutual boundaries do not show signs of reaction, (2) recognizing a group of phases replacing a single relict phase, (3) identifying incompatible phases, (4) identifying violations of the phase rule, and (5) observing systematic partitioning of elements among phases. It was not assumed that coexistence of minerals A + B in one sample and B + C in another sample implied that A + B + C were stable together. Many samples contain high-variance assemblages that are not useful for constraining conditions of metamorphism or identifying reaction relations among assemblages. Some samples, however, do contain low-variance assemblages, and these are emphasized in the following discussion.

Mineral parageneses reflect three periods of recrystallization in the Stuart Fork terrane (Goodge, 1989a): M_1, a prehnite-actinolite to greenschist-facies, low-P/T metamorphism; M_2, an eclogite- and blueschist-facies metamorphism; and M_3, a low-P/T greenschist- to amphibolite-facies contact metamorphism. The Red Ant terrane also contains M_1 and M_2 assemblages, but it does not contain eclogites or contact metamorphic rocks.

M_1 low-pressure metamorphism

Assemblages indicative of early low-pressure metamorphism are locally preserved within metavolcanic rocks in the Stuart Fork terrane (Goodge, 1989a). The common M_1 assemblage is chlorite + actinolite + epidote + albite ± biotite + quartz (Fig. 4). Some samples contain relict igneous diopside to augite phenocrysts (Goodge, 1990) that are partially replaced by chlorite + actinolite ± biotite (Fig. 5). Patches of mimetic actinolite and chlorite crystals are oriented randomly, indicating that M_1 recrystallization was not accompanied by penetrative deformation.

Calcic amphibole cores in metavolcanic rocks in the Red Ant terrane indicate an early phase of metamorphic recrystallization that predates growth of sodic-calcic amphibole parageneses. Although no other phases are observed in textural equilibrium with these early calcic amphibole cores, it is probable that the amphiboles grew in equilibrium with phases such as plagioclase, chlorite, epidote, quartz, and a Ti-phase (Fig. 4). Metasedimentary rocks in both terranes do not contain any phases that are demonstrably products of M_1.

M_2 high-pressure metamorphism

Minerals formed during high-pressure metamorphism (Fig. 4) grew over relict igneous or M_1 minerals, and constitute the characteristic assemblages of the Stuart Fork and Red Ant terranes. The Stuart Fork terrane contains widespread crossite-epidote assemblages in the central part of the terrane. Glaucophane-lawsonite zone rocks occur locally in the Salmon River area, but they predominate in the northern part of the terrane. Eclogites crop out in the northern part of the Stuart Fork terrane, overprinting glaucophane-lawsonite assemblages; they are believed to be in situ relics from high-temperature subduction-zone metamorphism (Borns, 1984). The Red Ant terrane contains prevalent pumpellyite-actinolite parageneses that overprint locally developed crossite-epidote assemblages, and are herein described as M_2 (Fig. 4). Eclogites do not occur in the Sierra Nevada.

Diagnostic blueschist-facies assemblages such as phengite + chlorite + quartz + sodic amphibole + lawsonite occur locally within metasedimentary rocks in both terranes (Fig. 6), but most commonly the nondiagnostic assemblage phengite + chlorite + quartz + albite ± stilpnomelane ± microcline is present. Lawson-

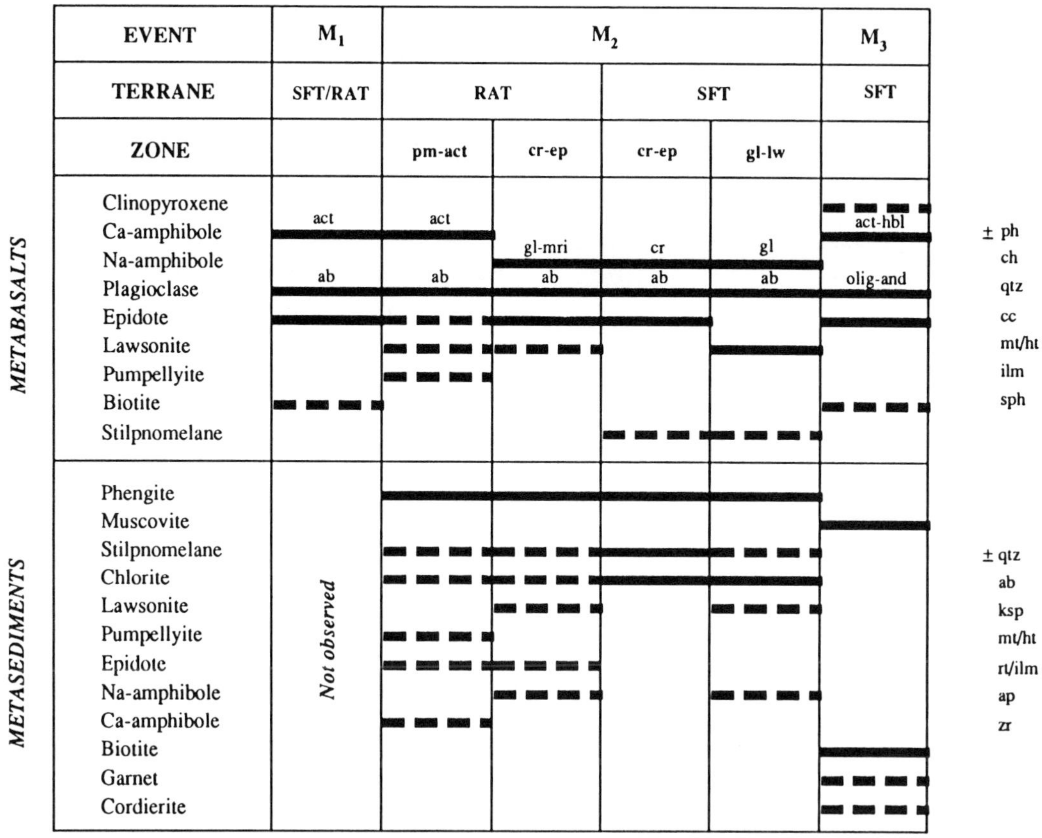

Figure 4. Mineral assemblages in the Red Ant and Stuart Fork terranes.

ite is rimmed by epidote in some samples. The common blueschist-facies assemblage in metavolcanic lithologies in the central Stuart Fork terrane is crossite + epidote + chlorite + phengite + magnetite ± stilpnomelane ± albite + quartz ± sphene (Fig. 7), whereas glaucophane-lawsonite zone rocks are predominant in the Yreka area (Hotz, 1973a, b; Borns, 1980). Blueschist-facies metavolcanic rocks in the Red Ant terrane contain epidote + sodic amphibole + hematite + chlorite + albite + quartz, and pumpellyite-actinolite facies rocks contain subassemblages of epidote + pumpellyite + actinolite + hematite/magnetite ± lawsonite + chlorite + albite + quartz.

M_3 contact metamorphism

Middle to Late Jurassic plutons intrude the Stuart Fork terrane and are surrounded by 1- to 2-km-wide aureoles containing muscovite + biotite ± garnet ± cordierite in semipelitic rocks, and hornblende + plagioclase ± epidote ± diopside ± biotite in mafic rocks (Fig. 4). In many samples, euhedral micas and amphiboles, designated as M_3, are randomly oriented and undeformed, indicating a lack of penetrative deformation during recrystallization. Contact metamorphism is unknown in the Red Ant terrane. Clumps of radiating stilpnomelane and actinolite are present on M_2 minerals in the Red Ant terrane, but their significance is unknown.

Figure 5. Back-scattered electron images of Stuart Fork metabasalts showing M_1 assemblages formed by replacement of primary igneous clinopyroxene and plagioclase assemblage. Scale bar is 100 μm. Sample GC84-13, containing nearly completely reacted clinopyroxene (cpx) phenocryst enclosed by an assemblage of actinolite (act), chlorite (ch), albite (ab), and biotite (not in view). Matrix phases (top, bottom, and left margins of photograph) surrounding relict phenocryst (center of photograph) include epidote (mottled), albite, quartz, chlorite, and fibrous M_2 crossite (cr). Crossite appears to be growing across actinolite-chlorite assemblage.

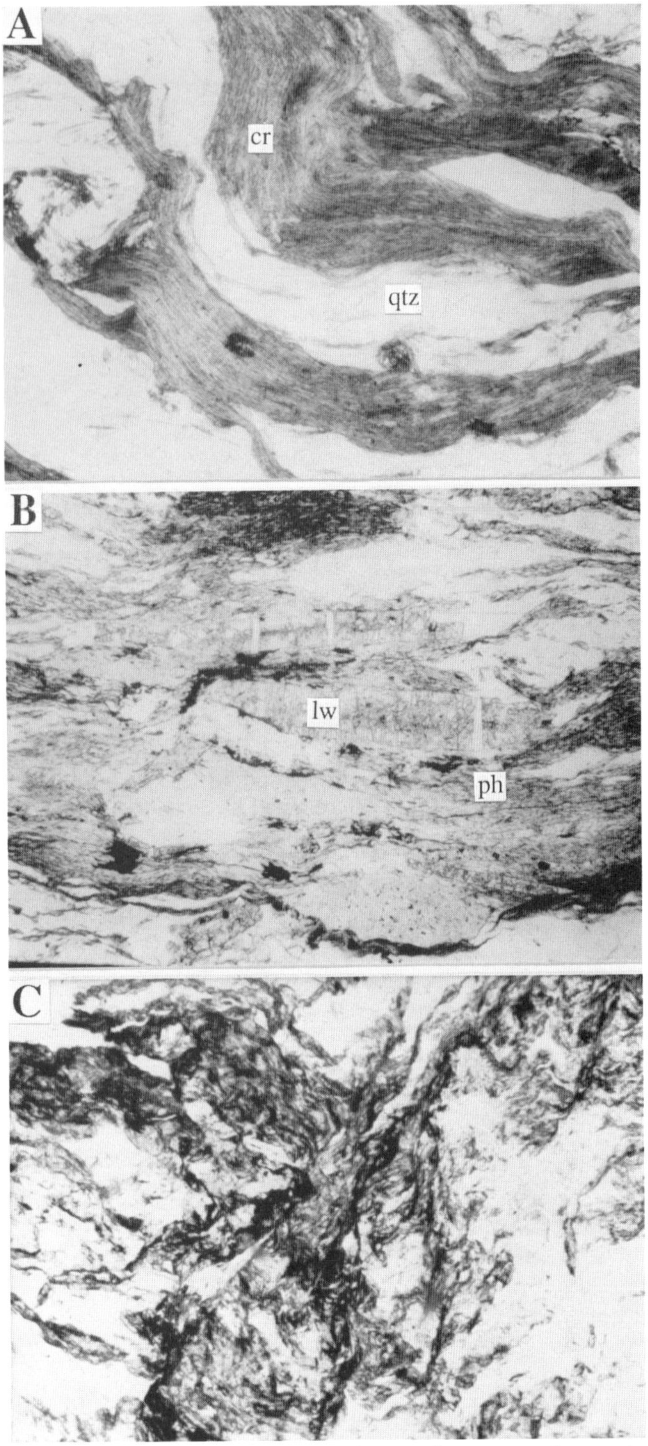

Figure 6. Optical photomicrographs of metasedimentary rocks containing M_2 assemblages. Field of view in all is approximately 1 mm. A, Fold in amphibole-mica quartzite from Stuart Fork terrane containing intergrown crossite (cr) and phengite, in addition to quartz (qtz); kink bands of amphibole in fold hinges indicate that the deformation was syn- to postmetamorphic. B, Lawsonite tablets (lw) intergrown with phengite (ph) in phyllitic quartzite from Stuart Fork terrane; note boudinage of lawsonite and growth of quartz between segments. C, Contorted folia of intergrown phengite and chlorite in Red Ant terrane phyllitic quartzite.

MINERAL CHEMISTRY

As described above, metasedimentary and metavolcanic rocks from the Stuart Fork and Red Ant terranes contain numerous phases related to each metamorphic event. Of these, the compositions of amphibole and mica provide some constraints on the conditions of metamorphism, and these are discussed here. Compositions of all phases are discussed in greater detail by Goodge (1989a) and Hacker (in review).

M_1 assemblages in the Stuart Fork terrane include chlorite, actinolite, epidote, albite, biotite, and quartz, whereas only amphibole is demonstrably an M_1 phase in the Red Ant terrane. For this reason, only M_1 amphiboles are compared, whereas M_2 amphiboles and phengites are discussed.

Amphibole

M_1 calcic amphiboles in metavolcanic rocks are uniformly actinolitic (Table 1), and M_2 sodic-calcic amphiboles are primarily winchite in both metasedimentary and metavolcanic rocks (Table 2). Sodic amphiboles span compositions from magnesioriebeckite through glaucophane. M_2 blueschist assemblages in the Stuart Fork terrane contain only one amphibole, rather than coexisting sodic or sodic-calcic amphibole and actinolite as in some high-pressure terranes (Fig. 4).

Amphibole compositions plotted in Figures 8 and 9 (after Laird and Albee, 1981b) illustrate that calcic amphiboles in M_1 assemblages are distinctly different from sodic-calcic amphiboles in M_2 assemblages. M_1 Ca-amphiboles are uniformly low in M4-site Na ($\leqslant$0.25 atoms per formula unit, pfu), and they show variable but significant A-site occupancy and tetrahedral Al; these amphiboles plot in the medium- to low-pressure facies-series field of Laird and Albee (1981b) and Hynes (1982). Ca-amphiboles in these assemblages also show strong tschermakite substitution, and minor edenite substitution. The weakly tschermakitic compositions of M_1 actinolites indicate conditions well below those of the garnet isograd in pelitic rocks (Laird and Albee, 1981b).

In contrast, sodic amphiboles from M_2 assemblages contain much Na in the M4 site (0.5 to 2.0 atoms pfu) but generally low A-site occupancy and tetrahedral A1; amphiboles from these assemblages are compositionally similar to those from the Sanbagawa and Franciscan high-pressure metamorphic belts.

Phengite

White mica in M_2 assemblages is phengite with only minor paragonite ($\leqslant$0.02 Na atoms pfu) or margarite (<0.02 Ca atoms pfu) components (Table 3). Phengite in metasedimentary rocks containing chlorite ± stilpnomelane ± lawsonite contains high Si (3.4 to 3.8 atoms pfu in the Stuart Fork terrane and 3.3 to 3.6 pfu in the Red Ant terrane). Phengite in metavolcanic rocks also contains high Si (3.6 to 3.9 pfu in the Stuart Fork terrane and 3.3 to 3.6 pfu in the Red Ant terrane), and octahedral cation substitution for Al (0.8 to 1.0 atoms in the Stuart Fork terrane and 0.2 to 0.6 in the Red Ant terrane).

CONDITIONS AND TIMING OF METAMORPHISM

M_1 low-pressure event

Stuart Fork M_1 assemblages contain actinolite, epidote, and albite, but lack zeolites, pumpellyite, hornblende, or intermediate-composition plagioclase. Temperatures inferred from phase equilibria are 300° to 450°C, and pressures are unconstrained (Goodge, 1989a); we term M_1 "low-pressure" metamorphism principally to distinguish it from the blueschist-facies metamorphism. M_1 amphiboles in both terranes (Fig. 8) are similar to medium- to low-pressure facies-series rocks metamorphosed below amphibolite grade (Laird and Albee, 1981a; Hynes, 1982). Based on the geobarometer of Brown (1977b), less than 0.20 Na in the amphibole M4 site indicates a pressure of approximately 200 MPa; because of metamorphic overprinting, we are unable to evaluate whether these amphiboles coexisted with an iron oxide, hence these may be minimum pressures. These conditions are compatible with the lowest pressure region of the greenschist facies, and overlap in part with the prehnite-actinolite facies (Liou and others, 1985). Thus, M_1 in both terranes is interpreted as a low-P/T greenschist-facies event. M_1 recrystallization may have been attended by moderate LIL-element enrichment in the Stuart Fork terrane suggestive of weak hydrothermal alteration or metasomatism (Goodge, 1990), which likely occurred near an oceanic spreading ridge or transform fault.

M_2 high-pressure event

Phase equilibria relevant to the high-pressure metamorphism are shown in Figure 10. Metamorphic pressures and temperatures can be estimated from coexisting minerals, as follows. The conditions of formation for the lawsonite + sodic amphibole rocks are limited by five of the experimentally determined equilibria shown in Figure 10 (reactions 1, 3, 5, 7, 10). The absence of jadeite and the presence of lawsonite and glaucophane indicate pressures less than 800 to 1,000 MPa, and more than 400 MPa (Newton and Smith, 1967; Liou, 1971; Maresch, 1977), and temperatures less than 350° to 400°C (Heinrich and Althaus, 1980). Brown (1977b) and Brown and Ghent (1983) have estimated the P-T stability of lawsonite + sodic amphibole + epidote assemblages as approximately 600 to 1,000 MPa and 200° to 300°C, using natural assemblages whose pressures and temperatures were inferred from experimentally determined reactions and independent thermobarometers such as sphalerite composition.

The conditions of formation for the epidote + sodic amphibole rocks are limited by two experimentally determined equilibria shown in Figure 10 (reactions 1, 5). The absence of jadeite and the presence of glaucophane indicate pressures less than 800 to 1000 MPa, and greater than 400 MPa (Newton and Smith, 1967; Maresch, 1977), and temperatures less than 450°C (Maresch, 1977). Brown (1977b), Brown and Ghent (1983), Brown and Blake (1987) have estimated the P-T stability of epidote +

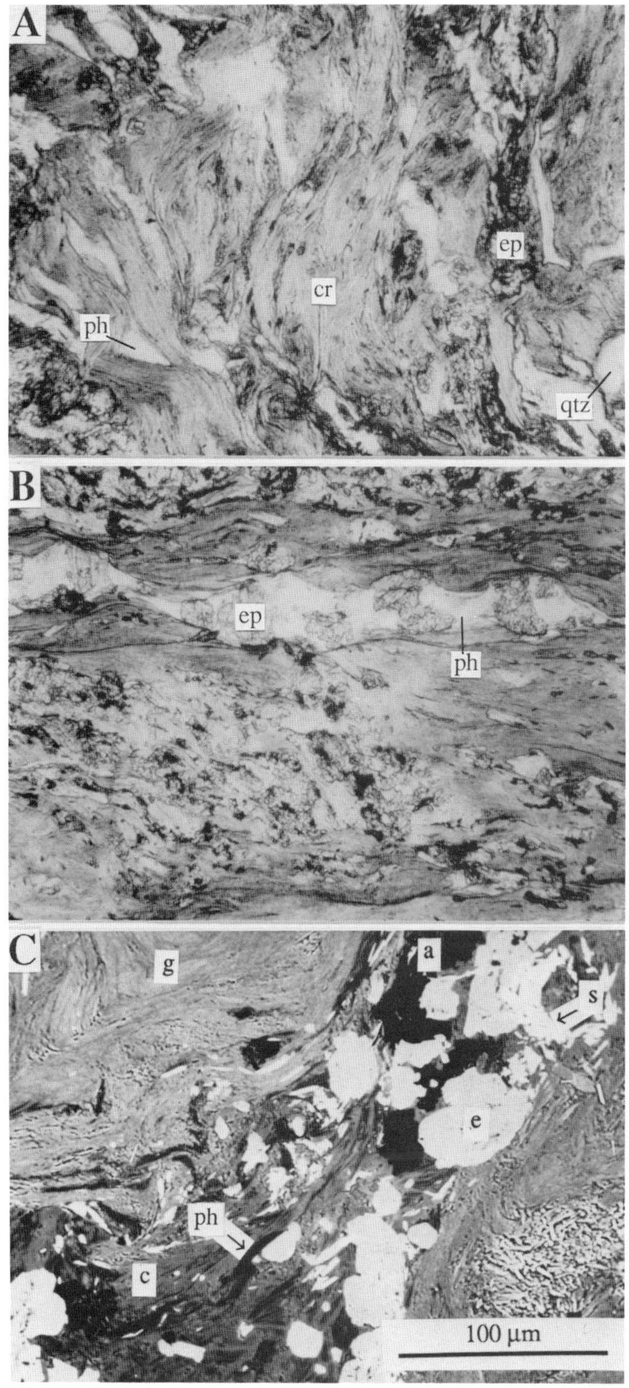

Figure 7. Photomicrographs of characteristic crossite-epidote metabasaltic assemblages in the Stuart Fork and Red Ant terranes. Field of view in (A) and (B) is approximately 1 mm; scale bar in (C) is 100 μm. A, Optical photomicrograph of foliated Stuart Fork crossite-epidote blueschist containing fibrous crossite (cr), epidote (ep), phengite (ph), and quartz (qtz). Kinks in crossite in lower left and lower right corners. B, Optical photomicrograph of foliated Stuart Fork crossite-epidote blueschist containing boudinaged epidote (ep) segments separated by phengite (ph). C, Back-scattered electron micrograph of vein in Red Ant volcaniclastic rock, showing elongate sphene (s), porphyroblastic epidote (e), phengite (ph), chlorite (c), sodic amphibole (g), and albite (a).

sodic amphibole + hematite assemblages as approximately 600 to 900 MPa and 250° to 400°C. Maruyama and others (1986) bracketed the equilibrium epidote + magnesio-riebeckite = tremolite + hematite at 300°C between 310 and 470 MPa. This bracket was derived from unreversed experiments, and may be in error because the run products were not fully characterized. If this determination is correct, the stability field of epidote + sodic amphibole extends downward to at least 470 MPa, and this provides a minimum crystallization pressure of the Red Ant terrane.

The pumpellyite-actinolite facies metamorphism did not occur at very low pressures (less than ~200 MPa), because the rocks lack characteristic low-pressure assemblages such as prehnite + epidote + actinolite or oligoclase + actinolite (e.g., Liou and others, 1987). Assemblages containing lawsonite and lacking sodic amphibole are limited to temperatures greater than 150°C and less than 350°C (Nitsch, 1968; Heinrich and Althaus, 1980; Schiffman and Liou, 1980) and to pressures greater than 300 MPa (Liou, 1971), by reactions 3, 6, 7, and 10, and bracket 4 in Figure 10. Maximum pressures are less than ~600 to 900 MPa

TABLE 1. REPRESENTATIVE CALCIC AMPHIBOLE ANALYSES FROM STUART FORK AND RED ANT TERRANE M_1 METABASALTIC ROCKS

	Stuart Fork terrane					Red Ant terrane					
Sample	84-14	8	11B	13	214	22	23	28	20	15	19
SiO_2	54.34	53.15	54.62	53.55	53.64	42.98	45.08	49.23	43.16	40.58	47.87
TiO_2	0.60	0.09	0.04	0.13	0.04	2.86	2.18	1.03	2.98	0.03	0.06
Al_2O_3	1.46	2.28	1.70	2.12	2.15	9.55	8.46	4.38	9.20	12.73	6.46
Cr_2O_3	0.05	0.08	0.35	b.d.	b.d.	b.d	0.05	0.06	0.05	b.d.	0.06
FeO*	10.87	13.22	9.37	13.65	13.54	19.20	18.43	17.83	20.77	21.09	22.55
MnO	0.24	0.26	0.25	0.27	0.28	0.35	0.29	0.36	0.30	0.50	0.35
MgO	16.54	14.29	17.73	14.99	14.71	8.50	9.76	12.34	8.50	6.80	7.94
CaO	12.35	11.93	12.83	12.50	11.66	11.15	11.02	10.29	10.81	11.62	11.31
Na_2O	0.41	0.37	0.25	0.33	0.73	2.11	1.91	0.94	2.50	1.70	1.26
K_2O	0.05	0.06	0.06	0.09	0.13	0.67	0.66	0.28	0.66	1.20	0.38
Total	96.91	95.73	97.20	97.63	96.88	97.37	97.84	96.74	98.93	96.25	98.24
	Cations based on 23 oxygen atoms										
Si	7.81	7.81	7.78	7.72	7.78	6.54	6.74	7.21	6.48	6.30	7.19
Al^{IV}	0.19	0.19	0.22	0.28	0.23	1.46	1.26	0.76	1.52	1.71	0.81
Σ	8.00	8.00	8.00	8.00	8.00	8.00	8.00	7.96	8.00	8.00	8.00
Al^{VI}	0.06	0.20	0.07	0.08	0.14	0.26	0.23	0.00	0.11	0.62	0.34
Ti	0.07	0.01	0.00	0.01	0.00	0.33	0.25	0.11	0.34	0.00	0.01
Fe^{3+}	0.08	0.10	0.13	0.19	0.22	0.15	0.32	1.05	0.40	0.46	0.38
Cr	0.00	0.00	0.01	b.d.	b.d.	b.d.	0.01	0.01	0.00	b.d.	0.00
Fe^{2+}	1.23	1.53	0.99	1.46	1.42	2.29	1.98	1.13	2.21	2.27	2.46
Mn	0.03	0.03	0.03	0.03	0.03	0.05	0.04	0.05	0.04	0.07	0.05
Mg	3.54	3.13	3.77	3.22	3.18	1.93	2.18	2.69	1.90	1.57	1.78
Σ	5.01	5.00	5.00	4.99	4.99	5.01	5.01	5.04	5.00	4.99	5.02
Ca	1.90	1.89	1.96	1.93	1.81	1.82	1.77	1.61	1.74	1.93	1.82
Na^{M4}	0.10	0.11	0.04	0.07	0.19	0.18	0.23	0.27	0.26	0.07	0.18
Σ	2.00	2.00	2.00	2.00	2.00	2.00	2.00	1.88	2.00	2.00	2.00
Na^A	0.02	0.00	0.03	0.02	0.02	0.44	0.32	0.00	0.47	0.44	0.19
K	0.01	0.01	0.01	0.02	0.02	0.13	0.13	0.05	0.13	0.24	0.07
Σ	0.03	0.01	0.04	0.04	0.04	0.57	0.45	0.05	0.60	0.68	0.26
Total	15.04	15.01	15.04	15.03	15.03	15.58	15.46	14.93	15.60	15.67	15.28

*All Fe as FeO.
b.d. = below detection.

TABLE 2. REPRESENTATIVE SODIC AND SOCIC-CALCIC AMPHIBOLE ANALYSES FROM STUART FORK AND RED ANT TERRANE M_2 ASSEMBLAGES

	Stuart Fork terrane								Red Ant terrane							
	Metavolcanics						Metasediments		Metavolcanics							
Sample	12A	13	18	34D	35C	V6D	268A	V6A	39	47	35	28	73	76/79	33	51
SiO_2	55.68	54.61	54.99	55.40	54.00	55.89	56.43	55.35	56.50	56.68	55.43	56.32	55.62	54.42	53.46	54.45
Al_2O_3	3.53	2.75	3.15	3.65	5.70	3.13	6.40	4.04	10.24	9.10	5.13	6.42	1.60	2.53	3.83	5.66
TiO_2	0.03	0.05	0.07	0.07	0.08	0.05	0.07	0.10	0.16	0.16	0.11	0.17	0.09	0.07	0.05	0.10
Cr_2O_3	b.d.	b.d.	0.06	0.03	0.02	0.03	0.04	0.02	0.03	b.d.	b.d.	b.d.	b.d.	0.02	0.04	b.d.
FeO*	16.64	19.14	17.68	13.39	16.31	18.58	15.16	20.66	16.10	16.37	20.83	19.70	20.40	24.39	19.68	18.19
MnO	0.20	0.24	0.27	0.26	0.22	0.19	0.37	0.48	0.19	0.16	0.39	0.36	0.30	0.22	0.30	0.20
MgO	11.77	10.06	10.3	13.66	10.28	10.74	10.54	8.11	6.81	7.13	7.20	7.16	10.40	6.68	10.34	8.50
CaO	6.45	3.91	3.71	5.91	3.01	3.03	2.85	1.00	0.28	0.73	0.95	0.69	1.99	1.30	6.83	4.53
Na_2O	3.33	4.83	5.05	3.70	4.86	5.49	5.41	6.73	7.31	6.84	6.67	6.82	5.99	6.55	3.54	4.61
K_2O	0.06	0.05	0.09	0.10	0.31	b.d.	b.d.	b.d.	b.d.	b.d.	b.d.	b.d.	b.d.	b.d.	b.d.	0.05
Total	97.69	95.64	59.70	96.17	94.77	97.13	97.27	96.49	97.62	97.17	96.71	97.64	96.39	96.18	98.07	96.29
	Cations based on 23 oxygen atoms															
Si	7.96	7.98	7.98	7.91	7.90	7.97	7.96	8.00	7.92	8.02	8.00	8.01	8.01	8.00	7.69	7.93
Al^{IV}	0.04	0.02	0.03	0.09	0.10	0.03	0.04	0.00	0.08	0.00	0.00	0.00	0.00	0.00	0.32	0.07
Σ	8.00	8.00	8.01	8.00	8.00	8.00	8.00	8.00	8.00	8.02	8.00	8.01	8.01	8.00	8.01	8.00
Al^{VI}	0.56	0.45	0.51	0.53	0.88	0.49	1.03	0.69	1.62	1.52	0.87	1.08	0.28	0.43	0.33	0.90
Ti	0.00	0.01	0.01	0.01	0.01	0.01	0.01	0.01	0.02	0.02	0.01	0.02	0.01	0.01	0.01	0.01
Fe^{3+}	0.39	0.92	0.90	0.55	0.53	1.03	0.46	1.09	0.41	0.29	0.98	0.74	1.36	1.41	0.95	0.45
Cr	b.d.	b.d.	0.01	0.00	0.00	0.00	0.00	0.00	0.00	b.d.	b.d.	b.d.	b.d.	0.00	0.10	b.d.
Fe^{2+}	1.61	1.42	1.25	1.05	1.47	1.18	1.32	1.41	1.48	1.65	1.53	1.61	1.10	1.58	1.42	1.77
Mn	0.03	0.03	0.03	0.03	0.03	0.02	0.05	0.06	0.02	0.02	0.05	0.04	0.04	0.03	0.04	0.02
Mg	2.51	2.19	2.30	2.91	2.24	2.28	2.22	1.75	1.42	1.50	1.55	1.52	2.23	1.46	2.22	1.85
Σ	5.10	5.02	5.01	5.08	1.16	5.01	5.09	5.01	4.97	5.00	4.99	5.01	5.02	4.92	5.07	5.00
Ca	0.99	0.61	0.56	0.90	0.47	0.46	0.43	0.16	0.04	0.11	0.15	0.11	0.31	0.20	1.05	0.71
Na^{M4}	0.92	1.37	1.42	1.03	1.38	1.52	1.48	1.84	1.99	1.88	1.87	1.88	1.67	1.87	0.99	1.30
Σ	1.91	1.98	1.98	1.93	1.85	1.98	1.91	2.00	2.03	1.99	2.02	1.99	1.98	2.07	2.04	2.01
Na^{A}	0.00	0.00	0.00	0.00	0.00	0.00	0.00	0.04	0.00	0.00	0.00	0.00	0.00	0.00	0.00	0.00
K	0.01	0.01	0.02	0.02	0.06	b.d.	b.d.	b.d.	b.d.	b.d.	b.d.	b.d.	b.d.	b.d.	b.d.	0.01
Σ	0.01	0.01	0.02	0.02	0.06	b.d.	b.d.	0.04	b.d.	b.d.	b.d.	b.d.	b.d.	b.d.	b.d.	0.01
Total	15.02	15.01	15.02	15.03	15.07	14.99	15.00	15.05	15.00	15.01	15.01	15.01	15.01	14.99	15.12	15.02

*All Fe as FeO.
b.d. = below detection.

TABLE 3. REPRESENTATIVE PHENGITE ANALYSES FROM STUART FORK AND RED ANT TERRANE M_2 ASSEMBLAGES

	STUART FORK TERRANE											
	Metabasalts				Metasediments							
Sample	17B	18	35C	V6D	11A	22	33A	35D	219C	268A	V6A	V9
SiO_2	55.47	56.34	55.30	57.42	52.54	53.92	51.43	53.50	50.70	54.62	57.00	53.02
TiO_2	0.04	0.04	0.05	0.06	0.06	0.10	0.11	0.05	0.12	0.03	0.03	0.06
Al_2O_3	16.58	18.21	19.71	18.54	24.74	25.87	26.67	21.65	24.51	22.59	18.24	25.01
Cr_2O_3	0.06	0.16	b.d.	b.d.	b.d.	0.04	0.02	b.d.	b.d.	0.06	b.d.	b.d.
FeO*	5.14	4.30	4.45	4.28	2.49	2.78	2.84	3.08	4.44	3.01	3.97	3.23
MnO	0.09	0.07	0.03	0.05	b.d.	0.12	0.05	0.12	0.18	b.d.	0.09	0.03
MgO	5.72	6.55	5.81	6.93	3.68	3.84	3.68	4.42	3.39	5.23	6.29	3.84
CaO	b.d.	0.03	0.06	0.29	b.d.	b.d.	0.04	b.d.	0.05	0.03	0.11	0.04
Na_2O	b.d.	b.d.	b.d.	0.09	b.d.	b.d.	b.d.	b.d.	b.d.	b.d.	0.30	b.d.
K_2O	11.18	9.82	9.26	7.77	9.86	7.84	8.47	10.80	9.31	9.78	8.03	8.30
Total	94.28	95.52	94.67	95.43	93.37	94.51	93.31	93.62	92.70	95.35	94.06	93.53
	Cations based on 11 oxygen atoms											
Si	3.80	3.76	3.71	3.78	3.53	3.49	3.40	3.63	3.43	3.58	3.75	3.50
Al^{IV}	0.20	0.24	0.29	0.22	0.47	0.51	0.60	0.37	0.57	0.42	0.25	0.51
Σ	4.00	4.00	4.00	4.00	4.00	4.00	4.00	4.00	4.00	4.00	4.00	4.00
Al^{VI}	1.14	1.19	1.27	1.22	1.49	1.47	1.47	1.37	1.39	1.32	1.16	1.44
Ti	0.00	0.00	0.00	0.00	0.00	0.01	0.01	0.00	0.01	0.00	0.00	0.00
Fe^{3+}	0.00	0.00	0.00	0.00	0.00	0.03	0.12	0.00	0.17	0.10	0.09	0.06
Cr	0.00	0.01	b.d.	b.d.	b.d.	0.00	0.00	b.d.	b.d.	0.00	b.d.	b.d.
Fe^{2+}	0.30	0.24	0.25	0.24	0.16	0.12	0.04	0.18	0.09	0.07	0.13	0.12
Mn	0.01	0.00	0.00	0.00	b.d.	0.01	0.00	0.01	0.01	b.d.	0.01	0.00
Mg	0.59	0.65	0.58	0.68	0.37	0.37	0.36	0.45	0.34	0.51	0.62	0.38
Σ	2.03	2.10	2.10	2.14	2.02	2.00	2.00	2.01	2.01	2.00	2.01	2.00
Ca	b.d.	0.00	0.01	0.02	b.d.	b.d.	0.00	b.d.	0.00	0.00	0.01	0.00
Na	b.d.	.b.d	b.d.	0.01	b.d.	b.d.	b.d.	b.d.	b.d.	b.d.	0.04	b.d.
K	0.98	0.84	0.79	0.65	0.85	0.65	0.71	0.94	0.80	0.82	0.67	0.70
Σ	0.98	0.84	0.80	0.68	0.85	0.65	0.71	0.94	0.80	0.82	0.72	0.70
Total	7.01	6.94	6.90	6.82	6.87	6.65	6.71	6.95	6.81	6.82	6.73	6.70

*All Fe as FeO.
b.d. = below detection.

(Brown, 1977b). If the experiments of Maruyama and others (1986) represent the minimum pressure stability of epidote + sodic amphibole, then assemblages containing lawsonite and lacking sodic amphibole may be stable to pressures as high as 470 MPa. Maresch's (1977) experiments delineate a maximum possible stability field for glaucophane, and do not provide an upper pressure limit to assemblages lacking sodic amphibole. Note, however, that Maresch's lower pressure limit is consonant with the bracket of Maruyama and others (1986).

Pumpellyite-bearing assemblages lacking lawsonite and sodic amphibole are stable up to maximum temperatures of 350°C (reaction 6; Schiffman and Liou, 1980), and to pressures at least as high as 400 MPa (reaction 5; Maresch, 1977). The lawsonite-out curve of Liou (1971) is displaced an unknown amount to higher pressures in the presence of other phases, and does not effectively limit the maximum pressure of these assemblages. Schiffman and Liou (1983) have suggested that Fe-bearing pumpellyite breaks down at temperatures perhaps 100° to 150°C lower than the thermal stability limit of Mg-pumpellyite such that ~250°C may be a more realistic upper temperature limit. Liou and others (1985) reported preliminary results for the reaction pumpellyite + chlorite = clinozoisite + tremolite, indicating that the upper thermal stability of pumpellyite + chlorite is about 325° to 350°C, but details of their experiments have not been presented.

Red Ant terrane samples contain the four-phase reaction

TABLE 3. REPRESENTATIVE PHENGITE ANALYSES FROM STUART FORK AND RED ANT TERRANE M_2 ASSEMBLAGES (continued)

	Red Ant terrane											
	Metabasalts								Metasediments			
Sample	26	25	76	3	5	9	37	55/56	92	63	61/62	57/58
SiO_2	52.24	51.30	52.91	54.03	53.94	53.70	54.63	51.87	49.31	50.35	50.97	51.27
TiO_2	0.07	0.05	0.08	0.14	0.07	b.d.	0.04	0.09	b.d.	0.21	0.35	0.08
Al_2O_3	25.49	27.90	21.63	22.59	19.50	22.20	21.38	22.88	30.24	26.61	25.04	25.38
Cr_2O_3	0.28	0.03	0.05	0.04	0.03	b.d.	b.d.	0.05	0.03	0.05	0.07	0.03
FeO*	3.21	2.70	6.06	5.39	7.63	5.60	3.14	4.59	3.19	2.53	3.11	2.85
MnO	b.d.	0.03	0.12	0.03	0.08	0.10	0.27	0.07	0.14	0.03	0.09	0.04
MgO	4.01	3.45	5.33	4.51	5.00	4.10	5.69	4.08	1.98	3.85	3.05	3.18
CaO	0.06	0.08	0.05	0.07	0.08	b.d.	b.d.	0.05	0.04	0.04	0.05	b.d.
Na_2O	0.09	b.d.	b.d.	b.d.	b.d.	0.48	b.d.	b.d.	0.11	0.11	b.d.	b.d.
K_2O	10.40	10.57	10.06	10.50	10.16	9.88	10.82	10.33	10.29	9.53	10.11	10.26
Total	95.85	96.11	96.29	97.30	96.49	96.06	95.97	94.01	95.34	93.31	92.84	93.09
	Cations based on 11 oxygen atoms											
Si	3.44	3.36	3.47	3.52	3.56	3.55	3.59	3.50	3.26	3.36	3.47	3.48
Al^{IV}	0.56	0.64	0.53	0.48	0.44	0.45	0.41	0.50	0.74	0.64	0.53	0.52
Σ	4.00	4.00	4.00	4.00	4.00	4.00	4.00	4.00	4.00	4.00	4.00	4.00
Al^{VI}	1.41	1.51	1.13	1.26	1.08	1.28	1.25	1.32	1.62	1.46	1.49	1.51
Ti	0.00	0.00	0.00	0.01	0.00	b.d.	0.00	0.00	b.d.	0.01	0.02	0.00
Fe^{3+}	0.14	0.13	0.33	0.20	0.35	0.17	0.15	0.18	0.12	0.14	0.04	0.01
Cr	0.01	0.00	0.00	0.00	0.00	b.d.	b.d.	0.00	0.00	0.00	0.00	0.00
Fe^{2+}	0.04	0.02	0.00	0.09	0.07	0.14	0.02	0.08	0.06	0.00	0.14	0.15
Mn	b.d.	0.00	0.01	0.00	0.00	0.01	0.01	0.00	0.01	0.00	0.00	0.00
Mg	0.39	0.34	0.52	0.44	0.49	0.40	0.56	0.41	0.20	0.38	0.31	0.32
Σ	1.99	2.00	2.00	2.00	2.00	2.00	1.99	1.99	2.01	2.00	2.00	1.99
Ca	0.00	0.01	0.00	0.00	0.01	b.d.	b.d.	0.00	0.00	0.00	0.00	b.d.
Na	0.01	b.d.	b.d.	b.d.	b.d.	0.06	b.d.	b.d.	0.01	0.01	b.d.	b.d.
K	0.87	0.88	0.84	0.87	0.86	0.83	0.91	0.89	0.87	0.81	0.88	0.89
Σ	0.88	0.88	0.84	0.87	0.87	0.89	0.91	0.89	0.88	0.82	0.88	0.89
Total	6.87	6.88	6.84	6.87	6.87	6.89	6.90	6.88	6.89	6.82	6.89	6.88

*All Fe as FeO.
b.d. = below detection.

assemblage pumpellyite + hematite + epidote + actinolite. Nitsch (1971) bracketed the equilibrium pumpellyite = epidote + actinolite at 700 MPa between 345° and 435°C (bracket 4 in Fig. 10)—and not at any other pressures or temperatures as has been implied by numerous authors; the bracket was derived from unreversed experiments on unbuffered natural materials, and may be in error because the experiments are truly only studies of reaction kinetics and not equilibrium. Unfortunately, this bracket does not provide information on the high temperature limit of pumpellyite beyond that provided by Schiffman and Liou (1980). Pumpellyite and actinolite do not coexist in low-pressure rocks, but an experimental low-pressure limit for their stability has not been determined (Schiffman and Liou, 1980; Liou and others, 1985).

The inferred conditions of recrystallization based solely on experimental determinations of reactions shown in Figure 10 are 150° to 350°C and 200 to 400(?) MPa for pumpellyite-actinolite facies rocks lacking lawsonite, 150° to 350°C and 300 to 400(?) MPa for pumpellyite-actinolite facies rocks containing lawsonite, and 150° to 400°C and 400(?) to 1,000 MPa for the blueschist-facies metamorphism. The pressures separating the blueschist and pumpellyite-actinolite facies are uncertain because there is only one experimental bracket on the stability of epidote and sodic amphibole. Empirical calibrations of Brown (1977a), Brown and

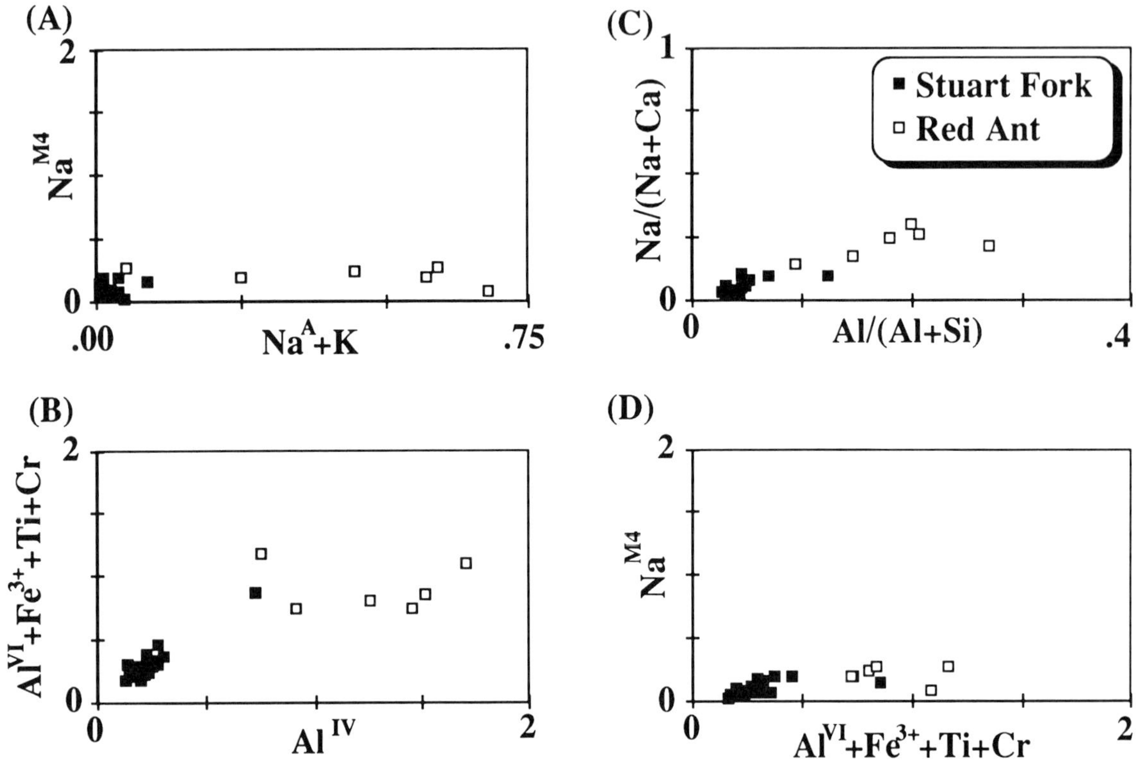

Figure 8. Compositions of M_1 calcic amphiboles. Common assemblage consists of, or is assumed to consist of, amphibole + epidote + chlorine + plagioclase + quartz. In each diagram, M_1 calcic amphiboles plot in fields defined by Laird and Albee (1981b) for the Abukuma low-pressure terrane of Japan. M_1 calcic amphiboles are low in Na^{M4} and show variable contents of A-site cations and tetrahedral Al. Ca amphiboles of M_1 assemblages show tschermak substitution (b), a coupled substitution represented by $(Al^{VI}+Fe^{3+}+Ti+Cr)$, $Al^{IV} \leftrightarrow (Fe^{2+}+Mg+Mn)$, Si.

Ghent (1983), and Brown and Blake (1987) suggest conditions of approximately 275° to 375°C and 200 to 600 MPa for pumpellyite-actinolite facies rocks, 250° to 400°C and 600 to 900 MPa for epidote + sodic amphibole, and 200° to 300°C and 600 to 900 MPa for lawsonite + sodic amphibole.

Na occupancy of the M4 site in amphibole coexisting with iron oxide, albite, chlorite, epidote, and H_2O is a function of pressure (Brown, 1977b). Brown's empirical calibration yields pressures of 500 to 700 MPa for the sodic and sodic-calcic amphiboles in the Red Ant and Stuart Fork terranes.

Phengite compositions are also compatible with high-pressure crystallization. Phengite white mica can indicate high pressure (Ernst, 1963; Guidotti and Sassi, 1976), and the Si content of phengite coexisting with K-feldspar, quartz, and biotite is apparently a function of pressure in unreversed experiments in the K_2O-MgO-Al_2O_3-SiO_2-H_2O (KMASH) system conducted in the temperature range 350° to 600°C (Massone and Schreyer, 1987). Two brackets for this same reaction were previously obtained on natural materials by Velde (1965). Assemblages from the Stuart Fork and Red Ant terranes do not contain biotite, and the biotite-absent reaction relating chlorite + microcline = celadonite + muscovite has a steep dP/dT slope (Bucher-Nurminen, 1987), rendering it a poor choice for barometry at low temperatures.

In summary, estimated metamorphic temperatures and pressures are roughly 250° to 400°C and 500 to 700 MPa for crossite-epidote assemblages, 200° to 300°C and 500 to 700 MPa for glaucophane-lawsonite assemblages, and 150° to 350°C and 200 to 400 MPa for pumpellyite-actinolite assemblages (Fig. 11). Note that although the pumpellyite-actinolite facies assemblages have been described with the blueschists, metamorphic conditions of the pumpellyite-actinolite facies are poorly constrained and include low P/T conditions.

Timing

Known ages of rock-forming, deformational, and metamorphic events place few constraints on the petrotectonic evolution of the Stuart Fork or Red Ant terranes. The ages of sedimentary and volcanic protoliths in both terranes are unknown, as is the age of M_1 metamorphism.

Phengite from Stuart Fork lawsonite-glaucophane blueschist in the Yreka area yields K-Ar and $^{40}Ar/^{39}Ar$ ages of approximately 220 Ma (i.e., Late Triassic) (Hotz and others, 1977). K-Ar geochronology was performed on the Red Ant terrane by Schweickert and others (1980); two whole-rock and three muscovite-separate dates range from 157 ± 5 to 190 ± 8 Ma. Because the Red Ant blueschists were overprinted by pump-

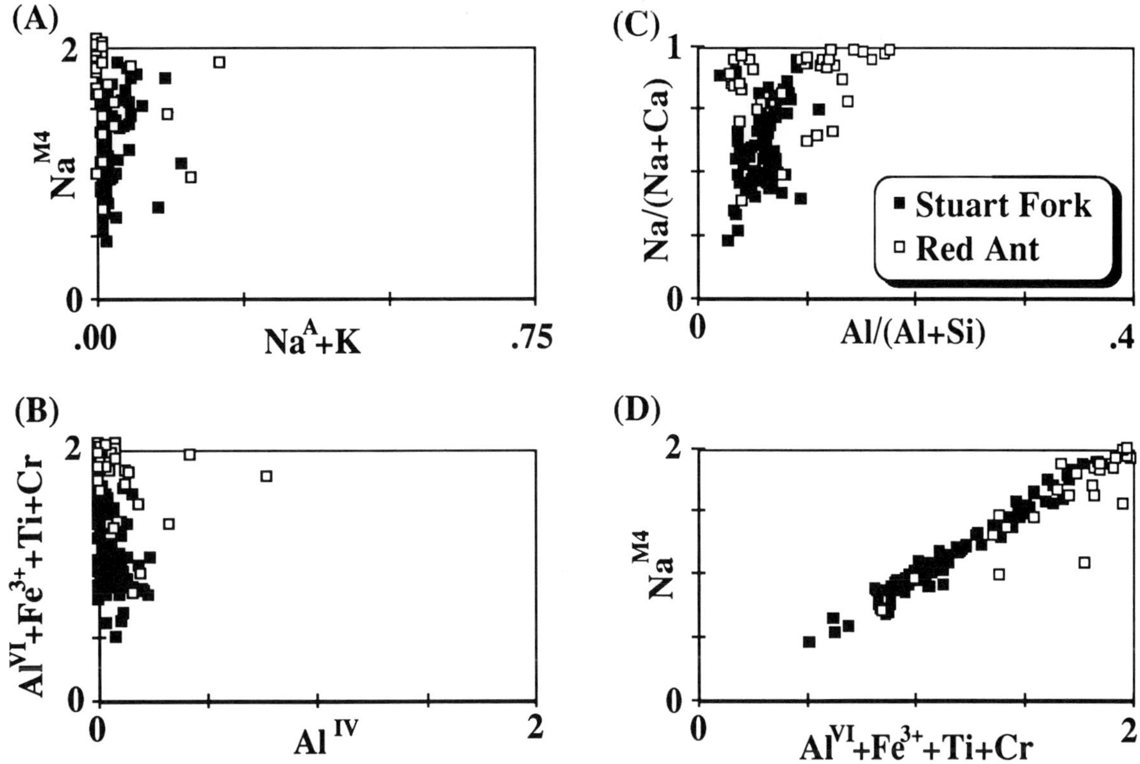

Figure 9. Compositions of amphiboles in M_2 assemblages (after Laird and Albee, 1981a, b). Common assemblage consists of amphibole + epidote + chlorite + plagioclase + quartz. M_2 sodic amphiboles are similar in composition to those from the Sanbagawa and Franciscan high-pressure metamorphic belts. M_2 sodic amphiboles are low in A-site cations and tetrahedral Al; they show a range of Na^{M4} to nearly 2.0 pfu. Sodic amphiboles in M_2 assemblages show glaucophane substitution (d), represented by Na^{M4},(Al^{VI}+Fe^{3+}+Ti+Cr) ↔ Ca, (Fe^{2+}+Mg+Mn).

ellyite-actinolite recrystallization, which may have influenced the K-Ar ages, the meaning of these dates is unclear. We tentatively infer that M_2 in both units is of a roughly comparable Late Triassic(?) age. Consequently, D_1 may have been concurrent in the Red Ant and Stuart Fork terranes, but D_2 need not have been.

P-T EVOLUTION

Rocks of the Stuart Fork and Red Ant terranes were metamorphosed during two or three distinct episodes, each of which was characterized by different P-T conditions (Fig. 11). The relative timing of these events was established by textural relations, and the P-T conditions of metamorphism were inferred from phase relations and mineral chemistry.

Protoliths consisted chiefly of pelagic radiolarian-bearing muds and oozes, and mafic volcanic rocks. We consider it likely that the Stuart Fork and the Red Ant protoliths represent crust from the same ocean basin, although they may be somewhat diachronous.

The first episode of metamorphism, M_1, formed assemblages at relatively high temperatures and low pressures, reflecting shallow crustal levels and high geothermal gradients (more than 50°C/km). Although no ages are available that directly date M_1, ocean crust formation and M_1 metamorphism must have occurred prior to M_2, providing a minimum age of Late Triassic. Randomly oriented M_1 minerals indicate that M_1 was not accompanied by penetrative deformation. We suggest that M_1 metamorphism occurred during hydrothermal alteration near an ocean ridge, intra-oceanic arc, or other shallow oceanic heat source.

The cooling oceanic crust then passed after an unknown period into a Late Triassic or older subduction zone at the western edge of North America. Rocks in the subducting slab were recrystallized (M_2) during deformation (D_1) at high pressures and moderate to low temperatures before or during Late Triassic time. The subduction zone probably dipped eastward, as indicated by west-vergent D_1 structures and the position of the subduction complex with respect to contemporaneous Early Triassic to Middle Jurassic arc volcanism and plutonism in the eastern Klamath Mountains and Sierra Nevada (e.g., Miller, 1989; Harwood, 1988). M_2 blueschist assemblages developed at relatively low temperatures and high pressures (500 to 1,100 MPa), reflecting deep crustal levels (on the order of 17 to 38 km).

Blueschist-facies rocks of the Sierra Nevada are overprinted by pumpellyite-actinolite facies minerals, whereas the Stuart Fork

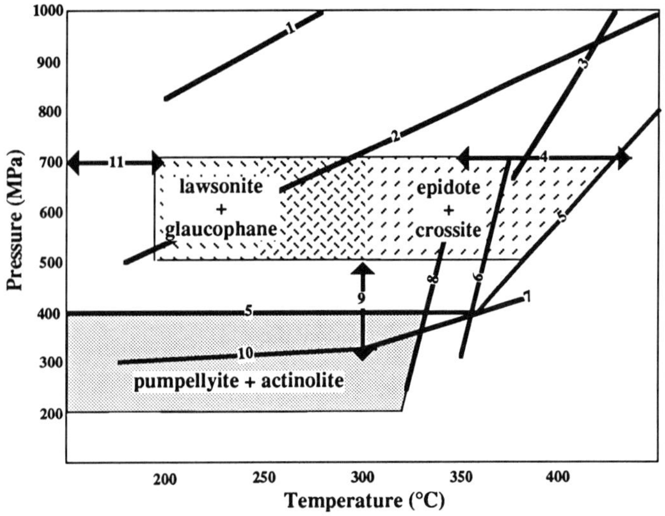

Figure 10. Phase equilibria relations of M_2 assemblages in metabasalt. Two bracketed reactions, 4 and 9, are shown by two-headed arrows. Reactions shown are (1) jadeite + quartz = albite (Newton and Smith, 1967); (2) aragonite = calcite (Johannes and Puhan, 1971); (3) lawsonite + albite = zoisite P paragonite (Heinrich and Althaus, 1980); (4) pumpellyite = epidote + actinolite (Nitsch, 1971); (5) maximum glaucophane stability field (Maresch, 1977); (6) Mg-pumpellyite = clinozoisite + grossulaire (Schiffman and Liou, 1980); (7) lawsonite = wairakite (Liou, 1971); (8) Mg-pumpellyite + chlorite = clinozoisite + grossularite (Liou and others, 1985); (9) epidote + magnesioriebeckite = tremolite + hematite (Maruyama and others, 1986); (10) laumontite = lawsonite (Liou, 1971); (11) heulandite = lawsonite (Nitsch, 1968). The position of the reaction as determined by Heinrich and Althaus (1980) is essentially coincident with a position calculated by Holland (1979) from thermodynamic data. In general, the P-T limits of blueschist and pumpellyite-actinolite facies are poorly known and the shaded regions in this figure should be regarded with caution.

terrane lacks any pervasive overprint. Either the Stuart Fork terrane returned toward the surface along a path with a high-P/T slope, or the kinetics of recrystallization were sufficiently slow to preserve the higher pressure parageneses.

By Middle Jurassic time, Stuart Fork rocks reached significantly lower pressure and higher temperature conditions when they were intruded by granitoid plutons of the widespread western Hayfork magmatic arc (Harper and Wright, 1984; Wright and Fahan, 1988).

CONCLUSIONS

The Stuart Fork terrane of the central Klamath Mountains and the Red Ant terrane of the Sierra Nevada have important similarities and differences. They represent either (1) unrelated rocks that formed independently under closely similar depositional and metamorphic conditions, or (2) separate fragments of the same crustal element that were physically contiguous during formation and at least two episodes of metamorphism. We consider the latter to be most likely because of the similarities between the Stuart Fork and Red Ant terranes, including:

1. The Stuart Fork and Red Ant terranes formed in pre–Late Triassic(?) time in a back-arc or larger ocean basin near a detrital volcanic source.

2. Prior to Late Triassic time, the ocean crust of these terranes were metamorphosed under static conditions at low pressures in an ocean-ridge, intraoceanic arc, or other oceanic hydrothermal environment.

3. The oceanic lithosphere, composed in part by the Stuart Fork and Red Ant terranes, was subducted eastward beneath an upper plate containing the Eastern Klamath–Northern Sierra magmatic arcs, and metamorphosed at depths of 17 to 38 km. By Late Triassic time, the rocks had ascended and cooled. This high-P/T metamorphism represents the characteristic metamorphic episode of both terranes in their evolution as subduction complexes. Strong folding and thrust imbrication occurred after and during the high-pressure metamorphism.

4. Regional shortening formed possible southwest-directed thrust faults and/or west-vergent, kilometer-scale folds by mid-Jurassic time. In the Klamath Mountains, this tectonic event is associated with contraction across the western Hayfork magmatic arc of the Middle Jurassic convergent margin (Harper and Wright, 1984; Coleman and others, 1988; Wright and Fahan, 1988).

5. The Stuart Fork terrane is bounded on the east by the central metamorphic belt, which contains mafic igneous rocks metamorphosed during Devonian time (Cashman, 1980; Peacock and Norris, 1980). Likewise, the Red Ant terrane lies in thrust contact with the Feather River terrane to the east which is

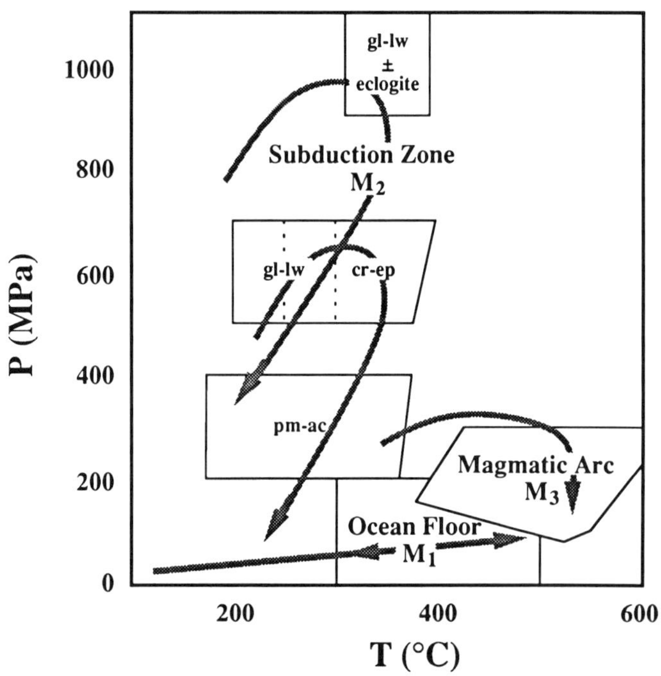

Figure 11. Schematic P-T-time diagram showing sequence of M_1–M_3 metamorphic events. Constraints on metamorphic conditions provided by data presented in this study, and from Borns (1980). Arrows show possible P-T paths for metamorphic events.

also composed of mafic metaigneous rocks metamorphosed in the Devonian (Edelman and others, 1989). The central metamorphic belt and the Feather River terrane share a number of petrologic characteristics, and they have been considered relicts of the same metamorphic belt (Hacker and Peacock, this volume).

6. The Stuart Fork terrane is bounded immediately to the west by the Late Paleozoic to Early Jurassic North Fork–Salmon River terrane, a relatively intact ophiolitic sequence (Ando, 1979; Ando and others, 1983). Farther west and tectonically beneath the North Fork–Salmon River terrane lies the Late Triassic to Early Jurassic siliceous melange of the eastern Hayfork terrane (Wright, 1982). The Red Ant terrane is bounded on the west by the Calaveras terrane, a package of disrupted metasedimentary and minor metavolcanic rocks that may be tectonically equivalent to the North Fork–Salmon River and eastern Hayfork terranes of the Klamath Mountains. These assemblages probably are oceanic materials accreted outboard of the Stuart Fork–Red Ant subduction complexes.

Important differences between the early Mesozoic high-pressure rocks in the Klamath Mountains and Sierra Nevada are probably representative of the spatial heterogeneity inherent in an arc-trench system that may have stretched for hundreds of kilometers:

1. The Middle to Late Jurassic M_3 contact metamorphism recognized in the Stuart Fork terrane is absent from the Red Ant terrane. This difference may simply be a result of the small outcrop of the Red Ant terrane; Cretaceous and Jurassic plutons were emplaced north, east, west, and south of the Red Ant terrane, but apparently none intruded close enough to cause recrystallization.

2. High-pressure parageneses of the Stuart Fork terrane consist of eclogite- and blueschist-facies minerals, whereas the Red Ant terrane contains only blueschists. Pumpellyite-actinolite facies parageneses in the Red Ant terrane may belong to this high-pressure event or to a later unrelated metamorphism.

In this chapter we have attempted to show the lithologic, structural, and petrologic equivalence of the Stuart Fork and Red Ant terranes. Our new data are a confirmation of the correlation between these metamorphic belts proposed earlier by Davis (1969) and expanded upon by others. Evidence presented herein suggests that these blueschist belts are relict pieces of a more broadly continuous convergent-margin trench complex. As such, they provide a tectonic and paleogeographic link between the early Mesozoic history of the Klamath Mountains and Sierra Nevada. Although we are certain that these rocks signify the existence of a convergent plate boundary, future studies must address the critical problem of where these belts fit in early Mesozoic paleogeographic reconstructions.

ACKNOWLEDGMENTS

Support for this research was provided by grants from the National Science Foundation (EAR83-12702, to W. G. Ernst), the Geological Society of America, Sigma Xi, and UCLA Graduate Fellowships. This work was stimulated by the 1988 Penrose Conference on Paleozoic–Early Mesozoic Paleogeographic Relations between the Klamath Mountains, Northern Sierra Nevada, and North America, sponsored by the Geological Society of America. Helpful reviews were provided by E. H. Brown, D. S. Harwood, and R. Haugerud.

REFERENCES CITED

Ando, C. J., 1979, Structural and petrologic analysis of the North Fork terrane, central Klamath Mountains, California [Ph.D. thesis]: Los Angeles, University of Southern California, 197 p.

Ando, C. J., Irwin, W. P., Jones, D. J., and Saleeby, J. B., 1983, The ophiolitic North Fork terrane in the Salmon River region, central Klamath Mountains, California: Geological Society of America Bulletin, v. 94, p. 236–252.

Beard, J. S., and Day, H. W., 1987, The Smartville intrusive complex, Sierra Nevada California; The core of a rifted volcanic arc: Geological Society of America Bulletin, v. 99, p. 779–791.

Blake, M. C., Jr., Howell, D. G., and Jones, D. L., 1982, Preliminary tectonostratigraphic terrane map of California: U.S. Geological Survey Open-File Report 82-593, scale 1:750,000.

Borns, D. J., 1980, Blueschist metamorphism of the Yreka–Ft. Jones area, Klamath Mountains, northern California [Ph.D. thesis]: Seattle, University of Washington, 155 p.

——— , 1984, Eclogites in the Stuart Fork terrane, Klamath Mountains, California: Geological Society of America Abstracts with Programs, v. 16, p. 271.

Brouxel, M., Lapierre, H., Michard, A., and Albarede, F., 1988, Geochemical study of an early Paleozoic island-arc–back-arc basin system; Part 2, Eastern Klamath, early to middle Paleozoic island-arc volcanic rocks (northern California): Geological Society of America Bulletin, v. 100, p. 1120–1130.

Brown, E. H., 1977a, Phase equilibria among pumpellyite, lawsonite, epidote, and associated minerals in low-grade metamorphic rocks: Contribution to Mineralogy and Petrology, v. 64, p. 123–136.

——— , 1977b, The crossite content of Ca-amphibole as a guide to pressure of metamorphism: Journal of Petrology, v. 18, p. 53–72.

Brown, E. H., and Blake, M. C., Jr., 1987, Correlation of Early Cretaceous blueschists in Washington, Oregon, and northern California: Tectonics, v. 6, p. 795–806.

Brown, E. H., and Ghent, E. D., 1983, Mineralogy and phase relations in the blueschist facies of the Black Butte and Ball rock areas, northern California Coast Ranges: American Mineralogist, v. 68, p. 365–372.

Bucher-Nurminen, K., 1987, A recalibration of the chlorite-biotite-muscovite geobarometer: Contributions to Mineralogy and Petrology, v. 96, p. 519–522.

Burchfiel, B. C., and Davis, G. A., 1981, Traissic and Jurassic tectonic evolution of the Klamath Mountains–Sierra Nevada geologic terrane, *in* Ernst, W. G., ed., The geotectonic development of California; Rubey Volume 1: Englewood Cliffs, New Jersey, Prentice-Hall, p. 50–70.

Cashman, S. M., 1980, Devonian metamorphic event in the northeastern Klamath Moutnains: Geological Society of America Bulletin, part 1, v. 91, p. 453–459.

Cebull, S. E., and Russell, L. R., 1979, Role of the Melones fault zone in the structural chronology of the North Yuba River area, western Sierra Nevada, California: Geological Society of America Bulletin, v. 90, part 1, p. 225–227, part 2, p. 528–544.

Coleman, R. G., Mortimer, N., Donato, M. M., Manning, C. E., and Hill, L. B., 1988, Tectonic and regional metamorphic framework of the Klamath

Mountains and adjacent Coast Ranges, California and Oregon, *in* Ernst, W. G., ed., Metamorphism and crustal evolution of the western United States; Rubey Volume 7, Englewood Cliffs, New Jersey, Prentice-Hall, p. 1061–1096.

D'Allura, J. A., Moore, E. M., and Robinson, L., 1977, Paleozoic rocks of the northern Sierra Nevada; Their structural and paleogeographic implications, *in* Stewart, J. H., Stevens, C. H., and Fritsche, A. E., eds., Paleozoic paleogeography of the western United States: Pacific Section, Society of Economic Paleontologists and Mineralogists Pacific Paleogeography Symposium p. 395–408.

Davis, G. A., 1968, Westward thrusting in the south-central Klamath Mountains, California: Geological Society of America, Bulletin, v. 82, p. 911–933.

——, 1969, Tectonic correlations, Klamath Mountains and western Sierra Nevada, California: Geological Society of America Bulletin, v. 8, p. 1095–1108.

Davis, G. A., Holdaway, M. J., Lipman, P. W., and Romey, W. D., 1965, Structure, metamorphism, and plutonism in the south-central Klamath Mountains, California: Geological Society of America Bulletin, v. 76, p. 933–965.

Davis, G. A., Monger, J.W.H., and Burchfiel, B. C., 1978, Mesozoic construction of the Cordilleran "collage," central British Columbia to central California, *in* Howell, D. G., and McDougall, K. A., eds., Mesozoic paleogeography of the western United States: Pacific Coast Section, Society of Economic Paleontologists and Mineralogists Pacific Coast Paleogeography Symposium 2, p. 1–32.

Day, H. W., Moores, E. M., and Tuminas, A. C., 1985, Structure and tectonics of the northern Sierra Nevada: Geological Society of America Bulletin, v. 96, p. 436–450.

Day, H. W., Schiffman, P., and Moores, E. M., 1988, Metamorphism and tectonics of the northern Sierra Nevada, *in* Ernst, W. G., ed., Metamorphism and crustal evolution of the western United States: Englewood Cliffs, New Jersey, Prentice-Hall, p. 737–763.

Donato, M. M., 1987, Evolution of an ophiolitic tectonic melange, Marble Mountains, northern California Klamath Mountains: Geological Society of America Bulletin, v. 98, p. 448–464.

Donato, M. M., Coleman, R. G., and Kays, M. A., 1980, Geology of the Condrey Mountain Schist, northern Klamath Mountains, California and Oregon: Oregon Geology, v. 42, no. 7, p. 125–129.

Donato, M. M., Barnes, C. G., Coleman, R. G., Ernst, W. G., and Kays, M. A., 1982, Geologic map of the Marble Mountain Wilderness, Siskiyou County, California: U.S. Geological Survey Miscellaneous Field Studies Map MF-1452-A, scale 1:48,000.

Edelman, S. H., Day, H. W., Moores, E. M., Zigman, S. M., Murphy, T. P., and Hacker, B. r., 1989, Structure across a Mesozoic ocean-continent suture zone in the northern Sierra Nevada, California: Geological Society of America Special Paper 224, 64 p.

Ehrenberg, S. N., 1975, Feather River ultramafic body, northern Sierra Nevada, California: Geological Society of America Bulletin, v. 86, p. 1235–1243.

Ernst, W. G., 1963, Significance of phengitic micas from low-grade schists: American Mineralogist, v. 48, p. 1357–1373.

——, 1976, Petrologic phase equilibria: San Francisco, California, W. H. Freeman, 333 p.

Garcia, M. O., 1979, Petrology of the Rogue and Galice Formations, Klamath Mountains, Oregon; Identification of a Jurassic island arc sequence: Journal of Geology, v. 87, p. 29–41.

Girty, G. H., and Wardlaw, M. S., 1985, Petrology and provenance of pre-Late Devonian sandstones, Shoo Fly complex, Sierra Nevada, California: Geological Society of America Bulletin, v. 96, p. 516–521.

Goodge, J. W., 1989a, Polyphase metamorphic evolution of a Late Triassic subduction complex, Klamath Mountains, northern California: American Journal of Science, v. 289, p. 874–943.

——, 1989b, Evolving early Mesozoic convergent margin deformation, central Klamath Mountains, northern California: Tectonics, v. 8, p. 845–863.

——, 1990, Convergent margin setting and petrotectonic evolution of a coherent late Triassic subduction complex, Stuart Fork terrane, Klamath Mountains, northern California: Geological Society of America Bulletin, v. 102, p. 86–101.

Guidotti, C. V., and Sassi, F. P., 1976, Muscovite as petrogenetic indicator mineral in pelitic schists: Neues Jahrbuchs für Mineralogie Abhandlungen, v. 127, no. 2, p. 97–142.

Hannah, J. L., and Moores, E. M., 1986, Age relationships and depositional environments of Paleozoic strata, northern Sierra Nevada, California: Geological Society of America Bulletin, v. 97, p. 787–797.

Hanson, R. E., and Schweickert, R. A., 1986, Stratigraphy of mid-Paleozoic island-arc rocks in part of the northern Sierra Nevada, Sierra and Nevada Counties, California: Geological Society of America Bulletin, v. 97, p. 986–998.

Harper, G. D., and Wright, J. E., 1984, Middle to Late Jurassic tectonic evolution of the Klamath Mountains, California-Oregon: Tectonics, v. 3, p. 759–772.

Harper, G. D., Bowman, J. R., and Kuhns, R., 1988, A field, chemical, and stable isotope study of subseafloor metamorphism of the Josephine ophiolite, California-Oregon: Journal of Geophysical Research, v. 93, p. 4625–4656.

Harwood, D. S., 1988, Tectonism and metamorphism in the northern Sierra terrane, northern California, *in* Ernst, W. G., ed., Metamorphism and crustal evolution of the western United States, Rubey Volume 7: Englewood Cliffs, New Jersey, Prentice-Hall, p. 764–788.

Heinrich, W., and Althaus, E., 1980, Die obere Stabilitätsgrenze von Lawsonit plus Albit bzw. Jadeit: Fortschritte für Mineralogie, v. 58, p. 49–50.

Helper, M. A., 1986, Deformation and high P/T metamorphism in the central part of the Condrey Mountain window, north-central Klamath Mountains, California and Oregon, *in* Evans, B. W., and Brown, E. H., Blueschists and eclogites: Geological Society of America Memoir 164, p. 125–142.

Hietanen, A., 1981, Geology west of the Melones fault between the Feather and North Yuba Rivers, California: U.S. Geological Survey Professional Paper 920, 30 p.

Holdaway, M. J., 1963, Petrology and structure of metamorphic and igneous rocks of parts of northern Coffee Creek and Cecilville Quadrangles [Ph.D. thesis]: Berkeley, University of California, 180 p.

Holland, T.J.B., 1979, Experimental determination of the reciton paragonite = jadeite + kyanite + H_2O, and internally consistent thermodynamic data for part of the system Na_2O-Al_2O_3-SiO_2-H_2O, with applications to eclogites and blueschists: Contributions to Mineralogy and Petrology, v. 68, p. 293–301.

Hotz, P. E., 1971, Geology of lode gold districts in the Klamath Mountains, California and Oregon: U.S. Geological Survey Bulletin 1290, 91 p.

——, 1973a, Blueschist metamorphism in the Yreka–Ft. Jones area, northeastern Klamath Mountains, California: Geological Society of America Abstracts with Programs, v. 5, p. 59–60.

——, 1973b, Blueschist metamorphism in the Yreka–Ft. Jones area, Klamath Mountains, California: U.S. Geological Survey Journal of Research, v. 1, no. 1, p. 53–61.

——, 1977a, Blueschist facies rocks in the Yreka–Ft. Jones area, Klamath Mountains, California, *in* Lindsley-Griffin, N., and Kramer, J. C., Guidebook to the geology of the Klamath Mountains, northern California; Geological Society of America Cordilleran Section 73rd Annual Meeting: p. 26–33.

——, 1977b, Geology of the Yreka Quadrangle, Siskiyou County, California: U.S. Geological Survey Bulletin 1436, 72 p.

Hotz, P. E., Lanphere, M. A., and Swanson, D. A., 1977, Triassic blueschist from northern California and north-central Oregon: Geology, v. 5, p. 659–663.

Hynes, A., 1982, A comparison of amphiboles from medium- and low-pressure metabasites: Contributions to Mineralogy and Petrology, v. 81, p. 119–125.

Irwin, W. P., 1966, Geology of the Klamath Mountain province, *in* Bailey, E. H., ed., Geology of northern California: California Division of Mines and Geology Bulletin 190, p. 19–38.

——, 1981, Tectonic accretion of the Klamath Mountains, *in* Ernst, W. G., ed., The geotectonic development of California; Rubey Volume 1: Englewood Cliffs, New Jersey, Prentice-Hall, p. 29–49.

Johannes, W., and Puhan, D., 1971, The calcite-aragonite transition, revisited:

Contributions to Mineralogy and Petrology, v. 31, p. 28–38.

Laird, J., and Albee, A. L., 1981a, High-pressure metamorphism in mafic schist from northern Vermont: American Journal of Science, v. 281, p. 97–126.

——, 1981b, Pressure, temperature, and time indicators in mafic schist; Their application to reconstructing the polymetamorphic history of Vermont: American Journal of Science, v. 281, p. 127–175.

Lanphere, M. A., Irwin, W. P., and Hotz, P. E., 1968, Isotopic age of the Nevadan orogeny and older plutonic and metamorphic events in the Klamath Mountains, California: Geological Society of America Bulletin, v. 79, p. 1027–1052.

Liou, J. G., 1971, Synthesis and stability relations of prehnite, $Ca_2Al_2Si_3O_{10}(OH)_2$: American Mineralogist, v. 56, p. 507–531.

Liou, J. G., Maruyama, S., and Cho, M., 1985, Phase equilibria and mineral parageneses of metabasites in low-grade metamorphism: Mineralogical Magazine, v. 49, p. 321–333.

——, 1987, Very low-grade metamorphism of volcanic and volcaniclastic rocks; Mineral assemblages and mineral facies, *in* Frey, M., ed., Low temperature metamorphism: Glasgow, Blackie, p. 59–112.

Maresch, W. V., 1977, Experimental studies on glaucophane; An analysis of present knowledge: Tectonophysics, v. 43, p. 109–125.

Maruyama, S., Cho, M., and Liou, J. G., 1986, Experimental investigations of blueschist-greenschist transition equilibria; Presssure dependenc eof Al_2O_3 contents in sodic amphiboles; A new geobarometer, *in* Evans, B. W., and Brown, E. H., eds., Blueschists and eclogites: Geological Society of America Memoir 164, p. 1–16.

Massone, H.-J., and Schreyer, W., 1987, Phengite geobarometry based on the limiting assemblage with K-feldspar, phlogopite, and quartz: Contributions to Mineralogy and Petrology, v. 96, p. 212–224.

Miller, M. M., 1989, Intra-arc sedimentation and tectonism; Late Paleozoic evolution of the eastern Klamath terrane, California: Geological Society of America Bulletin, v. 101, p. 170–187.

Mortimer, N., 1984, Geology of the Klamath Mountains near Yreka, California, *in* Nilsen, T. H., ed., Geology of the Upper Cretaceous Hornbrook Formation, Oregon and California: Pacific Section, Society of Economic Paleontologists and Mineralogists, v. 42, p. 167–178.

Newton, R. C., and Smith, J. V., 1967, Investigations concerning the breakdown of albite at depth in the earth: Journal of Geology, v. 75, p. 268–286.

Nitsch, K.-H., 1968, Die Stabilität von Lawsonit: Naturwissenschaften, v. 55, p. 388.

——, 1971, Stabilitätsbeziehungern von Prehnite- und Pumpellyite-haltigen Paragenesen: Contributions to Mineralogy and Petrology, v. 30, p. 240–260.

Passaglia, E., and Gottardi, G., 1973, Crystal chemistry and nomenclature of pumpellyites and julgoldites: Canadian Mineralogist, v. 12, p. 219–223.

Peacock, S. M., and Norris, P. J., 1989, Metamorphic evolution of the CMB, Klamath province, California; An inverted metamorphic gradient beneath the Trinity peridotite: Journal of Metamorphic Geology, v. 7, p. 191–209.

Rock, N.M.S., and Leake, B. E., 1984, The International Mineralogical Association amphibole nomenclature scheme; Computerization and its consequences: Mineralogical Magazine, v. 48, p. 211–227.

Saleeby, J. B., Harper, G. D., Snoke, A. W., and Sharp, W. D., 1982, Time relations and structural-stratigraphic patterns in ophiolite accretion, west-central Klamath Mountains, California: Journal of Geophysical Research, v. 87, p. 3831–3848.

Saleeby, J. B., Shaw, H. F., Moores, E. M., and Edelman, S. H., 1989, U/Pb, Sm/Nd, and Rb/Sr geochronological and isotopic study of northern Sierra Nevada ophiolitic assemblages, California: Contributions to Mineralogy and Petrology, v. 102, p. 205–220.

Schiffman, P., and Liou, J. G., 1980, Synthesis and stability of Mg-Al pumpellyite: Journal of Petrology, v. 21, p. 441–474.

——, 1983, Synthesis of Fe-pumpellyite and its stability relations with epidote: Journal of Metamorphic Geology, v. 1, p. 91–101.

Schweickert, R. A., 1981, Tectonic evolution of the Sierra Nevada range, *in* Ernst, W. G., ed., The geotectonic development of California; Rubey Volume 1: Englewood Cliffs, New Jersey, Prentice-Hall, p. 132–181.

Schweickert, R. A., Armstrong, R. L., and Harakal, J. E., 1980, Lawsonite blueschist in the northern Sierra Nevada: Geology, v. 8, p. 27–31.

Sharp, W. D., 1988, Pre-Cretaceous crustal evolution in the Sierra Nevada region, California, *in* Ernst, W. G., ed., Metamorphism and crustal evolution of the western United States; Rubey, Volume 7: Englewood Cliffs, New Jersey, Prentice-Hall, p. 823–864.

Standlee, L. A., 1978, Middle Paleozoic ophiolite in the Melones fault zone, northern Sierra Nevada, California: Geological Society of America Abstracts with Programs, v. 10, p. 148.

Strand, R. G., 1962, Geologic map of California, Redding sheet: California Division of Mines and Geology, scale 1:250,000.

Trexler, D. T., 1968, Geology of the northwest quarter of the Cecilville Quadrangle, Siskiyou County, California [M.S. thesis]: Los Angeles, University of Southern California, 133 p.

Velde, B., 1965, Phengite micas; Synthesis, stability, and natural occurrence: American Journal of Science, v. 263, p. 886–913.

Weisenberg, C. W., and Ave Lallement, H. G., 1977, Permo-Triassic emplacement of the Feather River peridotite body, northern Sierra Nevada mountains, California: Geological Society of America Abstracts with Programs, v. 9, p. 525.

Wright, J. E., 1982, Permo-Triassic accretionary subduction complex, southwestern Klamath Mountains, northern California: Journal of Geophysical Research, v. 87, no. 85, p. 3805–3818.

Wright, J. E., and Fahan, M. R., 1988, An expanded view of Jurassic orogenesis in the western United States Cordillera; Middle Jurassic (pre-Nevadan) regional metamorphism and thrust faulting within an active arc environment, Klamath Mountains, California: Geological Society of America Bulletin, v. 100, p. 859–876.

Wright, J. E., and Wyld, S. J., 1985, Multi-stage serpentinite matrix melange development; Rattlesnake Creek terrane, southwestern Klamath Mountains: Geological Society of America Abstracts with Programs, v. 17, p. 419.

MANUSCRIPT ACCEPTED BY THE SOCIETY MARCH 27, 1990

Printed in U.S.A.

Geological Society of America
Special Paper 255
1990

Accretionary terrane in the Sawyers Bar area of the Western Triassic and Paleozoic Belt, central Klamath Mountains, northern California

W. G. Ernst*
Department of Earth and Space Sciences, University of California, Los Angeles, California 90024-1567

ABSTRACT

Detailed mapping of greenschist-facies rocks in the central part of the Western Triassic and Paleozoic Belt near Sawyers Bar, California, demonstrates the stratal continuity and interlayered nature of intricately folded but unbroken, chemically distinctive mafic metavolcanic and metaturbiditic units. This section passes southward without break into the contiguous, weakly recrystallized Salmon River/North Fork amalgamated terrane, and northward, apparently along depositional contacts, into structurally lower, more intensely metamorphosed eastern(?) Hayfork metaclastic units. Although previously regarded as portions of several disparate terranes, these supracrustal rocks all belong to a single lithostratigraphic belt, tectonically emplaced beneath the more easterly Stuart Fork terrane. Because of the controversial nature of the correlations, this complex is referred to informally as the Sawyers Bar terrane. Contacts in the mapped area are depositional rather than faulted, except for the east-dipping thrust fault bringing the Stuart Fork over the Sawyers Bar terrane. No other terrane boundaries are recognized in the mapped area. This conclusion is supported by petrologic, structural/stratigraphic, and geochemical studies that demonstrate the local continuity of mineralogic, metamorphic, tectonic/stratigraphic, and bulk-rock compositional trends. Evidently this portion of the Western Triassic and Paleozoic Belt formed and evolved as a single, possibly parautochthonous accretionary complex, rather than as a collage of unrelated, exotic crustal entities.

INTRODUCTION TO THE WESTERN TRIASSIC AND PALEOZOIC BELT

The Western Triassic and Paleozoic Belt of the Klamath Mountains consists of three major lithotectonic units (plus the Stuart Fork terrane) juxtaposed along steeply to gently east-dipping thrust faults, and invaded by calc-alkaline plutons of approximately Middle and Late Jurassic age (Irwin, 1960, 1966, 1985; Wright and Fahen, 1988). Regional petrotectonic relations in northern and central California are illustrated in Figure 1. The Western Triassic and Paleozoic Belt contains fossils of latest Paleozoic and early Mesozoic age; some, especially the older macrofossils, apparently are confined to olistolithic and/or tectonic blocks (Irwin and others, 1978; Blome and Irwin, 1983). On the east, the structurally highest unit is the Stuart Fork terrane, metamorphosed under blueschist-facies conditions during Late Triassic time (Hotz, 1973; Hotz and others, 1977; Borns, 1984; Goodge, 1985, 1989a, b). To the west, the North Fork, Hayfork, and Rattlesnake Creek terranes of the Western Triassic and Paleozoic Belt are disposed in progressively more oceanward, structurally lower positions (Irwin, 1972). Yet farther to the west lies the lowest member of the Klamath Mountains stack, the feebly recrystallized Western Jurassic Belt (Coleman and others, 1988). Similar to the inboard Stuart Fork terrane, all Western Triassic and Paleozoic Belt terranes were deformed and overprinted by

*Present address: School of Earth Sciences, Stanford University, Stanford, California 94305-2210.

Ernst, W. G., 1990, Accretionary terrane in the Sawyers Bar area of the Western Triassic and Paleozoic Belt, central Klamath Mountains, northern California, *in* Harwood, D. S., and Miller, M. M., eds., Paleozoic and early Mesozoic paleogeographic relations; Sierra Nevada, Klamath Mountains, and related terranes: Boulder, Colorado, Geological Society of America Special Paper 255.

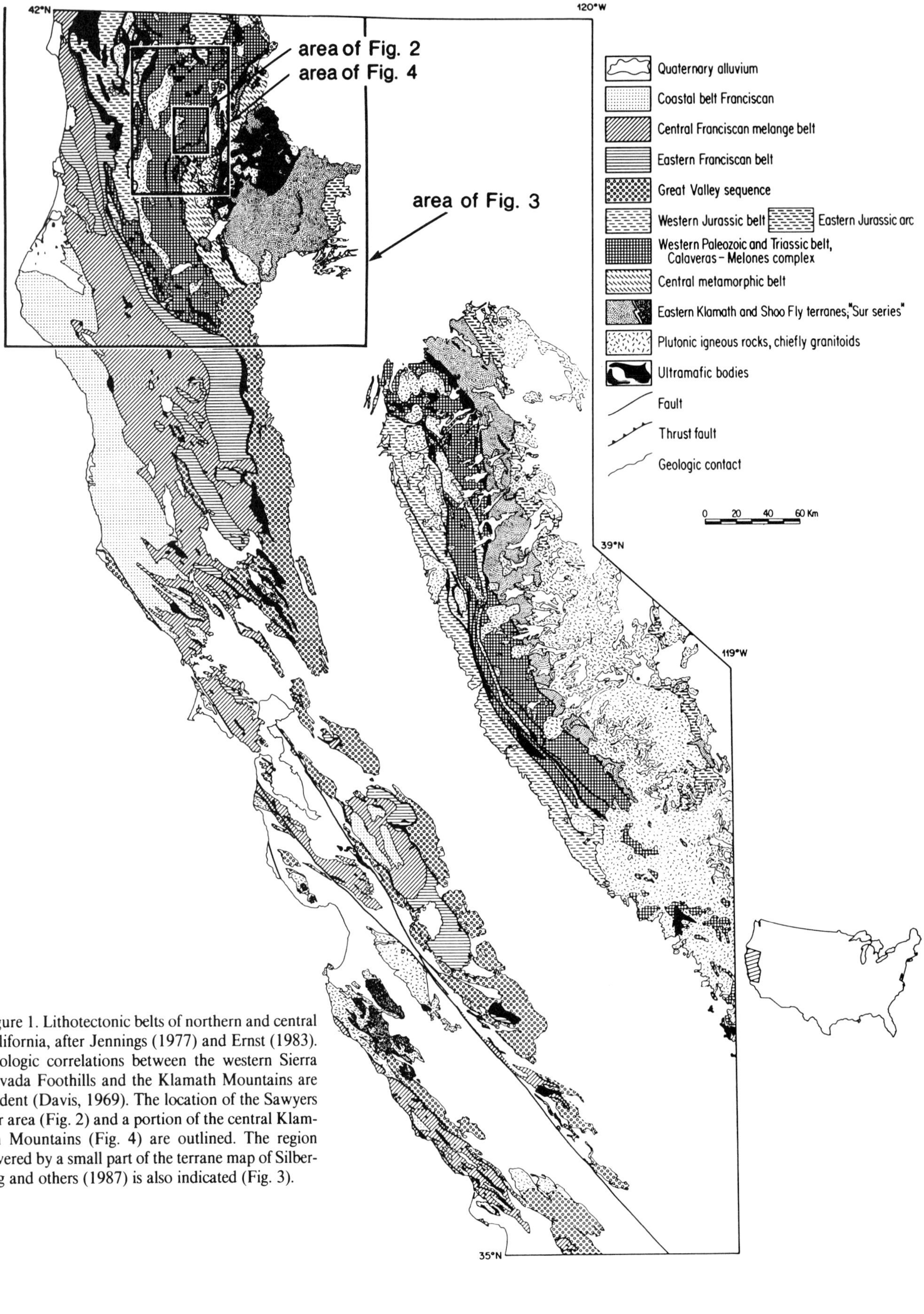

Figure 1. Lithotectonic belts of northern and central California, after Jennings (1977) and Ernst (1983). Geologic correlations between the western Sierra Nevada Foothills and the Klamath Mountains are evident (Davis, 1969). The location of the Sawyers Bar area (Fig. 2) and a portion of the central Klamath Mountains (Fig. 4) are outlined. The region covered by a small part of the terrane map of Silberling and others (1987) is also indicated (Fig. 3).

greenschist- to low amphibolite-facies metamorphism during their accretion (the Siskiyou event of Coleman and others, 1988; see also Mortimer, 1985; and Barnes and others, 1986), preceding and accompanying emplacement of the mid-Jurassic granitoids (Lanphere and others, 1968; Wright, 1982; Wright and Fahan, 1988; Harper and Wright, 1984; Hill, 1985).

Protoliths of the Western Triassic and Paleozoic Belt included widespread serpentinites, greenstones, and associated cherts, as well as abundant, locally chaotic argillites, minor limestones, and interlayered mafic and intermediate volcanic rocks. The ophiolitic metagabbros, metabasalts, serpentinized peridotites, and metacherts are dominantly of oceanic origin (Irwin, 1981, 1985); however, the source of the voluminous terrigenous debris was more probably the nearby eastern Klamath provinces (Wright, 1982; Gray, 1986). The Stuart Fork terrane is interpreted as a section of oceanic crust and distal fan deposits carried down a subduction zone in the vinicity of the central Klamath region and metamorphosed approximately 227 Ma[1] (Borns, 1980; Ernst, 1984; Wagner and Saucedo, 1987). The geologic site and manner of assembly of the ophiolitic North Fork, andesitic Hayfork, and ophiolitic Rattlesnake Creek terranes are less clear (e.g., see Burchfiel and Davis, 1981; Irwin, 1981; Wright, 1982; Ando and others, 1983; Mortimer, 1985). Based on mapping within a small area of the central part of the Western Triassic and Paleozoic Belt in the vicinity of Sawyers Bar, California (Donato and others, 1982; Ernst, 1987), I argue in the present paper that at least the structurally uppermost portion of the Hayfork and the Nork Fork/Salmon River terranes constitute the fragmentary remains of a single, evolving oceanic arc complex.

GEOLOGY OF THE SAWYERS BAR AREA

Local geologic relations for the investigated portion of the Western Triassic and Paleozoic Belt are shown in Figure 2. A small segment of the Stuart Fork thrust sheet, intruded by the Russian Peak pluton, is situated in the southeastern part of the map area (Goodge, 1989a, b); the English Peak pluton crops out along the western margin of the studied area and appears to have invaded chiefly the eastern part of the Hayfork terrane (Wright, 1982; Barnes, 1987; Coleman and others, 1988). Both granitoids were emplaced during mid-Jurassic time subsequent to, or at the terminal stages of, the Western Triassic and Paleozoic Belt regional dynamothermal Siskiyou event (Irwin, 1985; Coleman and others, 1988; Wright and Fahan, 1988).

The major portion of the mapped region consists of massive to weakly foliated greenstones (metabasalts and metadiabases), and less voluminous interlayered, stratigraphically coherent, predominantly metaclastic schists. Protoliths for the fine-grained metasedimentary rocks consist of pelitic units, argillites, micrograywackes, calcareous quartzofeldspathic rocks, rare limey units, and minor cherty layers; good marker horizons are exceedingly rare. Primary compositional layering and indistinct foliation in the greenstones, and more obvious stratification in the metaclastic units, coincide in attitude, strike north-northeast, and dip steeply, mostly to the east. Parallelism of primary layering is here interpreted to indicate that the metavolcanic-metasedimentary contacts are depositional. Although the greenstones are relatively massive, rare in-place pillows provide unequivocal facing indicators (right-side-up-section, with tops to the east-southeast) in the northeast part of the map area. Pillows have also been noted as float in the southern part of the region. Although the metavolcanic units contain chiefly massive flows, minor cross-cutting and concordant tabular bodies, most of which display coarser grained, relict diabasic texture, reflect the presence of hypabyssal metamorphosed intrusives and/or slowly cooled flows distributed throughout the volcanic pile. The entire greenstone assemblage, including associated diabasic dikes and sills, is referred to informally as the Yellow Dog greenstones, after the mountain northeast of Sawyers Bar.

Metavolcanic and metasedimentary rocks are intimately interlayered in the mapped area. The Yellow Dog greenstones dominate the exposed section near Sawyers Bar, and in general, appear to overlie the fine-grained metaclastic unit. Original interstratification and tight folding are evident from the intricately interdigitated outcrop patterns of the metasedimentary and metaigneous units shown in Figure 2.

The structural interpretation is based on the in situ occurrence of pillows, the broad, open, basinal fold on the east, and especially the stratigraphic continuity displayed by the several mapped lithologies. The region evidently consists of a major, intact, greenstone-cored synform on the east, and to the west, very gently south- and north-plunging, overturned and tightly appressed, west-verging metasediment-cored antiforms and greenstone-cored synforms. A relatively mafic, thin, schistose metavolcanic unit occurs systematically on both eastern and western margins of the investigated supracrustal section (Fig. 2). Interpreted as lying stratigraphically beneath the main mass of Yellow Dog metavolcanics, this P_2O_5 and iron-rich (fractionated), amygdaloidal flow breccia is distinctly more titaniferous ($TiO_2 = 2.5 \pm 0.6$ wt%) than the overlying, relatively more felsic, monolithic metabasalts ($TiO_2 = 0.9 \pm 0.2$ wt %). The areal disposition of this titaniferous greenstone accords with the general bilateral symmetry of the region about NNE–trending axial traces. Because Yellow Dog dikes and sills belong exclusively to the Ti-poor series, and intrude the more titaniferous flow breccia near the eastern margin of the fold, the mafic, schistose member appears to predate and stratigraphically underlie the main mass of greenstones, in accord with the mapped structure.

Tectonic discontinuities such as faults and shear zones between metasedimentary and metavolcanic lithologic packages were not recognized in the Sawyers Bar area. Scattered lenses of serpentinite are present, however, and may mark faults in the monotonous section. These ultramafic rocks are especially numerous and voluminous near the southern boundary of the stud-

[1]K-Ar mineral ages (Hotz and others, 1977) may represent the time of cooling through the blocking temperature, hence apparently providing a minimum age for the subduction/recrystallization event.

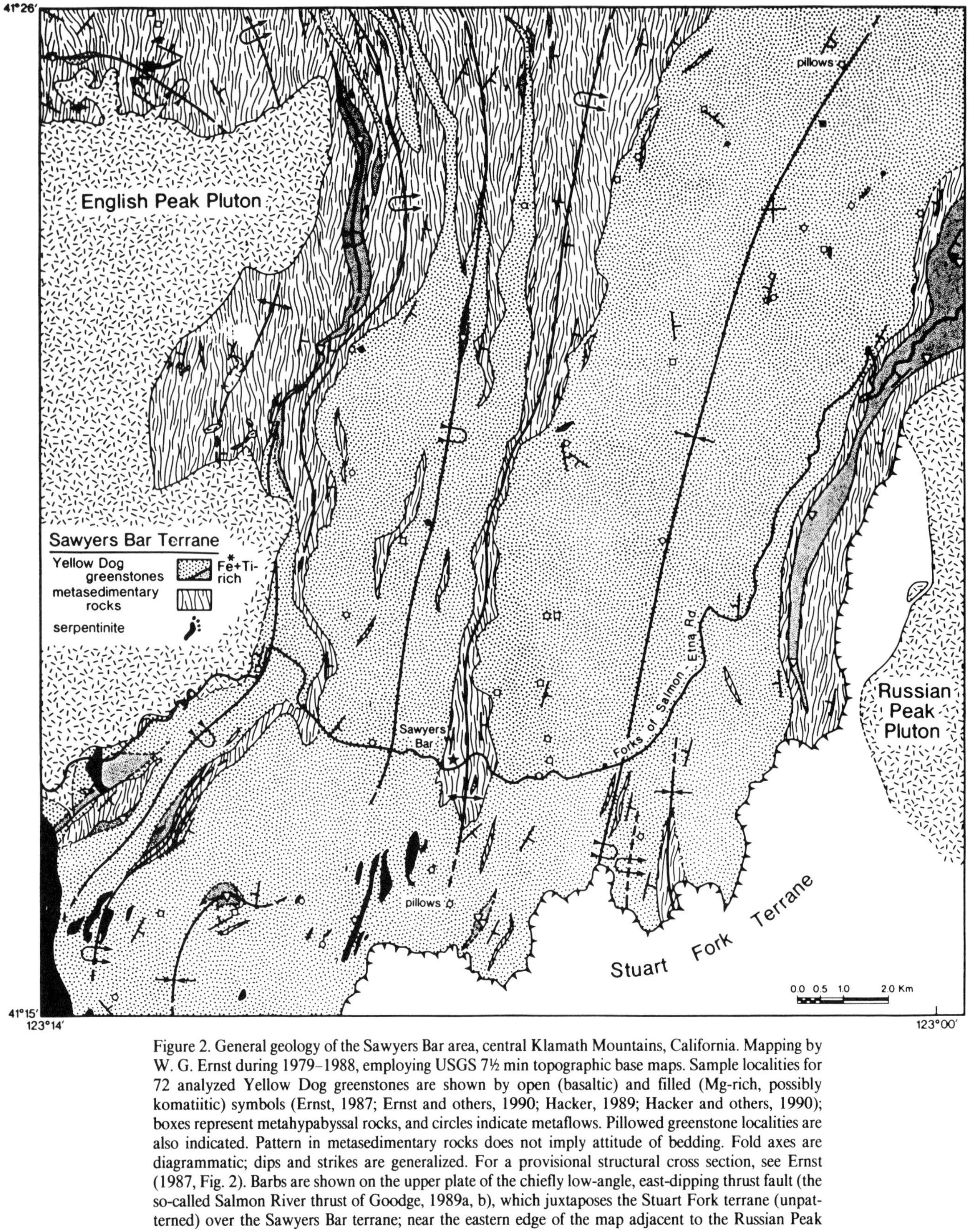

Figure 2. General geology of the Sawyers Bar area, central Klamath Mountains, California. Mapping by W. G. Ernst during 1979–1988, employing USGS 7½ min topographic base maps. Sample localities for 72 analyzed Yellow Dog greenstones are shown by open (basaltic) and filled (Mg-rich, possibly komatiitic) symbols (Ernst, 1987; Ernst and others, 1990; Hacker, 1989; Hacker and others, 1990); boxes represent metahypabyssal rocks, and circles indicate metaflows. Pillowed greenstone localities are also indicated. Pattern in metasedimentary rocks does not imply attitude of bedding. Fold axes are diagrammatic; dips and strikes are generalized. For a provisional structural cross section, see Ernst (1987, Fig. 2). Barbs are shown on the upper plate of the chiefly low-angle, east-dipping thrust fault (the so-called Salmon River thrust of Goodge, 1989a, b), which juxtaposes the Stuart Fork terrane (unpatterned) over the Sawyers Bar terrane; near the eastern edge of the map adjacent to the Russian Peak pluton, this fault dips steeply to the east.

ied region, where they apparently intrude, and are interlayered with, the greenstone host rocks. Directly to the south of the mapped area, Ando and others (1983) interpreted the somewhat disrupted mafic/ultramafic metaigneous rocks of the North Fork terrane as constituting the ophiolitic core of a major antiform, flanked by metasedimentary cover units. These authors based the inferred structure on the attitudes of bedding chiefly along the less disrupted eastern side of the belt, as well as on the fact that most identified microfossils from the metasedimentary section have nominal ages younger than the apparent Permo-Triassic (265 to 310 Ma) time of formation of the ophiolitic complex.[2] However, as is evident from the Geologic Map of California (Jennings, 1977), ultramafic material, in general, borders the mafic igneous rocks on both west and east, rather than being concentrated in the axis of the fold, which suggests the topology of a synform in the east-central segment of the Western Triassic and Paleozoic Belt.

The Yellow Dog greenstones include highly magnesian (komatiitic?) metabasalts + metadiabases, and the igneous suite exhibits a tholeiitic fractionation trend combined with marked light rare earth enrichment (Ernst, 1987). The associated metasedimentary section probably represents a distal turbiditic continental slope- or rise-type deposit. The interlayered association with distal turbidites, the light rare earth enrichment of the metalavas, and the inference that the extrusive section apparently largely overlies the sediments argue for submarine volcanism in a continental rise or immature island-arc setting (Ernst, 1987). This relation is corroborated by the occurrence of metahypabyssal rocks coeval with the Yellow Dog greenstones that are intrusive into the interlayered metasedimentary section (dikes and sills are too small to show at the scale of Fig. 2).

Except for the preservation of diabasic texture in some of the coarser grained intrusives, pillows in a few of the flows, and rare relict clinopyroxene ± igneous hornblende in several of the Yellow Dog hypabyssal rocks, recrystallization has obliterated most original features in both metaigneous and metasedimentary units. Petrographic studies demonstrate that these rocks were recrystallized under physical conditions of the upper greenschist facies (Donato and others, 1982; Mortimer, 1985; Coleman and others, 1988). Biotite is an essential phase in the metasedimentary rocks, especially in the pelitic and quartzofeldspathic units, and is a widespread accessory in the mafic metavolcanics. The typical upper greenschist-facies assemblage of the latter is albite + chlorite + actinolite hornblende + epidote + sphene. Average modes for 297 pre-Nevadan metamorphic rocks examined petrographically were presented by Ernst (1987), and 161 additional specimens have since been investigated. Low-variance, *P-T* diagnostic assemblages are rare in the studied area, and geothermometric/geobarometric investigations are still in progress. Judging from experimental *P-T* studies of the greenschist-epidote amphibolite facies boundary (Liou and others, 1974; Apted and Liou, 1983), and mineral parageneses in higher grade, low amphibolite-facies rocks directly to the north (Donato, 1987, 1989), estimated physical conditions for the upper greenschist-facies metamorphism in the Sawyers Bar area are 400 ± 75°C, 3 ± 1 kbar (Ernst, 1987).

TERRANES AND TERRANE BOUNDARIES IN THE SAWYERS BAR AREA

Nearly 20 years ago, Irwin (1972) recognized contrasting tectonostratigraphic entities in the southern, weakly metamorphosed, subgreenschist portion of the Western Triassic and Paleozoic Belt, namely the easterly, ophiolitic North Fork terrane, the central, calc-alkaline, in part chaotic Hayfork terrane, and the westerly, ophiolitic Rattlesnake Creek terrane. Since then, workers have attempted to extend these subdivisions northward into the more intensely recrystallized central Klamath Mountains. In the process, further complications and refinements of the Western Triassic and Paleozoic Belt lithotectonic units have been introduced. The central portion of the Western Triassic and Paleozoic Belt has been divided into a younger, mafic volcanic/volcaniclastic western Hayfork terrane, and an older, structurally disrupted eastern Hayfork terrane (Wright, 1982). The North Fork terrane is considered by some to be composite, consisting of a western, volcaniclastic, metaandesitic(?) Salmon River terrane, and an eastern, ophiolitic basement + metasedimentary cover North Fork sensu stricto terrane (Blake and others, 1982; Silberling and others, 1987). A northeastern, intensely metamorphosed, melanged, apparent equivalent of the Rattlesnake Creek belt, named the Marble Mountain terrane (Blake and others, 1982; Donato and others, 1982; Donato, 1987), has also been defined. Terrane boundaries after Silberling and others are illustrated for the Klamath Mountains in Figure 3.

Western Triassic and Paleozoic Belt rocks of the Sawyers Bar area are tectonically juxtaposed on the east beneath the Stuart Fork terrane. The latter is an allochthonous slab that dips gently to the east (Hotz, 1973; Goodge, 1985, 1989a, b). The Marble Mountain terrane (Donato, 1987) lies north of the mapped area, and to the south the Sawyers Bar terrane adjoins the Salmon River/North Fork amalgamated terrane assembly (Ando and others, 1983; Silberling and others, 1987). Lithic units in the mapped area pass northwestward into tectonically underlying rocks of the Hayfork terrane, and are extensively invaded by granitoids at this latitude; directly to the north, Donato (1987, p. 455) reported a gradational, interleaved transition between the structurally lower Marble Mountain terrane and what she recognized as the western Hayfork terrane. The question is, can Western Triassic and Paleozoic Belt terrane boundaries be identified within the Sawyers Bar area? And, to what terrane or terranes do the Yellow Dog greenstones and interfingered metasedimentary rocks belong? Some previous investigators have regarded the entire mapped area shown in Figure 2—except for the Stuart Fork—as Hayfork (Mortimer, 1985; Donato, 1987; Goodge, 1989a, b; Wyld and Wright, 1988; Irwin, 1989). Others have placed the Salmon River/North Fork–Hayfork tectonic contact

[2]The U-Pb mineral age reported by Ando and others (1983) is highly discordant, and thus does not provide tight constraints on the igneous formational age of the North Fork ophiolite.

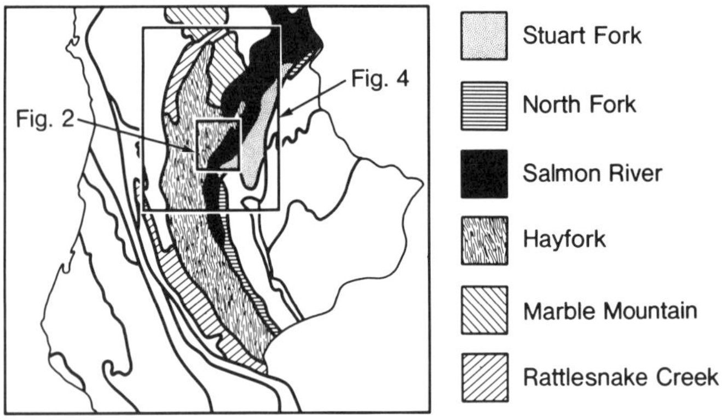

Figure 3. Terranes of the Californian sector of the Klamath Mountains, according to Silberling and others (1987); terrane designations in most cases follow the nomenclature of these authors. The column shows lithostratigraphic units roughly from east to west, and in general, from older to younger and from tectonically higher to lower. The areas of Figures 2 and 4 are also indicated.

along the eastern edge of the English Peak pluton near the western limit of the mapped area (Barnes, 1987), or along the eastern margin of the north-south–trending metasedimentary belt in the center of the Sawyers Bar area (Silberling and others, 1987; Coleman and others, 1988). Because of the debatable tectonostratigraphic correlations, the Yellow Dog greenstones and spatially associated metasedimentary strata in this chapter are informally referred to as constituting the Sawyers Bar terrane.

Other than the low-angle thrust contact between the tectonically overlying Stuart Fork terrane on the east and the underlying Western Triassic and Paleozoic Belt on the west, major faults and profoundly allochthonous entities have not been recognized in the region depicted in Figure 2. I regard all studied supracrustal units in the Sawyers Bar terrane, including hypabyssal rocks, to be equivalent to the North Fork terrane as described by Ando and others (1983), but shown as the Salmon River/North Fork amalgamated terrane by Silberling and others (1987), and as the western Hayfork terrane by Donato (1987).

A comparison of the bulk-rock chemistry of the Yellow Dog greenstones with the North Fork metavolcanic rocks to the south is presented in Table 1; also listed are average bulk-rock compositions for metabasaltic dikes and sills directly north of the Sawyers Bar area. The three analyzed samples of recrystallized North Fork ophiolitic intrusive (Ando, 1979, p. 122) compare favorably with the Yellow Dog diabasic rocks in the Sawyers Bar area. Except for the higher K_2O contents of the latter (but wide ranges in potash occur in both groups of metaigneous rocks), the two suites are chemically indistinguishable. Although broadly variable in composition, the North Fork metalavas are also compositionally similar to the Yellow Dog greenstones. No highly magnesian extrusives were analyzed, but the three North Fork samples associated with metasediments are quite rich in TiO_2, analogous perhaps to the titaniferous, amygdaloidal greenstone flow breccia that crops out apparently near the base of the section in the vicinity of Sawyers Bar. Metahypabyssal rocks in the Sawyers Bar terrane exhibit bulk-rock chemistries that are virtually identical, both in average values and in compositional ranges, to greenschist- and amphibolite-facies dikes and sills transecting the volcaniclastic unit directly north of the mapped area (Donato, 1985, Table 1.6). Although petrochemical similarity does not constitute proof for the proposed lithostratigraphic correlations, the similar basaltic nature of the analyzed rocks is evident from the data presented in Table 1.

The boundary with the Hayfork belt, if it passes through the Sawyers Bar area, appears to be depositional rather than tectonic (Wright, 1982; Donato, 1987), and is perhaps sited near the western edge of Figure 2, where transitional geologic relations are thoroughly obscured by baking associated with the English Peak pluton. Most lithologic sections in the mapped area are comparable to those of the type North Fork assemblage of Irwin (1972), as detailed by Ando and others (1983). No fault locally separates the northern extension of the so-called Salmon River terrane from the sediment-rich North Fork sensu stricto terrane of Blake and others (1982). In the Sawyers Bar area, metabasalts (on-strike equivalents of the Salmon River terrane) and distal turbidites + cherty layers (the northern extension of the North Fork sensu stricto terrane) are intimately interlayered and thrown into tight, concordant folds. Whether the Yellow Dog greenstones overlie or underlie the associated recrystallized distal turbidites is problematic, but is less important for the terrane analysis than the fact that the original sediments and basalts are depositionally interleaved.

The conclusion that the Sawyers Bar terrane represents a single, evolving supracrustal entity is based on the following observations of lithotectonic and geochemical relations displayed in the mapped area (Fig. 2).

1. The Yellow Dog greenstones pass southward without structural break into more feebly recrystallized lithologies of the North Fork ophiolite described by Ando and others (1983). Concomitantly, the primary grain size of many of the mafic protoliths gradually increases, imparting relict diabasic and gabbroic textures to the southern equivalents of the finer grained, more thoroughly reconstituted Yellow Dog greenstones.

2. The metavolcanic and metasedimentary rocks are interstratified, concordantly layered, and isoclinally folded; hence, the igneous-sedimentary contacts appear to be depositional, not tectonic, in nature. Volcanism and largely terrigenous sedimentation took place alternately, and probably simultaneously.

3. Although Al contents of calcic amphiboles, and $\delta^{18}O$ values for bulk metamafic supracrustal rocks are systematically higher in proximity to the English Peak pluton, no regional mineralogic/petrologic/geochemical discontinuity has been discovered in the map area (Ernst, 1987; Ernst and others, 1990; Hacker, 1989; Hacker and others, 1990). Instead, metamorphic mineral parageneses, phase compositions, and rock chemistries—major, minor, and trace-element concentrations, and values of ϵ_{Nd} and $\delta^{18}O$—show similar ranges and monotonic trends over

TABLE 1. COMPARISON OF AVERAGE BULK-ROCK CHEMISTRIES FOR YELLOW DOG GREENSTONES AND NEARBY, SIMILAR LITHOLOGIES, EASTERN PART OF WESTERN TRIASSIC AND PALEOZOIC BELT, CENTRAL KLAMATH MOUNTAINS, CALIFORNIA

	Yellow Dog†			North Fork§		Western Hayfork**
	Hypabyssals	Ti-rich greenstones	"Normal" greenstones	Gabbros and diabase	Mafic volcanics	Mafic dikes and sills
No. of Analyses	(18)	(16)	(38)	(3)	(5)	(8)
Oxide wt %						
SiO_2	53.26 ± 2.98	47.88 ± 2.47	50.78 ± 1.92	53.3 ± 2.7	51.7 ± 2.2	51.4 ± 2.0
Al_2O_3	14.04 ± 1.78	12.86 ± 1.85	12.85 ± 1.68	15.76 ± 0.68	15.39 ± 0.37	13.7 ± 1.6
TiO_2	0.86 ± 0.17	2.47 ± 0.60	0.91 ± 0.21	0.91 ± 0.60	1.84 ± 1.16	0.77 ± 0.19
$Fe_2O_3^*$	9.94 ± 1.21	13.87 ± 1.20	10.73 ± 0.98	8.23 ± 2.47	11.00 ± 1.53	9.41 ± 0.99
MgO	9.19 ± 3.20	9.95 ± 3.09	10.72 ± 2.96	8.14 ± 3.25	5.86 ± 1.94	9.07 ± 2.85
CaO	8.15 ± 1.39	8.44 ± 2.62	10.21 ± 2.23	10.44 ± 4.22	7.85 ± 1.68	10.09 ± 1.51
Na_2O	2.43 ± 0.70	2.06 ± 1.00	2.34 ± 0.77	2.65 ± 1.38	3.97 ± 0.72	2.77 ± 0.83
K_2O	1.42 ± 0.62	0.71 ± 0.60	0.48 ± 0.44	0.38 ± 0.48	1.19 ± 1.95	0.69 ± 0.27
MnO	0.17 ± 0.02	0.18 ± 0.02	0.17 ± 0.03	-	-	0.17 ± 0.02
P_2O_5	0.22 ± 0.08	0.34 ± 0.17	0.11 ± 0.05	-	-	0.17 ± 0.08
ppm						
Rb	35 ± 19	15 ± 17	11 ± 11	7 ± 9	58 ± 82	13 ± 5
Sr	383 ± 210	382 ± 204	179 ± 102	129 ± 35	418 ± 127	315 ± 224
Zr	113 ± 50	193 ± 57	76 ± 27	-	-	75 ± 21

†Ernst, 1987; Ernst and others, 1990.
§Ando, 1979; Ando and others, 1983.
**Donato, 1985, 1987.

the portion of the Western Triassic and Paleozoic Belt exposed in the Sawyers Bar area.

4. Structural and stratigraphic coherence are suggested by the areal disposition of the titaniferous, schistose greenstone flow breccia that crops out in both eastern and western parts of the mapped region.

5. Chemically distinctive, high-Mg metabasalts and chemically similar hypabyssal rocks of komatiitic(?) affinities are sporadically distributed in the Western Triassic and Paleozoic Belt throughout the mapped area (Ernst, 1987).

6. Except for the allochthonous Stuart Fork terrane, which exhibits the effects of two regional metamorphic events, not including contact aureoles surrounding mid-Jurassic-age calc-alkaline plutons (Goodge, 1989a, b; Hacker and Goodge, this volume), all rocks share a common metamorphic grade and deformational style, with axial planes dipping steeply to the east. These Western Triassic and Paleozoic Belt units show a single stage of regional folding about nearly horizontal NNE-trending axes throughout the area, evidently produced during the Siskiyou metamorphic event (Coleman and others, 1988).

The Yellow Dog greenstones and interstratified distal turbidites extend southward into the feebly recrystallized North Fork ophiolite terrane (Ando and others, 1983), and north-northeast into metavolcanic rocks of the Salmon River terrane (Silberling and others, 1987); the latter border the more intensely metamorphosed Marble Mountain terrane on the southeast (Donato and others, 1982; Donato, 1987). Regional relations indicating lithologic continuity are displayed on the 1:750,000 scale Geologic Map of California (Jennings, 1977) and the 1:250,000 scale Geologic Map of the Weed Sheet (Wagner and Saucedo, 1987); subsidiary occurrences of bedded limestone and larger proportions of ultramafic units crop out in both directions from the mapped area along the strike of this belt. Lithostratigraphic relations are shown in Figure 4. Donato (1987) described an intimately interleaved—possibly sedimentary interfingering—contact between her western(?) Hayfork and the underlying Marble Mountain terrane, which she correlated with the Rattlesnake Creek terrane. Alternatively, it seems possible that the Marble Mountain complex might represent an amphibolite-grade equivalent of the eastern(?) Hayfork tectonostratigraphic melange, rather than being an eastern salient of the Rattlesnake Creek terrane.

If portions of a petrotectonic continuum exist in the Sawyers Bar area, the extreme rarity of serpentinized peridotite and limestone may reflect the local exposure of volcanogenic upper stratigraphic/structural levels of a relatively coherent part of this zone. To the south, less intensely metamorphosed but tectonically disturbed and differentially uplifted units are exposed, whereas to the north, the deepest, most intensely recrystallized portion of the terrane has been exhumed adjacent to the Condrey Mountain

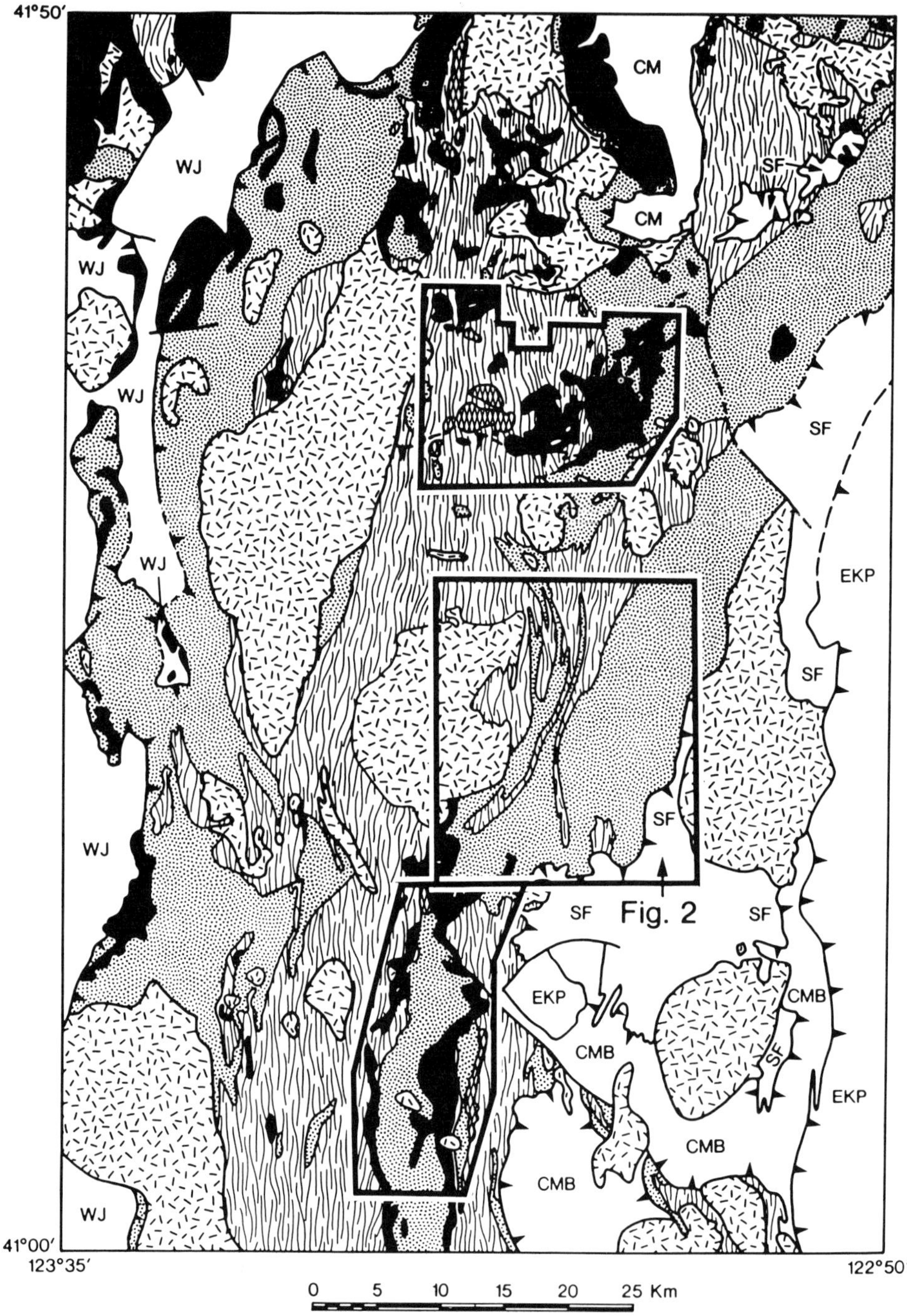

Figure 4. Regional lithotectonic relations in the central Klamath Mountains, California, emphasizing a portion of the Western Triassic and Paleozoic Belt (patterned), simplified after Wagner and Saucedo (1987). Rock types are as follows: black = serpentinized peridotite; fine dots = mafic metaigneous rocks; wavy lines = chiefly fine-grained metaclastic units (strike of layering not implied); gridiron pattern = marble blocks and layers. Mid-Jurassic granitic plutons are indicated by random dashes. Note that the stratigraphic package consisting of varying proportions of interlayered metabasaltic units, metasedimentary strata, and lenses of marble characterizes the entire eastern portion of the Western Triassic and Paleozoic Belt. Abbreviations for other belts (unpatterned) are: Central Metamorphic Belt = CMB; Condrey Mountain Schist = CM; Eastern Klamath terrane = EKP; Stuart Fork terrane = SF; and Western Jurassic Belt = WJ. The Sawyers Bar area is labeled Fig. 2; other studied regions referred to in the text include those mapped by Donato (1987) and by Ando and others (1983), north and south, respectively, of Fig. 2.

dome, floored by the Condrey Mountain Schist (Donato and others, 1980; Helper, 1986; Barnes, 1987; Coleman and others, 1988).

The intergradational and apparently depositional natures of the boundaries between the several postulated Western Triassic and Paleozoic Belt terranes in the mapped area and structural/stratigraphic extensions of the belt both north and south of Sawyers Bar have been partly obscured by subsequent metamorphism, deformation, and differential uplift. Nevertheless, primary extrusive depositional contacts are recognized among the eastern(?) Hayfork, Salmon River, and North Fork sensu stricto terranes. Similar relations directly to the northeast have recently been recognized by Elder and Cashman (1990). Accordingly, during growth of this sector of the North American margin, none of these terranes appears to have been tectonically introduced, and therefore exotic relative to the others. All three of these units, at least locally, thus may be regarded as portions of a single, Late Triassic–Early Jurassic convergent-margin or island-arc–type accretionary complex. Provenance studies suggest that this immature oceanic arc terrane probably evolved in the neighborhood of inboard lithotectonic belts constituting portions of the eastern Klamath Mountains (Wright, 1982; Gray, 1986). This conclusion is in accord with a recent paleomagnetic study reported by Mankinen and others (1989). These authors documented profound early/mid-Mesozoic clockwise rotation of the Eastern Klamath terrane but failed to find remnant magnetic data supporting significant latitudinal displacement for any Klamath Mountains terrane relative to the North American craton.

ACKNOWLEDGMENTS

This work is the result of field-based research and geochemical bulk-rock and mineralogic thermobarometric investigations in the Sawyers Bar area, which were supported by the U.S. Geological Survey and the University of California, Los Angeles, Research Committee. Most recently, substantial funding has been provided for ongoing studies by the U.S. Department of Energy through Grant DE-FG03-87ER13806. This manuscript, reporting some of the ongoing work, has been reviewed and materially improved by M. C. Blake, Jr., Mary Donato, R. G. Coleman, and John Goodge. I am indebted to the above institutions and researchers for their help.

REFERENCES CITED

Ando, C. J., 1979, Structured and petrologic analysis of the North Fork terrane, central Klamath Mountains [Ph.D. thesis]: Los Angeles, University of Southern California, 197 p.

Ando, C. J., Irwin, W. P., Jones, D. L., and Saleeby, J. B., 1983, The ophiolitic North Fork terrane in the Salmon River region, central Klamath Mountains, California: Geological Society of America Bulletin, v. 94, p. 236–252.

Apted, M. J., and Liou, J. G., 1983, Phase relations among greenschist, epidote amphibolite, and amphibolite in a basaltic system: American Journal of Science, v. 283A, p. 328–354.

Barnes, C. G., 1987, Mineralogy of the Wooley Creek batholith, Slinkard pluton, and related dikes, Klamath Mountains, northern California: American Mineralogist, v. 72, p. 879–901.

Barnes, C. G., Rice, J. M., and Gribble, R. F., 1986, Tilted plutons in the Klamath Mountains of California and Oregon: Journal of Geophysical Research, v. 91, p. 6059–6071.

Blake, M. C., Jr., Howell, D. G., and Jones, D. L., 1982, Preliminary tectonostratigraphic terranes map of California: U.S. Geological Survey Open-File Report 82-593, scale 1:750,000.

Blome, C. D., and Irwin, W. P., 1983, Tectonic significance of late Paleozoic to Jurassic radiolarians from the North Fork terrane, Klamath Mountains, California, *in* Stevens, C. H., ed., Pre-Jurassic rocks in western North American suspect terranes: Pacific Section, Society of Economic Paleontologists and Mineralogists, Los Angeles, p. 77–87.

Borns, D. J., 1980, Blueschist metamorphism of the Yreka–Fort Jones area, Klamath Mountains, northern California [Ph.D. thesis]: Seattle, University of Washington, 155 p.

——, 1984, Eclogites in the Stuart Fork terrane, Klamath Mountains, California: Geological Society of America Abstracts with Programs, v. 16, p. 271.

Burchfiel, B. C., and Davis, G. A., 1981, Triassic and Jurassic tectonic evolution of the Klamath Mountains–Sierra Nevada geologic terrane, *in* Ernst, W. G., ed., The geotectonic development of California; Rubey Volume 1: Englewood Cliffs, New Jersey, Prentice-Hall, p. 50–70.

Coleman, R. C., Mortimer, N., Donato, M. M., Manning, C. E., and Hill, L. B., 1988, Tectonic and regional metamorphic framework of the Klamath Mountains and adjacent Coast Ranges, California and Oregon, *in* Ernst, W. G., ed., Metamorphism and crustal evolution of the western United States; Rubey Volume 7: Englewood Cliffs, New Jersey, Prentice-Hall, p. 1061–1097.

Davis, G. A., 1969, Tectonic correlations, Klamath Mountains and western Sierra Nevada, California: Geological Society of America Bulletin, v. 80, p. 1095–1108.

Donato, M. M., 1985, Metamorphic and structural evolution of an ophiolitic tectonic melange, Marble Mountain, northern California [Ph.D. thesis]: Stanford, California, Stanford University, 258 p.

——, 1987, Evolution of an ophiolitic tectonic melange, Marble Mountains, northern California Klamath Mountains: Geological Society of America Bulletin, v. 98, p. 448–464.

——, 1989, Metamorphism of an ophiolitic tectonic melange, northern California Klamath Mountains, U.S.A.: Journal of Metamorphic Geology, v. 7, p. 515–525.

Donato, M. M., Coleman, R. G., and Kays, M. A., 1980, Geology of the Condrey Mountain schist, California and Oregon: Oregon Geology, v. 42, p. 125–129.

Donato, M. M., Barnes, C. G., Coleman, R. G., Ernst, W. G., and Kays, M. A., 1982, Geologic map of the Marble Mountain Wilderness, Siskiyou County, California: U.S. Geological Survey Map MF-1452A, scale 1:48,000.

Elder, D. R., and Cashman, S. M., 1990, Redefinition of early Mesozoic terranes, central Klamath Mountains, California: Geological Society of America Abstracts with Programs, v. 22, no. 3, p. 21.

Ernst, W. G., 1983, Phanerozoic continental accretion and the metamorphic evolution of northern and central California: Tectonophysics, v. 100, p. 287–320.

——, 1984, Californian blueschists, subduction, and the significance of tectonostratigraphic terranes: Geology, v. 12, p. 436–440.

——, 1987, Mafic meta-igneous arc rocks of apparent komatiitic affinities, Sawyers Bar area, central Klamath Mountains, northern California, *in* Mysen, B. O., ed., Magmatic processes; Physico-chemical principles: Geochemistry Society Special Publication 1, p. 191–208.

Ernst, W. G., Barton, M. D., Hacker, B. R., and Sen, G., 1990, Igneous petrogenesis of magnesian metavolcanic rocks from the central Klamath Mountains, northern California: Geological Society of America Bulletin (in press).

Goodge, J. W., 1985, Widespread blueschist assemblages in the Stuart Fork terrane, central Klamath Mountains, northern California: Geological Society of America Abstracts with Programs, v. 17, p. 357.

——, 1989a, Polyphase metamorphic evolution of a Late Triassic subduction complex, Klamath Mountains, northern California: American Journal of Science, v. 289, p. 874–943.

——, 1989b, Evolving early Mesozoic convergent margin deformation, central Klamath Mountains, northern California: Tectonics, v. 8, p. 845–864.

Gray, G. G., 1986, Native terranes of the central Klamath Mountains, California: Tectonics, v. 5, p. 1043–1054.

Hacker, B. R., 1989, Conditions and sequence of metamorphism of high-MgO volcanic rocks in the Hayfork/Northfork terrane, Sawyers Bar area, central Klamath Mountains: Geological Society of America Abstracts with Programs, v. 21, p. 87.

Hacker, B. R., Ernst, W. G., and Barton, M. D., 1990, Conditions of regional and contact metamorphism of the high-MgO Yellow Dog greenstones, central Klamath Mountains, northern California: Journal of Metamorphic Geology (in press).

Harper, G. D., and Wright, J. E., 1984, Middle to Late Jurassic tectonic evolution of the Klamath Mountains, California-Oregon: Tectonics, v. 3, p. 759–772.

Helper, M., 1986, Deformation and high P/T metamorphism in the central part of the Condrey Mountain window, north-central Klamath Mountains, California and Oregon, *in* Evans, B. W., and Brown, E. H., eds., Blueschists and eclogites: Geological Society of America Memoir 164, p. 125–141.

Hill, L. B., 1985, Metamorphic, deformational, and temporal constraints on terrane assembly, northern Klamath Mountains, California, *in* Howell, D. G., ed., Tectonostratigraphic terranes of the Circum-Pacific region: Houston, Texas, Circum-Pacific Council Energy and Mineral Resources, Earth Science Series 1, p. 173–186.

Hotz, P. E., 1973, Blueschist metamorphism in the Yreka–Fort Jones area, Klamath Mountains, California: U.S. Geological Survey Journal of Research, v. 1, p. 53–61.

Hotz, P. E., Lanphere, M. A., and Swanson, D. A., 1977, Triassic blueschists from northern California and north-central Oregon: Geology, v. 5, p. 659–663.

Irwin, W. P., 1960, Geologic reconnaissance of the northern Coast Ranges and the southern Klamath Mountains, California, with a summary of the mineral resources: California Division of Mines and Geology Bulletin 179, 80 p.

——, 1966, Geology of the Klamath Mountains province, *in* Bailey, E. H., ed., Geology of northern California: California Division of Mines and Geology Bulletin 190, p. 19–38.

——, 1972, Terranes of the western Paleozoic and Triassic belt in the southern Klamath Mountains, California: U.S. Geological Survey Professional Paper 800-C, p. 103–111.

——, 1981, Tectonic accretion of the Klamath Mountains, *in* Ernst, W. G., ed., The geotectonic development of California; Rubey Volume 1: Englewood Cliffs, New Jersey, Prentice-Hall, p. 29–49.

——, 1985, Age and tectonics of plutonic belts in accreted terranes of the Klamath Mountains, California and Oregon, *in* Howell, D. G., ed., Tectonostratigraphic terranes of the Circum-Pacific Region: Houston, Texas, Circum-Pacific Council for Energy and Mineral Resources, Earth Science Series 1, p. 187–199.

——, 1989, Terranes of the Klamath Mountains, California and Oregon, *in* Blake, M. C., Jr., and Harwood, D. S., eds., Tectonic evolution of northern California; Field Trip Guidebook T108: Washington, D.C., American Geophysical Union, p. 19–32.

Irwin, W. P., Jones, D. L., and Kaplan, T. A., 1978, Radiolarians from pre-Nevadan rocks of the Klamath Mountains, California and Oregon, *in* Howell, D. G., and McDougall, K. A., eds., Mesozoic paleogeography of the western United States: Pacific Section, Society of Economic Paleontologists and Mineralogists Pacific Coast Paleogeography Symposium 2, p. 303–310.

Jennings, C. W., 1977, Geologic map of California: California Division of Mines and Geology, scale 1:750,000.

Lanphere, M. A., Irwin, W. P., and Hotz, P. E., 1968, Isotopic age of the Nevadan orogeny and older plutonic and metamorphic events in the Klamath Mountains, California: Geological Society of America Bulletin, v. 79, p. 1027–1052.

Liou, J. G., Kuniyoshi, S., and Ito, K., 1974, Experimental studies of the phase relations between greenschist and amphibolite in a basaltic system: American Journal of Science, v. 274, p. 613–632.

Mankinen, E. A., Irwin, W. P., and Grommé, C. S., 1989, Paleomagnetic study of the eastern Klamath terrane, California, and implications for the tectonic history of the Klamath Mountains province: Journal of Geophysical Research, v. 94, p. 10444–10472.

Mortimer, N., 1985, Structural and metamorphic aspects of Middle Jurassic terrane juxtaposition, northeastern Klamath Mountains, California, *in* Howell, D. G., ed., Tectonostratigraphic terranes of the Circum-Pacific region: Houston, Texas, Circum-Pacific Council for Energy and Mineral Resources, Earth Science Series 1, p. 201–214.

Silberling, N. J., Jones, D. L., Blake, M. C., Jr., and Howell, D. G., 1987, Lithotectonic terrane map of the western conterminous United States: U.S. Geological Survey Miscellaneous Field Studies Map MF-1847-C, scale 1:2,500,000.

Wagner, D. L., and Saucedo, G. J., 1987, Geologic map of the Weed Quadrangle: California Division of Mines and Geology, scale 1:250,000.

Wright, J. E., 1982, Permo-Triassic accretionary subduction complex, southwestern Klamath Mountains, northern California: Journal of Geophysical Research, v. 87, p. 3805–3818.

Wright, J. E., and Fahan, M. R., 1988, An expanded view of Jurassic orogenesis in the western United States Cordillera; Middle Jurassic (pre-Nevadan) regional metamorphism and thrust faulting within an active arc environment, Klamath Mountains, California: Geological Society of America Bulletin, v. 100, p. 859–876.

Wyld, S. J., and Wright, J. E., 1988, The Devils Elbow ophiolite remnant and overlying Galice Formation; New constraints on the Middle to Late Jurassic evolution of the Klamath Mountains, California: Geological Society of America Bulletin, v. 100, p. 29–44.

Manuscript Accepted by the Society March 27, 1990

Printed in U.S.A.

Geological Society of America
Special Paper 255
1990

Stratigraphy and tectonics of Paleozoic arc-related rocks of the northernmost Sierra Nevada, California; The eastern Klamath and northern Sierra terranes

A. S. Jayko
U.S. Geological Survey, MS 975, 345 Middlefield Road, Menlo Park, California 94025

ABSTRACT

The Eastern Klamath and Northern Sierra terranes of northern California consist of Devonian to Jurassic arc-related rocks that structurally and/or stratigraphically overlie Devonian(?) or older complexes that consist of quartzite, quartzofeldspathic sandstone, chert, and mafic and ultramafic rocks. These terranes lie within a regional belt of Paleozoic arc-related rocks that can be recognized from the Sierra Nevada to British Columbia. The Eastern Klamath and Northern Sierra terranes are geographically separate and have paleontologic linkage, but lack direct stratigraphic ties in strata of Late Devonian to Early Permian age.

This chapter describes new stratigraphic correlations and structural interpretations of rocks that lie in the northernmost part of the Sierra Nevada. These rocks include the Butt Valley block of the Northern Sierra terrane, and rocks herein interpreted as dismembered Eastern Klamath terrane. The rocks of Eastern Klamath terrane affinity, Soda Ravine block, occur as a tectonic sliver that lies to the west of the Butt Valley block.

The Soda Ravine block is about 2 to 4 km by 20 km and includes limestone lenses equivalent to zone A of the McCloud Limestone of the Eastern Klamath terrane, and upper Middle and Upper Triassic limestone, slate, siltstone, and pebble conglomerate. Similar Permian limestone with McCloud faunal affinities and slate-bearing slivers have been noted by previous workers in this area and to the south.

The Butt Valley block had previously been interpreted as part of the west limb of a regional anticline, the Almanor anticline. New mapping suggests that the block is not contiguous with rocks that lie to the east. The Butt Valley block includes the Devonian and Mississippian Taylor Formation, which is unconformably overlain by an Upper Triassic (Carnian and Norian) basal conglomerate and sandstone, which in turn is overlain by Upper Triassic limestone. The Upper Triassic limestone is overlain by a sparsely fossiliferous, lithic-volcaniclastic sequence containing poorly preserved Triassic or Early Jurassic, and probable Early Jurassic, ammonites and clams. This sequence is similar to conglomerate, Triassic limestone, and volcaniclastic rocks of the Jurassic Sailor Canyon Formation, which overlie a regional Late Triassic unconformity west of Lake Tahoe, thus expanding the known distribution of the unconformity in the Sierra Nevada. The Triassic unconformity overlies Carboniferous and older rocks around the North Fork American River area to the south, and overlies Devonian and older rocks on the Butt Valley block, but overlies Permian rocks east and southeast of the Butt Valley block. These relations suggest that either the Butt Valley block was part of a highland or

Jayko, A. S., 1990, Stratigraphy and tectonics of Paleozoic arc-related rocks of the northernmost Sierra Nevada, California; The eastern Klamath and northern Sierra terranes,, *in* Harwood, D. S., and Miller, M. M., eds., Paleozoic and early Mesozoic paleogeographic relations; Sierra Nevada, Klamath Mountains, and related terranes: Boulder, Colorado, Geological Society of America Special Paper 255.

uplifted block along the western part of the northern Sierra terrane or that the Butt Valley block previously lay closer to the North Fork American River area and has been translated northward by right-slip displacement.

Northern Sierra terrane east of the Butt Valley block consists of the Hough and Genesse blocks, which are separated by the Grizzly Mountain fault zone. The Grizzly Mountain fault zone is here interpreted to be a transpressional right-slip fault.

INTRODUCTION

The Northern Sierra and Eastern Klamath terranes are both recognized as representing allochthonous or parallochthonous volcanic arc sequences that formed west of the Paleozoic passive margin of North America and were accreted to the continental margin probably during the Permo-Triassic Sonoma orogeny (Hamilton, 1969; Burchfiel and Davis, 1972, 1975, 1981; Schweickert, 1981; Miller, 1987). These terranes are also recognized as part of a more regionally extensive belt of Paleozoic arc-related rocks that extends from California to British Columbia (see, for example, Davis and others, 1977; Miller, 1987). Upper Triassic and Jurassic strata that are found in both the Northern Sierra and Eastern Klamath terranes, in particular the Hosselkus Limestone and overlying clastic rocks (Irwin, 1981; Harwood, 1988), probably constitute overlap units deposited after the terranes were accreted to the North American continental margin.

The Northern Sierra and Eastern Klamath terranes of northern California (terminology of Irwin, 1966, 1977; Blake and others, 1982; Silberling and Jones, 1984) (Fig. 1) consist of generally continuous stratigraphic sequences that range in age from early Paleozoic to at least Middle Jurassic (Sanborn, 1960; Irwin, 1966, 1977; McMath, 1966). The terranes are similar in that they are associated with Ordovician or older mafic and ultramafic rocks (Lindsley-Griffin, 1977; Schweickert, 1981), and with Ordovician and/or Silurian quartzites and quartzofeldspathic rocks (Potter and others, 1977; Bond and DeVay, 1980, D'Allura and others, 1977), and contain Devonian through Jurassic arc-related volcanic and volcaniclastic rocks, and inter-

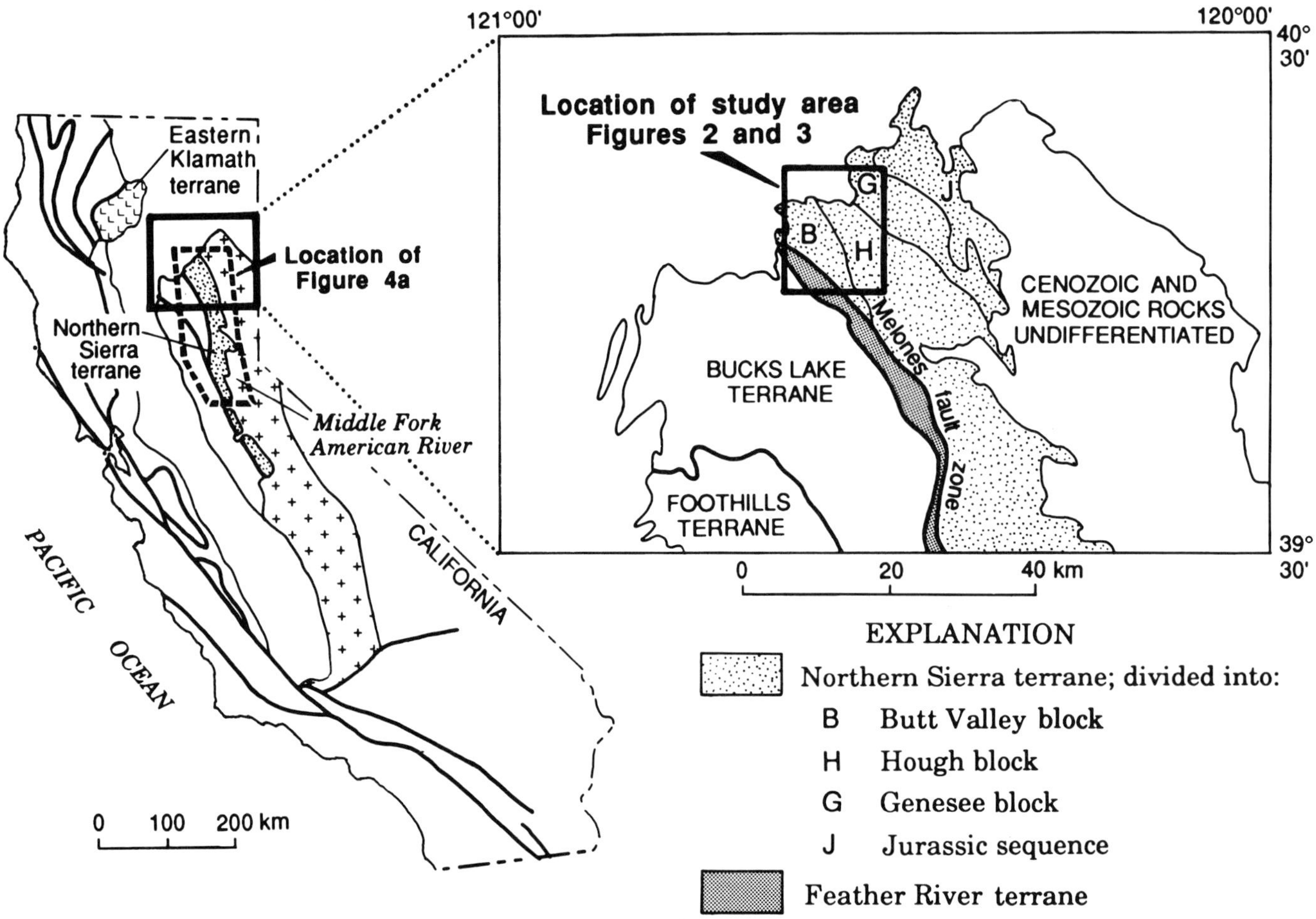

Figure 1. Map of northern California showing location of major tectonic units discussed in text, including Eastern Klamath and Northern Sierra terranes. Map also shows location of the Lake Almanor study area in the northern Sierra Nevada.

bedded basinal deposits, including chert and flysch deposits (Irwin, 1966; D'Allura and others, 1977; Durrell and D'Allura, 1977; Day and others, 1985, Harwood, 1983, 1988). However, stratigraphic units of a given time period from the two terranes commonly reflect different depositional environments, particularly during the Late Devonian to Early Permian (see for example Miller and Harwood, this volume).

The goals of this study were to examine the northwesternmost exposures of rocks associated with, and adjacent to, the Northern Sierra terrane, in an attempt to establish more specific stratigraphic ties with either the Northern Sierra terrane or possibly with the Eastern Klamath terrane. An additional goal was to determine whether there are structural or stratigraphic relations that could reveal the nature of the contact between the Northern Sierra and Eastern Klamath terranes, which mainly lies buried beneath Tertiary cover, and to examine major structures within the northwestern part of the Northern Sierra terrane to better understand the local structural history of the terrane.

The study was primarily field-oriented and focused on an area in the northernmost Sierra Nevada that lies east of the Feather River terrane, south of the Cenozoic volcanic rocks of the Cascade Range and Modoc Plateau, and west of the continuous structural blocks defined within the Northern Sierra terrane, the Hough and Genesee blocks (McMath, 1966; D'Allura and others, 1977; Day and others, 1985). This area includes the northwesternmost exposures of the Northern Sierra terrane and rocks of Northern Sierran terrane affiliation that lie within the Sierra Nevada. The next exposures of rock with Paleozoic arc affinity similar to the Northern Sierra terrane are part of the Eastern Klamath terrane, which lies in the Klamath Mountains about 90 km to the northwest (Fig. 1). The study consisted of (1) geologic mapping at a scale of 1:24,000 to establish the stratigraphic succession and structural relation of units (Jayko, 1988); (2) the sampling of chert and limestone for conodonts and radiolarians to establish stratigraphic control of the section; and (3) petrologic studies of clastic and volcanic rocks to verify stratigraphic association, to determine metamorphic grade and fabric development and their relation to faults and folds, and in the case of the clastic rocks, to study detritus for clues to the provenance and the nature of uplifted units.

Mapping indicates that (1) a tectonic sliver of Eastern Klamath terrane is present in the Sierra Nevada, west of the Northern Sierra terrane, suggesting that the Eastern Klamath terrane once lay to the west, or outboard of the Northern Sierra terrane, or that there has been left-lateral displacement of the sliver from the north along a zone that borders the Melones fault zone; (2) part of a regional Triassic unconformity extends into the western part of the Northern Sierra terrane, which helps delineate a new tectonic block in the Northern Sierra terrane, and which may help constrain the reconstruction of tectonic blocks of Northern Sierra terrane; and (3) the displacement along the Grizzly Mountain fault zone cannot be explained by a simple east-verging thrust mechanism, but requires a more complicated scenario.

Previous field studies in the Lake Almanor area. The first geologic studies of the Paleozoic and Mesozoic rocks from the Lake Almanor area were published around the turn of the century by Diller (1892, 1895, 1908) and Turner (1898). Subsequent work by McMath (1958, 1966), Moores and Wise (1970), Robinson (1975), D'Allura and others (1977), and Hannah (1980), established much of the geologic framework for the area, although some significant structural and stratigraphic revisions are described in the following text.

NORTHERN SIERRA TERRANE

The Northern Sierra terrane consists of continentally derived rocks of the Ordovician to Devonian Shoo Fly Complex, which are overlain by: (1) Upper Devonian to Pennsylvanian volcanic rocks, volcaniclastic rocks, and radiolarian chert of the Taylorsville sequence (McMath, 1966); (2) Permian volcanic and volcaniclastic rocks, referred to as the Permian volcanic sequence (Harwood, 1988); (3) Triassic slate, conglomerate, and limestone; and (4) Triassic(?) and Jurassic volcanic rocks, clastic rocks, and phyllite (McMath, 1966; Durrell and D'Allura, 1977; D'Allura and others, 1977; Varga and Moores, 1981; Day and others, 1985; Harwood, 1983, 1988; Jayko, 1988) (Table 1).

The Northern Sierra terrane consists of three lithotectonic blocks separated by west-dipping faults or fault zones that repeat the sections (Moores and Wise, 1970): the Genesee, Hough, and Butt Valley blocks (Figs. 2, 3). The two easternmost blocks, the

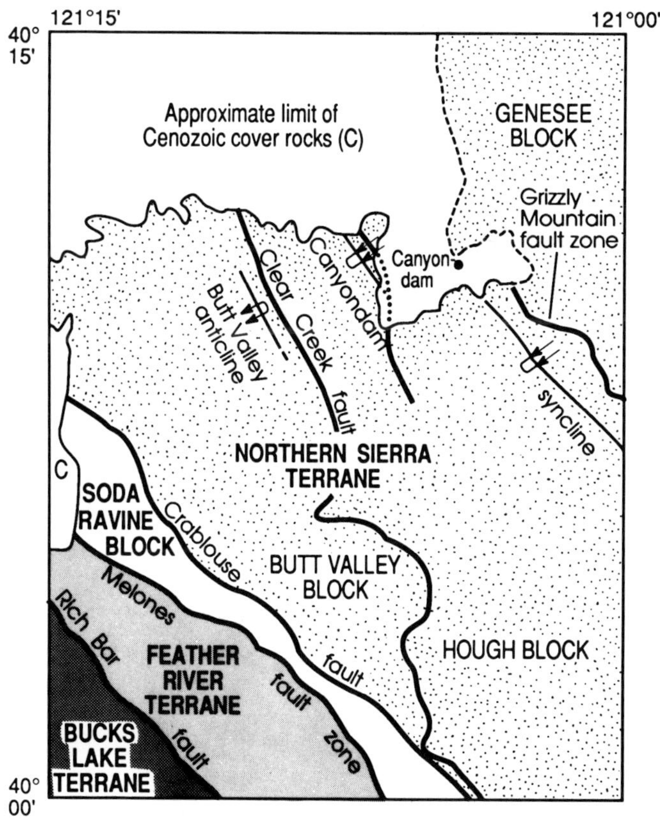

Figure 2. Simplified map showing major tectonic units of the Lake Almanor area.

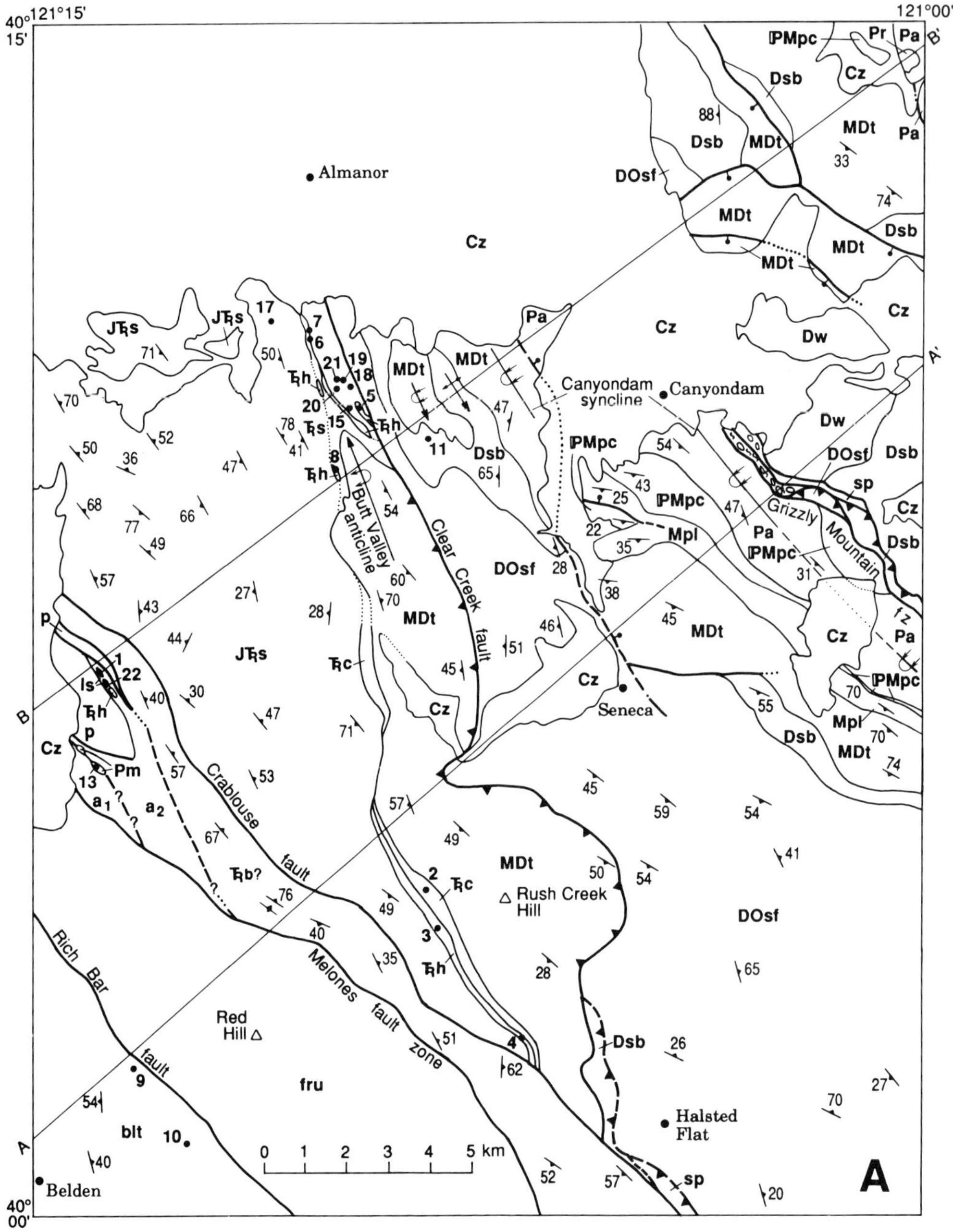

Figure 3. A (above), Geologic map of the Lake Almanor Quadrangle, modified from Jayko (1988). Explanation on facing page. Figure 3B (facing page), Structure sections of the Lake Almanor area, modified from Jayko (1988). Pattern in JTRs unit of sections A-A′ and B-B′, and in TRb unit of section A-A′ used to schematically show kink folds.

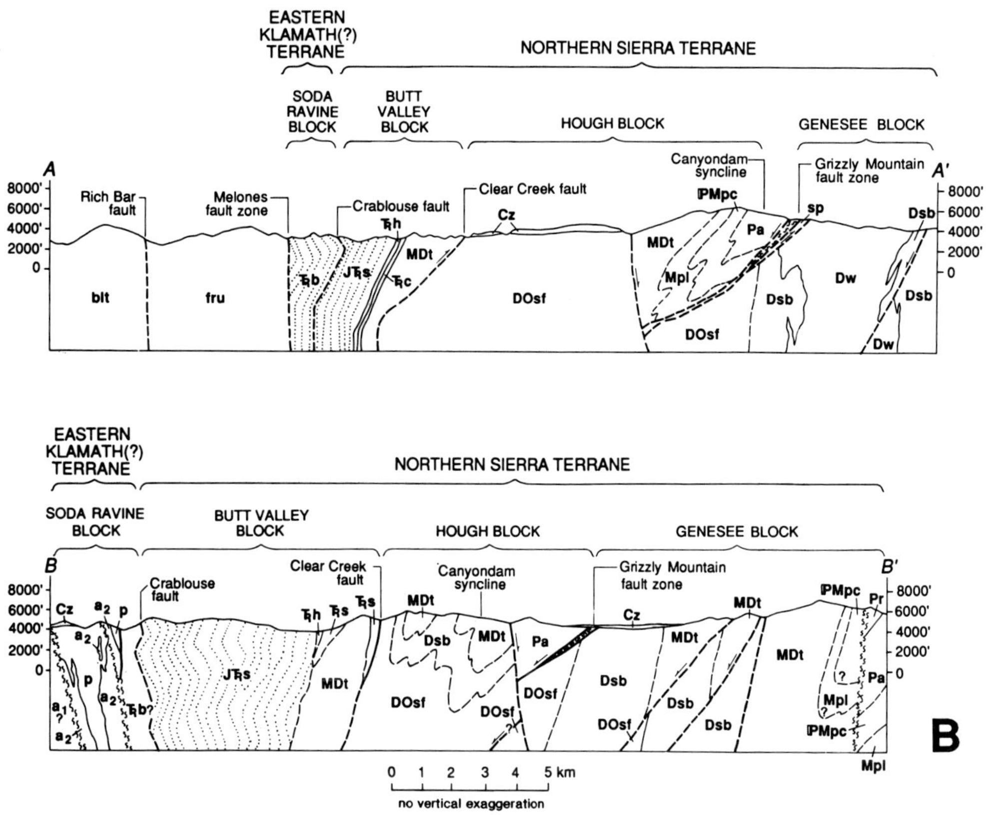

EXPLANATION

Cz Cenozoic deposits

fru Ultramafic and mafic rock (Devonian to Triassic?)

blt Bucks Lake terrane (Paleozoic and Mesozoic)

SODA RAVINE BLOCK (EASTERN KLAMATH(?) TERRANE)

TRb Brock(?) Shale (Triassic)

TRh Hosselkus Limestone (Upper Triassic)

a_2 Upper argillite and volcanic rocks, in part equivalent to the Pit(?) Formation (Triassic), includes limestone (ls)

Pm Mc Cloud(?) Limestone (Permian)

a_1 Lower argillite; volcanic and intrusive rocks; and quartzite sandstone and grits in part equivalent to the Baird(?) Formation (contact between a_1 and a_2 shown schematically)

p Porphyritic intrusive and extrusive rocks (Mesozoic)

NORTHERN SIERRA TERRANE

JTRs Metasedimentary rocks (Jurassic and Upper Triassic?)

TRh Hosselkus Limestone (Upper Triassic)

TRc Conglomerate (Upper Triassic)

TRs Sandstone and siltstone (Upper Triassic(?))

Permian volcanic sequence:

Pr Reeve Formation (Permian)

Pa Arlington Formation (Permian)

Taylorsville sequence:

Peale Formation (Pennsylvanian and Mississippian)--divided into:

PMpc Chert member (Pennsylvanian and Mississippian)

Mpl Lower member (Mississippian)

MDt Taylor Formation (Mississippian and(or) Devonian)

Dsb Sierra Buttes Formation (Upper Devonian)

Dw Wolf Creek Stock (Devonian)

DOsf Shoo Fly Complex (Devonian? to Ordovician)

sp Serpentinized ultramafic rock

SYMBOLS

Contact— Dotted where covered

Fault— Dashed where approximately located; queried where inferred; showing dip where measured

Thrust fault— Dashed where approximately located; sawteeth on upper plate

Normal fault— Dashed where approximately located; dotted where concealed; ball and bar on downthrown block of high-angle normal fault

Overturned syncline—Axial trace approximately located

Overturned anticline—Axial trace approximately located

Minor syncline— Axial trace approximately located

Minor anticline— Axial trace approximately located

Strike and dip of foliation:
Inclined
Vertical

Zone of sheared rocks, ellipses indicate the location of tectonic blocks

New fossil localities— modified from Jayco (1988):

Map#	Age	CAI*
1	late Middle to early Late Triassic	250
2	Late Triassic	350
3	Late Triassic	300-350
4	Late Triassic	300-350
5	Late Triassic	350
6	Late Triassic	350
7	Late Triassic	350
8	Ordovician to Triassic	300?
9	Late Pennsylvanian to EarlyPermian	
10	Mississippian to Early Pennsylvanian	
11	Middle? to Late Devonian	
13	Early Permian	
17	probably early Jurassic	
18-22	late Paleozoic to early Mesozoic	

* CAI— Conodont color alteration index, temperature in degrees C.

TABLE 1. GENERALIZED STRATIGRAPHY, NORTHERN SIERRA TERRANE, LAKE ALMANOR AREA

	Age	Formation	Unit Description
Triassic Unconformity	Jurassic and Upper Triassic(?)	Sedimentary rocks	Dominantly volcaniclastic sandstone, argillite, and conglomerate with minor tuff, and pyroxene andesite dikes or sills.
	Upper Triassic	Hosselkus Limestone	Schistose crinoid-bearing limestone.
	Upper Triassic	Sedimentary rocks	Basal conglomerate, sandstone, and argillite.
			unconformity
Permian Volcanic Sequence	Permian	Reeve Formation	Characterized by andesitic, plagioclase phyric volcanic rocks.
	Permian	Arlington Formation	Dominated by fine-grained lithic sandstone and siltstone with minor grit, pebble- and cobble-conglomerate.
			paraconformity
Taylorsville Sequence	Lower Mississippian to Middle(?) Pennsylvanian	Peale Formation	Chert member: consists predominantly of siliceous tuff, chert, and siliceous argillite. Lower (volcanic) member: characterized by dacitic commonly potassium-feldspar–bearing extrusive rocks (McMath, 1958, 1966). Consists of platy and massive, pastel-weathering, siliceous tuffs, welded tuffs, and localized lava flows.
	Upper Devonian and (or) Lower Mississippian	Taylor Formation	Characterized by pyroxene-bearing andesitic extrusive and volcaniclastic rocks. Includes brick-red–weathering, pyroxene-bearing lithic sandstone, with lesser andesitic flows, minor pillow lava, pillow breccias, tuff, and very minor intercalated calcareous tuff and limestone lenses.
	Upper Devonian	Sierra Buttes Formation	Characterized by siliceous quartz-phyric extrusive and volcaniclastic rocks (McMath, 1958, 1966). Includes slate, tuff, tuffaceous, or volcaniogenic lithic sandstone and siltstone, quartzose sandstone, conglomerate containing quartz-porphyry clasts, radiolarian chert, small lenses of black phosphatic chert, and conglomerate with abundant quartz-phyric volcanic clasts. Chert at one locality has yielded Middle(?) or Late Devonian radiolarians (Bonnie Murchey, 1987, oral communication).
	Devonian	Wolf Creek Stock	Granitic intrusive rock inferred to be syngenetic with Upper Devonian Sierra Buttes Formation. Occurs along the western part of the Genesee block east of the Grizzly Mountain fault zone (Hannah, 1980; Saleeby and others, 1987).
			unconformity
	Ordovician	Shoo Fly Complex	Consists of several thrust plates containing a variety of rocks including ultramafic rocks, mafic rocks, chert, rare limestone, and clastic rocks dominated by quartzite and quartzo-feldspathic sandstones (Bond and DeVay, 1980; Girty and Schweickert, 1984; Girty and Wardlaw, 1985).

Genesee and Hough blocks, were first recognized by McMath (1966) and further described by D'Allura and others (1977) and Hannah and Moores (1986). The Butt Valley block was subdivided from the Hough block by Jayko (1988) and is described in greater detail below.

Rocks of the Northern Sierra terrane in the study area have been regionally metamorphosed to lower greenschist facies. Schistosity is well developed in metasedimentary rocks, but is sporadic in massive lava flows. Metamorphic fabric is generally weaker in rocks in the Genesee block than in those of the Hough and Butt Valley blocks. Large-scale structures, including the Butt Valley anticline, Clear Creek fault, and Canyondam syncline, are generally east-verging, as reported by Day and others (1985). Alteration indices from conodonts within the Butt Valley block limestones suggest metamorphic temperatures ranging from 300° to 350°C (Fig. 3). (A. G. Harris, written communication, 1987). The metamorphism postdates deposition of Upper Triassic(?) and Lower Jurassic strata that occur in the western part of the terrane.

BUTT VALLEY AND SODA RAVINE BLOCKS

The Eastern Klamath and Northern Sierran terrane affinities of rocks in the study area have previously been recognized (Diller, 1908; Robinson, 1975; D'Allura and others, 1977); however, there have been only weakly established stratigraphic ties between rocks that lie in the wedge bounded on the west by the Melones fault zone and on the east by the Clear Creek fault (Figs. 2, 3) with those of the northern Sierra terrane. Rocks that lie within this wedge have been subdivided into two blocks, the Soda Ravine block on the west, which has Eastern Klamath terrane affinities, and the Butt Valley block on the east, which has Northern Sierra terrane affinities (Jayko, 1988).

Rocks of the study area that lie west of the Clear Creek fault (Fig. 3) were previously interpreted as lying within the west limb of the inferred Almanor anticline (D'Allura and others, 1977; Day and others, 1985); however, differences in stratigraphy between the east and west limbs of the inferred fold (Fig. 4), and structural constraints described below, suggest that rocks west of the Clear Creek fault occur within a fault-bounded structural block rather than on the west limb of a regional fold. The anticline had been inferred, in part, because of an errant correlation of clastic rocks of the Butt Valley block with the Arlington Formation of the Hough block. The clastic rocks of the Butt Valley block are now dated as Upper Triassic(?) and Lower Jurassic (Jayko, 1988), whereas the Arlington is Lower Permian (Harwood, 1988)—thus, these units are not correlative. The hinge west of Canyondam (Fig. 3), previously interpreted as part of the Almanor anticline, is interpreted as the northward continuation of the Canyondam syncline (Jayko, 1988). A major anticlinal fold seems to be confined to the eastern part of the Butt Valley block, where the nose and overturned limb of a west-dipping, overturned anticline (Robinson, 1975; D'Allura and others, 1977), the Butt Valley anticline, are truncated in the upper plate of the Clear Creek fault that cuts up-section toward the north (Fig. 3A, B).

Butt Valley block, Northern Sierra terrane

The Butt Valley block consists of minor serpentinite and the Sierra Buttes Formation at the base, overlain by pyroxene-bearing metaandesite assigned to the Taylor Formation (D'Allura and others, 1977). The serpentinite and Sierra Buttes Formation are present in the southern part of the block and faulted out toward the north. The Peale Formation and units of the Permian volcanic sequence are inferred to have been eroded prior to the Late Triassic.

Triassic unconformity. The Butt Valley block includes Late Triassic and Jurassic strata that overlie the Mississippian and Devonian Taylor Formation along a major regional Triassic unconformity. The oldest fossils from conglomerate that overlies the unconformity are Late Carnian and Norian (Jayko, 1988). The Upper Triassic sedimentary rocks are overlain by Upper Triassic limestone herein referred to as the Hosselkus Limestone. The Hosselkus Limestone is overlain by poorly dated Upper Triassic(?) and Lower Jurassic argillite, volcaniclastic, and volcanic rocks.

The regional unconformity can be recognized at several localities within the Northern Sierra terrane, including the North Fork of the American River, about 50 km south of the study area (Fig. 5), where Carboniferous and older rocks of the terrane are overlain by conglomerate and Upper Triassic limestone similar to rocks of the Butt Valley block. The conglomerate and limestone are in turn overlain by the volcaniclastic rocks of the Lower and Middle Jurassic Sailor Canyon Formation (Clark and others, 1962; Harwood, 1983, 1988). Near Genesee Valley, about 10 km southeast of the study area, Middle Triassic slate and Upper Triassic Hosselkus Limestone at the limestone's type locality (Diller, 1908) overlie the Permian rocks of the terrane (McMath, 1966). This unconformity may provide an important regional datum for structural reconstruction.

Isolated limestone lenses had previously been correlated with the Hosselkus Limestone of the Taylorsville area (Robinson, 1975), although the continuity and lateral extent of limestone were not recognized. The Upper Triassic(?) and Lower Jurassic volcaniclastic and volcanic rocks, together with the underlying Upper Triassic limestone and conglomerate from the Butt Valley area, were originally named the Cedar Formation by Diller (1895). Robinson (1975), D'Allura and others (1977), and Hannah and Moores (1986) subsequently assigned these rocks of the Butt Valley and Soda Ravine blocks to the Arlington Formation and concluded that they were Permian and Triassic in age. As noted above, the Arlington Formation and related rocks of the Permian volcanic sequence subsequently have been shown to be exclusively Permian in age (Harwood, 1983, 1988; oral communication, 1989); therefore, the volcaniclastic rocks of the Butt Valley block cannot be correlated with the Arlington Formation despite lithologic similarities. Thus the rocks that overlie andesitic rocks of the Taylor Formation in the Butt Valley block appear to be Late Triassic and younger in age.

Upper Triassic rocks. Conglomerate, volcanic breccia, tuff, slate, and sandstone unconformably overlie the Taylor Formation and underlie the Upper Triassic limestone in the Butt Valley block. The conglomerate and volcanic breccia overlie the unconformity in the southern part of the Butt Valley block, whereas tuff, slate, and sandstone overlie it on the northern part, near the axis of the Butt Valley anticline (Fig. 3). The conglomerate consists predominantly of subrounded and subangular clasts of pale greenish, pinkish, and gray chert, siliceous volcanic rock, pyroxene-bearing volcanic rock, and minor quartzite. These lithologies are consistent with derivation from the Paleozoic section of the Northern Sierra terrane. Rare lenses or clasts of Upper Triassic crinoidal limestone are also present locally. The finer grained sedimentary rocks that occur to the north of the conglomerate include a green- and purple-weathering platy tuff that locally underlies the Triassic limestone.

Upper Triassic limestone of the Butt Valley block consists of schistose, whitish and pale gray–weathering crinoidal limestone

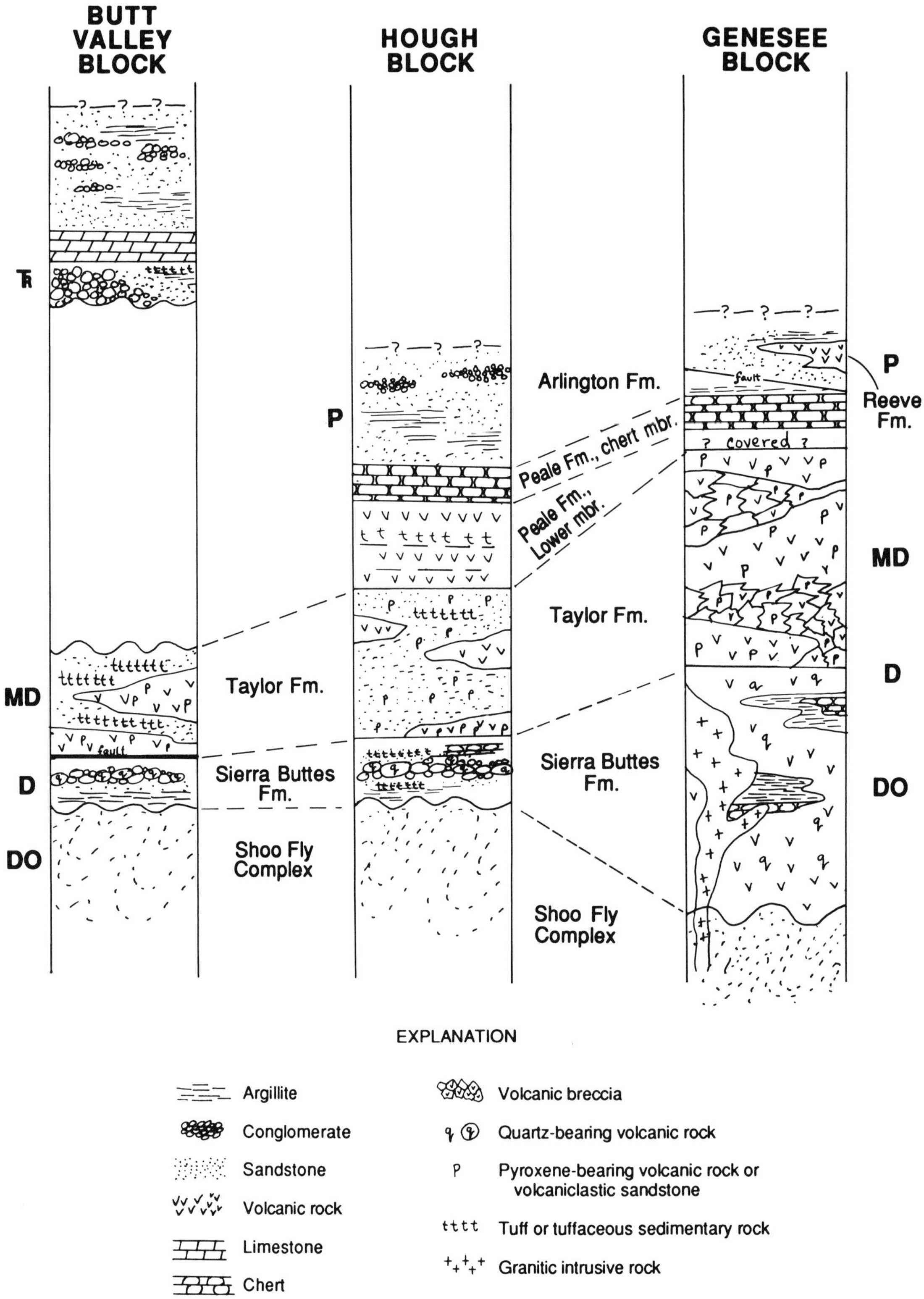

Figure 4. Schematic stratigraphic column showing Northern Sierra terrane units present in the study area.

(Jayko, 1988). The limestone occurs as a thin, discontinuous horizon as long as 8 km, and several meters to tens of meters thick (Jayko, 1988). The limestone has yielded Late Triassic conodonts from six localities (Jayko, 1988) (Fig. 3A). Based on the conodont data, this Upper Triassic limestone is here assigned to the Hosselkus Limestone. The Hosselkus Limestone is the only unit of the Northern Sierra terrane that has previously been recognized within the Eastern Klamath terrane of the Klamath Mountains (see, for example, Irwin, 1981).

Upper Triassic(?) and Lower Jurassic strata. The Hosselkus Limestone is overlain by a variety of rock types, including a lithic, crinoid-bearing sandstone near the base of the overlying section. Molds of pentagonal crinoid columnals are preserved, sometimes remarkably so, considering the schistose nature of the rocks; they are common in the fine-grained siltstones and sandstones, but only within a few hundred meters above the limestone. The overlying section, here referred to as Upper Triassic(?) and Lower Jurassic strata, includes sparsely fossiliferous sandstone, grit and conglomerate, lesser slate, and very minor tuff, dikes or sills(?) of pyroxene andesite, and extremely rare limestone lenses or blocks. Black slate that weathers buff color has yielded two ammonite localities (reported in Robinson, 1975; Jayko, 1988), as well as flat clams and ammonites tentatively identified as "most likely" Early Jurassic in age (N. J. Silberling, written communication, 1989).

Conglomerate within the Upper Triassic(?) and Lower Jurassic strata is similar to the basal Triassic conglomerate in that it consists of clasts of chert, siliceous volcanic rock, and quartzite typical of the underlying Paleozoic section. It also includes clasts of metamorphic rock, including epidote–plagioclase–biotite–white mica schist. The conglomerate occurs as discontinuous lenses with grit- to cobble-sized clasts. Pebble conglomerate is the most common.

Sandstone of the Upper Triassic(?) and Lower Jurassic strata is generally lithic or lithic-feldspathic in composition, greenish in outcrop, and grayish on fresh surfaces. Detrital clasts in the sandstone include abundant plagioclase, biotite, white mica, and clinopyroxene, minor green and brown amphibole, volcanic rock (some with hornblende phenocrysts), foliated quartz-albite-epidote schist, zoned and fritted plagioclase feldspar, quartzite, rare tourmaline, and chert.

All of the fossil localities from the Upper Triassic(?) and Lower Jurassic strata occur in the northeasternmost exposures of the unit (unit JTRs, Fig. 3A) near the axis of the Butt Valley anticline. These localities include *Posidonia*-like clams (N. J. Silberling, written communication, 1989), ammonite impressions that "could be at home in the upper Hettangian to Pleinsbachian" (Early Jurassic) (N. J. Silberling, written communication, 1989), as well as the ammonites previously reported as Late Triassic or Early Jurassic in age by Robinson (1975) and D'Allura and others (1977).

Constraints on foliation dip, Butt Valley block. All the metasedimentary rocks of the Butt Valley block have a pronounced metamorphic foliation that is most commonly very steeply dipping. The metamorphic fabric has previously been interpreted as resulting from a fan-shaped series of upright and overturned folds (D'Allura and others, 1977; Day and others, 1985) because the foliation is generally north-striking and mainly west-dipping in the eastern part of the block, and east-dipping in the western part of the block.

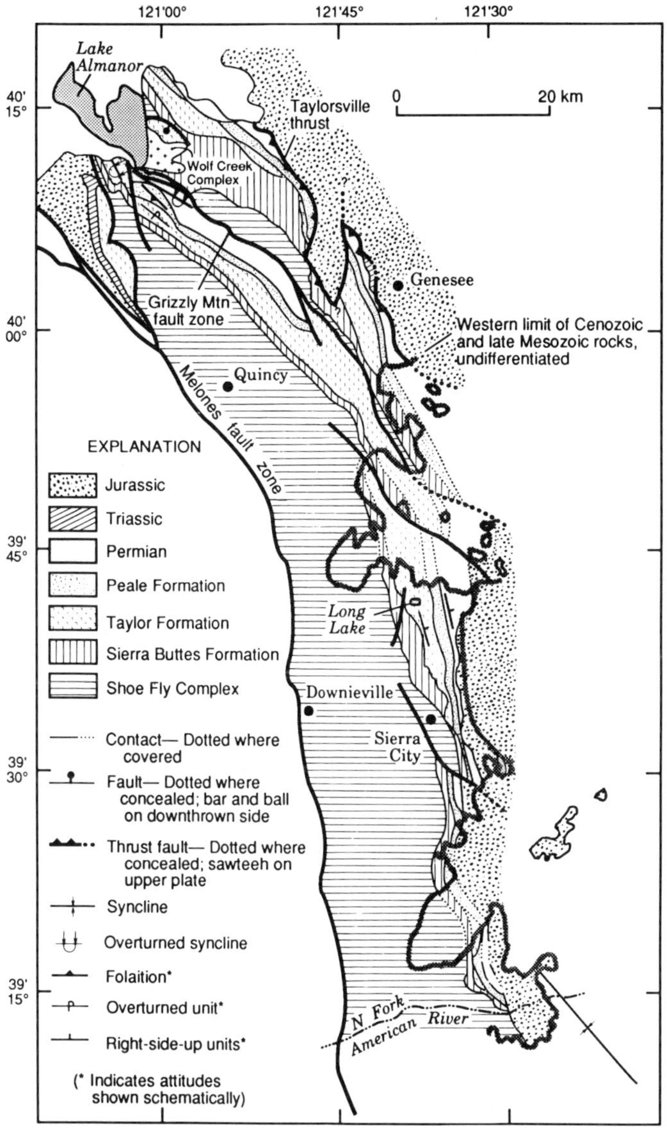

Figure 5. Simplified geologic map showing most of the northern Sierra terrane (modified from Harwood, 1988; D'Allura and others, 1977; Jayko, 1988).

An alternative interpretation to a series of upright fan-shaped folds is that the foliation dip direction is strongly controlled by large-scale, gently plunging kink folds with subhorizontal axial planes (see, for example, Fig. 3B). The foliation thus alternately dips east or west, above or below the axial planes of the kink folds. The foliation dip direction commonly changes with changes in elevation because the axial planes of the kink folds are gently dipping to subhorizontal (Fig. 6).

This kink fold interpretation is supported by several features: (1) structural domains that are controlled by changes in the foliation dip direction coincide with changes in elevation (Fig. 7); (2) numerous outcrop-scale kink folds have gently dipping axial planes; and (3) the dip direction of the early foliation changes along strike within continuous stratigraphic horizons. The axial trend of the large-scale kink folds tends to be subparallel to the strike direction of the foliation.

Soda Ravine block, Eastern Klamath(?) terrane

The Soda Ravine block is a tectonic sliver that lies along the westernmost edge of the Northern Sierra terrane (Jayko, 1988) (Fig. 3) and consists of slate, minor chert-pebble conglomerate, lithic and quartzose sandstone, and intercalated blocks of Permian and Triassic bioclastic limestone. The block has stronger stratigraphic affinities to the Eastern Klamath terrane of the Klamath Mountains than to the Northern Sierra terrane, and thus is interpreted as a sliver of Eastern Klamath terrane. Rocks of the Soda Ravine block were previously included in the Cedar Formation by Diller (1895) and in the Arlington Formation of the Hough block by D'Allura and others (1977).

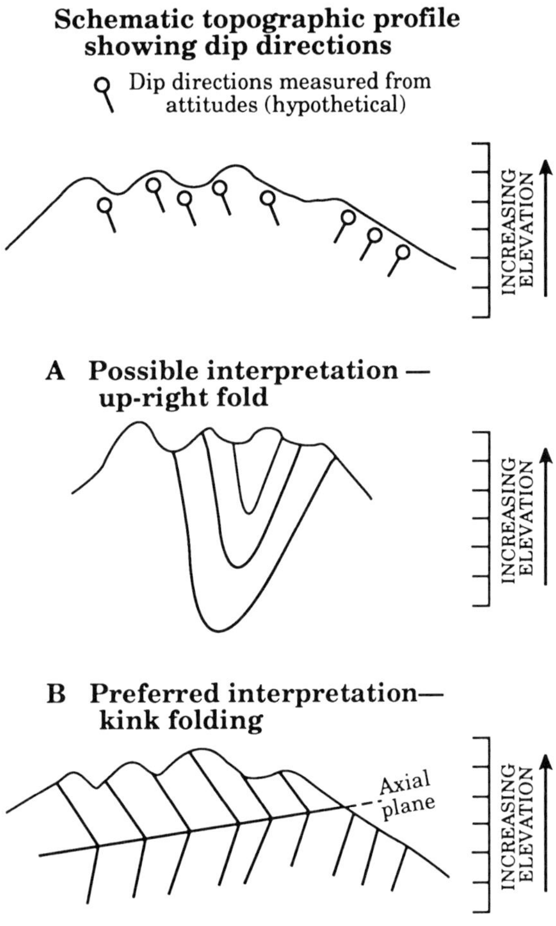

Figure 6. Schematic diagram showing steeply dipping foliation folded by kink fold with gently dipping axial plane vs. upright fold.

The Eastern Klamath terrane of the Klamath Mountains is perhaps best known for its unique Permian faunal assemblage, the McCloud fauna, which is distinct from North American and Tethyan assemblages (Skinner and Wilde, 1965, 1966; Wilson, 1984). Permian fusulinids of McCloud affinity were first reported from the Soda Ravine block by Robinson (1975) and D'Allura and others (1977). The gross stratigraphic similarity between rocks of the Soda Ravine and Butt Valley blocks with parts of the Eastern Klamath terrane was noted by Diller (1895), Robinson (1975), and D'Allura and others (1977). Tectonic slivers of Permian and Triassic(?) rock with Eastern Klamath terrane affinities have been reported in the same structural position along the Melones fault, south of the Almanor Quadrangle (McMath, 1958; D'Allura and others, 1977; Standlee, 1978; Harwood and others, 1988). The presence of Eastern Klamath terrane rocks along the western border of the Northern Sierra terrane in the Sierra Nevada implies either (1) that the Eastern Klamath terrane may have been a western facies of the Northern Sierra terrane that has since been displaced northward or that structurally overlies the northern Sierra terrane, or (2) that the slivers were offset from the north by left-lateral faulting along a zone marginal to the Melones fault zone.

The Soda Ravine block includes lithologies similar to, or in part identical to, parts of the McCloud Limestone, Pit Formation, Hosselkus Limestone, and Brock Shale (Jayko and others, 1987), and may also include clastic rocks similar to the Baird Formation (Fig. 8). These Eastern Klamath terrane–like lithologies in the Soda Ravine block are generally steeply dipping and young toward the northeast. The stratigraphic section, however, is disrupted and incomplete.

McCloud(?) Limestone. The McCloud limestone of the Eastern Klamath terrane consists of fossiliferous, shallow-water limestone (Watkins, 1985; Miller, 1989) with a distinctive faunal assemblage characterized by corals and fusulinids that differ from North American and Tethyan assemblages (Skinner and Wilde, 1966; Wilson, 1984; Miller, 1987).

A limestone body from the Soda Ravine block has yielded Early Permian fusulinids of probable McCloud Zone A affinity (Robinson, 1975; C. H. Stevens, written communication, 1985). This limestone is here referred to as the McCloud(?) Limestone. Although much of this limestone is recrystallized, some weathered surfaces are covered with fusulinids. The Northern Sierra terrane does not include any equivalent limestone lithofacies, although some younger fauna of McCloud zone H affinity occur in conglomerate of the Permian Reeve Formation (Harwood, 1988; D. S. Harwood and B. L. Murchey, this volume).

The limestone occurs as two isolated fault-bound blocks structurally underlain by slate, felsic porphyritic volcanic rocks, lithic-quartzofeldspathic sandstone and siltstone. The underlying sandstone contains abundant detrital felsic volcanic clasts, chert, quartz, and minor white mica and biotite, and may be equivalent to part of the Baird Formation, which would be expected to

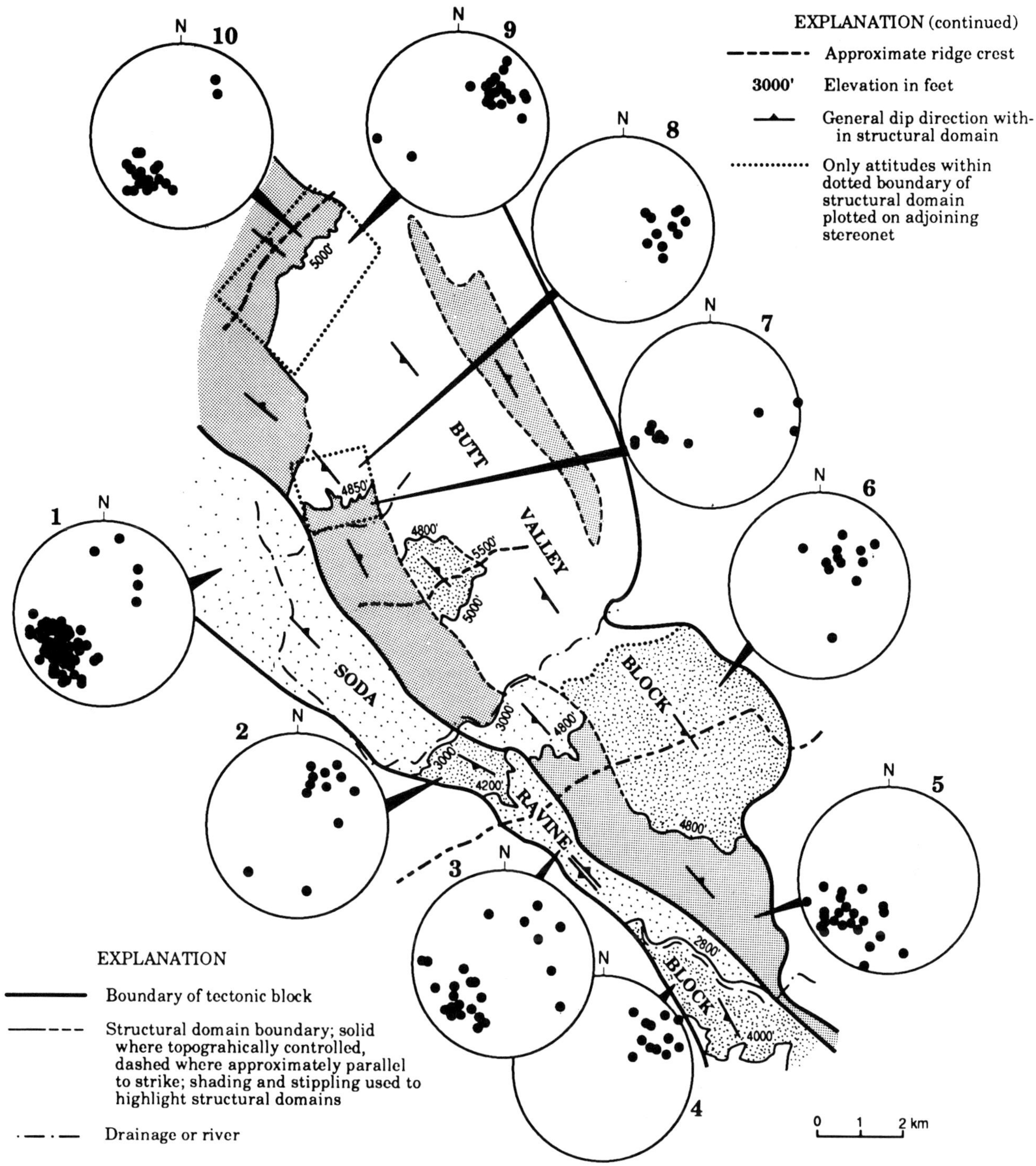

Figure 7. Lower-hemisphere, equal-area stereonet plots of poles to foliation from the Butt Valley and Soda Ravine blocks. The foliation dip direction clusters either to the east or west within the structural domains. The dip direction commonly changes along strike, coinciding with changes in elevation. Structural domains are bounded by northwest-trending boundaries parallel to strike (dashed lines); the boundaries approximately normal to the strike direction are drawn where the foliation dip direction changes from mainly east to west or vice versa, and are commonly subparallel to the topography (solid lines). The hinge lines of kink folds that control the foliation dip direction are inferred to approximately parallel the strike direction of the foliation. The change in foliation dip can be closely constrained in the field because abundant logging roads typically contour the hillsides at several levels on the slopes. Foliation symbols highlight the general orientation of foliation from each domain.

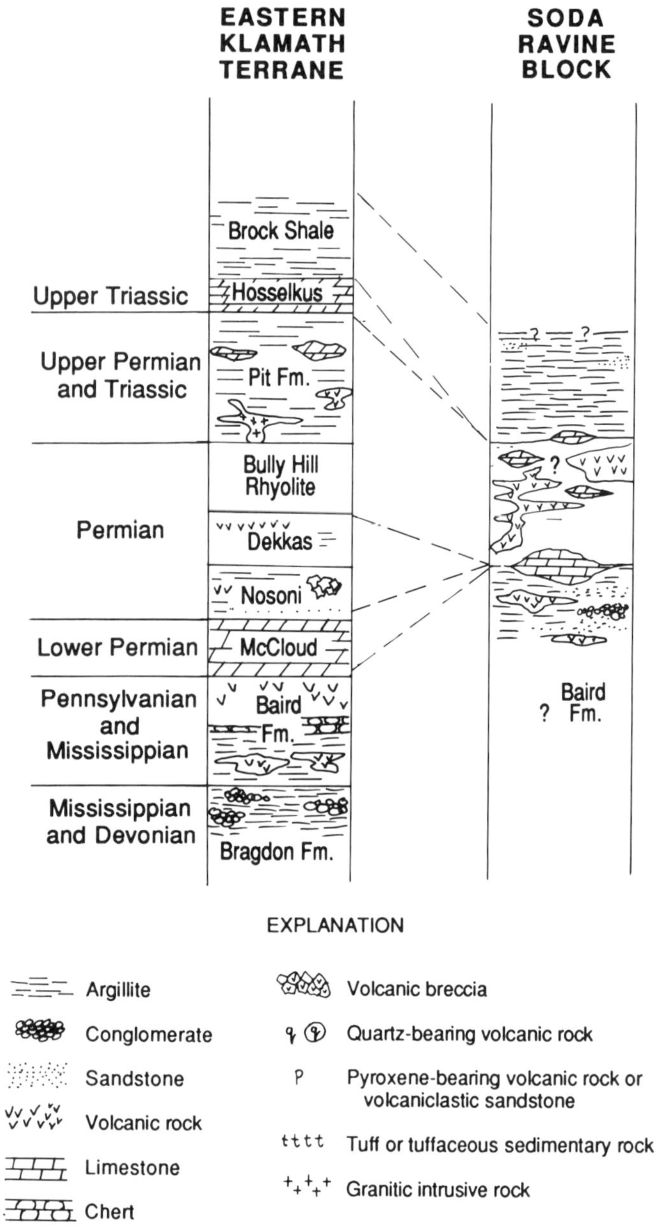

Figure 8. Schematic upper Paleozoic and lower Mesozoic part of the Eastern Klamath terrane, Klamath Mountains and similar units from the Soda Ravine block, Sierra Nevada (not drawn to scale). See previous work and this volume for description of these units in the Klamath Mountains (Sanborn, 1960; Albers and Robertson, 1961; Irwin, 1966; Watkins, 1985; Renne, 1988; Miller, 1989; Miller and Harwood, this volume.)

underlie the McCloud Limestone in the Klamath Mountains. The McCloud(?) Limestone is structurally overlain by slate with lesser siltstone and thin-bedded, mainly plagioclase crystal-rich volcanic rock.

Pit(?) Formation. The Pit Formation of the Eastern Klamath terrane consists predominantly of dark shale and siltstone, siliceous argillite, silicic tuffaceous and crystal-rich turbidites, and lenses of plagioclase porphyry and quartz keratophyre (Sanborn, 1960; Coogan, 1960; Albers and Robinson, 1961; Irwin, 1966; Miller and Harwood, this volume). The Pit Formation of the Eastern Klamath terrane includes fossiliferous limestone lenses of Middle Triassic age (Irwin, 1966; Albers and Robertson, 1961). Similar rock types are present in the Soda Ravine block where a sequence of Mesozoic dikes and flows (including the Humbug intrusives unit of Robinson, 1975) are interlayered with argillite, siltstone, and rare thin lenses of chert pebble conglomerate and limestone (Jayko, 1988); these rocks are here referred to as the Pit(?) Formation. The limestone block has yielded late Middle Triassic conodonts and ammonites (Ladinian to Carnian) (N. J. Silberling, written communication, 1987). The hypabyssal intrusive rocks include quartz porphyry, hornblende porphyry, and plagioclase porphyry. The intrusive layers are commonly a meter or two thick with sharp margins that lack evidence of shearing; some are bedding parallel. The contacts of the hypabyssal rocks are inferred to be primary. The intrusive rocks are characterized by weakly developed solution cleavage and a generally widespread overprint of low-grade metamorphic minerals, including chlorite, albite, epidote, green amphibole, and white mica. A penetrative fabric in the Mesozoic intrusive rocks indicates that intrusion occurred prior to the regional deformation, which is generally interpreted as Nevadan (Late Jurassic) in age (Day and others, 1985).

Hosselkus Limestone. Limestone bodies in the Soda Ravine block contain Late Triassic (late Carnian and Norian) fossils equivalent to those in the Hosselkus Limestone (Diller, 1892; Robinson, 1975; Jayko, 1988); the limestone is here assigned to the Hosselkus Limestone. The bioclastic limestone contains a large variety of fossils, including brachiopods, ichthyosaur vertebrae, conodonts, clams, bryozoans, and ammonites (Robinson, 1975; N. J. Silberling, oral communication, 1987). As noted earlier, the Hosselkus Limestone is the only lithologic unit that has been recognized in both the Eastern Klamath terrane (Sanborn, 1960) and the Northern Sierra terrane (see, for example, Diller, 1908; Irwin, 1966).

Brock(?) Shale. A thick sequence of slate and phyllite, with subordinate siltstone and minor sandstone occurs east of the Triassic Hosselkus Limestone in the Soda Ravine block (Jayko, 1988), and thus occupies the same stratigraphic position as the sparsely fossiliferous Triassic Brock Shale (see, for example, Sanborn, 1960) of the Eastern Klamath terrane; these sedimentary rocks are here referred to as the Brock(?) Shale. This unit in the Soda Ravine block is as yet undated. The Brock(?) Shale forms a continuous horizon subparallel to the trend of the Feather River terrane and Melones fault zone, but oblique to the Upper Triassic(?) and Lower Jurassic strata of the Butt Valley block. The Brock(?) Shale is interpreted as juxtaposed against the Butt Valley block along a steep fault that predates the regional metamorphism.

Tectonic implications of Eastern Klamath terrane slivers in the Sierra Nevada. The regional tectonic significance of the Soda Ravine block and related rocks of the Eastern Klam-

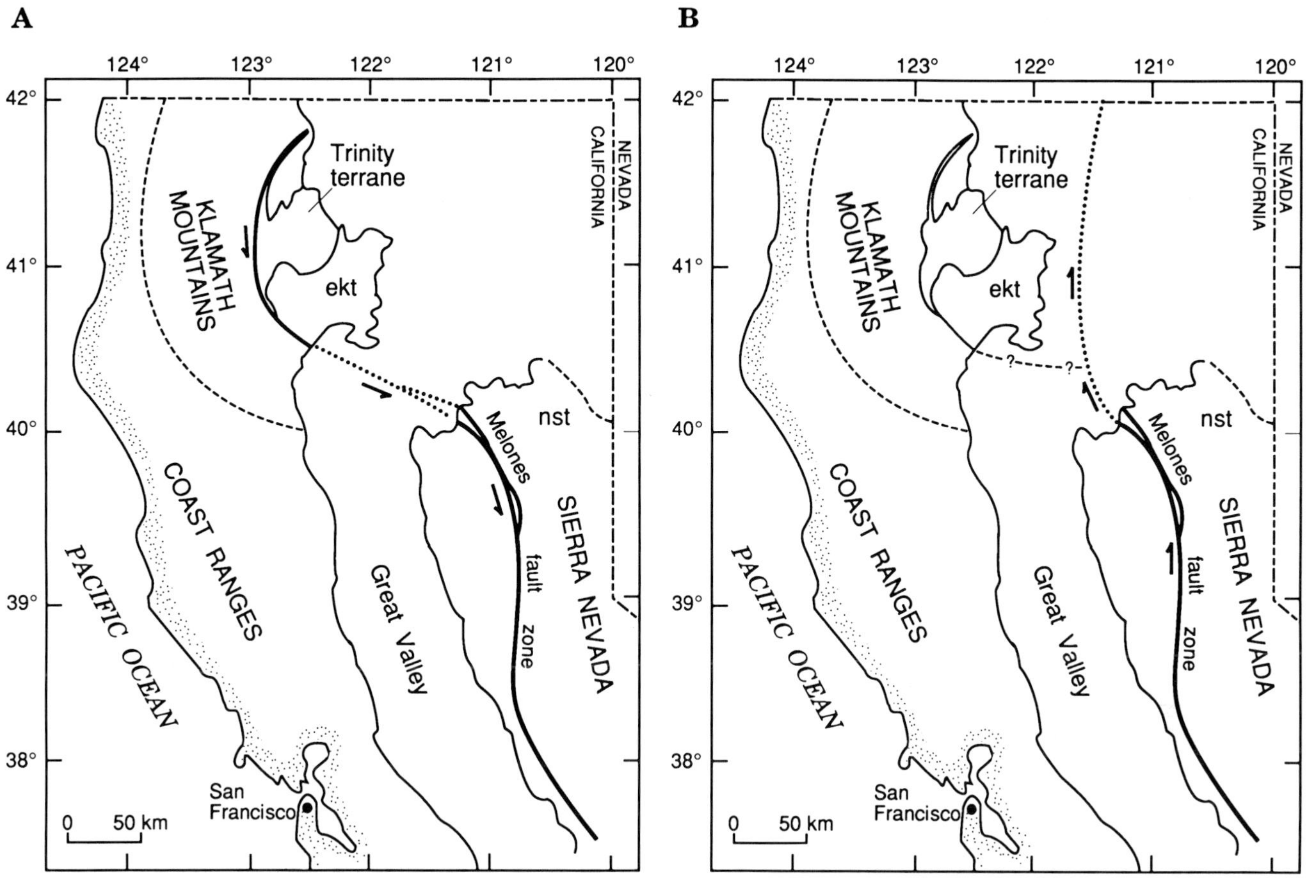

Figure 9. Schematic map. A, Northward continuation of the Melones fault zone to the west of the Eastern Klamath terrane (ekt), with inferred left-lateral displacement of tectonic slivers of Eastern Klamath terrane affinity. In this scenario the slivers and their bounding faults are considered to be part of the Melones fault zone. B, Northward continuation of the Melones fault zone east of the Klamath terrane, with inferred right-lateral displacement of the Eastern Klamath terrane relative to the northern Sierra terrane (nst). This scenario implies that the Eastern Klamath terrane was juxtaposed with the Northern Sierra terrane prior to northward displacement of the Eastern Klamath terrane.

ath(?) terrane is enigmatic. The simplest interpretation is that the Eastern Klamath terrane initially lay to the west and outboard of the Northern Sierra terrane and was emplaced along the Crablouse fault (Fig. 2) and similar structures to the south probably by thrust faulting. However, as the Soda Ravine block and slivers to the south (Harwood and others, 1988) occur along the eastern edge of the Melones fault zone, their position may be related to displacement along the fault zone. The slices could have been displaced from the north by left-lateral faulting if the northward continuation of the Melones fault zone lies west of the Eastern Klamath terrane in the Klamath Mountains (Hietanen, 1981) and the Crablouse fault is merely another splay within the Melones fault zone (Fig. 9A). Alternatively, if the Melones fault zone lies to the east of the Eastern Klamath terrane, beneath Cenozoic cover, the slices may represent remnants that were left behind, suggesting the Eastern Klamath terrane was displaced northward by right-lateral displacement (Fig. 9B). This interpretation also implies that the Eastern Klamath terrane may have lain to the west, and possibly structurally above, the Northern Sierra terrane prior to truncation of units east of the Melones fault zone.

EMPLACEMENT OF THE HOUGH BLOCK—IMPLICATIONS FOR THE GRIZZLY MOUNTAIN FAULT ZONE

Two major tectonic blocks of Northern Sierra terrane, the Genesee and Hough blocks, lie east of the Butt Valley block and are separated along the Grizzly Mountain fault zone (Fig. 5). The strata in the two blocks form east-facing, subparallel belts. The strata in the Hough block are overturned (east-facing and west-dipping) in the northern part of the block near the Canyondam syncline (McMath, 1966; D'Allura and others, 1977; Jayko, 1988), but upright (east-facing and east-dipping) along strike to the south in the Long Lake area (Durrell and D'Allura, 1977). Strata of the Genesee block are generally very steeply east and west dipping (Hannah, 1980; Hannah and Moores, 1986; Har-

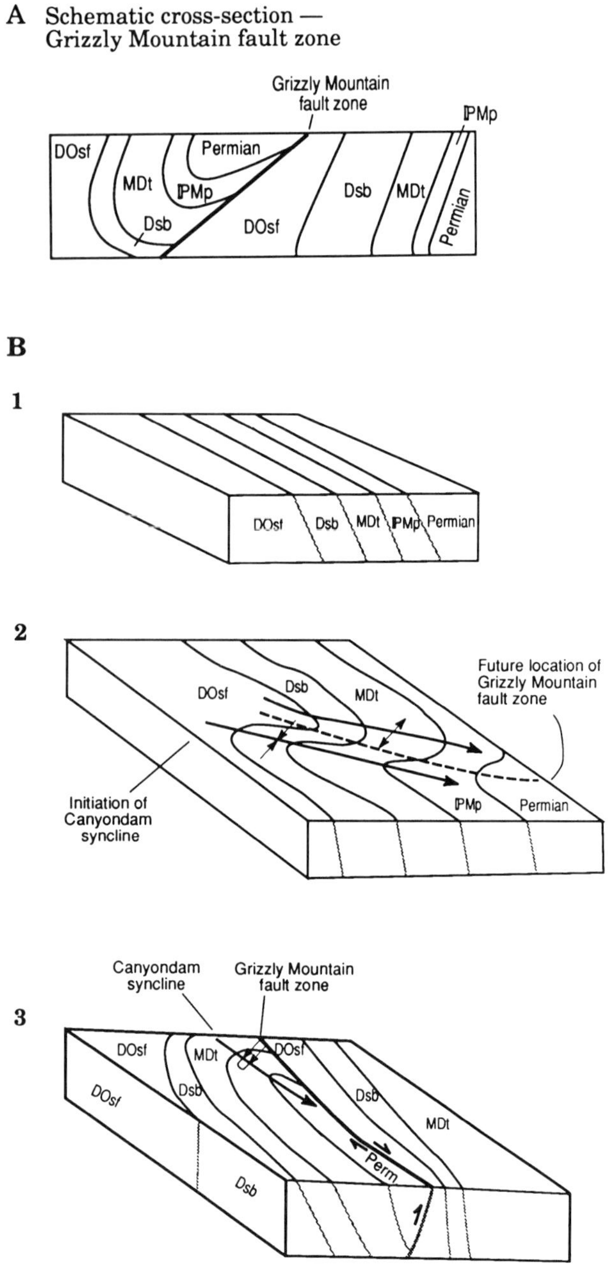

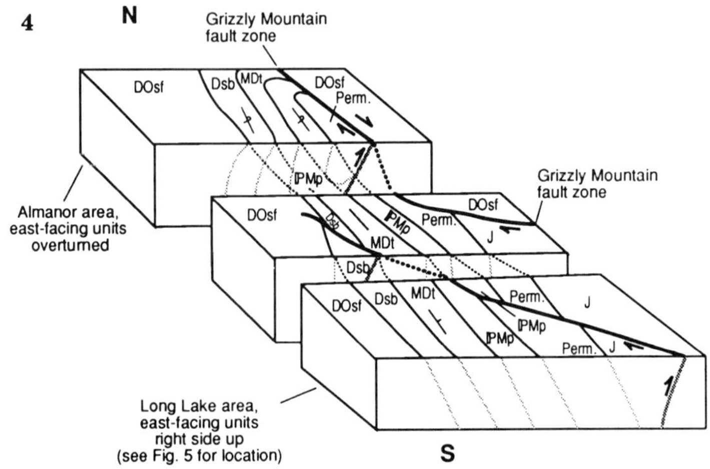

Figure 10. A, Schematic structural geometry along the Grizzly Mountain fault zone. The Grizzly Mountain fault zone was previously interpreted as a thrust fault (D'Allura and others, 1977; Day and others, 1985; Hannah and Moores, 1986; Harwood, 1988). DOsf = Shoo Fly Complex; Dsb = Sierra Buttes Formation; MDt = Taylor Formation; PMp = Peale Formation; Perm = Permian rocks undifferentiated; J = Jurassic rocks undifferentiated. B. Oblique strike-slip model for the Grizzly Mountain fault zone. Diagram 1 shows eastward-dipping strata prior to imposition of transpression. Diagram 2 shows the development of the Canyondam syncline and future trace of the Grizzly Mountain fault zone. Diagram 3 shows tightening and overturning of the Canyondam syncline and offset along the Grizzly Mountain fault zone, a west-dipping, oblique-slip fault. Diagram 4 schematically shows the structure of part of the Northern Sierra terrane west of the Grizzly Mountain fault zone. General dip directions shown by attitudes. Note that the units are east facing and overturned toward the north, but east facing and right-side-up along strike toward the south.

wood, 1988). The Canyondam syncline that lies in the northern part of the Hough block is the only major fold in the two blocks and occurs in the hanging wall block of the Grizzly Mountain fault zone (Figs. 3, 5).

The Grizzly Mountain fault zone in the study area is locally more than a kilometer wide, and is characterized by bodies of serpentinite and ultramafic igneous rocks. It is also characterized by a zone of melange-like deformation to the south (D. S. Harwood, oral communication, 1989). The fault zone has previously been interpreted as an eastward-verging thrust fault (D'Allura and others, 1977; Day and others, 1985; Hannah and Moores, 1986). The sense of displacement is not well constrained, however, and an overturned syncline in the hanging wall, combined with juxtaposition of younger over older rocks, is inconsistent with a simple thrust interpretation (Figs. 3B, 10A). Although a normal sense of offset could be inferred from juxtaposition of the Arlington Formation of the Hough block with the Shoo Fly Complex of the Genesee block (Fig. 10A), the greater development of schistosity in the Hough block and stratigraphic differences between the two blocks as described below are not consistent with a normal fault model. Likewise, the idea that structural geometry resulted from thrusting of a previously folded terrain is difficult to reconcile with the virtual absence of any other large-scale folds in the Hough and Genesee blocks (Fig. 5). Therefore, it seems reasonable to infer that the folding is related to faulting.

Stratigraphic contrasts across the Grizzly Mountain fault zone. There is only about five kilometers of separation between exposures of the base of the Taylorville sequence on the Hough block with equivalent units on the Genesee block in the study area; however, there are significant differences in the rocks that make up several of these units. Strata of the Taylorsville sequence within the Hough block have a higher percentage of sedimentary to volcanic rocks for a given unit than the Genesee block; the sedimentary rocks tend to be finer grained than in the Genesee block, and with respect to the Sierra Buttes Formation, unit thickness decreases considerably with respect to the Genesee block (Fig. 4). More specifically, the Sierra Buttes Formation consists predominantly of massive quartz-phyric flows and welded tuff in the Genesee block, whereas in the Hough block it consists mainly of tuffaceous sedimentary rocks, chert, and minor conglomerate with quartz-phyric clasts. The Taylor Formation consists of abundant pyroxene andesite flows and very coarse breccia in the Genesee block but predominantly pyroxene-bearing volcanogenic sandstone, tuff, and very minor flows in the Hough block. Similarly, the Permian of the Genesee block includes both clastic and volcanic rocks, whereas the Permian of the Hough block consists exclusively of clastic rocks. Therefore, although a normal-fault model for the Grizzly Mountain fault zone is geometrically appealing, it is not possible to simply restore the Hough block by removing down-dip displacement without advocating profound facies changes in much of the section across the fault.

Alternative interpretation of the Grizzly Mountain fault zone. An oblique-slip or transpressional model, with approximately 60 to 75 km of right-lateral slip, is consistent with observed structural and stratigraphic relations. These include the juxtaposition of different facies of the Taylorsville sequence, the juxtaposition of two predominantly east-facing stratigraphic sections, the presence of a tight overturned syncline in the hanging wall of the thrust and lack of other major folds, and the greater fabric development locally present in the upper plate rocks.

The model in Figure 10B shows the initial development of an anticline-syncline pair followed by the development of a right-oblique fault east of the syncline. As displacement on the fault progresses, the fold (Canyondam syncline) is tightened and translated northward.

Restoration of right-lateral slip of the Hough block along the Grizzly Mountain fault zone forms a single continuous belt of Northern Sierra terrane. The Canyondam syncline is interpreted as a transpressive drag fold associated with right-lateral slip. Smaller scale faults with apparent right-slip occur to the south (Harwood, 1988) (Fig. 5), where they are well exposed or can be inferred from isolated exposures. The Taylorsville thrust, which lies to the east, is inferred to be syntectonic with respect to displacement on the Grizzly Mountain fault. Deformation postdates Lower and Middle Jurassic rocks of the Mt. Jura section, which underlie the Taylorsville thrust (Day and others, 1985).

The Grizzly Mountain fault zone is also inferred to have been modified by later Cenozoic(?), high-angle normal faults (McMath, 1958; D'Allura and others, 1977; Day and others, 1985; Jennings, 1977). These northwest-trending normal faults are interpreted to represent incipient encroachment of the Basin and Range deformation on the northeastern Sierra Nevada.

CONCLUSIONS

Rocks of the northernmost Sierra Nevada that occur in a wedge lying east of the Feather River terrane and west of the Hough block have lithologic affiliation with both the Eastern Klamath and Northern Sierra terranes. These rocks have been divided into the Butt Valley and Soda Ravine blocks. Rocks of the Butt Valley block are considered to be part of the Northern Sierra terrane and are part of a structural block bounded on the east by the Clear Creek fault.

Rocks of the Soda Ravine block have a closer lithologic affiliation with the upper Paleozoic and lower Mesozoic part of the Eastern Klamath terrane than with the Northern Sierra terrane, as was recognized in part by Robinson (1975) and D'Allura and others (1977). Limestone similar to the McCloud Limestone occurs in the Soda Ravine block and also in similar tectonic slivers reported along the Melones fault zone to the south (D'Allura and others, 1977; Standlee, 1978; Harwood and others, 1988).

There are several tectonic scenarios that can be proposed to explain the presence of the Soda Ravine block and similar tectonic slices to the south; however, there are not yet enough data to choose one scenario over another. The spatial occurrence of these tectonic slices close to the Melones fault zone suggests a relation between the sense of offset along the fault and the juxtaposition of the tectonic slices. One interpretation is that the slivers were left behind as the Eastern Klamath terrane was displaced northwestward by right-lateral displacement along the Melones fault zone. The Soda Ravine block was juxtaposed with the Northern Sierra terrane prior to regional greenschist-facies metamorphism, which is post–early Jurassic and probably Middle Jurassic or Late Jurassic (Nevadan) in age.

In the Butt Valley block, Upper Triassic strata unconformably overlie the Devonian and/or Mississippian Taylor Formation; to the south in the Tahoe area, Upper Triassic conglomerate and limestone overlie Ordovician and Devonian rocks (Clark and others, 1962; Harwood, 1983; 1988). East of the study area, in the Genesee block, Middle and Upper Triassic strata overlie Permian rocks (Diller, 1908; McMath, 1966; Hannah and Moores, 1986). These differences in the oldest ages of the overlying strata, as well as differences in the ages of the underlying rocks, suggest that either there was uplift in the south and west (Tahoe and Butt Valley areas) relative to the east (Genesee area), or that the rocks of the Butt Valley block were displaced along a regional thrust or oblique-slip fault to the north or northeast.

Stratigraphic variations between the Hough and Genesee blocks in the Almanor area, combined with structural constraints, are consistent with repetition of the stratigraphic sequence by a major lateral fault. Available evidence suggests that the Grizzly Mountain fault zone may represent an oblique strike-slip fault rather than a thrust fault.

ACKNOWLEDGMENTS

I have benefited greatly from discussions with D. S. Harwood throughout the course of this study and acknowledge helpful discussions with Norm Silberling and Judy Hannah. Bruce Wardlaw and Anita Harris identified the conodonts in Table 1, Bonnie Murchey identified the radiolarians, Norm Silberling looked at the megafossils, and Cal Stevens reexamined fusulinids from the McCloud(?) Limestone. I thank John Bartley, Clark Blake, Reid Fisher, Dave Harwood, King Huber, and David L. Smith for helpful reviews.

REFERENCES CITED

Albers, J. P., and Robertson, J. F., 1961, Geology and ore deposits of East Shasta copper-zinc district, Shasta County, California: U.S. Geological Survey Professional Paper 338, 107 p.

Blake, M. C., Jr., Howell, D. G., and Jones, D. L., 1982, Preliminary tectono-stratigraphic terrane map of California: U.S. Geological Survey Open-File Report 82-593, 9 p., scale 1:750,000.

Bond, G. C., and DeVay, J. C., 1980, Pre-Upper Devonian quartzose sandstone in the Shoo Fly Formation, northern California; Petrology, provenance, and implications for regional tectonics: Journal of Geology, v. 88, p. 285–308.

Burchfiel, B. C., and Davis, G. A., 1972, Structural framework and evolution of the southern part of the Cordilleran Orogen, western United States: American Journal of Science, v. 272, p. 97–118.

—— , 1975, Nature and controls of Cordilleran orogenisis,, western United States; Extensions of an earlier synthesis: American Journal of Science, v. 275-A, p. 363–396.

—— , 1981, Triassic and Jurassic tectonic evolution of the Klamath Mountains–Sierra Nevada geologic terrane, *in* Ernst, W. G., ed., The geotectonic development of California; Rubey Volume 1: Englewood Cliffs, New Jersey, Prentice-Hall, p. 87–131.

Clark, L. D., Imlay, R. W., McMath, V. E., and Silberling, N. J., 1962, Angular unconformity between Mesozoic and Paleozoic rocks in the northern Sierra Nevada: U.S. Geological Survey Professional Paper 450-B, p. B15–B19.

Coogan, A. H., 1960, Stratigraphy and paleontology of the Permian Nosoni and Dekkas Formations (Bollibokka Group): University of California Publications in Geological Sciences, v. 5, p. 243–316.

D'Allura, J. A., Moores, E. M., and Robinson, L., 1977, Paleozoic rocks of the northern Sierra Nevada; Their structural and paleogeographic significance, *in* Stewart, J. H., Stevens, C. H, and Fritsche, A. E., eds., Paleozoic paleogeography of the western United States: Pacific Section, Society of Economic Paleontologists and Mineralogists Pacific Coast Paleogeograph Symposium 1, p. 394–408.

Day, H. W., Moores, E. M., and Tuminas, A. C., 1985, Structure and tectonics of the northern Sierra Nevada: Geological Society of America Bulletin, v. 96, p. 436–450.

Diller, J. S., 1892, Geology of the Taylorsville region of California: Geological Society of America Bulletin, v. 3, p. 370–394.

—— , 1895, Lassen Peak Folio, California: U.S. Geological Survey Geological Atlas, Folio 15.

—— , 1908, Geology of the Taylorsville region, California: U.S. Geological Survey Bulletin 353, 128 p.

Durrell, C., and D'Allura, J. A., 1977, Upper Paleozoic section in eastern Plumas and Sierra Counties, northern Sierra Nevada, California: Geological Society of America Bulletin, v. 88, p. 844–852.

Girty, G. H., and Schweickert, R. A., 1984, The Culbertson Lake allochthon; A newly identified structure within the Shoo Fly Complex, California; Evidence for four phases of deformation and extension of the Antler orogeny to the northern Sierra Nevada: Modern Geology, v. 8, p. 181–198.

Girty, G. H., and Wardlaw, M. S., 1985, Petrology and provenance of pre-Late Devonian sandstones, Shoo Fly Complex, northern Sierra Nevada, California: Geological Society of America Bulletin, v. 96, p. 516–521.

Hamilton, W., 1969, Mesozoic California and the underflow of the Pacific mantle: Geological Society of America Bulletin, v. 80, p. 2409–2430.

Hannah, J. L., 1980, Stratigraphy, petrology, paleomagnetism, and tectonics of Paleozoic arc complexes, northern Sierra Nevada [Ph.D. thesis]: Davis, University of California, 323 p.

Hannah, J. L., and Moores, E. M., 1986, Age relationships and depositional environments of Paleozoic strata, northern Sierra Nevada, California: Geological Society of America Bulletin, v. 97, p. 787–797.

Harwood, D. S., 1983, Stratigraphy of upper Paleozoic volcanic rocks and regional unconformities in part of the northern Sierra terrane: Geological Society of America Bulletin, v. 94, p. 413–422.

—— , 1988, Tectonism and metamorphism in the northern Sierra terrane, northern California, *in* Ernst, W. G., ed., Metamorphism and crustal evolution of the western United States; Rubey Volume 7: Englewood Cliffs, New Jersey, Prentice-Hall, p. 764–788.

Harwood, D. S., Jayko, A. S., Harris, A. G., Silberling, N. J., and Stevens, C. H., 1988, Permian-Triassic rocks slivered between the Shoo Fly Complex and the Feather River Peridotite belt, northern Sierra Nevada, California: Geological Society of America Abstracts with Programs, v. 20, p. 167–168.

Hietanen, A., 1981, The Feather River area as a part of the Sierra Nevada suture system in California: U.S. Geological Survey Professional Paper 1226-B, p. 1–13.

Irwin, W. P., 1966, Klamath Mountains Province: California Division of Mines and Geology Bulletin 190, p. 19–38.

—— , 1977, Review of Paleozoic rocks of the Klamath Mountains, *in* Stewart, J. H., Stevens, C. H., and Fritsch, A. E., eds., Paleozoic paleogeography of the western United States: Pacific Section, Society of Economic Paleontologists and Mineralogists Pacific Coast Paleogeography Symposium 1, p. 441–454.

—— , 1981, Tectonic accretion of the Klamath Mountains, *in* Ernst, W. G., ed., Metamorphism and crustal evolution of the western United States; Rubey Volume 7: Englewood Cliffs, New Jersey, Prentice-Hall, p. 764–788.

Jayko, A. S., 1988, Paleozoic and Mesozoic rocks of the Almanor 15′ Quadrangle, Plumas County, California: U.S. Geological Survey Open-File Report 88-757, scale 1:62,500.

Jayko, A. S., Murchey, B. L., Wardlaw, B. R., and Stevens, C. H., 1987, Revised interpretation of the pre-Cenozoic geology of the Butt Lake area, northern Sierra Nevada, California: Geological Society of America Abstracts with Programs, v. 19, p. 391.

Jennings, C. A., 1977, Geologic map of California: California Division of Mines and Geology, scale 1:750,000.

Lindsley-Griffin, N., 1977, Paleogeographic implications of ophiolites; The Ordovician Trinity Complex, Klamath Mountains, California, *in* Stewart, J. H., Stevens, C. H., and Fritsch, A. E., eds., Paleozoic paleogeography of the western United States: Pacific Section, Society of Economic Paleontologists and Mineralogists Pacific Coast Paleogeography Symposium 1, p. 409–420.

McMath, V. E., 1958, Geology of the Taylorsville area, Plumas County, California [Ph.D. thesis]: Los Angeles, University of California 199 p.

—— , 1966, Geology of the Taylorsville area, northern Sierra Nevada, California: California Division of Mines and Geology Bulletin 190, p. 173–183.

Miller, M. M., 1987, Dispersed remnants of a northeast Pacific fringing arc; Upper

Paleozoic terranes of Permian McCloud faunal affinity, western U.S.: Tectonics, v. 6, p. 807–830.

——, 1989, Intra-arc sedimentation and tectonism; Late Paleozoic evolution of the eastern Klamath terrane, California: Geological Society of America Bulletin, v. 101, p. 170–187.

Moores, E. M., and Wise, W. S., 1970, Prebatholithic structure and stratigraphy of the Lake Almanor and Greenville Quadrangles, northern Sierra Nevada; Progress report: Geological Society of America Abstracts with Programs, v. 2, p. 121.

Potter, A. W., Hotz, P. E., and Rohr, D. M., 1977, Stratigraphy and inferred tectonic framework of Lower Paleozoic rocks in the eastern Klamath Mountains, northern California, *in* Stewart, J. H., Stevens, C. H., and Fritsch, A. E., eds., Paleozoic paleogeography of the western United States: Pacific Section, Society of Economic Paleontologists and Mineralogists Pacific Coast Paleogeography Symposium 1, p. 421–440.

Renne, P. R., 1988, Structural chronology, oroclinal deformation, and tectonic evolution of the southeastern Klamath Mountains, California: Tectonics, v. 7, p. 1223–1242.

Robinson, L., 1975, Geology of the Arlington Formation, Butt Lake [M.S. thesis]: Davis, University of California, 77 p.

Saleeby, J. B., Hannah, J. L., and Varga, R. J., 1987, Isotopic age constraints on middle Paleozoic deformation in the northern Sierra Nevada, California: Geology, v. 15, p. 757–760.

Sanborn, A. F., 1960, Geology and paleontology of the southwest quarter of the Big Bend Quadrangle, Shasta County, California: California Division of Mines and Geology Special Report 63, 26 p.

Schweickert, R. A., 1981, Tectonic evolution of the Sierra Nevada Range, *in* Ernst, W. G., ed., The geotectonic development of California; Rubey Volume 1: Englewood Cliffs, New Jersey, Prentice-Hall, p. 87–131.

Silberling, N. J., and Jones, D. L., 1984, Lithotectonic terrane maps of the North American Cordillera: U.S. Geological Survey Open-File Report 84-523, scale 1:2,500,000.

Skinner, J. W., and Wilde, G. L., 1965, Permian biostratigraphy and fusulinid faunas of the Lake Shasta area, northern California: Lawrence, University of Kansas Paleontological Contributions, Protozoa, article 6, p. 1–98.

——, 1966, Permian fusulinids from Pacific Northwest and Alaska: Lawrence, University of Kansas Paleontologic Contribution Paper 4, p. 1–16.

Standlee, L. A., 1978, Middle Paleozoic ophiolite in the Melones Fault zone, northern Sierra Nevada, California: Geological Society of America Abstracts with Programs: v. 10, p. 148.

Turner, H. W., 1898, Bidwell Bar Folio: U.S. Geological Atlas Folio 43, 14 p.

Varga, R. J., and Moores, E. M., 1981, Age, origin, and significance of an unconformity that predates island-arc volcanism in the northern Sierra Nevada: Geology, v. 9, p. 512–518.

Watkins, R., 1985, Volcaniclastic and carbonate sedimentation in late Paleozoic island-arc deposits, Eastern Klamath Mountains, California: Geology, v. 13, p. 709–713.

Wilson, E. C., 1984, Stratigraphic range extensions and coral province affiliations of stony corals in the Lower Permian McCloud limestone of California [abs.], *in* Oliver, W. A., Jr., ed., Recent advances in paleobiology and geology of the Cnidaria: Paleontologica Americana, v. 54, p. 437.

Manuscript Accepted by the Society March 27, 1990

Printed in U.S.A.

Geological Society of America
Special Paper 255
1990

The Triassic-Jurassic magmatic arc in the Mojave-Sonoran Deserts and the Sierran-Klamath region; Similarities and differences in paleogeographic evolution

C. J. Busby-Spera, J. M. Mattinson, N. R. Riggs, and E. R. Schermer
Department of Geological Sciences, University of California, Santa Barbara, California 93106

ABSTRACT

We recognize two important paleogeographic links between the Klamath-Sierran segment and Mojave-Sonoran segment of the early Mesozoic arc in the southwest Cordillera: (1) both segments of the arc occupied an extensional graben-depression in Late Triassic to early Middle Jurassic time, and (2) the graben-depression acted as a long-lived (>40 m.y.) trap for Early and Middle Jurassic craton-derived quartz arenites deposited within both segments of the arc. The Early Mesozoic arc formed at a high angle to the truncated continental margin across cratonal, miogeoclinal, and eugeoclinal settings inherited from Paleozoic time. Intra-arc extension in the cratonal setting (Mojave-Sonoran Deserts) resulted in accumulation and preservation of thick terrestrial volcanic and sedimentary sequences. Intra-arc extension in the miogeoclinal setting (southern and central Sierra Nevada) resulted in deposition of very thick shallow marine and lesser deep marine successions. Intra-arc subsidence is also documented from deep marine to shallow marine successions in the eugeoclinal setting (northern Sierra and adjacent northwest Nevada).

The Klamath-Sierran segment and Mojave-Sonoran segment of the arc evolved dissimilarly in late Middle Jurassic to Late Jurassic time. Contractional deformation documented in the Klamath-Sierran segment was not nearly as widespread or intense in the central Mojave Desert, and probably did not affect the southern Mojave to Sonoran Desert areas at all. Extensional tectonics, instead, continued to dominate the Mojave-Sonoran segment of the arc, probably due to continental rifting associated with the opening of the Gulf of Mexico.

INTRODUCTION

A distinctive characteristic of the early Mesozoic magmatic arc in the Sierra Nevada and Mojave-Sonora Desert lies in the unusual association of volcanic and volcaniclastic rocks with supermature sandstones. A growing body of field and geochronologic evidence suggests that craton-derived quartz sands buried vent-proximal regions of the arc for at least 40 m.y. Rather than acting as a high-standing topographic barrier to the migration of craton-derived sands, the arc apparently occupied a deep fault-bounded depression (Fig. 1), similar to the modern magmatic arc of Central America (Busby-Spera, 1988). High calculated rates of subsidence, the presence of syndepositional normal faults and fault-talus breccias, the abundance of calderas and high-level intrusions, and interstratification of calc-alkalic and alkalic volcanic rocks provide support for this model. The arc graben depression thus acted as a long-lived trap for craton-derived quartz sands (Busby-Spera, 1988).

The early Mesozoic magmatic arc formed at a high angle to Paleozoic structural and stratigraphic trends, probably due to tectonic truncation of the continental margin in late Paleozoic time (Burchfiel and Davis, 1972; Miller and Cameron, 1982;

Busby-Spera, C. J., Mattinson, J. M., Riggs, N. R., and Schermer, E. R., 1990, The Triassic-Jurassic magmatic arc in the Mojave-Sonoran Deserts and the Sierran-Klamath region; Similarities and differences in paleogeographic evolution, *in* Harwood, D. S., and Miller, M. M., eds., Paleozoic and early Mesozoic paleogeographic relations; Sierra Nevada, Klamath Mountains, and related terranes: Boulder, Colorado, Geological Society of America Special Paper 255.

Stone and Stevens, 1988; Walker, 1988). The arc thus developed across cratonal, miogeoclinal, and eugeoclinal settings inherited from Paleozoic time. Intra-arc extension resulted in accumulation of thick terrestrial volcanic and sedimentary sequences in the cratonal setting, which extended from the Sonoran Desert (northern Sonora–southern Arizona) through the southern Mojave Desert (southeast California–southwest Arizona) to the eastern and central Mojave Desert (Figs. 1, 2-1 through 2-5). Thick, predominantly shallow marine volcanic and sedimentary sequences accumulated in the miogeoclinal setting, now preserved as roof pendants in the southern and central Sierra Nevada (columns 6, 7, Figure 2). Thick, deep marine and lesser shallow marine volcanic and sedimentary sequences accumulated on accreted island-arc basement in the northern Sierra (Fisher, 1989; this volume) and adjacent west-central Nevada (Dilles and Wright, 1988; Wyld, 1989). Entrapment of craton-derived quartz arenites occurred in all these paleogeographic settings, along the length of the arc graben-depression.

We propose here that the Klamath-Sierran and Mojave-Sonoran segments of the arc became dissimilar in their overall tectonic setting in late Middle to Late Jurassic time (Fig. 3). Shortening events of this age in the Klamath-Sierran segment probably had little effect on the Mojave-Sonoran segment; instead, rifting associated with the opening of the Gulf of Mexico resulted in continued regional extension of the arc.

This chapter synthesizes new field and geochronologic data with published data to discuss paleogeographic and paleotectonic similarities and differences between the Klamath-Sierran and Mojave-Sonoran segments of the Triassic-Jurassic arc. This synthesis is presented largely as a review of stratigraphy by geographic region, from south to north along the arc.

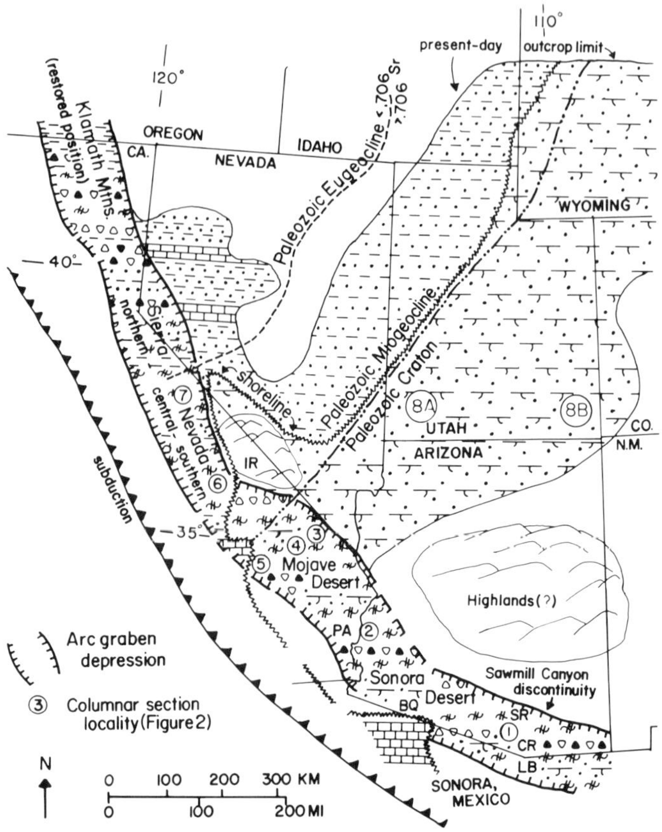

Figure 1. Tectonic model for the Late Triassic to early Middle Jurassic magmatic arc in Nevada, California, Arizona, and Sonora (present-day coordinates given). Several lines of evidence suggest that the arc occupied an extensional graben-depression more than 1,200 km long (Busby-Spera, 1988). Locations of tectono-stratigraphic columns (see Fig. 2) are plotted here. Key to patterning given in Figure 2. BQ = Baboquivari Mountains, CR = Cobre Ridge, IR = Inyo Range, LB = Las Borregas, PA = Palen Mountains, SR = Santa Rita Mountains.

CRATONAL ARC STRATIGRAPHY AND STRUCTURE

Fossils are absent from the nonmarine strata in the Triassic-Jurassic cratonal arc setting, making it necessary to rely on isotopic age data to establish stratigraphic correlations and to document patterns of volcanism, subsidence, and faulting. Zircon is highly resistant to isotopic resetting and is the most suitable mineral for dating Triassic and Jurassic rocks of this region, which were metamorphosed in Cretaceous and Tertiary time. The disadvantage of this resistance to melting and resetting is that zircon may retain a "memory" of its earlier history if it is inherited from a continental magma source area, or if it becomes entrained in a rising magma or in a pyroclastic flow moving over loose sediment.

Inheritance/entrainment is common in some zircon samples from the cratonal arc setting, but magmatic crystallization ages have been obtained by concordia interpretation (lower intercept) of multiple fractions of zircon from a single sample (Busby-Spera and others, 1987; Graubard and others, 1988; Reynolds and others, 1988). Similarly, the original crystallization ages of disturbed zircon from this region have been obtained by projecting back, through multiple fractions that have experienced varying degrees of disturbance, to the upper intercept on a concordia diagram (Wright and others, 1981). Our work in progress shows that some samples from this region combine inheritance or entrainment of older zircon components with later Pb loss (Riggs and others, 1986; Busby-Spera and others, 1989a; J. M. Mattinson and C. J. Busby-Spera, unpublished data). In these cases, we hand sort apparently magmatic zircon grains into several fractions according to size, shape, and color, and then perform partial digestion of each fraction to remove the more disturbed parts of the zircon grains in one or more steps. Use of these approaches is resulting in much more refined estimates of crystallization ages of the hitherto poorly dated metavolcanic rocks shown in Figure 2.

Southern Arizona–northern Sonora

The oldest U-Pb zircon ages from arc magmatic rocks in the southern Arizona–northern Sonora region come from metavolcanic rocks of the Mount Wrightson Formation in the Santa Rita Mountains south of Tucson (column 1, Fig. 2; locality plotted in Fig. 1) and date from approximately 205 Ma (Riggs and others, 1986, and in progress; Asmerom, 1988; Asmerom and others, 1988). Reconnaissance work suggests that the Mount Wrightson Formation may have originally extended as far south as the Las Borregas region of northern Sonora (Fig. 1).

The nature and distribution of volcanic lithofacies in the Mount Wrightson Formation in the Santa Rita Mountains record the evolution of a multivent volcanic complex that formed within a subsiding cratonal intra-arc graben (Riggs and Busby-Spera, 1990a). These lithofacies are grouped into the following facies assemblages: (1) the near-vent facies assemblage, consisting of silicic to intermediate hypabyssal intrusive bodies and peperites, as well as dacite domes and lava flows; (2) the proximal facies assemblage, containing intermediate lava flows, ignimbrites, and debris flow deposits; (3) the medial facies assemblage, consisting of ignimbrites, debris flow deposits, and reworked pyroclastic deposits; and (4) the distal facies assemblage, including nonwelded ignimbrites and fluvial deposits. Some of the thick, densely welded ignimbrites in the Mount Wrightson Formation probably represent outflow from a more distant caldera or calderas within the intra-arc basin, rather than being products of the vents within the complex.

Intra-arc subsidence resulted in a complex juxtaposition of vent, proximal, medial, and distal facies as vent areas were buried by volcanic/volcaniclastic material derived from adjacent vents in the multi-vent complex and by ignimbrites from calderas outside the complex (Riggs and Busby-Spera, 1990a). An overall evolution from dominantly effusive to dominantly explosive to waning volcanism is, however, recorded in the Mount Wrightson Formation. Rapid intra-arc subsidence also resulted in repeated burial of vent areas (and other subenvironments) by eolian quartz arenites, probably age-correlative with the Sinemurian-age Wingate Sandstone on the Colorado Plateau (columns 1 and 8, Fig. 2).

The Fresnal Canyon sequence in the Baboquivari Mountains and Sil Nakya Hills (Fig. 2-1) has yielded upper intercept U-Pb zircon ages of 190 Ma on metavolcanic rocks (Wright and others, 1981), indicating that interstratified eolian quartz arenites may be age-correlative with the Navajo sandstone on the Colorado Plateau (Haxel and others, 1985; Fig. 2-8). The Fresnal Canyon sequence comprises 7 to 8 km (composite stratigraphic thickness) of terrestrial volcanic and sedimentary rocks deposited within a short time (Haxel and others, 1985; Tosdal and others, 1989). Subsidence was dominantly controlled by regional extension, although local caldera collapse may have been a contributing factor (Riggs and others, 1990).

Our preliminary U-Pb zircon ages from ignimbrites of the Cobre Ridge group (Fig. 2-1) along the central Arizona–Sonora border suggest that these strata are 175 m.y. old (Riggs and others, 1990, and unpublished data); interstratified eolian quartz arenites may thus be age-correlative with eolianites of the Middle Jurassic Carmel Formation of the Colorado Plateau (Fig. 2-8). The Cobre Ridge group (Riggs and Busby-Spera, 1990b) records the collapse, filling, and resurgent doming of a very large (60×25 km) caldera that formed during the eruption of the Pajarito Tuff, one of the largest volume ignimbrites in the geologic record (estimated volume between 1,000 and 3,000 km^3). The intracaldera ignimbrite is as much as 3 km thick. The northwest half of the caldera subsided to greater depths than the southeast half, providing a depocenter for a $\geqslant$1.5-km-thick section of postcollapse volcanic and sedimentary strata. The postcollapse strata record the early stages of resurgent doming and attendant moat filling, and were intruded during the later stages of resurgence. The postcollapse strata include rhyolite dome talus cones, accidentally ponded ignimbrites (thick but areally restricted), water-lain tuffs, and a thick, localized conglomerate-breccia deposit that represents material eroded from the caldera walls (Riggs and Busby-Spera, 1990b).

Latest Triassic to Middle Jurassic intra-arc extension in the Arizona-Sonora segment of the arc appears to have gradually given way to Late Jurassic extension, volcanism, and sedimentation related to the opening of the Gulf of Mexico, shown as the Glance conglomerate of the Bisbee Group in Figure 2-1 (Dickinson and others, 1986). We suggest that the southern Mojave Desert to Sonoran Desert segment of the arc-graben depression was exploited, while still active, by a Gulf of Mexico–related rift (Busby-Spera and others, 1989a). We suspect that controversies regarding the paleotectonic setting (arc vs. rift) of specific stratigraphic units have resulted from this gradual transition. These units include: (1) the Canelo Hills volcanics (Fig. 2-1), the Gardner Canyon Formation, and "unnamed volcanic and sedimentary rocks" of the Huachuca and Patagonia Mountains, all in southern Arizona (Dickinson and others, 1987; Tosdal and others, 1989); and (2) the McCoy Mountains Formation (Fig. 2-2) of southeast California–southwest Arizona (Harding and Coney, 1985; Dickinson and others, 1986; Tosdal and others, 1989), discussed below. There is no evidence for late Jurassic compressional deformation in the southern Mojave to Sonoran segment of the arc, even though such an event was important in the Klamath-Sierran segment and may have affected some localities in the eastern to central Mojave Desert, as discussed below.

Southeast California–southwest Arizona region

Lower Mesozoic strata and plutonic rocks of the southeast California–southwest Arizona region record a transition from Paleozoic and Triassic(?) cratonal sedimentation to Jurassic continental arc magmatism, although the details of their age and correlation remain rather poorly constrained (Tosdal and others, 1989). Jurassic volcanic rocks locally occur in close spatial proximity to inferred deep-level plutons of probable Triassic age (Haxel and others, 1985; Barth and others, 1988). This may be the result of late Triassic uplift described by Reynolds and others (1989), or it may be due to middle or late Jurassic crustal extension proposed by Tosdal and Haxel (1987). In the Palen Mountains (Figs. 2-2, 4), a thick sequence of rhyodacitic to rhyolitic

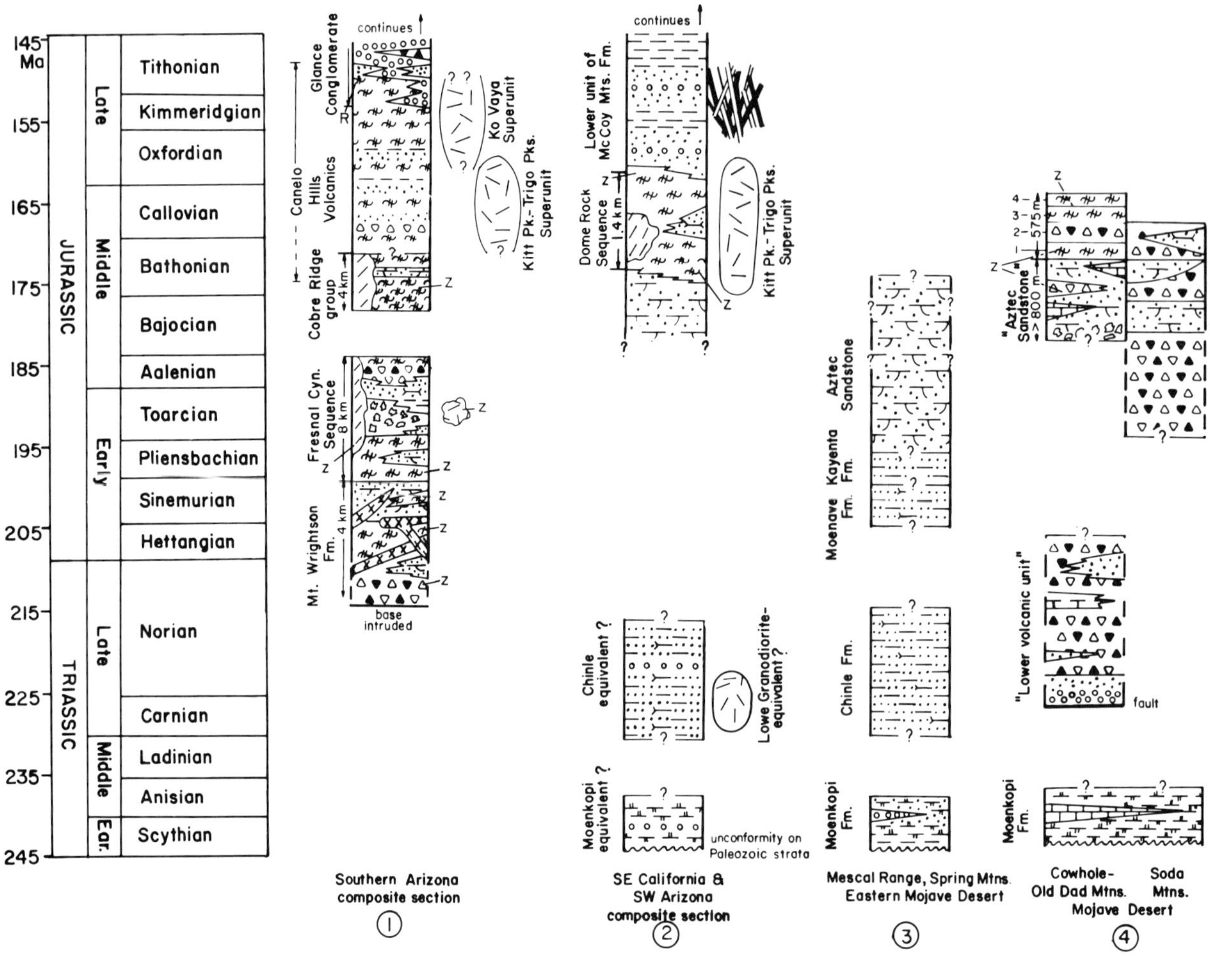

Figure 2. Composite, generalized tectonostratigraphic columns for Triassic and Jurassic time (localities plotted in Fig. 1). Stratigraphic thickness given where known. Data sources: column 1, Haxel and others, 1985; Dickinson and others, 1986; Tosdal and others, 1989; Riggs, 1987; Riggs and Busby-Spera, 1990. Column 2, largely from Tosdal and others, 1989; also Pelka, 1973; Harding and Coney, 1985; Stone and others, 1987. Column 3, Hewett, 1956; Marzolf, 1983. Column 4, Grose, 1959; Dunne, 1977; Novitsky-Evans, 1978; Marzolf, 1983. Column 5, Miller, 1978; Miller and Carr, 1978; Karish and others, 1987; Walker, 1988. Column 6, Busby-Spera, 1984a, b, 1985; Busby-Spera and Saleeby, 1987. Column 7, Fiske and Tobisch, 1978. Column 8, Petersen and Pipiringos, 1979; Imlay, 1980; Blakey and others, 1983. Additional references for all localities are given in text. Our new field and geochronologic data, discussed throughout text, are incorporated into columns 1 through 6.

metavolcanic rocks (Dome Rock sequence of Tosdal and others, 1989) overlies and intrudes Aztec-equivalent(?) eolian quartz arenites (Stone and Kelly, 1988; Stone and Pelka, 1989) and is overlain by the McCoy Mountains Formation of Pelka (1973). Our field studies in progress in the Palen Mountains suggest that the Aztec-equivalent strata are conformably overlain by the Dome Rock sequence, and that the Dome Rock sequence is in turn conformably overlain by the McCoy Mountains Formation. The interstices of lava flow-top breccias at the top of the Dome Rock sequence were infilled and draped by quartzose sandstone of the McCoy Mountains Formation, apparently before enough time had elapsed for weathering to occur. This observation, in conjunction with new, highly precise U-Pb zircon data from the upper part of the Dome Rock sequence, constrains the age of the base of the previously poorly dated McCoy Mountains Formation to be early Late Jurassic.

U-Pb isotopic analyses of five zircon fractions from a tuff near the top of the Dome Rock sequence (Fig. 4) show evidence for a combination of inheritance of older zircon components plus significant later Pb loss (Fig. 5, Table 1). We applied a two-step dissolution technique (Mattinson, 1984) consisting of a 20-day, 80°C digestion step to remove the more disturbed parts of the zircon grains, followed by total digestion of the residue, to all five zircon fractions. This dissolution technique allows not only sepa-

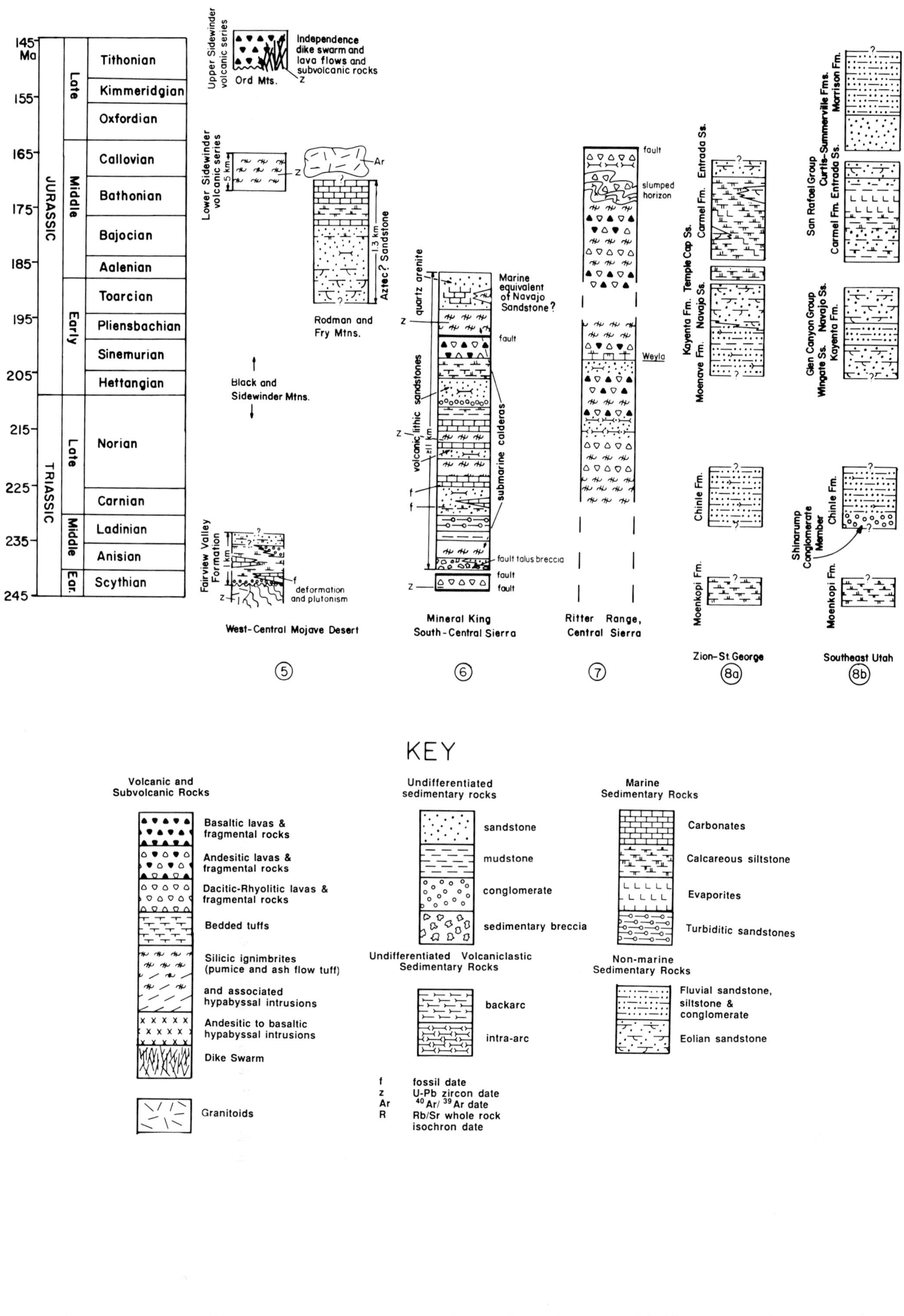
145 Ma
155
165
175
185
195
205
215
225
235
245
JURASSIC
TRIASSIC
Late
Middle
Early
Ear.
Tithonian
Kimmeridgian
Oxfordian
Callovian
Bathonian
Bajocian
Aalenian
Toarcian
Pliensbachian
Sinemurian
Hettangian
Norian
Carnian
Ladinian
Anisian
Scythian
Upper Sidewinder volcanic series
Independence dike swarm and lava flows and subvolcanic rocks
Ord Mts.
Lower Sidewinder volcanic series
~5 km
Ar
1.3 km
Aztec? Sandstone
Rodman and Fry Mtns.
Black and Sidewinder Mtns.
Fairview Valley Formation
1 km
deformation and plutonism
West-Central Mojave Desert
5
quartz arenite
Marine equivalent of Navajo Sandstone?
fault
volcanic lithic sandstones
≥11 km
submarine calderas
fault talus breccia
fault
fault
Mineral King South-Central Sierra
6
fault
slumped horizon
Weyla
Ritter Range, Central Sierra
7
Entrada Ss.
Carmel Fm.
Temple Cap Ss.
Navajo Ss.
Kayenta Fm.
Moenave Fm.
Chinle Fm.
Moenkopi Fm.
Zion–St. George
8a
Morrison Fm.
Curtis-Summerville Fms.
Entrada Ss.
Carmel Fm.
San Rafael Group
Navajo Ss.
Kayenta Fm.
Wingate Ss.
Glen Canyon Group
Chinle Fm.
Shinarump Conglomerate Member
Moenkopi Fm.
Southeast Utah
8b
KEY
Volcanic and Subvolcanic Rocks
Basaltic lavas & fragmental rocks
Andesitic lavas & fragmental rocks
Dacitic-Rhyolitic lavas & fragmental rocks
Bedded tuffs
Silicic ignimbrites (pumice and ash flow tuff)
and associated hypabyssal intrusions
Andesitic to basaltic hypabyssal intrusions
Dike Swarm
Granitoids
Undifferentiated sedimentary rocks
sandstone
mudstone
conglomerate
sedimentary breccia
Undifferentiated Volcaniclastic Sedimentary Rocks
backarc
intra-arc
f fossil date
z U-Pb zircon date
Ar 40Ar/39Ar date
R Rb/Sr whole rock isochron date
Marine Sedimentary Rocks
Carbonates
Calcareous siltstone
Evaporites
Turbiditic sandstones
Non-marine Sedimentary Rocks
Fluvial sandstone, siltstone & conglomerate
Eolian sandstone

rate analysis of the most soluble parts of the zircon grains and the less soluble residues, it also permits mathematical recombination of these results to give "bulk" or total fraction data equivalent to conventional zircon analyses. The four coarser recombined total fractions (open circles in Fig. 5) define a discordia trajectory with a lower intercept of about 155 Ma. However, the finest fraction clearly falls well below this trajectory, indicating significant Pb loss from this and possibly the other fractions. This is confirmed by results of the two step dissolution. Removal of the most easily dissolved and most discordant zircon domains (small solid diamonds in Fig. 5) leaves residues (solid circles in Fig. 5) that, in every case, show less Pb loss than do the total zircon fractions. The finest fraction still appears to show some Pb loss, but the four residues of the coarser fractions define a discordia trajectory with a lower intercept age (best crystallization age for the sample rock) of 157.7 ± 1.8 Ma (2σ error). The upper intercept age is poorly determined, owing to the leverage involved in the long projection to the upper intercept. However, at 1,470 ± 140, it clearly indicates the Precambrian age of the inherited zircons.

The first dissolution step removed most, if not all, of the common Pb from the zircon samples, as indicated by relatively low $^{206}Pb/^{204}Pb$ ratios (130 to 700), and hence, poorly determined Pb/Pb ages for this step. In contrast, $^{206}Pb/^{204}Pb$ ratios for the residues range from 17,000 to 50,000, permitting highly precise Pb/Pb determinations. In cases where the first step has a moderate error (e.g., the 200 to 300 and <300 mesh sizes, Table 1), an estimate of the time of isotopic disturbance might be obtained from lower concordia intercepts from residue-leach pairs. This yields mid–late Cretaceous ages, but uncertainties are too large to rule out a recent disturbance.

The geochronologic data and contact relations described here, along with gradational contacts between undated volcanic

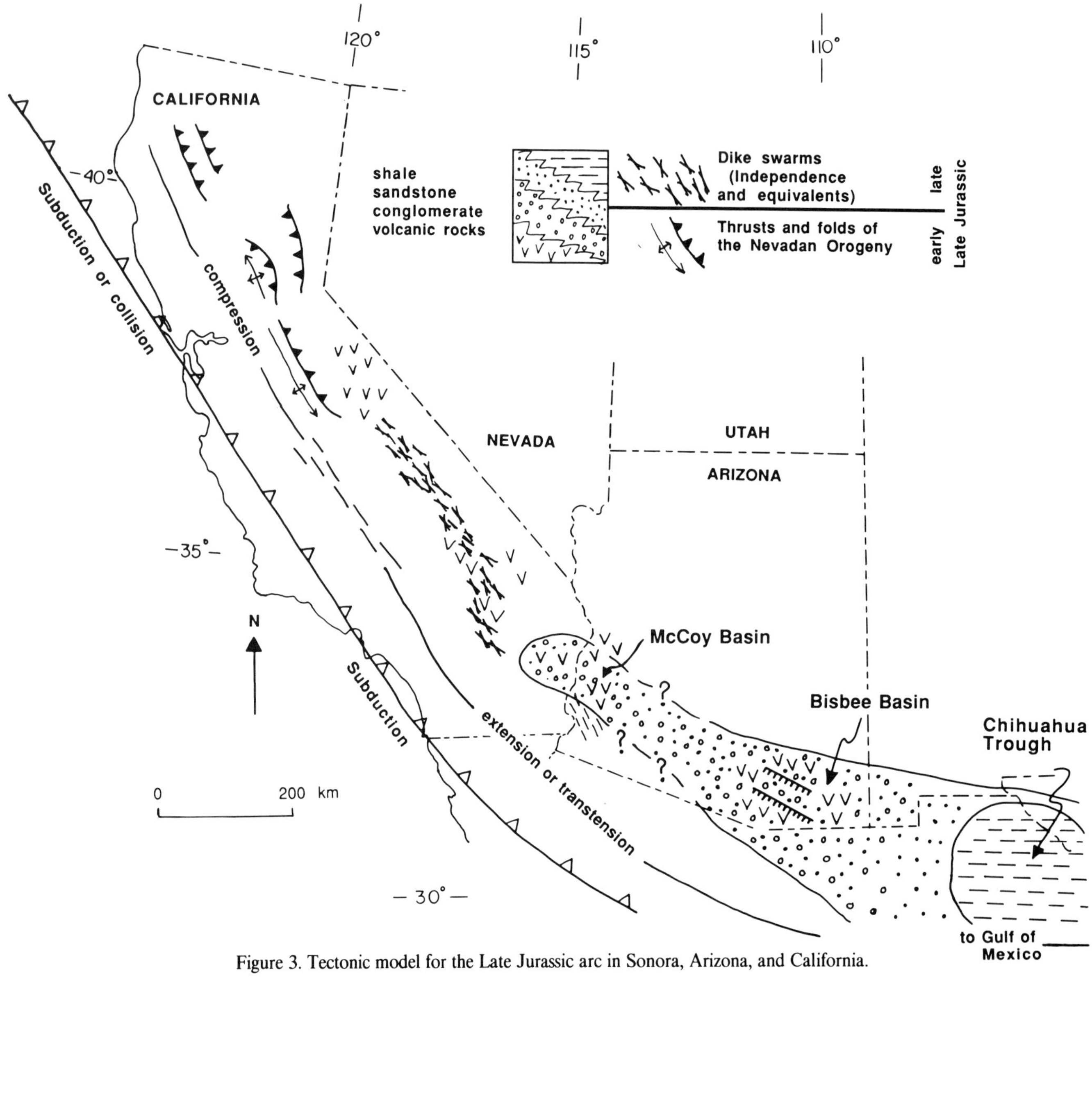

Figure 3. Tectonic model for the Late Jurassic arc in Sonora, Arizona, and California.

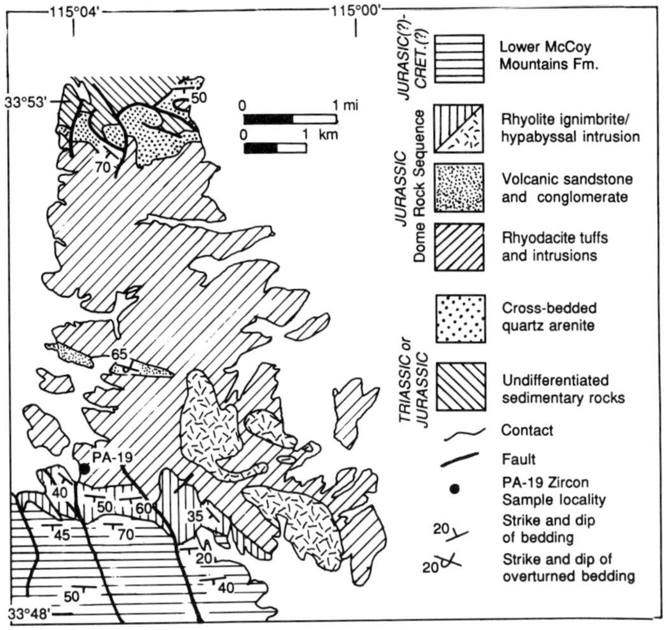

Figure 4. Geologic map of part of the Palen Mountains, southeastern California, simplified from Stone and Pelka (1989). Location plotted in Figure 1 (present-day coordinates given).

rocks and the overlying McCoy Mountains Formation (and equivalent units) in several other ranges (Tosdal and others, 1989), indicate that the lower unit of the McCoy Mountains Formation is Late Jurassic in age. This is similar in age to volcanic rocks interstratified with Glance Conglomerate in the Huachuca Mountains (U-Pb zircon age estimated at 160 to 170 Ma; Asmerom, 1988), at Canelo Pass (Rb-Sr whole-rock isochron of 151 ± 2 Ma; Kluth and others, 1982) and in the Santa Rita Mountains (Rb-Sr whole-rock isochron of 151 ± 5 Ma on Temporal Formation; Asmerom, 1988), all in southern Arizona. Our new data thus strengthen the correlation between the lower McCoy Mountains Formation and the Bisbee Basin fill (Fig. 3), and support the interpretation that rift sedimentation gradually replaced arc volcanism in Late Jurassic time. The tectonic setting of the upper unit of the McCoy Mountains Formation remains uncertain. This unit has yielded late Cretaceous fossil and isotopic ages (Stone and others, 1987; Tosdal and others, 1989) and has a provenance that is markedly less volcanic and more plutonic than that of the lower unit (Harding and Coney, 1985).

Mojave Desert Region

Hewett (1931) first applied the name Aztec Sandstone to 600 m of quartz arenite that overlies Triassic strata in the Spring Mountains of southern Nevada, and suggested it is correlative with the Navajo sandstone of the Colorado Plateau. He later extended this correlation to the Mescal Range in California, near the Nevada border (Fig. 2-3; Hewett, 1956). The name Aztec sandstone has since been applied to quartz arenites in several other ranges in the eastern to west-central Mojave Desert (Fig. 2-4, 2-5); these arenites were tentatively correlated with the Lower Jurassic Navajo sandstone. New isotopic age data, however, indicate that eolian quartz arenites in the Cowhole Mountains (Figs. 1, 2-4) are age-equivalent to the Carmel Formation of the Middle Jurassic San Rafael Group (Fig. 2-8). The presence of eolianites at some localities in the Carmel Formation supports this correlation (Petersen and Pipiringos, 1979; Kocurek and Dott, 1983; F. Petersen, 1989, personal communication). U-Pb zircon ages in the Cowhole Mountains are 172 ± 5 Ma on a lava flow within the Aztec sandstone, and 167 ± 1 Ma in the conformably overlying volcanic section, which contains sandstone with an eolian sand component (Busby-Spera and others, 1989b, and unpublished data). These ages overlap with $^{40}Ar/^{39}Ar$ single crystal laser probe age dates on bentonites from several horizons in the Carmel Formation, which range from 170 to 162 Ma (Everett and others, 1989; B. Kowallis, personal communication). Age controls on eolian quartz arenites in all other ranges in the Mojave Desert are too poor to determine whether they are Wingate-, Navajo-, Carmel-, or Entrada-equivalent in age.

Middle Jurassic syndepositional faulting in the Cowhole Mountains resulted in erosional stripping of a 125-m-thick ignimbrite from the upthrown side of a high-angle fault, leaving behind a horizon of ignimbrite cobbles that overlaps the fault and was deposited above the ignimbrite on the downthrown block. An immediately overlying megabreccia contains slabs that were shed from an inferred fault beyond the limits of present-day exposures. The megabreccia is encased within an ignimbrite that is

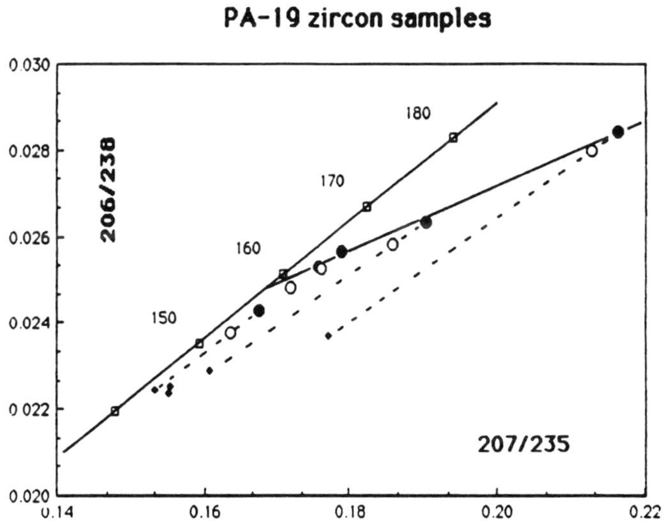

Figure 5. Enlarged portion of a concordia diagram showing isotopic data for leach (solid triangles), residue (solid circles), and total (open circles) fractions of zircon from the upper part of the Dome Rock sequence in the Palen Mountains. The coarsest fraction plots on the upper right side of the diagram; the finest, at the lower left. The concordia curve is plotted through small open squares. The dashed lines connect residue-total-leach data for individual fractions (only three shown to avoid crowding). The solid discordia line is a best fit to the residues (solid circles), which yields a crystallization age of 157.7 ± 1.8 Ma (2 σ error).

TABLE 1. ZIRCON ISOTOPIC AGE DATA

Sample	Fraction* (μm)		Amount (mg)	Concentration ^{238}U (ppm)	$^{206}Pb^*$ (ppm)	Atomic Ratios ^{206}Pb§/^{204}Pb	$^{206}Pb^*/^{238}U$	$^{207}Pb^*/^{238}U$	$^{207}Pb^*/^{206}Pb^*$	Isotopic Ages** $^{206}Pb^*/^{238}U$	$^{207}Pb^*/^{235}U$	$^{207}Pb^*/^{206}Pb^*$
PA-19-I	120-165	Res	3.6	427.8	10.52	16,600	0.02842	0.2164	0.05523	180.7	198.9	422 ± 2
		Lch	3.6	43.20	0.887	123.3	0.02368	0.1764	0.05402	151	165	372 ± 90
		Tot	3.6	471.0	11.41	1,267	0.02798	0.2128	0.05516	177.9	195.9	419 ± 10
PA-19-II	120-165	Res	11.7	407.6	9.29	18,023	0.02634	0.1903	0.05239	167.6	176.9	302 ± 2
		Lch	11.7	71.9	1.425	131.1	0.02289	0.1600	0.05069	146	151	227 ± 90
		Tot	11.7	479.5	10.72	829	0.02581	0.1858	0.05219	164.3	173.0	294 ± 15
PA-19-III	80-120	Res	10.6	451.6	10.02	34,540	0.02565	0.1788	0.05055	163.3	167.0	220 ± 2
		Lch	10.6	58.1	1.124	353.2	0.02237	0.1549	0.05022	143	146	205 ± 30
		Tot	10.6	509.7	11.14	3,067	0.02527	0.1760	0.05051	160.9	164.6	219 ± 5
PA-19-IV	45-80	Res	10.5	504.5	11.05	55,330	0.02531	0.1755	0.05030	161.1	164.2	209 ± 2
		Lch	10.5	111.5	2.173	697.7	0.02252	0.1553	0.05001	144	147	195 ± 15
		Tot	10.5	616.0	13.22	3,906	0.02480	0.1718	0.05025	157.9	161.0	206 ± 4
PA-19-V	<45	Res	6.7	553.2	11.63	50,960	0.02428	0.1674	0.05001	154.7	157.2	195 ± 2
		Lch	6.7	203.7	3.955	656.7	0.02243	0.1533	0.04958	143	145	175 ± 16
		Tot	6.7	756.9	15.58	2,439	0.02378	0.1637	0.04991	151.5	153.5	191 ± 6

*Radiogenic Pb; correction used for common Pb is $^{206}Pb/^{204}Pb = 19.0$, $^{207}Pb/^{204}Pb = 15.6$.
†Numbers refer to standard mesh sizes for zircon fractions. Res = residue after 20 days, 80°C dissolution step; Lch = data from 20 day step; Tot = Res + Lch = original bulk zircon.
§Observed 206/204 ratio, corrected for mass fractionation.
**Calculated ages in Ma. Uncertainties in 207/206 ages based on Mattinson, 1987.

nonwelded and thus probably does not represent an intracaldera accumulation. Instead, we infer that catastrophic landslides were triggered along a graben boundary fault by seismicity during the eruption of the ignimbrite (Busby-Spera and others, 1989b).

Normal faults that juxtapose the lower Sidewinder volcanic series against older rocks in the Black Mountain–Sidewinder Mountain area of the central Mojave Desert may also be Middle Jurassic in age (Schermer and Busby-Spera, 1989, 1990). The lower Sidewinder volcanic series in this area has yielded a preliminary U-Pb zircon age of 170 Ma (Graubard and others, 1988; and our unpublished data). Quartz monzonites of the Middle Jurassic plutonic suite (Fig. 2-5), as well as dikes that are probably part of the Late Jurassic Independence dike swarm, cross-cut the normal faults. Normal faulting thus probably occurred during and/or immediately after deposition of the lower Sidewinder volcanic series. Regional subsidence during lower Sidewinder volcanism may be indicated by outflow sequences greater than 1 km thick, but about 80 percent of the approximately 4.5-km-thick section at Sidewinder Mountain consists of 800- to 1,400-m-thick ignimbrites that probably represent intracaldera accumulations (Schermer and Busby-Spera, 1990).

We suggest that in Late Jurassic time the central to eastern Mojave segment of the arc occupied a transitional tectonic position along the plate margin (Fig. 3). To the north, in the Klamath-Sierran segment of the arc, thrust faulting and folding occurred from about 160 to 150 Ma (Nevadan Orogeny), probably in response to strongly coupled subduction or terrane collision (Saleeby and Busby-Spera, 1990). During the same time frame, to the south (Fig. 3), the McCoy and Bisbee Basins formed along the axis of the arc, probably in response to the opening of the Gulf of Mexico, as discussed above. We speculate that extensional tectonics continued to be important in the Mojave Desert through Jurassic time and that Late Jurassic shortening was not as widespread or intense in the Mojave Desert as it was in the Klamath-Sierran region. New evidence supports this interpretation. One piece of evidence is the concordant bedding above and below the contact between Early or Middle Jurassic eolianites and probable Cretaceous volcanic rocks in the eastern Mojave Desert. Eolian quartz arenite of Late Jurassic or Cretaceous age does not occur on the Colorado Plateau (Fig. 2-8), suggesting that climatic conditions that favor their development had ended by that time (Kocurek and Dott, 1983).In the Soda Mountains (Fig. 2-4), eolianites are conformably overlain by an ignimbrite that has yielded an early Cretaceous U-Pb zircon age (Busby-Spera and others, 1990). In the Mescal Range (Fig. 2-3), the eolianites are conformably overlain by the Delfonte volcanics; our preliminary U-Pb zircon data from the Delfonte volcanics are consistent with the interpretation of Fleck and Carr (1990) that these rocks are 98 m.y. old. Other evidence is that, in the central Mojave Desert, the late Late Jurassic Independence dike swarm and upper Sidewinder volcanic series, as well as plutonic rocks of the Middle Jurassic suite, all contain a locally well-developed foliation with the same orientation as that developed in the lower Sidewinder volcanics and the Fairview Valley Formation (Fig. 2-5; Schermer and Busby-Spera, 1989, 1990). This suggests that shortening in this part of the arc may not have begun until after Jurassic time, although previous workers have inferred a Middle Jurassic synplutonic contractional event in this region (Miller, 1978; Karish and others, 1987). Our new stratigraphic and structural data from the eastern to central Mojave Desert, in summary, suggest that >140- to 150-Ma metamorphism in the Halloran Hills of the eastern Mojave Desert (De Wit and others, 1984), as well as thrust faulting and folding between 166 and 155 Ma in the Cronese Hills of the east-central Mojave Desert (Walker and others, 1989, 1990), were not widespread events.

MIOGEOCLINAL ARC STRATIGRAPHY AND STRUCTURE

The segment of the Triassic-Jurassic magmatic arc now preserved in the southern and central Sierra Nevada and in the Inyo Range was constructed on the Paleozoic–Early Triassic miogeoclinal continental margin, in a dominantly shallow-marine environment (Fig. 2-6, 2-7). Local uplift to a terrestrial setting, however, accompanied the Early(?) or Middle(?) Triassic onset of arc volcanism in the Inyo Range (Dunne and others, 1978), shown as an island chain in Figure 1. The stratigraphic evolution of the Jurassic-Triassic arc is best documented in the Sierra Nevada, where marine fossils, as well as U-Pb zircon dates on metavolcanic rocks, provide age controls.

Southern and central Sierra Nevada

Upper Triassic through Lower Jurassic sequences in the Mineral King and Ritter Range pendants of the High Sierra represent the northern extension of the arc graben depression into the marine realm (Busby-Spera, 1988; Figs. 1, 2-6, 2-7). During this time interval, rapid subsidence kept pace with high sedimentation rates, resulting in accumulation of approximately 10 km of dominantly shallow-marine strata in the region (Fiske and Tobisch, 1978; Busby-Spera, 1984a). Additional, localized subsidence occurred at Mineral King as the result of four different caldera-forming eruptions (Busby-Spera, 1984b), and a syndepositional normal fault and associated fault talus breccia near the base of the section (Fig. 2-6) may record localized incipient ring fracturing rather than regional extension. Texturally mature quartz arenites occur in the upper part of the Mineral King section (Fig. 2-6), conformably above an ignimbrite dated at approximately 190 Ma (U-Pb zircon, Busby-Spera and Saleeby, 1987). This indicates that craton-derived sands age-equivalent to the Navajo sandstone (Fig. 2-8) gained access to the arc-graben depression in the miogeoclinal setting.

Triassic and Lower Jurassic strata of the Mineral King pendant record periods of volcanic activity characterized by large-volume caldera-forming eruptions of rhyolite ash-flow tuff (Busby-Spera, 1984b), construction of small-volume andesite stratocones (Busby-Spera, 1986) and growth of submarine fans and debris aprons (Busby-Spera, 1985). These periods alternated with periods of local volcanic and tectonic quiescence, characterized by deposition of fine-grained sediment and ash, or progradational shelf sequences. Water depths fluctuated from approximately 10 to >500 m. During periods of local volcanic quiescence, subsidence was even and continuous, whereas the parts of the section that contain proximal volcanic facies (such as calderas) record extreme fluctuations in relative sea level. Nearshore to outer shelf sediment accumulated along a high-energy ocean-facing coastline, and deposition occurred largely during storms (Busby-Spera, 1984a). Basinal metasedimentary rocks at Mineral King lack wave-generated sedimentary structures and include: (1) a quiescent basinal facies of parallel-laminated distal silicic tuffs that rained through the water column and were variably contaminated by carbonate mud; (2) a debris apron facies of polymictic breccias and pyroclastic flows shed from an incipient "leaky" ring fracture prior to a major caldera-forming eruption (Busby-Spera, 1984b); (3) a sand-rich turbidite facies that formed by downslope reworking of silicic pyroclastic debris supplied by contemporaneous volcanic eruptions; and (4) a starved basin facies, composed of pyritic shales and rare distal turbidites deposited in a restricted environment.

Triassic-Jurassic intra-arc facies patterns in the southern Sierra are complex due to original, rapid lateral variation (Saleeby and Busby-Spera, 1990). The Mineral King pendant preserves the proximal volcanic facies of the Jurassic-Triassic Kings sequence of Saleeby and Busby-Spera (1978). Pendants to the north, west, and south of Mineral King contain shallow-water quart arenite and carbonate, as well as deep-water quartzose to volcanic lithic turbidites and olistostromes, interstratified with distal subaqueous fallout tuff and tuffaceous quartzose sandstone (Saleeby and Busby-Spera, 1990). The unusual mixture of volcanic arc and mature continental detritus and interstratified shallow-water carbonates is consistent with accumulation of the Kings sequence along the truncated edge of the miogeocline (Saleeby and others, 1978; Saleeby and Busby-Spera, 1990).

EUGEOCLINAL ARC STRATIGRAPHY AND STRUCTURE

We recognize two important paleogeographic-paleotectonic links between the southern continental (cratonal and miogeoclinal) arc segment described in this chapter and the northern, eugeoclinal arc segment that formed on island-arc basement (Fig. 1): (1) Lower to Middle(?) Jurassic quartz arenites were deposited within the arc in the Northern Sierra terrane (Fisher, 1989, and this volume) and in adjacent western Nevada (Dilles and Wright, 1988), and (2) accumulation and preservation of thick Upper Triassic and Lower to Middle(?) Jurassic marine arc vol-

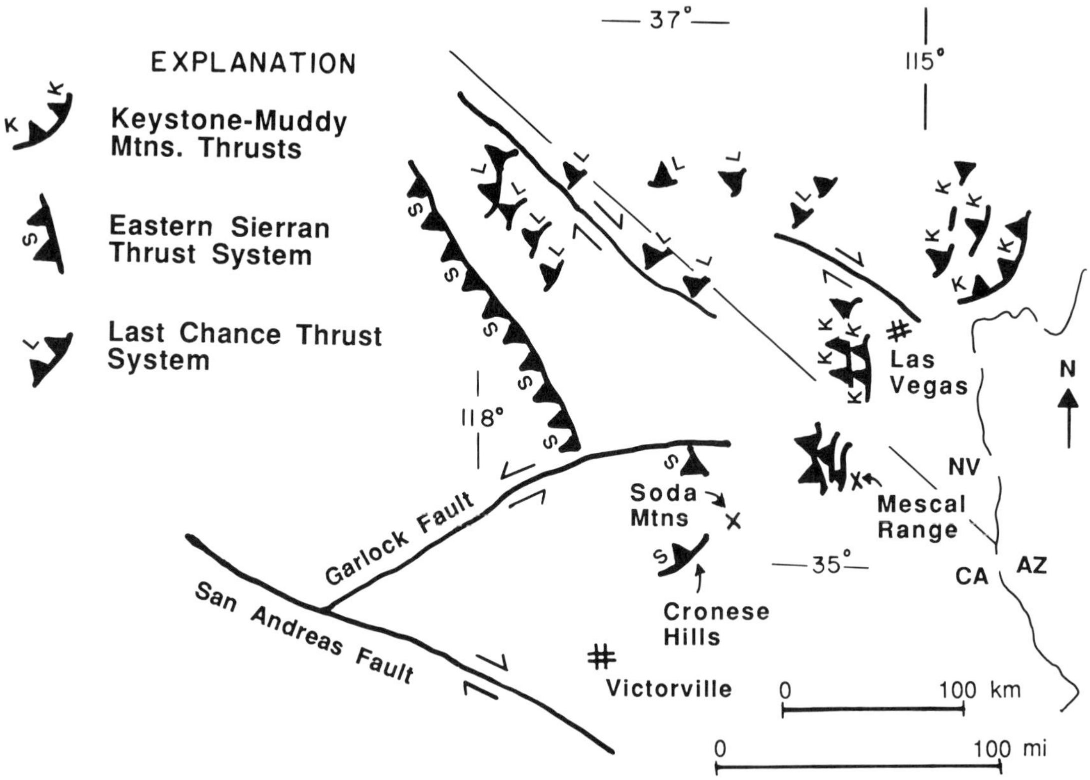

Figure 6. Generalized map of known and inferred Permian(?) to Mesozoic thrust faults in east-central California and southern Nevada, modified from Burchfiel and Davis (1981) and Dunne (1986). (Present-day coordinates given.)

canic and sedimentary sequences in the northern Sierra Nevada and adjacent Nevada, as well as in the Klamath Mountains, is consistent with an extensional arc setting (Wright and Miller, 1988). We thus infer that the arc-graben depression of Figure 1 extended more or less continuously from northern Sonora to the California-Oregon border, and that it acted as a trap for craton-derived quartz arenites along most of its length.

Lower Jurassic quartz arenite in the northern Sierra occurs in the Mt. Tallac roof pendant southwest of Lake Tahoe (Fisher, 1989, and this volume) and in an inlier surrounded by Cenozoic volcanic rocks in Blackwood Canyon west of Lake Tahoe (D. S. Harwood, personal communication, 1988). At both localities the quartz arenite conformably overlies, in gradational contact, a black slate-limestone unit with "probable Pliensbachian" ammonites. This black slate-limestone unit also occurs in the type area of the Sailor Canyon Formation (American River to Interstate 80), where it has yielded Pliensbachian ammonites. The quartzarenites at Mt. Tallac and Blackwood Canyon are considered to be intra-arc deposits because the black slate-limestone unit that underlies them is interstratified with arc volcanic rocks in the type area of the Sailor Canyon Formation, and because tuffs occur in the transition between the black slate-limestone unit and the quartz arenite unit in the Mt. Tallac pendant (D. S. Harwood, personal communication, 1988; Fisher, this volume).

Fisher (this volume) infers that regional subsidence in the northern Sierra Nevada occurred in Late Triassic to Early Jurassic time, and that, in Middle Jurassic time, extensional fault troughs formed. Regional extension also affected the inboard side of the arc in western Nevada where Early to Middle(?) Jurassic northwest-trending half grabens formed; quartz arenites of the Dunlap Formation, and possibly the Boyer Ranch Formation, were deposited in these half-grabens (Oldow and Bartel, 1987). Subsidence is also inferred to have controlled intra-arc sedimentation in the Black Rock Desert area of northwestern Nevada in Late Triassic time (Wyld, 1989; this volume). Lower to Middle(?) Jurassic quartz arenite of the Preachers Formation in the Yerrington District of west-central Nevada occurs in an 1,800-m-thick section of Late Triassic through Early Jurassic shallow marine clastic rocks, carbonate, evaporite, tuff, and tuff breccia (Dilles and Wright, 1988). The quartz arenite occurs above a Toarcian fossil locality and is thus similar in age to quartz arenite of the northern Sierra and to the Navajo Sandstone of the Colorado Plateau.

DISCUSSION AND CONCLUSIONS

Stratigraphic, structural and geochronologic data summarized here support the interpretation that the Late Triassic to early

Middle Jurassic arc of the Klamath-Sierran and Mojave-Sonoran regions occupied a low-lying structural depression that acted as a trap for craton-derived sands. Modern analogs to the arc-graben depression include the Taupo Volcanic Zone (Cole, 1984), the Central American arc (Burkart and Self, 1985), the Trans-Mexican Volcanic Belt (Allan and others, 1990), and the Kamchatka arc (Erhlich, 1979). All of these modern arcs are under regional extension, and fold and thrust belts are absent from the arc to back-arc region.

Other workers have suggested that the arc to back-arc region of the southwest Cordillera was undergoing shortening during Late Triassic to Early Jurassic time (Burchfiel and Davis, 1981; Dunne, 1986), but we know of no firm evidence for compressional deformation during this time frame. The first indications of thrust faulting, foreland subsidence patterns, and attendant synorogenic sedimentation documented in Idaho, Wyoming, Utah, and Nevada are Middle to Late Jurassic in age (see references in Heller, 1986). Thrust faults in southern Nevada and east-central to southeastern California (Fig. 6), interpreted by Burchfiel and Davis (1981) and Dunne 91986) as Late Triassic to Early Jurassic, can be reinterpreted as either older or younger structures for the following reasons:

1. The Last Chance thrust system involves strata as young as Lower Permian and is intruded by Lower Jurassic plutons (Dunne, 1986). Two thrust faults documented as Permian in age by Stevens and Stone (1988) may be related to this system. Structural correlations and stratigraphic interpretations by Snow (1990) suggest that the Last Chance thrust system predates a 228 ± 2 Ma intrusion and is middle Permian in age. Permian to Early Triassic thrusting is also well documented in the San Bernardino Mountains and the west-central Mojave Desert, and the effects of this shortening event may be traceable to the east-central Mojave Desert (summarized by Walker, 1988).

2. The Contact and Keystone thrust plates of the Clark Mountain thrust complex (Fig. 6), previously interpreted as late Triassic to early Jurassic in age (Carr and others, 1986), are now known to override synorogenic deposits of mid-Cretaceous age (Fleck and Carr, 1990).

3. The eastern Sierran thrust system (Fig. 6) generally cuts Middle to Late Jurassic plutons, as well as the 148-Ma Independence dike swarm (Dunne, 1986), and is thus largely latest Jurassic or younger. The Late Jurassic (165 to 155 Ma) Cronese Hills shear zone of the central Mojave Desert (Fig. 6) may be part of this system (Walker and others, 1989, 1990).

In summary, the available data from California, Nevada, and Arizona support our interpretation that shortening along the continental margin ended before subduction began, and that the length of the arc–back-arc region was dominated by extension from its Triassic inception until some time in the Middle Jurassic. Further work may determine exactly how synchronous or time-transgressive extensional structural provinces were along the length of the Late Triassic to Middle Jurassic arc, but tectonic regimes appear to have been broadly similar for both the Klamath-Sierran and Mojave-Sonoran segments in that time interval. By Late Jurassic time, the two major segments of the arc had diverged in character, with the Mojave-Sonoran segment of the arc coming under the influence of Atlantic margin extensional-transtensional tectonics (Dickinson and others, 1986; Busby-Spera and others, 1989a), while the Klamath-Sierran segment responded to collision or strongly coupled subduction along the Pacific margin (Saleeby and Busby-Spera, 1990).

ACKNOWLEDGMENTS

This research was supported by National Science Foundation Grants EAR-8519124 and EAR-8803769 (both to C.J.B.-S. and J.M.M.). We gratefully acknowledge stimulating discussions over the past several years with W. Dickinson, G. Dunne, D. S. Harwood, G. Haxel, E. Miller, F. Petersen, J. Saleeby, P. Stone, R. Tosdal, D. Walker, and J. Wright. We thank E. Miller and M. Carr for their reviews.

REFERENCES CITED

Allan, J. F., Nelson, S. A., Luhr, J. F., Carmichael, I.S.E., and Wopat, M., 1990, Pliocene-Recent rifting in southwest Mexico and associated alkaline volcanism, *in* Dauphin, J. P., ed., The Gulf and Peninsular province of the Californias: American Association of Petroleum Geologists Memoir (in press).

Asmerom, Y., 1988, Mesozoic igneous activity in the southern Cordillera of North America; Implication for tectonics and magma genesis [Ph.D. thesis]: Tucson, University of Arizona, 232 p.

Asmerom, Y., Zartman, R. E., Damon, P. E., and Shafiqullah, M., 1988, U-Th-Pb zircon ages from Santa Rita Mountains, southeast Arizona; Implications for timing of early Mesozoic magmatism: Geological Society of America Abstracts with Programs, v. 20, p. 140.

Barth, A. P., Tosdal, R. M., and Wooden, J. L., 1988, Characteristics and implications of Triassic and Jurassic granotoids in the San Gabriel, Mule, and Trigo Mountains, southern California and southwestern Arizona: Geological Society of America Abstracts with Programs, v. 20, p. 141.

Blakey, R. C., Peterson, F., Caputo, M. V., Geesaman, R. C., and Voorhees, B. J., 1983, Paleogeography of Middle Jurassic continental, shoreline, and shallow marine sedimentation, southern Utah, *in* Reynolds, M. W., and Dolly, E. D., eds., Mesozoic paleogeography of west-central United States: Rocky Mountain Section, Society of Economic Paleontologists and Mineralogists Rocky Mountain Geology Symposium 2, p. 77–100.

Burchfiel, B. C., and Davis, G. A., 1972, Structural framework and evolution of the southern part of the Cordilleran orogen, western United States: American Journal of Science, v. 272, p. 97–118.

—— , 1981, Mojave Desert and environs, *in* Ernst, W. G., ed., The geotectonic development of California; Rubey Volume 1: Englewood Cliffs, New Jersey, Prentice-Hall, p. 217–252.

Burkart, B., and Self, S., 1985, Extension and rotation of crustal blocks in northern Central America and effect on the volcanic arc: Geology, v. 13, p. 22–26.

Busby-Spera, C. J., 1984a, The lower Mesozoic continental margin and marine intra-arc sedimentation at Mineral King, California, *in* Bachman, S., and

Crouch, J., eds., Tectonics and sedimentation along the California margin; Field Trip Guidebook: Pacific Section, Society of Economic Paleontologists and Mineralogists, p. 135–156.

——, 1984b, Large-volume rhyolite ash-flow eruptions and submarine caldera collapse in the lower Mesozoic Sierra Nevada, California: Journal of Geophysical Research, v. 89, no. B10, p. 8417–8427.

——, 1985, A sand-rich submarine fan in the lower Mesozoic Mineral King caldera complex, Sierra Nevada, California: Journal of Sedimentary Petrology, v. 55, p. 376–391.

——, 1986, Depositional features of rhyolitic and andesitic volcaniclastic rocks of the Mineral King submarine caldera complex, Sierra Nevada, California: Journal of Volcanology and Geothermal Research, v. 27, p. 43–76.

——, 1988, Speculative tectonic model for the Lower Mesozoic arc of the southwest Cordilleran United States: Geology, v. 16, p. 1121–1125.

Busby-Spera, C. J., and Saleeby, J. B., 1987, Geologic guide to the Mineral King area, Sequoia National Park, California: Pacific Section, Society of Economic Paleontologists and Mineralogists Field Trip Guidebook, 44 p.

Busby-Spera, C. J., Mattinson, J. M., and Riggs, N. R., 1987, Lower Mesozoic extensional continental arc, Arizona and California; A depocenter for craton-derived quartz arenites: Geological Society of America Abstracts with Programs, v. 19, p. 607.

Busby-Spera, C. J., Mattinson, J. M., and Adams, B., 1989a, The Jurassic transition from continental intra-arc extension to Gulf of Mexico; Related rifting in the southwest Cordillera: EOS Transactions of the American Geophysical Union, v. 70, no. 43, p. 1300.

Busby-Spera, C. J., Schermer, E. R., and Mattinson, J. M., 1989b, Volcanotectonic controls on sedimentation in an extensional continental arc; A Jurassic example from the eastern Mojave Desert, California; International Association of Volcanology and Chemistry of the Earth's Interior, Santa Fe, New Mexico, June 1981: New Mexico Bureau of Mines and Mineral Resources Bulletin 131, p. 34.

Busby-Spera, C. J., Mattinson, J. M., and Schermer, E. R., 1990, Stratigraphic and tectonic evolution of the Jurassic arc; New field and U-Pb zircon geochronological data from the Mojave Desert: Geological Society of America Abstracts with Programs, v. 22, p. 11.

Carr, M. D., Evans, K. V., Fleck, R. J., Frizzell, V. A., Ort, K. M., and Zartman, R. E., 1986, Early middle Jurassic upper limit for movement on the Keystone Thrust, southern Nevada: Geological Society of America Abstracts with Programs, v. 18, no. 5, p. 345.

Cole, J. W., 1984, Taupo-Rotorua depression; An ensialic marginal basin of the North Island, New Zealand, *in* Kokelaar, B., and Howells, M., eds., Marginal basin geology: Geological Society of London Special Publication 16, p. 109–120.

De Witt, E., Armstrong, R. L., Sutter, J. F., and Zartman, R. E., 1984, U-Th-Pb, Rb-Sr, and Ar-Ar mineral and whole-rock isotopic systematics in a metamorphosed granitic terrane, southeastern California: Geological Society of America Bulletin, v. 95, p. 723–739.

Dickinson, W. R., Klute, M. A., and Swift, P. N., 1986, The Bisbee Basin and its bearing on Late Mesozoic paleogeographic and paleotectonic relations between the Cordilleran and Caribbean regions, *in* Abbott, P. L., ed., Cretaceous stratigraphy, western North America: Pacific Section, Society of Economic Paleontologists and Mineralogists Field Trip Guidebook 46, p. 51–62.

Dickinson, W. R., Klute, M. A., and Bilodeau, W. L., 1987, Tectonic setting and sedimentological features of upper Mesozoic strata in southeastern Arizona; Geological Society of America Cordilleran Section Field Trip Guidebook: Arizona Bureau of Geology and Mineral Technology Bulletin, p. 166–279.

Dilles, J. H., and Wright, J. E., 1988, The chronology of early Mesozoic arc magmatism in the Yererington district of western Nevada and its regional implications: Geological Society of America Bulletin, v. 100, p. 644–652.

Dunne, G. C., 1977, Geology and structural evolution of Old Dad Mountain, Mojave Desert, California: Geological Society of America Bulletin, v. 88, p. 737–748.

——, 1986, Geologic evolution of the southern Inyo Range Darwin Plateau, and Argus and Slate Ranges, east-central California—An overview, *in* Dunne, G. C., eds., Mesozoic and Cenozoic structural evolution of selected areas, east-central California; Geological Society of America Cordilleran Section Field Trip Guide, Field Trip 2: California State University, Los Angeles, p. 3–21.

Dunne, G. C., Gulliver, R. M., and Sylvester, A. G., 1978, Mesozoic evolution of rocks of the White, Inyo, Argus, and Slate Ranges, eastern California, *in* Howell, D. G., and McDougall, K. A., eds., Mesozoic paleogeography of the western United States: Pacific Section, Society of Economic Paleontologists and Mineralogists Pacific Coast Paleogeography Symposium 2, p. 189–208.

Erlich, E. N., 1979, Recent structure of Kamchatka and position of Quaternary volcanoes: Bulletin Volcanologique, v. 42, p. 13–42.

Everett, B. H., Kowallis, B. J., and Christiansen, E. H., 1989, Correlation of Jurassic sediments of southern Utah using bentonite characteristics: Geological Society of America Abstracts with Programs, v. 21, p. 77.

Fisher, R., 1989, The Mt. Tallac roof pendant, California; Mesozoic stratigraphic link between the northern Sierra terrane and west-central Nevada: Geological Society of America Abstracts with Programs, v. 21, p. 78.

Fiske, R. S., and Tobisch, O. T., 1978, Paleogeographic significance of volcanic rocks of the Ritter Range pendant, central Sierra Nevada, California, *in* Howell, G., and McDougall, K. A., eds., Mesozoic paleogeography of the western United States: Pacific Section, Society of Economic Paleontologists and Mineralogists Pacific Coast Paleogeography Symposium 2, p. 209–222.

Fleck, R. J., and Carr, M. D., 1990, The age of the Keystone thrust; Laser-fusion $^{40}Ar/^{39}Ar$ dating of foreland basin deposits, southern Spring Mountains, Nevada: Tectonics, v. 20, p. 467–476.

Graubard, C. M., Mattinson, J. M., and Busby-Spera, C. J., 1988, Age of the lower Sidewinder volcanics and reconstruction of the early Mesozoic arc in the Mojave Desert, California: Geological Society of America Abstracts with Programs, v. 20, p. A274–A275.

Grose, L. T., 1959, Structure and petrology of the northeast part of the Soda Mountains, San Bernardino County, California: Geological Society of America Bulletin, v. 70, p. 1509–1548.

Harding, L. E., and Coney, P. J., 1985, The geology of the McCoy Mountains Formation, southeastern California and southwestern Arizona: Geological Society of America Bulletin, v. 96, p. 755–769.

Haxel, G., Briskey, J., May, D., Tosdal, R., and Wright, J., 1985, Reconnaissance petrochemistry of Jurassic igneous rocks, Baboquivari and Comobabi Mountains area, southern Arizona: Geological Society of America Abstracts with Programs, v. 17, p. 361.

Heller, P. L., and 7 others, 1986, Timing of initial thrusting in the Sevier orogenic belt, Idaho-Wyoming and Utah: Geology, v. 14, p. 388–391.

Hewett, D. F., 1931, Geology and ore deposits of the Goodsprings Quadrangle, Nevada: U.S. Geological Survey Professional Paper 162, 172 p.

——, 1956, Geology and mineral resources of the Ivanpah Quadrangle, California and Nevada: U.S. Geological Survey Professional Paper 275, 172 p.

Imlay, R. W., 1980, Jurassic paleobiogeography of the coterminus United States in its continental setting: U.S. Geological Survey Professional Paper 1062, 134 p.

Karish, C. R., Miller, E. L., and Sutter, J. F., 1987, Mesozoic tectonic and magmatic history of the central Mojave Desert, *in* Dickinson, W., and Klute, M., eds., Mesozoic rocks of southern Arizona and adjacent regions: Arizona Geological Society Digest, v. 18, p. 15–32.

Kluth, C. F., Butler, R. F., Harding, L. E., Shafiqullah, M., and Damon, P. E., 1982, Paleomagnetism of Late Jurassic rocks in northern Canelo Hills, southeastern Arizona: Journal of Geophysical Research, v. 87, p. 7079–7086.

Kocurek, G., and Dott, R. H., Jr., 1983, Jurassic paleogeography and paleoclimate of the central and southern Rocky Mountains region, *in* Reynolds, M. W., and Dolly, E. D., eds., Mesozoic paleogeography of west-central United States: Rocky Mountain Section, Society of Economic Paleontologists and Mineralogists Rocky Mountain Paleogeography Symposium 2, p. 101–116.

Marzolf, J. E., 1983, Early Mesozoic eolian transition from cratonal margin to orogenic-volcanic arc: Utah Geological and Mineral Survey Special Studies 60, p. 39–46.

Mattinson, J. M., 1984, Labile Pb in young, high-uranium zircons: Geological Society of America Abstracts with Programs, v. 16, p. 559.

—— , 1987, U-Pb ages of zircons; A basic examination of error propagation: Chemical Geology, v. 66, p. 151–162.

Miller, E. L., 1978, Geology of the Victorville region, California [Ph.D. thesis]: Houston, Texas, Rice University, 226 p.

Miller, E. L., and Cameron, C. S., 1982, Late Precambrian to Late Cretaceous evolution of the southwestern Mojave Desert, California, *in* Cooper, J. P., Troxel, B. W., and Wright, L. A., eds., Geology of selected areas in the San Bernardino Mountains, western Mojave Desert and southern Great Basin, California; Geological Society of America Cordilleran Section Field Trip Guide, Field Trip 9: p. 21–34.

Miller, E. L., and Carr, M., 1978, Recognition of possible Aztec-equivalent sandstones and associated Mesozoic metasedimentary deposits within the Mesozoic magmatic arc in the southwestern Mojave Desert, California, *in* Howell, D. G., and McDougall, K. A., eds., Mesozoic paleogeography of the western United States: Pacific Section, Society of Economic Paleontologists and Mineralogists Pacific Coast Paleogeography Symposium 2, p. 283–289.

Novitsky-Evans, J. M., 1978, Geology of the Cowhole Mountains, southern California; Structural, stratigraphic, and geochemical studies [Ph.D. thesis]: Houston, Texas, Rice University, 95 p.

Oldow, J. S., and Bartel, R. L., 1987, Early to Middle(?) Jurassic extensional tectonism in the western Great Basin; Growth faulting and synorogenic deposition of the Dunlap Formation: Geology, v. 15, p. 740–743.

Pelka, G. J., 1973, Geology of the McCoy and Palen Mountains, southeastern California [Ph.D. thesis]: Santa Barbara, University of California, 162 p.

Petersen, F., and Pipiringos, G. N., 1979, Stratigraphic relationships of the Navajo Sandstone to Middle Jurassic formations, southern Utah and northern Arizona: U.S. Geological Survey Professional Paper 1036-B, 43 p.

Reynolds, S. J., Richard, S. M., Haxel, G. B., Tosdal, R. M., and Laubach, S. E., 1988, Geologic setting of Mesozoic and Cenozoic metamorphism in Arizona, *in* Ernst, W. G., ed., Metamorphism and crustal evolution of the western United States; Rubey Volume 7: Englewood Cliffs, New Jersey, Prentice-Hall, p. 466–501.

Reynolds, S. J., Spencer, J. E., Asmerom, Y., DeWitt, E., and Laubach, S. E., 1989, Early Mesozoic uplift in west-central Arizona and southeastern California: Geology, v. 17, p. 207–211.

Riggs, N. R., 1987, Stratigraphy, structure, and geochemistry of Mesozoic rocks of the Poyarito Mountains, Santa Cruz County, Arizona: Arizona Geological Society Digest, v. 18, p. 165–175.

Riggs, N. R., and Busby-Spera, C. J., 1990a, Evolution of a multivent volcanic complex in a subsiding intra-arc basin; The Early Jurassic Mount Wrightson Formation, southern Arizona: Geological Society of America Bulletin (in press).

—— , 1990b, Volcanotectonic controls on subsidence and preservation of arc facies; Jurassic magmatic arc, southern Arizona (USA) to northern Sonora (Mexico): International Association of Sedimentologists, 13th International Congress, Nottingham, England (abstract) p. 459–460.

Riggs, N. R., Mattinson, J. M., and Busby-Spera, C. J., 1986, U-Pb ages of the Mount Wrightson Formation, southern Arizona, and possible correlation with the Navajo sandstone: EOS Transactions of the American Geophysical Union, v. 67, p. 1249.

Riggs, N. R., Haxel, G. B., and Busby-Spera, C. J., 1990, Paleogeography and tectonic setting of the Jurassic magmatic arc in southern Arizona; Progress and problems: Geological Society of America Abstracts with Programs, v. 22, p. 78.

Saleeby, J. B., Goodin, S. E., Sharp, W. E., and Busby-Spera, C. J., 1978, Early Mesozoic paleogeographic-paleotectonic reconstruction of the southern Sierra Nevada region, *in* Howell, D. G., and McDougall, K. A., eds., Mesozoic paleogeography of the western United States: Pacific Section: Society of Economic Paleontologists and Mineralogists Pacific Coast Paleogeography Symposium 2, p. 311–336.

Saleeby, J. B., and Busby-Spera, C. J., 1990, Early Mesozoic evolution of the western U.S. Cordillera, *in* Burchfiel, B. C., Lipman, P. W., and Zoback, M. L., eds., The Cordilleran orogen; Conterminous U.S.: Boulder, Colorado, Geological Society of America, The Geology of North America, v. G-3 (in press).

Schermer, E. R., and Busby-Spera, C. J., 1989, Changing tectonic regimes in the Mesozoic Cordilleran arc; Evidence from the central Mojave Desert: EOS Transactions of the American Geophysical Union, v. 70, no. 43, p. 1300.

—— , 1990, Mesozoic volcanic stratigraphy and structure, west-central Mojave Desert; Implications for arc paleogeography and tectonics: Geological Society of America Abstracts with Programs, v. 22, p. 80.

Snow, J. K., 1990, A Permian age for the Last Chance thrust, California and Nevada?: Geological Society of America Abstracts with Programs, v. 22, no. 3, p. 85.

Stevens, C. H., and Stone, P., 1988, Early Permian thrust faults in east-central California: Geological Society of America Bulletin, v. 100, p. 552–562.

Stone, P., and Kelly, M. M., 1988, Preliminary geologic map of the Palen Pass Quadrangle, Riverside County, California: U.S. Geological Survey Open-File Report 88-503, scale 1:24,000.

Stone, P., and Pelka, G. J., 1989, Geologic map of the Palen-McCoy Wilderness Study area and vicinity, Riverside County, California: U.S. Geological Survey Miscellaneous Field Study Map MF-2092, scale 1:62,500.

Stone, P., and Stevens, C. H., 1988, Pennsylvanian and Early Permian paleogeography of the east-central California; Implications for the shape of the continental margin and the timing of continental truncation: Geology, v. 16, p. 330–333.

Stone, P., Page, V. M., Hamilton, W., and Howard, K. A., 1987, Cretaceous age of the upper part of the McCoy Mountains Formation, southeastern California and southwestern Arizona, and its tectonic significance; Reconciliation of paleobotanical and paleomagnetic evidence: Geology, v. 15, p. 561–564.

Tosdal, R. M., and Haxel, G. B., 1987, Late Jurassic regional metamorphism and deformation, southeastern California and southwestern Arizona: Geological Society of America Abstracts with Programs, v. 19, p. 870–871.

Tosdal, R. M., Haxel, G. B., and Wright, J. E., 1989, Jurassic geology of the Sonoran Desert region, southern Arizona, southeastern California, and northernmost Sonora; Construction of a continental-margin magmatic arc, *in* Jenny, J. P., and Reynolds, S. J., eds., Geologic evolution of Arizona: Arizona Geological Society Digest, v. 17, p. 397–434.

Walker, J. D., 1988, Permian and Triassic rocks of the Mojave Desert and their implications for timing and mechanism of continental truncation: Tectonics, v. 7, no. 3, p. 685–709.

Walker, J. D., Bartley, J. M., Martin, M. W., and Coleman, D. S., 1989, Implications of Late Jurassic deformation in the Cronese Hills, Mojave Desert, California, for Mesozoic development of the southern Cordillera: Geological Society of America Abstracts with Programs, v. 21, p. 268.

Walker, J. D., Martin, M. W., Bartley, J. M., and Coleman, D. S., 1990, Timing and kinematics of deformation in the Cronese Hills, California, and implications for Mesozoic structure of the southwestern Cordillera: Geology, v. 18, p. 554–557.

Wright, J. E., and Miller, E. L., 1988, The Early Mesozoic APW path of North America and its correspondence to western U.S. Cordilleran continental margin tectonics: Geological Society of America Abstracts with Programs, v. 20, p. A350.

Wright, J. E., Haxel, G., and May, D. J., 1981, Early Jurassic uranium-lead isotopic ages for Mesozoic supracrustal sequences, Papago Indian Reservation, southern Arizona: Geological Society of America Abstracts with Programs, v. 13, p. 115.

Wyld, S. J., 1989, Paleozoic-Triassic history of the Pine Forest Range, NW Nevada; Implication for the development of arc terranes of the U.S. Cordillera: Geological Society of America Abstracts with Programs, v. 21, p. 161.

MANUSCRIPT ACCEPTED BY THE SOCIETY MARCH 27, 1990

Printed in U.S.A.

Geological Society of America
Special Paper 255
1990

Middle Jurassic syntectonic conglomerate in the Mt. Tallac roof pendant, northern Sierra Nevada, California

G. Reid Fisher*
U.S. Geological Survey, 345 Middlefield Road, Menlo Park, California 94025

ABSTRACT

Middle Jurassic marine conglomeratic and debris-flow deposits in the Mt. Tallac roof pendant are interpreted to be syntectonic fault trough deposits. Similar deposits in similar stratigraphic successions throughout the northern Sierra Nevada demonstrate that the Middle Jurassic continental-margin arc north of latitude 39° was essentially marine, and possibly extensional or transtensional in nature. Recognition of Middle Jurassic syntectonic deposits in the northern Sierra Nevada establishes continuity of structural style between the arc in the northern Sierra Nevada and a portion of the arc south of latitude 39°, which has been interpreted as an ancient analog of the modern extensional or transtensional Central American arc. The Middle Jurassic marine conglomeratic deposits in the northern Sierra Nevada form one of three post-Late Triassic stratigraphic elements shared by the northern Sierra terrane and rocks of probable cratonal affinity in west-central Nevada. Upper Triassic limestone in the northern Sierra Nevada is linked faunally to correlative limestone of west-central Nevada. While correlation is speculative, Lower to Middle(?) Jurassic intervals of quartzose sandstone are common to parts of the Sailor Canyon Formation in the northern Sierra terrane and to rocks of the Auld Lang Syne Group, Boyer Ranch Formation, and Dunlap Formation of west-central Nevada. Middle Jurassic syntectonic conglomeratic deposits, such as in the Dunlap Formation of west-central Nevada, are now recognized in the northern Sierra terrane as well.

INTRODUCTION

Recent mapping in the northern Sierra Nevada has documented widespread Middle Jurassic polymict conglomeratic deposits (Schweickert and others, 1984; Jayko, 1988; Harwood and others, 1990; Fisher, 1989; Harwood, unpublished data; Harwood and Fisher, unpublished data; R. A. Schweickert and others, unpublished data). The mixed siliciclastic and volcaniclastic rocks contain marine fossils, and overlie an Upper Triassic to lower Middle Jurassic marine succession. While their distribution is incompletely known, similar coarse orogenic deposits have been identified in the Mt. Tallac roof pendant, along the North Fork of the American River, near Cisco Grove, on English Mountain and in Perazzo Canyon (Fig. 1). Wherever they occur, the deposits are intimately associated with volcanic rocks of the developing early Mesozoic marginal arc. The conglomeratic rocks are interpreted as syntectonic deposits correlative with Cordilleran rocks south of latitude 39° inferred to have been deposited in an extensional or transtensional continental arc setting (Busby-Spera, 1988, and this volume).

The recognition of Middle Jurassic syntectonic deposits in the northern Sierra Nevada suggests continuity of structural style for the portion of the marginal arc between the northern Sierra

*Present address: Harlan Tait Associates, 1269 Howard Street, San Francisco, California 94103.

Fisher, G. R., 1990, Middle Jurassic syntectonic conglomerate in the Mt. Tallac roof pendant, northern Sierra Nevada, California, *in* Harwood, D. S., and Miller, M. M., eds., Paleozoic and early Mesozoic paleogeographic relations; Sierra Nevada, Klamath Mountains, and related terranes: Boulder, Colorado, Geological Society of America Special Paper 255.

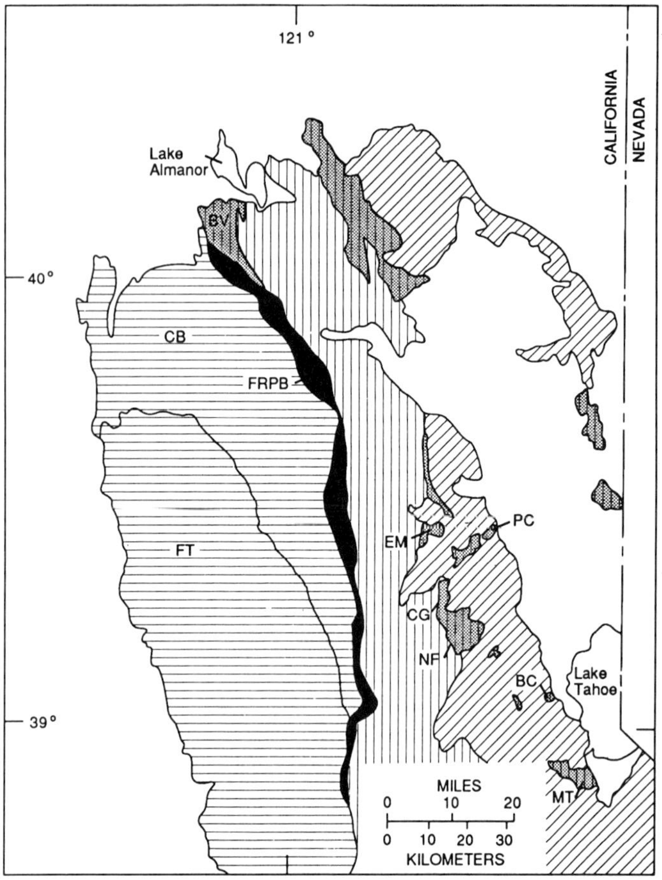

Figure 1. Location map of the northern Sierra Nevada. Vertical ruling indicates northern Sierra terrane, with Mesozoic rocks shaded; diagonal ruling, Cretaceous Sierra Nevada batholith; horizontal ruling, accreted Mesozoic terranes not discussed in text; unshaded, Cenozoic volcanic and alluvial cover. BC = Blackwood Creek, BV = Butt Valley structural block, CB = Central Belt terrane, CG = Cisco Grove, EM = English Mountain, FT = Foothills terrane, FRPB = Feather River peridotite belt, MT = Mt. Tallac roof pendant, NF = North Fork of American River, PC = Perazzo Canyon. Area of figure shown in regional context in Figure 5.

Nevada and southwestern Arizona (Busby-Spera, 1988). Much of the marginal arc is known to have been marine from sedimentologic and fossil evidence (Miller, 1978; Busby-Spera, 1984; Karish and others, 1987; Busby-Spera, 1988; Harwood, 1990). Busby-Spera (1988) has suggested that the Jurassic arc south of latitude 39° was similar to the modern extensional arc of Central America, in which subsiding graben(s) characterize the arc, rather than emergent areas of compressional uplift. Rocks of the northern Sierra Nevada suggest a similar extensional or transtensional arc setting. Lower and Middle Jurassic largely volcaniclastic sandstone of the Sailor Canyon Formation is entirely marine, deposited during a time of sea-level rise and gradual regional subsidence. Subsequent marine conglomerate, chaotic deposits, and mafic volcanic rocks of the Middle Jurassic Tuttle Lake Formation represent fault trough fill, associated with an episode of localized extension.

The Middle Jurassic conglomeratic deposits constitute one of the biostratigraphic and lithologic elements shared by post–Upper Triassic rocks of the northern Sierra terrane (Coney and others, 1980; Silberling and others, 1987) and coeval rocks of continental affinity in western Nevada (Ferguson and Muller, 1949; Lupe and Silberling, 1985; Speed and Jones, 1969; Stanley, 1971; Speed, 1978b; Stewart, 1980; Busby-Spera, 1988). Upper Triassic limestone in the northern Sierra Nevada is linked faunally to similar lithologies in western Nevada (Pessagno and Blome, 1986). Lower and Middle Jurassic faunal assemblages likewise are closely linked (Taylor and others, 1984). Lower and Middle Jurassic quartz arenites of cratonal derivation reported in intra-arc basins of western Nevada, southeastern California, and southeastern and southern Arizona (Lupe and Silberling, 1985; Miller and Carr, 1978; Marzolf, 1980; Tosdal and others, 1990; Busby-Spera, 1988) have more subdued lithologic counterparts in the northern Sierra Nevada. Finally, Middle Jurassic syntectonic conglomerate, while never a laterally continuous lithosome, is observed in the northern Sierra Nevada and west-central Nevada (Ferguson and Muller, 1949; Stanley, 1971; Stewart, 1980; Oldow and Bartel, 1987).

This chapter focuses first on the occurrence of diamictite and conglomerate in the Mt. Tallac roof pendant, then discusses correlative deposits of the northern Sierra Nevada. These deposits are then viewed in light of the Mesozoic stratigraphic elements shared by both the northern Sierra Nevada and rocks of known continental affinity in west-central Nevada. Finally, the issue of how the recognition of conglomerate and associated deposits in the northern Sierra Nevada affects evolving models for the Jurassic continental-margin arc is explored.

STRATIGRAPHY OF THE MT. TALLAC ROOF PENDANT

Conglomeratic deposits in the Mt. Tallac roof pendant occur roughly in the middle of the preserved sedimentary and volcanic section (Fig. 2). Stratigraphic units in ascending order are: an unnamed Upper Triassic(?) limestone, the Lower and Middle Jurassic Sailor Canyon Formation, and the Middle Jurassic Tuttle Lake Formation (Fisher, 1989). Conglomerate and diamictite lie disconformably on principally volcaniclastic sandstone of the Sailor Canyon Formation and form the base of the largely volcanic Tuttle Lake Formation. The succession observed in the Mt. Tallac roof pendant is reviewed below, for comparison with similar successions containing conglomeratic deposits in other roof pendants of the northern Sierra Nevada (Fig. 1).

Many mesoscopic primary structures and textures remain intact in the Mt. Tallac roof pendant because of the general lack of penetrative deformation, although metamorphic recrystallization associated with weak cleavage development has obscured some original textures, particularly at the microscopic scale. The masking effect of metamorphic recrystallization has been greatest on fine-grained sedimentary rocks (both quartz-rich and volcaniclastic), and least on volcanic, hypabyssal, and coarse volcaniclastic rocks. Although rocks in the Mt. Tallac roof pendant have been contact metamorphosed from albite-epidote hornfels to locally pyroxene hornfels (Loomis, 1966, 1981), protoliths are still discernible, and the prefix meta- is omitted from rock descriptions.

Many of the original sedimentary structures and stratigraphic relations are preserved, in part because of relatively mild tectonic deformation. The principal structural features of the Mt. Tallac roof pendant are an early thrust fault with poorly developed associated folds; two weak, mainly domainal metamorphic fabrics, the second one coeval with well-developed upright folds; zones of oblique, west-side-down ductile faulting; and finally, east-side-down brittle normal faulting associated with Basin and Range extension. The main effect of tectonism has been to slice up the stratigraphy, rather than fold it extensively (Fig. 3).

The stratigraphically lowest unit exposed in the Mt. Tallac roof pendant is an unnamed limestone (unit a in Fig. 2). Recrystallization has obscured primary textures, but the protolith is interpreted as calcarenite in beds of variable thickness, a few of them dolomitic. The calcarenite is locally faintly laminated, and interbedded with thin horizons of locally parallel-laminated siliceous sandstone and siltstone. Tabular blocks or intraclasts of sandstone and mudstone are present. The original geometry of preserved lamination is unclear because of plastic flow during deformation.

The age of the limestone has not been established directly, but the unit is here correlated with the Upper Triassic (late

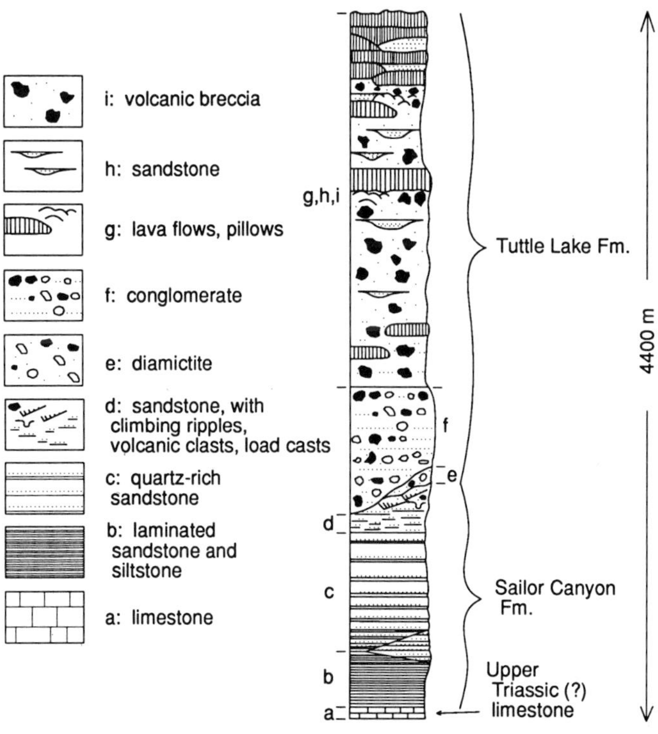

Figure 2. Stratigraphic column for the Mt. Tallac roof pendant. Lithologies shown schematically, but relative thicknesses of units are correct. Lithologic units discussed in text are lettered. Solid clasts are volcanic, open clasts are sedimentary.

Karnian to early Norian) Hosselkus Limestone (Harwood, 1983) exposed along the North Fork of the American River (Fig. 1). The unnamed limestone of the Mt. Tallac roof pendant and the biostratigraphically dated limestone on the American River appear to have similar, probably shallow-water sedimentary structures, are both overlain by the Sailor Canyon Formation, and display the same apparently conformable contact with correlative overlying strata (Fig. 4).

Although the transition between the units is not well exposed, rocks of the Sailor Canyon Formation appear to lie conformably on the unnamed Upper Triassic(?) limestone (Fig. 2). The lower part of the Sailor Canyon (unit b in Fig. 2) consists of interbedded parallel-laminated quartzofeldspathic sandstone, carbonaceous siltstone, and minor calcarenite. A death assemblage of Early Jurassic (most probably late Pliensbachian) ammonites has been recovered from this interval (Fisher, 1989). These laminated, carbonaceous deposits are succeeded by an interval marked by slightly calcareous, quartzose sandstone beds, some of them as much as 2 m in thickness (unit c in Fig. 2). The sandstone beds are interpreted as turbidites, and are interbedded with thin intervals of slightly calcareous, laminated dark siltstone. An interval of parallel-laminated, locally carbonaceous and pyritic feldspathic sandstone similar to the lowermost Sailor Canyon in the roof pendant, overlies the quartzose turbidites, and grades upward into a more volcanogenic interval at the top of the formation (unit d in Fig. 2). This uppermost part of the Sailor Canyon consists of laterally extensive and lenticular bodies of feldspathic turbidites, ripple-drift laminated tuffaceous(?) sandstone, and tuffaceous(?) lithic sandstone with rare volcanic pebbles.

The age of fossils recovered from the Sailor Canyon Formation in the Mt. Tallac roof pendant fit well with the age of the Sailor Canyon Formation near its type area, where it is known to be Early and Middle Jurassic (Clark and others, 1962; Imlay, 1968). The oldest fossils recovered from the formation to date are Early Jurassic (Late Sinemurian to early Pliensbachian) ammonites, found from 50 to 300 m above the base of the formation near the North Fork of the American River (Clark and others, 1962; Imlay, 1968). The fossils there are enclosed in a laterally persistent and widely fossiliferous interval of carbonaceous slate and feldspathic sandstone correlative with the lowermost part of the Sailor Canyon Formation exposed in the Mt. Tallac roof pendant. The youngest fossils found thus far in the Sailor Canyon are early Middle Jurassic (early Bajocian) ammonites that occur about 3,000 m above the base of the formation, near the North Fork of the American River (Clark and others, 1962; Imlay, 1968).

The Tuttle Lake Formation in the Mt. Tallac roof pendant consists of diamictite and conglomerate overlain by intercalated andesitic lava flows, volcanic breccia, and volcaniclastic sandstone (Fig. 3). The basal contact of the Tuttle Lake Formation is discordant with respect to the underlying Sailor Canyon Formation. Although the discordance is difficult to establish since layering is rare and of uncertain origin within the basal diamictite of the Tuttle Lake Formation, the discordance between bedding in

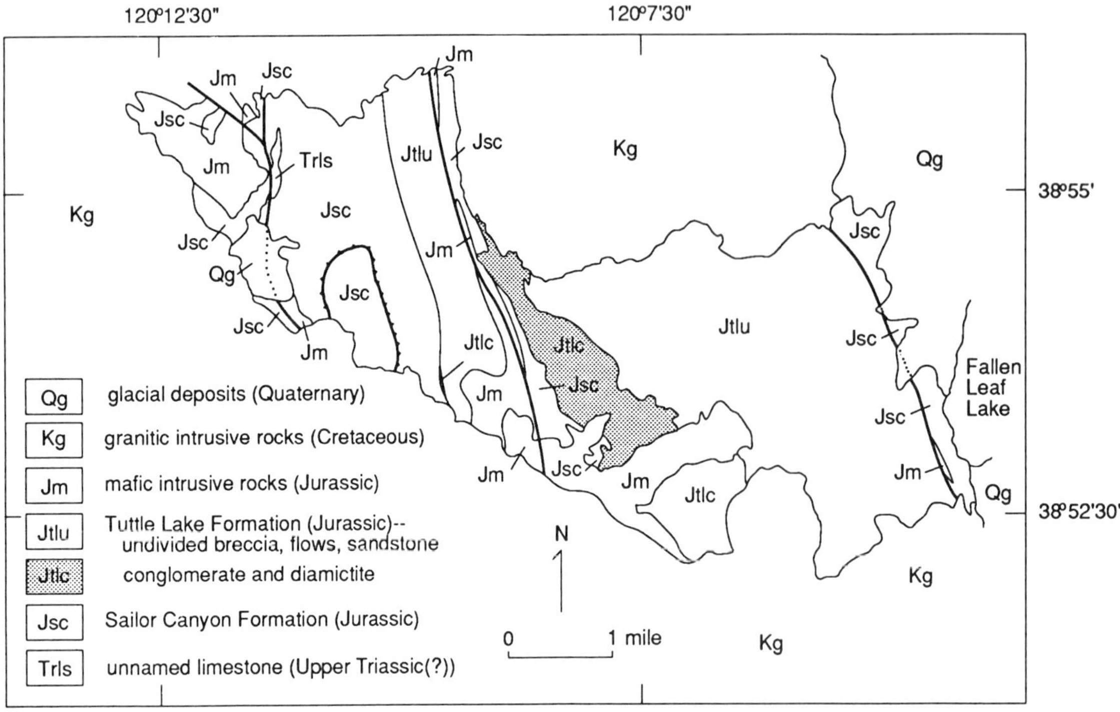

Figure 3. Generalized geologic map of the Mt. Tallac roof pendant showing major units.

the Sailor Canyon and the basal contact of the diamictite is approximately 25°.

Diamictite and conglomerate define a tapering, wedge-shaped map unit at the base of the Tuttle Lake Formation in the Mt. Tallac roof pendant (Fig. 3) (Fisher, 1989). Volumetrically, the diamictite (unit e in Fig. 2) consists mainly of matrix-supported pebbly sandstone, although the suspended clasts range up to 14 m in length. Many clasts are round, and some appear to have been semilithified or unlithified at the time of transport and deposition. Clast types include intermediate to mafic volcanic and hypabyssal rocks, very coarse-grained quartzofeldspathic sandstone; gray, fine-grained quartzite; massive gray siltstone; dark gray pelite chips; scarce laminated sandstone-pelite clasts; rare clear to slightly bluish, fine- to medium-grained quartzite, and still rarer white weathering metachert or very fine-grained quartzite (Fisher, 1989). Most of these lithologies appear to have been derived from the underlying Sailor Canyon section; some of them were clearly transported as soft masses. Quartzite and metachert were not derived from the Sailor Canyon Formation, but suitable sources occur in the lower Paleozoic Shoo Fly Complex, the upper Paleozoic arc sequence or the mid-Paleozoic part of the Lake Tahoe sequence, all of the northern Sierra terrane (Harwood, 1990). The matrix sandstone is of mixed volcanogenic and siliceous provenance.

The conglomerate (unit f in Fig. 2) above the diamictite contains poorly organized polymict conglomerate and subordinate grit beds, sandstone interlayered with pebble- and cobble-conglomerate, and structureless blankets of pebble-, cobble-, and rare boulder-breccia (Fisher, 1989). Both matrix- and clast-supported deposits are present. The matrix consists of medium- or coarser grained sandstone, of mixed volcanogenic and siliceous provenance. Clast lithologies in the conglomerate are essentially the same as those in the underlying diamictite, and were derived principally from two sources—the Sailor Canyon Formation and a nearby mafic to intermediate volcanic center (Fisher, 1989). Volcanic clasts increase upward, and locally compose greater than 85 percent of clasts.

Bedding thickness in the conglomerate is usually proportional to mean clast size, and inversely proportional to the degree of sorting. Lenses 15 cm thick of pebble- and granule-conglomerate commonly show diffuse bedding-parallel internal banding defined by small changes in mean clast size. The largest clasts in these lenses are usually matrix-supported. As bedding thickness increases, the mean clast size and the range in clast size increase, and the degree of organization falls off such that a single layer may be up to 3 m thick with no sign of internal stratification. The proportion of clast and matrix, though widely variable, also shifts gradually with bedding thickness so the thickest, coarsest beds appear to be clast-supported.

Intercalated lenses of volcanic breccia, andesitic lava flows, and tuffaceous sandstone overlie the diamictite and conglomerate in the Tuttle Lake Formation (Fig. 2). Volcanic breccias (unit i Fig. 2) are monolithologic or polylithologic and generally unstratified. Clasts range up to 1 m across. Almost all igneous clasts are intermediate to mafic in composition and reflect volcanic and hypabyssal sources. Extrusive clasts texturally and compositionally resemble the commonly clinopyroxene-phyric lava flows within the Tuttle Lake Formation. Monolithologic breccia fragments are surrounded by a matrix of the same modal composition that was apparently derived from a mixture of tuff, lithic fragments, and pyrogenic crystals of clinopyroxene, plagioclase, and minor hornblende. Monolithologic breccia deposits include pyroclastic breccia and epiclastic tuffaceous breccia (Fisher and Schmincke, 1984). Polymict breccia contains volcanic clasts of variable texture and mineralogy, and sparse sedimentary clasts.

Lava flows (unit g in Fig. 2) intercalated in the Tuttle Lake Formation on Mt. Tallac are basaltic to andesitic, with characteristic relict phenocrysts of augite, calcic plagioclase, and subordinate black amphibole. Massive flows lacking amygdules dominate over amygduloidal and pillowed flows. Flows commonly appear to be separated by basal flow breccia. Pillows, where observed, range from 65 to 135 cm in intermediate dimension, with radially elongated, quartz-filled amygdules. In one locale, hyaloclastite breccia formed where a pillowed flow intruded wet, unconsolidated sediment.

Volcaniclastic or tuffaceous sandstone (unit h in Fig. 2) is volumetrically much less abundant than volcanic breccia and lava flows in the Tuttle Lake Formation (Fig. 2). Most occurrences are small lenses of tuffaceous(?) or fine-grained volcaniclastic, feldspathic sandstone interbedded with thinly laminated siltstone and sandstone. Discontinuous, diffuse bands defined by mafic crystal concentrations outline scours, faint foreset beds, and grain-flow deposits. Sparse zones contain associated planar-parallel laminae, planar foresets, low-angle trough cross-beds, and grain-flow deposits commonly interlayered with polymict pebble breccia.

Depositional environments

Although the conglomeratic deposits accompanying fault trough formation are the focus of this chapter, the succession of environments and events inferred for the Mt. Tallac roof pendant is shared by other localities in the northern Sierra Nevada and demonstrates general contemporaneity of events and continuity of tectonic history for the northern part of the range.

The unnamed Upper Triassic(?) limestone at the base of the Mt. Tallac roof pendant section is inferred to represent a shallow marine setting. Locally dolomitic calcarenite is interbedded with quartzofeldspathic sandstone, suggesting current activity was energetic enough to allow deposition mainly of sand-size detritus. The presence of intraclasts or small tabular blocks of sandstone supports this interpretation. The original geometry of faintly preserved laminae is unclear. The closest correlative Upper Triassic (late Karnian to early Norian) limestone, on the North Fork of the American River (Figs. 1, 4), contains large-scale, low-angle trough cross-bedding (Reineck and Singh, 1980), possibly implying a water depth of less than 15 m (C. J. Busby-Spera, 1988, oral communication). Near the North Fork of the American River, the limestone is locally bioclastic, and is associated with limestone breccia and chert-pebble conglomerate (Harwood, 1990).

By contrast, the lowest part of the Sailor Canyon Formation, which overlies the shallow-water limestone in the Mt. Tallac roof pendant with apparent conformity, is interpreted to reflect deep-water conditions. This lithic assemblage of parallel-laminated feldspathic sandstone, carbonaceous siltstone, and minor calcarenite is widely represented in Sailor Canyon exposures throughout the northern Sierra Nevada, and is consistently fossiliferous (Fig. 4). Ammonites preserved as fossils settled after death. No bioturbation is evident, and the commonly pyritic, carbonaceous strata imply that conditions were reducing. The depositional environment is tentatively interpreted as a restricted euxinic basin, such as an adjacent sea with a strongly layered water column (Seibold, 1970; Reineck and Singh, 1980; Ghibaudo, 1980).

Minimum water depth is not well constrained for this interval of the Sailor Canyon Formation in the Mt. Tallac roof pendant, but the sudden, apparently conformable shift from shallow-water to deep-water facies implies either initially rapid tectonic subsidence, rapid sea-level rise, or a combined tectonic-eustatic apparent sea-level rise. The shift in environment is constrained by fossils to the period between the early Norian and late Sinemurian (Clark and others, 1962; Imlay, 1968; Harwood, 1983). Global sea-level curves (Vail and Mitchum, 1979; Vail and others, 1984) document a worldwide sea-level low in the latest Triassic (Norian), which may have persisted until the early Sinemurian (Vail and others, 1984) or the early Hettangian (Hallam, 1988). A second-order sea-level rise, beginning possibly in the early Sinemurian (Vail and Mitchum, 1979; Vail and others, 1984; Hallam, 1988), may have begun quite rapidly (Hallam, 1988). Such periods of rapid sea-level rise may result initially in a condensed or starved section as marine facies shift rapidly shoreward (Vail and others, 1984); this may have been the reason for the apparent lack of a documentable Hettangian and early Sinemurian section in the Sailor Canyon Formation throughout the northern Sierra Nevada. The section shows a rapid, conformable shift to deep-water conditions, with deposition not documented by fossils until at least the late Sinemurian.

The total thickness of deep-water strata in the Mt. Tallac roof pendant exceeds the amount of global sea-level rise during deposition, requiring some component of tectonic subsidence. The second-order cycle of sea-level rise initiated in the early Sinemurian (Jurassic supercycle) persisted until approximately the close of the Jurassic (Vail and Mitchum, 1979; Vail and others, 1984). However, total global sea-level rise during the Jurassic is estimated at less than 200 m (Vail and others, 1984). The Sailor Canyon Formation in the Mt. Tallac roof pendant totals 1,500 m in stratigraphic thickness (Fisher, 1989). All of it is marine; the lower 88 percent of the section reflects deep-water conditions, with possible shallowing only at the top, immediately beneath the coarse, chaotic, volcaniclastic and siliciclastic deposits of the Tuttle Lake Formation (Fisher, 1990). Regional tectonic subsidence must have occurred to accommodate the thickness of deep-water sediment in excess of the amount of sea-level rise. Deep-water conditions may have persisted until the early

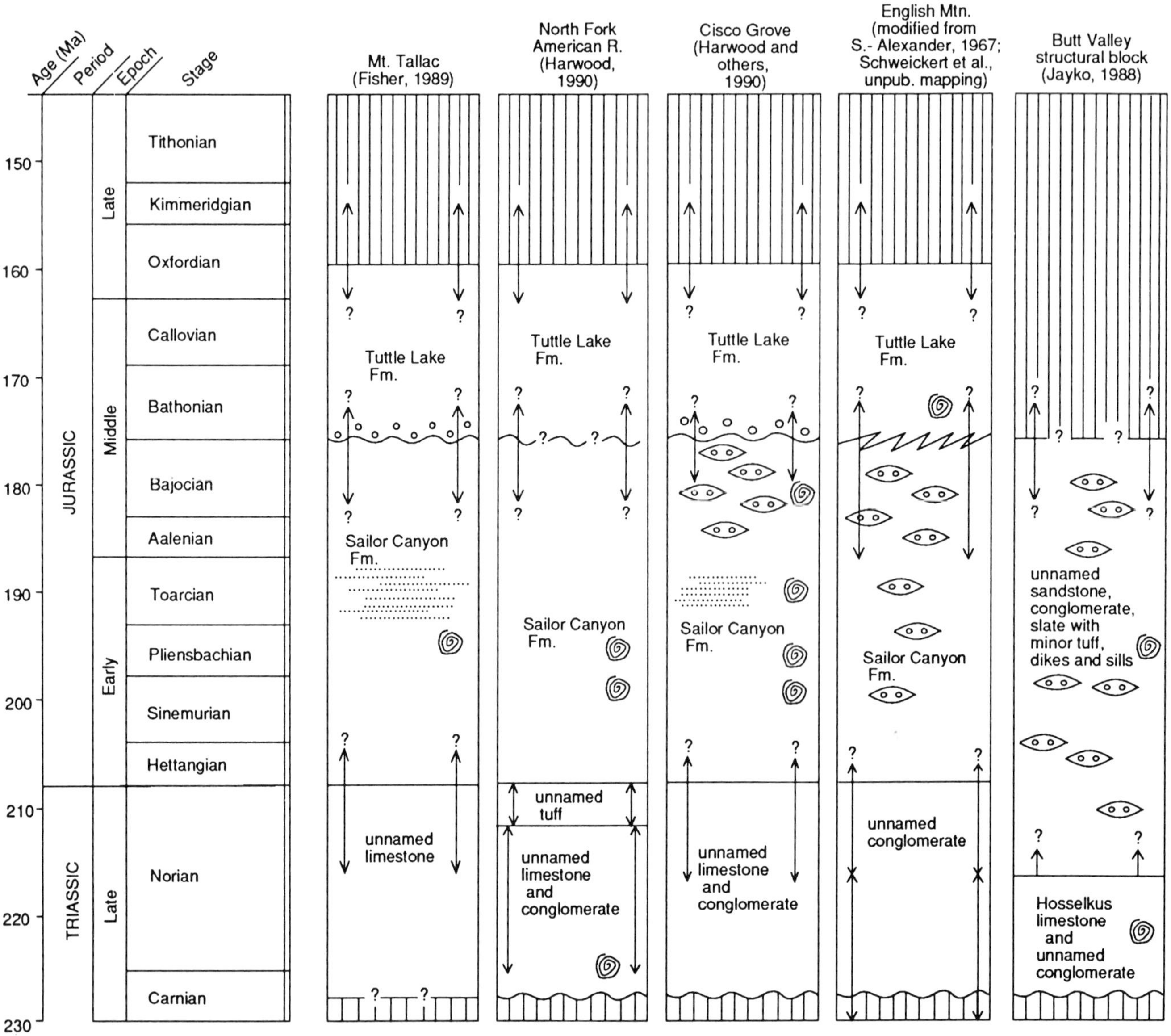

Figure 4. Generalized columnar sections for selected localities discussed in text, showing conglomerate (open circles, lenses with open circles), fossil control (spirals), quartz-rich sand intervals (horizontal dotted lines). Approximate maximum time range for placement of contacts indicated by arrows. Vertical lines signify missing section. Wavy lines indicate erosional unconformity. Queried horizontal lines indicate nature of contact unknown.

Bajocian, the age of the youngest fossils yet found in the Sailor Canyon Formation (Imlay, 1968). The average sedimentation rate for the Sailor Canyon Formation (uncorrected for compaction and tectonic flattening) in the Mt. Tallac roof pendant equals 50 m/m.y., a figure in line with high rates of thermal subsidence (Schwab, 1976) as along a newly formed passive margin, but significantly less than most tectonically driven subsidence rates.

Depositional conditions for overlying slightly calcareous, quartzose turbidites (lithology c in Fig. 2) are interpreted to reflect both turbidity and traction currents. Thick graded beds deposited by density currents were locally scoured and reworked by traction currents, leaving cross-laminated intervals (Stanley, 1988; Fisher, 1989). Intervals containing multiple thin sandstone and siltstone beds apparently record more quiescent conditions dominated by deposition of suspended sediment with minor current reworking in a continental rise or outermost shelf setting (Reineck and Singh, 1980; Stanley, 1988). The influx of quartzose sediment may have temporarily increased the rate of sediment accumulation; laminated pyritic sandstone overlying this interval appears to imply a return to relatively slow deposition in an oxygen-poor environment of restricted circulation.

Map pattern, sand composition, and bedding style suggest that the uppermost part of the Sailor Canyon Formation in the Mt. Tallac roof pendant reflects the development of the volcanic system responsible for generation of the Tuttle Lake Formation. Mappable units are lenticular and show rapid lateral variation (Fisher, 1989). Thin turbidites, ripple-drift cross-laminated sandstone, slump deposits, and possible lenses of hummocky cross-stratified sandstone supplant the varve-like laminae of underlying pyritic sandstone and siltstone, implying more rapid sedimentation and increased water circulation. Stratigraphically higher sandstone contains a greater component of tuffaceous and epiclastic volcanic grains. Pebble-size volcanic clasts are found first in tuffaceous, lithic sandstone very near the top of the Sailor Canyon Formation.

Several lines of evidence suggest that diamictite and conglomerate, which form the lower part of the Tuttle Lake Formation in the Mt. Tallac roof pendant, were derived locally from fault-generated relief and deposited in a fault-bounded trough:

1. The appearance of coarse clastic material in the section is abrupt. No hint of a prograding alluvial system is evident, nor is there any indication of subaerial exposure in the clasts. In fact, although the basal contact of the Tuttle Lake Formation is discordant with respect to the Sailor Canyon Formation and most clasts appear to be derived from the Sailor Canyon, many clasts appear to have been only semilithified at the time.

2. The lowest mappable unit (diamictite) of coarse clastic material is entirely chaotic. Stratification is essentially absent, and sorting is extremely poor. Detritus ranges from sand-size in the matrix up to slide blocks 14 m in length (Fisher, 1989). The bulk of the deposit consists of matrix-supported pebbly sandstone. The wide range in clast size, the lack of organization, and the matrix support strongly suggest the deposits originated by collapse of unstable slopes cut into the underlying section.

3. Map and thickness patterns of the diamictite and conglomerate (Fig. 3) suggest later ductile faulting reactivated the bounding fault of a depositional trough (Fig. 3) (Fisher, 1989). Two exposed sections through the diamictite and conglomerate are separated by the inferred bounding fault, and lie only 1,200 m apart. The total thickness of the diamictite and conglomerate in one section is approximately 575 m; the other exposure consists of thin, barely mappable lenses. While rapid lateral thickness changes in a conglomeratic deposit commonly reflect a channel system, there are none of the large- or small-scale sedimentary features normally associated with a channel setting. The trough, once filled with debris, was overlapped by volcanic deposits.

4. Several observations suggest that the clasts were transported a short distance. Superficially, the presence of stratification suggests the conglomerate deposits most closely resemble the graded-stratified deposits described by Walker (1975). According to this model, the deposits might have accumulated in a mid-fan setting, having been transported a significant distance from their source. However, clast support is often confined to only the coarsest beds, whereas graded-stratified conglomerates studied by Walker (1975) were primarily clast supported. The transport process was probably one of hyperconcentrated flow or debris flow, rather than fluid transport of bed load (Johnson, 1970). The grading described by Walker (1975) for graded-stratified deposits is essentially absent from conglomerate in the Mt. Tallac roof pendant. Sorting is poorly developed, and stratification is generally shown only by change in maximum clast size with no corresponding change in grain-size mode. More accurately, many conglomerate beds in the Mt. Tallac roof pendant resemble the "disorganized bed" deposits of Walker (1975) when considered individually. The deposits observed in the Mt. Tallac roof pendant are better interpreted as a sequence of internally disorganized beds with observed stratification the by-product of repeated deposition of internally chaotic deposits. Such deposits are inferred to be of proximal, steep slope origin with minimal transport.

The apparent lack of preferred clast orientation, transport of semilithified clasts, common subangular and inequant clasts, and derivation chiefly from underlying lithologies all suggest local derivation and minimal transport. The lack of preferred orientation implies either a debris flow or hyperconcentrated flow mechanism (Johnson, 1970). The common inequant and semilithified clasts are intact, suggesting minimal clast interaction and minimal transport. The derivation of most clasts from underlying lithologies means that transport from distant sources is minimal.

Volcanism was associated with syntectonic deposits of diamictite and conglomerate. Volcanic clasts larger than sand size first appear in tuffaceous volcaniclastic sandstone near the top of the Sailor Canyon Formation, less than 100 m stratigraphically below massive, chaotic diamictite of the Tuttle Lake Formation. The percentage of volcanic clasts increases upsection in the diamictite and conglomerate, eventually dominating. Pyrogenic clinopyroxene grains occur in the matrix sand of both units. Some pyroxene grains are euhedral, implying minimal transport and reworking in a sedimentary environment before final deposition. Feeder dikes and hypabyssal intrusions related to volcanic phases mapped as the Tuttle Lake Formation are in part syntectonic (Fisher, 1989). The map pattern of these intrusive bodies clearly indicates they intruded zones of structural weakness marked by ductile shearing and postdate at least some of the movement accommodated by these zones. The intrusive bodies themselves bear the same but less well-developed ductile shear fabric, implying post-intrusive shearing.

Summary of interpretations in the Mt. Tallac area

The argument for syndepositional, extensional arc-related tectonism in the Mt. Tallac roof pendant rests on several observations and interpretations. A history of regional subsidence beginning in Late Triassic(?) time is demonstrated by the facies succession and thickness of the Sailor Canyon Formation. There is no stratigraphic evidence to suggest compressional uplift or subaerial exposure. The rapid shift from shallow-water to deep-water conditions in Late Triassic to late Pliensbachian time is in part eustatic, but largely the result of prolonged subsidence, which ultimately allowed approximately 1,500 m of deep-water deposits to accumulate in a deep quiescent environment.

By contrast, the conglomeratic clastic deposits owe their existence to local fault-generated relief in post–early Bajocian time. The deposits appear abruptly in the section, with no evidence for regional subaerial exposure. Basal deposits are entirely chaotic and contain large blocks, suggesting their origin on unstable slopes. Dramatic changes in thickness between fault-separated exposures of the conglomerate and diamictite suggest a portion of a fault-bounded trough is preserved. Channel features are absent from the conglomeratic deposits filling the trough. The nature of the deposits themselves suggests minimal transport. Disorganized matrix-rich deposits contain many mechanically unstable clasts, most of which could have been derived locally. Submarine volcanism is associated with the syntectonic conglomerate. Volcanic clasts occur in the section only slightly beneath large-scale chaotic deposits, and increase in proportion upward through the section. Pyrogenic crystals in matrix sandstone suggest contemporary volcanism. Feeder dikes and hypabyssal intrusions are coeval with shearing.

CORRELATIVE DEPOSITS IN THE NORTHERN SIERRA NEVADA

Marine conglomerate deposits I interpret as syntectonic are reported at several other localities in the northern Sierra Nevada. The deposits occur in similar stratigraphic position, in similar tripartite stratigraphic successions (Fig. 4). Although Late Triassic shallow-water limestone is not present in all localities, the association of Lower and Middle Jurassic deep-water sedimentary strata with localized thick accumulations of polymict diamictite, conglomerate, and/or breccia is repeatedly observed.

Cisco Grove area

The tripartite lithologic association is best dated by fossils in the Cisco Grove area (Figs. 1, 4), where it consists of fossiliferous Upper Triassic shallow-water deposits, Lower and Middle Jurassic Sailor Canyon Formation containing deep-water strata, and Middle Jurassic Tuttle Lake Formation with coarse clastic and volcanic deposits. The Cisco Grove area also contains the type locality for the Tuttle Lake Formation (Harwood, 1990).

In the Cisco Grove area, Late Triassic (late Karnian to early Norian) conglomerate and limestone clast breccia deposited in a shallow-water setting is abruptly but conformably overlain by deep-water carbonaceous slate and feldspathic sandstone of the Sailor Canyon Formation (Harwood, 1983). As noted in the Mt. Tallac pendant, rapid sea-level rise and regional subsidence appear to have contributed to the rapid facies change and may explain the seeming lack of a Hettangian section. The rapid facies shift must predate late Sinemurian to early Pliensbachian ammonites found within 50 to 300 m above the base of the formation (Clark and others, 1962; Imlay, 1968; Harwood, 1990).

In the Cisco Grove area, the upper third of the Sailor Canyon Formation contains lenses of polylithologic conglomerate and pebbly mudstone enclosed in deep-water strata (Harwood and others, 1990). Lithologically, the lenses resemble more laterally extensive debris flow or conglomerate blankets mapped at the contact of the Sailor Canyon and Tuttle Lake Formations (Harwood and others, 1990). The lenses are typically internally structureless or crudely stratified, matrix- and clast-supported, and contain pebble- to cobble-size clasts derived mainly from the Sailor Canyon Formation. Conglomerate lenses resembling both the disorganized bed and graded-stratified deposits of Walker (1975) are present. Although their origin is unclear, I infer the lenses to have originated by slumping of tectonically oversteepened submarine slopes, and interpret them as precursors of later, more extensive accumulations of fault-generated debris mapped in the Tuttle Lake Formation.

In the type area, the Tuttle Lake Formation (Harwood, 1990) consists principally of andesitic breccia, but locally contains polymict debris flow and unorganized conglomerate deposits. Breccia deposits consist of pyroclastic tuff breccia and epiclastic tuffaceous breccia (Fisher, 1966). The basal contact of the Tuttle Lake Formation is discordant, indicating a disconformity between the Sailor Canyon and Tuttle Lake Formations. Since conglomerate lenses resembling those in the Tuttle Lake Formation occur in the upper part of the Sailor Canyon Formation, the disconformity between the formations probably does not indicate a profound break. The youngest fossils found in the Sailor Canyon Formation are early Middle Jurassic (early Bajocian) (Clark and others, 1962; Imlay, 1968). Rocks correlated with the Tuttle Lake Formation on English Mountain have produced a Middle Jurassic marine pelecypod (Stuart-Alexander, 1967). Stratigraphic columns presented here (Fig. 4) provisionally show the base of the Tuttle Lake Formation at the end of the Bajocian, for want of better constraint.

Classification of the deposits in the Tuttle Lake Formation as breccia rather than conglomerate is partly a semantic division based on clast angularity, lack of internal structure, and the dominant volcanic lithology. The co-occurrence of deposits containing monolithologic volcanic and mixed volcanic and sedimentary clasts documents unstable slopes in the tectonically and volcanically active environment. The clast angularity and lack of internal structure within the disorganized beds reflect minimal transport.

North Fork of the American River

Exposures along the North Fork of the American River (Figs. 1, 4) demonstrate a rapid post–Late Triassic facies shift as well. Shallow-water Upper Triassic marine limestone is there gradationally overlain by Upper Triassic and/or Lower Jurassic volcaniclastic sandstone and tuff (Harwood, 1990; Harwood and others, 1990). Lenses of shallow-water polymict pebble conglomerate are locally mapped at the top of the tuffaceous unit, and the contact with overlying deep-water slate and feldspathic sandstone of the Sailor Canyon Formation is abrupt. The sudden, apparently conformable shift to overlying deep-water deposits implies either rapid subsidence, rapid sea-level rise, or a combination of both tectonic and eustatic movement.

The nature of the contact between the Sailor Canyon Formation and the Tuttle Lake Formation in the North Fork of the American River vicinity is poorly known. While extensive areas of monolithologic, andesitic pyroclastic, and epiclastic breccia have been mapped within the Tuttle Lake Formation, conglomeratic siliciclastic polymict deposits like those in the Cisco Grove area to the north and the Mt. Tallac area to the southeast have not been identified.

English Mountain

In the vicinity of English Mountain (Figs. 1, 4), Jurassic conglomerate is found as lenses in deep-water strata overlying Late Triassic(?) shallow-water deposits. The stratigraphic section of English Mountain is very similar to that at Cisco Grove. Stuart-Alexander (1967) reported limestone and chert-rich conglomerate east of Bowman Lake on Prairie Creek at the base of what is now interpreted as the Sailor Canyon Formation. R. A. Schweickert and others (1984, unpublished data) did not confirm this occurrence, but the units may be correlative with Late Triassic limestone and conglomerate in the Cisco Grove area and on the North Fork of the American River. D. Davis (1989, unpublished mapping, oral communication) has reported the same conglomerate along strike, south of Sawmill Lake. Inferred depositional conditions were shallow marine (D. Davis, 1989, oral communication).

The Late Triassic(?) shallow-water deposits are overlain by deep-water deposits of the Sailor Canyon Formation (Schweickert and others, 1984, unpublished data; D. Davis, unpublished mapping). Fine-grained rock types include feldspathic and tuffaceous, parallel- and cross-laminated, thin-bedded sandstone, black pyritic argillite, slate, and calcarenite (Stuart-Alexander, 1967; R. A. Schweickert and others, unpublished mapping). I infer the environment of deposition to be deep water, as at Cisco Grove. Significantly, the deep-water strata enclose lenses of polymict conglomerate (R. A. Schweickert and others, unpublished mapping), and polymict debris flow or diamictite (Stuart-Alexander, 1967), particularly in the upper part of the Sailor Canyon Formation. I interpret the deposits to be syntectonic like those in the Cisco Grove and Mt. Tallac areas.

Lithologies of the Sailor Canyon interfinger with volcanic rocks of the Tuttle Lake Formation on English Mountain itself (Stuart-Alexander, 1967). Both monolithologic tuff breccia and polylithologic (epiclastic?) breccia units are reported in the Tuttle Lake Formation on English Mountain (Stuart-Alexander, 1967; Schweickert and others, unpublished data). Breccia occurrences are unsorted and chaotic, with rare indications of stratification (Stuart-Alexander, 1967). Such disorganized bed deposits are unlikely to have undergone significant transport. Volcaniclastic sandstone and siltstone lenses are reported but not mapped (Stuart-Alexander, 1967; R. A. Schweickert and others, unpublished data). One of these lenses near the base of the Tuttle Lake Formation has yielded a Middle Jurassic pelecypod.

If the basal conglomerate and limestone is Upper Triassic, the English Mountain section almost certainly records the same events as at Cisco Grove. A Late Triassic shallow-marine depositional setting was succeeded rapidly by Early Jurassic deep-water conditions, with syntectonic conglomerate deposits presaging the development of the volcanic system that finally became dominant by Middle Jurassic time.

Perazzo Canyon

Incompletely mapped exposures of crudely stratified, polylithologic, quartz-rich conglomerate and dioritic breccia in Perazzo Canyon (Fig. 1) are in fault contact with Sailor Canyon Formation strata (R. A. Schweickert and others, unpublished mapping). Sedimentary clast lithologies imply local derivation primarily from the Sailor Canyon Formation with other possible northern Sierra Nevada sources as well. Coarse, as yet undated, carbonate clasts are abundant. Intermediate to mafic volcanic clasts texturally and compositionally resemble mafic breccia mapped elsewhere as Tuttle Lake Formation. Larger blocks of possible slide origin are contained within pebble conglomerate intervals. Internally, some beds are graded, and are separated by intervals of moderately sorted medium-grained sandstone. The deposits closely resemble the crudely stratified deposits at the top of the Sailor Canyon Formation in the Cisco Grove and Mt. Tallac areas, and I also interpret them to reflect syndepositional faulting.

Butt Valley

Ongoing mapping by Jayko (1988, and this volume) indicates that the Butt Valley block southwest of Lake Almanor contains a coherent, west-facing stratigraphic sequence spanning the Late Triassic through Early(?) Jurassic (Figs. 1, 4). Upper Triassic limestone and conglomerate, shown by Robinson (1975) to correlate with the type Hosselkus Limestone, are also correlated by Jayko (1988) with (Upper Triassic) Hosselkus limestone and conglomerate at the North Fork of the American River (Clark and others, 1962; Harwood, 1983) on the basis of similar conodont faunas.

Mapped by Jayko (1988) as part of the Butt Valley structural block, the Upper Triassic limestone and conglomerate are interpreted by Jayko to be depositionally overlain by volcanogenic metagraywacke, metasiltstone, and conglomerate. Conglomerate and grit deposits are interlayered with the dominant sandstone and siltstone. They contain abundant chert and siliceous volcanic clasts, and minor quartzite clasts (Jayko, 1988). Ammonites recovered from volcaniclastic(?) sandstone indicate a probable Early Jurassic age (A. S. Jayko, 1989, oral communication). Previous mapping assigned this sequence to the Triassic Cedar Formation (Diller, 1895) or to the Arlington Formation, then considered Permian and Triassic (Robinson, 1975; D'Allura and others, 1977). Descriptions of the deposits imply the entire sequence was marine, with submarine conglomerate deposits I infer to have been shed from tectonically oversteepened slopes in an active volcanic environment.

TIES TO THE CRATON AND REGIONAL IMPLICATIONS

Middle Jurassic conglomeratic deposits in the northern Sierra Nevada will help to constrain models for the evolution of the continental-margin arc, and help to tie in northern Sierra terrane stratigraphy with cratonally influenced stratigraphy in west-central Nevada. Biostratigraphic ties for post–Late Triassic time have been recognized for some time but do not provide enough information for paleogeographic reconstruction. Refinement of Mesozoic stratigraphy in the northern Sierra Nevada is beginning to permit tentative lithologic correlations, in part on the basis of quartz-rich sand, and conglomeratic deposits.

Faunal data indicate affinity or continuity, but are of limited use in paleogeographic reconstructions since they cannot resolve questions of lateral displacement of less than many hundreds of kilometers (Pessagno and Blome, 1986; Taylor and others, 1984). Telescoping in the Luning Fencemaker thrust system (Fig. 5) is this order of magnitude, but is not resolved by faunal data. Upper Triassic, largely shallow-water limestone is overlain by Lower Jurassic basinal sedimentary rocks in widely distributed localities through the northern Sierra terrane and western Nevada (Fig. 5). Carbonate depositional cycles are synchronous (Lupe and Silberling, 1985), and fauna in Upper Triassic limestone imply the two areas were contiguous (Pessagno and Blome, 1986). The largely ammonite fauna of the Lower and Middle Jurassic marine clastic sequence also suggest continuity between the northern Sierra terrane and western Nevada (Imlay, 1968; Taylor and others, 1984; Pessagno and Blome, 1986; Callomon, 1984).

Occurrences of quartz-rich sandstone in some cases may approximate laterally persistent sheets useful for correlation. The earliest pulses of quartzose sand occurred in post–Late Triassic time, and link the Chinle Formation of east-central Utah to the Auld Lang Syne Group (Burke and Silberling, 1973) of west-central Nevada (Lupe and Silberling, 1985). As envisaged by Lupe and Silberling (1985), cratonally derived sediment was swept westward (present coordinates) across the shelf into the Auld Lang Syne basin ("basinal terrane" of Speed, 1978a, b) (Fig. 5). Quartz-rich sandstone of this age has not been reported to the west of the Auld Lang Syne Group, either because of a profound structural boundary with the continental-margin arc ("volcanic arc terrane" of Speed, 1978a, b) (Fig. 5), or simple lateral facies change behind the continental-margin arc.

Influx of quartz sand into parts of west-central Nevada occurred again in late Early and Middle Jurassic time, when poorly dated orogenic activity affected portions of the continental-margin arc and Auld Lang Syne Group basin ("orogenic terrane" of Speed, 1978a). The Boyer Ranch Formation (Speed and Jones, 1969) and Dunlap Formation (Ferguson and Muller, 1949; Stanley, 1971) are interpreted as syntectonic units characterized in part by once-widespread but possibly discontinuous deposits of quartz sandstone. Oldow and Bartel (1987) have demonstrated that the Dunlap Formation accumulated in discrete extensional basins, and hence was never continuous.

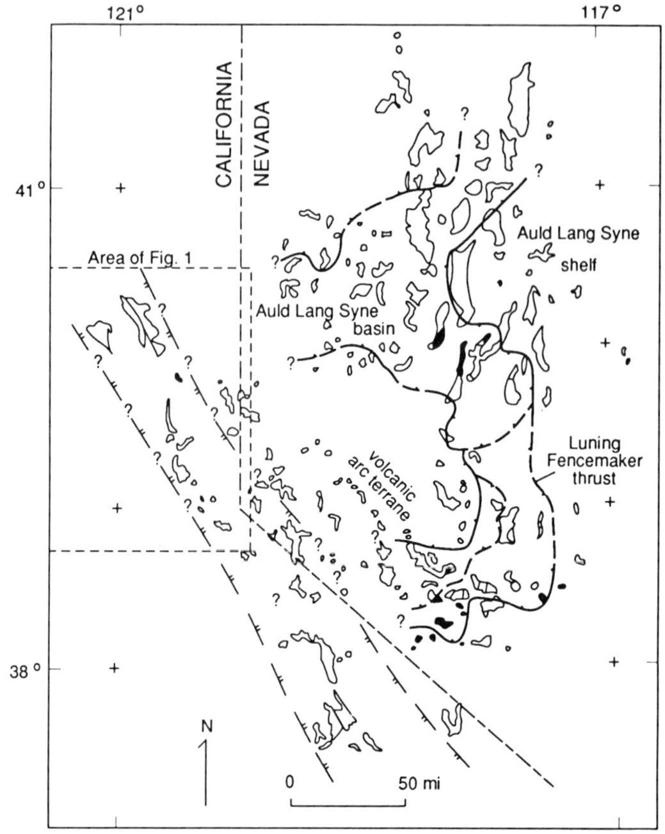

Figure 5. Regional map showing distribution of Mesozoic rocks (outlined areas). Area between dashed hachured lines approximates arc graben depression (adapted from Busby-Spera, 1988). Auld Lang Syne shelf and basin modified from Burke and Silberling (1973). Approximate combined area of Dunlap (south) and Boyer Ranch (north) Formations in black (modified from Stewart, 1980).

Localities in the northern Sierra Nevada exhibiting an Early Jurassic pulse of quartzose detritus include the Mt. Tallac roof pendant, Blackwood Creek, and to a lesser extent, the Cisco Grove area (Fig. 1). In the Mt. Tallac roof pendant, an interval 600 m thick of quartzose turbidites punctuated the deposition of more feldspathic, lithic sandstone (Fig. 2). Thick quartzose turbidites partially reworked by bottom currents occur near Blackwood Creek (Fig. 1), where they are mapped as the Ellis Peak Formation of Harwood (1990). The similarity of the deposits and the similarity of underlying strata suggest the two occurrences are correlative, and are here regarded as part of the Sailor Canyon Formation. The Sailor Canyon Formation section near Cisco Grove contains a less pronounced influx of quartzose sand in the same stratigraphic position (D. Davis, 1989, oral communication).

The timing for deposition of these quartzose intervals appears to be equivalent. Quartzose sedimentation in the Mt. Tallac roof pendant, Blackwood Creek, and Cisco Grove areas is reliably constrained by fossils as post–late Pliensbachian and pre–early Bajocian. Tenuous correlation along strike from the Cisco

Grove area suggests the interval may be stratigraphically very near strata bearing Toarcian ammonites (D. Davis, 1989, oral communication; Harwood, 1990). Influx of quartz sand into the northern Sierra Nevada appears to correlate with Toarcian deposition of quartz sandstone in the Navajo Sandstone of the Colorado Plateau, and the Boyer Ranch (Speed and Jones, 1969) and Dunlap Formations of Nevada (Ferguson and Muller, 1949; Stanley, 1971). Distribution of the Dunlap Formation is patchy, and corresponds approximately with the southern extent of tectonism recognized by Speed (1978a) (Fig. 5). Oldow and Bartel (1987) have shown that this tectonism was extensional, at least in areas where the Dunlap Formation is preserved.

The presence of quartz sandstone of cratonal origin and similar age has already been established in the Jurassic continental-margin arc south of latitude 39° (Miller, 1978; Miller and Carr, 1978; Karish and others, 1987; Busby-Spera, 1984, 1988; Tosdal and others, 1990). The northern Sierra terrane until now has not been known to contain quartz sandstone similar to that in the southern Sierra Nevada (Busby-Spera, 1988) or west-central Nevada.

Middle Jurassic conglomeratic deposits, while never a laterally persistent lithosome, demonstrate lateral continuity of structural style in the evolving continental-margin arc. In the northern Sierra Nevada, morphology of the deposits suggests locally derived material was deposited in a marine fault trough with no evidence for compressional uplift, subaerial erosion, or channel development. The occurrence of such a setting within an entirely marine sequence supports and extends the contention that the Jurassic continental-margin arc was extensional or transtensional (Busby-Spera, 1988).

The same extensional style is demonstrated inboard of the main arc terrane by subaerial extensional fault troughs containing the heterolithic Dunlap Formation (Fig. 5). The conglomeratic deposits reflect local, fault-generated relief and are syntectonic. Formation of extensional basins is loosely constrained as Toarcian in west-central Nevada. Conglomeratic troughs in the northern Sierra Nevada are post–early Bajocian, suggesting time-transgressive growth of the extensional structural province.

The marginal arc appears to have been largely or entirely submerged at the latitude of the northern Sierra Nevada, much like the modern Central American extensional or transtensional arc (Burkart and Self, 1985; Busby-Spera, 1988) (Fig. 5). The low positive relief inherent to an extensional arc setting would facilitate the dispersal of cratonally derived sand sheets (Busby-Spera, 1988). Although quartz-rich sandstone found locally in the Jurassic section of the northern Sierra Nevada is not unequivocally cratonally derived, the occurrence of such sand in west-central Nevada and along the length of the continental-margin arc south of latitude 39° suggests this may be the case.

ACKNOWLEDGMENTS

I have gained immensely from ongoing discussion with many people throughout this study, in particular D. S. Harwood, R. A. Schweickert, and J. C. Yount. Norm Silberling identified the ammonites, and Anita Harris examined the limestone for conodonts. I gratefully acknowledge U.S. Geological Survey field and research support arranged by D. S. Harwood, and a Geological Society of America research grant. This work constitutes part of a Ph.D. dissertation undertaken at the University of Nevada, Reno. Thanks go to Dave Harwood for tireless, careful reviews.

REFERENCES CITED

Burkart, B., and Self, S., 1985, Extension and rotation of crustal blocks in northern Central America and effect on the volcanic arc: Geology, v. 13, p. 22–26.

Burke, D. B., and Silberling, N. J., 1973, The Auld Lang Syne of Late Triassic and Jurassic(?) age, north-central Nevada: U.S. Geological Survey Bulletin 1394-E, 14 p.

Busby-Spera, C. J., 1984, The lower Mesozoic continental margin and marine intra-arc sedimentation at Mineral King, California, *in* Bachman, S., and Crouch, J., eds., Tectonics and sedimentation along the California margin: Pacific Section, Society of Economic Paleontologists and Mineralogists, p. 135–156.

——, 1988, Speculative tectonic model for the early Mesozoic arc of the southwest Cordilleran United States: Geology, v. 16, p. 1121–1125.

Callomon, J. H., 1984, A review of the biostratigraphy of the post–lower Bajocian Jurassic ammonites of western and northern North America, *in* Westermann, G.E.G., ed., Jurassic-Cretaceous biochronology and paleogeography of North America: Geological Association of Canada Special Paper 27, p. 143–174.

Clark, L. D., Imlay, R. W., McMath, V. E., and Silberling, N. J., 1962, Angular unconformity between Mesozoic and Paleozoic rocks in the northern Sierra Nevada, California: U.S. Geological Survey Professional Paper 450-B, p. B-15–B-19.

Coney, P. J., Jones, D. L., and Monger, J.W.H., 1980, Cordilleran suspect terranes: Nature, v. 288, no. 5789, p. 329–333.

D'Allura, J. A., Moores, E. M., and Robinson, L., 1977, Paleozoic rocks of the northern Sierra Nevada; Their structural and paleogeographic implications, *in* Stewart, J. H., and others, eds., Paleozoic paleogeography of the western United States: Pacific Section, Society of Economic Paleontology and Mineralogy, Pacific Coast Paleogeography Symposium 1, p. 395–408.

Diller, J. S., 1895, Lassen Peak Folio, California: U.S. Geological Survey Geological Atlas, Folio 15.

Ferguson, H. G., and Muller, S. W., 1949, Structural geology of the Hawthorne and Tonopah Quadrangles, Nevada: U.S. Geological Survey Professional Paper 216, 53 p.

Fisher, G. R., 1989, Geologic map of the Mt. Tallac roof pendant, El Dorado County, California: U.S. Geological Survey Miscellaneous Field Studies Map, MF-1943, scale 1:24,000.

Fisher, R. V., 1966, Rocks composed of volcanic fragments: Earth and Science Letters, v. 1, p. 287–298.

Fisher, R. V., and Schmincke, H.-U., 1984, Pyroclastic rocks: New York, Springer-Verlag, 472 p.

Ghibaudo, G., 1980, Deep-sea fan deposits in the Macigno Formation (middle–upper Oligocene) of the Gordana Valley, northern Apennines, Italy; Journal of Sedimentary Petrology, v. 50, no. 3, p. 0723–0742.

Hallam, A., 1988, A reevaluation of Jurassic eustasy in the light of new data and the revised Exxon curve, *in* Wilgus, C. K. Posamentier, H., Ross, C. A., and Kendall, C.G.S.C., eds., Sea level changes; An integrated approach: Society of Economic Paleontologists and Mineralogists Special Publication 42, p. 261–273.

Harwood, D. S., 1983, Stratigraphy of upper Paleozoic volcanic rocks and regional unconformities in part of the northern Sierra terrane: Geological Society of America Bulletin, v. 94, p. 413–422.

——, 1990, Stratigraphy of Paleozoic and Lower Mesozoic rocks in the northern Sierra terrane: U.S. Geological Survey Bulletin (in press).

Harwood, D. S., Fisher, G. R., and Waugh, B. J., 1990, Geologic map of the Duncan Peak and southern part of the Cisco Grove Quadrangles, Placer and Nevada Counties, California: U.S. Geological Survey Miscellaneous Investigations Map, scale 1:24,000 (in press).

Imlay, R. W., 1968, Lower Jurassic (Pliensbachian and Toarcian) ammonites from eastern Oregon and California: U.S. Geological Survey Professional Paper 593-C, p. C1–C51.

Jayko, A. S., 1988, Paleozoic and Mesozoic rocks of the Almanor 15′ Quadrangle, Plumas County, California: U.S. Geological Survey Open-File Report 88-757, 12 p.

Johnson, A. M., 1970, Physical processes in geology: San Francisco, California, Freeman and Cooper, 577 p.

Karish, C. R., Miller, E. L., and Sutter, J. F., 1987, Mesozoic tectonic and magmatic history of the central Mojave Desert, *in* Dickinson, W., and Klute, M., eds., Mesozoic rocks of southern Arizona and adjacent regions: Arizona Geological Society Digest, v. 18, p. 15–32.

Loomis, A. L., 1966, Contact metamorphic reactions and processes in the Mt. Tallac roof remnant, Sierra Nevada, California: Journal of Petrology, v. 7, no. 2, p. 221–245.

——, 1981, Geology of the Fallen Leaf Lake 15′ Quadrangle, El Dorado County, California: California Division of Mines and Geology Map Sheet 32, scale 1:62,500, 24 p.

Lupe, R., and Silberling, N. J., 1985, Genetic relationship between Lower Mesozoic continental strata of the Colorado Plateau and marine strata of the western Great Basin; Significance for accretionary history of Cordilleran lithotectonic terranes, *in* Hull, D. G., ed., Tectonostratigraphic terranes of the circum-Pacific region: Proceedings, Circum-Pacific Council for Energy and Mineral Resources, Earth Science Series, no. 1, p. 263–271.

Marzolf, J. E., 1980, the Aztec Sandstone and stratigraphically related rocks in the Mojave Desert, *in* Fife, D. L., and Brown, G. R., eds., Geology and mineral wealth of the California desert: Santa Ana, California, South Coast Geographical Society, p. 215–220.

Miller, E., 1978, The Fairview Valley Formation; A Mesozoic intraorogenic deposit in the southwestern Mojave Desert, *in* Howell, D. G., and McDougall, K. A., eds., Mesozoic paleogeography of the western United States: Pacific Section, Society of Economic Paleontologists and Mineralogists Pacific Coast Paleogeography Symposium 2, p. 277–282.

Miller, E. M., and Carr, M. D., 1978, Recognition of possible Aztec-equivalent sandstones and associated Mesozoic metasedimentary deposits within the Mesozoic magmatic arc in the southwestern Mojave desert, California, *in* Howell, D. G., and McDougall, K. A., eds., Mesozoic paleogeography of the western United States: Pacific Section, Society of Economic Paleontologists and Mineralogists Pacific Coast Paleogeography Symposium 2, p. 285–289.

Oldow, J. S., and Bartel, R. L., 1987, Early to Middle(?) Jurassic extensional tectonism in the western Great Basin; Growth faulting and synorogenic deposition of the Dunlap Formation: Geology, v. 15, p. 740–743.

Pessagno, E. A., and Blome, C. D., 1986, Faunal affinities and tectonogenesis of Mesozoic rocks in the Blue Mountains province of eastern Oregon and western Idaho, *in* Vallier, T. L. and Brooks, H. C., eds., Geology of the Blue Mountains region of Oregon, Idaho, and Washington; Geologic implications of Paleozoic and Mesozoic paleontology and biostratigraphy, Blue Mountains province, Oregon and Idaho: U.S. Geological Survey Professional Paper 1435, p. 65–78.

Reineck, H. E., and Singh, I. B., 1980, Depositional sedimentary environments, 2nd ed.: New York, Springer-Verlag, 549 p.

Robinson, L., 1975, Geology of the Arlingon Formation, Butt Lake [M.S. thesis]: Davis, University of California, 77 p.

Schwab, F. L., 1976, Modern and ancient sedimentary basins; Comparative accumulation rates: Geology, v. 4, p. 723–727.

Schweickert, R. A., Harwood, D. S., Girty, G. H., and Hanson, R. E., 1984, Tectonic development of the northern Sierra terrane; An accreted Late Paleozoic island arc and its basement, *in* Lintz, J., Jr., ed., Western geological excursions; Geological Society of America Guidebook, v. 4: Reno, University of Nevada Department of Geology, p. 1–65.

Seibold, E., 1970, Nebenmeere im humiden und ariden Klimabereich: Geologische Rundschau, v. 60, p. 73–105.

Silberling, N. J., Jones, D. L., Blake, M. C., Jr., and Howell, D. J., 1987, Lithotectonic terrane map of the western conterminous United States: U.S. Geological Survey Miscellaneous Field Studies Map MF-1874-C, scale 1:2,500,000.

Speed, R. C., 1978a, Paleogeographic and plate tectonic evolution of the early Mesozoic marine province of the western Great Basin, *in* Howell, D. G., and McDougall, K. A., eds., Mesozoic paleogeography of the western United States: Pacific Section, Society of Economic Paleontologists and Mineralogists Pacific Coast Paleogeography Symposium 2, p. 253–270.

——, 1978b, Basinal terrane of the early Mesozoic marine province of the western Great Basin, *in* Howell, D. G., and McDougall, K. A., eds., Mesozoic paleogeography of the western United States: Pacific Section, Society of Economic Paleontologists and Mineralogists Pacific Coast Paleogeography Symposium 2, p. 237–252.

Speed, R. C., and Jones, D. L., 1969, Synorogenic quartz sandstone in the Jurassic mobile belt of western Nevada; Boyer Ranch Formation: Geological Society of America Bulletin, v. 80, p. 2551–2584.

Stanley, D. J., 1988, Turbidites reworked by bottom currents; Upper Cretaceous examples from St. Croix, U.S. Virgin Islands: Smithsonian Contributions to the Marine Sciences, no. 33, Smithsonian Institution Press, 74 p.

Stanley, K. O., 1971, Tectonic and sedimentological history of Lower Jurassic Sunrise and Dunlap Formations, west-central Nevada: American Association of Petroleum Geologists Bulletin, v. 55, no. 3, p. 454–477.

Stewart, J. H., 1980, Geology of Nevada; A discussion to accompany the geologic map of Nevada: Nevada Bureau of Mines and Geology Special Publication 4, 136 p.

Stuart-Alexander, D., 1967, Contrasting deformation of Paleozoic and Mesozoic rocks near Sierra City, northern Sierra Nevada, California [Ph.D. thesis]: Stanford, California, Stanford University, 91 p.

Taylor, D. G., and 5 others, 1984, Jurassic ammonite biogeography of western North America; The tectonic implications, *in* Westermann, G.E.G., ed., Jurassic-Cretaceous biochronology and paleogeography of North America: Geological Association of Canada Special Paper 27.

Tosdal, R. M., Haxel, G. B., and Wright, J. E., 1990, Jurassic geology of the Sonoran Desert region, southern Arizona, southeastern California, and northernmost Sonora; Construction of a continental-margin magmatic arc, *in* Jenny, J. P., and Reynolds, S. J., eds., Geologic evolution of Arizona: Arizona Geological Society Digest, v. 17 (in press).

Vail, P. R., and Mitchum, R. M., Jr., 1979, Global cycles of relative changes of sea level from seismic stratigraphy *in* Watkins, J. S., Montadert, L., and Dickerson, P. W., eds., Geological and geophysical investigations of continental margins: American Association of Petroleum Geologists Memoir 29, p. 469–472.

Vail, P. R., Hardenbol, J., and Todd, R. G., 1984, Jurassic unconformities, chronostratigraphy, and sea-level changes from seismic stratigraphy and biostratigraphy, *in* Schlee, J. S., ed., Interregional unconformities and hydrocarbon accumulation: American Association of Petroleum Geologists Memoir 36, p. 129–144.

Walker, R. G., 1975, Generalized facies models for resedimented conglomerates of turbidite association: Geological Society of America Bulletin, v. 86, p. 737–748.

MANUSCRIPT RECEIVED BY THE SOCIETY MARCH 27, 1990

Printed in U.S.A.

Geological Society of America
Special Paper 255
1990

Late Paleozoic–early Mesozoic oceanic basement of a Jurassic arc terrane in the northwestern Sierra Nevada, California

Y. Dilek*
Department of Geology, University of California, Davis, California 95616
P. Thy
NASA, Johnson Space Center, SN2, Houston, Texas 77058
E. M. Moores
Department of Geology, University of California, Davis, California 95616
S. Grundvig
Geologisk Institut, Aarhus Universitet, DK-8000 Arhus C, Denmark

ABSTRACT

The Jurassic Smartville arc sequence in the northern Sierra Nevada foothills is bounded both in the north and east by ophiolitic and marine rocks of a disrupted oceanic terrane. The Jarbo Gap ophiolite in the north consists of serpentinized harzburgite and dunite intruded by gabbro and diorite and overlain by volcanic and sedimentary rocks, all of which are metamorphosed to upper greenschist-amphibolite facies. Contacts between the ophiolitic subunits, which commonly have lenticular shapes, are generally tectonic and discontinuous along-strike. Ophiolitic rocks display a penetrative fabric of a west-northwest–striking and steeply to moderately north-northeast–dipping foliation. Locally, mafic to intermediate volcanic rocks within the ophiolite are cut by fault-bounded and/or diapiric intrusions of serpentinized peridotites. Contact relations and the rock types indicate that the Jarbo Gap ophiolite is part of a multiply deformed, heterogeneous oceanic crust.

Marine rocks exposed south of and structurally beneath the Jarbo Gap ophiolite form a chert-argillite unit that is composed mainly of blocks of sedimentary, volcanic, plutonic, and ultramafic rocks in an argillite matrix. The matrix displays a northwest-trending, northeast-dipping penetrative foliation, whereas some of the blocks in the matrix show little or no deformation. Olistostromal origin of these undeformed blocks suggests that gravity debris sliding in addition to tectonic mixing was a significant agent for development of the generally chaotic internal structure of the chert-argillite unit. Ophiolitic blocks and detritus in the matrix indicate an older ophiolitic provenance, whereas blocks and clasts of relatively undeformed and fresh volcanic rocks, which occur as olistoliths in a mudstone-sandstone sequence, suggest volcanic activity adjacent to the depocenter(s) of the chert-argillite unit.

Age relations, reconstructions of stratigraphic, structural, and contact relations, and regional correlations suggest that the Jarbo Gap ophiolite may have been originally

*Present address: Department of Geology and Geography, Vassar College, Poughkeepsie, New York 12601.

Dilek, Y., Thy, P., Moores, E. M., and Grundvig, G., 1990, Late Paleozoic–early Mesozoic oceanic basement of a Jurassic arc terrane in the northwestern Sierra Nevada, California, *in* Harwood, D. S., and Miller, M. M., eds., Paleozoic and early Mesozoic paleogeographic relations; Sierra Nevada, Klamath Mountains, and related terranes: Boulder, Colorado, Geological Society of America Special Paper 255.

overlain by the chert-argillite unit. Thus the Jarbo Gap ophiolite and the overlying chert-argillite unit are interpreted to have constituted a late Paleozoic–early Mesozoic oceanic basement. The existence of similar structural and stratigraphic relations between stratigraphically and chronologically correlative ophiolitic and marine rocks exposed in the Central Belt east of the Smartville complex suggests that much of the Central Belt is composed of this late Paleozoic–early Mesozoic oceanic basement. Both the Jarbo Gap ophiolite and the chert-argillite unit are intruded by dikes of a basaltic andesite composition that display low FeO and high MgO contents and enrichment in compatible as well as some LIL elements. These dikes, which are Lower Jurassic in age and chemically unrelated to the ophiolitic sequence, show petrographic and petrologic characteristics typical of boninitic affinity and are interpreted as products of primary melts generated by high degrees of melting of a metasomatically enriched source during early stages of subduction. We infer that this subduction during Early Jurassic time may have underplated the preexisting oceanic basement and produced a submarine volcanic arc terrane. Part of the Smartville complex and the entire Slate Creek complex are interpreted as fragments of this arc terrane, which was built on and across the preexisting oceanic basement. This model suggests that the Western Belt and much of the Central Belt in the northern Sierra Nevada are part of a single ensimatic and polygenetic arc terrane that evolved in early Mesozoic time. This interpretation precludes tectonic scenarios suggesting the existence in the northern Sierra Nevada metamorphic belt of a number of discrete, exotic island-arc systems that were presumably accreted into the North American continental margin during early Mesozoic time. The current structural grain defined by northwest-striking and steeply northeast-dipping faults and associated structures is a relatively late-stage feature that modified the primary contact relations during a period of regional contraction in Late Jurassic to Early Cretaceous time.

INTRODUCTION

The northern Sierra Nevada metamorphic belt is subdivided into four main tectonic belts that are separated by nearly north-trending and steeply east-dipping, late-stage faults (Figs. 1, 2; Day and others, 1985). Variably metamorphosed, Paleozoic-Mesozoic marine, ophiolitic, and arc rocks compose a significant part of these tectonic belts. Whereas the Western and Eastern Belts are composed mainly of volcanic arc rocks, the intervening Central and Feather River Belts consist of marine and ophiolitic rocks interleaved with each other in a way reminiscent of melange terranes (Central Belt) and high-grade oceanic rocks tectonically mixed with blueschist and amphibolite rocks (Feather River Belt) characteristic of subduction complexes (Fig. 1; Day and others, 1985; Edelman and others, 1989). The Eastern, Feather River, and much of the Central Belt rocks have been interpreted as part of an oceanward-building active margin of western North America during late Paleozoic to early Mesozoic time (Burchfiel and Davis, 1981; Day and others, 1985; Edelman and others, 1989; Schweickert, 1981).

The Western Belt, composed of Jurassic plutonic, volcanic, hypabyssal, and volcaniclastic rocks of the Smartville complex, has received special attention because of its tectonic evolution and the paleogeography of the Smartville complex. Various tectonic models have been proposed addressing these problems (e.g., Moores and Day, 1984; Saleeby, 1981; Schweickert and Cowan, 1975). Some workers have suggested that the Smartville complex represents an exotic island-arc sequence that evolved above a west-dipping subduction zone (in present coordinates) and was accreted to the North American continental margin during an arc-continent collision (e.g., Moores, 1972, Moores and Day, 1984; Schweickert and Cowan, 1975; Ingersoll and Schweickert, 1986). Other workers have suggested that the Smartville complex is a volcanoplutonic arc that developed over a subduction zone dipping east beneath the continent, and that it was in a proximal position to North America throughout its evolution (e.g., Burchfiel and Davis, 1981; Saleeby, 1981). In all models the age of the Smartville complex has been considered as Late Jurassic based on U/Pb zircon ages from its sheeted dikes, felsic plutons, and upper volcanic units (McJunkin and others, 1979; Saleeby, 1982), and thus its accretion and deformation age has been interpreted to correspond to the Late Jurassic Nevadan orogeny (Day and others, 1985). Recent studies have shown, however, that the Smartville complex does not represent a cogenetic arc-ophiolite suite as previously thought (Beard and Day, 1987; Dilek, 1989a), and that it comprises plutonic and hypabyssal rocks as old as Lower Jurassic (Bickford and Day, 1988; Saleeby and others, 1989) in addition to the Upper Jurassic plutonic and volcanic rocks intruding and overlying, respectively, these older arc rocks. It is therefore interpreted that the older part of the Smartville complex is coeval with and correlative to the Early Jurassic Slate Creek complex, a volcanoplutonic arc sequence to the east, and

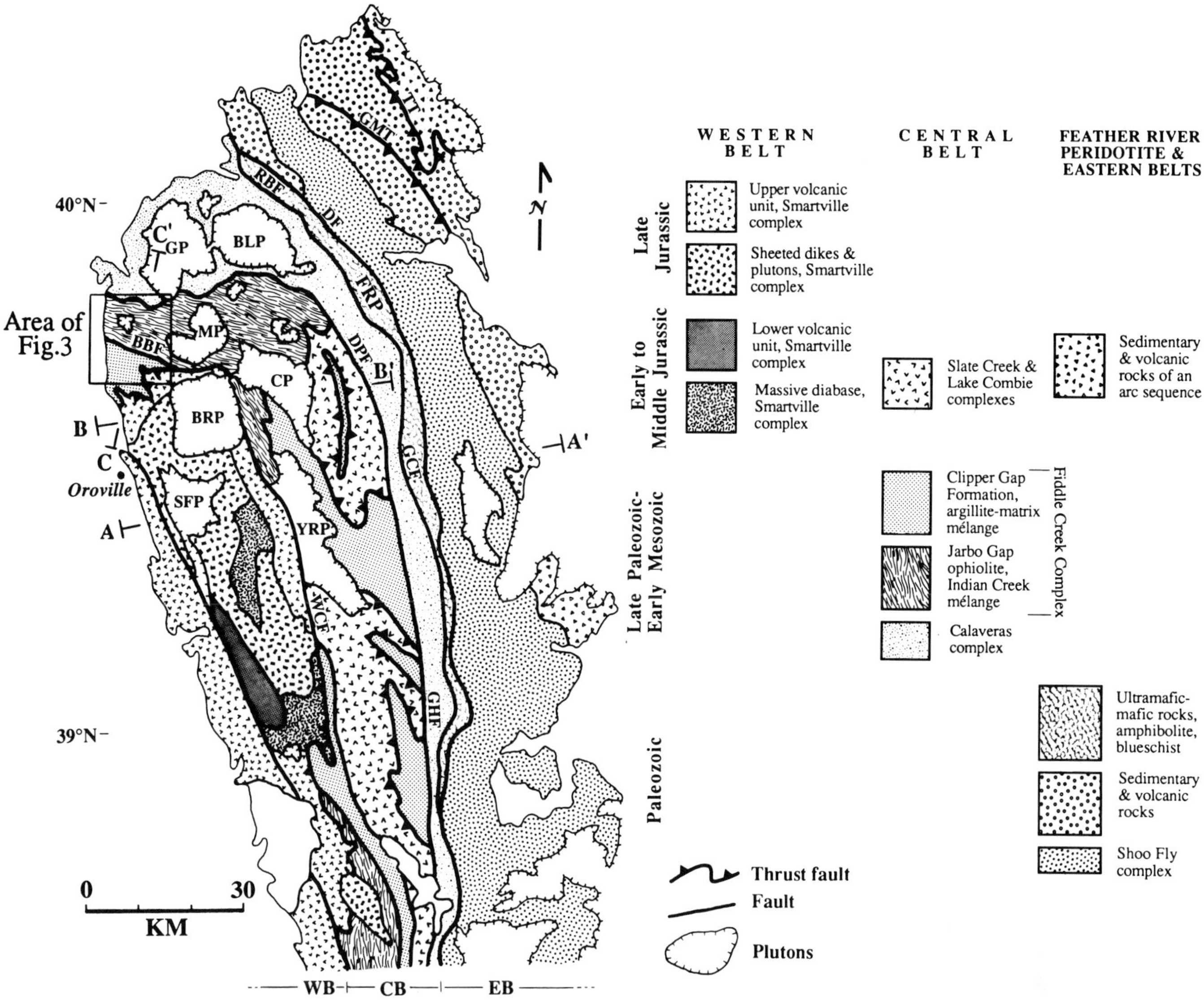

Figure 1. Simplified tectonic map of the northern Sierra Nevada (modified from Dilek, 1989b). Major tectonic belts include: EB = Eastern Belt; CB = Central Belt; FRP = Feather River peridotite; WB = Western Belt. Major faults include: BBF = Big Bend fault zone; DF = Downieville fault; DPF = Dogwood Peak fault; GCF = Goodyear's Creek fault; GHF = Gillis Hill fault; GMT = Grizzly Mountain thrust; RBF = Rich Bar fault; TT = Taylorsville thrust; WCF = Wolf Creek fault. Major plutons include: BLP = Bucks Lake pluton; BRP = Bald Rock pluton; CP = Cascade pluton; GP = Grizzly pluton; MP = Merrimac pluton; SFP = Swedes Flat pluton; YRP = Yuba Rivers pluton; area of Figure 3 is outlined.

that both arc sequences may be fragments of the same arc terrane (Bickford and Day, 1988; Dilek, 1989a, b).

The Smartville arc complex is bounded in the north by an ophiolitic assemblage (Jarbo Gap ophiolite; Dilek, 1989a) and a chert-argillite unit (Fig. 3). The chert-argillite unit has been previously interpreted as the tectonic basement of the Smartville complex, implying an allochthonous origin of the arc sequence (Day and others, 1985; Ricci and others, 1985). Contact relations between the Jarbo Gap ophiolite and the chert-argillite unit have been less well understood, mostly because of the chaotic internal structure of these tectonic units as well as the late-stage structures that overprinted and modified the primary structures and contacts within and between them. Both the Jarbo Gap ophiolite and the chert-argillite unit are significant components of an enigmatic geologic province bounding the Smartville and Slate Creek arc fragments, and therefore, understanding their primary structural and stratigraphic relations with each other and with the arc fragments is important to reconstruct the tectonic evolution of the northern Sierra Nevada metamorphic belt.

In this chapter, we describe the Jarbo Gap ophiolite and the chert-argillite unit and the contact relations within and between them. On the basis of structural and chronologic relations and the

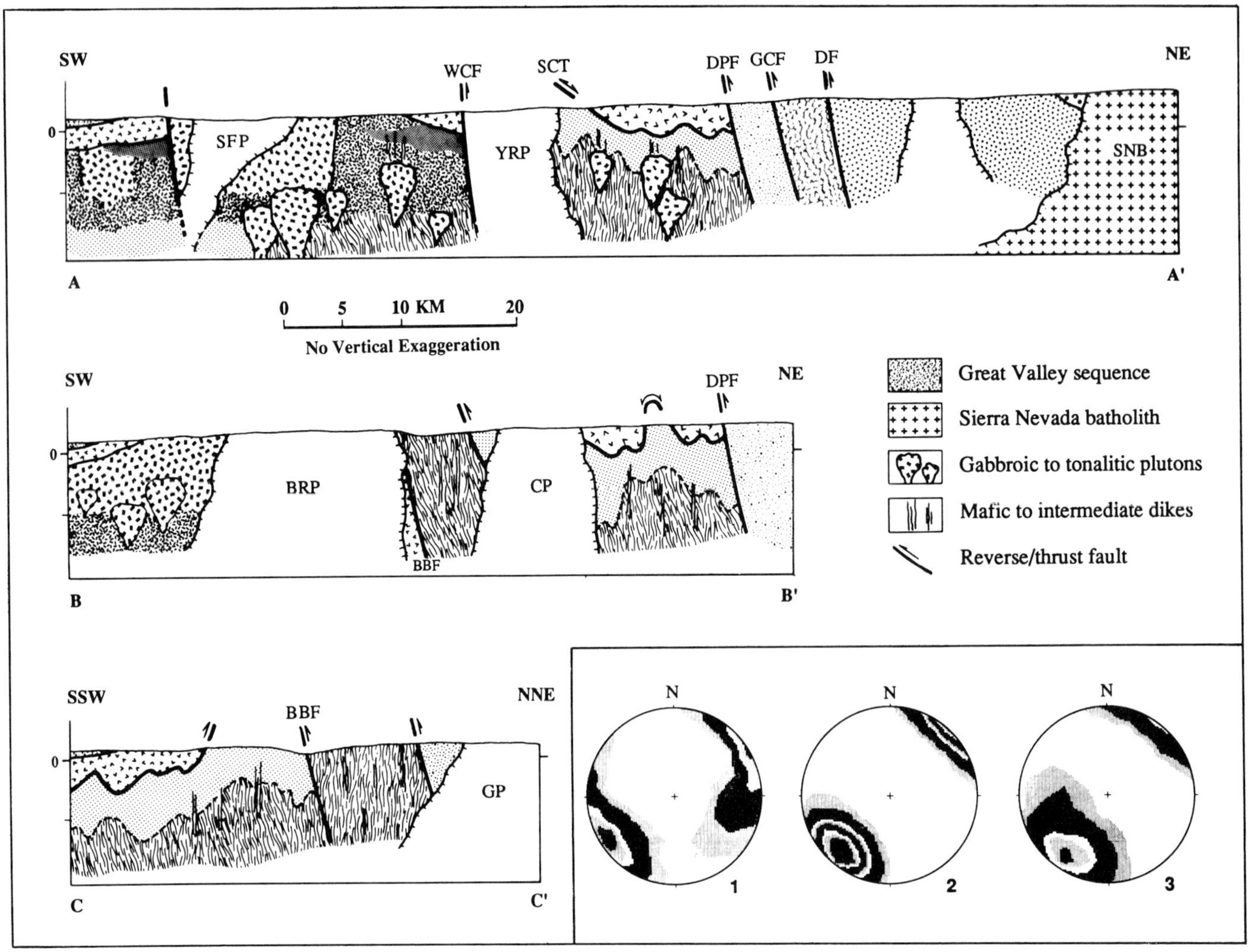

Figure 2. Interpretive geologic cross sections from the northern Sierra Nevada metamorphic belt (see Fig. 1 for orientations). Symbols are the same as in Figure 1; dashed line represents the inferred depositional/stratigraphic contact between the chert-argillite unit and the ophiolitic basement. Inset figure shows lower-hemisphere equal-area contour diagrams for poles to foliation in (1) the upper volcanic unit of the Smartville complex, (2) argillite matrix of the chert-argillite unit, and (3) metavolcanic and metasedimentary rocks of the Jarbo Gap ophiolite within the area outlined in Figure 1. Contour interval is 2 σ.

regional correlations, we interpret the chert-argillite unit as a disrupted marine sequence that once depositionally overlay the Jarbo Gap ophiolite. Both the Jarbo Gap ophiolite and the chert-argillite unit are intruded by basaltic andesite dikes of incipient island-arc character, suggesting that the inferred oceanic basement was underplated by a subducting oceanic slab in early Mesozoic time (Dilek, 1989a, b; Thy and Dilek, 1987). Radiometric dates from the dikes indicate an Early Jurassic age for the inferred arc volcanism (U/Pb zircon age of 196 +23/−11 Ma for the youngest dikes; Saleeby and others, 1989), which is coeval with the igneous ages of the Slate Creek arc complex in the Central Belt and with the older part of the Smartville complex in the Western Belt (Bickford and Day, 1988). These relations suggest that the Smartville and Slate Creek complexes may have evolved as part of a Jurassic arc terrane on and across the late Paleozoic–early Mesozoic oceanic basement described in this study. This interpretation implies that the contact relations between the volcanic arc fragments and the inferred oceanic basement were originally intrusive and depositional, although they are currently tectonic. We describe the Jarbo Gap ophiolite and the chert-argillite unit in the first part of the chapter and then summarize the petrology and geochemistry of cross-cutting dikes in the second part. Following the description of the contact, age, and correlation relations in the third part, we present our local and regional interpretations, which we hope will help refine and constrain tectonic models for geology of the northern Sierra Nevada.

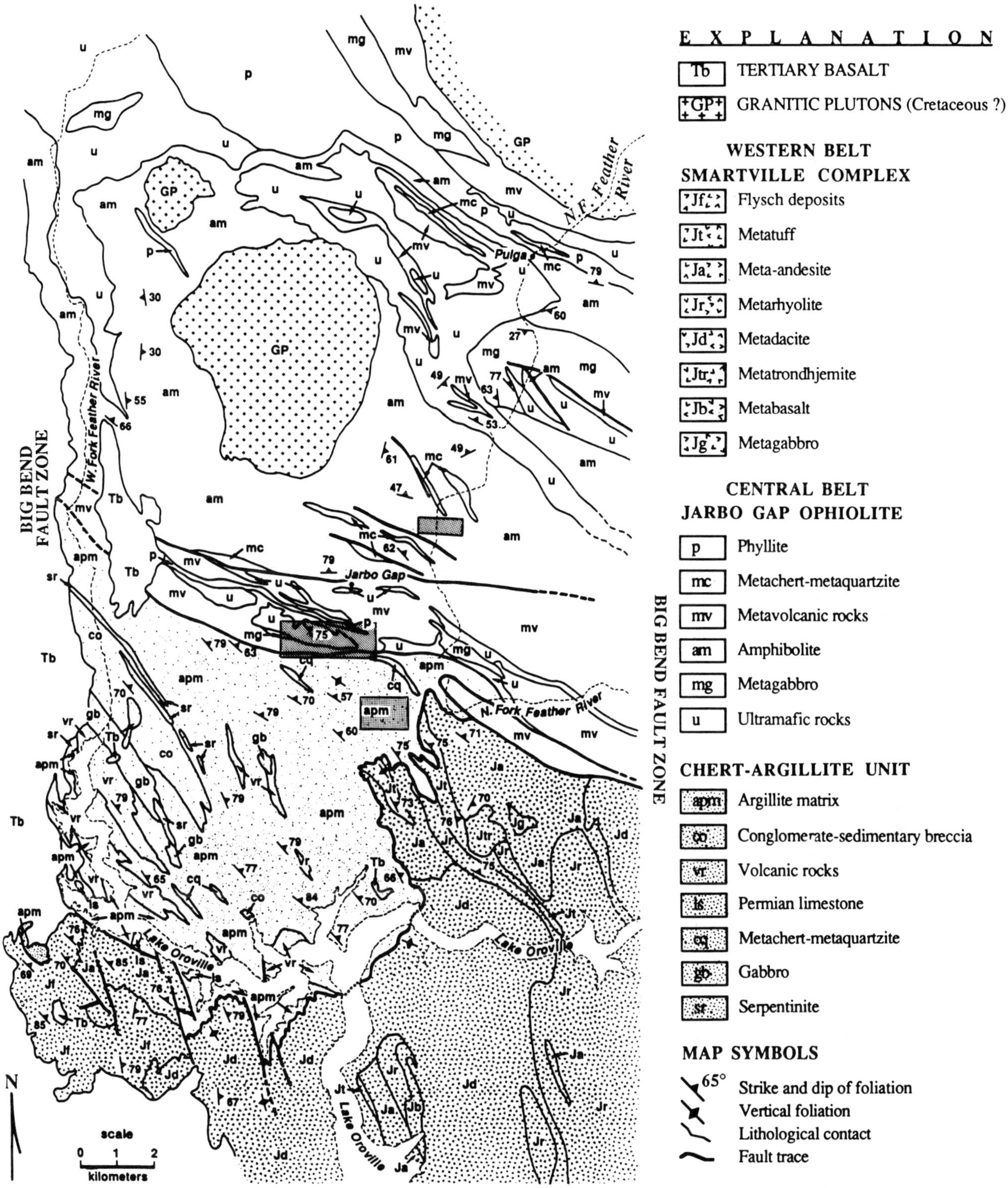

Figure 3. Geologic map of the Jarbo Gap area, northwestern Sierra Nevada foothills showing distribution of rock units and structures in Jarbo Gap ophiolite, chert-argillite unit, and Smartville complex (after Dilek, 1989a). Outlined areas show the areal extent of widespread dike intrusions within the Jarbo Gap ophiolite and the chert-argillite unit.

JARBO GAP OPHIOLITE

The Jarbo Gap ophiolite shows a progression from tectonized harzburgite and dunite to metagabbro, metadiorite, and metavolcanic and metasedimentary rocks. The contacts between these subunits are commonly tectonic and discontinuous along strike (Fig. 3). The ophiolite is intruded at all stratigraphic levels by mafic to silicic dikes (Jarbo Gap dikes) that are relatively undeformed compared to the host rocks (Dilek, 1989b). The main outcrop of mafic-ultramafic rocks consists of serpentinized and tectonized harzburgite, dunite, troctolite, metagabbro, diorite, and metavolcanic rocks, and is exposed in a northwest-trending belt south of Pulga (Fig. 3). The tectonized harzburgite occupies stratigraphically and structurally the lowest position within the ophiolite and consists of relict orthopyroxene grains and folded spinel layers (Fig. 4A). Structurally upward within the harzburgite, layers of dunite arc common, suggesting a transition between the harzburgite and the overlying dunite. The dunite is entirely serpentinized and includes rare olivine grains surrounded with antigorite, tremolite, and chlorite, and cut by veins of lizardite and chrysotile.

The harzburgite-dunite sequence is intruded and overlain by a metagabbro and metamorphosed plutonic and volcanic rocks (Figs. 3, 4A). Locally, layers of troctolite are present between the metagabbro and the underlying dunite. The metagabbro crops out in a southeast-plunging synform, which is nearly 2.5 km wide and approximately 500 m thick and, in turn, is cut by small dioritic to tonalitic intrusions (Fig. 4A). The metagabbro displays a distinct magmatic layering defined by an apparent change of color index and grain-size variation. It is locally strongly foliated and displays shear zones of flaser gabbro. The foliation in the flaser gabbro is associated with a blastomylonitic texture characterized by augen-shaped hornblende and plagioclase porphyroclasts, which are commonly parallel or slightly oblique to the magmatic layering. Development of this foliation is associated with deformation of the magmatic clinopyroxene and formation of synkinematic amphibole. Fine-grained mylonitic bands form shear zones 30 to 50 cm thick in the flaser gabbro. These shear zones are slightly oblique (~20°) or at high angles (70°) to the mineral layering and are composed of thin layers of amphibole, recrystallized plagioclase, and opaque minerals.

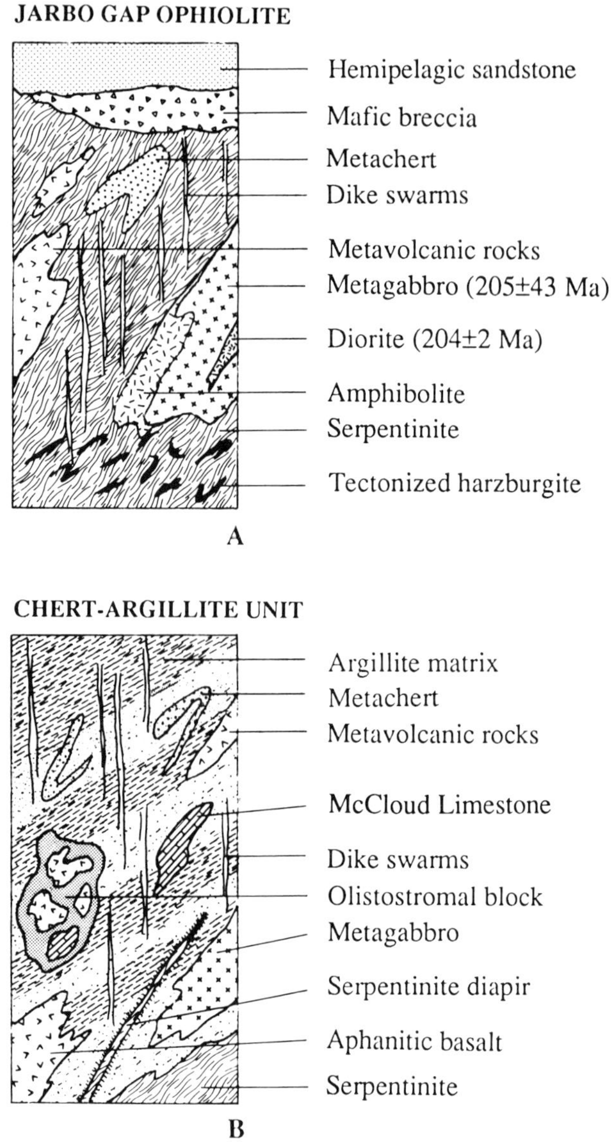

Figure 4. Schematic columnar sections showing the internal structure of (A) the Jarbo Gap ophiolite and (B) the chert-argillite unit.

The northwest-trending belt of mafic-ultramafic rocks south of Pulga structurally overlies an areally extensive amphibolite body exposed along a northwest-striking, moderately northeast-dipping fault surface (Fig. 3). The amphibolite is medium to coarse grained and displays a pronounced northwest-striking foliation and gneissic banding. The foliation plane in the amphibolite is defined by elongated and parallel crystals of green to blue-green amphibole with alternating layers of albite and quartz. This foliation is subsequently deformed by kink bands that form a crenulation cleavage, which is associated with the growth of actinolite, chlorite, and white mica crystals. Lenses and faulted blocks of metachert and metaquartzite are common within the amphibolite. Relatively undeformed basaltic andesite dikes intrude the amphibolite at an angle parallel or slightly oblique to the main foliation.

The amphibolite unit is juxtaposed against metavolcanic rocks along a nearly east-trending and steeply north-dipping northern strand of the Big Bend fault zone (Fig. 3). Metavolcanic rocks include fine-grained lavas interlayered with tuff, epiclastic volcanic rocks, mafic breccia, and tuff breccia. The mafic breccia unit consists of angular and unsorted clasts of gabbro, basalt, andesite, and dacite in a fine-grained siliceous matrix, and it is faulted against the serpentinite. Metasedimentary rocks consist of phyllite, metaquartzite, and metachert, and commonly form fault-bounded blocks and lens-shaped bodies within the metavolcanic unit (Fig. 3).

Metavolcanic rocks within the Big Bend fault zone are locally cut by fault-bounded blocks and diapiric intrusions of serpentinite. The serpentinite is commonly schistose and exhibits northwest-striking and steeply northeast-dipping penetrative foliation. Protolith textures and the primary mineralogy in serpentinized peridotite(s) are entirely obliterated; the common metamorphic minerals in the serpentinite include antigorite, chrysotile, tremolite, chlorite, talc, magnesite, and opaque minerals. Pipelike intrusions of rodingitized plutons, varying from dolerite to gabbro and ranging in size from several meters to tens of meters, are widespread within the serpentinite. Most rodingitized intrusions are elongated parallel to the foliation in the serpentinite and display fine-grained, dark-colored reaction zones ("blackwall") around them. Rodingitic rocks typically include epidote, clinozoisite, tremolite, actinolite, chlorite, sphene, and calcite, and display a granoblastic texture. Both the serpentinite and the rodingitic intrusions are cut by local shear zones and faults that display pyrite-chalcopyrite concentrations.

The serpentinite and metavolcanic rocks are intruded by isolated dikes and local dike complexes (Jarbo Gap dikes), which show multiple intrusive relations (Dilek, 1989b). At one locality south of Jarbo Gap (Fig. 3), cross-cutting relations and lithologic and compositional differences between the dikes indicate at least four separate generations of dike intrusions (Dilek, 1989b). The oldest dikes intruding the serpentinite and the rodingite are medium-grained dacite to quartz andesite and are commonly folded and faulted. The youngest dikes intrude both serpentinite and metavolcanic rocks and cross-cut all the existing dikes. These dikes are relatively undeformed and are composed of plagiogranite-leucotonalite. A U/Pb zircon date from one of these plagiogranite dikes gives an igneous age of 196 +23/−11 Ma (Saleeby and others, 1989). The other generations of dikes include dike swarms of basaltic andesite, andesite, and dacite in the serpentinite, and individual andesite and andesitic basalt dikes in metavolcanic rocks. The individual andesite and basaltic andesite dikes have a characteristic porphyritic texture with yellow to brown, euhedral phenocrysts of igneous amphibole. The first three dike generations strike northwest and dip moderately to steeply to the northeast, whereas the youngest dike generation strikes northeast and dips to the northwest at shallow angles.

CHERT-ARGILLITE UNIT

The chert-argillite unit contains various rock types as millimeter- to kilometer-long blocks in a fine-grained argillite matrix (Fig. 3). Dominant rock types in the matrix include chert, turbiditic sandstone, conglomerate, sedimentary breccia, pebbly mudstone, fossiliferous limestone, graywacke, volcaniclastic rocks, massive volcanic rocks, gabbro, and serpentinite (Fig. 4B; Dilek, 1989a; Vaitl, 1980).

The argillite matrix displays a pervasive northwest-striking and northeast-dipping foliation (Fig. 2) and lacks, in general, a coherent internal stratigraphy (Fig. 3). However, the argillite matrix is locally deprived of exotic blocks or fragments and forms a broken formation consisting of phyllite, pebbly phyllite, and fine-grained phacoidal chert (Ricci and others, 1985). Bedding is rarely preserved in the broken formation; common structures include rootless fold hinges, intrafolial folds, discrete crenulation cleavages, and folded and dismembered quartz veins (Dilek, 1989a; Ricci and others, 1985). All these structures are overprinted by the pervasive northwest-striking foliation, with the exception of the least deformed quartz veins that cut across the foliation at high angles.

Despite the well-pronounced deformation of the argillite matrix, some of the blocks and clasts of various rock types display evidence for little or no deformation. Large outcrops of debris flow containing blocks and clasts of basaltic lava, dolerite, gabbro, and serpentinite are present as olistostromes in the matrix. Highly angular volcanic clasts in these olistostromes are vesicular and aphyric and are commonly associated with turbiditic sandstone, black mudstone, and pebbly mudstone. Chert cobbles in black siliceous mudstone contain fine-grained limestone clasts that are elongated and broken due to soft-sediment deformation.

In contrast to the relatively undeformed olistostromal blocks in the matrix, some of the blocks that are bounded by steeply dipping faults are highly deformed, displaying structures generally subparallel to the main foliation in the matrix (Fig. 4B). Fine-grained chert outcrops in some of these fault blocks are isoclinally folded around vertically plunging fold axes, whereas nearly vertically inclined pillow-lavas are sheared and elongated parallel to their long dimensions. Similarly, some fault blocks that are composed of successions of hemipelagic sandstone-shale sequence display isoclinal folds with vertically plunging fold axes.

Relatively undeformed mafic lavas occur as kilometers-long blocks in the argillite matrix north and south of the western branch of Lake Oroville (Fig. 3). Dark green, fine-grained, aphanitic volcanic rocks include massive splitic basalt (sample LR-1; see Tables 2 through 4) and basalt breccias with undeformed amygdules filled with calcite, quartz, and/or chlorite. The massive splitic basalt is a TiO_2-enriched lava chemically similar to basalts recovered from modern propagating rifts (Sinton and others, 1983) and is clearly different from any of the other volcanic rocks in the Jarbo Gap ophiolite and the chert-argillite unit. Blue-gray, fine- to medium-grained limestones are locally intercalated with these basaltic lavas. Watkins and others (1987) demonstrated that some of these limestone blocks contain distinctive Permian faunas found in the McCloud Limestone of the eastern Klamath Mountains.

Farther north toward the Big Bend fault zone, deformed blocks in the argillite matrix include sheared metagabbro, serpentinite, metabasalt, folded chert, and metasedimentary rocks. Blocks and clasts of sheared metagabbro and metabasalt are commonly enveloped by fine-grained sandstone and mudstone. These relations imply that a relatively older and deformed oceanic basement was exposed on the sea floor during deposition of the sandstone-mudstone sequence, which was subsequently incorporated into the argillite matrix as olistoliths. A northwest-striking serpentinite septum, approximately 3 m in width,

TABLE 1. REPRESENTATIVE MICROPROBE ANALYSES OF PRIMARY AND SECONDARY AMPHIBOLES

	JG-5	JG-8	JG-12	DC-4	DC-8	JG-20	JG-16	JG-19	JG-7
SiO_2	41.57	40.49	41.24	43.39	40.24	42.43	48.43	52.80	48.34
TiO_2	2.55	2.14	2.18	1.84	2.73	0.44	0.12	0.14	1.24
Al_2O_3	13.49	13.54	12.96	12.35	13.78	16.54	9.13	4.23	7.37
FeO*	10.64	12.72	11.31	9.56	11.08	6.14	11.85	11.11	10.32
MnO	0.00	0.18	0.20	0.24	0.17	0.19	0.28	0.38	0.00
MgO	15.78	12.95	14.76	16.17	14.83	16.52	14.56	16.58	16.41
CaO	11.32	11.92	11.41	11.27	11.13	12.02	11.80	12.23	12.10
Na_2O	2.81	2.05	2.87	2.82	2.98	3.39	1.25	0.33	1.50
K_2O	0.42	0.57	0.41	0.39	0.55	0.00	0.00	0.00	0.34
Total	98.58	96.56	97.50	98.03	97.49	97.67	97.42	97.80	97.62
Si	5.885	5.970	5.962	6.155	5.808	5.982	6.908	7.442	6.901
Ti	0.272	0.237	0.254	0.196	0.296	0.047	0.013	0.015	0.133
Al	2.253	2.355	2.210	2.066	2.346	2.751	1.536	0.703	1.241
Fe^{3+}	1.153	0.771	0.943	0.961	1.068	0.633	0.671	0.599	0.513
Fe^{2+}	0.106	0.796	0.424	0.173	0.269	0.091	0.741	0.710	0.719
Mn	0.000	0.022	0.025	0.029	0.021	0.023	0.034	0.045	0.000
Mg	3.332	2.848	3.182	3.420	3.192	3.474	3.097	3.485	3.494
Ca	1.717	1.883	1.767	1.713	1.721	1.816	1.803	1.847	1.851
Na	0.771	0.586	0.804	0.775	0.834	0.927	0.346	0.090	0.415
K	0.076	0.107	0.076	0.071	0.101	0.000	0.000	0.000	0.062

JG-5 to DC-9 are cores of magnesio-hastingsite phenocrysts and microphenocrysts in dikes and breccia.
JG-20 is massive pargasite, JG-16 is fibrous magneisio-hornblende, and JG-19 is fibrous actinolitic hornblende in metagabbros.
JG-7 is fibrous magnesio-hornblende in microphenocrystic dike.
Classification and distribution of Fe^{2+} and Fe^{3+} are according to IMA recommendations (Leake, 1978; Rock, 1987).

intrudes the same sandstone-mudstone sequence along a subvertical plane and displays a several-centimeter-thick, dark-colored baked zone. The baked zone is composed of metamorphosed sedimentary host rock, suggesting that the serpentinized peridotite was hot when it was injected.

The argillite matrix and the locally developed broken formation south of the Big Bend fault zone are intruded by commonly northwest-striking and northeast-dipping individual dikes and dike swarms (Dark Canyon dikes; see Fig. 3 for the location). Some of these dikes are highly altered and folded, faulted, and sheared, suggesting that they were probably injected prior to deformation of the matrix. The primary texture and mineralogy of these dikes are entirely obliterated. Some other dikes are relatively undeformed and injected subparallel to the northwest-striking and steeply northeast-dipping foliation. These dikes are aphyric and are altered to the extent that their primary mineralogy is not preserved. Finally, there exist dikes that are oblique to and/or cut the main foliation in the matrix, suggesting their post-foliation injection. These dikes, commonly several tens of centimeters thick, are relatively fresh basaltic andesites with green to brown hornblende phenocrysts in a fine-grained groundmass. There are no radiometric dates yet available from these intrusions. Petrographically and geochemically, however, these dikes are similar to Lower Jurassic basaltic andesite dikes that intrude the Jarbo Gap ophiolite discussed in the next section.

MINERALOGY AND GEOCHEMISTRY

Both the Jarbo Gap and the Dark Canyon basaltic andesite dikes are relatively mafic and contain igneous amphiboles as phenocrysts and microphenocrysts. In addition to these primary phenocrysts, albite, presumably pseudomorphed after calcic plagioclase, and secondary amphiboles are present in the groundmass. Some fine-grained dikes contain devitrification textures. Andesitic and dacitic dikes and the mafic breccia unit in the ophiolite contain amphibole and albite as phenocrysts and orthoclase and quartz in the groundmass, whereas plagiogranite dikes have quartz phenocrysts. The relatively undeformed but amphibolized gabbros in the ophiolite locally contain relict calcic plagioclases.

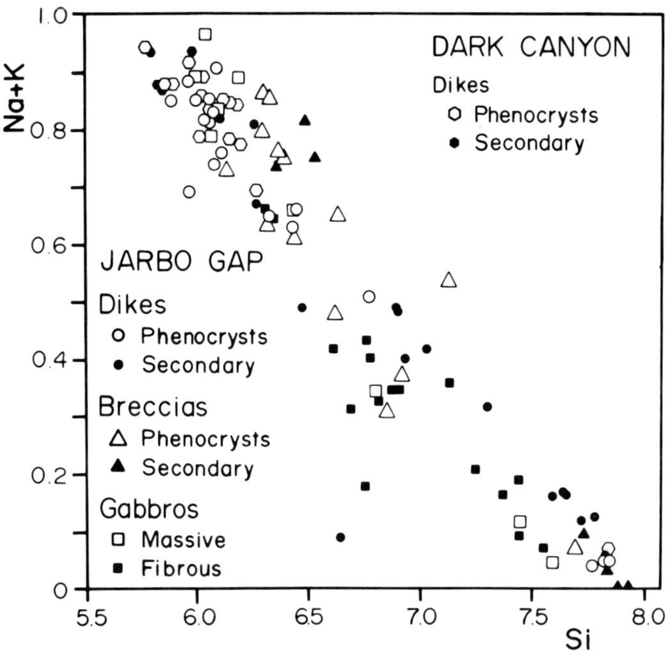

Figure 5. Variation diagram showing Na + K vs. Si for amphiboles (calculated to 23 oxygens and without correcting for the presence of Fe^{3+}). Filled symbols are secondary or fibrous amphiboles; unfilled symbols are primary or massive amphiboles.

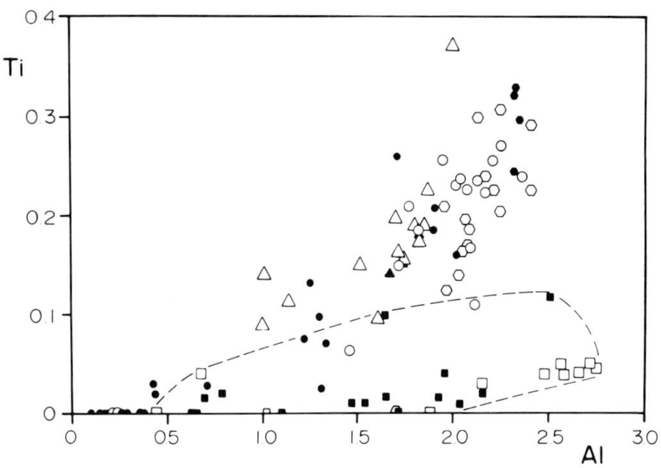

Figure 6. Ti vs. Al for amphiboles (calculated to 23 oxygens). Dashed line encloses amphiboles from the gabbros in the Jarbo Gap ophiolite. Plotting symbols are the same as in Figure 5.

The primary amphiboles in the dikes are mainly magnesio-hastingsitic hornblende with a high estimated Fe^{3+} (Table 1; nomenclature and calculation of Fe^{3+} according to IMA recommendations; Leake, 1978). The secondary amphiboles in the dikes are dominantly actinolitic. In contrast, the massive amphiboles of relatively undeformed gabbros in the Jarbo Gap ophiolite are dominantly paragasite or tschermakite. There is a linear trend between high Na + K for the hastingsitic and paragasitic amphiboles and high Si for the secondary and fibrous actinolitic amphiboles (Fig. 5). Although there is no distinct difference between the amphiboles of the dikes and those of the gabbros on this diagram, most of the primary amphiboles of the dikes have relatively high Ti, whereas all analyses of amphiboles from the gabbros show a systematic lower Ti content (Fig. 6). This latter phenomenon probably reflects the lower Ti-content of the parental pyroxenes in the gabbros. In addition, massive as well as fibrous amphiboles of the gabbros, when compared to the amphiboles of the dikes, have significantly higher Al, which primarily appears as higher octahedral coordinated Al (Fig. 7). These relations suggest that the dikes and the gabbros are genetically unrelated. The high Fe^{3+} of the hastingsitic amphibole points to a relatively oxidized melt. On the other hand, the secondary amphibole of the dikes does not reflect a high Fe^{3+} regime during their formation. Both albite ($An_{1-5}Or_{0-7}$) and orthoclase ($An_{0-1}Ab_{5-15}$) show limited solid solutions. The majority of the relict plagioclases of the gabbros range compositionally from An_{50} to An_{92}.

Despite extensive alteration and hydration of many of the samples, primary igneous features can be identified in relatively unaltered samples (Fig. 8; Table 2). The majority of the analyzed dikes have relatively high MgO (5 to 10 wt %) and SiO_2 (50 to 57 wt %) characteristic of basaltic andesites and andesites. In addition to the basaltic andesite and andesite dikes, there are dacite dikes with a SiO_2 of 60 to 62 wt % and plagiogranite dikes with 76 wt % SiO_2. The volcanic compositions of the breccias appear to be relatively evolved and are andesitic.

Compared to oceanic basalts, the Jarbo Gap and Dark Canyon dikes have systematic low FeO and high TiO_2 contents. The Al_2O_3 content appears relatively high in many samples. The variable CaO, Na_2O, K_2O, and H_2O contents may in part reflect extensive alteration and metasomatism. The general freshness and the lack of correlation between CaO and H_2O contents suggest, however, that the generally high H_2O and K_2O contents of the dikes may be primary. Compared to typical calc-alkalic evolution

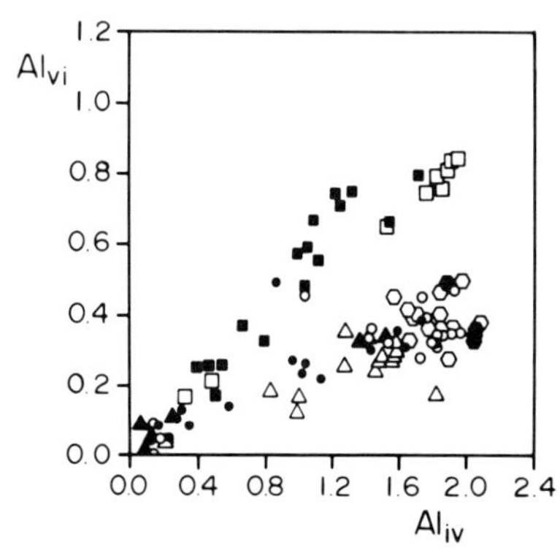

Figure 7. Al_{vi} vs. Al_{iv} for amphiboles calculated to 23 oxygens and assuming 2 tetrahedral cations in the structural formula. Plotting symbols are the same as in Figure 5.

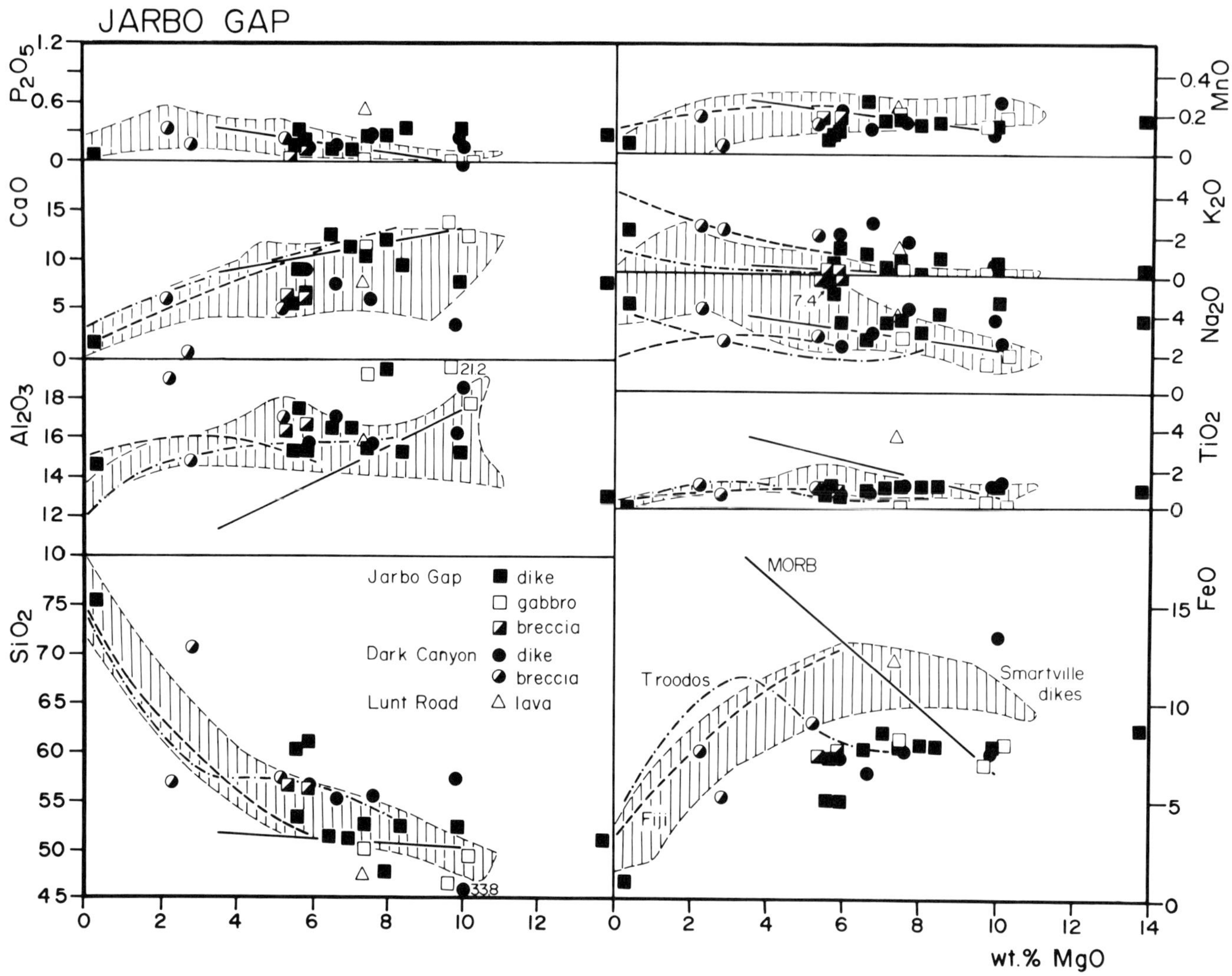

Figure 8. Major-element variation of the analyzed Jarbo Gap and Dark Canyon dikes, mafic breccia unit and gabbros from the Jarbo Gap ophiolite, and a basaltic lava (Lunt Road) from the chert-argillite unit. For the purpose of comparison the general fields/trends of the Smartville dikes (Xenophontos, 1984), Troodos volcanic glasses (Robinson and others, 1983), volcanic glasses of the Fiji Plateau (Jezek, 1975), and mid-ocean ridge basalts (MORB) are shown.

trends, both the Jarbo Gap and Dark Canyon dikes are significantly higher in MgO and have systematic lower FeO contents (Fig. 9). These features differ from the Upper Jurassic Smartville dikes, which follow a typical calc-alkalic trend interpreted as indicating an intra-arc/back-arc origin (Xenophontos, 1984; Dilek, 1989a). The Jarbo Gap and Dark Canyon dikes chemically seem to compare well to basaltic andesites from upper pillow lavas of the Troodos ophiolite (Robinson and others, 1983) and some low-TiO_2 basalts and basaltic andesite lavas of arc volcanoes (Johnson and others, 1985; Tatsumi and Ishizaka, 1982). Except for the low FeO and high MgO, there is no other major-element evidence for a strong source depletion and high degrees of melting.

Of the trace elements (Fig. 10, Table 3), Cr (80 to 500 ppm) and Ni (50 to 200 ppm) show systematic high values and a positive correlation with MgO. This enrichment may reflect high degrees of source melting. Nevertheless, high-field-strength elements such as Zr and Ti are not systematically depleted as would be expected in the case of extensive melting. Incompatible trace elements, such as Ba and Rb, as well as K_2O, of unaltered rocks are systematically enriched and may reflect source metasomatism. These features are typical of many basaltic andesites and boninites believed to have formed by melting of a depleted mantle wedge above a subducting slab (Hickey and Frey, 1982; Cameron and others, 1983; Johnson and others, 1985). The Zr/Y ratios correspond to those observed in many basaltic andesites and boninitic lavas and appear slightly higher than those observed in typical mid-ocean ridge basalts and in the Smartville dikes. In addition, the Ti/Zr ratios of the Jarbo Gap and Dark Canyon dikes are just on the low side of those of mid-ocean

TABLE 2. MAJOR-ELEMENT CHEMISTRY OF ANALYZED SAMPLES

Sample	Type*	SiO_2	TiO_2	Al_2O_3	Fe_2O_3	FeO	MnO	MgO	CaO	Na_2O	K_2O	P_2O_5	LOI
wt %													
JG-2	D	49.60	1.05	15.97	2.94	5.56	0.16	6.81	10.97	3.46	0.40	0.13	2.00
JG-3	D	59.15	0.79	15.01	0.00	4.95	0.08	5.39	4.80	7.20	0.30	0.17	1.38
JG-4	D	51.75	1.24	16.94	2.98	4.34	0.10	5.49	8.68	4.82	0.55	0.32	2.22
JG-5	D	48.98	1.02	12.57	1.50	6.86	0.16	13.18	7.55	3.62	0.34	0.28	3.01
JG-6	D	45.75	1.11	18.68	4.57	3.50	0.14	7.64	11.55	3.05	0.11	0.28	3.72
JG-7	D	50.63	1.17	14.69	1.21	6.45	0.14	9.50	7.54	4.48	0.61	0.35	2.58
JG-8	D	49.79	1.13	14.48	1.17	6.31	0.15	7.99	8.91	3.85	0.85	0.33	3.21
JG-11	D	48.75	0.93	15.64	2.78	4.80	0.26	6.18	12.00	2.60	1.08	0.14	3.55
JG-12	D	50.68	1.19	14.85	1.73	5.99	0.17	7.18	9.92	3.62	0.77	0.26	2.71
JG-13	D	56.02	0.69	14.02	0.91	3.82	0.10	5.41	5.91	3.31	1.29	0.21	5.93
JG-15	D	74.66	0.07	14.41	0.13	0.73	0.06	0.36	1.56	4.41	2.20	0.07	1.01
JG-9	B	53.28	0.90	15.39	0.93	6.06	0.18	5.02	6.31	5.34	0.28	0.12	5.20
JG-10	B	53.21	0.96	15.69	1.02	6.29	0.18	5.53	5.76	5.47	0.30	0.12	5.08
JG-16	G	47.87	0.14	17.22	0.73	7.07	0.17	9.91	11.94	1.84	0.10	0.01	2.01
JG-18	G	45.25	0.37	20.69	0.69	6.04	0.14	9.43	13.64	1.36	0.11	0.01	2.20
JG-19	G	48.78	0.15	18.74	1.04	7.10	0.19	7.22	11.09	2.74	0.14	0.01	1.72
JG-1	S	29.35	1.47	20.24	2.78	11.20	0.38	20.61	2.75	0.13	0.06	0.07	10.47
JG-14	S	38.85	0.34	11.61	0.14	9.13	0.19	21.97	7.80	0.23	0.07	0.02	8.64
JG-17	S	39.39	0.04	0.86	5.10	2.50	0.09	37.65	1.49	0.08	0.04	0.01	12.53
JG-20	S	45.77	0.15	10.12	0.68	6.43	0.13	20.63	10.10	0.85	0.07	0.01	3.92
DC-1	D	50.82	0.69	14.21	0.37	6.15	0.20	5.27	8.00	2.07	1.81	0.12	9.92
DC-4	D	53.56	0.86	16.58	1.48	5.01	0.13	6.42	7.39	3.03	2.52	0.17	1.97
DC-6	D	53.46	1.07	15.16	0.11	6.89	0.09	9.13	3.09	3.47	0.57	0.23	6.29
DC-7	D	28.23	1.17	15.45	0.59	10.62	0.22	8.30	16.47	2.13	0.16	0.13	16.21
DC-9	D	52.85	1.17	14.84	0.73	6.61	0.16	7.18	5.73	4.09	1.63	0.28	2.61
DC-2	B	55.08	1.12	18.38	1.80	5.78	0.18	2.16	5.81	4.13	2.30	0.34	2.40
DC-3	B	54.89	1.01	16.28	0.70	7.99	0.14	4.96	4.85	2.84	1.90	0.22	5.62
DC-5	B	67.88	0.66	14.19	0.63	4.39	0.05	2.68	0.64	2.46	2.13	0.17	3.16
DC-8	S	41.21	0.06	1.04	3.52	3.80	0.10	31.08	4.79	0.04	0.04	0.01	14.33
LE-1	L	44.33	3.42	14.88	1.81	9.74	0.21	6.85	7.21	3.56	1.41	0.50	5.19

*D = dike; B = breccia; G = gabbro; S = serpentinite or mafic gabbro; L = lava.

basalts and the Smartville dikes. This phenomenon may be a reflection of a relative Zr enrichment of their sources.

Relatively unaltered and some altered samples were selected for rare earth element analyses (Table 4). When normalized to chondrite, three of the Jarbo Gap dikes show systematic trends with enrichments in the LREE and high La_N/Yb_N ratios of 4 to 5 (Fig. 11A). These trends are similar to the patterns observed for basaltic andesites of the Setouchi volcanic belt of Japan described by Tatsumi and Ishizaka (1982) but clearly differ from the LREE-depleted patterns of the Troodos basaltic andesites (Rautenschlein and others, 1985). A rodingitized dolerite intrusion (JG-2) in the ophiolite shows a fairly flat trend with an La_N/Yb_N ratios of about 1 (Fig. 11B). Only one of the Dark Canyon dikes was analyzed for REE, and this sample (DC-4) shows the same general pattern as the Jarbo Gap dikes (Fig. 11B). It has lower

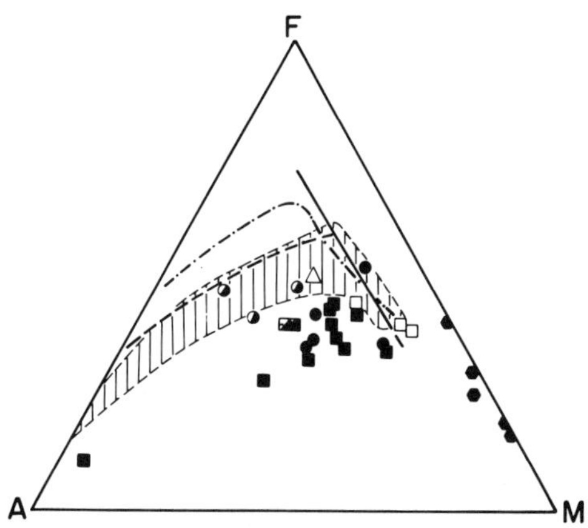

Figure 9. Alkali (Na_2O + K_2O) - FeO - MgO) variation of the Jarbo Gap and Dark Canyon dikes. Data and symbols are the same as in Figure 8. Filled hexagonal symbols near the MgO corner represent ultramafic rocks (serpentinized peridotites and serpentinites) from the Jarbo Gap ophiolite.

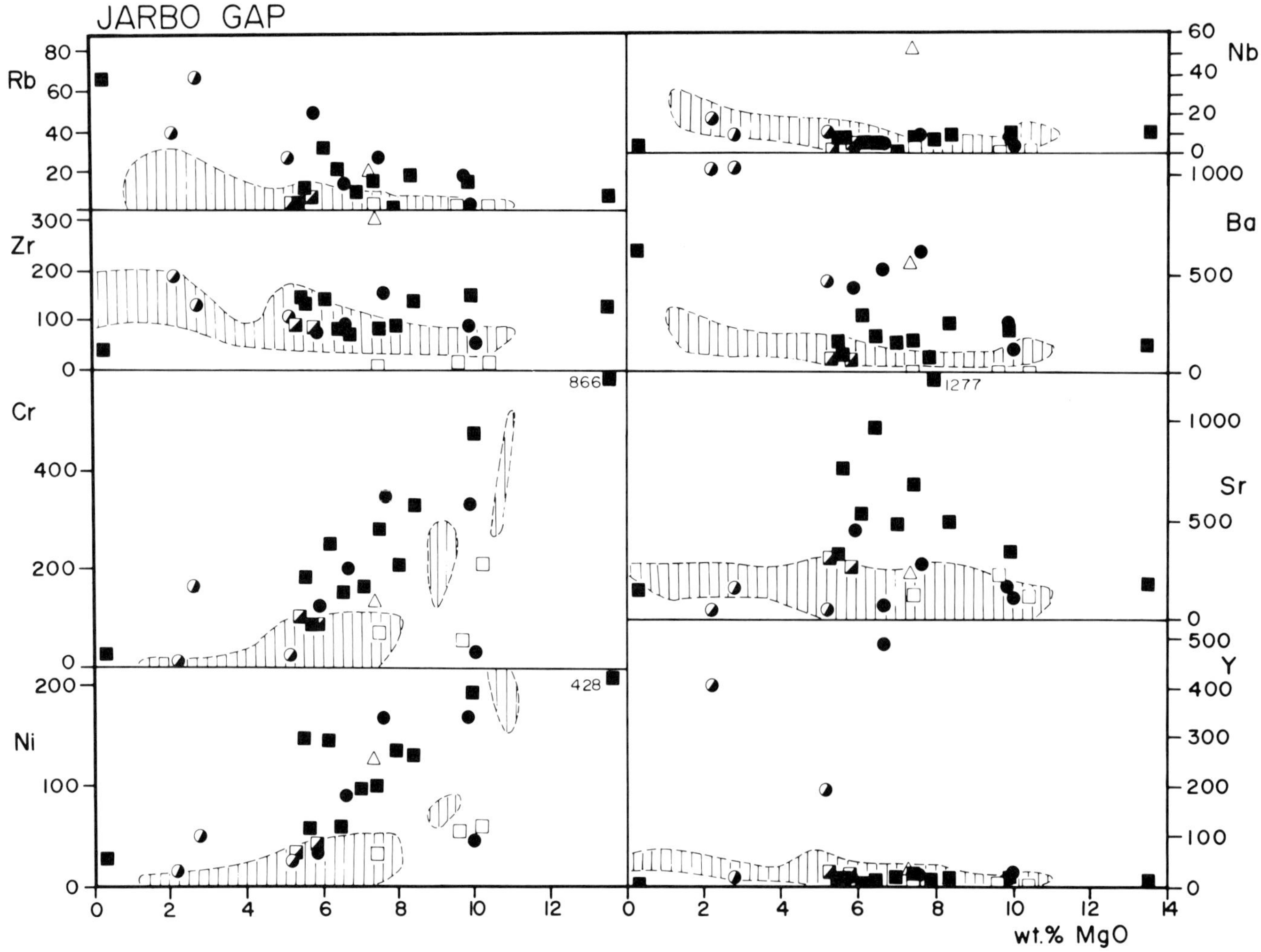

Figure 10. Trace-element (ppm) variation vs. MgO (wt%) for the Jarbo Gap and Dark Canyon rocks. Symbols are the same as in Figure 8. The ruled field represents the Smartville dikes (Smartville data from Xenophontos, 1984).

relative abundances, however, compared to the Jarbo Gap dikes. Basaltic lava in the chert-argillite unit (LR-1) shows a strong enrichment in LREE (Fig. 11b; La_N/Yb_N ~8 to 9). It is significant to note that, except for the plagiogranite dike of the Jarbo Gap ophiolite (JG-15), none of the analyzed samples shows negative Eu anomalies. This phenomenon may be a reflection of the lack of plagioclase phenocrysts and/or of the highly oxidized parental melts, but it may also indicate pre-melting depletion in source.

Figure 12 summarizes the trace-element variations for fresh Jarbo Gap dikes normalized to chondrite and a Yb_N of 10 according to Thompson (1982). With some notable exceptions, the systematic increase going in general from Yb to La can be related to increasing incompatibility during partial melting. However, Sr shows significant enrichment, whereas Nd and Ta (and Th) display marked depletions typical of island-arc series.

CONTACT RELATIONS

The Jarbo Gap ophiolite is juxtaposed against the chert-argillite unit along a WNW-trending, steeply NNE-dipping reverse fault, which forms the southern strand of the Big Bend fault zone (Figs. 2, 3; Dilek, 1989a, b; Hietanen, 1981). The dip angle of the fault changes along strike, emplacing various rock types of the ophiolite over the broken formation and the argillite matrix of the chert-argillite unit. The trace of the fault is rarely exposed, however, and the location of the fault is inferred from a variety of indirect evidence, such as the existence of a well-pronounced and linear topographic depression, highly sheared and brecciated metavolcanic and metasedimentary rocks near the inferred fault, truncation of the foliation and the contacts, and a sharp change from dominantly mafic and ultramafic rocks in the north to mainly phyllitic and sedimentary rocks in the south (Fig. 3).

TABLE 3. X-RAY FLUORESCENCE TRACE-ELEMENT ANALYSES

Sample	Type*	Ba	Cr	Ni	Nb	Zr	Y	Sr	Rb
ppm									
JG-2	D	150	163	99	0	70	22	488	8
JG-3	D	153	183	139	6	146	20	323	3
JG-4	D	137	94	57	7	130	19	769	11
JG-5	D	141	866	428	9	125	16	189	6
JG-6	D	69	209	137	6	88	16	1,277	1
JG-7	D	215	476	196	10	151	19	344	13
JG-8	D	248	332	131	9	138	20	494	16
JG-11	D	186	158	48	5	78	16	971	20
JG-12	D	161	280	104	8	92	15	677	14
JG-13	D	291	250	146	5	144	12	542	30
JG-15	D	728	27	25	4	42	8	149	65
JG-9	B	62	106	46	2	91	25	324	4
JG-10	B	58	87	41	5	85	23	260	6
JG-16	G	0	116	60	0	8	4	118	1
JG-18	G	0	55	22	0	15	6	226	1
JG-19	G	0	69	32	0	16	6	128	1
JG-1	S	30	232	322	5	104	31	183	2
JG-14	S	0	2,453	935	0	22	6	31	1
JG-17	S	6	2,083	1,894			2	6	1
JG-20	S	0	1,023	391	2	18	5	8	2
DC-1	D	424	124	30	4	77	18	447	49
DC-4	D	535	201	90	5	90	488	69	13
DC-6	D	244	338	170	7	93	16	171	16
DC-7	D	111	29	43	3	55	27	111	2
DC-9	D	613	346	171	9	157	20	279	26
DC-2	B	1,042	12	15	17	188	408	46	39
DC-3	B	458	24	15	9	103	197	48	26
DC-5	B	1,040	163	48	9	131	21	149	66
DC-8	S	0	1,288	973	0	15	3	41	1
LR-1	L	547	127	96	53	304	31	239	19

*D = dike; B = breccia; G = gabbro; S = serpentinite or mafic gabbro; L = lava.

The contact between the Jarbo Gap ophiolite and the Smartville complex is the east-southeastward continuation of the Big Bend fault (Figs. 1, 2). This steeply NNE-dipping reverse fault juxtaposes upper greenschist to amphibolite grade mafic and ultramafic rocks of the Jarbo Gap ophiolite against the lower greenschist facies volcanic, volcaniclastic, and epiclastic rocks of the Smartville complex. The Big Bend fault zone turns sharply to a north-south orientation around the Bald Rock pluton (Fig. 1), which apparently intrudes and truncates the fault (M. McLeod, personal communication, 1989). Farther south, the north-trending, steeply east-dipping Big Bend fault zone separates the Indian Creek ophiolitic melange in the east from volcanic rocks of the Smartville complex in the west (Edelman and others, 1989). The Big Bend fault zone and its southern continuation, the Wolf Creek fault (Fig. 1), are interpreted to have developed around 160 Ma, based on cross-cutting relations with the 160-Ma Yuba Rivers pluton (Bickford and Day, 1988; Day and others, 1985).

The chert-argillite unit is juxtaposed against volcanic, volcaniclastic, and flysch-type sedimentary rocks of the Smartville complex in the south and east. The nature of the contact is not well constrained because of its poor exposure and tectonic disruption by closely spaced shear zones striking northwest and dipping steeply to the northeast. On the basis of map patterns and gravity data that indicate a folded, sheet-like geometry for the Smartville rocks, Ricci and others (1985) suggested that the contact may be a deformed thrust fault, placing the Smartville complex over the chert-argillite unit. However, coarse sandstones and sedimentary breccias in the flysch sequence (Pentz sandstone; Vaitl, 1980), which depositionally overlie the volcanic and volcaniclastic rocks of the Smartville complex, contain clasts derived from both the Smartville complex and the chert-argillite unit (Vaitl, 1980). This

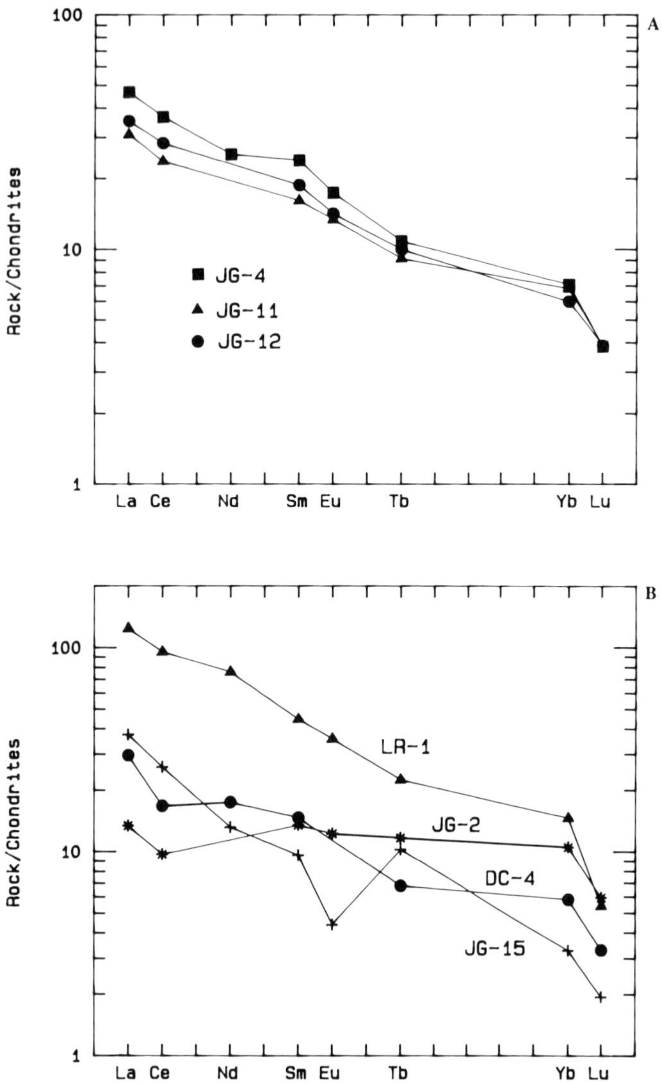

Figure 11. Rare earth elements normalized to chondrite. A, Jarbo Gap dikes (JG-4, JG-11, JG-12); B, Rogingitized dolerite (JG-2) and plagiogranite dike (JG-15) from the Jarbo Gap ophiolite, Dark Canyon dike (DC-4, an anomalous high Eu anomaly not shown), and spilitic basalt (LR-1) from the chert-argillite unit.

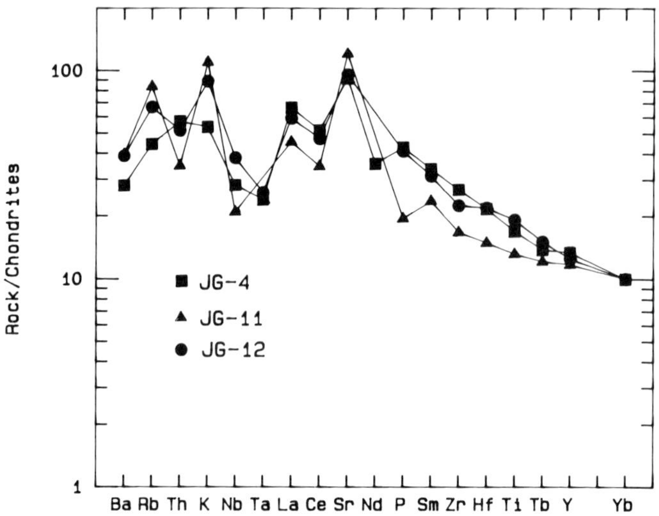

Figure 12. Trace-element variation diagram for the Jarbo Gap dikes (JG-4, JG-11, JG-12) normalized to chondrite and to Yb_N of 10 (Thompson, 1982).

observation suggests that the flysch sequence may be an "overlap sequence" overlying both the Smartville complex and the chert-argillite unit and that both tectonic entities were in contact prior to deposition of this Late Jurassic "overlap sequence."

The northwest-trending, northeast-dipping steep faults and shear zones that deform the original contact between the Smartville complex and the chert-argillite unit possibly are Late Jurassic structures. A tuffaceous sandstone in the turbiditic sequence of the flysch deposit contains *Buchia Concentrica,* a pelecypod of middle Oxfordian to upper Kimmeridgian age (Marlette and others, 1978; Vaitl, 1980), suggesting that the deformation of this turbiditic sequence with the rest of the Smartville complex occurred during and/or after Late Jurassic time.

AGE AND CORRELATION

Jarbo Gap ophiolite

Several radiometric dates from intrusive rocks within the Jarbo Gap ophiolite constrain the minimum igneous age of the ophiolite as early Mesozoic. A deformed leucodiorite intrusion in the flaser gabbro gives a U/Pb zircon age of 204 ± 2 Ma, which is consistent with a Sm/Nd isotopic age of 205 ± 43 Ma from a suite of rocks from the flaser gabbro outcrop (Saleeby and others, 1989). These age relations suggest that the tectonized harzburgite-dunite sequence intruded by the metagabbro-flaser gabbro unit is probably older than the Lower Jurassic. The youngest dike generation intruding the serpentinized peridotite and mafic breccia in the ophiolite reveals a U/Pb zircon age of 196 +23/−11 Ma (Saleeby and others, 1989) supporting this interpretation. While these intrusive and geochronologic relations set the minimum igneous age of the ophiolite as the early Mesozoic, the possibility of an older age (late Paleozoic?) cannot be ruled out. Discontinuous exposures of late Paleozoic and early Mesozoic ophiolitic fragments crop out along strike within the entire length of the Sierra Nevada metamorphic belt (Saleeby, 1982).

The Jarbo Gap ophiolite contains the Horseshoe Bend Formation of Hietanen (1981), which continues farther southeast between the Cascade and Bald Rock plutons (Fig. 1). In this area these ophiolitic rocks are described by Edelman and others (1989) as components of the Indian Creek ophiolitic melange, which constitutes the ophiolitic basement of the Fiddle Creek Complex (Fig. 1). The Indian Creek ophiolitic melange crops out beneath the Clipper Gap Formation in the cores of several anticlines and consists of blocks of ophiolitic rocks in a serpentinite matrix (Fig. 13). The blocks are composed of tectonized harzbur-

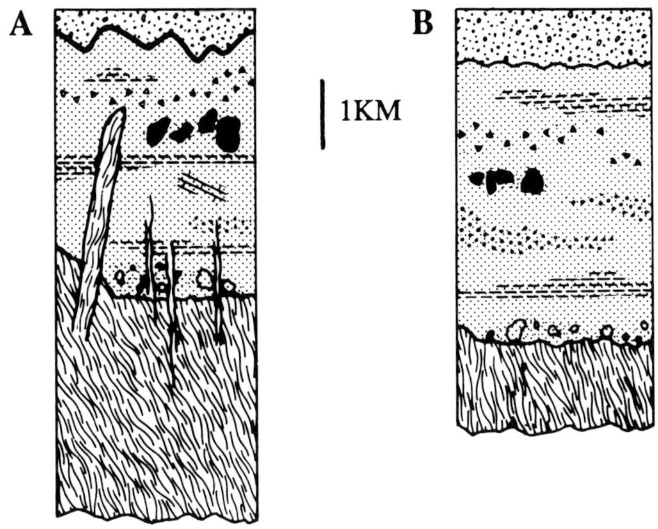

Pentz sandstone: turbiditic sandstone, conglomerate, argillite, pebbly mudstone.

Chert-argillite unit: argillite, chert, sandstone, volcanic and volcaniclastic rocks, olistostromes, limestone blocks, serpentinite diapir, dike intrusions.

Jarbo Gap ophiolite: serpentinized peridotite, gabbro, diorite, amphibolite, volcanic rocks, chert, dike intrusions.

Colfax sequence: lithic and quartzose sandstone, conglomerate, siltstone, argillite.

Clipper Gap Formation: argillite, ribbon chert, sandstone, volcanic and volcaniclastic rocks, olistostromes, marble lenses.

Indian Creek ophiolitic mélange: blocks of harzburgite, pyroxenite, gabbro, diorite, volcanic rocks, and chert in serpentinite matrix.

Figure 13. Constructed stratigraphic column sections. A, The late Paleozoic–early Mesozoic oceanic basement north of the Smartville complex. B, The Fiddle Creek complex of Edelman and others (1989) east of the Smartville complex. The Pentz sandstone and the Colfax sequence represent flysch deposits of an overlap sequence that structurally (A) and unconformably (B) overlie both basements.

TABLE 4. INSTRUMENTAL NEUTRON ACTIVATION TRACE-ELEMENT ANALYSES

Sample	Type*	Sc	La	Ce	Nd	Sm	Eu	Tb	Yb	Lu	Hf	Ta	Th
ppm													
JG-2	D	36	4.4	8.4		2.74	0.943	0.55	2.31	0.202	2.02	0.6	0.4
JG-4	D	20	15.5	31.7	16	4.86	1.341	0.51	1.56	0.131	3.07	0.3	1.7
JG-11	D	24	10.2	20.5		3.27	1.032	0.43	1.50	0.131	2.03		1.0
JG-12	D	27	11.6	24.5		3.80	1.092	0.47	1.32	0.132	2.62	0.3	1.3
JG-15	D	2	12.4	22.5	8	1.95	0.338	0.48	0.72	0.066	1.68	0.3	3.8
JG-10	B	20	7.1	11.0		3.17	0.727	0.42	1.60	0.204	1.60		0.6
JG-18	G	62				0.31	0.215		0.45	0.043			
DC-4	D	14	9.8	14.5	11	2.98	6.840	0.32	1.28	0.112	1.78	0.2	0.8
DC-2	B	20	24.1	46.5	31	7.07	1.744	1.41	4.03	0.346	5.15	1.1	5.3
LR-1	L	20	41.2	82.6	48	9.08	2.750	1.06	3.21	0.184	7.38	3.2	4.7

*D = dike; B = breccia; G = gabbro; L = lava.

gite, pyroxenite, metamorphosed gabbro, diorite, metavolcanic rocks, and chert, ranging in length from a few centimeters to several kilometers. There are no isotopic age data from the Indian Creek ophiolite melange; however, a U/Pb zircon date from a felsic tuff in the overlying Clipper Gap Formation gives an eruption age of 193 ± 7 Ma (Saleeby and others, 1989), suggesting that the ophiolitic melange is Early Jurassic or older. This interpretive age of the Indian Creek ophiolitic melange is consistent with the inferred age of the Jarbo Gap ophiolite, suggesting that the Jarbo Gap ophiolite and the Indian Creek ophiolitic melange may in part be correlative units (Fig. 13).

Chert-argillite unit

An age for the chert-argillite unit is not well constrained because of its generally chaotic (melange) character, intense deformation, and tectonic mixing in the matrix. The ages of the blocks in the matrix suggest a maximum age of the formation of the chert-argillite unit. Large outcrops of fossiliferous limestone, with faunal affinity to the McCloud Limestone of the eastern Klamath Mountains (Watkins and others, 1987), indicate a maximum age of Permian for the chert-argillite unit.

The northwest-striking, steeply northeast-dipping deformational fabric involving the contact between the chert-argillite unit and the Smartville complex (Fig. 2) suggests a minimum Late Jurassic age for the deformation of the chert-argillite unit and may indicate that part of the chert-argillite unit evolved in Late Jurassic time. Thus, these interpretations and age relations suggest that the evolution of the chert-argillite unit may have occurred during late Paleozoic to early Mesozoic time.

The chert-argillite unit was originally correlated with the Calaveras Complex in the central Sierra Nevada foothills metamorphic belt based on lithologic similarities and an inferred Paleozoic age (Creely, 1965). However, the redefinition of the Calaveras Complex by different workers (e.g., Hietanen, 1981; Schweickert, 1981) indicates that the type Calaveras crops out farther east between the Dogwood Peak–Gillis Hill and the Goodyear's Creek faults (Fig. 1). Therefore, the chert-argillite unit is not considered to be correlative with the Calaveras Complex.

A chert-argillite unit similar to that described in this study crops out in the cores of several anticlines between the Cascade and Bald Rock plutons northeast of the Smartville complex (Fig. 1). Described by Edelman and others (1989) as the Clipper Gap Formation of the Fiddle Creek Complex, this chert-argillite unit contains black siliceous (cherty) argillite with black ribbon chert, volcaniclastic sandstone, diamictite, sedimentary breccia, conglomerate, rare marble lenses, and detrital serpentinite (Fig. 13). Clasts of amphibolite, chert, and metavolcanic rocks are common in sandstones, and thin layers of volcanic rocks occur interbedded and/or tectonically interleaved with the sedimentary rocks. The Clipper Gap Formation in this area depositionally overlies the Indian Creek ophilitic melange and is in turn unconformably overlain by the Colfax sequence (Fig. 13), which contains clasts of quartz-mica schist possibly derived from the Shoo Fly Complex to the east (Edelman and others, 1989).

Farther south in the Central tectonic belt the chert-argillite unit is best correlated with broken formation in the Grass Valley area south of the Yuba Rivers pluton (Fig. 1; Day and others, 1985, and references therein). The broken formation in this area includes fault-bounded blocks of polymictic conglomerate as well as pyroclastic and volcanic rocks in a chert-argillite matrix and is unconformably overlain by flysch deposits of the Late Jurassic Colfax sequence (Day and others, 1985).

DISCUSSION

Structural and intrusive relations, as well as rock types, suggest that the Jarbo Gap ophiolite is a fragment of heterogeneous and multiply deformed oceanic crust (Dilek, 1989b), part of which may be as old as late Paleozoic. The chert-argillite unit displays characteristics of a progressively disrupted sequence deposited at an active convergent margin (Cowan, 1985). Blocks of metagabbro, serpentinite, metavolcanic rocks, and deformed chert, enveloped by the sandstone-mudstone sequence, suggest a provenance of older, deformed ophiolitic basement exposed on the sea floor during deposition of the chert-argillite unit. The serpentinite diapir intruding these supracrustal rocks indicates remobilization of this oceanic basement during and after deposition of the sedimentary rocks (Fig. 13). On the basis of these relations the Jarbo Gap ophiolite is interpreted to have been part of the ophiolitic basement of the chert-argillite unit. Similarities between the lithologic, petrographic, and geochemical characteristics of dikes intruding both the ophiolite and the chert-argillite unit support this interpretation and suggest that a common magma source underlay both the ophiolitic basement and the chert-argillite unit during Early Jurassic time. Additional evidence for a stratigraphic relation between the chert-argillite unit and the Jarbo Gap ophiolite comes from the correlation of these two units with the Clipper Gap Formation and the Indian Creek ophiolitic melange (Fig. 13), respectively, which are in depositional contact (Edelman and others, 1989). These interpretations imply that much of the Central tectonic belt in the early Mesozoic northern Sierra Nevada consisted of a multiply deformed and metamorphosed ophiolitic basement depositionally overlain by a progressively disrupted marine sequence (Fig. 13; Dilek and Moores, 1989).

The dikes intruding both the Jarbo Gap ophiolite and the chert-argillite unit are basalts and basaltic andesites. The low FeO and high MgO contents of these dikes point to primary melts and high degrees of melting. The enrichment in compatible as well as some LIL elements suggests a metasomatized depleted source. These characteristics of the dikes are similar to those of high-magnesian andesites and boninites derived from primary magmas that are generated as second-stage melts from hydrous melting of a peridotite source during early stages of subduction (Cameron and others, 1983; Hickey and Frey, 1982; Tatsumi and Ishizaka, 1982). The lack of felsic phenocrysts in the dikes is typical of basaltic andesites of boninitic affinity. The existence of the hastingsitic amphibole phenocrysts in these dikes indicates water-rich

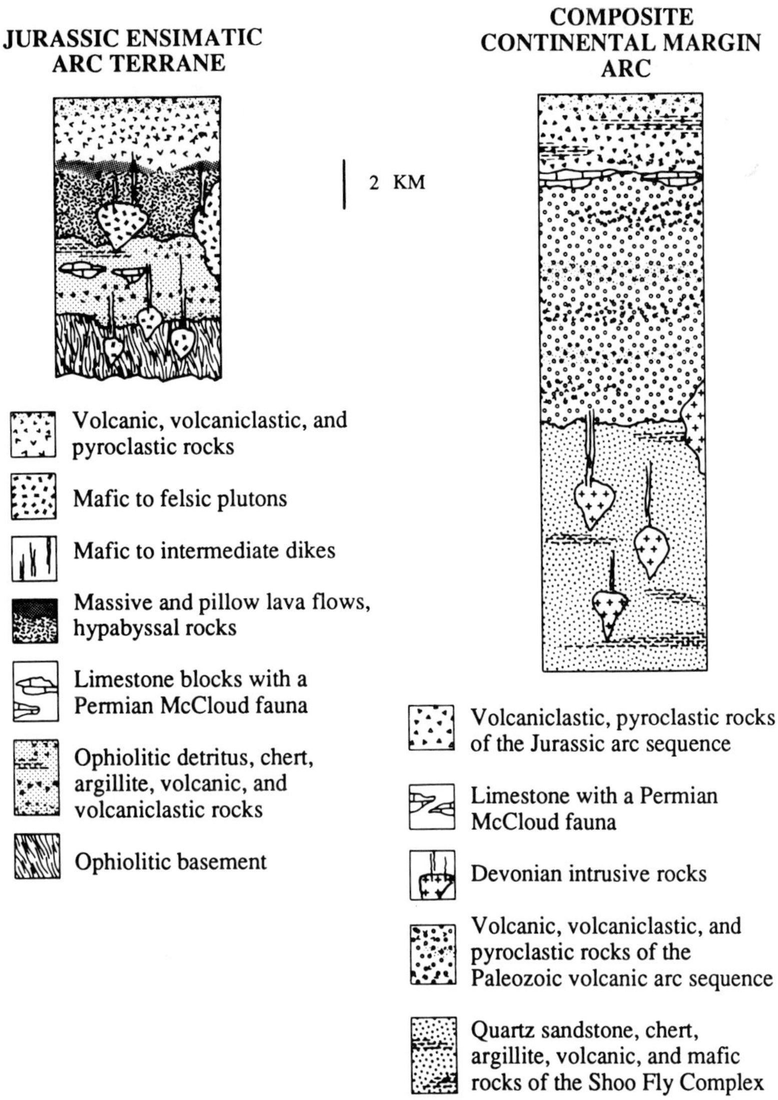

Figure 14. Reconstructed stratigraphic column sections of the ensimatic Jurassic arc terrane and the composite continental margin arc in the northern Sierra Nevada.

and oxidized parental melts. The lower calculated Fe^{3+} content of the secondary amphiboles of the dikes suggests that alteration occurred in a more reduced environment.

The inferred origin of basaltic andesite dikes during early stages of subduction suggests that these intrusions may have been the conduits for Early Jurassic arc volcanism that evolved through and across the preexisting oceanic basement. The basaltic andesite dikes are coeval with the hypabyssal, volcanic, and plutonic rocks in the Slate Creek complex and with some of the hypabyssal and plutonic rocks in the Smartville complex (Bickford and Day, 1988; Dilek, 1989a, b; Saleeby and others, 1989). Based on lithologic and chronologic correlations, Bickford and Day (1988) and Dilek (1989a) suggested that the Smartville and Slate Creek complexes are part of a single arc terrane. It is inferred that this Early Jurassic arc terrane, consisting of the Smartville and Slate Creek complexes, may have developed on the preexisting oceanic basement composed of ophiolitic fragments and the overlying chert-argillite unit (Fig. 14). If this interpretation is correct, most of the Western and Central tectonic belts are then part of a single, polygenetic arc terrane that evolved in early Mesozoic time (Dilek and Moores, 1989). This interpretation also rules out the possibilities, as previously suggested (Edelman, 1987), of the existence in the northern Sierra Nevada metamorphic belt of several genetically unrelated, allochthonous arc terranes that were presumably accreted to the continent in early Mesozoic time.

The polygenetic, ensimatic arc terrane in the northern Sierra Nevada is separated by the Dogwood Peak–Gillis fault system from the continental margin rocks in the east (Fig. 1). The Dogwood Peak–Gillis fault system comprises curvilinear and steeply east-dipping reverse faults that are associated with strongly sheared and foliated rocks in both upper and lower blocks (Edelman and others, 1989). The continental margin rocks east of this fault system include, from west to east, the Calaveras Complex of a possible late Paleozoic(?)–early Mesozoic accretionary prism, the Feather River Belt as a remnant of an early Mesozoic subduction complex, and the Eastern Belt consisting of the Lower Paleozoic Shoo Fly Complex successively overlain by Devonian and Mississippian, mid-Permian, and Jurassic arc sequences (Schweickert, 1981; Edelman and others, 1989). The Jurassic arc sequence in the Eastern Belt is clearly constructed over a composite and mainly continental basement, as opposed to the polygenetic and single arc terrane that was built on and across the oceanic basement described in this study (Fig. 14). The significant difference in the character of the arc basements suggests that the continental margin arc and the ensimatic arc terrane were probably genetically unrelated. The absence in the continental margin arc and in the subduction complex of Lower Jurassic dike intrusions, which in turn exist in the ensimatic arc terrane, supports this interpretation and further suggests that the two arc systems were probably separate tectonic entities during the earliest Mesozoic time.

The prevailing structural grain across the northern Sierra Nevada metamorphic belt that is defined by steeply east-dipping reverse faults and associated cleavage must have evolved following the juxtaposition of the two arc systems. These structures are truncated by Upper Jurassic–Lower Cretaceous granitic batholiths that intrude deformed rocks in both arc systems in the northern Sierra Nevada (Fig. 1). We think that these structures are shallow features that developed as a result of regional contraction and plate boundary reorganization(s) along the active margin of western North America during Late Jurassic to Early Cretaceous time.

ACKNOWLEDGMENTS

This study, as part of our ongoing research project in the northwestern Sierra Nevada foothills, has been supported in part by Sigma Xi, R. C. Baker, C. Durrell, and University of California, Davis, Department of Geology research grants (to Y.D.), and by National Science Foundation Grants EAR-7803640 and EAR-8010697 (to E.M.M. and H. W. Day). Communications and discussions with H. W. Day, S. H. Edelman, J. B. Saleeby, R. A. Schweickert, and R. K. Springer at various stages of the study were most helpful. P. Thy acknowledges a National Science Research Council-NASA Research Associateship. REE analyses (INA) were supported by the Danish Natural Research Council and were made at Forsøgsanlæg Risø, Roskilde. Critical and constructive reviews by D. S. Harwood, J. B. Saleeby, and an anonymous referee significantly improved the paper.

APPENDIX 1

Petrographic description and intrusive relations of the analyzed samples from the Jarbo Gap ophiolite:

JG-2. Rodingitized dolerite pipe within serpentinite.

JG-3. Highly altered, aphyric, quartz andesite dike that represents first episode of dike injection in serpentinite.

JG-4. Basaltic andesite dike with a microphyric groundmass of brown hornblende and albite that represents the second episode dike injection in serpentinite.

JG-5. Basaltic andesite dike intruding serpentinite.

JG-6. Basaltic andesite dike with igneous hornblende phenocrysts and a microphyric texture that intrudes serpentinite.

JG-7. Aphyric, relatively altered andesitic dike with hornblende and albite in the groundmass that intrudes serpentinite.

JG-8. Fine-grained, hornblende-phyric andesite-andesitic basalt dike with an altered matrix that intrudes mafic breccia.

JG-9, 10. Mafic talus breccia composed mainly of angular fragments of dacitic-andesitic rocks containing albite phenocrysts.

JG-11, 12. Fine-grained andesite dikes intruding mafic breccia; they contain brown hornblende phenocrysts and microcrysts and show devitrification textures.

JG-13. Highly altered andesitic dike intruding mafic breccia.

JG-15. Medium- to fine-grained plagiogranite dike with quartz phenocrysts intruding serpentinite.

JG-16, 18, 19. Relatively less-deformed metagabbros with relict plagioclase ($An_{<92}$).

JG-1, 14, 17, 20. Various mafic gabbros and serpentinites without primary mineralogy.

Petrographic description and intrusive relations of the analyzed samples from the chert-argillite unit:

Dark Canyon

DC-1. Mafic, highly altered dike intruding argillite matrix.

DC-2, 3. Various clasts in volcaniclastic breccia.

DC-4. Basaltic andesite dike containing brown hornblende phenocrysts in altered, fine-grained groundmass; crosscuts foliation in the argillite matrix.

DC-5. Composite sample of an agglomerate composed of angular quartz, albite, and fragments of volcanic rocks.

DC-6. Extensively altered, basaltic andesite dike parallel to foliation in argillite matrix.

DC-7. Highly altered mafic dike crosscutting metagabbro unit that intrudes argillite matrix.

DC-8. Highly altered metaggabro intruding argillite matrix; its primary mineralogy and texture are obliterated.

DC-9. Basaltic andesite dike containing microphenocrysts and phenocrysts of brown hornblende and intruding argillite matrix.

Lunt Road:

LR-1. Dark green, fine-grained, aphanitic and amygduloidal splitic basalt as blocks in argillite matrix.

REFERENCES CITED

Beard, J. S., and Day, H. W., 1987, The Smartville intrusive complex, Sierra Nevada, California; The core of a rifted volcanic arc: Geological Society of America Bulletin, v. 99, p. 779–791.

Bickford, M. E., and Day, H. W., 1988, Jurassic ages of arc-ophiolite complexes, northern Sierra Nevada; Implications for duration of the Nevadan "Orogeny": Geological Society of America Abstracts with Programs, v. 20, p. A274.

Burchfiel, B. C., and Davis, G. A., 1981, Triassic and Jurassic tectonic evolution of the Klamath Mountains–Sierra Nevada geologic terrane, *in* Ernst, W. G., ed., The geotectonic development of California; Rubey Volume 1: Englewood Cliffs, New Jersey, Prentice-Hall, p. 50–70.

Cameron, W. E., McCulloch, M. T., and Walker, D. A., 1983, Boninite petrogenesis; Chemical and Nd-Sr isotopic constraints: Earth and Planetary Science Letters, v. 65, p. 75–89.

Cowan, D. S., 1985, Structural styles in Mesozoic and Cenozoic melanges in the western Cordillera of North America: Geological Society of America Bulletin, v. 96, p. 451–462.

Creely, R. S., 1965, Geology of the Oroville Quadrangle, California: California Division of Mines and Geology Bulletin 184, 86 p.

Day, H. W., Moores, E. M., Tuminas, A. C., 1985, Structure and tectonics of the northern Sierra Nevada: Geological Society of America Bulletin, v. 96, p. 436–450.

Dilek, Y., 1989a, Tectonic significance of post-accretion rifting of a Mesozoic island-arc terrane in the northern Sierra Nevada, California: Journal of Geology, v. 97, p. 503–518.

——, 1989b, Structure and tectonics of an Early Mesozoic oceanic basement in the northern Sierra Nevada metamorphic belt, California; Evidence for transform faulting and ensimatic arc evolution: Tectonics, v. 8, p. 999–1014.

Dilek, Y., and Moores, E. M., 1989, Island-arc evolution and fracture zone tectonics in the Mesozoic Sierra Nevada, California, and implications for transform offset of the Sierran/Klamath convergent margins, *in* Bartholomew, M. J., Hyndman, D. W., Mogk, D. W., and Mason, R., eds., Characterization and comparison of Precambrian through Mesozoic continental margins; Proceedings of the 8th International Conference on Basement Tectonics: Salt Lake City, Utah, International Basement Tectonics Association (in press).

Edelman, S. H., 1987, Evidence for two Mesozoic arc-continent collisions in the western Sierra Nevada metamorphic belt, California; Plate kinematic interpretation of a terrane analysis: Geological Society of America Abstracts with Programs, v. 19, p. 651.

Edelman, S. H., and 5 others, 1989, Structure across a Mesozoic ocean-continent suture zone in the northern Sierra Nevada, California: Geological Society of America Special Paper 224, 56 p.

Hickey, R., and Frey, F. A., 1982, Geochemical characteristics of boninite series volcanics: Implications for their source: Geochimica et Cosmochimica Acta, v. 46, p. 2099–2115.

Hietanen, A., 1981, Petrologic and structural studies in the northwestern Sierra Nevada, California: U.S. Geological Survey Professional Paper 1226, 59 p.

Ingersoll, R. V., and Schweickert, R. A., 1986, A plate-tectonic model for ophiolite genesis, Nevadan orogeny and forearc initiation, northern California: Tectonics, v. 5, p. 901–912.

Jezek, P. A., 1975, Geological studies in the area of the Fiji plateau, southwest Pacific [Ph.D. thesis]: Amherst, University of Massachusetts, 407 p.

Johnson, R. W., Jaques, A. L., Hickey, R. L., McKee, C. O., and Chappell, B. W., 1985, Manam Island, Papua New Guinea; Petrology and geochemistry of a low-TiO_2 basaltic island-arc volcano: Journal of Petrology, v. 26, p. 283–323.

Leake, B. E., 1978, Nomenclature of amphiboles: Minerals Magazine, v. 42, p. 533–563.

Marlette, J. W., Akers, R. J., Cole, K. A., and McJunkin, R. D., 1978, Geologic investigations, *in* The August 1, 1975, Oroville earthquake investigation: California Department of Water Resources Bulletin 204-278, p. 15–103.

McJunkin, R. D., Davis, T. E., and Criscione, J. J., 1979, An isotopic age for Smartville ophiolite and the obduction of metavolcanic rocks in the northern Sierran Foothills, California: Geological Society of America Abstracts with Programs, v. 11, p. 91.

Moores, E. M., 1972, Model for Jurassic island arc-continental margin collision in California: Geological Society of America Abstracts with Programs, v. 4, p. 202.

Moores, E. M., and Day, H. W., 1984, An overthrust model for the Sierra Nevada: Geology, v. 12, p. 416–419.

Rautenschlein, M., and 6 others, 1985, Isotopic and trace element composition of volcanic glasses from the Akaki canyon, Cyprus; Implications for the origin of the Troodos ophiolite: Earth and Planetary Science Letters, v. 75, p. 369–383.

Ricci, M. P., Moores, E. M., Verosub, K. L., and McClain, J. S., 1985, Geologic and gravity evidence for thrust emplacement of the Smartville ophiolite: Tectonics, v. 3, p. 133–155.

Robinson, P. T., Melson, W. G., O'Hearn, T., and Schmincke, H.-U., 1983, Volcanic glass compositions of the Troodos ophiolite, Cyprus: Geology, v. 11, p. 400–404.

Rock, N.M.S., 1987, A FORTRAN program for tabulating and naming amphibole analyses according to the International Mineralogical Association scheme: Mineralogy and Petrology, v. 37, p. 79–88.

Saleeby, J. B., 1981, Ocean floor accretion and volcano-plutonic arc evolution in the Mesozoic Sierra Nevada, California, *in* Ernst, W. G., ed., The geotectonic development of California; Rubey Volume 1: Englewood Cliffs, New Jersey, Prentice-Hall, p. 132–181.

——, 1982, Polygenetic ophiolite belt of the California Sierra Nevada; Geochronological and tectonostratigraphic development: Journal of Geophysical Research, v. 87, p. 1803–1824.

Saleeby, J. B., Shaw, H. F., Niemeyer, S., Moores, E. M., and Edelman, S. H., 1989, U/Pb, Sm/Nd, and Rb/Sr geochronological and isotopic study of northern Sierra Nevada ophiolitic assemblages, California: Contributions to Mineralogy and Petrology, v. 102, p. 205–220.

Schweickert, R. A., 1981, Tectonic evolution of the Sierra Nevada range, *in* Ernst, W. G., ed., The geotectonic development of California; Rubey Volume 1: Englewood Cliffs, New Jersey, Prentice-Hall, p. 87–131.

Schweickert, R. A., and Cowan, D. S., 1975, Early Mesozoic tectonic evolution of the western Sierra Nevada, California: Geological Society of America Bulletin, v. 86, p. 1329–1336.

Sinton, J. M., Wilson, D. S., Christie, D. M., Hey, R. N., and Delaney, J. R., 1983, Petrologic consequences of rift propagation on oceanic spreading ridges: Earth and Planetary Science Letters, v. 62, p. 193–207.

Tatsumi, Y., and Ishizaka, K., 1982, Origin of high-magnesian andesites in the Setouchi volcanic belt, southwest Japan; 1, Petrological and chemical characteristics: Earth and Planetary Science Letters, v. 60, p. 293–304.

Thompson, R. N., 1982, Magmatism of the British Tertiary volcanic province: Scottish Journal of Geology, v. 18, p. 49–107.

Thy, P., and Dilek, Y., 1987, Boninitic dikes from a fracture zone basement of the Smartville complex, NW Sierra Nevada, California: EOS Transactions of the American Geophysical Union, v. 68, p. 1519.

Vaitl, J., 1980, Geology and structure of a northern Sierra Nevada foothills melange and adjacent areas, Butte County, California [M.S. thesis]: Davis, University of California, 93 p.

Watkins, R., Reinheimer, C. E., Wallace, J. W., and Nestell, M. K., 1987, Paleogeographic significance of a Permian sedimentary megamictite in the Central Belt of the northern Sierra Nevada: Geological Society of America Bulletin, v. 99, p. 771–778.

Xenophontos, C., 1984, Geology, petrology, and geochemistry of part of the Smartville complex, northern Sierra Nevada, California [Ph.D. thesis]: Davis, University of California, 446 p.

Manuscript Received by the Society March 27, 1990

Printed in U.S.A.

Geological Society of America
Special Paper 255
1990

Cross section and Mesozoic paleogeography of the northern Sierra Nevada

Steven H. Edelman*
Department of Geological Sciences, University of California, Davis, California 95616

ABSTRACT

Pre–Middle Jurassic stratigraphic assemblages in the northern Sierra Nevada are contained in tightly folded but macroscopically shallowly dipping nappes. Four shallowly east-dipping f_1 faults juxtapose, in descending structural order, the Northern Sierra terrane, the Feather River peridotite/Devils Gate ophiolite, the Red Ant Schist, the Calaveras Complex, and the Fiddle Creek Complex. The f_2 Slate Creek thrust dips shallowly west; it places the Slate Creek Complex over the Red Ant Schist, Calaveras Complex, and Fiddle Creek Complex, and thus cuts two f_1 faults. As many as three Early Jurassic magmatic arcs may be represented by the stratigraphic assemblages in the f_1-f_2 nappe pile. A continental-margin arc is represented by Late Triassic–Middle Jurassic volcanogenic rocks in the upper part of the Northern Sierra terrane stratigraphy and by coeval plutons in the eastern part of the Sierra Nevada batholith; the associated subduction complex is represented by blueschist-facies metamorphism in the Red Ant Schist and melange structure in oceanic rocks of the Calaveras Complex. An intraoceanic arc is represented by the Fiddle Creek Complex, which contains ophiolitic basement overlain and intruded by Early Jurassic volcanic and plutonic rocks; the volcanic strata are intercalated with cherty deposits, olistostromes, and volcaniclastic rocks devoid of continental detritus. A more westerly Early Jurassic intraoceanic arc is represented by the Slate Creek Complex, which has ophiolitic structure. These units were juxtaposed along the f_1 and f_2 faults before Callovian-Kimmeridgian (Middle to Late Jurassic) deposition of volcanic and epiclastic strata and coeval intrusion of mafic-intermediate plutons; this postamalgamation arc was extensional in its western part, as indicated by sheeted dikes in the Smartville Complex. The Callovian-Kimmeridgian rocks and their basement of f_1-f_2 nappes are cut by two sets of oppositely dipping f_3 faults, a shallowly west-dipping set (Taylorsville and related thrust faults) and a steeply east-dipping set (Foothills fault system); f_3 faults are the only major Late Jurassic (Nevadan) or younger faults in the northern Sierra Nevada. This structural chronology rules out models for Late Jurassic (Nevadan) collisions, but allows collision(s) of the Fiddle Creek and Slate Creek arcs in the Early or Middle Jurassic.

INTRODUCTION

The upper continental crust in the northern Sierra Nevada (Fig. 1) is composed of metamorphosed igneous and sedimentary rocks, many of which formed in oceanic environments. These rocks were deformed during the post–early Kimmeridgian (middle Late Jurassic) Nevadan orogeny by steep dip-slip faults, upright folds, and steep cleavages indicative of horizontal shortening and crustal thickening. This tectonically thickened oceanic material forms the wall rocks of the main latest Jurassic-Cretaceous part of the Sierra Nevada batholith (gr in Fig. 1). Whereas the Sierra Nevada batholith formed essentially in situ with respect to

*Present address: Department of Geological Sciences, University of California, Santa Barbara, California 93106.

Edelman, S. H., 1990, Cross section and Mesozoic paleogeography of the northern Sierra Nevada, *in* Harwood, D. S., and Miller, M. M., eds., Paleozoic and early Mesozoic paleogeographic relations; Sierra Nevada, Klamath Mountains, and related terranes: Boulder, Colorado, Geological Society of America Special Paper 255.

other parts of the North American continent, probably with some strike-slip and extensional translations, the kinematic history of the metamorphosed wall rocks is not as well constrained.

Several of the older wall-rock units represent Paleozoic to early Mesozoic magmatic arcs, and arc magmatism is a primary crust-forming process (Hamilton, 1981). An important problem that can be addressed in the northern Sierra Nevada is determination of the relative volumes of arc rocks that were tectonically transported to the North American continent versus those that intruded the continental crust like the Sierra Nevada batholith. A second related problem is to determine the emplacement mechanisms of transported rock assemblages and the kinematic relations of these mechanisms to Nevadan crustal thickening.

This chapter attempts to place some constraints on the solutions to these problems by presenting a crustal-scale cross section across the northern Sierra Nevada. Emphasis is placed on the configuration of the major stratigraphic assemblages in the uppper and middle crust, and the cross-cutting relations and ages of major faults. Major stratigraphic assemblages and their paleographic settings are first summarized, with emphasis on those that represent Mesozoic magmatic arcs.

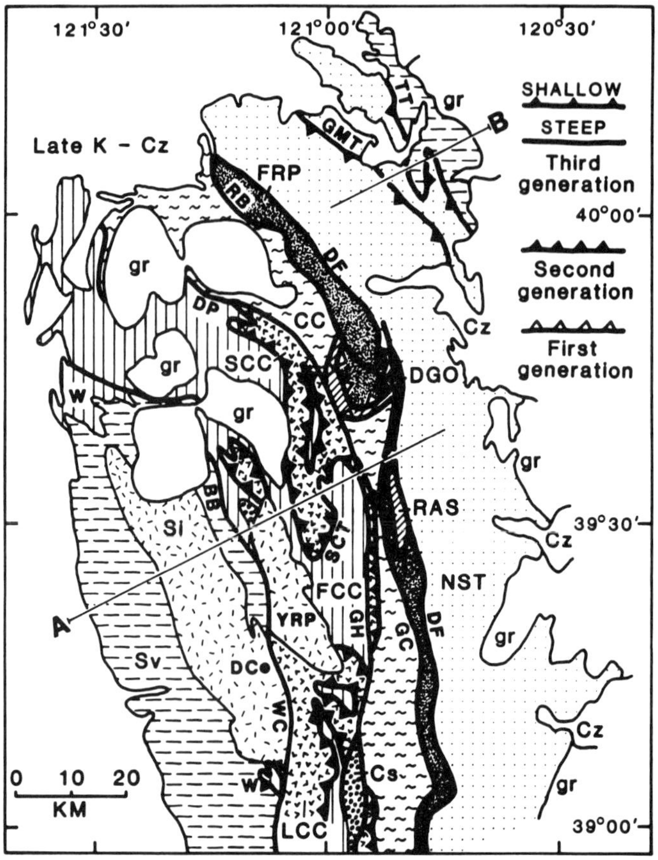

Figure 1. General geologic map of the northern Sierra Nevada showing line of cross section A-B (Fig. 3). Plutonic rocks that are about 160 m.y. old are shown by random dash; younger plutons (gr) are unpatterned. BB = Big Bend fault; CC = Calaveras Complex; Cs = Colfax sequence; Cz = Cenozoic cover; DC = Deer Creek locality where 160-Ma dike cuts 200-Ma tonalite; DF = Downieville fault; DGO = Devils Gate ophiolite; DP = Dogwood Peak fault; FCC = Fiddle Creek Complex; FRP = Feather River peridotite; GC = Goodyears Creek fault; GH = Gillis Hill fault; GMT = Grizzly Mountain thrust; gr = latest Jurassic–Cretaceous granites associated with the Sierra Nevada batholith; K = Cretaceous cover, LCC = Lake Combie Complex; NST = Northern Sierra terrane, including Jurassic rocks northeast of TT; RAS = Red Ant Schist; RB = Rich Bar fault; SCC = Slate Creek Complex; SCT = Slate Creek thrust; Si and Sv = Smartville Complex intrusive and volcanic rocks, respectively; TT = Taylorsville thrust; W = window of Smartville Complex over Fiddle Creek Complex; WC = Wolf Creek fault; YRP = Yuba Rivers pluton. Compiled from McMath (1966), Hietanen (1973, 1976, 1981), D'Allura and others (1977), Jennings (1977), Xenophontos and Bond (1978), Day and others (1985), Beard and Day (1987), and Edelman and others (1989a).

STRATIGRAPHIC ASSEMBLAGES

Northern Sierra terrane

The Northern Sierra terrane is the thickest and oldest stratigraphic assemblage in the northern Sierra Nevada (Fig. 1; D'Allura and others, 1977; Schweickert, 1981; Silberling and others, 1987; Harwood, 1988; Edelman and Sharp, 1989). It contains the Shoo Fly Complex, an internally deformed and imbricated assemblage of latest Proterozoic–Silurian subarkosic sandstone, chert, mudstone, limestone, and mafic rocks (Saleeby and others, 1987, 1989), overlain unconformably by an intact succession of Late Devonian to Middle Jurassic, chiefly volcanogenic, strata. The volcanic section contains three major unconformity-bounded sequences of Late Devonian–Pennsylvania, Permian, and Late Triassic–Callovian ages (Harwood, 1988). The third sequence is coeval with the oldest, eastern components of the Sierra Nevada batholith (Schweickert and Cowan, 1975; Stern and others, 1981).

The long and complex paleogeographic history of the Northern Sierra terrane is beyond the scope of this chapter. The important point here is its location in the early Mesozoic, which was probably at the edge of the North American continent (Davis and others, 1978; Saleeby, 1983). The Late Triassic–Callovian volcanogenic strata in the uppermost part of the Northern Sierra terrane stratigraphy and coeval plutons in the eastern part of the Sierra Nevada batholith represent an early Mesozoic Andean-style magmatic arc (Schweickert and Cowan, 1975).

Feather River peridotite and Devils Gate ophiolite

The Feather River peridotite (Fig. 1) contains harzburgite and lherzolite tectonite, dunite, pyroxenite, layered and massive gabbro, and amphibolite (Hietanen, 1973; Ehrenberg, 1975; Edelman and others, 1989a). Mafic rocks and felsic differentiates in the Feather River peridotite have yielded a range of middle and late Paleozoic radiometric ages (references cited *in* Edelman and Sharp, 1989). The Devils Gate ophiolite (Fig. 1) is an intact succession of gabbro, sheeted mafic dikes, pillow basalt, and metachert (Edelman and others, 1989a). Parts of the volcanic unit are greenstone, and the remainder of the mafic rocks are amphibolite. Several late Paleozoic radiometric ages have been reported for the Devils Gate ophiolite (references cited *in* Edelman and Sharp, 1989). The Feather River peridotite and Devils Gate ophiolite are remnants of a single polygenetic or cogenetic ophiolite, as indicated by their shared late Paleozoic igneous and/or metamorphic ages, their map continuity, and their ophiolitic characteristics, and have thus been included in a single terrane (Edelman and others, 1989a; Edelman and Sharp, 1989; Saleeby and others, 1989).

The Feather River peridotite and Devils Gate ophiolite represent the upper parts of oceanic or back-arc lithosphere. Their place of origin is unknown, and they cannot be paleogeographically tied with certainty to the Northern Sierra terrane until the Middle Jurassic (Edelman and Sharp, 1989). However, earlier juxtaposition with the Northern Sierra terrane may be indicated by Devonian felsic dikes in the Feather River peridotite, which may be correlative with Devonian felsic intrusions in the Northern Sierra terrane (Saleeby and others, 1989).

Red Ant Schist

The Red Ant Schist (Fig. 1) is a chaotic assemblage of quartz-mica schist, metaclastic rocks, and metavolcanic rocks with blueschist-facies metamorphic minerals, including glaucophane and lawsonite plus quartz (Schweickert and others, 1980; Hietanen, 1981; Edelman and others, 1989a). The blueschist-facies metamorphism is Middle Jurassic or older (Schweickert and others, 1980). The Red Ant Schist is interpreted to be a remnant of an early Mesozoic, subcontinental subduction complex associated with the early Mesozoic magmatic arc in the Northern Sierra terrane (Schweickert and others, 1980).

Calaveras Complex

The Calaveras Complex (Fig. 1) is a mesoscopically chaotic assemblage of argillite, chert, limestone, and volcanic rocks. The complex contains two regionally mapped melange units, an eastern chert-rich unit and a western argillite-rich unit (Hietanen, 1981; Edelman and others, 1989a). Fossils of Mississippian, Pennsylvanian, Permian, and Triassic ages have been recovered from the Calaveras Complex in the northern Sierra Nevada (references cited in Edelman and Sharp, 1989). The Calaveras Complex is interpreted to have formed in the same general regime of early Mesozoic, continental-margin subduction as the Red Ant Schist (Schweickert and others, 1980). The wide range of depositional ages may reflect subduction of an expansive, Mississippian-Triassic ocean.

Fiddle Creek Complex

The Fiddle Creek Complex (Fig. 1) consists of a disrupted ophiolitic basement overlain depositionally by volcanic rocks, volcaniclastic rocks, distal tuff, chert, and olistostromes (Ohlin, 1986; Edelman and others, 1989a). The chert contains Middle Triassic–Early Jurassic radiolaria, and the ophiolitic basement is locally intruded by Early Jurassic (ca. 200 Ma) mafic-intermediate plutons (Saleeby and others, 1989). The complex is part of the Tuolomne River terrane of Edelman and Sharp (1989) and is probably correlative with the Rattlesnake Creek terrane in the Klamath Mountains. The Fiddle Creek Complex is interpreted as an Early Jurassic, intraoceanic magmatic arc because of its ophiolitic basement, associated chert-rich sedimentary rocks, paucity of continental detritus, and Early Jurassic igneous rocks.

Slate Creek and Lake Combie complexes

The Slate Creek Complex (Fig. 1) is a pseudostratigraphic sequence of serpentinized cumulate ultramafic rocks, gabbroic-tonalitic plutonic rocks, and volcanogenic rocks. The cumulate origin of the ultramafic rocks and mutually intrusive relations between plutons and feeder dikes along plutonic-volcanic contacts indicate that the entire sequence is comagmatic (Edelman and others, 1989a. U-Pb zircon ages of 200 to 210 Ma for the plutonic unit (Edelman and others, 1989b; Saleeby and others, 1989) probably also represent the ages of the ultamafic and volcanic rocks. In addition, tonalite exposed on Deer Creek in the Smartville Complex (DC in Fig. 1) is about 200 m.y. old (Edelman and others, 1989b). Although most of the Smartville Complex dates from about 160 Ma (see below), this older tonalitic basement in the Smartville Complex is correlative with the Slate Creek Complex. The Lake Combie Complex (Fig. 1) is undated, but similar pseudostratigraphic sequence and structural position suggest that it is correlative with the Slate Creek Complex (Edelman and Sharp, 1989).

The Slate Creek and Lake Combie complexes are interpreted as remnants of an Early Jurassic magmatic arc. They are ophiolitic, but continuous sheeted dike units are absent. Generation of these complexes in an arc rather than a mid-ocean ridge is indicated by the abundance of tonalitic, pyroclastic, and volcaniclastic rocks, and geochemical similarity of the Lake Combie Complex to modern arcs (Rogers and others, 1984).

Post-amalgamation intrusive and cover rocks

Callovian-Kimmeridgian volcanic and sedimentary rocks and coeval intrusive rocks depositionally overlie and intrude the six units described above and postdate their amalgamation. This younger "overlap sequence" (Fig. 2; Edelman and Sharp, 1989) includes volcanogenic strata in the Northern Sierra terrane northeast of the Taylorsville thrust, the Colfax sequence, Smartville volcanic rocks, the Smartville intrusive complex with its sheeted dikes, and other ca. 160-Ma intrusive bodies such as the Yuba Rivers pluton (Fig. 1; Snoke and others, 1982; Edelman and others, 1989b; Saleeby and others, 1989). Callovian and possibly younger strata northeast of the Taylorsville thrust stratigraphically overlie the older units in the Northern Sierra terrane (McMath, 1966; Harwood, 1988). The Colfax sequence depositionally overlies the Fiddle Creek Complex, contains clastic detri-

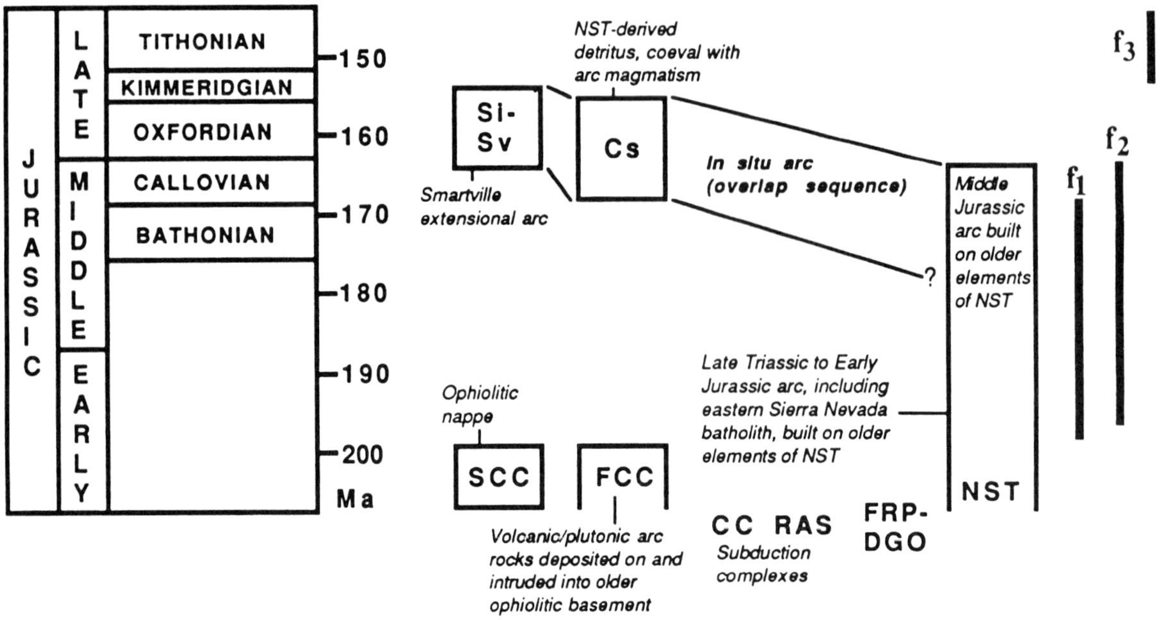

Figure 2. Jurassic time diagram summarizing ages of arc rocks and bracketed ages of faulting events. Early Jurassic arc fragments are Slate Creek Complex (SCC), Fiddle Creek Complex (FCC), and Early Jurassic part of Northern Sierra terrane stratigraphy, including eastern Sierra Nevada batholith. Smartville Complex (Si-Sv), Colfax sequence (Cs), and Middle Jurassic part of NST are parts of a single, in situ arc that overlaps previously amalgamated basement units. Abbreviations as in Figure 1. Time scale from Palmer (1983).

tus derived in part from the northern Sierra terrane, and ranges in age from early Callovian to Oxfordian. The Colfax sequence is partly coeval with, and is inferred to interfinger with, Smartville volcanic strata, in a fashion similar to the interfingering of the Mariposa (Colfax equivalent) and Logtown Ridge (Smartville equivalent) Formations south of the area of Figure 1 (Behrman and Parkison, 1978). The 200-Ma tonalite at Deer Creek in the Smartville Complex is cut by a 160-Ma dike, so the 160-Ma Smartville intrusive complex intrudes Slate Creek Complex basement (Edelman and others, 1989b).

Although direct correlation of geographically separated Callovian-Kimmeridgian units, particularly the Smartville volcanic rocks and volcanic rocks northeast of the Taylorsville thrust, has not been demonstrated, they have been interpreted as a single magmatic arc built across an older basement amalgam (Davis and others, 1978; Saleeby, 1981, 1983). This interpretation was supported by independent evidence that the basement terranes were mutually juxtaposed before the Callovian (Edelman and others, 1989a; Edelman and Sharp, 1989). The Callovian-Kimmeridgian units were thus built across a previously amalgamated basement. This configuration rules out models in which two Callovian-Kimmeridgian arcs, represented in the area of Figure 1 by the Smartville Complex and the volcanic strata northeast of the Taylorsville thrust, were associated with two different subduction zones that collided during the Late Jurassic (Moores, 1970, 1972; Schweickert and Cowan, 1975; Hietanen, 1981; Edelman and others, 1983; Ingersoll and Schweickert, 1986).

FAULT CHRONOLOGY

Three major generations of faults f_{1-3} are recognized in the northern Sierra Nevada (Figs. 1, 2). These generations are different from the four generations recognized in the Colfax area by Day and others (1985); their F_1 corresponds to f_2 here, and their F_2, F_3, and F_4 are grouped into f_3 here. First-generation f_1 faults juxtapose parts of the Northern Sierra terrane, Feather River peridotite, Devils Gate ophiolite, Red Ant Schist, and Calaveras Complex (Fig. 1). These faults are tightly folded or transposed along steeply dipping, northwest-striking hinge planes, but have shallowly east-dipping enveloping surfaces (Fig. 3; Edelman and others, 1989a). These faults are remnants of a major system of mostly cryptic, originally east-dipping faults that juxtapose, in descending structural order, the Northern Sierra terrane, Feather River peridotite/Devils Gate ophiolite, Red Ant Schist, Calaveras Complex, and Fiddle Creek Complex (Fig. 3; Edelman and others, 1989a; Edelman and Sharp, 1989).

The Slate Creek thrust and equivalent faults constitute f_2 (Fig. 1). The Slate Creek thrust is an isoclinally folded, west-dipping fault that places the Slate Creek Complex over the Fiddle Creek Complex, Calaveras Complex, and Red Ant Schist (Fig. 3; Edelman and others, 1989a). Thus, the Slate Creek thrust is younger than the two f_1 faults that juxtapose the Fiddle Creek Complex, Calaveras Complex, and Red Ant Schist, and is inferred to be younger than the other members of the f_1 fault system. The Slate Creek thrust is cut by the Scales pluton (SP in Fig. 3), which is at least 163 m.y. old, and by the Yuba Rivers

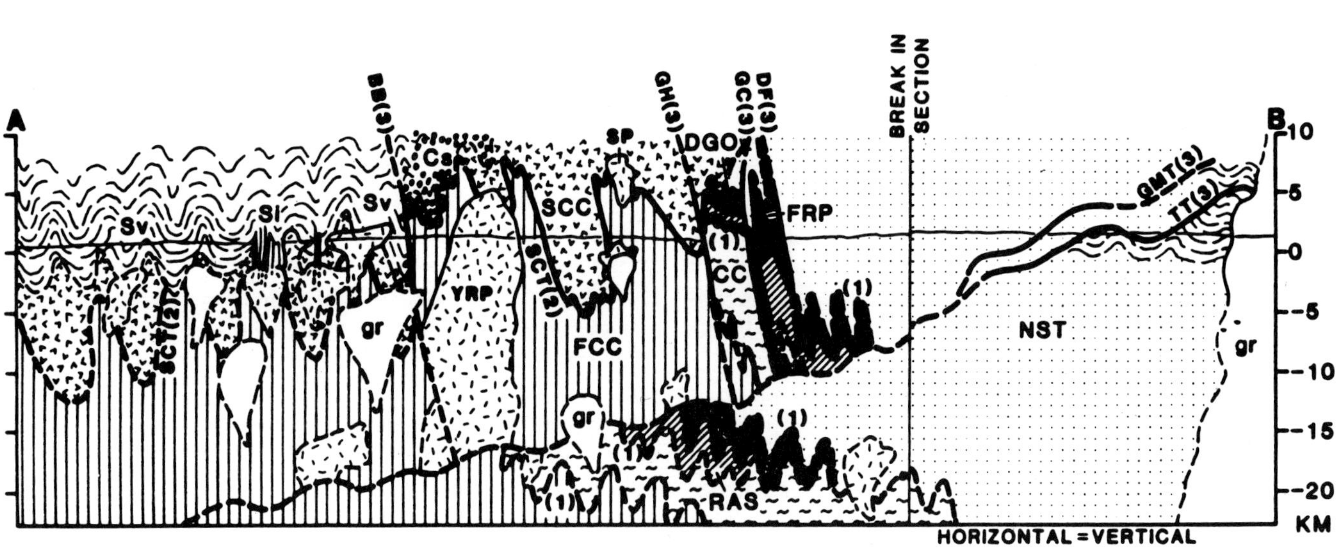

Figure 3. Cross section across the northern Sierra Nevada; see Figure 1 for location. Numbers in parentheses indicate fault generations. Dashed contacts are inferred or show observed relations schematically at depth. SP = 163-Ma Scales pluton; other abbreviations as in Figure 1.

pluton (YRP in Fig. 3), which is about 160 m.y. old (Edelman and others, 1989a, 1989b; Saleeby and others, 1989). Neither the Lake Combie Complex nor the low-angle faults that place it over the Fiddle Creek Complex are dated, but they are tentatively correlated with the Slate Creek Complex and Slate Creek thrust, respectively (Day and others, 1985; Edelman and Sharp, 1989).

The Fiddle Creek Complex is overlain by the Smartville Complex in two windows (W in Fig. 1). These low-angle contacts were formerly correlated with the Slate Creek thrust and were interpreted as thrust faults that cut ca. 160-Ma Smartville rocks (Moores and Day, 1984; Day and others, 1985). This interpretation is precluded by the evidence that the Slate Creek thrust is pre-163 Ma (Edelman and others, 1989b). The significance of these highly deformed, low-angle contacts is now an unresolved problem (Edelman and Sharp, 1989); I speculate that the southern window (Higgins Corner window of Day and others, 1985) is an f_2 fault remnant framed by undated 200-Ma rocks like those at the Deer Creek locality, and that the northern window (Ricci and others, 1985) is an unconformity along which Oxfordian-Kimmeridgian Smartville volcanogenic rocks overlie Fiddle Creek basement (Fig. 1). This unconformity may reflect pre-Oxfordian erosion through the Slate Creek thrust sheet and deposition of Oxfordian strata across a topographic surface of Slate Creek rocks and tectonic windows framing Fiddle Creek rocks. Depositional contacts along which the Colfax sequence, which is coeval with Smartville volcanic rocks, overlies Fiddle Creek rocks, demonstrate erosional breaching of the Slate Creek thrust sheet before deposition of the Colfax sequence. This depositional contact, which is itself now folded, is shown schematically in Figure 3.

The f_1-f_2 nappes juxtapose the major pre–Middle Jurassic stratigraphic packages, and only these faults can represent sutures. Senses of slip and magnitudes of displacement along the f_1 and f_2 faults are unconstrained. The f_1-f_2 nappes are interpreted as accretion-related contractional faults (Edelman, 1988), but origins as extensional faults or low-angle strike-slip faults cannot be ruled out.

Whereas f_2 faults are cut by ca. 160-Ma plutons and predate deposition of Callovian-Kimmeridgian strata, two sets of oppositely dipping faults cut these rocks and are assigned to f_3. The shallowly west-dipping Taylorsville thrust cuts fossiliferous Callovian rocks and overlying unfossiliferous strata; the Grizzly Mountain thrust is inferred to be the same age. The conspicuous, throughgoing faults of the northern Sierra Nevada, including the Big Bend, Wolf Creek, Dogwood Peak, Gillis Hill, Rich Bar, Goodyears Creek, and Downieville faults (Fig. 1), belong to the post–early Kimmeridgian Foothills fault system of Clark (1960). In the northern Sierra Nevada, the Big Bend, Wolf Creek, and Gillis Hill faults demonstrably cut Callovian-Kimmeridgian strata and related plutons (Fig. 1). These throughgoing faults dip steeply east; they are marked by mylonitic fabrics with down-dip stretching lineations and display generally east-side-up separations of earlier surfaces. These faults are thus dip-slip, generally east-side-up reverse faults (Edelman and others, 1989a). They are folded in map view and probably were steepened during upright folding reflected by folds in Callovian-Kimmeridgian units (Fig. 3; Edelman, 1989). These steep faults, the Taylorsville and Grizzly Mountain thrusts, the major upright folds, and pervasive steep cleavages, are permissively the same age (post–early Kimmeridgian; Fig. 2). The cross-cutting relations of the two oppositely dipping fault sets at depth are not known, but Edelman (1989) suggested that the steep faults terminate against the Taylorsville thrust (Fig. 3).

RELATIONS OF FAULTS TO METAMORPHISM

Each f_1 and f_2 fault marks a major metamorphic break as well as a stratigraphic break; thus prograde metamorphism in the

f_1-f_2 nappes predates these faults. The f_1 faults separate rocks displaying maximum metamorphic conditions in the lower greenschist facies (Northern Sierra terrane), epidote-amphibolite facies (Feather River peridotite/Devils Gate ophiolite), blueschist facies (Red Ant Schist), lower greenschist facies (Calaveras Complex), and lowermost of subgreenschist facies (Fiddle Creek Complex; Day and others, 1988; Edelman and others, 1989a). The f_2 Slate Creek thrust places greenschist- and amphibolite-facies rocks of the Slate Creek Complex, in which metamorphic grade is controlled by pseudostratigraphic position, over lower grade Fiddle Creek and Calaveras rocks (Edelman and others, 1989a).

Postamalgamation (post-f_1-f_2) regional metamorphism is lowermost or subgreenschist facies, as indicated by prehnite-pumpellyite metamorphism of Smartville and Colfax rocks, and preservation of prehnite in Fiddle Creek rocks (Day and others, 1988; Edelman and others, 1989a). This very low-grade regional metamorphism is the only candidate for Late Jurassic (Nevadan) regional metamorphism (Day and others, 1988). Higher grade contact metamorphism is present near plutons.

MESOZOIC TECTONIC HISTORY

As many as three different magmatic arcs may have been active in the Early Jurassic: a continental-margin arc represented by Early Jurassic volcanic rocks in the Northern Sierra terrane and coeval plutonic rocks in the eastern part of the Sierra Nevada batholith, an intraoceanic arc to the west represented by the Fiddle Creek Complex, and an even more westerly intraoceanic arc represented by the Slate Creek Complex and equivalent units. The Calaveras Complex and Red Ant Schist represent the continental-margin subduction complex. The Fiddle Creek arc accreted sometime between 200 Ma (age of arc magmatism) and the early Callovian (age of Northern Sierra terrane-derived detritus in the unconformably overlying Colfax sequence). The Fiddle Creek arc was thrust beneath the Calaveras Complex along an f_1 fault; because the other f_1 faults appear to form a subparallel set (Fig. 3), all the f_1 faults may have formed during this collision. The Slate Creek arc was thrust eastward along f_2 fault(s) sometime later and in the 200- to 163-Ma interval (Fig. 2). Whether the Fiddle Creek and Slate Creek arcs were separate arcs associated with separate subduction zones or fragments of a single arc is not known.

The assembled f_1-f_2 nappe pile formed the basement for a complicated continental-margin arc. This Callovian-Kimmeridgian arc formed across the entire northern Sierra Nevada and was extensional in its western part, as reflected by the Smartville sheeted dikes (Figs. 2, 3; Beard and Day, 1987). The subduction complex associated with this arc is presumably represented by the oldest parts of the Franciscan Complex located west of the area of Figure 1. Deformation, uplift, and erosion during Callovian-Kimmeridgian arc magmatism is indicated by syn-f_3 intrusion of the Yuba Rivers pluton (Bobbitt and others, 1986), sedimentologic evidence for deposition of the Colfax sequence in actively deforming basins (Tuminas and Moores, 1982), erosion of the Northern Sierra terrane and other rocks to supply detritus to the Colfax sequence, and erosion through the Slate Creek thrust sheet indicated by deposition of the Colfax sequence and Smartville volcanic rocks on both Fiddle Creek and Slate Creek basement. During the Late Jurassic, the arc evolved from this mafic-intermediate regime to a regime in which granitic plutons related to the Sierra Nevada batholith were intruded (gr in Figs. 1, 3). During this transition the crust, consisting of the f_1-f_2 nappe pile and the younger mafic-intermediate arc rocks, was shortened and thickened by f_3 faults and related Nevadan (post–early Kimmeridgian) structures, which are cut by Cretaceous granitic plutons. The origin of the f_3 deformation is an outstanding problem; suggestions include transpressional deformation (Saleeby, 1981) and a continuation of collisional deformation associated with accretion of the Slate Creek arc (Edelman, 1989).

CONCLUSIONS

The upper crust in the northern Sierra Nevada is composed of a stack of shallowly dipping nappes. The five nappes west of the Northern Sierra terrane contain stratigraphically unrelated assemblages of oceanic rocks. The f_1-f_2 faults that bound these nappes formed in Early and/or Middle Jurassic time. The f_1 system of shallowly east-dipping faults is geometrically and temporally similar to the major west-vergent thrust faults of the Klamath Mountains (Irwin, 1981; Wright and Fahan, 1988). The structure of the northern Sierra Nevada differs from the Klamath Mountains chiefly by f_2 eastward emplacement of the Early Jurassic Slate Creek Complex and by apparently stronger and more pervasive f_3 horizontal shortening (Nevadan orogeny).

The Fiddle Creek Complex is the structurally lowest unit in the f_1-f_2 nappe pile (Fig. 3). Unless unrecognized, structurally lower units are present, the middle and lower crust in the northern Sierra Nevada may be predominantly disrupted ophiolitic rocks belonging to the Fiddle Creek Complex basement. The structurally higher units lie in relatively thin f_1-f_2 tectonic sheets. Although sense of displacement is unconstrained on f_1 and f_2 faults, the stratigraphic and metamorphic contrasts across these faults suggest they are thrust faults (rather than normal or strike-slip faults). New continental crust apparently formed by stacking of both oceanward- and continentward-directed thrust sheets of oceanic material, in situ arc magmatism, and shortening and thickening of the nappe pile and the early (Callovian-Kimmeridgian) phases of the in situ arc by reverse faulting and upright folding.

ACKNOWLEDGMENTS

The field work was partially supported by National Science Foundation Grant EAR-80-19697 (to H. W. Day and E. M. Moores). The radiometric dating was supported by NSF Grant EAR-85-16262 (to H. W. Day and M. E. Bickford). Day kindly read an early version of the mansucript. Reviews by G. H. Girty, A. Griscom and M. M. Miller were very helpful.

REFERENCES CITED

Beard, J. S., and Day, H. W., 1987, Smartville intrusive complex, Sierra Nevada, California; Core of a rifted volcanic arc: Geological Society of America Bulletin, v. 99, p. 779–791.

Behrman, P. G., and Parkison, G. A., 1978, Paleogeographic significance of the Callovian to Kimmeridgian strata, central Sierra Nevada foothills, California, *in* Howell, D. G., and McDougall, K. A., eds., Mesozoic paleogeography of the western United States: Pacific Section, Society of Economic Paleontologists and Mineralogists Pacific Coast Paleogeography Symposium 2, p. 349–360.

Bobbitt, J. B., Day, H. W., and Eddy, C. A., 1986, Syntectonic crystallization of the Yuba Rivers pluton, northern Sierra Nevada: Geological Society of America Abstracts with Programs, v. 18, p. 87.

Clark, L. D., 1960, Foothills fault system, western Sierra Nevada, California: Geological Society of America Bulletin, v. 71, p. 483–496.

D'Allura, J. A., Moores, E. M., and Robinson, L., 1977, Paleozoic rocks of the northern Sierra Nevada; Their structural and paleogeographic implications, *in* Stewart, J. H., and others, eds., Paleozoic paleogeography of the western United States: Pacific Section, Society of Economic Paleontologists and Mineralogists Pacific Coast Paleogeography Symposium 1, p. 395–408.

Davis, G. A., Monger, J.W.H., and Burchfiel, B. C., 1978, Mesozoic construction of the Cordilleran "collage," central British Columbia to central California, *in* Howell, D. G., and McDougall, K. A., eds., Mesozoic paleogeography of the western United States: Pacific Section, Society of Economic Paleontologists and Mineralogists Pacific Coast Paleogeography Symposium 2, p. 1–32.

Day, H. W., Moores, E. M., and Tuminas, A. C., 1985, Structure and tectonics of the northern Sierra Nevada: Geological Society of America Bulletin, v. 96, p. 436–450.

Day, H. W., Schiffman, P., and Moores, E. M., 1988, Metamorphism and tectonics of the northern Sierra Nevada, *in* Ernst, W. G., ed., Metamorphism and crustal evolution of the western United States; Rubey Volume 4: Englewood Cliffs, New Jersey, Prentice-Hall, p. 737–763.

Edelman, S. H., 1988, Continental growth by stacking, folding, and reverse faulting of oceanic tectonic flakes, Sierra Nevada metamorphic belt, California: Geological Society of America Abstracts with Programs, v. 20, p. 184.

—— , 1989, Relationships between kinematics of arc-continent collision and kinematics of thrust faults, folds, shear zones, and foliations in the Nevadan orogen, northern Sierra Nevada, California: Tectonophysics (in press).

Edelman, S. H., and Sharp, W. D., 1989, Terranes, early faults, and pre–Late Jurassic amalgamation of the western Sierra Nevada metamorphic belt, California: Geological Society of America Bulletin (in press).

Edelman, S. H., Day, H. W., and Moores, E. M., 1983, The Nevadan suture, northern Sierra Nevada, California; A Late Jurassic arc-continent collision: Geological Society of America Abstracts with Programs, v. 15, p. 565.

Edelman, S. H., and 5 others, 1989a, Structure across a Mesozoic ocean-continent suture zone in the northern Sierra Nevada, California: Geological Society of America Special Paper 224, 56 p.

Edelman, S. H., Day, H. W., and Bickford, M. E., 1989b, Implications of U-Pb zircon ages for the tectonic settings of the Smartville and Slate Creek complexes, northern Sierra Nevada, California: Geology, v. 17, p. 1032–1035.

Ehrenberg, S. N., 1975, Feather River ultramafic body, northern Sierra Nevada, California: Geological Society of America Bulletin, v. 86, p. 1235–1243.

Hamilton, W., 1981, Crustal evolution by arc magmatism: Philosophical Transactions of the Geological Society of London, v. A-301, p. 279–291.

Harwood, D. S., 1988, Tectonism and metamorphism in the Northern Sierra terrane, California, *in* Ernst, W. G., ed., Metamorphism and crustal evolution of the western United States; Rubey Volume 7: Englewood Cliffs, New Jersey, Prentice-Hall, p. 764–788.

Hietanen, A., 1973, Geology of the Pulga and Bucks Lake Quadrangles, Butte and Plumas Counties, California: U.S. Geological Survey Professional Paper 731, 66 p.

—— , 1976, Metamorphism and plutonism around the Middle and South forks of the Feather River, California: U.S. Geological Survey Professional Paper 920, 30 p.

—— , 1981, Petrologic and structural studies in the northwestern Sierra Nevada, California: U.S. Geological Survey Professional Paper 1226, 59 p.

Ingersoll, R. V., and Schweickert, R. A., 1986, A plate-tectonic model for ophiolite genesis, Nevadan orogeny and forearc initiation, northern California: Tectonics, v. 5, p. 901–912.

Irwin, W. P., 1981, Tectonic accretion of the Klamath Mountains, *in* Ernst, W. G., ed., The geotectonic development of California; Rubey Volume 7: Englewood Cliffs, New Jersey, Prentice-Hall, p. 29–49.

Jennings, C. W., 1977, Geologic map of California: California Division of Mines and Geology, California Geologic Data Map Series, scale 1:750,000.

McMath, V. E., 1966, Geology of the Taylorsville area, northern Sierra Nevada: California Division of Mines and Geology Bulletin 190, p. 173–183.

Moores, E. M., 1970, Ultramafics and orogeny, with models of the U.S. Cordillera and the Tethys: Nature, v. 228, p. 837–842.

—— , 1972, Model for Jurassic island arc-continental margin collision in California: Geological Society of America Abstracts with Programs, v. 4, p. 202.

Moores, E. M., and Day, H. W., 1984, Overthrust model for the Sierra Nevada: Geology, v. 12, p. 416–419.

Ohlin, H. N., 1986, Pre–Late Jurassic melanges, northern Sierra Nevada foothills, Butte County, California; Southeast extension of the Rattlesnake Creek terrane?: Geological Society of America Abstracts with Programs, v. 18, p. 165.

Palmer, A. R., 1983, The Decade of North American Geology 1983 geologic time scale: Geology, v. 11, p. 503–504.

Ricci, M. P., Moores, E. M., Verosub, K. L., and McClain, J. S., 1985, Geologic and gravity evidence for emplacement of the Smartville ophiolite: Tectonics, v. 4, p. 539–546.

Rogers, J.J.W., Suayah, I. B., and Edwards, J. M., 1984, Trace elements in continental margin magmatism; Part 4, Geochemical criteria for recognition of two volcanic assemblages near Auburn, western Sierra Nevadas, California: Geological Society of America Bulletin, v. 95, p. 1437–1445.

Saleeby, J. B., 1981, Ocean floor accretion and volcanoplutonic arc evolution of the Mesozoic Sierra Nevada, *in* Ernst, W. G., ed., The geotectonic development of California; Rubey Volume 7: Englewood Cliffs, New Jersey, Prentice-Hall, p. 132–181.

—— , 1983, Accretionary tectonics of the North American Cordillera: Annual Reviews of Earth and Planetary Science, v. 15, p. 45–73.

Saleeby, J. B., Hannah, J. L., and Varga, R. J., 1987, Isotopic age constraints on middle Paleozoic deformation in the northern Sierra Nevada, California: Geology, v. 15, p. 757–760.

Saleeby, J. B., Shaw, H. F., Moores, E. M., and Edelman, S. H., 1989, U/Pb, Sm/Nd, and Rb/Sr geochronological and isotopic study of northern Sierra Nevada ophiolitic assemblages, California: Contributions to Mineralogy and Petrology, v. 102, p. 205–220.

Schweickert, R. A., 1981, Tectonic evolution of the Sierra Nevada Range, *in* Ernst, W. G., ed., The geotectonic development of California; Rubey Volume 7: Englewood Cliffs, New Jersey, Prentice-Hall, p. 87–131.

Schweickert, R. A., and Cowan, D. S., 1975, Early Mesozoic tectonic evolution of the western Sierra Nevada, California: Geological Society of America Bulletin, v. 86, p. 1329–1336.

Schweickert, R. A., Armstrong, R. L., and Harakal, J. E., 1980, Lawsonite blueschist in the northern Sierra Nevada, California: Geology, v. 8, p. 27–31.

Silberling, N. J., Jones, D. L., Blake, M. C., Jr., and Howell, D. G., 1987, Lithotectonic terrane map of the western conterminous United States: U.S. Geological Survey Miscellaneous Field Studies Map MF-1874-C, scale 1:2,500,000.

Snoke, A. W., Sharp, W. D., Wright, J. E., and Saleeby, J. B., 1982, Significance of mid-Mesozoic peridotitic to dioritic intrusive complexes, Klamath Mountains–western Sierra Nevada, California: Geology, v. 10, p. 160–166.

Speed, R. C., and Moores, E. M., 1980, Geologic cross section of the Sierra

Nevada and the Great Basin along 40°N. lat., northeastern California and northern Nevada: Geological Society of America Map and Chart Series MC-28L, scale 1:250,000.

Stern, T. W., Bateman, P. C., Morgan, B. A., Newell, M. F., and Peck, D. L., 1981, Isotopic U-Pb ages of zircon from the granitoids of the central Sierra Nevada, California: U.S. Geological Survey Professional Paper 1185, 17 p.

Tuminas, A. C., and Moores, E. M., 1982, Sedimentology and possible paleotectonic setting of a Late Jurassic flysch sequence in the western foothills of the northern Sierra Nevada, California: Geological Survey of America Abstracts with Programs, v. 14, p. 241.

Wright, J. E., and Fahan, M. R., 1988, An expanded view of Jurassic orogenesis in the western United States Cordillera; Middle Jurassic (pre-Nevadan) regional metamorphism and thrust faulting within an active arc environment, Klamath Mountains, California: Geological Society of America Bulletin, v. 100, p. 859–876.

Xenophontos, C., and Bond, G. C., 1978, Petrology, sedimentation, and paleogeography of the Smartville terrane (Jurassic); Bearing on the genesis of the Smartville ophiolite, *in* Howell, D. G., and McDougall, K. A., eds., Mesozoic paleogeography of the western United States: Pacific Section, Society of Economic Paleontologists and Mineralogists Pacific Coast Paleogeography Symposium 2, p. 291–302.

MANUSCRIPT ACCEPTED BY THE SOCIETY MARCH 27, 1990

Printed in U.S.A.

Geological Society of America
Special Paper 255
1990

A structural study of a metamorphic sole beneath the Josephine ophiolite, western Klamath terrane, California-Oregon

Gregory D. Harper and Kristin Grady
Department of Geology, State University of New York, Albany, New York 12222
John Wakabayashi
Department of Geology, University of California, Davis, California 95616

ABSTRACT

The 162-Ma Josephine ophiolite was emplaced over an active mafic batholith (Chetco River complex) along the Madstone thrust in southwestern Oregon during the Nevadan orogeny, beginning at ~155 Ma. Strongly deformed amphibolite and minor quartzite occur between the ophiolite and the batholith and are interpreted to make up a metamorphic sole formed during thrusting. Retrograde metamorphism is ubiquitous, and amphibolite has been locally converted to greenschist-facies mafic phyllonite adjacent to the Madstone thrust. Pegmatite dikes locally cut the amphibolite but are also penetratively deformed, indicating syntectonic intrusion. A geochronologic study (Harper and others, 1989) indicates cooling from ~450°C at 153 Ma, intrusion of the pegmatite at 150 Ma, and cooling to ~350°C at 146 Ma. Geobarometry, using amphibole composition and phengite content of muscovite, indicates relatively low P/T metamorphism. The lower contact of the amphibolite sole with the Chetco River complex, as described by previous workers, is intrusive and syntectonic with deformation of the amphibolite sole.

In the hanging wall of the Madstone thrust, 20 to 40 m of high-T serpentinite mylonite occurs along the base of the Josephine Peridotite. The serpentinite apparently formed during ophiolite emplacement because it is structurally concordant with the underlying amphibolite and phyllonite. In addition, the serpentinite locally shows metasomatism, which probably resulted from interaction with fluids derived from the amphibolite sole.

The amphibolite shows two generations of folds having fold hinges parallel to a NNE-stretching lineation. These structures, along with grain-size reduction and asymmetric fabrics, indicate that the amphibolites are mylonites formed by progressive simple shear. The lineations and sense-of-shear criteria for the amphibolite and serpentinite mylonite indicate thrusting of the Josephine ophiolite toward the north-northeast, over the Chetco River complex. Continued north-northeast thrusting during greenschist retrograde metamorphism is indicated by lineations and microstructures in phyllonites and a pegmatite dike. A minimum displacement of 12 km is inferred from the outcrop pattern of the Madstone thrust.

The metamorphic sole and regional geologic setting of the Josephine ophiolite are distinct from other ophiolites. There is no inverted gradient, maximum temperatures were lower, syntectonic magmas were intruded into both the metamorphic sole and the

Harper, G. D., Grady, K., and Wakabaiashi, J., 1990, A structural study of a metamorphic sole beneath the Josephine ophiolite, western Klamath terrane, California-Oregon, *in* Harwood, D. S., and Miller, M. M., eds., Paleozoic and early Mesozoic paleogeographic relations; Sierra Nevada, Klamath Mountains, and related terranes: Boulder, Colorado, Geological Society of America Special Paper 255.

ophiolite, and the ophiolite was thrust over an active magmatic arc rather than a continental margin. In addition, the ophiolite and overlying Galice Formation were thrust beneath the North American continent by >40 km along the roof thrust (Orleans fault) and regionally metamorphosed to low grade.

Geochronologic and structural studies indicate that the basal Madstone thrust and the roof thrust were both active at 150 ± 1 Ma, but the thrusting direction along the roof thrust appears to have been west or northwest. The cause and tectonic significance of nearly orthogonal thrusting directions between the basal and roof thrusts of the ophiolite is enigmatic. One possibility is that thrusting occurred during sinistral oblique subduction, and the Josephine thrust sheet was effectively decoupled along the roof thrust due to high pore-fluid pressures in the Galice Formation.

INTRODUCTION

Regional thrust faults are one of the fundamental structural aspects of the Klamath Mountains (Burchfiel and Davis, 1981; Snoke, 1977). Determination of the direction of thrusting is important for fully understanding the tectonic evolution of the Klamath Mountains and for relating orogeny to plate motions (e.g., Engebretson and others, 1985; May and Butler, 1986). Thrust faults are a major feature in the Western Klamath Terrane, including the basal (Madstone) and roof (Orleans) thrusts of the Josephine ophiolite and overlying Galice Formation (Fig. 1). These thrusts were both active during the Late Jurassic Nevadan orogeny (Harper and others, 1989).

A tectonic model for the formation and emplacement of the Josephine ophiolite is shown in Figure 2 and provides a framework for the study of the basal thrust. It is modified from Harper and Wright (1984) with the addition of new geochronologic data (Wright and Wyld, 1986; Wright and Fahan, 1988; Harper and others, 1989). A Middle Jurassic magmatic arc was built on older rocks of the Klamath Mountains beginning at ∼177 Ma in response to eastward subduction beneath western North America. Following a major intra-arc orogeny at 165 to 170 Ma, rifting and formation of a back-arc basin resulted in generation of the Josephine ophiolite; the inferred east-west orientation of spreading centers in the Josephine ophiolite suggests that the back-arc basin may have been dominated by arc-parallel transforms formed in response to oblique subduction (Saleeby, 1983; Harper and others, 1985). The back-arc basin was short-lived, and the Josephine ophiolite was buried by flysch and thrust beneath the Klamath Mountains. The active arc to the west was also buried by flysch and thrust beneath the Josephine ophiolite. With continued shortening, the ophiolite became situated over the zone of arc magmatism, resulting in syntectonic intrusion of the ophiolite, Galice Formation, and thrust faults.

This paper focuses on the basal thrust of the Josephine ophiolite (Madstone thrust, Figs. 1, 3, 4), including an amphibolite sole. The main objectives of this chapter are to document that metamorphism and deformation of the amphibolite sole occurred during ophiolite emplacment, to determine the direction of thrusting, and to estimate pressure and temperature conditions during thrusting.

REGIONAL FRAMEWORK

The Klamath Mountains province consists of thrust slices of Paleozoic and Mesozoic marine volcanic arc and sedimentary strata, along with ophiolitic rocks (Irwin, 1966, 1981; Burchfiel and Davis, 1981). The major tectonic units are generally younger toward the west and are separated by regional east-dipping thrust faults (Davis and others, 1978; Wright and Fahan, 1988).

The Western Klamath Terrane (western Jurassic belt of Irwin, 1966) is a generally east-dipping belt of ophiolitic rocks, island-arc intrusive and volcanic rocks, and Upper Jurassic flysch (Galice Formation) which is bounded at the top and base by regional thrust faults (Irwin, 1966, 1981). Blake and others (1985) subdivided the Western Klamath Terrane into three subterranes (Fig. 1): the Smith River subterrane (Josephine ophiolite and overlying flysch), the Dry Butte subterrane (primarily intrusive rocks), and the Rogue River subterrane (mostly volcaniclastic rocks and flysch).

The structurally highest thrust sheet in the Western Klamath Terrane is the Smith River subterrane, which includes the Josephine ophiolite and overlying Galice Formation (Harper, 1984). The ophiolite has been dated in the area of Figure 1 at 162 ± 1 Ma by Pb/U dating of zircon (Saleeby, 1987), and latest Callovian by radiolaria in mudstone interbedded with lavas (E. Pessagno, written communication, 1989). A southern remnant of the Josephine ophiolite has yielded a Pb/U zircon age of 164 ± 1 Ma (Wright and Wyld, 1986). The Galice Formation conformably overlies the ophiolite and consists of ∼50 m of radiolarian argillite and chert, ranging from early to middle Oxfordian age, which grade upward into flysch of middle Oxfordian to Kimmeridgian age (Pessagno and Mizutani, 1988).

The Rogue River subterrane (Fig. 1) consists of marine volcanic-arc strata (Rogue Formation) overlain by turbidites (flysch) of the Galice Formation. The Rogue Formation has yielded Pb/U zircon ages of 157 Ma (Saleeby, 1984). The Galice Formation conformably overlies the Rogue Formation (Wood, 1987) and is the same age and petrographically similar to flysch of the "Galice" that overlies the Josephine ophiolite (Harper, 1983; Pessagno and Blome, 1988; Pessagno and Mizutani, 1988). The Rogue and Galice Formations were folded and regionally

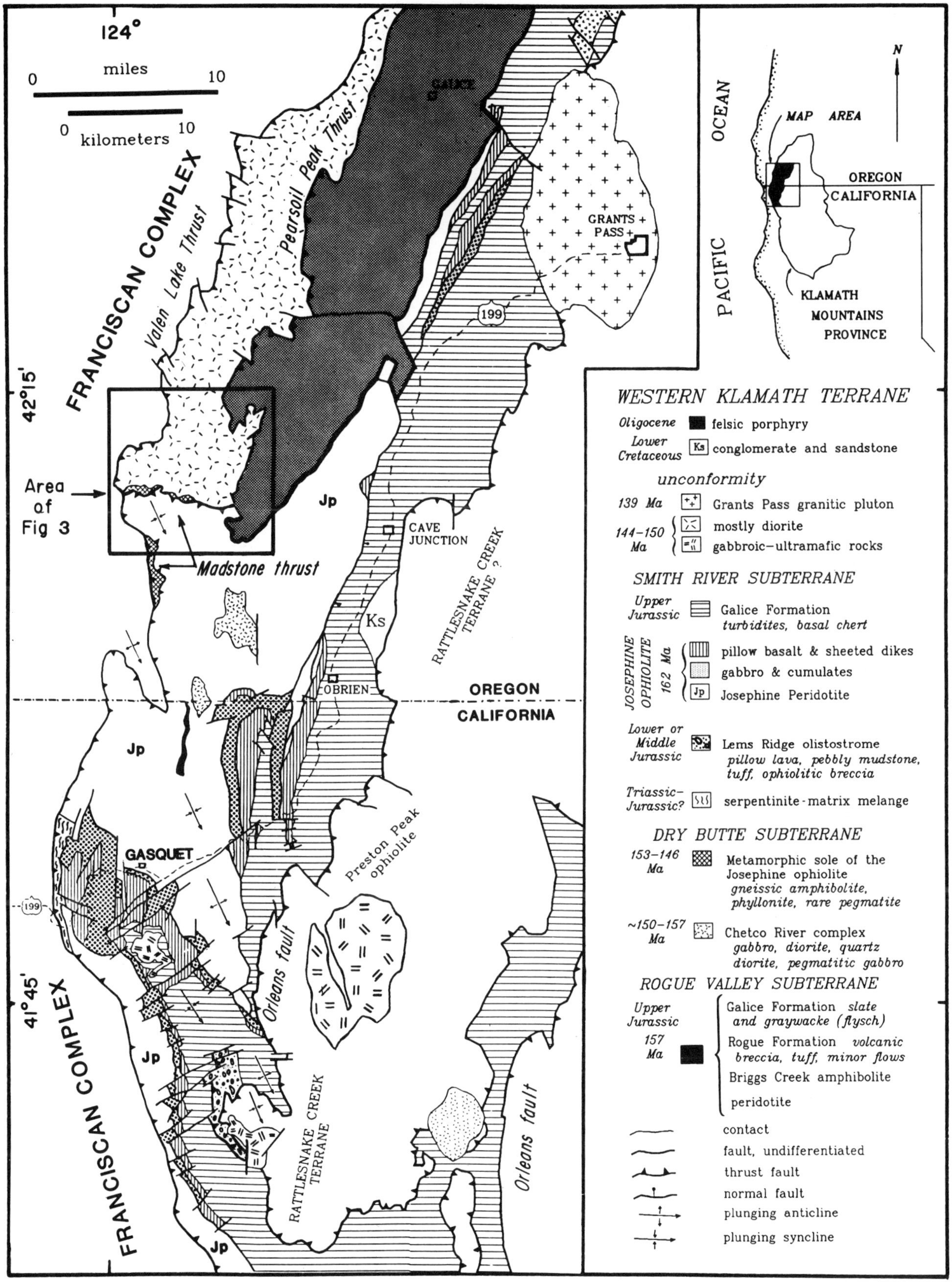

Figure 1. Generalized geologic map of the western Klamath terrane, modified from Harper (1989).

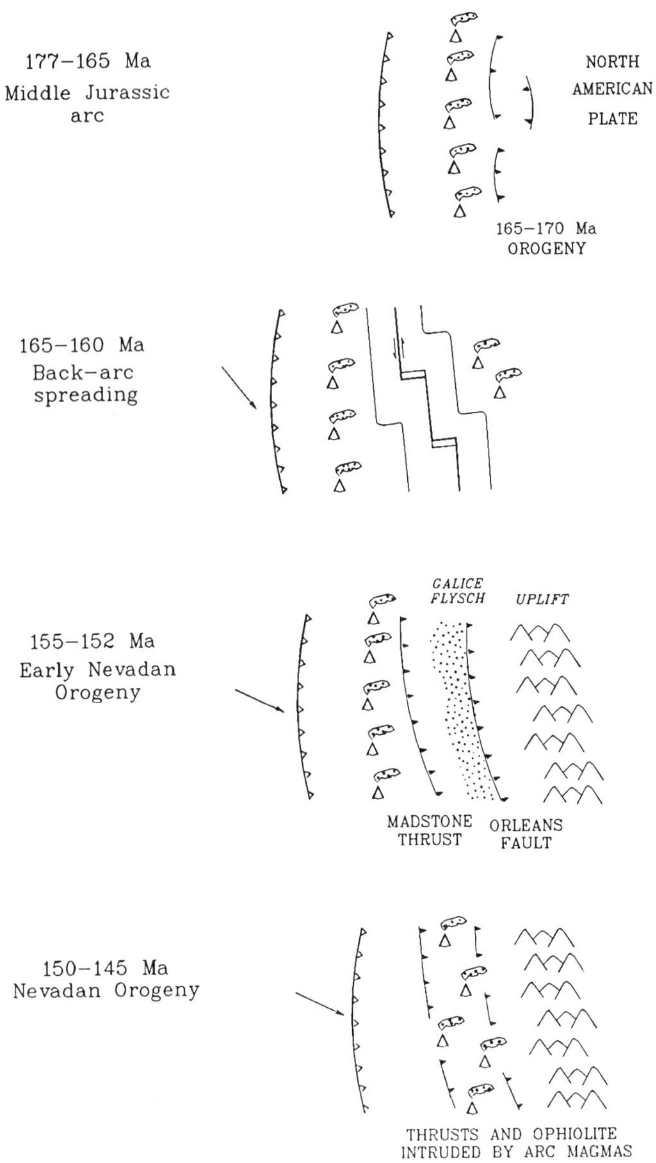

Figure 2. Tectonic model for the Middle to Late Jurassic tectonic evolution of the Klamath Mountains, modified from Harper and Wright, 1984.

metamorphosed during the Nevadan orogeny (Wells and Walker, 1953; Garcia, 1979; Riley, 1987; Park-Jones, 1987). The Briggs Creek amphibolite forms an elongate belt structurally beneath the Rouge Formation and was defined by Blake and others (1985) as a separate subterrane. The Briggs Creek amphibolite, along with a large ultramafic body to the west, are tentatively included in the Rogue River subterrane, consistent with the interpretation that they may represent ophiolitic basement upon which the Rogue arc was built (Garcia, 1982).

The Dry Butte subterrane (Fig. 1) consists of a Late Jurassic batholithic complex, known as the Chetco River complex or Illinois River gabbro (Hotz, 1971), and probably represents the plutonic roots of the Rogue volcanic arc (Dick, 1976, 1977; Garcia, 1982). The intrusive rocks range from gabbro through quartz diorite. The Chetco River complex is generally isotropic in the central part, whereas it becomes foliated and lineated near its margins, with the foliation dipping toward the east-southeast (Wells and others, 1949).

Gneissic amphibolite north of Vulcan Peak occurs adjacent to the Madstone thrust along the southern margin of the Dry Butte subterrane. These rocks, along with amphibolite near Chetco Lake to the south, form the metamorphic sole of the Josephine ophiolite, and also form the roof of the Chetco River complex (Fig. 3, cross section). This contact is intrusive, but the intrusive rocks are also deformed along with the amphibolite (Loney and Himmelberg, 1977; Dick, 1976), indicating syntectonic intrusion. Deformed ultramafic rocks appear to be associated with the amphibolite and locally form the roof of the batholith (Fig. 3).

The Josephine ophiolite of the Smith River subterrane is thrust over the Dry Butte subterrane along the Madstone thrust (Figs. 1, 3, 4). The nature of fault contacts between the other subterranes is uncertain, but mapping by Dick (1976) and Harding (1987), along with aeromagnetic interpretation (Harding, 1987), suggests that the Josephine ophiolite (Smith River subterrane) is thrust over the Rogue River subterrane, probably along an eastward and northeastward extension of the Madstone thrust. The boundary between the Rogue River and Dry Butte subterranes is tentatively taken to be a NNE-striking thrust fault that forms the eastern boundary of the Dry Butte subterrane (Ramp, 1984); we informally refer to this fault as the Pearsoll Peak thrust (Fig. 1). Mapping of the Pearsoll Peak thrust by Ramp (1984) shows similarities to the Madstone thrust: foliated intrusive rocks and local amphibolite are present along the thrust (Ramp, 1984). In addition, a locally foliated intermediate-granitic intrusive sheet extends along the thrust and locally intrudes the peridotite in the hanging wall (Hotz, 1971; Ramp, 1984); a 151 ± 4-Ma K/Ar age on muscovite from this unit is similar to K/Ar and $^{40}Ar/^{39}Ar$ muscovite ages of deformed pegmatite dikes that intrude the gneissic amphibolite below the Madstone thrust (Dick, 1976; Harper and others, 1989).

The roof thrust of the Smith River subterrane is the Orleans fault (Fig. 1). The outcrop pattern of the Orleans fault, including a large northeast-trending reentrant (Fig. 1), indicates the Western Klamath Terrane has been thrust >40 km beneath older terranes of the Klamath Mountains. Gravity data suggest that the fault extends much farther eastward, implying >110 km of horizontal displacement, assuming west-directed thrusting (Jachens and others, 1986). The Orleans fault is constrained to have been active during the Nevadan orogeny by 150- to 139-Ma plutons that cut the thrust (Fig. 1) and contact metamorphose the upper and lower plates (Snoke, 1977; Wright and Fahan, 1988; Harper and others, 1989).

Regional metamorphic grade in the Rogue River and Smith River subterranes ranges from prehnite-pumpellyite to lower greenschist facies (Irwin, 1966; Garcia, 1978; Harper, 1984;

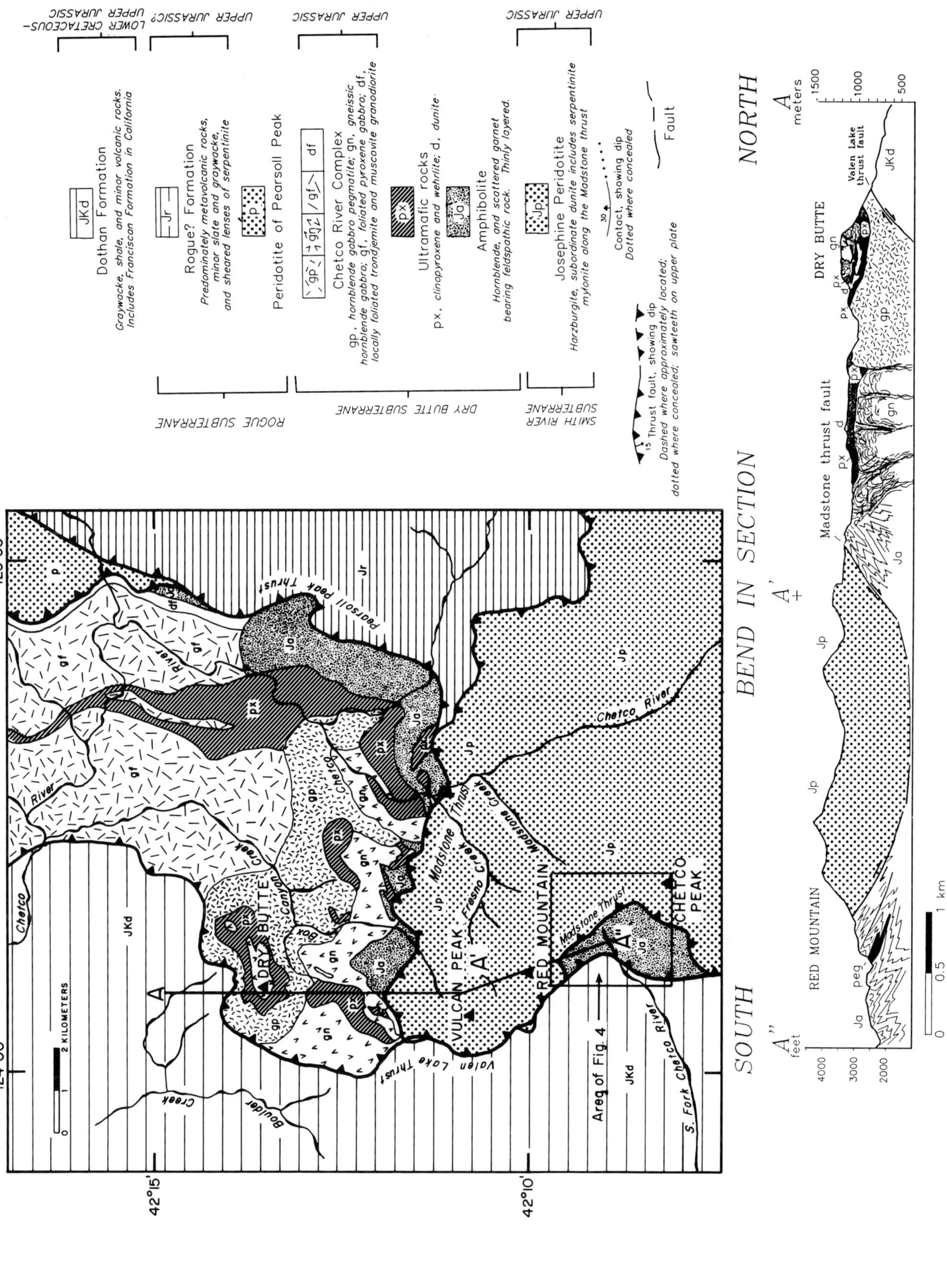

Figure 3. Geologic map in the region where basal (Madstone) thrust of the Josephine ophiolite is exposed; modified from Loney and Himmelberg, 1977).

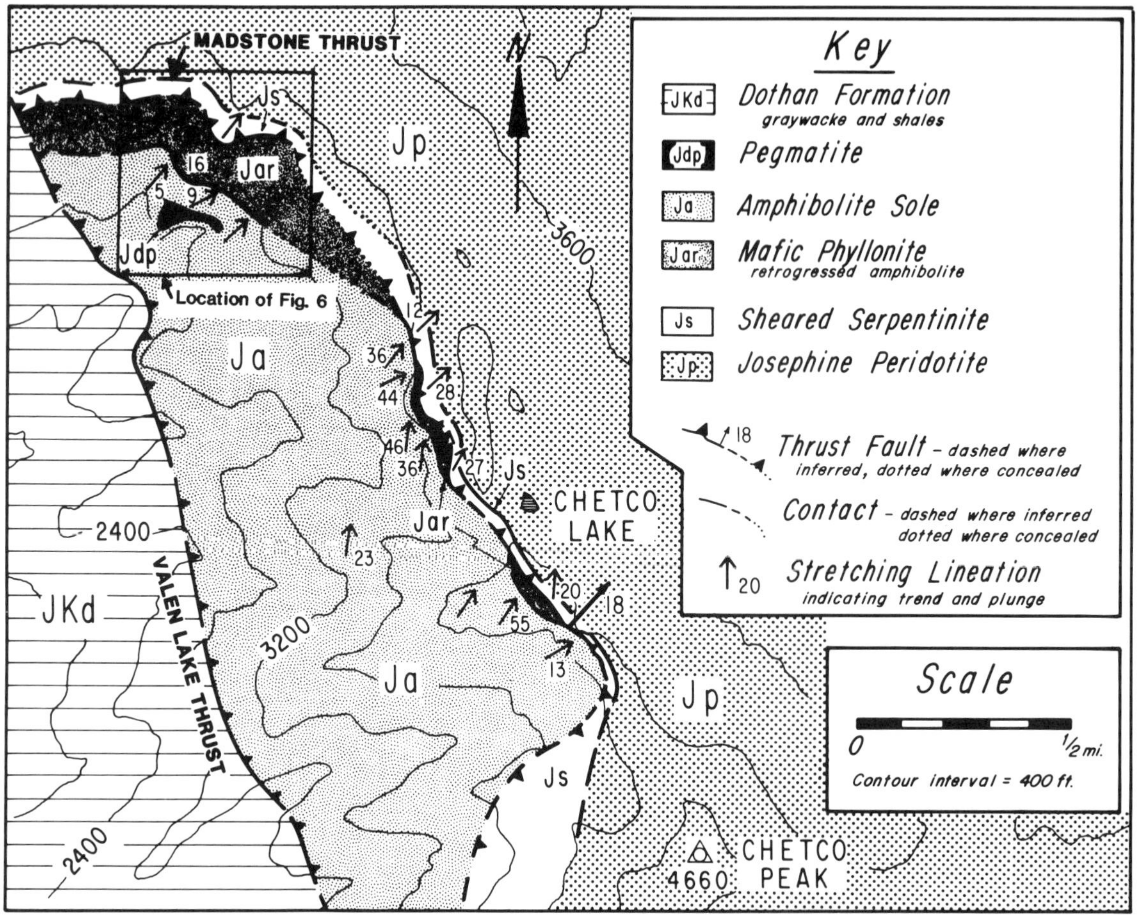

Figure 4. Map of the study area near Chetco Peak. The Josephine ophiolite is thrust over amphibolite along the Madstone thrust. The east dip of the thrust is the result of large-scale post-Nevadan folding.

Harper and others, 1988). As discussed in this study, Nevadan-age amphibolite-grade metamorphism occurred in the metamorphic sole of the Josephine ophiolite, but was followed by greenschist-facies retrograde metamorphism. The additional heat required for the amphibolite-grade metamorphism was probably provided by syntectonic intrusion of the underlying Chetco River complex. The regional metamorphism and deformation along thrusts show that the present exposure of the Western Klamath Terrane was at midcrustal depths during the Nevadan orogeny (~150 Ma). Uplift and exposure occurred by ~125 Ma, as indicated by a Lower Cretaceous angular unconformity south of Cave Junction, Oregon (Fig. 1).

The western boundary of the Western Klamath Terrane is an east-dipping thrust (Fig. 1) called the Valen Lake thrust in Oregon and the South Fork fault in California. The Valen Lake thrust is post-Nevadan in age and truncates the Madstone thrust (Fig. 1). Within the area of Figure 1, the footwall of the thrust consists of latest Jurassic to Early Cretaceous graywacke and shale (broken formation) and melange assigned to the Yolly Bolly terrane of the Franciscan complex (Aalto and Harper, 1982; Blake and others, 1985; Aalto, 1979).

The map pattern of the Western Klamath Terrane in the area of Figure 1 largely reflects post-Nevadan folding and faulting. The Josephine ophiolite, the Orleans thrust, the Madstone thrust, and the Valen Lake thrust are folded by large, open, southeast-plunging anticlines and synclines (Fig. 1). The post-Nevadan age of these folds is evident from folding of the Valen Lake thrust, as well as from folding of Nevadan-age foliations in the Josephine ophiolite and overlying Galice Formation.

MADSTONE THRUST AND METAMORPHIC SOLE

The Madstone thrust in southwestern Oregon forms the basal contact of the Josephine Peridotite. Along the northern edge of the peridotite, the trace of the Madstone thrust is approximately east-west (Figs. 1, 3). Along the western 10 km of this portion of the thrust, the Josephine Peridotite is thrust over intrusive rocks of the Chetco River complex, whereas farther east the peridotite is thrust over the Rogue River subterrane. The present outcrop pattern of the Madstone thrust is due to post-Nevadan folding: the east-west–trending segment is in the hinge of a southeast plunging syncline (Fig. 1, 3, cross section; Dick, 1976),

Figure 5. Photograph of area shown in Figure 4. View is toward the southwest and shows the trace of the Madstone thrust. Ja = amphibolite, Js = serpentinite mylonite, Jp = Josephine Peridotite.

whereas a north-south segment, exposed farther south near Chetco Peak, is situated on the west-dipping limb of the syncline (Figs. 4, 5).

The basal Josephine Peridotite adjacent to the contact with underlying amphibolite has been converted into serpentinite having a fabric structurally concordant with the amphibolite. The contact is well exposed on a ridge northeast of Vulcan Peak and in a canyon west of Chetco Lake (Figs. 5, 6). At both localities, the contact is sharp with no evidence of later faulting. As much as 1 m of coarse-grained talc-actinolite schist occurs along the contact and probably represents a metasomatic reaction zone between the serpentinite and amphibolite.

Hanging wall of the Madstone thrust

The upper plate of the thrust in the Chetco Peak area consists of nearly fresh harzburgite of the Josephine ophiolite, underlain by a 20- to 40-m-thick sole of serpentinite along the Madstone thrust (Figs. 4, 6). The serpentinites are coherent, often tough, and vary from weakly to strongly foliated and lineated. Carbonate is present in many of the serpentinites, and may constitute 15 percent or more of the rocks. Talc-rich serpentinite and chlorite veins occur locally near the thrust.

In thin section, two generations of serpentine are evident. An older serpentine is apparent from pseudomorphic textures after olivine (mesh and hourglass textures) and orthopyroxene (bastite), which are typical of lizardite (Wicks and Whittaker, 1977; Wicks and others, 1977); these textures are variably overprinted by coarser grained serpentine having a bladed habit characteristic of antigorite. In weakly foliated samples, antigorite occurs as randomly oriented blades overprinting pseudomorphic lizardite and within discrete shear zones. Strongly foliated samples consist mostly of aligned antigorite blades, patches of aligned chlorite, and streaks of magnetite; lozenges of relict finer grained lizardite partially overgrown by antigorite are common. The coherent nature, lineation, and microstructures of the antigorite-rich rocks suggest that the later serpentinization formed during relatively high-temperature ductile flow rather than by brittle faulting, and fall within the definition of "serpentinite mylonite," a new type of fault rock recently described by Norrell and others (1989).

Where the contact between the serpentinite and underlying gneissic amphibolite is exposed, as much as 1 m of coarse-grained, nonlineated tremolite + talc schist occurs. Talc-rich zones and chlorite seams also occur within the overlying serpentinite mylonite. In thin section, the chlorite is in textural equilibrium with antigorite and has a preferred orientation parallel to the antigorite, suggesting that metasomatism occurred during high-temperature serpentinization associated with emplacement of the ophiolite.

Footwall of the Madstone thrust

Gneissic amphibolite. The amphibolite beneath the Madstone thrust is gneissic and has a stretching lineation defined by aligned hornblende and aggregates of recrystallized plagioclase and quartz. The thickness of the amphibolite is ~200 m in the Chetco Lake area but is highly variable farther north near Vulcan Peak (Fig. 3). Ultramafic rocks exposed on and south of Dry Butte are strongly deformed and are probably related to the gneissic amphibolite (Fig. 3; Loney and Himmelberg, 1977; Cannat and Boudier, 1985). Amphibolites exposed north of Vulcan Peak (Fig. 3) have large plagioclase porphyroclasts (Fig. 7), suggesting a gabbro protolith. Some of the amphibolite is apparently metabasalt because it is associated with thinly layered micaceous quartzite (metachert) near Chetco Lake. A manganese deposit (rhodenite) occurs in the quartzite (Ramp, 1975); these metasediments are probably analogous to Fe-Mn sediments of hydro-

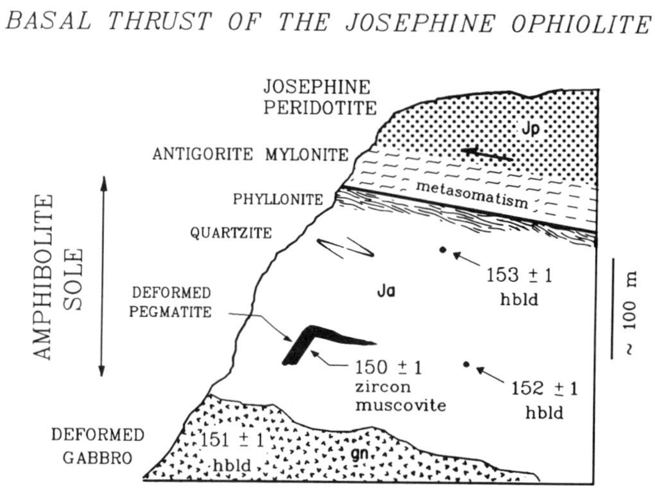

Figure 6. Geologic and age relations of the rocks along the Madstone thrust. U/Pb zircon age of pegmatite and $^{40}Ar/^{39}Ar$ ages of hornblende and muscovite are from Harper and others (1989).

Figure 7. Amphibolite (metagabbro) from the metamorphic sole north of Vulcan Peak showing strong foliation, apparent grain-size reduction, and asymmetric plagioclase porphyroclasts. The asymmetric porphyroclasts of plagioclase and offset vein indicate a dextral sense of shear.

Figure 8. Typical gneissic amphibolite in Chetco Peak region showing folded foliation. The sample is cut normal to the fold hinges that are parallel to a stretching lineation.

Figure 9. Photograph of pegmatite dike northwest of Chetco Lake that cuts folded gneissic layering in amphibolite. View is toward the northeast. White line on amphibolite is drawn parallel to lineation (L). The pegmatite is lineated and has an S-C fabric (Fig. 12).

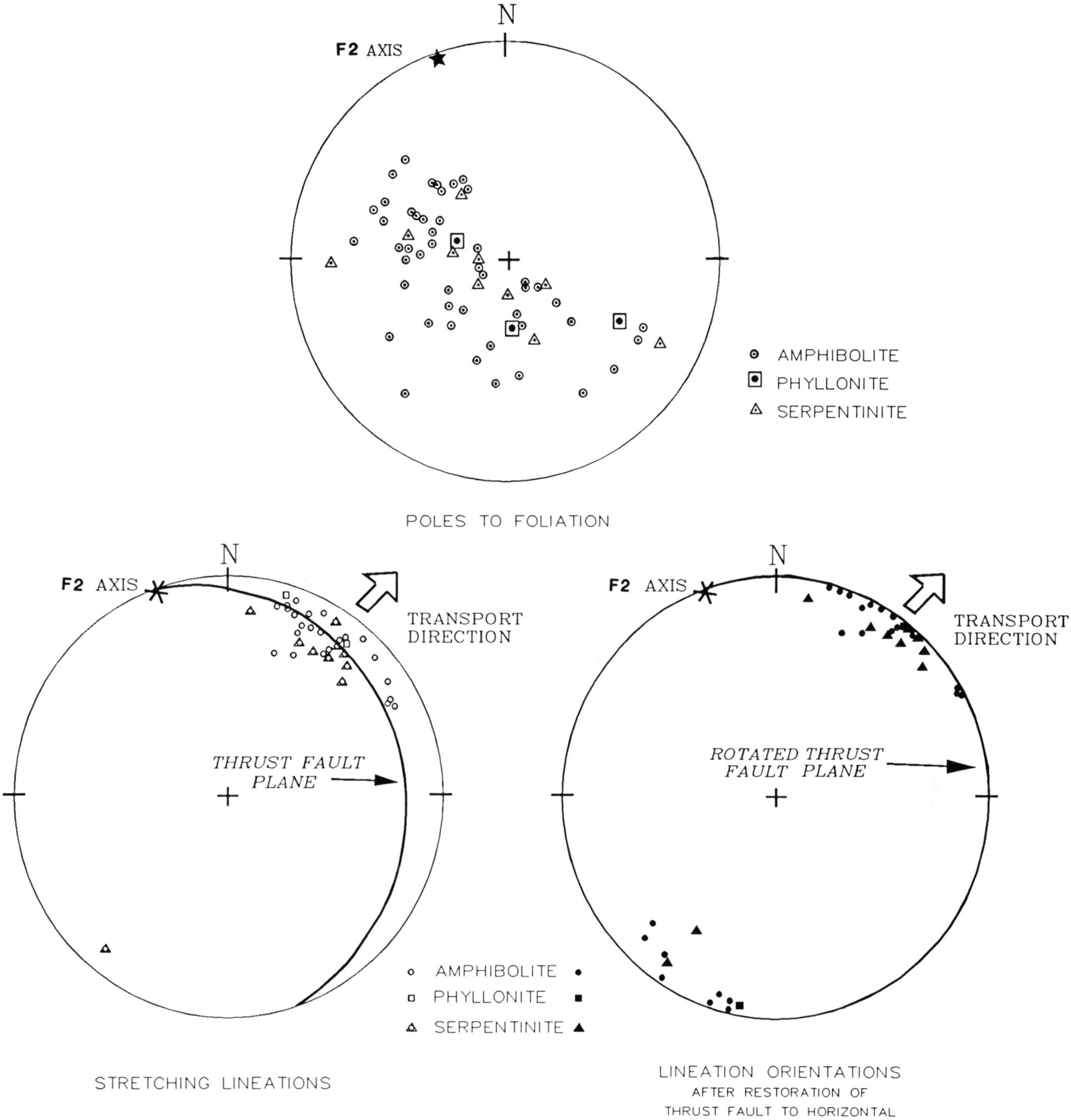

Figure 10. Equal-area projection of poles to foliations and lineations, and lineations after restoration of thrust to horizontal by rotation around F2 axis. $F1_a$ and $F1_b$ axes are parallel to lineations in amphibolite. Trend and plunge of F2 is based on two mesoscopic F2 axes and the axis defined by the orientation of the Madstone thrust in the two areas north and south of Vulcan Peak.

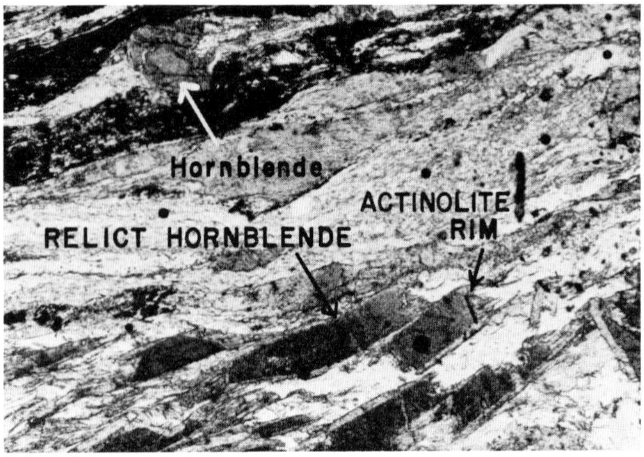

Figure 11. Photomicrograph of greenschist-facies mafic phyllonite showing porphyroclasts of relict brown hornblende rimmed by blue-green hornblende and actinolite (Table 2, sample SA102). Asymmetric fabric indicates dextral sense of shear (top to the northeast). The hornblende porphyroclasts show that the phyllonite is a retrogressed amphibolite. Field of view, 3 mm.

thermal origin, which commonly overlie ophiolites and are known to form from hot springs on mid-ocean ridges (e.g, Crerar and others, 1982). The protoliths of the gneissic amphibolite unit are thus typical of an ophiolitic assemblage, but their igneous age is only constrained to be pre-Nevadan.

The gneissic layering and quartzite are folded by two generations of tight to isoclinal, nearly similar folds (Figs. 8, 9), and sheath-like fold profiles have been observed. Fold hinges of both generations of folds are parallel to the hornblende lineation (Fig. 10). The geometry of these folds suggests that both generations formed during a single continuous deformation; the folds are thus termed F_{1a} and F_{1b}. Late open folds (F_2) are evident in some outcrops and have subhorizontal NNW-trending axes; these folds appear to be related to large-scale folding of the Madstone thrust and the younger Valen Lake thrust (Figs. 1, 10). In the Chetco Lake area, this late folding accounts for the northeast dip of the thrust (Figs. 3, 4), as well as for minor scatter in the plunge of lineations and broadening of the girdle defined by poles to foliation (Fig. 10).

In the Chetco Lake area, the gneissic amphibolite consists of hornblende + plagioclase + sphene or Fe-Ti oxide ± epidote ± quartz ± biotite. Retrograde metamorphism to greenschist or prehnite-pumpellyite facies is evident from veinlets of epidote, quartz, and rarely, pumpellyite. In addition, much of the primary plagioclase is cloudy due to alteration to epidote, muscovite, and albite(?), and hornblende is commonly partially altered to actinolite ± chlorite.

Mafic phyllonite locally occurs along the Madstone thrust (Fig. 4; Ramp, 1975; Dick, 1976). These rocks are highly retrogressed amphibolites, as indicated by the presence of relict hornblende as porphyroclasts (Fig. 11, Table 1). The foliated matrix consists of actinolite, chlorite, epidote, albite and quartz, and locally, muscovite.

Pegmatite dikes locally intrude the gneissic amphibolite (Figs. 4, 9). Of three samples examined in thin section, all contain abundant primary muscovite, two have garnet, and the third sample has minor biotite. Secondary minerals observed in one sample include prehnite, epidote, and chlorite. The pegmatite is lineated, and microstructures indicate the dikes are strongly deformed (Fig. 12): foliations are defined by aligned mica and aggregates of quartz, relict quartz grains have undulous extinctions and subgrains, some quartz is very fine-grained and is apparently recrystallized, micas are kinked and bent, and some plagioclase is bent and has deformation twins.

The gneissic amphibolite, quartzite, phyllonite, and pegmatite have mesoscopic and microscopic features characteristic of mylonites, such as stretching lineations, grain-size reduction, composite foliations, and rotated porphyroclasts (Figs. 7, 11, 12). Furthermore, multiple generations of coaxial folds and sheath folds, such as those observed in the gneissic amphibolite, are common in other high strain zones. The parallelism of F_{1a} and F_{1b} fold hinges with the stretching lineation suggests rotation of the hinges into the displacement direction during progressive simple shear (e.g., Bell and Hammond, 1984).

Chetco River complex. The Chetco River complex (Figs. 1, 3) has been mapped and described by Wells and others (1949), Ramp (1975), Dick (1976), and Loney and Himmelberg (1977), and the following summary is taken from this previous work. In the vicinity of the Madstone thrust, the Chetco River complex consists predominantly of hornblende metagabbro (Fig. 3), which has been partially retrogressed to fine-grained greenschist-facies assemblages. Late-stage undeformed gabbro pegmatite occurs north of the Madstone thrust (Fig. 3).

The contact between the Chetco River complex and the gneissic amphibolite is intrusive (Fig. 3); it is gradational and difficult to map because the gneissic amphibolite and the intrusive

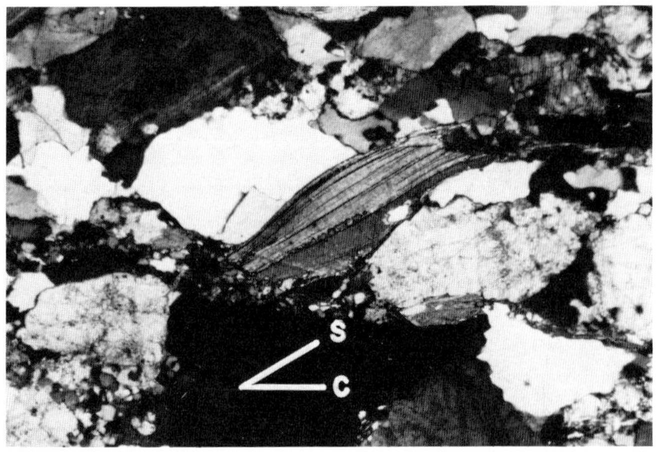

Figure 12. Photomicrograph of pegmatite dike (Fig. 9) that has a well-developed S-C fabric defined by muscovite. Shown here is a typical asymmetric muscovite showing bending into C surfaces. Field of view, 1.5 mm.

TABLE 1. SHEAR SENSE FOR FAULT ROCKS ALONG THE MADSTONE THRUST

SERPENTINITE

Sample No.	Magnetite Porphyroclasts	Shear Band C'	Antigorite Preferred Orientation
K10		NE*	NE
K14	NE		NE
K16	NE?		NE
K19A1			NE?
SA109	NE	unclear	
MKG2B	unclear		NE?
MKG2D			NE?
<LG2F			imc;ear
<LG2G			imc;ear

AMPHIBOLITE

Sample No.	Shape Preferred Orientation	SC fabric	Crystallographic Preferred Orientation	Retrograde: Vein	Retrograde: Crystallographic Preferred Orientation
Sa12	NE?	NE?	h† NE	NE	s§ NE
Sa13	SW?	NE	h NE?	NE	s NE
Sa34		NE?	h NE	NE	m** NE
Sa44		NE	h NE	unclear	
			q‡ SW?		
K12	NE	NE	h NE	NE	m NE
M2H		NE?			m NE?
MKG1F		NE	h NE		
K11-Quartzite		unclear	q NE		
Sa36-Pegmatite		NE			

MAFIC PHYLLONITE

Sample No.	Porphyroclasts	Shear Band C'	Retrograde Vein
SA100	NE		
SA102	NE	NE	
K21	NE		SW?
Sa105		NE	
Sa150	unclear	NE	
K2	NE?		
K5	NE		

*Top to the NE or SW
†Hornblende
§Sphene
**Muscovite
‡Quartz
Note: All determinations were made from thin sections cut perpendicular to foliation and parallel to lineation.

rocks are foliated, lineated, and have amphibolite-facies mineral assemblages (Wells and others, 1949; Ramp, 1975; Dick, 1976; Loney and Himmelberg, 1977). Thus, intrusion of the metagabbro, as well as the granitic pegmatite dikes in the gneissic amphibolite, was apparently syntectonic, a conclusion supported by geochronologic data (Fig. 6; Hotz, 1971; Dick, 1976; Harper and others, 1989).

Displacement direction on the Madstone thrust

Shear-sense indicators. The sense of shear in the mylonitic rocks associated with the Madstone thrust can be interpreted using criteria discussed by Berthe and others (1979), White and others (1980), Simpson and Schmid (1983), Lister and Snoke (1984), and Passchier and Simpson (1986). These criteria include composite planar fabrics (S-C), shear bands (C′), rotated porphyroclasts, and asymmetric pressure shadows. In addition, crystallographic fabrics evident in thin section were used for hornblende (Hall, 1984) and quartz (e.g., Lister and Hobbs, 1980).

The orientations of stretching lineations in the mylonite associated with the Madstone thrust are interpreted to be parallel to the bulk transport direction. They plunge gently toward the northeast in the Chetco Lake area for both the gneissic amphibolite and serpentinite mylonite (Figs. 4, 10). The phyllonite and pegmatite dikes also have northeast-trending lineations (Figs. 4, 9).

Hornblende in gneissic amphibolite has a strong preferred crystallographic and shape orientation parallel to the gneissic layering. The hornblende crystals have ragged edges and commonly show undulatory extinction. Quartz and plagioclase are intergrown and generally form relatively fine-grained equant grains in the lighter gneissic layers. The much finer grain size of quartz and feldspar, along with the local presence of plagioclase porphyroclasts (Fig. 7)—some of which have relict igneous zoning—suggests that these rocks have undergone significant grain-size reduction typical of mylonites. In some gneissic amphibolites, the sense of shear can be inferred from S-C fabrics. A foliation oblique to the gneissic layering, interpreted as an S-foliation, is defined by the shape and crystallographic preferred orientation of hornblende, sphene, plagioclase, and/or epidote. The shear sense in amphibolite was also determined by a technique developed by Hall (1984), who studied crystallographic fabrics of hornblende in thin ductile shear zones of known displacement: where c-axes of the hornblende are strongly aligned, an asymmetry may be evident from the preferred orientation of a-axes, which will be inclined to the foliation; for dextral shear, the a-axes are inclined in a clockwise sense with respect to the foliation in a thin section cut parallel to lineation and normal to foliation. This asymmetry can be identified in thin section using changes in pleochroism as the microscope stage is rotated. Quartzites have a strong asymmetric crystallographic preferred orientation evident in thin section; for intercrystalline slip on the basal plane of quartz, the trace of the c-axes should be inclined in a counterclockwise sense from the normal to the foliation.

The foliated, antigorite-rich serpentinites along the base of the Josephine Peridotite are structurally concordant with the amphibolite (Fig. 10). Many of these rocks, however, do not have well-developed shear-sense indicators and appear to be weakly strained. Sense-of-shear indicators are evident, however, in a few strongly foliated, antigorite-rich mylonites and include asymmetric porphyroclasts and shear bands (Fig. 13).

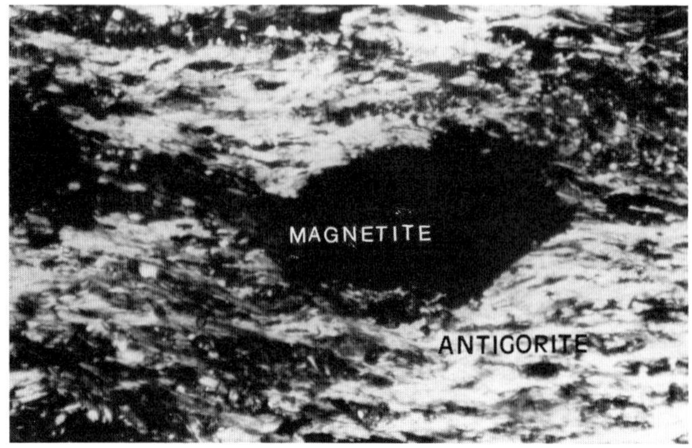

Figure 13. Thin section of serpentinite mylonite cut perpendicular to foliation and parallel to lineation. Dextral sense of shear (top to the northeast) is suggested by asymmetric tails on magnetite porphyroclast. The matrix consists of antigorite. Field of view is 1.5 mm.

Sense of shear has also been determined for the later stages of deformation using (1) microstructures in mafic phyllonite; and (2) microstructures in a pegmatite dike that is deformed but cuts gneissic layering in the amphibolite (Fig. 9). The sense of shear in the mafic phyllonite is indicated by porphyroclasts having asymmetric tails (Fig. 11) and by shear bands. The pegmatite dikes are lineated, and some have a well-developed S-C fabric defined by the preferred orientation of micas (Fig. 12) and by lenticular aggregates of recrystallized quartz. In addition, the sense of shear in some gneissic amphibolites during retrograde metamorphism is interpreted from very fine-grained muscovite that defines a planar fabric oblique to the gneissic layering, and from extension veins (some with fibers) that are at a high angle but oblique to the gneissic layering. Both of these criteria assume that the plane of shear is essentially the same during both amphibolite and later retrograde metamorphism; this assumption seems justified based on the similar orientation of foliations and lineations in the amphibolite, pegmatite, and mafic phyllonite (Figs. 4, 10).

Results. The results of the shear-sense determinations are given in Table 1. The lineations and sense of shear for amphibolite, quartzite, pegmatite, and serpentinite consistently indicate north-northeast thrusting of the Josephine ophiolite over the gneissic amphibolite and Chetco River complex. Continued thrusting in the same direction is indicated by similarly oriented lineations and sense of shear for phyllonite, late veins and minerals in amphibolite, and pegmatite. The same sense of shear was determined by Cannat and Boudier (1985) for quartzites in the

Chetco Lake area, who used asymmetric crystallographic fabrics and pressure shadows. A minimum displacement of 12 km is indicated by the current exposures of the Madstone thrust.

Determination of the actual displacement direction during the Nevadan orogeny requires that the structural data be corrected for post-Nevadan rotations. The east dip of the Madstone thrust near Chetco Lake is the result of late folding and was removed by rotation about the subhorizontal F_2 fold axis. Rotating the thrust to horizontal results in the average lineation still trending northeast, but the plunge becomes subhorizontal (Fig. 10). Thus, the inferred direction of thrusting is essentially unchanged.

A less certain structural correction involves possible Tertiary clockwise rotations, as suggested by paleomagnetic studies. Declination anomalies in the Klamath Mountains are generally compatible with Tertiary oroclinal bending of the Klamath Mountains from an initial structural trend similar to pre-Tertiary rocks in the Sierra Nevada foothills (~N40°W; Renne and Scott, 1988; Renne and others, 1988). If the regional change in structural grain from northwest to northeast in the Klamath Mountains (Fig. 1) is assumed to be the result of oroclinal bending, then the structural grain in the Western Klamath Terrane at the latitude of the Madstone thrust (~N20°E) would suggest ~60° of clockwise rotation. Paleomagnetic declination anomalies determined on rocks northeast of the study area are 78 ± 26° for the 139-Ma Grants Pass pluton (Fig. 1) and 99 ± 10° for probable Nevadan-age magnetization near Galice, Oregon (Schultz and Levi, 1983; Bogen, 1986). Removal of a 60 to 100° clockwise rotation would change the inferred north-northeast transport direction for the Madstone thrust to northwest. Additional paleomagnetic and structural studies are needed to better constrain possible clockwise rotations and evaluate the size of domains that have undergone rotations (e.g., Renne and Scott, 1988).

Relative ages and geochronology

Based on the geometry and structure of the gneissic amphibolite, several workers have interpreted these rocks as a metamorphic sole formed during emplacement of the Josephine ophiolite (Armstrong and Dick, 1974; Dick, 1976; Cannat and Boudier, 1985). Cannat and Boudier (1985) assumed that the amphibolite was a metamorphic sole and suggested that a thick basal section of peridotite mylonite, typically present along the base of other ophiolites, had been faulted away. They noted the presence of a few meters of "serpentinized peridotite mylonite" along the thrust, but these rocks are actually serpentinite mylonites (Fig. 13).

In contrast, Loney and Himmelberg (1977) considered the Madstone thrust to be younger than the earliest deformation in the amphibolite. These workers, along with Cannat and Boudier (1985) and the present study, recognized two generations of folds that have axes parallel to a stretching lineation. Loney and Himmelberg (1977) observed gabbro dikelets that cut "F_1" (F_{1a}) folds but which are folded by "F_2" (F_{1b}) folds. In the context of the present study, this observation is consistent with syntectonic dike intrusion during one continuous deformation involving progressive simple shear.

Additional relative age controls from the present study that indicate the amphibolite was deformed during thrusting and syntectonic intrusion of magmas include the following: (1) pegmatite dikes cut folded amphibolite but are also deformed (Figs. 4, 9, 12); (2) gneissic amphibolite and overlying serpentinite mylonite are structurally concordant (Figs. 4, 10); and (3) the serpentinite has undergone metasomatism along the thrust, apparently from fluids derived from the footwall.

A Nevadan age for thrusting, intrusion, and amphibolite-grade metamorphism is indicated by geochronologic studies. Dick (1976) obtained K/Ar ages of ~150 Ma on hornblende from both the amphibolite and Chetco River complex, as well as on muscovite from pegmatite dikes intruding the amphibolite. $^{40}Ar/^{39}Ar$ dating has yielded similar results (Fig. 6; Grady and others, 1989; Harper and others, 1989), but avoids some of the assumptions inherent in K/Ar dating. Two amphibolites gave $^{40}Ar/^{39}Ar$ hornblende cooling ages (~400 to 500°C) of 152 ± 1 and 153 ± 1 Ma, and a metagabbro north of Vulcan Peak yielded a hornblende cooling age of 151 ± 1 Ma. A pegmatite dike intruding the amphibolite (Figs. 6, 9) has a Pb/U zircon age of 150 ± 1 Ma, and $^{40}Ar/^{39}Ar$ dating of muscovite indicates slow cooling near its closure temperature (~350°C) from 150 to 146 Ma.

Thus, the relative and absolute ages, along with field relations, clearly demonstrate that ductile deformation and syntectonic magmatism along the Madstone thrust are Nevadan in age. Furthermore, the cooling history evident from hornblende and muscovite ages is consistent with the observed retrograde nature of metamorphism.

Pressure, temperature, and fluids

Mineral assemblages and compositions can provide constraints on the temperatures and pressures during emplacement of the Josephine ophiolite. Two gneissic amphibolites from the Chetco Lake area were studied in detail: they contain pargasitic hornblende + plagioclase (largely altered) + epidote + minor chlorite + quartz + sphene + Fe-Ti oxide. For these samples, no quantitative estimate of metamorphic temperature is possible, but the mineral assemblage suggests a temperature range of 400° to 540°C (Apted and Liou, 1983; Maruyama and others, 1983). Pressure was estimated using the NaM4 content of calcic amphibole (Brown, 1977). For this estimate, the amphibole formulae obtained from microprobe analysis were recalculated by normalizing the number of cations (excluding Ca, Na, and K) to 13, then adding ferric iron to balance a charge of 46; other normalization schemes do not change the calculated NaM4 content enough to significantly alter the pressure estimate. The resulting pressure estimate is <4 kbar for two samples (Table 1, SA13, SA44). Sample SA44 is rimmed by a less aluminous calcic amphibole having approximately the same NaM4 but less total Na than the amphibole core, suggesting temperature at constant or decreasing pressure (Raase, 1974; Brown, 1977).

TABLE 2. ANALYSES OF MINERALS IN METAMORPHIC SOLE OF JOSEPHINE OPHIOLITE*

	Amphibole†						Muscovite§
	Gneissic amphibolite			Phyllonite			Pegmatite
	SA13	SA44		SA 102			J1113-7
		rim	core	rim	intermediate	core	
SiO^2	42.10	48.56	44.94	57.17	49.45	43.43	45.15
TiO_2	0.80	0.48	0.76	....	1.74	3.45	0.39
Al_2O_3	11.86	8.28	11.73	1.11	6.86	11.16	32.16
Cr_2O_3	0.08	0.12	0.01	....	....	....	....
FeO	17.71	12.62	14.12	7.40	10.76	11.41	3.17
MgO	10.73	14.63	12.54	19.75	16.15	14.83	1.19
MnO	0.31	0.23	0.31	0.47	0.38	0.21	0.03
CaO	11.92	11.71	11.67	12.98	12.19	11.23	0.01
Na_2O	1.58	1.32	1.74	0.25	1.19	2.50	0.60
K_2O	1.29	0.18	0.39	0.02	0.16	0.16	10.05
Total	98.39	98.14	98.21	99.15	98.88	98.37	92.75
Si	6.23	6.91	6.48	7.87	6.96	6.20	Si 3.12
Al_{IV}	1.77	1.09	1.52	0.13	1.03	1.80	Al 2.62
Al_{VI}	0.30	0.30	0.48	0.05	0.11	0.08	Ti 0.02
Ti	0.09	0.05	0.08	....	0.19	0.37	Fe 0.18
Fe^{3+}(calc)	0.80	0.70	0.71	0.14	0.52	0.81	Mg 0.12
Cr	0.01	0.01	....	....	....	....	Na 0.08
Fe^{2+}	1.39	0.80	1.00	0.71	0.75	0.56	K 0.89
Mg	2.37	3.10	2.70	4.06	3.39	3.16	
Mn	0.04	0.03	0.04	0.06	0.05	0.03	
Ca	1.89	1.79	1.80	1.92	1.84	1.72	
NaM_4	0.11	0.21	0.20	0.07	0.16	0.28	
Na_A	0.35	0.15	0.29	....	0.17	0.41	
K	0.24	0.03	0.07	....	0.03	0.03	

*Analyses done with Cameca SX-50 electron microprobe, 1 μm spot size.
†Formula based on Σ cations-(Ca+Na+K) = 13.
§Formula based on 11 oxygens.

A mafic phyllonite from beneath the Madstone thrust (Fig. 11) contains porphyroclasts of relict brown tschermakitic amphibole + calcic plagioclase (altered). Although no quantitative pressure or temperature estimates can be made from these relict minerals, the assemblage of tschermakitic amphibole + plagioclase as well as the high TiO_2 content of the calcic amphibole (Table 2, SA102) are suggestive of higher temperatures than the other amphibolite samples. The brown hornblende is rimmed by a less aluminous hornblende, which is in turn rimmed by actinolite (Fig. 11, Table 2). The only other brown hornblende observed in samples from the Chetco Lake area was also in a phyllonite near the Madstone thrust. The zoning trend in amphiboles from the phyllonite (SA102) is consistent with retrograde metamorphism from amphibolite-facies to greenschist-facies conditions (Raase, 1974).

Pressure can be estimated from the phengite content of muscovite in a deformed pegmatite dike (Figs. 9, 12), because it has the assemblage muscovite + biotite + K-feldspar + quartz (Massonne and Schreyer, 1987). Because the assemblage in the pegmatite is igneous, the derived pressure corresponds to the time of the dike intrusion: the low-T alteration assemblage, field relations, and age of the dike indicate that it was intruded during greenschist-facies retrograde metamorphism. Temperature is needed to apply the geobarometer and can be estimated from the upper temperature stability of this assemblage, which melts at ~700°C in the presence of water at pressures >3 kbar (Massonne and Schreyer, 1987). For an equilibration temperature of 700°C, and assuming $a_{H_2O} = 1$, the measured Si content per formula for muscovite (J113-7, Table 2) gives a pressure of 5.5 kbar. Muscovites in this sample have a uniform composition, and no zoning was detected. The uncertainty in temperature has a relatively small effect on the estimated pressure. The effect of $a_{H_2O} < 1$ is also relatively small and causes a shift toward higher pressures at a given Si content. The quantitative effects of iron in muscovite on the geobarometer are not known, but addition of Fe^{2+} and especially Fe^{3+} result in a lower Si content, which would yield a pressure estimate that is somewhat low. The phengite barometer is sensitive to fluorine content: pressure will be overestimated by ~1.5 kbar for each increase of 0.1 in the F/(F + OH) ratio (Massonne and Schreyer, 1987). Because fluorine is likely to be high in a pegmatitic muscovite, the pressure estimate (5.5 kbar) should be considered a maximum.

A further constraint on pressure is given by regional and contact metamorphism in other rocks of the Western Klamath Terrane. Mineral assemblages formed during regional prehnite-pumpellyite facies metamorphism in northwestern California suggest conditions of ~300°C and 2 to 4 kbar (Harper and others, 1988). The presence of andalusite in contact aureoles around Nevadan plutons that intrude the Galice Formation (Harper, 1980; Wyld and Wright, 1988; Harper and others, 1989) also indicates low pressure (<4 kbar). These data, along with the pressure estimates from geobarometry for the metamorphic sole, indicate relatively low-P/T metamorphism during emplacement of the Josephine ophiolite.

The mineralogy of the serpentinite mylonite above the Madstone thrust gives some constraint on temperature. The serpentinite mylonite contains bladed, typically schistose antigorite (Fig. 13). In general terms, antigorite is the high-temperature serpentine mineral, whereas lizardite and chrysotile are the low-temperature minerals (Wenner and Taylor, 1974; Evans, 1977). For the Josephine samples (serpentinized harzburgite), the stability of antigorite at ~4 kbar can be estimated using the following reactions in the Al-poor system: lizardite = antigorite + brucite (~225°C), antigorite + brucite = forsterite (~450°C), and antigorite = fosterite + talc + H_2O (~530°C; O'Hanley and others, 1989). The carbonate associated with the antigorite suggests that CO_2 was present in the fluid phase and would reduce the maximum stability temperature of antigorite (~30°C for a fluid with 20 percent CO_2; Trommsdorff and Evans, 1977). Thus, the antigorite is estimated to have formed at temperatures between 250° to 550°C. The relict pseudomorphic low-T (<250°C) lizardite evident in some samples could be either oceanic or early Nevadan in age and indicates that the hanging wall of the Madstone thrust was heated during thrusting to form antigorite mylonites.

The presence of fresh peridotite within 40 m of the thrust suggests that the basal serpentinite mylonite formed an impermeable barrier to metamorphic fluids during metamorphism. Such a barrier may have resulted in channeling of fluids along the Madstone thrust, resulting in extensive hydration of amphibolite to mafic phyllonite. The high strains evident from the fabrics of the phyllonite suggest that shearing was localized in these rocks during retrograde metamorphism, probably due to strain softening as fluids allowed formation of abundant, relatively weak phyllosilicates (e.g., White and Knipe, 1978).

Infiltration of fluids from below the thrust into the antigorite serpentinites is indicated by metasomatism evident from talc, chlorite, and carbonate. Talc + actinolite schist as much as 1 m thick occurs along the contact with underlying amphibolite or phyllonite and probably represents a metasomatic reaction zone. These metasomatic effects are similar to those formed during infiltration of fluids along the basal thrust of the Paleozoic Trinity ultramafic sheet (Peacock, 1987).

Retrograde metamorphism is evident from amphibolite-facies assemblages overprinted by typical greenschist-facies assemblages. The presence of a pumpellyite vein in an amphibolite sample and prehnite in the syntectonic pegmatite in the Checto Lake region suggest that retrograde metamorphism extended down to the prehnite-pumpellyite facies. The regional metamorphic grade higher in the Josephine thrust sheet at this latitude is prehnite-pumpellyite facies (Harper and others, 1988), and the excess heat needed for the early amphibolite-grade metamorphism beneath the Madstone thrust may have been supplied by the syntectonic intrusions of the underlying Chetco River complex.

DISCUSSION AND CONCLUSIONS

This study has documented the presence of a metamorphic sole beneath the Josephine ophiolite. Emplacement of the ophiolite, however, was much different than other ophiolites, such as the Semail and Bay of Islands, which also have metamorphic soles (Spray, 1984; Moores, 1982):

1. The Josephine ophiolite is thrust over a magmatic arc complex (Figs. 1, 3) rather than over a continental shelf. The ophiolite and overlying Galice Formation were thrust *beneath* the continental margin and were metamorphosed to low grade during the Nevadan (Harper and others, 1989).

2. An inverted metamorphic gradient apparently is absent, and metamorphism was retrograde in nature and most extensive adjacent to the thrust.

3. A basal zone of very highly strained peridotite is absent in the Josephine ophiolite. Instead, a 20- to 40-m zone containing serpentinite mylonite occurs above the basal thrust. The presence of antigorite mylonite rather than peridotite mylonite, as well as the absence of granulite-facies metamorphism, suggests lower maximum temperatures (<~600°C) during thrusting, compared to those in metamorphic soles of other ophiolites (≥800°C; Nicolas and Le Pichon, 1980; Spray, 1984).

These differences are a reflection of tectonic setting of ophiolite emplacement. The Josephine ophiolite represents back-arc basin lithosphere thrust over an active arc (Fig. 2). In contrast, emplacement of other ophiolites appears to have involved collision in which the continental margin was carried beneath the ophiolite (Moores, 1982).

The gneissic amphibolite and quartzite, mafic phyllonite, pegmatite dikes, and antigorite serpentinite are all structurally concordant and show evidence of deformation by progressive simple shear. The concordant structures, along with metasomatism of the serpentinite, indicate that these rocks were deformed during north-northeast emplacement of the Josephine ophiolite over an active mafic batholith (Chetco River complex). Metamorphism was retrograde in nature, ranging from amphibolite through prehnite-pumpellyite facies, and occurred under low-P/T conditions. Geochronologic studies (Dick, 1976; Harper and others, 1989) demonstrate that thrusting and magmatism is Nevadan in age (Fig. 6) and indicate cooling from ~450°C at 153 Ma to ~350°C at 146 Ma (Harper and others, 1989). The minimum offset along the Madstone thrust is 12 km, which gives a minimum displacement of ~1.2 mm/yr.

In contrast to the NNE-directed thrusting on the basal (Madstone) thrust throughout the retrograde metamorphism, west- or northwest-directed thrusting on the roof thrust (Orleans fault) has been inferred from regional mapping and mesoscopic structures in the Galice Formation in the area of Figure 1 (Snoke, 1977; Harper, 1980; Norman, 1984; Gray, 1985; Jones, 1988; Harper and others, 1989). The offset along the fault is very large: >40 km based on its map pattern (Fig. 1) and probably ≥110 km based on gravity studies (Jachens and others, 1986).

The north-northeast emplacement of the ophiolite thus appears to have been at a high angle to synchronous thrusting along the roof thrust and is essentially parallel to the inferred paleogeographic trends at this latitude during the Late Jurassic. It is also subparallel to the spreading direction in the Josephine ophiolite, based on the approximately east-west strike of sheeted dikes. The differences in thrusting direction for the basal and roof thrusts may result partly from differing amounts of Tertiary clockwise rotation, especially since the size of rotated domains is not known. Different thrusting directions during the Nevadan orogeny might result from decoupling of the Josephine and overlying Galice Formation from the Klamath Mountains as a result of high pore-fluid pressures in the Galice; such high pressures are likely since the Galice flysch is syntectonic and was thrust beneath the Klamath Mountains shortly after deposition (Fig. 2; Harper and others, 1989).

North-northeast thrusting on the Madstone thrust may reflect sinistral oblique subduction during the Nevadan orogeny. A dramatic change in North American plate motion at ~150 Ma is suggested by the North American polar wander path and should have resulted in sinistral oblique convergence from ~150 to 135 Ma (May and others, 1989). Future structural studies of thrust faults and paleomagnetic data should provide additional quantitative constraints on the kinematics of thrusting and its relation to plate motions during the Nevadan orogeny.

ACKNOWLEDGMENTS

This research was funded by National Science Foundation Grants EAR-8518974 and EAR-8722425 (to G.D.H.) and by a Sigma Xi grant (to K.G.). H.J.B. Dick provided some of the samples used for dating and petrography. David O'Hanley assisted with the identification of serpentine minerals and textures. Susan Delay and Phil Hicks provided valuable assistance in the field, and Marilyn Peacock drafted many of the figures.

REFERENCES CITED

Aalto, K. R., 1979, Franciscan complex geology of the Crescent City area, northern California, *in* Aalto, K. R., and Harper, G. D., eds., Geologic evolution of the northernmost Coast Ranges and western Klamath Mountains, California; 28th Geologic Congress Field Trip Guidebook T308: Washington, D.C., American Geophysical Union, p. 21–46.

Aalto, K. R., and Harper, G. D., 1982, Geology of the Coast Ranges in the Klamath and Ship Mountain Quadrangles, Del Norte County, California: California Division of Mines and Geology Open-File Map, scale 1:62,500.

Apted, M., and Liou, J. G., 1983, Phase relations among greenschist, epidote amphibolite, and amphibolite in a basaltic system: American Journal of Science, v. 283A, p. 328–354.

Armstrong, R. L., and Dick, H.J.B., 1974, A model for the development of thin overthrust sheets of crystalline rock: Geology, v. 2, p. 35–40.

Bell, T. H., and Hammond, R. L., 1984, On the internal geometry of mylonite zones: Journal of Geology, v. 92, p. 667–686.

Berthe, D., Choukroune, P., and Jegouzo, P. E., 1979, Orthogneiss, mylonite, and non-coaxial deformation of granites; The example of the South Armorican Shear zone: Journal of Structural Geology, v. 1, p. 32–42.

Blake, M. C., Engebretson, D. C., Jayko, A. S., and Jones, D. L., 1985, Tectonostratigraphic terranes of the Circum-Pacific region: Circum-Pacific Council for Energy and Mineral Resources Earth Science Series 1, p. 147–157.

Bogen, N. L., 1986, Paleomagnetism of the Upper Jurassic Galice Formation, southwestern Oregon; Evidence for differential rotation of the eastern and western Klamath Mountains: Geology, v. 14, p. 335–338.

Brown, E. H., 1977, The crossite content of Ca amphibole as a guide to pressure of metamorphism: Journal of Petrology, v. 18, p. 53–72.

Burchfiel, B. C., and Davis, G. A., 1981, Triassic and Jurassic tectonic evolution of the Klamath Mountains–Sierra Nevada geologic terrane, *in* Ernst, W. G., ed., The geotectonic development of California; Rubey Volume 1: Englewood Cliffs, New Jersey, Prentice-Hall, p. 50–70.

Cannat, M., and Boudier, F., 1985, Structural study of intra-oceanic thrusting in the Klamath Mountains, northern California; Implications on accretion geometry: Tectonics, v. 4, p. 435–452 (and corrections v. 4, p. 598–601).

Crerar, D. A., Namson, J., Chyi, M. S., Williams, L., and Feigenson, 1982, Manganiferous cherts of the Franciscan assemblage; 1, General geology, ancient and modern analogures, and implications for hydrothermal convection at oceanic spreading centers: Economic Geology, v. 77, p. 519–522.

Davis, G. A., Monger, J.W.H., and Burchfiel, B. C., 1978, Mesozoic construction of the Cordilleran "collage," central British Columbia to central California, *in* Howell, D. G., and McDougall, K. A., eds., Mesozoic paleogeography of the western United States: Pacific Section, Society of Economic Paleontologists and Mineralogists Pacific Coast Paleogeography Symposium 2, p. 1–32.

Dick, H.J.B., 1976, The origin and emplacement of the Josephine Peridotite of southwestern Oregon [Ph.D. thesis]: Ne Haven, Connecticut, Yale University, 409 p.

—— , 1977, Partial melting in the Josephine Peridotite; 1, The effect on mineral composition and its consequences for geothermometry and geobarometry: American Journal of Science, v. 277, p. 801–832.

Engebretson, D. C., Cox, A., and Gordon, R. G., 1985, Relative motions between oceanic and continental plates in the Pacific basin: Geologic Society of America Special Paper 206, 59 p.

Evans, B. W., 1977, Metamorphism of alpine peridotites and serpentinites: Annual Review of Earth and Planetary Sciences, v. 5, p. 397–448.

Garcia, M. O., 1979, Petrology of the Rogue and Galice Formations, Klamath Mountains, Oregon; Identification of a Jurassic island arc sequence: Journal of Geology, v. 86, p. 29–41.

—— , 1982, Petrology of the Rogue River island-arc complex, southwest Oregon: American Journal of Science, v. 282, p. 783–807.

Grady, K. A., Harper, G. D., and Heizler, M., 1989, N-NE obduction of the Josephine ophiolite along an amphibolite sole, SW Oregon: Geological Society of American Abstracts with Programs, v. 21, p. 86.

Harper, G. D., 1980, Structure and petrology of the Josephine ophiolite and overlying metasedimentary rocks, northwestern California [Ph.D. thesis]: Berkeley, University of California, 260 p.

—— , 1983, A depositional contact between the Galice Formation and a Late Jurassic ophiolite in northwestern California and southwestern Oregon: Oregon Geology, v. 45, p. 3–9.

—— , 1984, The Josephine ophiolite, northwestern California: Geological Society of America Bulleti, v. 95, p. 1009–1026.

—— , 1989, Field guide to the Josephine ophiolite and coeval island arc complex, Oregon-California, *in* Aalto, K. R., and Harper, G. D., eds., Geologic evolution of the northernmost Coast Ranges and western Klamath Mountains, California; 28th Geologic Congress Field Trip Guidebook T308: American Geophysical Union, p. 2–21.

Harper, G. D., and Wright, J. E., 1984, Middle to Late Jurassic tectonic evolution of the Klamath Mountains, California-Oregon: Tectonics, v. 3, p. 759–772.

Harper, G. D., Saleeby, J. B., and Norman, E., 1985, Geometry and tectonic setting of sea-floor spreading for the Josephine ophiolite, and implications for Jurassic accretionary events along the California margin, *in* Howell, D., ed., Tectonostratigraphic terranes of the Circum-Pacific region: Circum-Pacific Council for Energy and Mineral Resources Earth Science Series 1, p. 239–257.

Harper, G. D., Bowman, J. R., and Kuhns, R., 1988, Field, chemical, and isotopic aspects of submarine hydrothermal metamorphism of the Josephine ophiolite, Klamath Mountains, California-Oregon: Journal of Geophysical Research, v. 93, p. 4625–4656.

Harper, G. D., Saleeby, J. B., Pessagno, E. A., and Heizler, M., 1989, The Josephine ophiolite; Generation, translation and emplacement of Late Jurassic parautochthonous back-arc basin lithosphere, Klamath Mountains, California-Oregon: Geological Society of America Abstracts with Programs, v. 21, p. A28.

Hotz, P. E., 1971, Plutonic rocks of the Klamath Mountains, California and Oregon: U.S. Geological Survey Bulletin 1290, 91 p.

Irwin, W. P., 1966, Geology of the Klamath Mountain province, *in* Bailey, E. H., ed., Geology of northern California: California Division of Mines and Geology Bulletin 190, p. 19–38.

—— , 1981, Tectonic accretion of the Klamath Mountains, *in* Ernst, W. G., ed., The geotectonic development of California; Rubey Volume 1: Englewood Cliffs, New Jersey, Prentice-Hall, p. 29–49.

Jachens, R. C., Barnes, C. G., and Donato, M. M., 1986, Subsurface configuration of the Orleans fault; Implications for deformation in the western Klamath Mountains, California: Geological Society of America Bulletin, v. 97, p. 388–395.

Jones, F. R., 1988, Structural geology of the northern Galice Formation, western Klamath Mountains, Oregon and California [M.S. thesis]: Albany, State University of New York, 211 p.

Lister, G. S., and Hobbs, B. E., 1980, The simulation of fabric development during plastic deformation and its application to quartzite; The influence of deformation history: Journal of Structural Geology, v. 2, p. 355–370.

Lister, G. S., and Snoke, A. W., 1984, S-C mylonites: Journal of Structural Geology, v. 6, p. 617–638.

Loney, R. A., and Himmelberg, G. R., 1977, Geology of the gabbroic complex along the northern border of the Josephine peridotite, Vulcan Peak area, southwestern Oregon: U.S. Geological Survey Journal of Research, v. 5, p. 761–781.

Maruyama, S., Suzuki, K., and Liou, J. G., 1983, Greenschist-amphibolite transition equilibria at low pressures: Journal of Petrology, p. 24, p. 583–604.

Massonne, H. J., and Schreyer, W., 1987, Phengite geobarometry based on the limiting assemblage with K-feldspar, phlogopite, and quartz: Contributions to Mineral and Petrology, v. 96, p. 212–224.

May, S. R., and Butler, R. R., 1986, North American Jurassic apparent polar wander; Implications for plate motion, paleogeography, and Cordilleran tectonics: Journal of Geophysical Research, v. 91, p. 11519–11544.

May, S. R., Beck, M. E., Jr., and Butler, R. R., 1989, North American APW, plate motions, and left oblique convergence; Late Jurassic/Early Cretaceous orogenic consequences: Tectonics, v. 8, p 443–452.

Moores, E. M., 1982, Origin and emplacement of ophiolites: Reviews of Geophysics and Space Physics, v. 20, p. 735–760.

Nicolas, A., and Le Pichon, X., 1980, Thrusting of young lithosphere in subduction zones with special reference to structures in ophiolitic peridotites: Earth and Planetary Science Letters, v. 46, p. 397–406.

Norrell, G. N., and Tiexell, A., and Harper, G. D., 1989, Microstructure of serpentinite mylonites from the Josephine ophiolite and serpentinization in retrogressive shear zones, California: Geological Society of America Bulletin, v. 101, p. 673–682.

Norman, E.A.S., 1984, Petrology and structure of the Summit Valley area, Klamath Mountains, California [M.S. thesis]: Salt Lake City, University of Utah, 148 p.

O'Hanley, D. S., Chernosky, J. V., Jr., and Wicks, F. J., 1989, The stability of lizardite and chrysotile: Canadian Mineralogist, v. 27, p. 483–493.

Park-Jones, R., 1987, Sedimentology, structure, and geochemistry of the Galice Formation; sediment fill of a back-arc basin and island arc in the western Klamath Mountains [M.S. Thesis]: Albany, State University of New York, 165 p.

Passchier, C. W., and Simpson, C., 1986, Porphyroclast systems as kinematic indicators: Journal Structural Geology, v. 8, p. 831–843.

Peacock, S. M., 1987, Serpentinization and infiltration metasomatism in the Trinity peridotite, Klamath province, northern California; Implications for subduction zones: Contributions to Mineralogy and Petrology, v. 95, p. 55–70.

Pessagno, E. A., Jr., and Blome, C. D., 1988, Biostratigraphic, chronostratigraphic, and U/Pb geochronometric data from the Rogue and Galice Formations, western Klamath Terrane (Oregon and California); Their bearing on the age of the Oxfordian-Kimmeridgian boundary and the Mirifusus first occurrence event: Proceedings of the 2nd International Symposium on Jurassic Stratigraphy, Lisbon, Portugal, p. 1–14.

Pessagno, E. A., Jr., and Mizutani, S., 1988, Correlation of radiolarian biozones of the eastern and western Pacific (North America and Japan), *in* Westermann, G.E.G., ed., Jurassic of the Circum-Pacific region (in press).

Raase, P., 1974, Al and Ti contents of hornblende, indicators of pressure and temperature of regional metamorphism: Contributions to Mineralogy and Petrology, v. 45, p. 231–236.

Ramp, L., 1975, Geology and mineral resources of the upper Chetco drainage area, Oregon: Oregon Department of Geology and Mineral Industries Bulletin 88, 47 p.

—— , 1984, Geologic map of the southeast quarter of the Pearsoll Peak Quadrangle, Curry and Josephine Counties, Oregon: Oregon Department of Geology and Mineral Industries Geological Map Series GMS-30, scale 1:24,000.

—— , 1988, Structural chronology, oroclinal deformation, and tectonic evolution of the southeastern Klamath Mountains, California: Tectonics, v. 7, p. 1223–1242.

Renne, P. R., Scott, G. R., and Bazard, D. R., 1988, Multicomponent paleomagnetic data from the Nosoni Formation, eastern Klamath Mountains, California; Cratonic Permian primary direction with Jurassic overprints: Journal of Geophysical Research, v. 93, p. 3387–3400.

Riley, T. A., 1987, The petrogenetic evolution of a Late Jurassic island arc; The Rogue Formation, Klamath Mountains, Oregon [M.S. thesis]: Stanford, California, Stanford University, 40 p.

Saleeby, J. B., 1983, Accretionary tectonics of the North American Cordillera: Annual Review of Earth and Planetary Sciences, v. 15, p. 45–73.

—— , 1984, Pb/U zircon ages from the Rogue River area, western Jurassic belt, Klamath Mountains, Oregon: Geological Society of America Abstracts with Programs, v. 16, p. 331.

—— , 1987, Discordance patterns in Pb/U zircon ages of the Sierra Nevada and Klamath Mountains: EOS Transactions of the American Geophysical Union, v. 68, p. 1514–1515.

Schultz, K. L., and Levi, S., 1983, Paleomagnetism of Middle Jurassic plutons of the north-central Klamath Mountains: Geological Society of America Ab-

stracts with Programs, v. 15, p. 427.
Simpson, C., and Schmid, S. M., 1983, An evaluation of criteria to deduce the sense of movement in sheared rocks: Geological Society of America Bulletin, v. 94, p. 1281–1288.
Snoke, A. W., 1977, A thrust plate of ophiolitic rocks in the Preston Peak area, Klamath Mountains, California: Geological Society of America Bulletin, v. 88, p. 1641–1659.
Spray, J. G., 1984, Possible causes and consequences of upper mantle decoupling and ophiolite displacement, *in* Gass, I. G., Lippard, S. J., and Shelton, A. W., eds., Ophiolites and oceanic lithosphere: Oxford, Blackwell Scientific Publications, p. 255–268.
Trommsdorff, V., and Evans, B. W., 1977, Antigorite-ophicarbonates; Phase relations in a portion of the system CaO-MgO-SiO_2-H_2O-CO_2: Contributions to Mineralogy and Petrology, v. 60, p. 39–56.
Wells, F. G., and Walker, G. W., 1953, Geology of the Galice Quadrangle, Oregon: U.S. Geological Survey, Geologic Quadrangle Map GO-25, scale 1:62,500.
Wells, F. G., Hotz, P. E., and Cater, F. W., 1949, Preliminary description of the geology of the Kerby Quadrangle, Oregon: Oregon Department of Geology and Mineral Industries Bulletin 40, 23 p.
Wenner, D. B., and Taylor, H. P., Jr., 1974, Temperatures of serpentinization of ultramafic rocks based on O^{18}/O^{16} fractionation between coexisting serpentine and magnetite: Contributions to Mineralogy and Petrology, v. 32, p. 165–185.
White, S. H., and Knipe, R. J., 1978, Transformation- and reaction-enhanced ductility in rocks: Journal of the Geological Society of London, v. 135, p. 513–516.
White, S. H., Burrows, S. E., Carreras, J., Shaw, N. D., and Humphreys, F. J., 1980, On mylonites in ductile shear zones: Journal of Structural Geology, v. 2, p. 175–187.
Wicks, F. J., and Whittaker, E.J.W., 1977, Serpentine textures and serpentinization: Canadian Mineralogist, v. 15, p. 459–488.
Wicks, F. J., Whittaker, E.J.W., and Zussman, J., 1977, An idealized model for serpentine textures after olivine: Canadian Mineralogist, v. 15, p. 446–358.
Wood, R. A., 1987, Geology and geochemistry of the Almeda mine, Josephine County, Oregon [M.S. thesis]: Los Angeles, California State University, 237 p.
Wright, J. E., and Fahan, M. R., 1988, An expanded view of Jurassic orogenesis in the western United States Cordillera; Middle Jurassic (pre-Nevadan) regional metamorphism and thrust faulting within an active arc environment, Klamath Mountains, California: Geological Society of America Bulletin, v. 100, p. 859–876.
Wright, J. E., and Wyld, S. J., 1986, Significance of xenocrystic Precambrian zircon contained within the southern continuation of the Josephine ophiolite; Devils Elbow ophiolite remnant, Klamath Mountains, northern California: Geology, v. 14, p. 671–674.
Wyld, S. J., and Wright, J. E., 1988, The Devils Elbow ophiolite remnant and overlying Galice Formation; New constraints on the Middle to Late Jurassic evolution of the Klamath Mountains, California: Geologic Society of America Bulletin, v. 100, p. 29–44.

Manuscript Accepted by the Society March 27, 1990

Printed in U.S.A.

Geological Society of America
Special Paper 255
1990

Review of paleomagnetic data from the Klamath Mountains, Blue Mountains, and Sierra Nevada; Implications for paleogeographic reconstructions

Edward A. Mankinen and William P. Irwin
U.S. Geological Survey, 345 Middlefield Road, Menlo Park, California 94025

ABSTRACT

Paleomagnetic studies of the Klamath Mountains, Blue Mountains, Sierra Nevada, and northwestern Nevada pertain mostly to Jurassic and Cretaceous rocks, but some data also are available for Permian and Triassic rocks of the region. Large vertical-axis rotations are indicated for rocks in many of the terranes, but few studies show statistically significant latitudinal displacements. The most complete paleomagnetic record is from the Eastern Klamath terrane, which shows large post-Triassic clockwise rotations and virtual cessation of rotation by Early Cretaceous time, when accretion to the continent was completed. Data from Permian strata of the Eastern Klamath terrane indicate no paleolatitude anomaly, in contrast to preliminary results from coeval strata of Hells Canyon in the Blue Mountains region, which are suggestive of some southward movement. If these Hells Canyon results are confirmed, some of the terranes in these two regions must have been traveling on separate plates during late Paleozoic time. Data from Triassic and younger strata in the Blue Mountains region indicate paleolatitudes that are concordant with North America. Results from Triassic rocks of the Koipato Formation in west-central Nevada also indicate southward transport, but when this movement ceased is unknown. The Nevadan orogeny may have occurred in the Sierra Nevada during Jurassic accretion of the ophiolitic and volcanic-arc terranes of that province to the continent, whereas what has been considered to be the same orogeny in the Klamath Mountains may have occurred before accretion. Using the concordance of observed and expected paleomagnetic directions as a guide, the allochthonous Sierra Nevada, Klamath Mountains, and Blue Mountains composite terranes seem to have accreted to the continent sequentially from south to north.

INTRODUCTION

The Sierra Nevada, Klamath Mountains, and Blue Mountains are isolated geologic provinces, each of which is subdivided into several terranes. Most of these terranes represent fragments of oceanic crust, island arcs, and melange of varying age, and in many respects they are not strikingly dissimilar from one province to another. Similar rocks are present also in western Nevada. Selected elements of this tectonic accumulation forming the western Cordillera are shown in Figure 1. The provinces lie north and west of the Golconda and Roberts Mountains terranes, which separate them from the craton.

Contained in the terranes shown in Figure 1 are many deformed assemblages of rocks that range from early Paleozoic to late Mesozoic in age. Volcanic and volcaniclastic rocks of Permian and Triassic age are common; they represent island arcs formed on oceanic crust. Plutons that intrude the various terranes are mostly Jurassic and Cretaceous in age. Some of these plutons

Mankinen, E. A., and Irwin, W. P., 1990, Review of paleomagnetic data from the Klamath Mountains, Blue Mountains, and Sierra Nevada; Implications for paleogeographic reconstructions, *in* Harwood, D. S., and Miller, M. M., eds., Paleozoic and early Mesozoic paleogeographic relations; Sierra Nevada, Klamath Mountains, and related terranes: Boulder, Colorado, Geological Society of America Special Paper 255.

represent the roots of the volcanic arcs, forming plutonic-volcanic pairs, whereas others intruded after the terranes had been sutured to one another. Despite certain similarities, correlation of the terranes from one province to another is controversial, as are the geographic origins of the terranes relative to North America. Some terranes may have originated near the early continental margin, whereas others may be exotic to North America. Summaries of the geology of the region and possible tectonic interpretations are found in such studies as Davis and others (1978), Hamilton (1978), and Speed and others (1988). In this chapter, we briefly review the paleomagnetic data that are now available from the pre-Tertiary rocks to determine what constraints, if any, they may impose on geographic reconstructions of the region.

PALEOMAGNETIC DATA

The terranes in the region are difficult to study paleomagnetically because the rocks are generally altered. Many of these rocks have been subjected to conditions of metamorphism due to hydrothermal conditions on the sea floor or to heating because of deep burial, and/or they have undergone additional alteration events related to their tectonic emplacement along the continental margin and to subsequent orogenesis. All of these circumstances tend to mask or completely replace the primary remanent magnetization of the rocks. Thus, paleomagnetic sampling commonly has been restricted to the youngest rocks within a sequence, which are the freshest, the least disrupted, and the most likely to retain a primary magnetization. Plutonic rocks within some of the terranes have been sampled, but the paleomagnetic data from them can be difficult to interpret because the rocks commonly lack the structural control necessary to detect postemplacement tilting. Where overprinting of a primary magnetization has occurred, the direction carried by the secondary magnetization occasionally can provide useful information if the age of remagnetization is known or can be reasonably inferred, even though indicators of paleohorizontal are generally indeterminate. Despite these difficulties, some paleomagnetic data are now available from the localities shown in Figure 1; these are discussed briefly below.

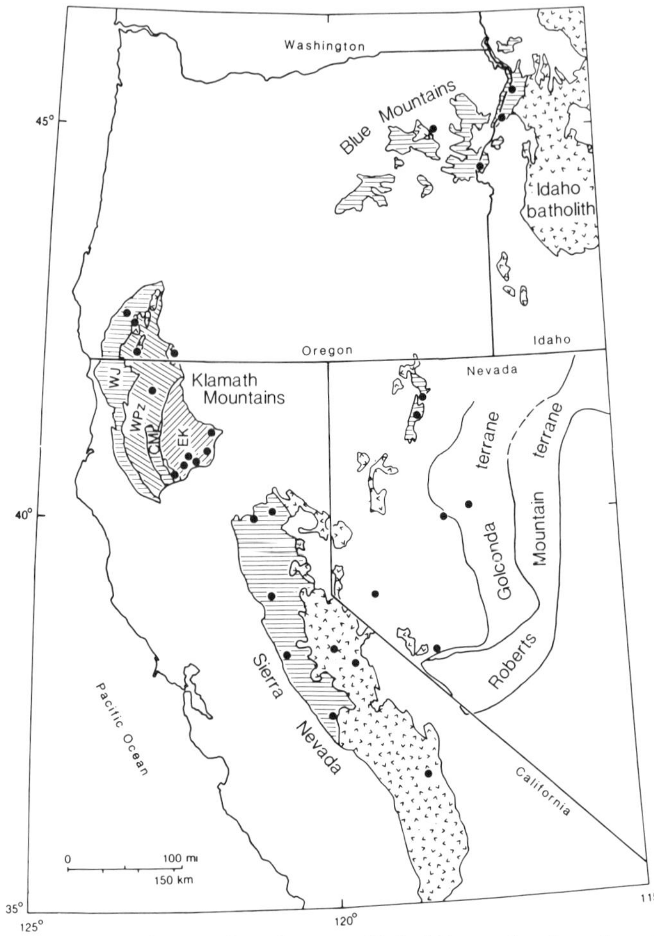

Figure 1. Sketch map of northwestern United States, showing selected elements of the Klamath Mountains, Blue Mountains, and Sierra Nevada region. Klamath Mountains province is subdivided into four accretionary plates (Irwin, 1981): WJ = western Jurassic; WPz = western Paleozoic and Triassic; CM = central metamorphic; EK = eastern Klamath. Ruled areas indicate pre-Tertiary rocks of accretionary terranes; v-pattern, Jurassic and Cretaceous granitoid plutons; dots, paleomagnetic-sample localities. Modified from Silberling and others (1987).

The laboratory and field tests commonly used to help determine whether a primary magnetization has been preserved are described in standard textbooks (e.g., McElhinny, 1973). Whether a suite of igneous rocks or a sedimentary sequence has been studied, the paleomagnetic data should represent a sampling of the geomagnetic field over several thousands of years, so that short-term variations of the field will have been averaged. When these conditions have been met, the data should represent the geomagnetic field at the sampling site, and a paleomagnetic pole can be calculated by assuming that the field is due to a geocentric dipole. This paleopole can then be compared with either a reference pole of the appropriate age or an apparent-polar-wander (APW) path for the continent (e.g., Irving and Irving, 1982; Gordon and others, 1984; May and Butler, 1986) to determine whether the results indicate discordance with cratonal North America. Uncertainties given for declination and paleolatitude anomalies are 95 percent confidence limits determined by the method of Demarest (1983). We emphasize that, because of these associated uncertainties, the smallest amount of latitudinal displacement that can be detected paleomagnetically is generally about 600 or 700 km. Conversely, some movement within these limits is allowed, but it can be neither proved nor disproved by the paleomagnetic method. A synthesis of all paleomagnetic data is presented in a later section.

Devonian and Carboniferous rocks

Few paleomagnetic data from Devonian and Carboniferous rocks in the region have been reported. A sequence of lava flows and volcanogenic sedimentary rocks of the Upper Devonian

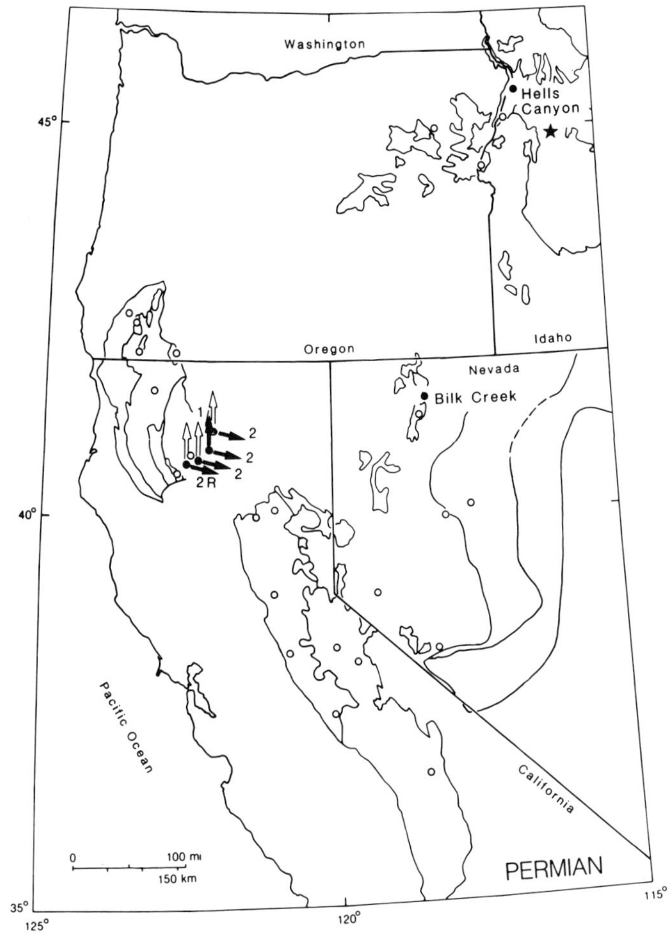

Figure 2. Sketch map of northwestern United States. Solid arrows denote vertical-axis rotations of magnetizations pertaining to Permian time; open arrows denote expected direction for cratonal North America, shown as true north. Dots, localities discussed in text where rocks of this age have been sampled. Star denotes study area indicating significant latitudinal displacement. R = remagnetization direction. Numbers are keyed to references in Table 1.

TABLE 1. SOURCES OF PALEOMAGNETIC DATA

Source Number in Figures 2 to 5	Reference
1.	Renne and others, 1988
2.	Mankinen and others, 1989
3.	Hillhouse and others, 1982
4.	Skalbeck and others, 1989
5.	Wilson and Cox, 1980
6.	Bogen, 1986a
7.	Schultz, 1983
8.	Achache and others, 1982
9.	Hannah and Verosub, 1980
10.	Gromme and others, 1967
11.	Bogen and others, 1985
12.	Hudson and Geissman, 1984
13.	Russell and others, 1982
14.	Mankinen and others, 1988
15.	Gromme and Merrill, 1965
16.	Frei and others, 1984
17.	Geissman and others, 1984
18.	Frei, 1986

and/or Lower Mississippian Taylor Formation in the northern Sierra Nevada were sampled by Hannah and Verosub (1980), who found that the rocks were remagnetized, probably during Late Jurassic time.

Samples of Middle Devonian limestone at 10 sites in the Kennett Formation of the Eastern Klamath terrane were found by Achache and others (1982) to contain only a postfolding magnetization, which they related to remagnetization by the intrusion of the Early Cretaceous Shasta Bally batholith. However, this batholith has a magnetization direction (Mankinen and others, 1988) markedly dissimilar to that recorded by the Kennett, so the pluton could not have caused the remagnetization. Overprinting of the Kennett most likely occurred during Late Jurassic time (Mankinen and others, 1989).

Paleomagnetic samples from clastic rocks of the Kennett Formation and the lower part of the Bragdon Formation of Carboniferous age were obtained by Fagin and Gose (1983). Only one site was sampled in each formation, and the remanence directions have large uncertainties (28.1° and 15.1°, respectively). Large clockwise rotations are indicated when the site-mean directions are compared with the APW path for cratonal North America. However, details of laboratory or field tests that could help determine whether these rocks have retained a record of their primary remanence were not provided by Fagin and Gose (1983), and thus the significance of their results is difficult to assess.

The Devonian Mule Mountain stock, which represents part of the root zone of the volcanic island arc that formed the early part of the Eastern Klamath terrane, was sampled by Mankinen and others (1989), who found that the pluton records a prefolding, secondary magnetization acquired during Permian time. Thus, none of these studies has provided data that can be used to deduce the pre-Permian tectonic history of the region.

Permian rocks

Permian rocks have been sampled for paleomagnetic analysis in the northern Sierra Nevada, the Bilk Creek Range of northwestern Nevada, along the Snake River in Oregon and Idaho, and in the Eastern Klamath terrane (Fig. 2). Preliminary results from Lower Permian limestone in the Bilk Creek Range indicate that those rocks may have been remagnetized (Russell and others, 1982). Volcaniclastic turbidites of the Permian Reeve Formation in the northern Sierra Nevada also were found to be remagnetized (Hannah and Verosub, 1980), probably during the Late Jurassic, along with the Taylor Formation.

Results from Lower Permian sedimentary rocks of the Hells Canyon fragment of the allochthonous terrane known as Wrangellia (Jones and others, 1977; Hillhouse and others, 1982) were presented in abstract form by Harbert and others (1988). Their preliminary results show a primary, reversed-polarity magnetization for these rocks and demosntrate that Wrangellia was in the Northern Hemisphere during Permian time, and indicate that this terrane had undergone a significant amount of southward transport relative to the North American craton by Middle Triassic time. A secondary overprint, with an inclination similar to that expected for North America during Early Cretaceous time but with a declination rotated substantially clockwise, was also isolated in these rocks (Harbert and others, 1988).

The Lower Permian McCloud Limestone in the central part of the Eastern Klamath terrane along the McCloud River arm of Shasta Lake was sampled by Fagin and Gose (1983), who considered the magnetization recorded by the McCloud to be a primary remanence that indicated considerable clockwise rotation for the terrane. Mankinen and others (1989) recently have shown that the paleomagnetically sampled outcrops of the McCloud Limestone were remagnetized by mafic quartz-diorite intrusive rocks that are closely associated with much of the McCloud. The intrusive rocks and, thus, the remagnetization, however, seem to be Permian in age (see summary by Mankinen and others, 1989). A Permian age is also supported by chemical-composition similarities between the mafic intrusions and Permian volcanic rocks that overlie the McCloud Limestone (Renne, 1987). The sites involved in the Permian remagnetization contain a postfolding, normal-polarity overprint that probably was acquired during latest Cretaceous or earliest Tertiary time.

The Permian Nosoni Formation, which conformably overlies the McCloud Limestone, was sampled along Nosoni Creek in the central part of the Eastern Klamath terrane by Renne and others (1988). Those workers isolated a prefolding, reversed-polarity remanence that yields a paleomagnetic pole which seems to be representative of cratonic North America. Only three cooling units (tuffs) were sampled in that study, however, and it is unlikely that geomagnetic secular variation could have been averaged. Therefore, Renne and others' (1988) conclusion that the resulting paleopole represents the Permian geomagnetic field is questionable. These Nosoni rocks also contain a postfolding, normal-polarity overprint that Renne and others (1988) considered to be Middle Jurassic in age. Fagin and Gose (1983) sampled a locality at the top of the Nosoni in the same area, where they considered the rocks to be remagnetized. Mankinen and others (1989) sampled the same outcrop sampled by Fagin and Gose and concurred that the rocks were remagnetized, probably during Late Jurassic time. Two other sites sampled by Mankinen and others within the Nosoni Formation at Nosoni Creek also were found to be remagnetized.

Pyroclastic rocks of the Permian Dekkas Andesite, which conformably overlie the Nosoni Formation, were sampled at eight sites along Hawkins Creek in the northern part of the Eastern Klamath terrane by Mankinen and others (1989). These sites span a stratigraphic thickness of about 3,000 m, and multiple cooling units are exposed at most sites. The rocks at seven of the sites were found to contain a primary, reversed-polarity remanence and a postfolding, normal-polarity overprint probably acquired during Late Jurassic time. The Pit Formation, which overlies the Dekkas Andesite at Nosoni Creek, is dominantly Triassic in age; however, the basal part of the Pit was sampled by Mankinen and others (1989), who found it to contain Permian radiolarians (Silberling and Jones, 1982). This locality contained a primary, normal-polarity remanence; combining its mean direction with those from Hawkins Creek provides a fold test that is statistically significant at the 95 percent confidence level. A paleopole calculated from these data indicates that the terrane has been rotated $101° \pm 10°$ clockwise but not substantially displaced latitudinally ($0° \pm 10°$) relative to cratonal North America.

The Permian Pit River stock and the mafic quartz diorite that intrudes the McCloud Limestone were sampled at two localities south of Lake Shasta (Mankinen and others, 1989). The Pit River stock was found to have been completely remagnetized during Late Jurassic time. The diorite locality also was overprinted but to a lesser degree than the stock; using principal-component analysis (Kirschvink, 1980), the primary remanence of the diorite was found to be similar to that recorded by the Permian strata sampled by Mankinen and others (1989) at both Hawkins Creek and Nosoni Creek.

Paleomagnetic data from Permian rocks of the Eastern Klamath terrane, including the remagnetization recorded by the Mule Mountain stock, are summarized in Figure 2. Most of these Permian data, except those of Renne and others (1988), indicate a large clockwise rotation for the Eastern Klamath terrane. Although Renne and others (1988) and Renne and Scott (1988) considered their data to indicate oroclinal deformation, Mankinen and others (1989) have shown Permian magnetization directions to be consistent over a broad area of the eastern Klamath Mountains. This consistency indicates that the terrane has generally behaved as a single rigid block. If the magnetization isolated by Renne and others (1988) proves to be primary (see discussion in Mankinen and others, 1989), and the paleopole they calculate can be shown to be representative of Permian time, differential rotation of a relatively small structural block may explain the observed differences in paleomagnetic declination.

Triassic rocks

Paleomagnetic samples have been obtained from Triassic rocks in the Snake River area, the Eastern Klamath terrane, and the Tobin Range of north-central Nevada (Fig. 3). Upper Triassic rocks of the Seven Devils Group and the Huntington Formation of Brooks (1979) were sampled at 29 sites along the Snake River in the Blue Mountains region by Hillhouse and others (1982). Each site consists of a single lava flow or a stratigraphic section of volcaniclastic sedimentary rocks from 2 to 100 m thick. Both normal and reversed polarities were obtained, and the magnetization at 11 of the sites appears to predate folding. In only one of the two groups of primary directions is there sufficient structural variation to provide a fold test that is statistically significant, but

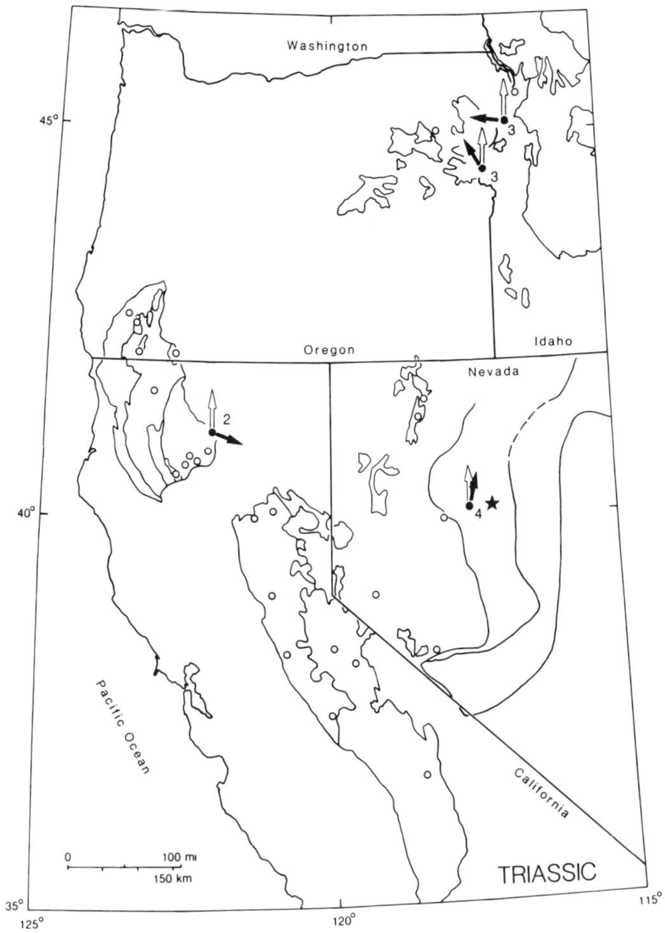

Figure 3. Sketch map of northwestern United States. Solid arrows denote vertical-axis rotations of magnetizations pertaining to Triassic time. Rotation of the Seven Devils area is calculated using Northern Hemisphere option for Wrangellia. See Figure 2 for explanation.

the corrected inclinations of both groups are consistent with those for rocks belonging to a fragment of Wrangellia.

We compare these paleomagnetic data with the Late Triassic reference pole calculated by Mankinen and others (1989) and assume that Wrangellia was in the Northern Hemisphere during Triassic time, as seems to be indicated by the data of Harbert and others (1988). The paleopole for the group 1 directions (reversed polarity) of Hillhouse and others (1982) is located 84° ± 15° counterclockwise from the Late Triassic pole, and that for the group 2 directions (normal polarity) is 29° ± 12° counterclockwise. These differences in rotation are explained by local rotations of three structural blocks within the area (Hillhouse and others, 1982). Because of these large variations in declination, Hillhouse and others used only the paleomagnetic inclinations to calculate a paleolatitude for the area at the time when the magnetization of the rocks was acquired. Comparison of the observed paleolatitude (18° ± 4°N) with that expected, based on the above reference pole, indicates insignificant transport (3° ± 7° southward) relative to the craton. Hillhouse and others (1982) also determined that the rocks at 17 sites contained only a postfolding magnetization (their group 3 directions) and that this remagnetization occurred during Late Jurassic and Early Cretaceous time.

Middle and Upper Triassic siliceous shale and siltstone of the Pit Formation were sampled in the vicinity of Hawkins Creek along a northern transect of the Eastern Klamath terrane by Mankinen and others (1989). Limestone and calcareous sandstone from the stratigraphically higher Upper Triassic Modin Formation also were sampled in the same area by Fagin and Gose (1983) and Mankinen and others (1989). Rocks at five stably magnetized localities possess a normal, presumably primary magnetic polarity, although a statistically significant fold test is precluded by the absence of structural variation. Mankinen and others (1989) calculated 110° ± 12° of clockwise rotation for these Triassic rocks, and the displacement indicated (7° ± 11° southward) is not statistically meaningful.

Volcanic rocks of the Lower Triassic Koipato Fórmation, which is the oldest unit of a sequence that unconformably overlies deformed rocks of the Golconda allochthon, were sampled in the southern Tobin Range by Skalbeck and others (1989). Both normal- and reversed-polarity lava flows were found, and the data pass a fold test at the 95 percent confidence level, indicating that the magnetization of these rocks is prefolding and probably primary. Although only 11 lava flows were sampled, Skalbeck and others (1989) noted that the statistical parameters of their data set are characteristic of those from other paleomagnetic studies that have averaged secular variation. Comparison of the paleopole calculated for these volcanic rocks (after correcting for tilt and plunge) with the 240-Ma reference pole of Irving and Irving (1982) indicates 13° ± 8° of clockwise rotation and a statistically significant 11° ± 7° of southward transport relative to the craton.

Triassic data from central Nevada and the Blue Mountains and Klamath Mountains provinces are shown in Figure 3.

Jurassic rocks

Jurassic strata and intrusive rocks were sampled throughout much of the region shown in Figure 1. Sheeted dikes and pillow basalt of the Middle to Late Jurassic Smartville ophiolite complex in the northwestern Sierra Nevada were sampled by Levi and Banerjee (1977). The rocks were found to contain a normal-polarity overprint that was probably acquired during Quaternary time.

Also in the northwestern Sierra Nevada, Jurassic rocks of the Penon Blanco Volcanics, the Sonora dike swarm, and Upper Jurassic rocks of the Logtown Ridge Formation were sampled by Bogen and others (1985). Samples from all three units were found to contain a postfolding, normal-polarity overprint whose direction is concordant with North American APW paths for Late Jurassic time. The Logtown Ridge Formation also has a reversed-polarity direction that is presumed to be prefolding but shows near-concordance with the Middle rather than Late Jurassic APW path (Bogen and others, 1985). Because this direction was found in a total of only nine samples from 3 of 12 sites, its significance is problematic. Similarly, the reversed-polarity direction reported by Bogen and others for four of the Sonora dikes may not have averaged geomagnetic secular variation, and the paleohorizontal at the time of intrusion is unknown.

Plutonic rocks at four sites within the Guadalupe igneous complex of Best (1963) and at 12 sites within the Bucks batholith in the western foothills of the Sierra Nevada were sampled by Gromme and others (1967) and Merrill and Gromme (1969), who found both normal and reversed polarities that were stable to alternating-field (AF) and thermal demagnetization. Potassium-argon ages reported by Curtis and others (1958) and Gromme and others (1967) range from 146 to 132 Ma (corrected to conform with currently accepted abundance and decay constants), indicating that intrusion of these plutons apparently spanned the Jurassic-Cretaceous boundary at 138 Ma (Sohl and Wright, 1980). Recent U-Pb determinations (Tobisch and others, 1989) yielded a date of 151 Ma for the Guadalupe complex.

Although no reference pole for latest Jurassic time for the craton is available, the mean magnetization direction recorded by these plutons indicates some rotation (≈20° counterclockwise) but no apparent latitudinal shift relative to most Late Jurassic to Early Cretaceous APW paths. However, these plutons are decoupled from the main part of the Sierra Nevada batholith by the Melones fault zone and may have been tilted (C. S. Gromme, oral communication, 1989). A steeply dipping igneous lamination and compositional layering in the Guadalupe complex led Best (1963) to consider that the plutonic rocks may have been tilted eastward, but possibly no more than about 10° since the final phase solidified. In contrast to the Guadalupe complex, layering in the Bucks batholith is generally horizontal, but sheared quartz diorite surrounding the eastern lobe of this intrusion indicated to Gromme and others (1967) that some postconsolidation movement may have occurred. Although conclusive evidence for tilting is absent, we note that restoration of as little as 10° of northeastward tilt is sufficient to bring these paleopoles into concordance with the APW path.

Late Jurassic plutons of the Yerington district, Nevada, yield normal- and reversed-polarity magnetizations whose in-situ directions are markedly anomalous in comparison with the APW path for the craton (Geissman and others, 1982). Those workers attributed the discrepant directions to late Cenozoic tilting of the plutons, ranging from 70° to 135° westward. The actual amount of tilting cannot be determined independently, however, so the data are not usable for determining paleoposition.

Layered and massive gabbro of the Jurassic Humboldt lopolith and its extrusive equivalents record both normal- and reversed-polarity directions (Hudson and Geissman, 1984). The paleomagnetic data from 15 sites, after structural correction using compositional layering in the gabbro and flow contacts in the lava flows, provide a fold test that is significant at the 99 percent confidence level. The resulting mean magnetization direction is discordant relative to the Jurassic APW paths for the craton. Although isotopic studies summarized by Hudson and Geissman (1984) indicate ages of 163 to 144 Ma for the Humboldt lopolith, these workers favored an earlier date for intrusion of the pluton and noted that this interpretation is not inconsistent with stratigraphic relations. Hudson and Geissman compared the paleopole they obtained with Early Jurassic reference poles and reported 37° of counterclockwise rotation (uncertainty limits range from 14° to 22°) and no paleomagnetically detectable latitudinal discordance (7° ± 7°) for the lopolith.

Jurassic andesite and related intrusive rocks of the Happy Creek igneous complex of Russell (1984) and Lower Cretaceous volcanogenic sedimentary rocks of the King Lear Formation, both of which are in the Jackson Mountains of the Black Rock Desert, were found by Russell and others (1982) to contain a postfolding remagnetization that was probably acquired during Late Cretaceous time. Additional sampling of the same units was undertaken by Mahar and Kirschvink (1987), and their preliminary results indicate large clockwise rotations and significant amounts of southward transport relative to the craton. However, the results from the King Lear Formation are much more anomalous in terms of both indicated rotation and latitudinal displacement than those from the older Happy Creek complex. A meaningful assessment of these data must await final publication of the results.

Granitoid intrusive rocks and their contact aureoles were sampled at 27 sites in the Blue Mountains province by Wilson and Cox (1980). Rocks at 18 sites did not yield internally consistent paleomagnetic data, but stably magnetized rocks at sites within the Wallowa batholith and Cornucopia stock yielded both normal and reversed polarities. Potassium-argon ages of the intrusive rocks were shown by Armstrong and others (1977) to range from 164 to 95 Ma; some of the youngest ages may reflect argon loss and/or protracted magma emplacement. These plutons, as well as some of those in the Sierra Nevada, spanned the Jurassic-Cretaceous boundary. Hillhouse and others (1982) compared the structurally uncorrected data of Wilson and Cox (1980) with a revised Jurassic-Cretaceous reference pole and calculated 66° ± 17° of clockwise rotation and no statistically meaningful translation (4° ± 11° northward) for those plutons.

Rocks at 37 sites within five calc-alkaline plutons of the composite Western Paleozoic and Triassic terrane in the northern part of the Klamath Mountains province were sampled by Schultz (1983). Ages summarized by Irwin (1985) and additional determinations by Wright and Fahan (1988) range from 170 to 147 Ma, indicating that the plutons are Middle to Late Jurassic in age. Stable magnetizations that include both normal and reversed polarities in some plutons generally seem mutually consistent and were considered by Schultz (1983) to indicate a primary remanence. Only nine of the sampled sites failed to yield a stable remanence direction. Although there is evidence that some of the plutons have been tilted (e.g., Barnes and others, 1986), actual amounts and timing for most plutons are not closely constrained, and the magnetic data from them are consistent with results from the Eastern Klamath terrane even if no structural corrections are applied (Mankinen and others, 1989). If the tilting is ig-

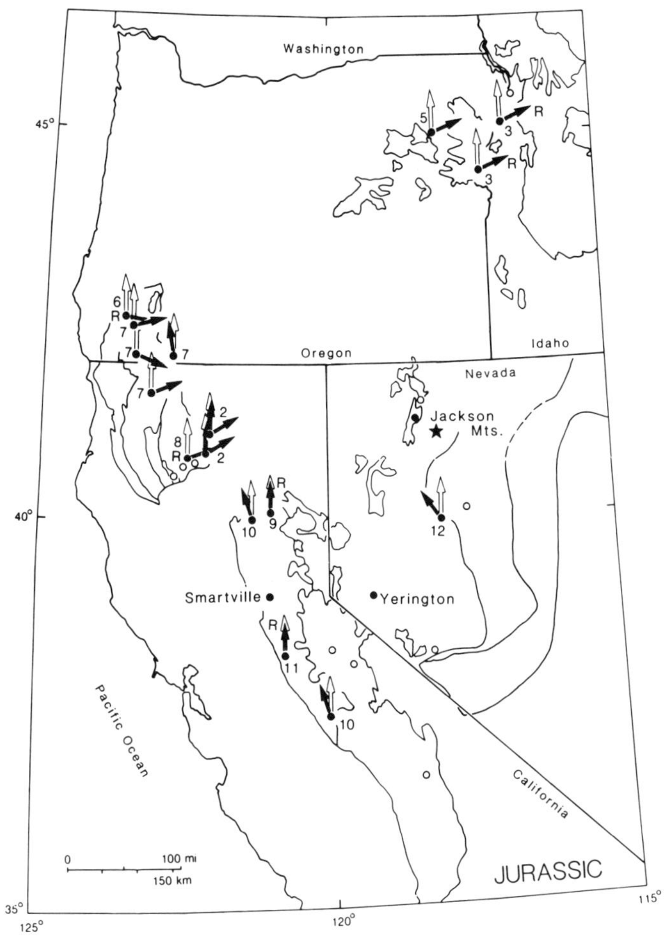

Figure 4. Sketch map of northwestern United States. Solid arrows denote vertical-axis rotations of magnetizations pertaining to Jurassic time. See Figure 2 for explanation.

nored, the direction for the Ashland pluton is nearly concordant with the Jurassic APW path, whereas those for the Grants Pass, Wooley Creek, Slinkard, and Greyback intrusions indicate clockwise rotations ranging from 67° ± 19° to 113° ± 16° (Schultz, 1983), but no paleolatitudinal movement.

Metasedimentary rocks of the Upper Jurassic Galice Formation in the Western Jurassic terrane of the Klamath Mountains province were found to contain a postfolding, reversed-polarity magnetization (Bogen, 1986a) at 13 of the 22 sites sampled. These data are difficult to analyze because the paleohorizontal at the time of remagnetization is uncertain, but considerable clockwise rotation seems likely. Bogen (1986a) calculated 99° ± 10° of clockwise rotation and 0° ± 7° of latitudinal displacement for the Galice Formation, assuming that remagnetization occurred during Late Jurassic time and that subsequent tilting has not occurred. The possibility of structural complexities affecting these paleomagnetic data was discussed by Renne and Scott (1986), Harper and Park (1986), and Bogen (1986b).

In the Eastern Klamath terrane, volcaniclastic strata of the Arvison Formation of Sanborn (1960) and the overlying Potem Formation, which range from Early to Middle Jurassic in age, were sampled by Mankinen and others (1989), who found two distinct groups of directions recorded at multiple sites. Fold tests for both normal-polarity directional groups are significant at the 75 and 95 percent confidence levels, strongly indicating a prefolding remanence for the rocks. The directional groups occur at the same stratigraphic level in the areas sampled and appear to be duplicated in the data from the Jurassic plutons sampled by Schultz (1983). The older direction (three sites) is nearly concordant with the Jurassic APW path, whereas the younger direction (four sites) is rotated approximately 60° clockwise. The postfolding overprint affecting the Permian sites sampled by Renne and others (1988) coincides with the older, concordant direction.

Paleomagnetic data pertaining to Jurassic time are summarized in Figure 4. Jurassic remagnetizations, most of which seem to be fairly well determined, are included.

Cretaceous rocks

The earliest paleomagnetic study of Cretaceous rocks in the region (Fig. 5) was undertaken in the Yosemite Valley area of the Sierra Nevada batholith by Currie and others (1963), who considered less than half of their samples to possess a reliable remanent magnetization. Additional sampling in the area was done by Gromme and Merrill (1965), who also reported K-Ar ages, which together with the results of Curtis and others (1958), established that these plutons were emplaced between 94 and 85 Ma. Gromme and Merrill (1965) reanalyzed the data from the earlier study and combined them with their own results. A paleomagnetic pole, based on 14 site-mean virtual geomagnetic poles (VGPs), was found to be concordant with others reported for the craton at that time. Comparison of Gromme and Merrill's paleopole with the Cretaceous reference pole of Globerman and Irving (1988) indicates a statistically insignificant rotation (3° ± 13° counterclockwise) and latitudinal displacement (1° ± 8° southward). However, not all of the samples used by Gromme and Merrill to calculate the paleopole were magnetically cleaned.

Three plutons of the central Sierra Nevada batholith in the vicinity of Tower Peak north of Yosemite were sampled at 49 sites by Frei and others (1984), who found that all were of normal polarity and that the samples exhibited a high magnetic stability to both AF and thermal demagnetization. Radiometric ages summarized by those workers range from 96 to 84 Ma for these intrusive rocks. Frei and others calculated a mean magnetization direction for each of the plutons after excluding data from 10 sites where only hand samples were collected or where the site-mean directions were poorly determined ($\alpha_{95} > 9°$). A paleopole based on the three pluton-means shows insignificant rotation (1° ± 12° counterclockwise) and translation (7° ± 8° northward) relative to the Cretaceous reference pole.

Rocks at 17 sites within four plutons in the area of Kings Canyon and Sequoia National Parks were sampled by Frei (1986), who summarized radiometric ages indicating that the rocks were emplaced between 102 and 97 Ma. All samples were

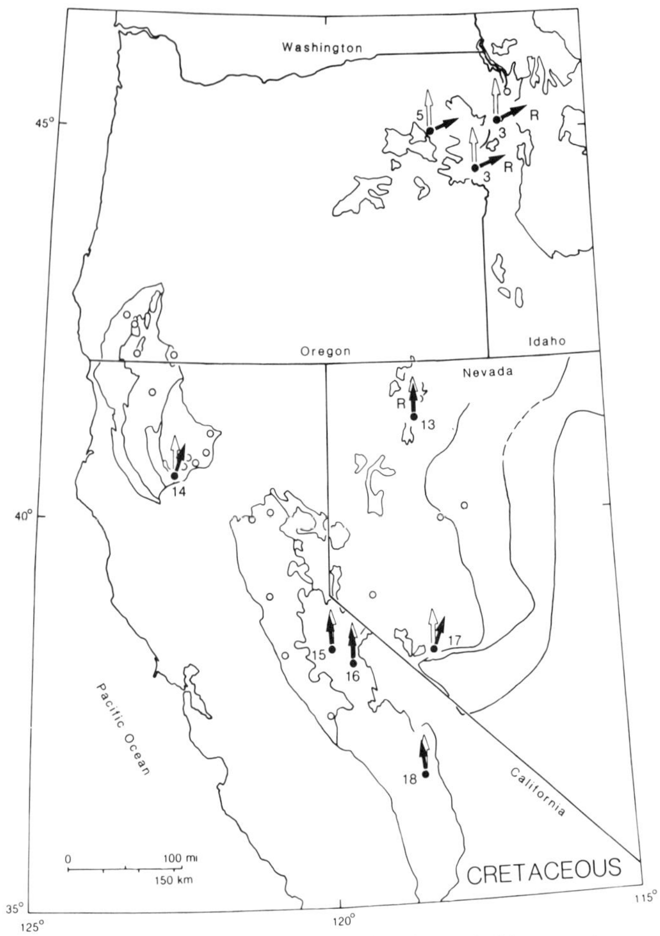

Figure 5. Sketch map of northwestern United States. Solid arrows denote vertical-axis rotations of magnetizations pertaining to Cretaceous time. See Figure 2 for explanation.

magnetically cleaned using thermal or AF demagnetization, and all were found to possess a normal-polarity remanence. However, the distribution of site VGPs was found to be markedly skewed, presumably due, in part, to an overprint parallel to the present geomagnetic field. Modal analysis of the data showed a circular concentration about nine VGPs that were considered by Frei to best characterize the data set. A paleomagnetic pole calculated for these nine sites shows insignificant rotation (7° ± 7° counterclockwise), but significant translation (6° ± 5° northward) relative to the Cretaceous reference pole of Globerman and Irving (1988).

Siliceous plutonic rocks in the Mina area of western Nevada were sampled by Geissman and others (1984), who also summarized available radiometric ages indicating that the plutons were emplaced during Late Cretaceous time. Stable normal- and reversed-polarity magnetizations were found, but the mean directions between plutons differed slightly and indicated to those workers that differential tilting and/or rotation had occurred. Because the differences are small and not systematic, we assume here that the observed dispersion could indicate the true dispersion within the plutonic suite as a whole. Accordingly, the paleomagnetic results of Geissman and others (1984) for the five plutons studied are averaged and yield a paleomagnetic pole at 84.7°N, 239.0°E (k = 32, α_{95} = 13.8°). This paleopole indicates 21° ± 16° of clockwise rotation and statistically insignificant latitudinal displacement (6° ± 11° northward) for these rocks relative to the craton.

Granitoid rocks from the Shasta Bally belt, the youngest belt of postamalgamation plutons in the Eastern Klamath terrane, were sampled by Mankinen and others (1988), who found normal- and reversed-polarity magnetizations that were stable to both AF and thermal demagnetization. Plutons within this intrusive belt are Early Cretaceous (ca. 136 Ma) in age (Irwin, 1985). Mankinen and others considered seven of the nine sites sampled as providing an accurate representation of the geomagnetic field recorded by these rocks. Assuming that tilting has not occurred, the data indicate 18° ± 14° of clockwise rotation and statistically insignificant latitudinal transport (5° ± 9° northward) for the plutons relative to the reference pole of Globerman and Irving (1988).

Paleomagnetic data indicating Cretaceous magnetizations are summarized in Figure 5, which includes the data of Wilson and Cox (1980) and Hillhouse and others (1982) from the Blue Mountains region because emplacement of the plutons in this area spanned the Jurassic-Cretaceous boundary. Note that the only Cretaceous rocks indicating latitudinal displacement are those studied by Frei (1986). Frei's paleopole is discordant in comparison with the latest version of the Cretaceous reference pole but shows insignificant rotations and translations in comparison with most previous determinations of the reference pole, such as those summarized by Globerman and Irving (1988). Caution should be used before considering these rocks to be latitudinally displaced, also because the uncertainty (3.7°) given by Frei (1986) seems improbably small, considering the highly skewed distribution of the data and the possibility of relatively recent overprinting.

Needed for the current summary is a paleopole representative of the northern and central Sierra Nevada batholith as a whole. A simple mean of all site VGPs would be inappropriate because the resulting paleopole would be biased toward the direction indicated by those individual plutons that were most heavily sampled. The large number of sites also would produce unrealistically small uncertainty limits. However, neither would averaging the pluton-mean VGPs be entirely correct, because some of the pluton means may be well determined (many sites), whereas others may be based on the data from a single site. Nonetheless, we believe that averaging pluton means is the better of the two alternatives because the paleopole would then be based on a more representative sampling of intrusions (and therefore, time) within the batholith and would have reasonable uncertainty limits. Combining the data of Gromme and Merrill (1965), Frei and others (1984), and Frei (1986) into 14 pluton means yields a paleopole at 70.5°N, 188.7°E, with k = 43 and α_{95} = 6.1°. Note that this paleopole does not differ significantly from the four-pluton paleopole for the Sierra Nevada calculated by Frei (1986). A comparison of the paleopole we calculate with the Cretaceous reference pole of Globerman and Irving (1988) indicates that the Sierra Nevada batholith has not been significantly rotated (2° ± 9° counterclockwise) or translated (2° ± 6° northward).

DISCUSSION AND CONCLUSIONS

The foregoing discussion indicates that many rocks within the region shown in Figure 1 have magnetization directions that are discordant in comparison with reference paleomagnetic poles for cratonal North America. To interpret these results, some assumptions must be made as to the style of the deformation that has occurred. Small amounts of relative rotation about vertical axes between sampling sites have been shown to be quite common in tectonically disrupted terranes; these rotations cause variations in magnetic declination, such as that mentioned above for Triassic rocks in the Snake River area. In general, however, most terranes as a whole can still be considered to have behaved as semi-rigid blocks during deformation.

Oroclinal bending has been proposed by some workers as a mode of deformation in certain areas, most recently by Renne and others (1988) and Renne and Scott (1988) to explain tectonic rotations in the Klamath Mountains. To support the oroclinal concept, Renne and Scott (1988, Fig. 10) summarized paleomagnetic data that seem to indicate that major rotations are more common in the northern than in the southern parts of the Klamath Mountains province. Because not all the available data were shown in their figure, and because the rotations shown range in age from Permian to Cretaceous, the significance of their summary is questionable. Renne and Scott (1988, Fig. 12) showed poles to bedding and measured paleomagnetic declination anomalies for three domains within the Eastern Klamath terrane. Their figure purports to show that the declination anomalies can be directly correlated to changes in the trend of major fold axes. Here, again, however, not all the data are represented, and not all the indicated rotations are coeval. Comparisons of magnetizations of different ages can be valid if no age-dependent differences in rotation exist, but the studies by Mankinen and Irwin (1982) and Mankinen and others (1988, 1989) indicate large differences in rotation between Permian, Jurassic, and Cretaceous rocks of the terrane. Because Permian magnetic declinations seem to be consistent over a broad area (Mankinen and others, 1989), we believe that the Eastern Klamath terrane generally has behaved as a semirigid block since Permian time, although differential rotation of small discrete blocks within this terrane may have occurred.

Similarity in magnetization directions between rocks of the Eastern Klamath terrane and some plutons in the Western Paleozoic and Triassic terranes indicates that a composite terrane containing the Eastern Klamath terrane and parts of the Western Paleozoic and Triassic, and intervening Central Metamorphic terranes also may have undergone rotation as a single block since Middle Jurassic time. We do not, however, discount that other types of deformation, such as oroclinal bending or dextral shear, may have been more important in other parts of the province (e.g., Wells and Heller, 1988). In general, disruption of the terranes in the region shown in Figure 1 seems to be adequately described by assuming rigid-block behavior, given the paleomagnetic data that are currently available. A thorough description of the processes by which such crustal blocks are disrupted has been given by Beck (1989).

Klamath Mountains province

The most complete paleomagnetic record to date from the Klamath Mountains province is from the Eastern Klamath terrane (Mankinen and others, 1989), but even here reliable data are unavailable from pre-Permian rocks. Because most of the indicated rotations from Permian and Triassic strata of the Eastern Klamath terrane are similar (Figs. 2, 3) and greater than those indicated by the overlying Jurassic strata (Fig. 4), major rotation of the Permian and Triassic rocks must have begun during latest Triassic or earliest Jurassic time. Once started, rotation continued as Lower and Middle Jurassic strata were being deposited. Some similarities in rotation indicated by the Jurassic plutons sampled by Schultz (1983) and the strata in the Eastern Klamath terrane (Fig. 4) show that the rocks of the Western Paleozoic and Triassic terrane, the Eastern Klamath terrane, and the intervening Central Metamorphic terrane may have joined by Middle to Late Jurassic time, as deduced earlier on the basis of various geological criteria (Irwin, 1981).

As summarized in this report and shown in Figure 4, Jurassic results from the Klamath Mountains are complex, and Mankinen and others (1989) have considered this complexity as due both to the rapid APW that has been documented for Jurassic time and to a rotational history that may have included some retrograde movement. The rapid rotation during the Jurassic seems to have ceased at about the time of emplacement of the Shasta Bally batholith (Early Cretaceous); during the succeeding ≈100 m.y. until middle Tertiary time, virtually no rotation occurred (Mankinen and others, 1988, 1989). The close correspondence between the waning stages of rotation and the inception of deposition of the Lower Cretaceous basal strata of the Great Valley sequence probably represents completion of accretion to the North American Continent. This timing is somewhat later than that suggested by some workers, most recently Goodge (1989), who has considered that accretion occurred during Middle Jurassic time, when major deformation and magmatism affected much of the province. Although the Middle and Late Jurassic were times of major consolidation of Klamath Mountains terranes, collision of the composite terrane with the craton may have occurred at a later time. It is difficult to envision how more than 60° of post–150-Ma rotation (see Mankinen and others, 1989) could be structurally accommodated if the province were firmly attached to the continental margin. We consider it more likely that most of the rotations indicated by rocks of the

Klamath Mountains occurred while the various terranes were parts of oceanic plates. The absence of available evidence for significant latitudinal displacement of any Klamath Mountains terranes relative to the craton indicates that any large post-Permian transport must have been nearly perpendicular to the continental margin. Because paleomagnetism cannot detect longitudinal displacements, the distance of the paleo-Klamath arcs from the craton cannot be determined by this method.

Blue Mountains region

Preliminary results from Permian strata in the Hells Canyon area indicate that the rocks originated in the Northern Hemisphere and subsequently moved southward relative to continental North America (Harbert and others, 1988). This relative motion is in contrast to that indicated by coeval strata from the Klamath Mountains, where no paleolatitude anomaly was observed. If these preliminary results are confirmed by the published data, the differences in paleolatitude show that the arc rocks of these two areas must have been transported on separate oceanic plates during late Paleozoic time. A secondary, postfolding overprint observed in the Triassic strata sampled by Hillhouse and others (1982) indicates that the rocks have undergone 66° of post-Jurassic clockwise rotation. Restoration of this amount of rotation indicates that the pre–Late Jurassic rotation of the Triassic rocks must have been more than 100° counterclockwise. Primary magnetic inclinations recorded by these strata (Hillhouse and others, 1982) indicate a paleolatitude concordant with North America by Late Triassic time. Because no paleolatitude anomaly is indicated by Triassic or later magnetizations (Wilson and Cox, 1980; Hillhouse and others, 1982), transport of the rocks of the Blue Mountains since Triassic time must have been perpendicular to the continental margin.

Intrusion of the Wallowa batholith and Cornucopia stock into both the Seven Devils Group and other terranes of the Blue Mountains province indicates that these rocks were amalgamated by Late Jurassic time (Wilson and Cox, 1980; Ave Lallemant and Oldow, 1988). Hillhouse and others (1982) and Gromme and others (1986) favored an Early Cretaceous arrival of the composite terrane near the continental margin, and accretion between 130 and 50 Ma, when rotation was nearly complete. Geologic constraints and $^{40}Ar/^{39}Ar$ age-spectrum data indicated to Lund and Snee (1988) that accretion of the composite terrane occurred between 118 and 93 Ma. Arrival of the Blue Mountains terranes, therefore, was considerably later than what we consider to be the time of accretion ($\simeq$136 Ma) of the Klamath Mountains terranes to the continent.

Northwestern Nevada

Paleomagnetic data that could help delineate the Paleozoic and early Mesozoic tectonic history of west-central Nevada are limited to those from the Koipato Formation within the Tobin Range. The anomalously shallow magnetic inclinations recorded by those rocks (Skalbeck and others, 1989) indicate that the Koipato Formation and, presumably, underlying rocks of the Golconda allochthon have been transported approximately 1,200 $\pm$ 800 km southward relative to the craton since Triassic time. Paleomagnetic data are unavailable to help determine when this southward transport ceased, but it must have occurred before overthrusting of the Fencemaker allochthon during Middle or Late Jurassic through Early Cretaceous time (Oldow, 1984). The Jurassic Humboldt lopolith, which intrudes Jurassic sedimentary rocks of the Fencemaker allochthon, shows no paleolatitude anomaly (Hudson and Geissman, 1984). Although Hudson and Geissman (1984) considered the counterclockwise rotation of the Humboldt lopolith to reflect active thrusting during the final stages of accretion of the Fencemaker allochthon, later work (Hudson and Geissman, 1987) indicates that at least 25° of this rotation occurred during middle Tertiary Basin and Range extension. If the Tertiary rotation of the Humboldt lopolith is removed, the data shown in Figure 4 generally indicate that the rocks of the Sierra Nevada and northwestern Nevada, with the possible exception of the Golconda allochthon, were in place and relatively stable by Late Jurassic time, in contrast to the terranes of the Klamath and Blue Mountains province.

Sierra Nevada

The Jurassic Nevadan orogeny of the northern Sierra Nevada is thought by some workers to reflect the collision and accretion of those volcanic-arc terranes to North America (e.g., Moores and Day, 1984). Sharp (1988), however, summarized geological data indicating that several accretionary events may have occurred between Triassic and Early Jurassic time. In either case, the northern Sierra Nevada terranes must have been accreted to the continent well before those in the Klamath and Blue Mountains. If the remagnetizations described by Hannah and Verosub (1980) and Bogen and others (1985) were acquired in Late Jurassic time, as seems likely, then the differences in rotation indicated by the Jurassic rocks in those three areas support this conclusion.

Summary

Rotation of the Late Cretaceous plutons of the Mina area of Nevada is not well constrained, although Geissman and others (1984) considered that it may be due to localized, late Cenozoic tectonics. This interpretation seems reasonable because of the absence of significant disruption of Late Cretaceous magnetizations to the west (Sierra Nevada) and north (Jackson Mountains). Because similar rotations were recorded by plutons of the Shasta Bally belt in the Eastern Klamath terrane, by Tertiary volcanic rocks in the Cascade Range immediately east of the Klamath Mountains (Beck and others, 1986), and by Cretaceous onlap sequences (Mankinen and Irwin, 1982), the area including these rocks probably rotated as a single block during the earliest stages of extension in the Basin and Range province (Mankinen and others, 1988). When the Tertiary rotation of the Klamath Mountains province is removed, most of the elements in the region (Fig. 5) appear to have been accreted early in Cretaceous time and to have remained relatively stable during most of that period. The date of major rotation in the Blue Mountains province has been shown to be post-130 and pre-50 Ma (Wilson and Cox, 1980; Hillhouse and others, 1982; Gromme and others, 1986), so this composite terrane was the last of those considered to be in place.

Significant latitudinal displacements seem to be indicated for Triassic strata of the Koipato Formation and underlying rocks of the Golconda allochthon, probably for Permian strata of Wrangellia, and questionably for Mesozoic rocks of the Jackson Mountains. Results indicating southward transport for the Koipato Formation are unusual because the terranes now situated farther westward (oceanward) from the Paleozoic continental margin show no evidence for latitudinal displacement. Results from the Koipato Formation, however, do seem compatible with the concept of transpressional terranes (Ave Lallemant and Oldow, 1988), which predicts that parts of the Cordilleran borderland should show evidence of southward transport relative to the craton during early Mesozoic time. Transport of the Koipato is not related to that of Wrangellia, where Triassic rocks indicate latitudinal concordance with the craton.

The other terranes studied in the region of Figure 1 show no evidence of significant latitudinal transport, at least while their youngest units were being deposited or emplaced. This result could indicate that none of these terranes was ever far from the continental margin. However, convergence between the Farallon plate and the western margin of North America, though poorly constrained, appears to have been predominantly east-west during Middle Jurassic to Early Cretaceous time (Engebretson and others, 1985), so considerable amounts of longitudinal transport are not precluded. Therefore, many of the terranes may have developed separately and far removed from each other. For example, the Eastern Klamath terrane must have originated sufficiently far offshore to have developed distinctive Permian fusulinid and coral faunas that differ from typical North American forms (Stevens, 1977), and must have been far enough from the Sierra Nevada terranes to show significant geochemical and petrologic differences (Lapierre and others, 1987; Rouer and others, 1989) between nearly contemporaneous Paleozoic and early Mesozoic magmatism in the two areas.

Few paleomagnetic data are available that indicate a common tectonic history for many of the elements shown in Figure 1, except for the youngest time frames illustrated in Figures 2 through 5. Some correlatives of the terranes may also exist outside of the region considered in this report. Paleolatitudes determined for volcanic rocks of the Lower Permian Asitka Group in the Stikinia terrane are similar to those for the Dekkas Andesite in the Eastern Klamath terrane, and Irving and Monger (1987) suggested that these two areas were much closer during Permian time. How these terranes became separated is unknown. In one possible scenario, Wernicke and Klepacki (1988) suggested that the rocks composing Stikinia became displaced from their original position as a result of the collision and penetration of Wrangellia into North America beginning in Middle Jurassic time. The paleomagnetic signature of Wrangellia is well established (e.g., Hillhouse and Gromme, 1984) and has been reported from several separate fragments. Evidently, Wrangellia was once a coherent, subcontinent-size terrane whose fragments have been distributed great distances along the continental margin (Jones and others, 1977).

ACKNOWLEDGMENTS

Reviews by J. W. Hillhouse, P. R. Renne, R. E. Wells, and informal comments by C. S. Gromme are appreciated. Thanks also to M. E. Beck, Jr., for providing us with a preprint of the Koipato Formation study.

REFERENCES CITED

Achache, J., Cox, A., and O'Hare, S., 1982, Paleomagnetism of the Devonian Kennett limestone and the rotation of the eastern Klamath Mountains, California: Earth and Planetary Science Letters, v. 61, p. 365–380.

Armstrong, R. L., Taubeneck, W. H., and Hales, P. O., 1977, Rb-Sr and K-Ar geochronometry of Mesozoic granitic rocks and their Sr isotopic composition, Oregon, Washington, and Idaho: Geological Society of America Bulletin, v. 88, p. 397–411.

Ave Lallemant, H. G., and Oldow, J. S., 1988, Early Mesozoic southward migration of Cordilleran transpressional terranes: Tectonics, v. 7, p. 1057–1075.

Barnes, C. G., Rice, J. M., and Gribble, R. F., 1986, Tilted plutons in the Klamath Mountains of California and Oregon: Journal of Geophysical Research, v. 91, p. 6059–6071.

Beck, M. E., Jr., 1989, Block rotations in continental crust; Examples from western North America, *in* Laj, C., and Kissel, C., eds., Paleomagnetic rotations and continental deformation: Boston, Massachusetts, Kluwer, p. 1–16.

Beck, M. E., Jr., Burmester, R. F., Craig, D. E., Gromme, C. S., and Wells, R. E., 1986, Paleomagnetism of middle Tertiary volcanic rocks from the Western Cascade Series, northern California: Journal of Geophysical Research, v. 91, p. 8219–8230.

Best, M. G., 1963, Petrology of the Guadalupe igneous complex southwestern Sierra Nevada foothills California: Journal of Petrology, v. 4, p. 223–259.

Bogen, N. L., 1986a, Paleomagnetism of the Upper Jurassic Galice Formation, southwestern Oregon; Evidence for differential rotation of the eastern and western Klamath Mountains: Geology, v. 14, p. 335–338.

——, 1986b, Reply to comments: Geology, v. 14, p. 1050–1051.

Bogen, N. L., Kent, D. V., and Schweickert, R. A., 1985, Paleomagnetism of Jurassic rocks in the western Sierra Nevada metamorphic belt and its bearing on the structural evolution of the Sierra Nevada block: Journal of Geophysical Research, v. 90, p. 4627–4638.

Brooks, H. C., 1979, Geologic map of Huntington and part of the Olds Ferry Quadrangles, Baker and Malheur Counties, Oregon: Oregon Department of Geology and Mineral Industries Geologic Map Series Map GMS-13.

Currie, R. G., Gromme, C. S., and Verhoogen, J., 1963, Remanent magnetization of some Upper Cretaceous granitic plutons in the Sierra Nevada, California: Journal of Geophysical Research, v. 68, p. 2263–2279.

Curtis, G. H., Evernden, J. F., and Lipson, J., 1958, Age determination of some granitic rocks in California by the potassium-argon method: California Division of Mines Special Report 54, 16 p.

Davis, G. A., Monger, J.W.H., and Burchfiel, B. C., 1978, Mesozoic construction of the Cordilleran "collage," central British Columbia to central California, *in* Howell, D. G., and McDougall, K. A., eds., Mesozoic paleogeography of the western United States: Pacific Section, Society of Economic Paleontologists and Mineralogists Pacific Coast Paleogeography Symposium 2, p. 1–32.

Demarest, H. H., Jr., 1983, Error analysis for the determination of tectonic rotation from paleomagnetic data: Journal of Geophysical Research, v. 88, p. 4321–4328.

Engebretson, D. C., Cox, A., and Gordon, R. G., 1985, Relative motions between oceanic and continental plates in the Pacific basin: Geological Society of America Special Paper 206, 59 p.

Fagin S. W., and Gose, W. A., 1983, Paleomagnetic data from the Redding section of the eastern Klamath belt, northern California: Geology, v. 11, p. 505–508.

Frei, L. S., 1986, Additional paleomagnetic results from the Sierra Nevada; Further constraints on Basin and Range extension and northward displacement in the western United States: Geological Society of America Bulletin, v. 97, p. 840–849.

Frei, L. S., Magill, J. R., and Cox, A., 1984, Paleomagnetic results from the central Sierra Nevada; Constraints on reconstructions of the western United States: Tectonics, v. 3, p. 157–177.

Geissman, J. W., Van der Voo, R., and Howard, K. L., Jr., 1982, A paleomagnetic study of the structural deformation in the Yerington district, Nevada; 2, Mesozoic "basement" units and their total and pre-Oligocene tectonism: American Journal of Science, v. 282, p. 1080–1109.

Geissman, J. W., Callian, J. T., Oldow, J. S., and Humphries, S. E., 1984, Paleomagnetic assessment of oroflexural deformation in the west-central Nevada and significance for emplacement of allochthonous assemblages: Tectonics, v. 3, p. 179–200.

Globerman, B. R., and Irving, E., 1988, Mid-Cretaceous paleomagnetic reference field for North America; Restudy of 100 Ma intrusive rocks from Arkansas: Journal of Geophysical Research, v. 93, p. 11721–11733.

Goodge, J. W., 1989, Evolving early Mesozoic convergent margin deformation, central Klamath Mountains, northern California: Tectonics, v. 8, p. 845–864.

Gordon, R. G., Cox, A., and O'Hare, S., 1984, Paleomagnetic Euler poles and the apparent polar wander and absolute motion of North America since the Carboniferous: Tectonics, v. 3, p. 499–537.

Gromme, C. S., and Merrill, R. T., 1965, Paleomagnetism of Late Cretaceous granitic plutons in the Sierra Nevada, California; Further results: Journal of Geophysical Research, v. 70, p. 3407–3420.

Gromme, C. S., Merrill, R. T., and Verhoogen, J., 1967, Paleomagnetism of Jurassic and Cretaceous plutonic rocks in the Sierra Nevada, California, and its significance for polar wandering and continental drift: Journal of Geophysical Research, v. 72, p. 5661–5684.

Gromme, C. S., Beck, M. E., Jr., Wells, R. E., and Engebretson, D. C., 1986, Paleomagnetism of the Tertiary Clarno Formation of central Oregon and its significance for the tectonic history of the Pacific Northwest: Journal of Geophysical Research, v. 91, p. 14089–14103.

Hamilton, W., 1978, Mesozoic tectonics of the western United States, *in* Howell, D. G., and McDougall, K. A., eds., Mesozoic paleogeography of the western United States: Pacific Section, Society of Economic Paleontologists and Mineralogists Pacific Coast Paleogeographic Symposium 2, p. 33–70.

Hannah, J. L., and Verosub, K. L., 1980, Tectonic implications of remagnetized upper Paleozoic strata in the northern Sierra Nevada: Geology, v. 8, p. 520–524.

Harbert, W., Hillhouse, J., and Vallier, T., 1988, Leonardian paleolatitude of Wrangellia [abs.]: EOS Transactions of the American Geophysical Union, v. 69, p. 1169.

Harper, G. D., and Park, R., 1986, Comment on a paper by Nicholas L. Bogen, "Paleomagnetism of the Upper Jurassic Galice Formation, southwestern Oregon; Evidence for differential rotation of the eastern and western Klamath Mountains": Geology, v. 14, p. 1049–1050.

Hillhouse, J. W., and Gromme, C. S., 1984, Northward displacement and accretion of Wrangellia; New paleomagnetic evidence from Alaska: Journal of Geophysical Research, v. 89, p. 4461–4477.

Hillhouse, J. W., Gromme, C. S., and Vallier, T. L., 1982, Paleomagnetism and Mesozoic tectonics of the Seven Devils volcanic arc in northeastern Oregon: Journal of Geophysical Research, v. 87, p. 3777–3794.

Hudson, M. R., and Geissman, J. W., 1984, Preliminary paleomagnetic data from the Jurassic Humboldt lopolith, west-central Nevada; Evidence for thrust belt rotation in the Fencemaker allochthon: Geophysical Research Letters, v. 11, p. 828–831.

——, 1987, Paleomagnetic and structural evidence for middle Tertiary counterclockwise block rotation in the Dixie Valley region, west-central Nevada: Geology, v. 15, p. 638–642.

Irving, E., and Irving, G. A., 1982, Apparent polar wander paths Carboniferous through Cenozoic and the assembly of Gondwana: Geophysical Surveys, v. 5, p. 141–188.

Irving, E., and Monger, J.W.H., 1987, Preliminary paleomagnetic results from the Permian Asitka Group, British Columbia: Canadian Journal of Earth Sciences, v. 24, p. 1490–1497.

Irwin, W. P., 1981, Tectonic accretion of the Klamath Mountains, *in* Ernst, W. G., ed., The geotectonic development of California; Rubey Volume 1: Englewood Cliffs, New Jersey, Prentice-Hall, p. 30–49.

——, 1985, Age and tectonics of plutonic belts in accreted terranes of the Klamath Mountains, California and Oregon, *in* Howell, D. G., ed., Tectonostratigraphic terranes of the circum-Pacific region: Circum-Pacific Council for Energy and Mineral Resources, v. 1, p. 187–199.

Jones, D. L., Silberling, N. J., and Hillhouse, J. W., 1977, Wrangellia; A displaced terrane in northwestern North America: Canadian Journal of Earth Sciences, v. 14, p. 2565–2577.

Kirschvink, J. L., 1980, The least-squares line and plane and the analysis of palaeomagnetic data: Geophysical Journal of the Royal Astronomical Society, v. 62, p. 699–718.

Lapierre, H., and 7 others, 1987, Paleozoic and lower Mesozoic magmas from the eastern Klamath Mountains (north California) and the geodynamic evolution of northwestern America: Tectonophysics, v. 140, p. 155–177.

Levi, S., and Banerjee, S. K., 1977, The effects of alteration on the natural remanent magnetization of three ophiolite complexes; Possible implications for the oceanic crust: Journal of Geomagnetism and Geoelectricity, v. 29, p. 421–439.

Lund, K., and Snee, L. W., 1988, Metamorphism, structural development, and age of the continent; Island arc juncture in west-central Idaho, *in* Ernst, W. G., ed., Metamorphism and crustal evolution of the western United States; Rubey Volume 7: Englewood Cliffs, New Jersey, Prentice-Hall, p. 296–331.

Mahar, K. A., and Kirschvink, J., 1987, Paleomagnetism and tectonics of the Jackson Mountains, NW Nevada [abs.]: EOS Transactions of the American Geophysical Union, v. 68, p. 1253.

Mankinen, E. A., and Irwin, W. P., 1982, Paleomagnetic study of some Cretaceous and Tertiary sedimentary rocks of the Klamath Mountains province, California: Geology, v. 10, p. 82–87.

Mankinen, E. A., Irwin, W. P., and Gromme, C. S., 1988, Paleomagnetic results from the Shasta Bally plutonic belt in the Klamath Mountains province, northern California: Geophysical Research Letters, v. 15, p. 56–59.

——, 1989, Paleomagnetic study of the Eastern Klamath terrane, California, and implications for the tectonic history of the Klamath Mountains province: Journal of Geophysical Research, v. 94, p. 10444–10472.

May, S. R., and Butler, R. F., 1986, North American Jurassic apparent polar wander; Implications for plate motion, paleogeography, and Cordilleran tectonics: Journal of Geophysical Research, v. 91, p. 11519–11544.

McElhinny, M. W., 1973, Palaeomagnetism and plate tectonics: New York, Cambridge University Press, 358 p.

Merrill, R. T., and Gromme, C. S., 1969, Nonreproducible self-reversal of magnetization in diorite: Journal of Geophysical Research, v. 74, p. 2014–2024.

Moores, E. M., and Day, H. W., 1984, Overthrust model for the Sierra Nevada: Geology, v. 12, p. 416–419.

Oldow, J. S., 1984, Evolution of a late Mesozoic back-arc fold and thrust belt, northwestern Great Basin, U.S.A.: Tectonophysics, v. 102, p. 245–274.

Renne, P. R., 1987, Permian to Jurassic tectonic evolution of the eastern Klamath Mountains, California [Ph.D. thesis]: Berkeley, University of California, 127 p.

Renne, P. R., and Scott, G. R., 1986, Comment on a paper by Nicholas L. Bogen, "Paleomagnetism of the Upper Jurassic Galice Formation, southwestern Oregon; Evidence for differential rotation of the eastern and western Klamath Mountains": Geology, v. 14, p. 1048–1049.

—— , 1988, Structural chronology, oroclinal deformation, and tectonic evolution of the southeastern Klamath Mountains, California: Tectonics, v. 7, p. 1223–1242.

Renne, P. R., Scott, G. R., and Bazard, D. R., 1988, Multicomponent paleomagnetic data from the Nosoni Formation, eastern Klamath Mountains, California; Cratonic Permian primary directions with Jurassic overprints: Journal of Geophysical Research, v. 93, p. 3387–3400.

Rouer, O., Lapierre, H., Mascle, G., Coulon, C., and Albers, J., 1989, Geodynamic implications of Devonian silicic arc magmatism in the Sierra Nevada and Klamath Mountains, California: Geology, v. 17, p. 177–180.

Russell, B. J., 1984, Mesozoic geology of the Jackson Mountains, northeastern Nevada: Geological Society of America Bulletin, v. 95, p. 313–323.

Russell, B. J., Beck, M. E., Jr., Burmester, R. F., and Speed, R. C., 1982, Cretaceous magnetizations in northwestern Nevada and tectonic implications: Geology, v. 10, p. 423–428.

Sanborn, A. F., 1960, Geology and paleontology of the southwest quarter of the Big Bend Quadrangle, Shasta County, California: California Division of Mines Special Report 63, 26 p.

Schultz, K. L., 1983, Paleomagnetism of Jurassic plutons in the central Klamath Mountains, southern Oregon and northern California [M.S. thesis]: Corvallis, Oregon State University, 153 p.

Sharp, W. D., 1988, Pre-Cretaceous crustal evolution in the Sierra Nevada region, California, *in* Ernst, W. G., ed., Metamorphism and crustal evolution of the western United States; Rubey Volume 7: Englewood Cliffs, New Jersey, Prentice-Hall, p. 823–864.

Silberling, N. J., and Jones, D. L., 1982, Tectonic significance of Permian–Triassic strata in northwestern Nevada and northern California: Geological Society of American Abstracts with Programs, v. 14, p. 234.

Silberling, N. J., Jones, D. L., Blake, M. C., Jr., and Howell, D. G., 1987, Lithotectonic terrane map of the western conterminous United States: U.S. Geological Survey Miscellaneous Field Studies Map MF-1874-C, scale 1:2,500,000.

Skalbeck, J. D., Burmester, R. F., Beck, M. E., Jr., and Speed, R. C., 1989, Paleomagnetism of the Late Permian–Early Triassic Koipato volcanics, Nevada; Implications for latitudinal displacement: Earth and Planetary Science Letters, v. 95, p. 403–410.

Sohl, N. F., and Wright, W. B., 1980, Changes in stratigraphic nomenclature by the U.S. Geological Survey, 1979: U.S. Geological Survey Bulletin 1502-A, 138 p.

Speed, R., Elison, M. W., and Heck, F. R., 1988, Phanerozoic tectonic evolution of the Great Basin, *in* Ernst, W. G., ed., Metamorphism and crustal evolution of the western United States; Rubey Volume 7: Englewood Cliffs, New Jersey, Prentice-Hall, p. 572–605.

Stevens, C. H., 1977, Permian depositional provinces and tectonics, western United States, *in* Stewart, J H., Stevens, C. H., and Fritsche, A. E., eds., Paleozoic paleogeography of the western United States: Pacific Section, Society of Economic Paleontologists and Mineralogists Pacific Coast Paleogeography Symposium 1, p, 113–135.

Tobisch, O. T., Paterson, S. R., Saleeby, J. B., and Geary, E. E., 1989, Nature and timing of deformation in the Foothills terrane, central Sierra Nevada, California; Its bearing on orogenesis: Geological Society of America Bulletin, v. 101, p. 401–413.

Wells, R. E., and Heller, P. L., 1988, The relative contribution of accretion, shear, and extension to Cenozoic tectonic rotation in the Pacific Northwest: Geological Society of America Bulletin, v. 100, p. 325–338.

Wernicke, B., and Klepacki, D. W., 1988, Escape hypothesis for the Stikine block: Geology, v. 16, p. 461–464.

Wilson, D., and Cox, A., 1980, Paleomagnetic evidence for tectonic rotation of Jurassic plutons in Blue Mountains, eastern Oregon: Journal of Geophysical Research, v. 85, p. 3681–3689.

Wright, J. E., and Fahan, M. R., 1988, An expanded view of Jurassic orogenesis in the western United States Cordillera; Middle Jurassic (pre-Nevadan) regional metamorphism and thrust faulting within an active arc environment, Klamath Mountains, California: Geological Society of America Bulletin, v. 100, p. 859–876.

Manuscript Accepted by the Society March 27, 1990

Printed in U.S.A.

Index

[Italic page numbers indicate major references]

Typeset by WESType Publishing Services, Inc., Boulder, Colorado
Printed in U.S.A. by JL Printing, Loveland, Colorado